Battery
Reference Book

Battery
Reference Book

T R Crompton
MSc, BSc

Butterworths
London . Boston . Singapore
Sydney . Toronto . Wellington

 PART OF REED INTERNATIONAL P.L.C.

First published 1990

Butterworth International Edition, 1990
  ISBN 0-408-00791-5

© **Butterworth & Co. (Publishers) Ltd, 1990**

British Library Cataloguing in Publication Data

Crompton, T. R. (Thomas Roy)
  Battery reference book.
  1. Batteries
  I. Title
  621.31'242

ISBN 0-408-00790-7

Library of Congress Cataloging-in-Publication Data

Crompton, T. R. (Thomas Roy)
  Battery reference book / T. R. Crompton
      p.   cm.
  Bibliography: p.
  Includes index.
  ISBN 0-408-00790-7
  1. Electric batteries. 2. Storage batteries.
  I. Title.
TK2921.C75   1989
621.31'242—dc20

Composition by Genesis Typesetting
Printed and bound in Great Britain by Courier
International Ltd, Tiptree, Essex

Preface

Primary (non-rechargeable) and secondary (rechargeable) batteries are an area of manufacturing industry that has undergone a tremendous growth in the past two or three decades, both in sales volume and in variety of products designed to meet new applications. Not so long ago, mention of a battery to many people brought to mind the image of an automotive battery or a torch battery and, indeed, these accounted for the majority of batteries being produced. There were of course other battery applications such as submarine and aircraft batteries, but these were of either the lead–acid or alkaline type. Lead–acid, nickel–cadmium, nickel–iron and carbon–zinc represented the only electrochemical couples in use at that time.

There now exist a wide range of types of batteries, both primary and secondary, utilizing couples that were not dreamt of a few years ago. Many of these couples have been developed and utilized to produce batteries to meet specific applications ranging from electric vehicle propulsion, through minute batteries for incorporation as memory protection devices in printed circuits in computers, to pacemaker batteries used in heart surgery. This book attempts to draw together in one place the available information on all types of battery now being commercially produced. It starts with a chapter dealing with the basic theory behind the operation of batteries. This deals with the effects of such factors as couple materials, electrolyte composition, concentration and temperature on battery performance, and also discusses in some detail such factors as the effect of discharge rate on battery capacity. The basic thermodynamics involved in battery operation are also discussed. The theoretical treatment concentrates on the older types of battery, such as lead–acid, where much work has been carried out over the years. The ideas are, however, in many cases equally applicable to the newer types of battery and one of the objectives of this chapter is to assist the reader in carrying out such calculations.

The following chapters discuss various aspects of primary and secondary batteries including those batteries such as silver–zinc and alkaline manganese which are available in both forms.

Chapter 2 is designed to present the reader with information on the types of batteries available and to assist him or her in choosing a type of battery which is suitable for any particular application, whether this be a digital watch or a lunar landing module.

Part 1 (Chapters 3–17) presents all available information on the performance characteristics of various types of battery and it highlights the parameters that it is important to be aware of when considering batteries. Such information is vital when discussing with battery suppliers the types and characteristics of batteries they can supply or that you may wish them to develop.

Part 2 (Chapters 18–29) is a presentation of the theory, as far as it is known, behind the working of all the types of battery now commercially available and of the limitations that battery electrochemistry might place on performance. It also discusses the ways in which the basic electrochemistry influences battery design. Whilst battery design has always been an important factor influencing performance and other factors such as battery weight it is assuming an even greater importance in more recently developed batteries.

Part 3 (Chapters 30 and 31) is a comprehensive discussion of practical methods for determining the performance characteristics of all types of battery. This is important to both the battery producer and the battery user. Important factors such as the measurement of the effect of discharge rate and temperature on available capacity and life are discussed.

Part 4 (Chapters 32–43) is a wide ranging look at the current applications of various types of battery and indicates areas of special interest such as vehicle propulsion, utilities loading and microelectronic and computer applications.

v

Part 5 (Chapters 44–49) deals with all aspects of the theory and practice of battery charging and will be of great interest to the battery user.

Finally, Part 6 (Chapters 50–63) discusses the massive amount of information available from battery manufacturers on the types and performance characteristics of the types of battery they can supply. The chapter was assembled from material kindly supplied to the author following a worldwide survey of battery producers and their products and represents a considerable body of information which has not been assembled together in this form elsewhere.

Within each Part, chapters are included on all available types of primary batteries, secondary batteries and batteries available in primary and secondary versions. The primary batteries include carbon–zinc, carbon–zinc chloride, mercury–zinc and other mercury types, manganese dioxide–magnesium perchlorate, magnesium organic, lithium types (sulphur dioxide, thionyl chloride, vanadium pentoxide, iodine and numerous other lithium types), thermally activated and sea-water batteries. Batteries available in primary and secondary forms include alkaline manganese, silver–zinc, silver–cadmium, zinc–air and cadmium–air. The secondary batteries discussed include lead–acid, the nickel types (cadmium, iron, zinc, hydrogen), zinc–chlorine, sodium–sulphur and other fast ion types.

The book will be of interest to battery manufacturers and users and the manufacturers of equipment using batteries. The latter will include manufacturers of domestic equipment, including battery-operated household appliances, power tools, TVs, radios, computers, toys, manufacturers of emergency power and lighting equipment, communications and warning beacon and life-saving equipment manufacturers. The manufacturers of medical equipment including pacemakers and other battery operated implant devices will find much to interest them, as will the manufacturers of portable medical and non-medical recording and logging equipment. There are many applications of batteries in the transport industry, including uses in conventional vehicles with internal combustion engines and in aircraft, and the newer developments in battery-operated automobiles, fork lift trucks, etc. Manufacturers and users of all types of defence equipment ranging from torpedoes to ground-to-air and air-to-air missiles rely heavily on having available batteries with suitable characteristics and will find much to interest them throughout the book; the same applies to the manufacturers of aerospace and space equipment, the latter including power and backup equipment in space vehicles and satellites, lunar vehicles, etc. Finally, there is the whole field of equipment in the new technologies including computers and electronics.

The teams of manufacturers of equipment who manufacture all these types of equipment which require batteries for their performance include the planners and designers. These must make decisions on the performance characteristics required in the battery and other relevant factors such as operating temperatures, occurrence of vibration and spin, etc., weight, volume, pre-use shelf life; these and many other factors play a part in governing the final selection of the battery type. It is a truism to say that in many cases the piece of equipment has to be designed around the battery.

Battery manufacturers will also find much to interest them, for it is they who must design and supply batteries for equipment producers and who must try to anticipate the future needs of the users, especially in the new technologies. Battery manufacturers and users alike will have an interest in charging techniques and it is hoped that Part 5 will be of interest to them. The development of new types of batteries usually demands new charger designs, as does in many instances the development of new applications for existing battery types.

Throughout the book, but particularly in Chapter 1, there is a discussion of the theory behind battery operation and this will be of interest to the more theoretically minded in the user and manufacturer industries and in the academic world. Students and postgraduates of electrical and engineering science, and design and manufacture will find much to interest them, as will members of the lay public who have an interest in power sources and technology.

Finally, it is hoped that this will become a source book for anyone interested in the above matters. This would include, among others, researchers, journalists, lecturers, writers of scientific articles, government agencies and research institutes.

Contents

Acknowledgements

Acknowledgements are hereby given to the companies listed under Suppliers at the end of the book for supplying information on their products and particularly to the following companies for permission to reproduce figures in the text.

Catalyst Research Corporation, 9.10, 24.14, 24.15, 27.1, 27.2–27.8, 27.10, 27.13–27.15, 56.21, 56.23–56.29

Chloride Batteries, 4.2, 4.3, 11.1, 18.2, 18.5–18.8, 19.8–19.10, 32.1, 32.2–32.6, 43.1–43.3, 48.1, 48.2, 49.3–49.8, 50.1, 50.2, 50.4–50.11, 50.14

Chloride Silent Power, 29.2

Crompton-Parkinson, 8.1, 8.2, 31.2, 50.16

Dryfit, 4.6–4.11, 20.2, 31.17, 31.45, 50.15

Duracell, 30.22, 30.30, 30.50, 30.53–30.55, 52.4, 56.11, 56.12, 59.1

Eagle Picher, 4.12–4.14, 10.1, 18.3, 19.12, 19.20, 20.1, 24.9, 24.10, 24.16, 25.1, 27.9, 30.5, 30.24–30.27, 30.32, 30.42, 30.52, 31.1, 31.11–31.14, 31.19, 31.29, 31.31, 31.32, 31.44, 43.4, 43.5, 45.5, 45.6, 50.17–50.19, 51.25, 51.26, 51.38, 51.42–51.44, 52.1, 52.2, 52.7, 52.11–52.13, 56.18–56.20, 56.22, 57.1–57.3, 60.1, 60.2

Energy Development Associates, 28.1

Ever Ready (Berec), 19.4, 19.5, 26.1

Ford Motors, 29.1

General Electric, 31.23, 31.24, 31.41, 31.47, 45.1, 47.1, 50.20, 51.23

W. R. Groce, 18.20–18.23

Honeywell, 9.1–9.9, 24.1–24.8, 24.11, 24.12, 30.14, 30.19–30.21, 30.33, 30.46, 56.1

Mallory, 8.3, 23.2, 23.3, 30.1, 30.10, 30.15–30.18, 30.28, 30.29, 30.31, 30.34, 30.35, 30.49, 53.1, 53.2, 55.3, 55.5, 55.6, 56.2–56.4

Marathon, 11.1, 25.2, 30.38–30.41, 30.51, 30.57, 31.25, 57.4, 57.5

McGraw Edison, 30.43–30.45, 59.2

Nife Jungner, 31.40, 31.48, 33.1, 47.7, 47.11, 47.15, 51.20–51.22, 51.30–51.32

SAFT, 4.5, 30.23, 30.56, 31.22, 31.26–31.28, 31.35, 31.40, 47.8–47.10, 47.12, 47.13, 51.1–51.3, 56.7–56.10, 56.13, 56.17, 59.3–59.8, 63.1–63.3

Silberkraft FRIWO, 56.5, 56.6

Swiss Post Office, Berne, 18.9–18.19

Union Carbide, 5.1, 5.2, 6.1–6.5, 7.1, 8.1, 19.6, 19.7, 19.11, 21.1, 21.2, 22.1–22.3, 23.1, 30.2–30.4, 30.6–30.9, 30.36, 30.37, 30.47, 31.4, 31.20, 31.21, 31.30, 31.33, 45.3, 46.1–46.5, 47.4–47.7, 47.17, 51.10–51.19, 52.3, 53.3–53.7, 55.1, 55.2

Varley, 31.16, 31.34, 50.21

Varta, 4.1, 4.4, 19.1–19.3, 31.5–31.10, 31.38, 31.39, 31.49, 40.1, 40.2, 47.3, 47.16, 50.12, 50.13, 51.4–51.9, 51.34–51.37, 56.14–56.16

Vidor, 30.11–30.13, 55.4

Yardney, 20.3, 31.42, 31.43, 33.2–33.5, 47.14, 51.39–51.41, 52.8–52.10

Yuasa, 18.4, 31.3, 31.18, 31.36, 31.37, 31.46, 31.50, 31.51, 45.2, 45.4, 51.27–51.29, 52.5, 52.6, 54.1

1

Introduction to battery technology

Contents

1.1 Electromotive force

A galvanic or voltaic cell consists of two dissimilar electrodes immersed in a conducting material such as a liquid electrolyte or a fused salt; when the two electrodes are connected by a wire a current will flow. Each electrode, in general, involves an electronic (metallic) and an ionic conductor in contact. At the surface of separation between the metal and the solution there exists a difference in electrical potential, called the electrode potential. The electromotive force (e.m.f.) of the cell is then equal to the algebraic sum of the two electrode potentials, appropriate allowance being made for the sign of each potential difference as follows. When a metal is placed in a liquid, there is, in general, a potential difference established between the metal and the solution owing to the metal yielding ions to the solution or the solution yielding ions to the metal. In the former case, the metal will become negatively charged to the solution; in the latter case, the metal will become positively charged.

Since the total e.m.f. of a cell is (or can in many cases be made practically) equal to the algebraic sum of the potential differences at the two electrodes, it follows that, if the e.m.f. of a given cell and the value of the potential difference at one of the electrodes are known, the potential difference at the other electrode can be calculated. For this purpose, use can be made of the standard calomel electrode, which is combined with the electrode and solution between which one wishes to determine the potential difference.

In the case of any particular combination, such as the following:

Zn/N ZnSO$_4$/Hg$_2$Cl$_2$ in N KCl/Hg

the positive pole of the cell can always be ascertained by the way in which the cell must be inserted in the side circuit of a slide wire potentiometer in order to obtain a point of balance on the bridge wire. To obtain a point of balance, the cell must be opposed to the working cell; and therefore, if the positive pole of the latter is connected with a particular end of the bridge wire, it follows that the positive pole of the cell in the side circuit must also be connected with the same end of the wire.

The e.m.f. of the above cell at 18°C is 1.082 V and, from the way in which the cell has to be connected to the bridge wire, mercury is found to be the positive pole; hence, the current must flow in the cell from zinc to mercury. An arrow is therefore drawn under the diagram of the cell to show the direction of the current, and beside it is placed the value of the e.m.f., thus:

Zn/N ZnSO$_4$/Hg$_2$Cl$_2$ in N KCl/Hg

$$\longrightarrow$$
1.082

It is also known that the mercury is positive to the solution of calomel, so that the potential here tends to produce a current from the solution to the mercury. This is represented by another arrow, beside which is placed the potential difference between the electrode and the solution, thus:

Zn/N ZnSO$_4$/Hg$_2$Cl$_2$ in N KCl/Hg

$$\longrightarrow$$
0.281

$$\longrightarrow$$
1.082

Since the total e.m.f. of the cell is 1.082 V, and since the potential of the calomel electrode is 0.281 V, it follows that the potential difference between the zinc and the solution of zinc sulphate must be 0.801 V, referred to the normal hydrogen electrode, and this must also assist the potential difference at the mercury electrode. Thus:

Zn/N ZnSO$_4$/Hg$_2$Cl$_2$ in N KCl/Hg

$\longrightarrow$ $\longrightarrow$
0.801   0.281

$$\longrightarrow$$
1.082

From the diagram it is seen that there is a tendency for positive electricity to pass from the zinc to the solution, i.e. the zinc gives positive ions to the solution, and must, therefore, itself become negatively charged relative to the solution. The potential difference between zinc and the normal solution of zinc sulphate is therefore −0.801 V. By adopting the above method, errors both in the sign and in the value of the potential difference can be easily avoided.

If a piece of copper and a piece of zinc are placed in an acid solution of copper sulphate, it is found, by connecting the two pieces of metal to an electrometer, that the copper is at a higher electrical potential (i.e. is more positive) than the zinc. Consequently, if the copper and zinc are connected by a wire, positive electricity flows from the former to the latter. At the same time, a chemical reaction goes on. The zinc dissolves forming a zinc salt, while copper is deposited from the solution on to the copper.

Zn + CuSO$_4$(aq.) = ZnSO$_4$(aq.) + Cu

This is the principle behind many types of electrical cell.

Faraday's Law of Electrochemical Equivalents holds for galvanic action and for electrolytic decomposition. Thus, in an electrical cell, provided that secondary reactions are excluded or allowed for, the current of chemical action is proportional to the quantity of electricity produced. Also, the amounts of different substances liberated or dissolved by the same amount of electricity are proportional to their chemical equivalents. The quantity of electricity required to produce one equivalent of chemical action (i.e. a quantity of chemical action equivalent to the liberation of 1 g of hydrogen from an acid) is known as the faraday (F). One faraday is equivalent to 96 494 ampere seconds or coulombs. The reaction quoted above involving the passage into solution of one equivalent of zinc and the deposition of one equivalent of copper is therefore accompanied by the production of 2 F (192 988 C), since the atomic weights of zinc and copper both contain two equivalents.

1.1.1 Measurement of the electromotive force

The electromotive force of a cell is defined as the potential difference between the poles when no current is flowing through the cell. When a current is flowing through a cell and through an external circuit, there is a fall of potential inside the cell owing to its internal resistance, and the fall of potential in the outside circuit is less than the potential difference between the poles at open circuit.

In fact if R is the resistance of the outside circuit, r the internal resistance of the cell and E its electromotive force, the current through the circuit is:

$$C = \frac{E}{R + r} \tag{1.1}$$

The potential difference between the poles is now only $E' = CR$, so that

$$E'/E = R/R + r$$

The electromotive force of a cell is usually measured by the compensation method, i.e. by balancing it against a known fall of potential between two points of an auxiliary circuit. If AB (Figure 1.1) is a uniform wire connected at its ends with a cell M, we may find a point X at which the fall of potential from A to X balances the electromotive force of the cell N. Then there is no current through the loop ANX, because the potential difference between the points A and X, tending to cause a flow of electricity in the direction ANX, is just balanced by the electromotive force of N which acts in the opposite direction. The point of balance is observed

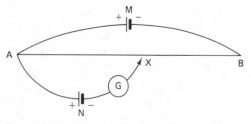

Figure 1.1 The Poggendorf method of determining electromotive force

by a galvanometer G, which indicates when no current is passing through ANX. By means of such an arrangement we may compare the electromotive force E of the cell N with a known electromotive force E' of a standard cell N'; if X' is the point of balance of the latter, we have:

$$\frac{AX}{AX'} = \frac{E}{E'} \tag{1.2}$$

1.1.2 Origin of electromotive force

It is opportune at this point to consider why it comes about that certain reactions, when conducted in galvanic cells, give rise to an electrical current. Many theories have been advanced to account for this phenomenon. Thus, in 1801, Volta discovered that if two insulated pieces of different metals are put in contact and then separated they acquire electric charges of opposite sign. If the metals are zinc and copper, the zinc acquires a positive charge and the copper a negative charge. There is therefore a tendency for negative electricity to pass from the zinc to the copper. Volta believed that this tendency was mainly responsible for the production of the current in the galvanic cell. The solution served merely to separate the two metals and so eliminate the contact effect at the other end.

It soon became evident that the production of the current was intimately connected with the chemical actions occurring at the electrodes, and a 'chemical theory' was formulated, according to which the electrode processes were mainly responsible for the production of the current. Thus there arose a controversy which lasted, on and off, for a century.

On the one hand the chemical theory was strengthened by Faraday's discovery of the equivalence of the current produced to the amount of chemical action in the cell and also by the discovery of the relation between the electrical energy produced and the energy change in the chemical reaction stated incompletely by Kelvin in 1851 and correctly by Helmholtz in 1882. Nernst's theory of the metal electrode process (1889) also added weight to the chemical theory.

On the other hand, the 'metal contact' theorists showed that potential differences of the same order of magnitude as the electromotive forces of the cells occur at the metal junctions. However, they fought a losing battle against steadily accumulating evidence on the 'chemical' side. The advocates of the chemical theory ascribed these large contact potential differences to the chemical action of the gas atmosphere at the metal junction at the moment of separating the metals. They pointed out that no change occurred at the metal junction which could provide the electrical energy produced. Consequently, for 20 years after 1800 little was heard of the metal junction as an important factor in the galvanic cell. Then (1912–1916) it was conclusively demonstrated by Richardson, Compton and Millikan, in their studies on photoelectric and thermionic phenomena, that considerable potential differences do occur at the junction of dissimilar metals. Butler, in 1924, appears to have been the first to show how the existence of a large metal junction potential difference can be completely reconciled with the chemical aspect.

Nernst's theory of the electrode process

In the case of a metal dipping into a solution of one of its salts, the only equilibrium that is possible is that of metal ions between the two phases. The solubility of the metal, as neutral metal atoms, is negligibly small. In the solution the salt is dissociated into positive ions of the metal and negative anions, e.g.

$$CuSO_4 = Cu^{2+} + SO_4^{2-}$$

and the electrical conductivity of metals shows that they are dissociated, at any rate to some extent, into metal ions and free electrons, thus:

$$Cu = Cu^{2+} + 2e$$

The positive metal ions are thus the only constituent of the system that is common to the two phases. The equilibrium of a metal and its salt solution therefore differs from an ordinary case of solubility in that only one constituent of the metal, the metal ions, can pass into solution.

Nernst, in 1889, supposed that the tendency of a substance to go into solution was measured by its solution pressure and its tendency to deposit from the solution by its osmotic pressure in the solution. Equilibrium was supposed to be reached when these opposing tendencies balanced each other, i.e. when the osmotic pressure in the solution was equal to the solution pressure.

In the case of a metal dipping into a solution containing its ions, the tendency of the metal ions to dissolve is thus determined by their solution pressure, which Nernst called the electrolytic solution pressure, P, of the metal. The tendency of the metal ions to deposit is measured by their osmotic pressure, p.

Consider what will happen when a metal is put in contact with a solution. The following cases may be distinguished:

1. $P > p$ The electrolytic solution pressure of the metal is greater than the osmotic pressure of the ions, so that positive metal ions will pass into the solution. As a result the metal is left with a negative charge, while the solution becomes positively charged. There is thus set up across the interface an electric field which attracts positive ions towards the metal and tends to prevent any

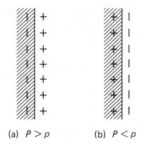

(a) $P > p$ (b) $P < p$

Figure 1.2 The origin of electrode potential difference

more passing into solution (Figure 1.2(a)). The ions will continue to dissolve and therefore the electric field to increase in intensity until equilibrium is reached, i.e. until the inequality of P and p, which causes the solution to occur, is balanced by the electric field.

2. $P < p$ The osmotic pressure of the ions is now greater than the electrolytic solution pressure of the metal, so that the ions will be deposited on the surface of the latter. This gives the metal a positive charge, while the solution is left with a negative charge. The electric field so arising hinders the deposition of ions, and it will increase in intensity until it balances the inequality of P and p, which is the cause of the deposition (Figure 1.2(b)).

3. $P = p$ The osmotic pressure of the ions is equal to the electrolytic solution pressure of the metal. The metal and the solution will be in equilibrium and no electric field will arise at the interface.

When a metal and its solution are not initially in equilibrium, there is thus formed at the interface an electrical double layer, consisting of the charge on the surface of the metal and an equal charge of opposite sign facing it in the solution. By virtue of this double layer there is a difference of potential between the metal and the solution. The potential difference is measured by the amount of work done in taking unit positive charge from a point in the interior of the liquid to a point inside the metal. It should be observed that the passage of a very minute quantity of ions in the solution or vice versa is sufficient to give rise to the equilibrium potential difference.

Nernst calculated the potential difference required to bring about equilibrium between the metal and the solution in the following way. He determined the net work obtainable by the solution of metal ions by means of a three-stage expansion process in which the metal ions were withdrawn from the metal at the electrolyte solution pressure P, expanded isothermally to the osmotic pressure p, and condensed at this pressure into the solution. The net work obtained in this process is

$$w' = RT \ln P/p \text{ per mol} \tag{1.3}$$

If V is the electrical potential of the metal with respect to the solution (V being positive when the metal is positive), the electrical work obtained when 1 mol of metal ions passes into solution is nVF, where n is the number of unit charges carried by each ion. The total amount of work obtained in the passage of 1 mol of ions into solution is thus

$$RT \ln P/p + nVF \qquad (1.4)$$

and for equilibrium this must be zero; hence

$$V = \frac{RT}{nF} \ln P/p \qquad (1.5)$$

Objection can be made to this calculation on the grounds that the three-stage process employed does not correspond to anything that can really occur and is really analogous in form only to the common three-stage transfer. However, a similar relation to which this objection does not apply has been obtained by thermodynamic processes.

In an alternative approach to the calculation of electrode potentials and of potential differences in cells, based on concentrations, it is supposed that two pieces of the same metal are dipping into solutions in which the metal ion concentrations are m_1 and m_2 respectively (Figure 1.3).

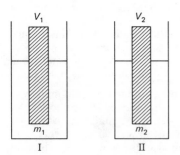

Figure 1.3 Calculation of electrode potential and potential difference

Let the equilibrium potential differences between the metal and the solutions be V_1 and V_2. Suppose that the two solutions are at zero potential, so that the electrical potentials of the two pieces of metal are V_1 and V_2.

We may now carry out the following process:

1. Cause one gram-atom of silver ions to pass into the solution from metal 1. Since the equilibrium potential is established at the surface of the metal, the net work of this change is zero.
2. Transfer the same amount (1 mol) of silver ions reversibly from solution 1 to solution 2. The net work obtained is

$$w' = RT \ln m_1/m_2 \qquad (1.6)$$

provided that Henry's law is obeyed.

3. Cause the gram-atom of silver ions to deposit on electrode 2. Since the equilibrium potential is established, the net work of this change is zero.
4. Finally, to complete the process, transfer the equivalent quantity of electrons (charge nF) from electrode 1 to electrode 2. The electrical work obtained in the transfer of charge $-nF$ from potential V_1 to potential V_2 (i.e. potential difference = $V_1 - V_2$), for metal ions of valency n when each gram-atom is associated with nF units of electricity, is

$$-nF(V_1 - V_2) \qquad (1.7)$$

The system is now in the same state as at the beginning (a certain amount of metallic silver has been moved from electrode 1 to electrode 2, but a change of position is immaterial).

The total work obtained in the process is therefore zero, i.e.

$$-AF(V_1 - V_2) + RT \ln (m_1/m_2) = 0 \qquad (1.8)$$

or the potential difference is

$$V = V_1 - V_2 = \frac{RT}{nF} \ln \left(\frac{m_1}{m_2}\right) \qquad (1.9)$$

i.e.

$$V_1 = \frac{RT}{nF} \ln m_1 \quad \text{and} \quad V_2 = \frac{RT}{nF} \ln m_2 \qquad (1.10)$$

i.e.

$$V = \frac{RT}{nF} \ln \frac{m_1}{m_2} \qquad (1.11)$$

Inserting values for R, T (25°C) and F and converting from napierian to ordinary logarithms,

$$V = \frac{2.303 \times 1.988 \times 298.1 \times 4.182}{n \times 96\ 490} \log \left(\frac{m_1}{m_2}\right)$$

$$= \frac{0.0591}{n} \log \left(\frac{m_1}{m_2}\right) \qquad (1.12)$$

From Equation 1.5 the electrical potential (V) of a metal with respect to the solution is given by

$$V = \frac{-RT}{nF} \ln \left(\frac{P}{p}\right) \qquad (1.5)$$

where P is the electrolytic solution pressure of the metal and p is the osmotic pressure of metal ions. For two different metal solution systems, 1 and 2, the electrical potentials V_1 and V_2 are given by

$$V_1 = \frac{-RT}{nF} \ln \left(\frac{P_1}{p}\right)$$

$$V_2 = \frac{-RT}{nF} \ln \left(\frac{P_2}{p}\right)$$

Therefore

$$V_1 - V_2 = \text{potential difference } (V)$$

$$= \frac{RT}{nF} \ln \left(\frac{P_1}{P_2}\right)$$

$$= \frac{0.059}{n} \log \left(\frac{P_1}{P_2}\right) \quad \text{at } 25°C \qquad (1.13)$$

Comparing Equations 1.12 and 1.13 it is seen that, as would be expected, $m_1 \propto P_1$ and $m_2 \propto P_2$, i.e. the concentrations of metal ions in solution (m) are directly proportional to the electolytic solution pressures of the metal (P).

Kinetic theories of the electrode process

A more definite physical picture of the process at a metal electrode was given by Butler in 1924. According to current physical theories of the nature of metals, the valency electrons of a metal have considerable freedom of movement. The metal may be supposed to consist of a lattice structure of metal ions, together with free electrons either moving haphazardly among them or arranged in an inter-penetrating lattice. An ion in the surface layer of the metal is held in its position by the cohesive forces of the metal, and before it can escape from the surface it must perform work in overcoming these forces. Owing to their thermal agitation the surface ions are vibrating about their equilibrium positions, and occasionally an ion' will receive sufficient energy to enable it to overcome the cohesive forces entirely and escape from the metal. On the other hand, the ions in the solution are held to the adjacent water molecules by the forces of hydration and, in order that an ion may escape from its hydration sheath and become deposited on the metal, it must have sufficient energy to overcome the forces of hydration.

Figure 1.4 is a diagrammatic representation of the potential energy of an ion at various distances from the surface of the metal. (This is not the electrical potential, but the potential energy of an ion due to

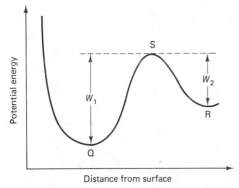

Distance from surface

Figure 1.4 Potential energy of an ion at various distances from the surface of a metal

the forces mentioned above.) The equilibrium position of an ion in the surface layer of the metal is represented by the position of minimum energy, Q. As the ion is displaced towards the solution it does work against the cohesive forces of the metal and its potential energy rises while it loses kinetic energy. When it reaches the point S it comes within the range of the attractive forces of the solution. Thus all ions having sufficient kinetic energy to reach the point S will escape into the solution. If W_1 is the work done in reaching the point S, it is easily seen that only ions with kinetic energy W_1 can escape. The rate at which ions acquire this quantity of energy in the course of thermal agitation is given by classical kinetic theory as $\theta_1 = k' \exp(-W_1/kT)$, and this represents the rate of solution of metal ions at an uncharged surface.

In the same way, R represents the equilibrium position of a hydrated ion. Before it can escape from the hydration sheath the ion must have sufficient kinetic energy to reach the point S, at which it comes into the region of the attractive forces of the metal.

If W_2 is the difference between the potential energy of an ion at R and at S, it follows that only those ions that have kinetic energy greater than W_2 can escape from their hydration sheaths. The rate of deposition will thus be proportional to their concentration (i.e. to the number near the metal) and to the rate at which these acquire sufficient kinetic energy. The rate of deposition can thus be expressed as $\theta_2 = k''c \exp(-W_2/kT)$.

θ_1 and θ_2 are not necessarily equal. If they are unequal, a deposition or solution of ions will take place and an electrical potential difference between the metal and the solution will be set up, as in Nernst's theory. The quantities of work done by an ion in passing from Q to S or R to S are now increased by the work done on account of the electrical forces. If V' is the electrical potential difference between Q and S, and V'' that between S and R, so that the total electrical potential difference between Q and R is $V = V' + V''$, the total work done by an ion in passing from Q to S is $W_1 - neV'$ and the total work done by an ion in passing from R to S is $W_2 + neV''$, where n is the valency of the ion and e the unit electronic charge. V' is the work done by unit charge in passing from S to Q and V'' that done by unit charge in passing from R to S. The rates of solution and deposition are thus:

$$\theta_1 = k' \exp \left[-(W_1 - nV')/kT\right]$$
$$\theta_2 = k''c \exp \left[-(W_2 + nV'')/kT\right]$$

For equilibrium these must be equal, i.e.

$$k' \exp \left[-(W_1 - nV')/kT\right] = k''c \exp \left[-(W_2 + nV'')/kT\right]$$

or

$$V' + V'' = \frac{W_1 - W_2}{ne} + \frac{kT}{ne} \ln c + \frac{kT}{ne} \ln \left(\frac{k''}{k'}\right)$$

If N_0 is the number of molecules in the gram-molecule, we may write:

$$N_0 (W_1 - W_2) = \Delta E$$
$$N_0 e = F$$
$$N_0 k = R$$

and we then have

$$V = \frac{\Delta E}{nF} + \frac{RT}{nF} \ln c + \frac{RT}{nF} \ln \left(\frac{k''}{k'} \right)$$

The final term contains some statistical constants which are not precisely evaluated, but it is evident that, apart from this, V depends mainly on ΔE, the difference of energy of the ions in the solution and in the metal.

Comparing this with the Nernst expression we see that the solution pressure P is

$$\ln P = \frac{\Delta E}{RT} + \ln \left(\frac{k''}{k'} \right) \qquad (1.14)$$

One of the difficulties of Nernst's theory was that the values of P required to account for the observed potential differences varied from enormously great to almost infinitely small values, to which it was difficult to ascribe any real physical meaning. This difficulty disappears when it is seen that P does not merely represent a concentration difference, but includes a term representing the difference of energy of the ions in the two phases, which may be large.

The electrode process has also been investigated using the methods of quantum mechanics. The final equations obtained are very similar to those given above.

Work function at the metal–metal junction

When two dissimilar metals are put in contact there is a tendency for negative electricity, i.e. electrons, to pass from one to the other. Metals have different affinities for electrons. Consequently, at the point of junction, electrons will tend to pass from the metal with the smaller to that with the greater affinity for electrons. The metal with the greater affinity for electrons will become negatively charged and that with the lesser affinity will become positively charged. A potential difference is set up at the interface which increases until it balances the tendency of electrons to pass from the one metal to the other. At this junction, as at the electrodes, the equilibrium potential difference is that which balances the tendency of the charged particle to move across the interface.

By measurements of the photoelectric and thermionic effects, it has been found possible to measure the amount of energy required to remove electrons from a metal. This quantity is known as its thermionic work function and is usually expressed in volts, as the potential difference through which the

electrons would have to pass in order to acquire as much energy as is required to remove them from the metal. Thus, if ϕ is the thermionic work function of a metal, the energy required to remove one electron from the metal is $e\phi$, where e is the electronic charge. The energy required to remove one equivalent of electrons (charge F) is thus ϕF or $96\,500\phi/4.182$ cal. The thermionic work functions of a number of metals are given in Table 1.1.

Table 1.1 The thermionic work functions of the metals

Metal	Thermionic work function (V)
Potassium	2.12
Sodium	2.20
Lithium	2.28
Calcium	3.20
Magnesium	3.68
Aluminium	4.1
Zinc	3.57
Lead	3.95
Cadmium	3.68
Iron	4.7
Tin	4.38
Copper	4.16
Silver	4.68
Platinum	6.45

The energy required to transfer an equivalent of electrons from one metal to another is evidently given by the difference between their thermionic work functions. Thus, if ϕ_1 is the thermionic work function of metal 1 and ϕ_2 that of metal 2, the energy required to transfer electrons from 1 to 2 per equivalent is

$$\Delta E = (\phi_1 - \phi_2) F \qquad (1.15)$$

The greater the thermionic work function of a metal, the greater is the affinity for electrons. Thus electrons tend to move from one metal to another in the direction in which energy is liberated. This tendency is balanced by the setting up of a potential difference at the junction. When a current flows across a metal junction, the energy required to carry the electrons over the potential difference is provided by the energy liberated in the transfer of electrons from the one metal to the other. The old difficulty that no apparent change occurred at the metal junction which could contribute to the electromotive force of a cell thus disappears.

It should be noted that the thermionic work function is really an energy change and not a reversible work quantity and is not therefore a precise measure of the affinity of a metal for electrons. When an electric current flows across a junction the difference between the energy liberated in the transfer of electrons and the electric work done in passing through the potential difference appears as heat liberated at the junction. This heat is a relatively small quantity, and the junction potential difference can be taken as approximately equal to the difference between the thermionic work functions of the metals.

Taking into account the above theory, it is now possible to view the working of a cell comprising two dissimilar metals such as zinc and copper immersed in an electrolyte. At the zinc electrode, zinc ions pass into solution leaving the equivalent charge of electrons in the metal. At the copper electrode, copper ions are deposited. In order to complete the reaction we have to transfer electrons from the zinc to the copper, through the external circuit. The external circuit is thus reduced to its simplest form if the zinc and copper are extended to meet at the metal junction. The reaction

$$Zn + Cu^{2+}(aq.) = Zn^{2+}(aq.) + Cu$$

occurs in parts, at the various junctions:

1. Zinc electrode:
   $$Zn = Zn^{2+}(aq.) + 2e(Zn)$$
2. Metal junction:
   $$2e(Zn) = 2e(Cu)$$
3. Copper electrode:
   $$Cu^{2+}(aq.) + 2e(Cu) = Cu$$

If the circuit is open, at each junction a potential difference arises which just balances the tendency for that particular process to occur. When the circuit is closed there is an electromotive force in it equal to the sum of all the potential differences. Since each potential difference corresponds to the net work of one part of the reaction, the whole electromotive force is equivalent to the net work or free energy decrease of the whole reaction.

1.2 Reversible cells

During the operation of a galvanic cell a chemical reaction occurs at each electrode, and it is the energy of these reactions that provides the electrical energy of the cell. If there is an overall chemical reaction, the cell is referred to as a chemical cell. In some cells, however, there is no resultant chemical reaction, but there is a change in energy due to the transfer of solute from one concentration to another; such cells are called 'concentration cells'. Most, if not all, practical commercial batteries are chemical cells.

In order that the electrical energy produced by a galvanic cell may be related thermodynamically to the process occurring in the cell, it is essential that the latter should behave reversibly in the thermodynamic sense. A reversible cell must satisfy the following conditions. If the cell is connected to an external source of e.m.f. which is adjusted so as exactly to balance the e.m.f. of the cell, i.e. so that no current flows, there should be no chemical or other change in the cell. If the external e.m.f. is decreased by an infinitesimally small amount, current will flow from the cell, and a chemical or other change, proportional in extent to the quantity of electricity passing, should take place. On the other hand, if the external e.m.f. is increased by a very small amount, the current should pass in the opposite direction, and the process occurring in the cell should be exactly reversed.

It may be noted that galvanic cells can only be expected to behave reversibly in the thermodynamic sense, when the currents passing are infinitesimally small, so that the system is always virtually in equilibrium. If large currents flow, concentration gradients arise within the cell because diffusion is relatively slow; in these circumstances the cell cannot be regarded as existing in a state of equilibrium. This would apply to most practical battery applications where the currents drawn from the cell would be more than infinitesimal. Of course, with a given type of cell, as the current drawn is increased the departure from the equilibrium increases also. Similar comments apply during the charging of a battery where current is supplied and the cell is not operating under perfectly reversible conditions.

If this charging current is more than infinitesimally small, there is a departure from the equilibrium state and the cell is not operating perfectly reversibly in the thermodynamic sense. When measuring the e.m.f. of a cell, if the true thermodynamic e.m.f. is required, it is necessary to use a type of measuring equipment that draws a zero or infinitesimally small current from the cell at the point of balance. The e.m.f. obtained in this way is as close to the reversible value as is experimentally possible. If an attempt is made to determine the e.m.f. with an ordinary voltmeter, which takes an appreciable current, the result will be in error.

In practical battery situations, the e.m.f. obtained is not the thermodynamic value that would be obtained for a perfectly reversible cell but a non-equilibrium value which for most purposes suffices and in many instances is, in fact, close to the value that would have been obtained under equilibrium conditions.

One consequence of drawing a current from a cell which is more than infinitesimally small is that the current obtained would not be steady but would decrease with time. The cell gives a steady current only if the current is very low or if the cell is in action only intermittently. The explanation of this effect, which is termed 'polarization', is simply that some of the hydrogen bubbles produced by electrolysis at the metal cathode adhere to this electrode. This results in a two-fold action. First, the hydrogen is an excellent insulator and introduces an internal layer of very high electrical resistance. Secondly, owing to the electric field present, a double layer of positive and negative ions forms on the surface of the hydrogen and the cell actually tries to send a current in the reverse direction or a back e.m.f. develops. Clearly, the two opposing forces eventually balance and the current falls to zero. These consequences of gas production at the electrodes are avoided, or at least considerably reduced, in practical batteries by placing between the positive and negative electrodes a suitable inert separator material. The separators perform the additional and, in many cases, more important function of preventing short-circuits between adjacent plates.

A simple example of a primary (non-rechargeable) reversible cell is the Daniell cell, consisting of a zinc electrode immersed in an aqueous solution of zinc sulphate, and a copper electrode in copper sulphate solution:

$$Zn \mid ZnSO_4(soln) \mid CuSO_4(soln) \mid Cu$$

the two solutions being usually separated by a porous partition. Provided there is no spontaneous diffusion through this partition, and the electrodes are not attacked by the solutions when the external circuit is open, this cell behaves in a reversible manner. If the external circuit is closed by an e.m.f. just less than that of the Daniell cell, the chemical reaction taking place in the cell is

$$Zn + Cu^{2+} = Zn^{2+} + Cu$$

i.e. zinc dissolves from the zinc electrode to form zinc ions in solution, while copper ions are discharged and deposit copper on the other electrode. Polarization is prevented. On the other hand, if the external e.m.f. is slightly greater than that of the cell, the reverse process occurs; the copper electrode dissolves while metallic zinc is deposited on the zinc electrode.

A further example of a primary cell is the well known Leclanché carbon–zinc cell. This consists of a zinc rod anode dipping into ammonium chloride paste outside a linen bag inside which is a carbon rod cathode surrounded by solid powdered manganese dioxide which acts as a chemical depolarizer.

The equation expressing the cell reaction is as follows:

$$2MnO_2 + 2NH_4Cl + Zn \rightarrow 2MnOOH + Zn(NH_3)_2Cl_2$$

The e.m.f. is about 1.4 V. Owing to the fairly slow action of the solid depolarizer, the cell is only suitable for supplying small or intermittent currents.

The two cells described above are primary (non-rechargeable) cells, that is, cells in which the negative electrode is dissolved away irreversibly as time goes on. Such cells, therefore, would require replacement of the negative electrode, the electro-

lyte and the depolarizer before they could be re-used. Secondary (rechargeable) cells are those in which the electrodes may be re-formed by electrolysis, so that, effectively, the cell gives current in one direction when in use (discharging) and is then subjected to electrolysis (recharging) by a current from an external power source passing in the opposite direction until the electrodes have been completely re-formed. A well known secondary cell is the lead–acid battery, which consists of electrodes of lead and lead dioxide, dipping in dilute sulphuric acid electrolyte and separated by an inert porous material. The lead dioxide electrode is at a steady potential of about 2 V above that of the lead electrode. The chemical processes which occur on discharge are shown by the following equations:

1. Negative plate:
   $$Pb + SO_4^{2-} \rightarrow PbSO_4 + 2e$$
2. Positive plate:
   $$PbO_2 + Pb + 2H_2SO_4 + 2e \rightarrow 2PbSO_4 + 2H_2O$$

or for the whole reaction on discharge:

$$PbO_2 + Pb + H_2SO_4 \rightarrow 2PbSO_4 + 2H_2O$$

The discharging process, therefore, results in the formation of two electrodes each covered with lead sulphate, and therefore showing a minimum difference in potential when the process is complete, i.e. when the cell is fully discharged. In practice, the discharged negative plate is covered with lead sulphate and the positive plate with compounds such as $PbO.PbSO_4$.

In the charging process, current is passed through the cell in such a direction that the original lead electrode is reconverted into lead according to the equation:

$$PbSO_4 + 2H^+ + 2e^- \rightarrow H_2SO_4 + Pb$$

while the lead peroxide is re-formed according to the equation:

$$PbSO_4 + 2H_2O \rightarrow PbO_2 + H_2SO_4 + 2e^- + 2H^+$$

Overall, the charge cell reaction is:

$$2PbSO_4 + 2H_2O \rightarrow Pb + PbO_2 + 2H_2SO_4$$

It is clear from the above equations that in the discharging process water is formed, so that the relative density of the acid solution drops steadily. Conversely, in the charging process the acid concentration increases. Indeed, the state of charge of an accumulator is estimated from the density of the electrolyte, which varies from about 1.15 when completely discharged to 1.21 when fully charged. Throughout all these processes the e.m.f. remains approximately constant at 2.1 V and is therefore useless as a sign of the degree of charge in the battery.

The electromotive force mentioned above is that of the charged accumulator at open circuit. During

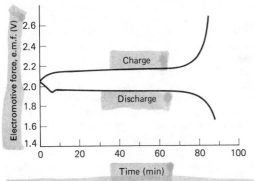

Figure 1.5 Charge and discharge curves for a lead–acid battery

the passage of current, polarization effects occur, as discussed earlier, which cause variations of the voltage during charge and discharge. Figure 1.5 shows typical charge and discharge curves. During the charge the electromotive force rises rapidly to a little over 2.1 V and remains steady, increasing very slowly as the charging proceeds. At 2.2 V oxygen begins to be liberated at the positive plates and at 2.3 V hydrogen at the negative plates. The charge is now completed and the further passage of current leads to the free evolution of gases and a rapid rise in the electromotive force. If the charge is stopped at any point the electromotive force returns, in time, to the equilibrium value. During discharge it drops rapidly to just below 2 V. The preliminary 'kink' in the curve is due to the formation of a layer of lead sulphate of high resistance while the cell is standing, which is soon dispersed. The electromotive force falls steadily during cell discharge; when it has reached 1.8 V the cell should be recharged, as the further withdrawal of current causes the voltage to fall rapidly.

The difference between the charge and discharge curves is due to changes of concentration of the acid in contact with the active materials of the plates. These are full of small pores in which diffusion is very slow, so that the concentration of the acid is greater during the charge and less during the discharge than in the bulk of the solution. This difference results in a loss of efficiency.

The current efficiency of the lead accumulator, i.e.

$$\text{Current efficiency} = \frac{\text{Amount of current taken out during discharge}}{\text{Amount of current put in during charge}}$$

is high, about 94–96%, but the charging process takes place at a higher electromotive force than the discharge, so that more energy is required for the former.

The energy efficiency measured by

$$\frac{\text{Energy obtained in discharge}}{\text{Energy required to charge}} = \frac{\sum (\text{Discharge voltage} \times \text{Quantity of electricity})}{\sum (\text{Charge voltage} \times \text{Quantity of electricity})}$$

is comparatively low, at 75–85%.

A further example of a rechargeable battery is the nickel–iron cell. In the discharged state the negative plate of this cell is iron with hydrated ferrous oxide, and the positive plate is nickel with hydrated nickel oxide. When charged, the ferrous oxide is reduced to iron, and the nickel oxide is oxidized to a hydrated peroxide. The cell reaction may thus be represented by

$$FeO + 2NiO \underset{\text{discharge}}{\overset{\text{charge}}{\rightleftharpoons}} Fe + Ni_2O_3$$

The three oxides are all hydrated to various extents, but their exact compositions are unknown. In order to obtain plates having a sufficiently large capacity, the oxides have to be prepared by methods which give particularly finely divided and active products. They are packed into nickel-plated steel containers, perforated by numerous small holes – an arrangement which gives exceptional mechanical strength. The electrolyte is usually a 21% solution of potash, but since hydroxyl ions do not enter into the cell reaction the electromotive force (1.33–1.35 V) is nearly independent of the concentration. Actually, there is a difference between the amount of water combined with the oxides in the charged and discharged plates. Water is taken up and the alkali becomes more concentrated during the discharge, but water is given out during the charge. The electromotive force therefore depends to a small extent on the free energy of water in the solution, which in turn is determined by the concentration of the dissolved potash. Actually 2.9 mol of water are liberated in the discharge reaction, as represented above, and the variation of the electromotive force between 1.0_N and 5.3_N potash is from 1.351 to 1.335 V. The potential of the positive plate is +0.55 and that of the negative plate −0.8 on the hydrogen scale.

The current efficiency, about 82%, is considerably lower than that of the lead accumulator. The voltage during the charge is about 1.65 V, rising at the end to 1.8 V, whereas during the discharge it falls gradually from 1.3 to 1.1 V. Hence the energy efficiency is only about 60%.

1.3 Reversible electrodes

The electrodes constituting a reversible cell are reversible electrodes, and three chief types of such electrodes are known. The combination of any two reversible electrodes gives a reversible cell.

The first type of reversible electrode involves a metal (or a non-metal) in contact with a solution of its own ions, e.g. zinc in zinc sulphate solution, or copper in copper sulphate solution, as in the Daniell cell. Electrodes of the first kind are reversible with respect to the ions of the electrode material, e.g. metal or non-metal; if the electrode material is a univalent metal or hydrogen, represented by M, the reaction which takes place at such an electrode, when the cell of which it is part operates, is

$$M \rightleftharpoons M^+ + e$$

where e indicates an electron, and M^+ implies a hydrated (or solvated) ion in solution. The direction of the reaction depends on the direction of flow of current through the cell. If the electrode material is a univalent non-metal A, the ions are negative and the corresponding reaction is

$$A^- \rightleftharpoons A + e$$

As will be seen later, the potentials of these electrodes depend on the concentration (or activity) of the reversible ions in the solution.

Electrodes of the second type involve a metal and a sparingly soluble salt of this metal in contact with a solution of a soluble salt of the same anion:

$$M \mid MX(s) \, HX(soln)$$

The electrode reaction in this case may be written as

$$M(s) + X^- \rightleftharpoons MX(s) + e$$

the ion X being that in the solution of the soluble acid, e.g. HX. These electrodes behave as if they were reversible with respect to the common anion (the ion X in this case).

Electrodes of the second type have been made with various insoluble halides (silver chloride, silver bromide, silver iodide and mercurous chloride) and also with insoluble sulphates, oxalates, etc.

The third important type of reversible electrode consists of an unattackable metal, e.g. gold or platinum, immersed in a solution containing both oxidized and reduced states of an oxidation–reduction system, e.g. Sn^{4+} and Sn^{2+}; Fe^{3+} and Fe^{2+}; or $Fe(CN)_6^{3-}$ and $Fe(CN)_6^{4-}$. The purpose of the unattackable metal is to act as a conductor to make electrical contact, just as in the case of a gas electrode. The oxidized and reduced states are not necessarily ionic. For example, an important type of reversible electrode involves the organic compound quinone, together with hydrogen ions, as the oxidized state, with the neutral molecule hydroquinone as the reduced state. Electrodes of the kind under consideration, consisting of conventional oxidized and reduced forms, are sometimes called oxidation–reduction electrodes; the chemical reactions taking place at these electrodes are either oxidation of the reduced state or reduction of the oxidized state of the metal ion M:

$$M^{2+} \rightleftharpoons M^{4+} + 2e$$

depending on the direction of the current. In order that the electrode may behave reversibly it is essential that the system contain both oxidized and reduced states.

The three types of reversible electrodes described above differ formally as far as their construction is concerned; nevertheless, they are all based on the same fundamental principle. A reversible electrode always involves an oxidized and a reduced state, using the terms 'oxidized' and 'reduced' in their broadest sense; thus, oxidation refers to the liberation of electrons while reduction implies the taking up of electrons. If the electrode consists of a metal M and its ions M^+, the former is the reduced state and the latter is the oxidized state; similarly, for an anion electrode, the A^- ions are the reduced state while A represents the oxidized state. It can be seen, therefore, that all three types of reversible electrode are made up from the reduced and oxidized states of a given system, and in every case the electrode reaction may be written in the general form

$$\text{Reduced state} \rightleftharpoons \text{Oxidized state} + ne$$

where n is the number of electrons by which the oxidized and reduced states differ.

A reversible electrode consists of an oxidized and a reduced state, and the reaction which occurs at such an electrode, when it forms part of an operating cell, is either oxidation (i.e. reduced state → oxidized state + electrons) or reduction (i.e. oxidized state + electrons → reduced state). It can be readily seen, therefore, that in a reversible cell consisting of two reversible electrodes, a flow of electrons, and hence a flow of current, can be maintained if oxidation occurs at one electrode and reduction at the other. According to the convention widely adopted, the e.m.f. of the cell is positive when in its normal operation oxidation takes place at the left-hand electrode of the cell as written and reduction occurs at the right-hand electrode. If the reverse is the case, so that reduction is taking place at the left-hand electrode, the e.m.f. of the cell, by convention, will have a negative sign.

The Daniell cell, represented by

$$Zn \mid \text{M} \, ZnSO_4 \, (soln) \mid \text{M} \, CuSO_4 \, (soln) \mid Cu$$

has an e.m.f. of 1.10 V, and by the convention its sign is positive. This means that when the cell operates oxidation occurs at the left-hand electrode; that is to say, metallic zinc atoms are being oxidized to form zinc ions in solution, i.e.

$$Zn = Zn^{2+} + 2e$$

At the right-hand electrode there must, therefore, be reduction of the cupric ions, from the copper sulphate solution, to copper atoms, i.e.

$$Cu^{2+} + 2e = Cu$$

The electrons liberated at the zinc electrode travel along the external connecting circuit and are available for the discharge (reduction) of the cupric ions at the copper electrode. The complete cell reaction, obtained by adding the separate electrode reactions, is consequently:

$$Zn + Cu^{2+} = Zn^{2+} + Cu$$

Since two electrons are involved for each zinc (or copper) atom taking part in the reaction, the whole process as written, with quantities in gram-atoms or gram-ions, takes place for the passage of 2 F of electricity.

The practical convention, employed in connection with cells for yielding current, is to call the 'negative' pole the electrode at which the process is oxidation when the cell is producing current; the 'positive' electrode is the one at which reduction is the spontaneous process. The reason for this is that oxidation is accompanied by the liberation of electrons, and so the electrode metal acquires a negative charge; similarly, the reduction electrode will acquire a positive charge, because electrons are taken up from it. According to the widely used convention, the e.m.f. of a cell is positive when it is set up in such a way that the negative (oxidation) electrode is to the left, and the positive (reduction) electrode is to the right.

1.4 Relationship between electrical energy and energy content of a cell

It may be asked what is the relation between the electrical energy produced in a cell and the decrease in the energy content of the system, as a result of the chemical reaction going on therein. Considering only cells working at constant (atmospheric) pressure, when a chemical reaction occurs at constant pressure, without yielding any electrical energy, the heat evolved is equal to the decrease in the heat content of the system. In 1851, Kelvin made the first attempt to answer the question, by assuming that in the cell the whole of the heat of reaction appeared as electrical energy, i.e. the electrical energy obtained is equal to the decrease in the heat content of the system. This was supported by measurement on the Daniell cell. When the reaction

$$Zn + CuSO_4(aq.) = Cu + ZnSO_4(aq.)$$

is carried out in a calorimeter, an evolution of heat of 50.13 kcal occurs, which agrees well with the value of 50.38 kcal obtained for the electrical energy yielded by the reaction. This agreement, however, has since proved to be a coincidence. In other cell reactions, the electrical energy is sometimes less, sometimes greater, than the difference in heat content of the system. In the former case, the balance must appear as heat evolved in the working of the cell; in the latter case heat must be absorbed by the cell from its surroundings and to maintain the conservation of energy it is necessary to have

$$w' = -H + q \tag{1.16}$$

where w' is the electrical energy yielded by the cell reaction, $-H$ the decrease in heat content of the system and q the heat absorbed in the working of the cell.

It is necessary, therefore, to determine the heat absorbed in the working of the cell before the electrical energy yield of the cell can be found.

In methods for the accurate measurement of the electromotive force of a cell, the electromotive force of the cell is balanced by an applied potential difference. If the applied potential difference is slightly decreased, the cell reaction will go forward and the cell will do electrical work against the applied potential difference. If the applied potential difference is slightly increased, the reaction will occur in the reverse direction and work will be done by the external electromotive force on the cell. The reaction thus occurs reversibly in the cell when its electromotive force is balanced by an outside potential difference. When a reaction goes forward under these conditions, i.e. when the tendency of the reaction to go is just balanced by an external force, the maximum work that the reaction can yield is obtained. In a reaction at constant pressure, work is necessarily done against the applied pressure if any volume change occurs and this work cannot be obtained as electrical energy. The electrical energy obtained under these conditions is, therefore, the net work of the reaction.

For n equivalents of chemical reaction, nF coulombs are produced. If E is the electromotive force of the cell, an applied potential difference E is required to balance it. The electrical work w' done when the reaction goes forward in a state of balance (or only infinitesimally removed from it) is thus nFE, and this is equal to the net work of the reaction. Thus

$$w' = nFE \tag{1.17}$$

It should be observed that w' is the electrical work done against the applied potential difference. If there is no opposing potential difference in the circuit, no work is done against an applied potential difference, and the electrical energy nFE is dissipated in the circuit as heat.

According to the Gibbs–Helmholtz equation,

$$\Delta H + w' = T \left(\frac{dw'}{dT}\right)_P \tag{1.18}$$

putting $w' = nFE$ and $dw'/dT = nF(dE/dT)$,

$$\Delta H + nFE = TnF \left(\frac{dE}{dT}\right)_P \tag{1.19}$$

i.e.

$$\Delta H\,(\mathrm{J}) \;=\; -nF\left[E - T\left(\frac{\mathrm{d}E}{\mathrm{d}T}\right)_P\right]$$

$$\Delta H\,(\mathrm{cal}) \;=\; -\frac{nF}{4.18}\left[E - T\left(\frac{\mathrm{d}E}{\mathrm{d}T}\right)_P\right]$$

(1.20)

where $\mathrm{d}E/\mathrm{d}T$ is the temperature coefficient of the electromotive force at constant pressure.

Comparing Equation 1.19 with Equation 1.16 it can be seen that

$$TnF\left(\frac{\mathrm{d}E}{\mathrm{d}T}\right)_P$$

(1.21)

corresponds with the heat absorbed in the working of the cell. Thus

$$q = w' - (-\Delta H)$$

$$= nFT\left(\frac{\mathrm{d}E}{\mathrm{d}T}\right)_P$$

(1.22)

where q is the heat absorbed in working the cell, w' is the electrical energy yielded by the cell reaction, and $-\Delta H$ is the decrease in heat content of the system.

The sign of q thus depends on the sign of the temperature coefficient of the electromotive force:

1. If $\mathrm{d}E/\mathrm{d}T$ is positive, heat is absorbed in the working of the cell, i.e. the electrical energy obtained is greater than the decrease in the heat content in the reaction.

   $w' - (-\Delta H)$ is positive

2. If $\mathrm{d}E/\mathrm{d}T$ is negative, heat is evolved in the working of the cell, i.e. the electrical energy obtained is less than the decrease in the heat content in the reaction.

   $w' - (-\Delta H)$ is negative

3. If $\mathrm{d}E/\mathrm{d}T$ is zero, no heat is evolved in the working of the cell, i.e. the electrical energy obtained is equal to the decrease in the heat content in the reaction.

   $w' - (-\Delta H) = 0$

1.5 Free energy changes and electromotive forces in cells

More recent work has regarded the processes occurring in a cell in terms of free energy changes. The free energy change accompanying a process is equal to the reversible work, other than that due to a volume change, at constant temperature and pressure. When a reversible cell operates, producing an infinitesimal current, the electrical work is thermodynamically reversible in character, and does not include any work due to a volume change.

Furthermore, since the temperature and pressure remain constant, it is possible to identify the electrical work done in a reversible cell with the free energy change accompanying the chemical or other process taking place in the cell. The work done in a cell is equal to the product of the e.m.f. and the quantity of electricity passing. The practical unit of electrical energy is defined as the energy developed when one coulomb is passed under the influence of an e.m.f. of one volt; this unit is called the volt-coulomb, and is equivalent to one international joule. The calorie defined by the US Bureau of Standards is equivalent to 4.1833 international joules, and hence one volt-coulomb is equivalent to 1/4.1833, i.e. 0.2390 (defined), calorie.

If the e.m.f. of a reversible cell is E volts, and the process taking place is associated with the passage of n faradays, i.e. nF coulombs, the electrical work done by the system is consequently nFE volt-coulombs or international joules. The corresponding increase of free energy (ΔF) is equal to the electrical work done on the system; it is therefore possible to write

$$\Delta F = -nFE$$

(1.23)

This is an extremely important relationship, which forms the basis of the whole treatment of reversible cells.

The identification of the free energy change of a chemical reaction with the electrical work done when the reaction takes place in a reversible cell can be justified experimentally in the following manner. By the Gibbs–Helmholtz equation,

$$\Delta F = \Delta H + T\left(\frac{\mathrm{d}(\Delta F)}{\mathrm{d}T}\right)_P$$

(1.24)

where ΔH is the heat change accompanying the cell reaction and T is temperature in kelvins. If ΔF is replaced by $-nFE$, the result is

$$-nFE = \Delta H - nFT\left(\frac{\mathrm{d}E}{\mathrm{d}T}\right)_P$$

$$\Delta H = nF\left[E - T\left(\frac{\mathrm{d}E}{\mathrm{d}T}\right)_P\right]$$

(1.25)

It can be seen from Equation 1.20 that if the e.m.f. of the reversible cell, i.e. E, and its temperature coefficient $\mathrm{d}E/\mathrm{d}T$, at constant pressure, are known, it is possible to evaluate the heat change of the reaction occurring in the cell. The result may be compared with that obtained by direct thermal measurement; good agreement would then confirm the view that $-nFE$ is equal to the free energy increase, since Equation 1.20 is based on this postulate.

Using Equation 1.20 it is possible, having the e.m.f. of a cell on open circuit at a particular temperature, the temperature coefficient of $\mathrm{d}E/\mathrm{d}T$

and the e.m.f., to calculate the heat change accompanying the cell reaction ΔH:

$$\Delta H = nF \left[E - T \left(\frac{dE}{dT} \right)_P \right] \quad V\,C$$

$$= \frac{-nF}{4.183} \left[E - T \left(\frac{dE}{dT} \right)_P \right] \quad \text{cal}$$

For example, the open circuit voltage of a lead–acid cell is 2.01 V at 15°C (288 K) and its temperature coefficient of resistance is $dE/dT = 0.000\,37$ V/K, $n = 2$. The heat change accompanying the cell reaction in calories is

$$\Delta H = \frac{-2 \times 96\,500}{4.18} \left(2.01 - 288 \times 0.000\,37 \right)$$

$$= -87\,500 \,\text{cal} = -87.5 \,\text{kcal}$$

which is in quite good agreement with the calorimetrically derived value of -89.4 kcal.

Similarly, in the Clark cell, the reaction

Zn(amalgam) + Hg_2SO_4(s) + $7H_2O$
= $ZnSO_4.7H_2O$(s) + 2Hg(l)

gives rise to 2 F of electricity, i.e. $n = 2$, the open circuit voltage is 1.4324 V at 15°C and the temperature coefficient is 0.000 19, hence:

$$\Delta H = \frac{-2 \times 96\,540}{4.18} \left(1.4324 - 288 \times 0.001\,19 \right)$$

$$= 81.92 \,\text{kcal}$$

which agrees well with the calorimetric value of 81.13 kcal.

1.6 Relationship between the energy changes accompanying a cell reaction and concentration of the reactants

It is important when studying the effect of concentrations of reactants in a cell on the e.m.f. developed by the cell to consider this in terms of free energies (ΔF).

Free energy (ΔF) is defined by the following expression:

$$-\Delta F = w - P\Delta V$$

at constant temperature and pressure.

In a reversible, isothermal process, w is the maximum work that can be obtained from the system in the given change.

The quantity $P\Delta V$ is the work of expansion done against the external pressure, and so $-\Delta F$ represents the maximum work at constant temperature and pressure other than that due to volume change. The quantity $w - P\Delta V$ is called the net work and so the decrease $-\Delta F$ in the free energy of a system is equal to the net work obtainable (at constant temperature and pressure) from the system under reversible conditions. An important form of net work, since it does not involve external work due to a volume change, is electrical work; consequently, a valuable method for determining the free energy change of a process is to carry it out electrically, in a reversible manner, at constant temperature and pressure.

By Equation 1.23,

$$\Delta F = -nFE$$

where ΔF is the free energy increase, E the e.m.f. of a reversible cell, nF the number of faradays associated with the process occurring ($F = 1$ F), and

$$E = \frac{-\Delta F}{nF} \tag{1.26}$$

The free energy change accompanying a given reaction depends on the concentrations or, more accurately, the activities, of the reactants and the products. It is evident, therefore, that the e.m.f. of a reversible cell, in which a particular reaction takes place when producing current, will vary with the activities of the substances present in the cell. The exact connection can be readily derived in the following manner. Suppose the general reaction

$$aA + bB + \ldots \rightleftharpoons lL + mM + \ldots$$

occurs in a reversible cell; the corresponding free energy change is then given by the following equation

$$\Delta F = \Delta F^0 + RT \ln \frac{a^l L \times a^m M \times \ldots}{a^a A \times a^b B \times \ldots} \tag{1.27}$$

where $aA, aB, \ldots, aL, aM, \ldots$ now represent the activities of A, B, $\ldots$, L, M, $\ldots$ as they occur in the reversible cell. If the arbitrary reaction quotient, in terms of activities, is represented by the symbol Q_a, Equation 1.27 may be written as

$$\Delta F = \Delta F^0 + RT \ln Q_a \tag{1.28}$$

As before, ΔF^0 is the free energy change when all the substances taking part in the cell reaction are in their standard states.

If E is the e.m.f. of the cell under consideration when the various substances have the arbitrary activities $aA, aB, \ldots, aL, aM, \ldots$, as given above, and the reaction as written occurs for the passage of n faradays, it follows from Equation 1.26 that ΔF, as given by Equation 1.27 or 1.28, is also equal to $-nFE$. Furthermore, if the e.m.f. of the reversible cell is E^0 when all the substances involved are in their standard states, the ΔF^0 is equal to $-nFE^0$. Substituting these values for ΔF and ΔF^0 into Equation 1.28 and dividing through by $-nF$, the result is

$$E = E^0 - \frac{RT}{nF} \ln Q_a \tag{1.29}$$

This expression is seen to relate the e.m.f. of a cell to the activities of the substances taking part; E^0, the standard e.m.f., is a constant for the given cell reaction, varying only with the temperature, at 1 atmosphere pressure.

The foregoing results may be illustrated by reference to the cell

$$\tfrac{1}{2}H_2(g) + AgCl(s) = H^+ + Cl^- + Ag(s)$$

for the passage of 1 F. The reaction quotient in terms of activities is

$$Q_a = \frac{a_{H^+} \times a_{Cl^-} \times a_{Ag}}{a_{H_2/2} \times a_{AgCl}}$$

but since the silver and the silver chloride are present in the solid state, their activities are unity; hence

$$Q_a = \frac{a_{H^+} \times a_{Cl^-}}{a_{H_2/2}}$$

Inserting this expression into Equation 1.29, with n equal to unity, the e.m.f. of the cell is given by

$$E = E^0 - \frac{RT}{F} \ln \left(\frac{a_{H^+} \times a_{Cl^-}}{a_{H_2/2}} \right) \qquad (1.30)$$

The e.m.f. is thus seen to be dependent upon the activities of the hydrogen and chloride ions in the solution of hydrochloric acid, and of the hydrogen gas in the cell. If the substances taking part in the cell behaved ideally, the activities in Equation 1.30 could be replaced by the corresponding concentrations of the hydrogen and chloride ions, and by the pressure of the hydrogen gas. The resulting form of Equation 1.30, namely

$$E = E^0 - \frac{RT}{F} \ln \left(\frac{c_{H^+} \times c_{Cl^-}}{p_{H_2/2}} \right) \qquad (1.31)$$

could be used if the solution of hydrochloric acid were dilute and the pressure of the hydrogen gas low.

1.7 Single electrode potentials

There is at present no known method whereby the potential of a single electrode can be measured; it is only the e.m.f. of a cell, made by combining two electrodes, that can be determined experimentally. However, by choosing an arbitrary zero of potential, it is possible to express the potentials of individual electrodes. The arbitrary zero of potential is taken as the potential of a reversible hydrogen electrode, with gas at 1 atm. pressure, in a solution of hydrogen ions of unit activity. This particular electrode, namely H_2 (1 atm.) H^+ ($a = 1$), is known as the standard hydrogen electrode. The convention, therefore, is to take the potential of the standard

hydrogen electrode as zero; electrode potentials based on this zero are said to refer to the hydrogen scale. If any electrode, M, M^+, is combined with the standard hydrogen electrode to make a complete cell, i.e.

$$M \mid M^+ (\text{soln}) \vdots H^+ (a = 1) \mid H_2 (1 \text{ atm.})$$
$$E \qquad\qquad\qquad\qquad 0$$

the e.m.f. of this cell, E, is equal to the potential of the M, M^+ electrode on the hydrogen scale.

When any reversible electrode is combined with a standard hydrogen electrode, as indicated above, and oxidation reaction takes place at the former, while the hydrogen ions are reduced to hydrogen gas at the latter. The electrode (oxidation) process may be written in the following general form:

$$\text{Reduced state} = \text{Oxidized state} + ne$$

and the corresponding hydrogen electrode reaction is

$$nH^+ + ne = \tfrac{1}{2}nH_2(g)$$

The complete cell reaction for the passage of n faradays is consequently

$$\text{Reduced state} + nH^+ = \text{Oxidized state} + \tfrac{1}{2}nH_2(g) \quad (1.32)$$

The e.m.f. of the cell, which is equal to the potential of the reversible electrode under consideration, is then given by Equation 1.29 as

$$E = E_{el}^0 - \frac{RT}{nF} \ln \left[\frac{(\text{Oxidized state}) \times a_{H_2}^{n/2}}{(\text{Reduced state}) \times a_{H^+}^n} \right] \quad (1.33)$$

where parentheses have been used to represent the activities of the oxidized and reduced states as they actually occur in the cell. In the standard hydrogen electrode, the pressure of the gas is 1 atm., and hence the activity a_{H_2} is unity; furthermore, by definition, the activity of the hydrogen ions a_{H^+} in the electrode is also unity. It can thus be seen that Equation 1.33 for the electrode potential can be reduced to the simple form

$$E = E_{el}^0 - \frac{RT}{nF} \ln \left[\frac{(\text{Oxidized state})}{(\text{Reduced state})} \right] \quad (1.34)$$

This is the general equation for the oxidation potential of any reversible electrode; E_{el}^0 is the corresponding standard electrode potential; that is, the potential of the electrode when all the substances concerned are in their standard states of unit activity. The qualification 'oxidation' is used because it describes the process taking place at the electrode; the corresponding 'reduction potentials' will be considered below.

The application of Equation 1.34 may be illustrated by reference to a few simple cases of different types. Consider, first, an electrode consisting of a metal in contact with a solution of its own cations, e.g. copper in copper (cupric) sulphate solution.

The electrode (oxidation) reaction is

Cu = Cu^{2+} + 2e

the Cu being the reduced state and Cu^{2+} the oxidized state; in this case n is 2, and hence by Equation 1.34

$$E = E_{el}^0 - \frac{RT}{2F} \ln \left(\frac{a_{Cu^{2+}}}{a_{Cu}} \right)$$

The activity of a_{Cu} of the solid metal is unity, by convention, and hence

$$E = E_{el}^0 - \frac{RT}{2F} \ln a_{Cu^{2+}} \qquad (1.35)$$

so that the electrode potential is dependent on the standard (oxidation) potential E_{el}^0 of the Cu, Cu^{2+} system, and on the activity $a_{Cu^{2+}}$ of the cupric ions in the copper sulphate solution. The result may be generalized, so that for any metal M (or hydrogen) in equilibrium with a solution of its ions M$^+$ of valence n, the oxidation potential of the M, M$^+$ electrode is given by

$$E = E_{el}^0 - \frac{RT}{nF} \ln a_{M^+}$$

At 25°C,

$$E = E_{el}^0 - \frac{0.059\ 15}{n} \log a_{M^+} \qquad (1.36)$$

where a_{M^+} is the activity of the M$^+$ ions in the solution. For a univalent ion (e.g. hydrogen, silver, cuprous), n is 1; for a bivalent ion (e.g. zinc, nickel, ferrous, cupric, mercuric), n is 2 and so on.

Similarly, the general equation for the oxidation potential of any electrode reversible to the anion A$^-$ of valence n is

$$E = E_{el}^0 + \frac{RT}{nF} \ln a_{A^-}$$

At 25°C,

$$E = E_{el}^0 + 0.059\ 15 \log a_{A^-} \qquad (1.37)$$

where a_{A^-} is the activity of the A$^-$ ions in the given electrode solution. It should be noted, in comparing Equations 1.36 and 1.37, that in the former, for cations, the second term is preceded by a negative sign, whilst in the latter, for anions, the sign is positive. To make practical use of the electrode potential equations it is necessary to insert values for R and F in the factor RT/nF which appears in all such equations. The potential is always expressed in volts, and since F is known to be 96 500 C, the value of R ($R = 1.998$ cal) must be in volt coulombs, i.e. in international joules; thus R (1.998×4.18) is 8.314 absolute joules or 8.312 international joules per degree per mole. Taking Equation 1.36 for the oxidation potential of an electrode reversible with respect to cations, that is

$$E = E_{el}^0 - \frac{RT}{nF} \ln a_{M^-}$$

inserting the values of R and F given above, and introducing the factor 2.303 to convert natural logarithms to common logarithms, i.e. to the base 10, the result is

$$E = E_{el}^0 - \frac{2.303 \times 8.312}{96\ 500} \frac{T}{n} \log a_{M^+}$$

$$= E_{el}^0 - 1.984 \times 10^{-4} \frac{T}{n} \log a_{M^{2+}} \qquad (1.38)$$

At 25°C, i.e. $T = 298.16$ K, which is the temperature most frequently employed for accurate electrochemical measurements, this equation becomes

$$E = E_{el}^0 - \frac{0.059\ 15}{n} \log a_{M^+} \qquad (1.39)$$

Similarly, for the oxidation potential of an anion electrode at 25°C,

$$E = E_{el}^0 + \frac{0.059\ 15}{n} \log a_{A^-} \qquad (1.40)$$

The general form of the equation at 25°C, which is applicable to all reversible electrodes (see Equation 1.34) is

$$E = E_{el}^0 - \frac{0.059\ 15}{n} \log \left[\frac{(\text{Oxidized state})}{(\text{Reduced state})} \right] \qquad (1.41)$$

where the parentheses are used to indicate activities.

It should be evident from the foregoing examples that it is not a difficult matter to derive the equation for the oxidation potential of any electrode; all that is necessary is to write down the electrode reaction, and then to insert the appropriate activities of the oxidized and reduced states in Equation 1.34. The result is then simplified by using the convention concerning the standard states of unit activity. Thus, for any metal present in the pure state, for any pure solid compound, for a gas at 1 atm. pressure, and for water forming part of a dilute solution, the activity is taken as unity. The corresponding activity factors may then be omitted from the electrode potential equation.

It has been seen that, in every galvanic cell, oxidation occurs at one electrode, but a reduction process takes place at the other electrode. The equations just derived give the potential of the electrode at which oxidation occurs, and now reference must be made to the potential of the electrode at which reduction is occurring. The situation is, fortunately, quite simple; the reduction potential of any electrode is equal to the oxidation potential for the same electrode but with the sign reversed. It is quite unnecessary, and in fact undesirable, to write out separate formulae for reduction potentials. The recommended procedure is to derive the oxidation potential for the given electrode and then merely to reverse the sign. For example, the reduction potential of the copper–cupric ion electrode, for which the reaction is

$$Cu^{2+} + 2e = Cu$$

would be given by an equation identical to Equation 1.35 but with the sign reversed.

To facilitate the representation of electrodes, a simple convention is adopted; when the electrode is a metal M, and the process is oxidation to M^+ ions, the reduced state of the system is written to the left and the oxidized state to the right, namely M, M^+, as in the electrochemical equation $M \rightarrow M^+ +$ electrons. Examples of oxidation electrodes are thus

Cu, Cu^{2+} (or Cu, $CuSO_4$ (soln))
Zn, Zn^{2+} (or Zn, $ZnSO_4$ (soln))

The potentials of such electrodes are given by Equation 1.34 or 1.36. On the other hand, if the electrodes are represented in the reverse manner, i.e. M^+, M, with the oxidized state to the left and the reduced state to the right, e.g.

Cu^{2+}, Cu (or $CuSO_4$ (soln), Cu)
Zn^{2+}, Zn (or $ZnSO_4$ (soln), Zn)

the electrode process is reduction, and the potentials are opposite in sign to those of the corresponding oxidation electrodes.

If two reversible electrodes are combined to form such cells as

$$Zn \mid ZnSO_4 \text{ (soln)} \vdots CuSO_4 \text{ (soln)} \mid Cu$$

then, in accordance with the convention given above, the reaction at the left-hand electrode is oxidation, while at the right-hand electrode a reduction process is taking place when the cell operates spontaneously to produce current upon closing the external circuit. Thus, the e.m.f. of the complete cell is equal to the algebraic sum of the potentials of the two electrodes, one being an oxidation potential and the other a reduction potential. An important point to which attention may be called is that since the e.m.f. of a cell is equal to the sum of an oxidation and a reduction electrode potential, it is equivalent to the difference of two oxidation potentials. As a consequence, the e.m.f. of a cell is independent of the arbitrary potential chosen as the zero of the potential scale; the actual value, whatever it may be, cancels out when taking the difference of the two oxidation potentials based on the same (e.g. hydrogen) scale.

According to the equations derived above, the potential of any electrode is determined by the standard potential E^0_{el}, and by the activity or activities of the ions taking part in the electrode process. These activities are variable, but the standard potential is a definite property of the electrode system, having a constant value at a given temperature. If these standard potentials were known, it would be a simple matter to calculate the actual potential of any electrode, in a solution of given concentration or activity, by using the appropriate form of Equation 1.34. The standard potentials of many electrodes have been determined, with varying degrees of accuracy, and the results have been tabulated. The principle of the method used to evaluate E^0_{el} for a given electrode system is to measure the potential E of the electrode, on the hydrogen scale, in a solution of known activity; from these two quantities the standard potential E^0_{el} can be calculated at the experimental temperature, using Equation 1.34. Actually the procedure is more complicated than this, because the activities are uncertain. The results obtained for the standard oxidation potentials of some electrodes at 25°C are recorded in Table 1.2; the appropriate electrode process is given in each case.

Table 1.2 Standard oxidation potentials at 25°C on the hydrogen scale

Electrode	Reaction	E^0_{el}
K, K^+	$K \rightarrow K^+ + e$	+2.924
Na, Na^+	$Na \rightarrow Na^+ + e$	+2.714
Zn, Zn^{2+}	$Zn \rightarrow Zn^{2+} + 2e$	+0.761
Fe, Fe^{2+}	$Fe \rightarrow Fe^{2+} + 2e$	+0.441
Cd, Cd^{2+}	$Cd \rightarrow Cd^{2+} + 2e$	+0.402
Co, Co^{2+}	$Co \rightarrow Co^{2+} + 2e$	+0.283
Ni, Ni^{2+}	$Ni \rightarrow Ni^{2+} + 2e$	+0.236
Sn, Sn^{2+}	$Sn \rightarrow Sn^{2+} + 2e$	+0.140
Pb, Pb^{2+}	$Pb \rightarrow Pb^{2+} + 2e$	+0.126
Pt, $\frac{1}{2}H_2$, H^+	$\frac{1}{2}H_2 \rightarrow H^+ + e$	±0.000
Pt, Sn^{2+}, Sn^{4+}	$Sn^{2+} \rightarrow Sn^{4+} + 2e$	−0.15
Pt, Cu^+, Cu^{2+}	$Cu^+ \rightarrow Cu^{2+} + e$	−0.16
Ag, AgCl(s), Cl^-	$Ag + Cl^- \rightarrow AgCl + e$	−0.2224
Cu, Cu^{2+}	$Cu \rightarrow Cu^{2+} + 2e$	−0.340
Pt, $Fe(CN)_6^{4-}$, $Fe(CN)_6^{2-}$	$Fe(CN)_6^{4-} \rightarrow Fe(CN)_6^{2-} + e$	−0.356
Pt, O_2, OH^-	$2OH^- \rightarrow \frac{1}{2}O_2 + H_2O + 2e$	−0.401
Pt, $I_2(s)$, I^-	$I^- \rightarrow \frac{1}{2}I_2 + e$	−0.536
Pt, Fe^{2+}, Fe^{3+}	$Fe^{2+} \rightarrow Fe^{3+} + e$	−0.771
Ag, Ag^+	$Ag \rightarrow Ag^+ + e$	−0.799
Hg, Hg_2^{2+}	$Hg \rightarrow \frac{1}{2}Hg_2^{2+} + e$	−0.799
Pt, Hg_2^{2+}, Hg^{2+}	$Hg_2^{2+} \rightarrow 2Hg^{2+} + 2e$	−0.906
Pt, $Br_3(l)$, Br	$Br^- \rightarrow \frac{1}{2}Br_2 + e$	−1.066
Pt, $Cl_2(g)$, Cl^-	$Cl^- \rightarrow \frac{1}{2}Cl_2 + e$	−1.358
Pt, Ce^{3+}, Ce^{4+}	$Ce^{3+} \rightarrow Ce^{4+} + e$	−1.61

It should be remembered that the standard potential refers to the condition in which all the substances in the cell are in their standard states of unit activity. Gases such as hydrogen, oxygen and chlorine are thus at 1 atm. pressure. With bromine and iodine, however, the standard states are chosen as the pure liquid and solid, respectively; the solutions are therefore saturated with these elements in the standard electrodes. For all ions the standard state of unit activity is taken as the hypothetical ideal solution of unit molality or, in other words, a solution for which the product $m\gamma$ is unity, where m is the molality of the ion and γ its activity coefficient.

The standard reduction potentials, corresponding to the oxidation potentials in Table 1.2 but involving the reverse electrode processes, would be obtained by reversing the sign in each case; thus, for example, for the zinc electrode,

$$Zn, Zn^{2+} \quad E^0_{el} = +0.761 \text{ V} \quad Zn = Zn^{2+} + 2e$$

$$Zn^{2+}, Zn \quad E^0_{el} = -0.761 \text{ V} \quad Zn^{2+} + 2e = Zn$$

whereas, for the chlorine electrode,

$$Pt, Cl_2(g), Cl^- \quad E^0_{el} = -1.358 \text{ V} \quad Cl^- = \tfrac{1}{2}Cl_2(g) + e$$

$$Cl^-, Cl_2(g), Pt \quad E^0_{el} = +1.358 \text{ V} \quad \tfrac{1}{2}Cl_2(g) + e = Cl^-$$

1.8 Activities of electrolyte solutions

The use of activities instead of concentrations in the types of thermodynamic calculations dealing with cells is of great significance. The extensive use of the activity term has been seen in the preceding equations. For an ideal solution, activity equals the concentrations of dissolved electrolytes. Very few solutions, in fact, behave ideally, although in some cases very dilute solutions approach ideal behaviour. By definition, however, cell electrolytes are not dilute and hence it is necessary when carrying out thermodynamic calculations to use activities rather than concentrations. Most electrolytes consist of a solute dissolved in a solvent, commonly water, although, in some types of cell, solutions of various substances in organic solvents are used. When a solute is dissolved in a liquid, the vapour pressure of the latter is lowered. The quantitative connection between the lowering of the vapour pressure and the composition of a solution was discovered by F. M. Raoult. If p^0 is the vapour pressure of pure solvent at a particular temperature, and p is the vapour pressure of the solution at the same temperature, the difference $p^0 - p$ is the lowering of the vapour pressure. If this is divided by p^0 the result, $(p^0 - p)/p^0$, is known as the relative lowering of the vapour pressure for the given solution. According to one form of Raoult's law, the relative lowering of the vapour pressure is equal to the mole fraction of the solute in the solution. If n_1 and n_2 are the numbers of moles of solvent and solute, respectively, the mole fraction x_2 of the solute is

$$x_2 = \frac{n_2}{n_1 + n_2}$$

and hence, by Raoult's law,

$$\frac{p^0 - p}{p^0} = x_2 = \frac{n_2}{n_1 + n_2} \tag{1.42}$$

This law, namely that

$$\frac{\text{Relative lowering of vapour pressure}}{\text{Mole fraction of solute}} = 1$$

is obeyed, at least approximately, for many solute–solvent systems. There are, however, theoretical reasons for believing that Raoult's law could only be expected to hold for solutions having a heat of dilution of zero, and for which there is no volume change upon mixing the components in the liquid state. Such solutions, which should obey Raoult's law exactly at all concentrations and all temperatures, are called ideal solutions. Actually very few solutions behave ideally and some deviation from Raoult's law is always to be anticipated; however, for dilute solutions these deviations are small and can usually be ignored.

An alternative form of Raoult's law is obtained by subtracting unity from both sides of Equation 1.42; the result is

$$\frac{p}{p^0} = 1 - x_2 \tag{1.43}$$

The sum of the mole fractions of solvent and solute must always equal unity; hence, if x_1 is the mole fraction of the solvent, and x_2 is that of the solute, as given above, it follows that

$$x_1 + x_2 = 1 \tag{1.44}$$

Hence Equation 1.43 can be reduced to

$$p = x_1 p^0 \tag{1.45}$$

Therefore, the vapour pressure of the solvent in a solution is directly proportional to the mole fraction of the solvent, if Raoult's law is obeyed. It will be observed that the proportionality constant is p^0, the vapour pressure of the solvent.

As, in fact, most cell electrolyte solutions are relatively concentrated, they are non-ideal solutions and Raoult's law is not obeyed. To overcome this problem the activity concept is invoked to overcome departure from ideal behaviour. It applies to solutions of electrolytes, e.g. salts and bases, and is equally applicable to non-electrolytes and gases. The following is a simple method of developing the concept of activity when dealing with non-ideal solutions.

Consider a system of two large vessels, one containing a solution in equilibrium with its vapour at the pressure p', and the other containing another solution, of the same solvent and solute at a different concentration, whose vapour pressure is p''. The external pressure, e.g. 1 atm., and the temperature, T, are the same for both vessels. One mole of solvent is then vaporized isothermally and reversibly from the first solution at constant pressure p'; the quantity of solution is supposed to be so large that the removal of 1 mol of solvent does not appreciably affect the concentration or vapour pressure. The vaporization has been carried out reversibly, and so every stage represents a state of equilibrium. Furthermore, the temperature and pressure have remained constant, and hence there is no change of free energy.

The mole of vapour at pressure p' is now removed and compressed or expanded at constant temperature until its pressure is changed to p'', the vapour pressure of the second solution. If the pressures are sufficiently low for the vapour to be treated as an ideal gas without incurring serious error, as is generally the case, the increase of free energy is given by

$$F = RT \ln \left(\frac{p''}{p'}\right) \tag{1.46}$$

Finally, the mole of vapour at the constant pressure p'' is condensed isothermally and reversibly into the second solution. The change of free energy for this stage, like that for the first stage, is again zero; the total free energy change for the transfer of 1 mol of solvent from the first solution to the second is thus given by Equation 1.46.

Let F' represent the actual free energy of 1 mol of solvent in the one solution and F'' the value in the other solution. Since the latter solution gains 1 mol while the former loses 1 mol, the free energy increase F is equal to $F'' - F'$; it is thus possible to write, from Equation 1.46,

$$F'' - F' = FT \ln \left(\frac{p''}{p'}\right) \tag{1.47}$$

If both solutions behave ideally, so that Raoult's law is applicable, the vapour pressure is proportional to the mole fraction of the solvent in the particular solution (Equation 1.45); hence, for ideal solutions, Equation 1.47 becomes

$$F'' - F' = RT \ln \left(\frac{x''}{x'}\right) \tag{1.48}$$

For non-ideal solutions this result is not applicable, but the activity of the solvent, represented by a, is defined in such a way that the free energy of transfer of 1 mol of solvent from one solution to the other is given exactly by:

$$F'' - F' = RT \ln \left(\frac{a''}{a'}\right) \tag{1.49}$$

This means, in a sense, that the activity is the property for a real solution that takes the place of the mole fraction for an ideal solution in the free energy equation.

Although the definition of activity as represented by Equation 1.49 has been derived with particular reference to the solvent, an exactly similar result is applicable to the solute. If F' is the free energy of 1 mol of solute in one solution, and F'' is the value in another solution, the increase of free energy accompanying the transfer of 1 mol of solute from the first solution to the second is then given by Equation 1.49, where a' and a'' are, by definition, the activities of the solute in the two solutions.

Equation 1.49 does not define the actual or absolute activity, but rather the ratio of the activities of the particular substances in two solutions. To express activities numerically, it is convenient to choose for each constituent of the solution a reference state or standard state, in which the activity is arbitrarily taken as unity. The activity of a component, solvent or solute in any solution is thus really the ratio of its value in the given solution to that in the chosen standard state. The actual standard state chosen for each component is the most convenient for the purpose, and varies from one to the other, as will be seen shortly. If the solution indicated by the single prime is taken as representing the standard state, a' will be unity, and Equation 1.49 may be written in the general form

$$F - F^0 = RT \ln a \tag{1.50}$$

the double primes being omitted, and a superscript zero used, in accordance with the widely accepted convention, to identify the standard state of unit activity. This equation defines the activity or, more correctly, the activity relative to the chosen standard state, of either solvent or solute in a given solution.

The deviation of a solution from ideal behaviour can be represented by means of the quantity called the activity coefficient, which may be expressed in terms of various standard states. In this discussion the solute and solvent may be considered separately; the treatment of the activity coefficient of the solute in dilute solution will be given first. If the molar concentration, or molarity of the solute, is c moles (or gram-ions) per litre, it is possible to express the activity a by the relationship

$$a = fc \quad \text{or} \quad f = \frac{a}{c} \tag{1.51}$$

where f is the activity coefficient of the solute. Inserting this into Equation 1.50 gives the expression

$$F - F_c^0 = RT \ln fc \tag{1.52}$$

applicable to ideal and non-ideal solutions. An ideal (dilute) solution is defined as one for which f is unity, but for a non-ideal solution it differs from unity. Since solutions tend to a limiting behaviour as they become more dilute, it is postulated that at the same time f approaches unity, so that, at or near infinite dilution, Equation (1.51) becomes

$$a_a = c_0 \tag{1.53}$$

that is, the activity of the solute is equal to its molar concentration. The standard state of unit activity may thus be defined as a hypothetical solution of unit molar concentration possessing the properties of a very dilute solution. The word 'hypothetical' is employed in this definition because a real solution at a concentration of 1 mol (or gram-ion) per litre will generally not behave ideally in the sense of having the properties of a very dilute solution.

Another standard state for solutes that is employed especially in the study of galvanic cells is that based on the relationships

$$a = \gamma m \quad \text{or} \quad \gamma = \frac{a}{m} \tag{1.54}$$

where m is the molality of the solute, i.e. moles (or gram-ions) per 1000 g solvent, and γ is the appropriate activity coefficient. Once again it is postulated that γ approaches unity as the solution becomes more and more dilute, so that at or near infinite dilution it is possible to write

$$a_0 = m_0 \tag{1.55}$$

the activity being now equal to the molality. The standard state of unit activity is consequently defined as a hypothetical solution of unit molality possessing the properties of a very dilute solution. The difference between the actual value of the activity coefficient γ and unity is a measure of the departure of the actual solution from an ideal solution, regarded as one having the same properties as at high dilution.

In view of Equations 1.53 and 1.55 it is evident that in the defined ideal solutions the activity is equal to the molarity or to the molality, respectively. It follows, therefore, that the activity may be thought of as an idealized molarity or molality), which may be substituted for the actual molarity (or molality) to allow for departure from ideal dilute solution behaviour. The activity coefficient is then the ratio of the ideal molarity (or molality) to the actual molarity (or molality). At infinite dilution both f and γ must, by definition, be equal to unity, but at appreciable concentrations the activity coefficients differ from unity and from one another. However, it is possible to derive an equation relating f and γ, and this shows that the difference between them is quite small in dilute solutions.

When treating the solvent, the standard state of unit activity almost invariably chosen is that of the pure liquid; the mole fraction of the solvent is then also unity. The activity coefficient f_x of the solvent in any solution is then defined by

$$a = f_x x \quad \text{or} \quad f_x = \frac{a}{x} \tag{1.56}$$

where x is the mole fraction of the solvent. In the pure liquid state of the solvent, a and x are both equal to unity, and the activity coefficient is then also unity on the basis of the chosen standard state.

Several methods have been devised for the determination of activities; measurements of vapour pressure, freezing point depression, etc., have been used to determine departure from ideal behaviour, and hence to evaluate activities. The vapour pressure method has been used particularly to obtain the activity of the solvent in the following manner.

Equation 1.49 is applicable to any solution, ideal or non-ideal, provided only that the vapour behaves as an ideal gas; comparison of this with Equation 1.49 shows that the activity of the solvent in a solution must be proportional to the vapour pressure of the solvent over a given solution. If a represents the activity of the solvent in the solution and p is its vapour pressure, then $a = kp$, where k is a proportionality constant. The value of this constant can be determined by making use of the standard state postulated above, namely that $a = 1$ for the pure solvent, i.e. when the vapour pressure is p^0; it follows, therefore, that k, which is equal to a/p, is $1/p^0$, and hence

$$a = \frac{p}{p^0} \tag{1.57}$$

The activity of the solvent in a solution can thus be determined from measurements of the vapour pressure of the solution, p, and of the pure solvent, p^0 at a given temperature. It is obvious that for an ideal solution obeying Raoult's law p/p^0 will be equal to x, the mole fraction of solvent. The activity coefficient as given by Equation 1.56 will then be unity. It is with the object of obtaining this result that the particular standard state of pure solvent was chosen. For a non-ideal solution the activity coefficient of the solvent will, of course, differ from unity, and its value can be determined by dividing the activity as given by Equation 1.57 by the mole fraction of the solvent.

Table 1.3 gives the activity coefficients at various concentrations of two typical liquids used as battery electrolytes, namely sulphuric acid and potassium hydroxide. It will be seen that the activity coefficients initially decrease with increasing concentrations. Subsequently at higher concentrations activity coefficients rise becoming greater than one at high

Table 1.3 Activity coefficients (γ) and activities (a) of strong electrolytes

Molality, m (mol *solute*/ 1000 g *solvent*)	Potassium hydroxide		Sulphuric acid	
	γ	a	γ	a
0.01	0.920	0.009 20	0.617	0.006 17
0.02	–	–	0.519	0.010 38
0.05	0.822	0.041 10	0.397	0.019 85
0.1	0.792	0.079 2	0.313	0.031 3
0.2	0.763	0.152 6	0.244	0.048 8
0.5	0.740	0.370	0.178	0.089 0
1	0.775	0.775	0.150	0.150
2	–	–	0.147	0.294
3	1.136	3.408	0.166	0.498
4	–	–	0.203	0.812
5	–	–	0.202	1.010

concentrations. The activity coefficient molality relationship for sulphuric acid is shown in Figure 1.6. Figure 1.7 shows the relationship between activity, a, and molality for sulphuric acid.

For many purposes, it is of more interest to know the activity, or activity coefficient, of the solute rather than that of the solvent as discussed above. Fortunately, there is a simple equation which can be derived thermodynamically, that relates the activity a_1 of solvent and that of solute a_2; thus

$$n_2 \ln a_2 = -n_1 \ln a_1 \qquad (1.58)$$

where n_1 and n_2 are the numbers of moles of solvent and solute. If the values of a_1 for the solvent are known at a series of concentrations n_1, the activity a_2 of the solute can be solved by graphical integration of the above equation.

The activity of a solution changes with the temperature. For many purposes in thermodynamic calculations on batteries this factor may be ignored but, nevertheless, it is discussed below.

The variation of the partial molar free energy of a solute is given by

$$\overline{F_1} = \overline{F_1^0} + RT \log a$$

i.e.

$$\log a = \frac{\overline{F_1}}{RT} - \frac{\overline{F_1^0}}{RT}$$

where $\overline{F_1^0}$ is the partial free energy of the solute in the solution for which the activity has been taken as unity. $\overline{F_1}$ may be termed the standard free energy under the conditions defined.

Let F be the free energy of a solution containing n_1 moles of S_1, n_2 moles of S_2, etc.; by definition,

$$F = E - TS + PV$$

differentiating with respect to n_1,

$$\frac{dF}{dn_1} = \frac{dE}{dn_1} - T\frac{dS}{dn_1} + p\frac{dV}{dn_1} \qquad (1.59)$$

or

$$\overline{F_1} = \overline{E_1} - T\overline{S_1} + P\overline{V_1}$$
$$= \overline{H_1} - T\overline{S_1} \qquad (1.60)$$

where

$$S_1 = \left(\frac{ds}{dn_1}\right)_{t,p,n_2} \text{ etc.}$$

is the partial molar entropy defined in the same way as the partial molar energy or volume, etc. We can now find the change of F with temperature and pressure. First, differentiating F with respect to T, we have

$$\left(\frac{dF}{dT}\right)_{p,n} = -S$$

and differentiating again with respect to n_1,

$$\left(\frac{d^2F}{dT\,dn_1}\right)_p = -\frac{dS}{dn_1} = \overline{S_1}$$

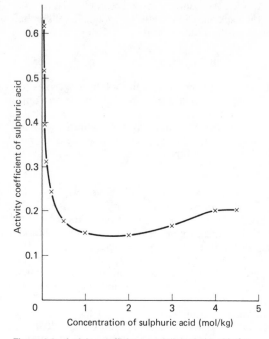

Figure 1.6 Activity coefficient–molality relationship for aqueous sulphuric acid

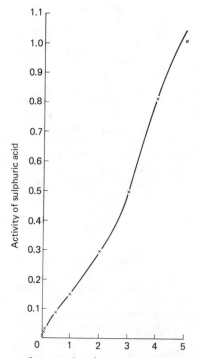

Figure 1.7 Activity–molality relationship for aqueous sulphuric acid

since

$$\left(\frac{d^2F}{dT\,dn_1}\right)_p = \frac{d^2F}{dn_1\,dT} = \frac{d\overline{F}_1}{dT}$$

Introducing the value of $\overline{S}_1$ given by Equation 1.60 we have

$$\frac{d\overline{F}_1}{dT} = \overline{S}_1 = \frac{\overline{F}_1 - \overline{E}_1}{T}$$

dividing through by T, we find

$$\frac{I}{T} \times \frac{d\overline{F}_1}{dT} - \frac{\overline{F}_1}{T^2} = -\frac{\overline{H}_1}{T^2}$$

or

$$\frac{d(\overline{F}_1/T)}{dT} = -\frac{\overline{H}}{T^2}$$

i.e.

$$\frac{d(\overline{F}_1)}{dT} = \frac{d(\overline{F}_1^0/T)}{dT} = -\frac{\overline{H}_1}{T^2} + \frac{\overline{H}_1^0}{T^2} \tag{1.61}$$

from Equations 1.59 and 1.61,

$$\frac{d(\log a)}{dT} = \frac{d(\overline{F}_1/T)}{R\,dT} - \frac{d(\overline{F}_1^0/T)}{R\,dT}$$

$$= \frac{-\overline{H}_1}{RT^2} + \frac{\overline{H}_1^0}{RT^2} = \frac{-L}{RT^2} \tag{1.62}$$

In this equation for the temperature coefficient of activity, $\overline{H}_1$ and H_1^0 are the partial heat contents of the component in a given solution and in the standard state to which F_1^0 refers and $L_1 = \overline{H}_1 - H_1^0$ is the heat content relative to the standard state.

The change in activity (and also in the activity coefficient, if the composition is expressed in a way which does not depend on the temperature) over a range of temperature can be obtained by integrating this equation. For a wide range of temperatures it may be necessary to give L_1 as a function of temperature as in the Kirchhoff equation.

1.9 Influence of ionic concentration in the electrolyte on electrode potential

The oxidation potential of a cation electrode in a solution of ionic activity a is given by the general equation

$$E_1 = E_M^0 - \frac{RT}{nF} \ln a \tag{1.63}$$

As has been seen from Equation 1.54,

$$a = \gamma m$$

where m is the molality of the solute (moles or

gram-ions per 1000 g solvent) and γ is the appropriate activity coefficient. Hence

$$E_1\,(V) = E_M^0 - \frac{RT}{nF} \ln \gamma m$$

$$= E_M^0 - \frac{0.059\,15}{n} \log \gamma m \tag{1.64}$$

If the solution is diluted to decrease the activity of the cations to one-tenth of its initial value, that is to say to $0.1a$, i.e. $0.1a = \gamma' m'$, the electrode potential becomes

$$E_2\,(V) = E_M^0 - \frac{RT}{nF} \ln 0.1a$$

$$= E_M^0 - \frac{0.059\,15}{n} \log 0.1a \tag{1.65}$$

i.e.

$$E_2\,(V) = E_M^0 - \frac{RT}{nF} \ln \gamma' m'$$

$$= E_M^0 - \frac{0.059\,15}{n} \log \gamma' m' \tag{1.66}$$

The resulting change of potential is obtained by subtracting E_1 from E_2:

$$E_2 - E_1\,(V) = \frac{RT}{nF} (\ln 0.1a - \ln a)$$

$$= \frac{RT}{nF} \ln 10$$

$$= \frac{0.059\,15}{n} \log 10$$

$$= \frac{0.059\,15}{n} \tag{1.67}$$

i.e.

$$E_2 - E_1\,(V) = \frac{RT}{nF} (\ln \gamma' m' - \ln \gamma m)$$

$$= \frac{RT}{nF} \ln \left(\frac{\gamma' m'}{\gamma m}\right)$$

$$= \frac{0.059\,15}{n} \log \left(\frac{\gamma' m'}{\gamma m}\right) \tag{1.68}$$

It can be seen, therefore, that at 25°C every ten-fold decrease in ionic activity or, approximately, in the concentration of the cations results in the oxidation potential becoming more positive by $0.059\,15/n$ V, where n is the valence of the ions. For bivalent ions, such as Zn^{2+}, Cd^{2+}, Fe^{2+}, Cu^{2+}, etc., the value of n is 2, and hence the electrode potential changes by $0.059\,15/2$, i.e. 0.0296 V, for every ten-fold change of ionic activity; a hundred-fold change, which is equivalent to two successive ten-fold changes, would mean an alteration of $0.059\,15$ V in the potential at 25°C. For univalent ions, n is 1 and hence ten-fold and hundred-fold changes in the activities of the reversible ions produce potential changes of 0.059 15 and 0.1183 V, respectively. The alteration of potential is not determined by the actual ionic concentrations or activities, but by the ratio of the two concentrations; that is, by the relative change of concentration. Thus, a change from 1.0 gram-ion to 0.1 gram-ion per litre produces the same change in potential as a decrease from 10^{-6} to 10^{-7} gram-ions per litre; in each case the ratio of the two concentrations is the same, namely 10 to 1.

An equation similar to Equation 1.67, but with a negative sign, can be derived for electrodes reversible with respect to anions; for such ions, therefore, a ten-fold decrease of concentration or activity, at 25°C, causes the oxidation potential to become 0.059 15/n V more negative. For reduction potentials, the changes are of the same magnitude as for oxidation potentials, but the signs are reversed in each case.

To quote a particular example, the concentration of sulphuric acid in a fully charged lead–acid battery is approximately 29% by weight (relative density 1.21) whilst that in a fully discharged battery is 21% by weight (relative density 1.15).

Weight concentrations of 29% and 21% of sulphuric acid in water, respectively, correspond to molalities (mmol H_2SO_4/1000 g water) of

$$m = \frac{21 \times 1000}{98 \times (100 - 21)} = 2.71$$

and

$$m' = \frac{29 \times 1000}{98 \times (100 - 29)} = 4.17$$

The activity coefficients (γ) corresponding to $m =$ 2.71 and $m' = 4.17$ molal sulphuric acid are respectively $\gamma = 0.161$ and $\gamma' = 0.202$ (obtained from standard activity tables, see Table 1.3).

Hence the activities (γm) are

$$a = 0.161 \times 2.71 = 0.436$$

and

$$2a' = 0.202 \times 4.17 = 0.842$$

Hence, from Equation 1.64,

$$E_1 = E_M^0 - \frac{0.059\ 15}{n} \log 0.436$$

From Equation 1.65,

$$E_2 = E_M^0 - \frac{0.059\ 15}{n} \log 0.842$$

From Equation 1.68,

$$E_2 - E_1 = \frac{0.059\ 15}{n} \log \left(\frac{0.842}{0.436}\right)$$

$$= \frac{0.059\ 15}{n} \log 1.931$$

if $n = 2$, $E_2 - E_1$, the potential change accompanying an increase in the concentration of sulphuric acid from 2.71 to 4.17 molal, is 0.008 45 V at 25°C. For relatively concentrated solutions of sulphuric acid, as in the case of the example just quoted, the ratio of the activities at the two acid concentrations is similar to the ratio of the molal concentrations:

Molality of H_2SO_4, m	2.71	4.17
Activity coefficient of H_2SO_4, γ	0.161	0.202
Activity, $a = \gamma m$	0.436	0.842

$$\text{Ratio of molalities} = \frac{4.17}{2.71} = 1.54$$

$$\text{Ratio of activities} = \frac{0.842}{0.436} = 1.93$$

i.e. the two ratios are fairly similar. For less concentrated sulphuric acid solutions (0.01 molal and 0.0154 molal, i.e. in the same concentrations ratio as the stronger solutions) these two ratios are not as similar, as the following example illustrates:

Molality of H_2SO_4, m	0.01	0.0154
Activity coefficient of H_2SO_4, γ	0.617	0.555
Activity, $a = \gamma m$	0.006 17	0.0855

$$\text{Ratio of molalities} = \frac{0.1}{0.01} = 10.00$$

$$\text{Ratio of activities} = \frac{0.0855}{0.006\ 17} = 1.38$$

Potential changes at 25°C resulting from the same two-fold change in concentration of sulphuric acid are as follows:

From 2.71 to 4.17 molal

$$E_2 - E_1 = \frac{0.059\ 15}{2} \log \left(\frac{0.600}{0.393}\right)$$

$$= 0.005\ 43 \text{ V}$$

From 0.01 to 0.0154 molal

$$E_2 - E_1 = \frac{0.059\ 15}{2} \log \left(\frac{0.0855}{0.006\ 27}\right)$$

$$= 0.0418 \text{ V}$$

That is, when more dilute solutions and stronger solutions are diluted by the same amount, the e.m.f. difference obtained with the former is greater than that obtained with the latter.

The greater the concentration difference of the two solutions, the greater the e.m.f. difference $E_2 - E_1$, as shown in Table 1.4.

Table 1.4 Change of e.m.f. with concentration of electrolyte

	Concentration, m (mol/kg)	Activity coefficient, γ	Activity $a = \gamma m$	Ratio of concentrations	Ratio of activities	$E_2 - E_1$ (V) $n = 2$
0.2 molal H_2SO_4 diluted two-fold	0.2 0.1	0.244 0.313	0.048 8 0.031 3 } 2		1.56	0.005 71
0.2 molal H_2SO_4 diluted ten-fold	0.2 0.02	0.244 0.519	0.048 8 0.010 38 } 10		4.70	0.019 9
2 molal H_2SO_4 diluted two-fold	2 1	0.147 0.150	0.294 0.150 } 2		1.96	0.008 64
2 molal H_2SO_4 diluted ten-fold	2 0.2	0.147 0.244	0.294 0.048 8 } 10		6.02	0.023 0

1.10 Effect of sulphuric acid concentration on e.m.f. in the lead–acid battery

For the reaction

$$Pb + PbO_2 + 2H_2SO_4 = 2PbSO_4 + 2H_2O$$

from the free energy (Equation 1.27),

$$\Delta F = \Delta F^0 + RT \ln \left(\frac{a^2_{PbSO_4} \times a^2_{H_2O}}{a_{Pb} \times a_{PbO_2} \times a^2_{H_2SO_4}} \right)$$

as, from Equation 1.23,

$$\Delta F = -nFE \quad \text{and} \quad \Delta F^0 = -nFE^0$$

we have

$$E = E^0 - \frac{RT}{nF} \log \left(\frac{a^2_{PbSO_4} \times a^2_{H_2O}}{a_{Pb} \times a_{PbO_2} \times a^2_{H_2SO_4}} \right) \quad (1.69)$$

where a is the activity of sulphuric acid solution ($m\gamma$), m the concentration in moles per kilogram, γ the activity coefficient, E^0 the e.m.f. in standard state, E the cell e.m.f., $F = 1\,F$ (96 500 C), n the number of electrons involved in the net chemical reaction, T the temperature in kelvins, and R the gas constant (8.312 international joules).

By definition, $a_{PbSO_4} = a_{H_2O} = a_{Pb} = a_{PbO_2} = 1$.

$$E = E^0 - \frac{RT}{nF} \ln \left(\frac{1}{a^2_{H_2SO_4}} \right) \quad (1.70)$$

at 25°C

$$E = E^0 - \frac{8.312\,(273 + 25) \times 2.303}{2 \times 96\,500} \log \left(\frac{1}{a^2_{H_2SO_4}} \right)$$

$$= E^0 - 0.029\,57 \, \log \left(\frac{1}{a^2_{H_2SO_4}} \right) \quad (1.71)$$

Consider the previously discussed case of a lead–acid battery at the start and the end of discharge. At the start of discharge the electrolyte contains 29% by weight of sulphuric acid, i.e. molality = 4.17 and activity coefficient = 0.202

(Table 1.3). Therefore, the activity is $0.202 \times 4.17 = 0.842$. Similarly, at the end of discharge, the acid strength is 21% by weight, i.e. $m = 2.71$ and $\gamma = 0.161$, i.e. $a = 0.436$.

If the cell potentials at the start and end of discharge are respectively by E_{29} and E_{21}, then

$$E_{29} = E^0 - 0.029\,57 \, \log \left(\frac{1}{(0.842)^2} \right)$$

$$= E^0 - 0.004\,40 \text{ V}$$

and

$$E_{21} = E^0 - 0.029\,57 \, \log \left(\frac{1}{(0.436)^2} \right)$$

$$= E^0 - 0.021\,32 \text{ V}$$

Therefore the potential decrease ($E_{29} - E_{21}$) during discharge is given by

$$E_{29} - E_{21} = -0.004\,40 + 0.021\,52 \text{ V}$$
$$= 0.0169 \text{ V at } 25°C$$

Alternatively

$$E_{29} = E^0 - 0.029\,57 \, \log \left(\frac{1}{a^2_{29}} \right)$$

$$E_{21} = E^0 - 0.029\,57 \, \log \left(\frac{1}{a^2_{21}} \right)$$

where E_{29} is the cell e.m.f. at 25°C when the sulphuric acid concentration is 29% by weight, i.e. activity $a_{29} = 0.842$, and E_{21} is the cell e.m.f. at 25°C when the sulphuric acid concentration is 21% by weight, i.e. activity $a_{21} = 0.436$.

$$E_{29} - E_{21} = 0.029\,57 \left[\log \left(\frac{1}{a^2_{21}} \right) - \log \left(\frac{1}{a^2_{29}} \right) \right]$$

$$= 2 \times 0.029\,57 \, \log \left(\frac{a_{29}}{a_{21}} \right)$$

$$= 0.059\,15 \, \log \left(\frac{a_{29}}{a_{21}} \right)$$

$$= 0.0169 \text{ V at } 25°C$$

or, in general,

$$E_{c_1} - E_{c_2} = \frac{2 \times 8.312 \times (273 + T) \times 2.303 \log (a_{c_1}/a_{c_2})}{2 \times 96\,500}$$

$$= 0.000\,198\,4\,(273 + T) \log (a_{c_1}/a_{c_2})$$

where E_{c_1} is the e.m.f. at $T°C$ when the acid has activity a_{c_1} at the start of discharge, E_{c_2} is the e.m.f. at $T°C$ when acid has activity a_{c_2} at the end of discharge, and T is the cell temperature in °C. That is,

$$E_{c_1} - E_{c_2} = \frac{2 \times R \times (273 + T)\,2.303}{nF} \log \left(\frac{a_{c_1}}{a_{c_2}}\right) \quad (1.72)$$

To ascertain the value of the standard potential E^0

According to the literature, the e.m.f. of a cell containing 21% by weight (2.71 molal) sulphuric acid at 15°C is 2.0100 V.

$$E = E^0 - \frac{2.303 \times 8.312 \times 288}{2 \times 96\,500} \log \left(\frac{1}{(0.436)^2}\right)$$

$$= E^0 - 0.028\,56 \times \log \left(\frac{1}{(0.436)^2}\right)$$

$$E^0 = E + 0.028\,56 \times \log \left(\frac{1}{(0.436)^2}\right)$$

$$= 2.0100 + 0.028\,56 = 2.030\,59\,\text{V}$$

$$E_{29}\,(25°C) = 2.030\,59 - 0.004\,40 = 2.0262\,\text{V}$$

and

$$E_{21}\,(25°C) = 2.030\,59 - 0.021\,32 = 2.0093\,\text{V}$$

These equations can be used to calculate the effect of sulphuric acid concentration (expressed as activity) and cell temperature on cell e.m.f. If, for example, the electrolyte consists of 29% by weight sulphuric acid at the start of discharge (i.e. activity $a_{29} = 0.842$) decreasing to 21% by weight sulphuric acid (i.e. activity $a_{21} = 0.436$) at the end of discharge, and if the temperature at the start of discharge, T_I, is 15°C increasing to 40°C (T_F) during discharge, then from Equation 1.71

$$E_{29} = E^0 - \frac{8.312 \times (273 + 15) \times 2.303}{2 \times 96\,500} \log \left(\frac{1}{(0.842)^2}\right)$$

and

$$E_{21} = E^0 - \frac{8.312 \times (273 + 40) \times 2.303}{2 \times 96\,500} \log \left(\frac{1}{(0.436)^2}\right)$$

If $E^0 = 2.0359\,\text{V}$,

$$E_{29} = 2.0263\,\text{V}$$

$$E_{21} = 2.0082\,\text{V}$$

During discharge, the e.m.f. decreases from 2.0263 V to 2.0082 V, i.e. by 0.0181 V. Both the temperature increase and the lowering of acid concentration contribute to a reduction in e.m.f. Alternatively,

$$E_{c_1} = E^0 - \frac{8.312\,(273 + T_I)\,2.303}{2 \times 96\,500} \log \left(\frac{1}{(a_{c_1})^2}\right)$$

$$E_{c_2} = E^0 - \frac{8.312\,(273 + T_F)\,2.303}{2 \times 96\,500} \log \left(\frac{1}{(a_{c_2})^2}\right)$$

$$E_{c_1} - E_{c_2} = \frac{8.312 \times 2.303}{96\,500} \left\{(273 + T_I) \log a_{c_1}\right.$$

$$\left. - (273 + T_F) \log a_{c_2}\right\}$$

$$= 0.000\,198\,4\,(288 \log 0.842 - 313 \log 0.436)$$

$$= 0.0181\,\text{V}$$

In general,

$$E_{c_1} - E_{c_2} = \frac{8.312 \times 2.303}{96\,500} \left[(273 + T_I) \log a_{c_1}\right.$$

$$\left. - (273 + T_F) \log a_{c_2}\right]$$

$$= \frac{R \times 2.303}{F} \left[(273 + T_I) \log a_{c_1}\right.$$

$$\left. - (273 + T_F) \log a_{c_2}\right] \quad (1.73)$$

and

$$E_{c_1} = E^0 - \frac{R \times 2.303}{2F}\,(273 + T_I) \quad (1.74)$$

$$E_{c_2} = E^0 - \frac{R \times 2.303}{2F}\,(273 + T_F)$$

The effect of a change of sulphuric acid concentration and consequently activity on cell e.m.f. at 25°C is shown in Table 1.5. The e.m.f. difference of 0.062 V obtained between acid concentrations of 47% and 32.9% is quite small. However, on a 500-cell 100-V lead–acid battery this difference will cause a difference in e.m.f. of approximately 3 V which, in some types of battery application, will be significant.

1.11 End-of-charge and end-of-discharge e.m.f. values

As seen in Table 1.5 the thermodynamically calculated e.m.f. of a lead–acid battery at the end of charge is about 2.03 V at 25°C, corresponding to a sulphuric acid concentration of about 29% by weight. The thermodynamic e.m.f. is that of the charged battery on open circuit. Correspondingly, the open circuit e.m.f. of a discharged battery (corresponding to 21% sulphuric acid concentration by weight) is about 2.00 V at 25°C. The actual, as

Table 1.5 Effect of activity of sulphuric acid on cell e.m.f. of a lead–acid battery at 25°C

Sulphuric acid concentration (% by weight)	Molality, m (g H$_2$SO$_4$/1000 g water)	Activity coefficient, γ (see Table 1.3)	Activity ($a = \gamma m$)	e.m.f. (V)*
0.097	0.01	0.617	0.006 17	1.8999
0.196	0.02	0.519	0.010 38	1.9133
0.487	0.05	0.397	0.019 85	1.9299
0.970	0.1	0.313	0.031 3	1.9417
1.92	0.2	0.244	0.048 8	1.9530
4.67	0.5	0.178	0.089	1.9685
8.92	1	0.150	0.150	1.9819
16.39	2	0.147	0.294	1.9992
22.72	3	0.166	0.498	2.0127
28.16	4	0.203	0.812	2.0253
32.88	5	0.202	1.010	2.0308

* $2.030\,59 - 0.029\,57 \log (1/a^2_{H_2SO_4})$

opposed to the thermodynamic, e.m.f. values differ from these values because of various charge and discharge effects. During passage of current, polarization effects occur which cause variations of the voltage from the thermodynamically calculated values during charge and discharge. Figure 1.8 shows typical charge and discharge curves for a lead–acid battery. During charge, the e.m.f. rises rapidly to a little over 2.1 V and remains steady, increasing very slowly as the charge proceeds. At 2.2 V, oxygen begins to be liberated at the positive plate, and at 2.3 V hydrogen is liberated at the negative plate. The charge is now completed and we enter the region of overcharge. Further passage of current leads to the free evolution of gases and a rapid rise in e.m.f. The overcharge state is not only wasteful in charging current, which is now entirely devoted to gas production rather than charging of the plates, but it causes mechanical damage to the battery. If the charge is stopped at any point during charge (i.e. open circuit), the e.m.f. returns in time, when free gases have become dislodged, to the equilibrium thermodynamic value for the particular state of charge (Table 1.5). During discharge, the cell e.m.f. drops rapidly to just below 2 V. The preliminary kink observed in Figure 1.8 is due to the formation of a high-resistance layer of lead sulphate on the surface of the plates which is soon dispersed.*

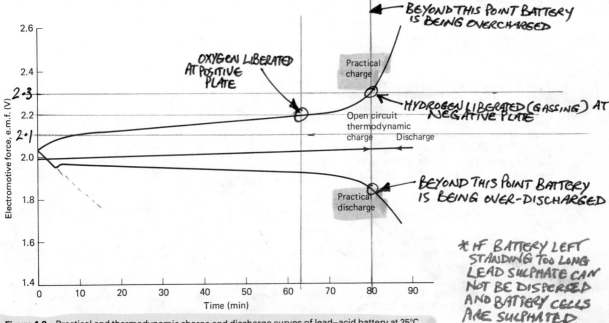

Figure 1.8 Practical and thermodynamic charge and discharge curves of lead–acid battery at 25°C

Table 1.6 Electromotive force and temperature coefficient of voltage for various concentrations of sulphuric acid electrolyte in the lead–acid battery

Sulphuric acid concentration (% by weight)	E			Temperature coefficient, dE/dT			Decrease in e.m.f. (V) with 25°C temperature rise
	15°C	25°C	40°C	15–25°C	25–40°C	15–40°C	
0.097	1.9043	1.8999	1.8934	0.000 44	0.000 43	0.000 44	0.0109
0.196	1.9172	1.9133	1.9074	0.000 39	0.000 39	0.000 39	0.0098
0.487	1.9333	1.9299	1.9249	0.000 34	0.000 33	0.000 34	0.0084
0.970	1.9446	1.9417	1.9372	0.000 29	0.000 30	0.000 30	0.0074
1.920	1.9556	1.9530	1.9491	0.000 26	0.000 26	0.000 26	0.0065
4.67	1.9706	1.9685	1.9654	0.000 21	0.000 20	0.000 21	0.0052
8.92	1.9835	1.9819	1.9794	0.000 16	0.000 17	0.000 16	0.0041
16.39	2.0002	1.9992	1.9976	0.000 10	0.000 11	0.000 10	0.0026
21	2.0100	2.0092	2.0082	0.000 07	0.000 07	0.000 07	0.0018
22.72	2.0133	2.0127	2.0118	0.000 06	0.000 06	0.000 06	0.0015
28.16	2.0254	2.0253	2.0250	0.000 01	0.000 02	0.000 02	0.0004
29	2.0263	2.0262	2.0260	0.000 01	0.000 01	0.000 01	0.0003
32.88	2.0308	2.0308	2.0308	0.000 00	0.000 00	0.000 00	0.0000

The e.m.f. falls steadily during the discharge, and when it has reached 1.8 V the cell should be recharged, as the further withdrawal of current causes the voltage to fall rapidly. The difference between the practically observed charge and discharge curves shown in Figure 1.8 is due to changes in concentration of the acid in contact with the active material of the plates. The plates are full of small pores in which acid diffusion is very slow, so that the concentration of acid is greater during the charge and less during the discharge than in the bulk of electrolyte. The thermodynamic treatment previously discussed does not take account of such effects as these, which occur in a practical battery. Figure 1.8 compares cell e.m.f. values during charge and discharge at 25°C with the thermodynamic curve at 25°C for charge and discharge.

1.12 Effect of cell temperature on e.m.f. in the lead–acid battery

Table 1.6 and Figure 1.9 illustrate the effect of cell temperature on thermodynamic cell e.m.f. at various sulphuric acid concentrations. Three temperatures likely to be encountered in battery operation are included: 15°C, 25°C and 40°C. Also calculated are the average temperature coefficients of voltage (dE/dT) for the three temperature ranges 15–25°C, 25–40°C and 15–40°C. It can be seen that dE/dT is independent of temperature range but depends on the electrolyte concentration. The temperature coefficient of resistance reduces to very low values, i.e. e.m.f. is relatively insensitive to temperature change, for stronger solutions of sulphuric acid. Figure 1.10 shows the relationship of dE/dT to acid concentration for the temperature range 15–40°C.

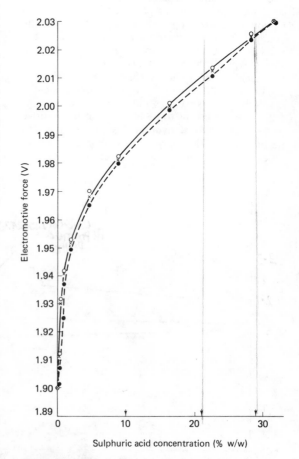

Figure 1.9 Effect of sulphuric acid concentration on thermodynamic cell e.m.f. of a lead–acid battery: ○, 15°C; ×, 25°C; ●, 40°C

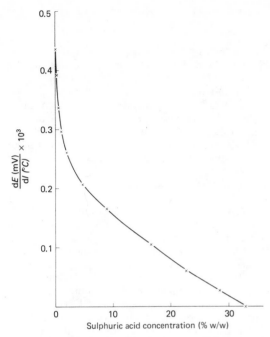

Figure 1.10 Relationship of temperature coefficient of e.m.f. with sulphuric acid concentration in lead–acid battery

1.13 Effect of temperature and temperature coefficient of voltage dE/dT on heat content change of cell reaction

According to the Gibbs–Helmholtz equation (Equation 1.20) the heat content change (ΔH calories) in a reaction is given by

$$\Delta H = -\frac{nFE}{4.18} + \frac{nFT}{4.18}\left(\frac{dE}{dT}\right)_p$$

where ΔH is the heat content change (in calories), n the number of electrons involved in the reaction (the number of equivalents of chemical action), $F = 1\,F$ (96 500 C), T the temperature (in kelvins), $(dE/dT)_p$ the temperature (T) coefficient of cell e.m.f. (E) at constant pressure, p, and 4.18 the conversion factor from international joules to calories.

Table 1.7 shows heat content changes occurring at various sulphuric acid concentrations in the lead–acid battery. The ΔH values with 21% by weight of sulphuric acid at 15–40°C of 91 915 to 91 749 calories for the reaction

$$Pb + PbO_2 + 2H_2SO_4 = 2PbSO_4 + 2H_2O$$

are in fair agreement with the calorimetrically determined value of 89 400 cal at the same acid concentration. In general, heat content changes are higher at higher acid concentrations and lower cell temperatures.

A linear relationship would be expected, and is indeed found, between cell e.m.f. (E) and heat content change (ΔH), at a particular cell temperature (T) as shown in Figure 1.11 for results obtained at 25°C.

1.14 Derivation of the number of electrons involved in a cell reaction

Consider the following cell reaction:

$$A^{n+} + ne = A$$

or

$$B + ne = B^{n-}$$

The number of equivalents of chemical action or the number of electrons involved in the reaction is denoted by n (i.e. $n \times F = n \times 96\,500$ amp-seconds). If the heat content change (ΔH) of this cell reaction

Table 1.7 Heat content changes in the lead–acid cell as a function of cell temperature, temperature coefficient of e.m.f. and acid concentration

% H_2SO_4 (w/w)	E (V)			$(dE/dT)_p$			$T(dE/dT)_p$			$\Delta H = 46191\,(T(dE/dT)_p - E)$ cal		
	15°C	25°C	40°C	15–25°C	25–40°C	15–40°C	288 K	298 K	313 K	288 K	298 K	313 K
0.097	1.9043	1.8999	1.8934	0.000 44	0.000 43	0.000 43	0.1267	0.1281	0.1346	−82 109	−81 795	−81 241
0.196	1.9172	1.9133	1.9074	0.000 39	0.000 39	0.000 39	0.1123	0.1162	0.1221	−83 370	−83 010	−82 465
0.487	1.9333	1.9299	1.9249	0.000 34	0.000 33	0.000 34	0.0950	0.0983	0.1064	−84 913	−84 603	−83 998
0.970	1.9446	1.9417	1.9372	0.000 29	0.000 30	0.000 30	0.0864	0.0894	0.0939	−85 832	−85 559	−85 144
1.92	1.9556	1.9530	1.9491	0.000 26	0.000 26	0.000 26	0.0759	0.0775	0.0814	−86 825	−86 631	−86 271
4.67	1.9706	1.9685	1.9654	0.000 21	0.000 20	0.000 21	0.0576	0.0626	0.0657	−88 363	−88 035	−87 749
8.92	1.9835	1.9819	1.9794	0.000 16	0.000 17	0.000 16	0.0490	0.0507	0.0501	−89 356	−89 204	−89 116
16.39	2.0002	1.9992	1.9976	0.000 10	0.000 11	0.000 10	0.0289	0.0328	0.0313	−91 056	−90 830	−90 825
21	2.0100	2.0093	2.0082	0.000 70	0.000 70	0.000 70	0.0201	0.0209	0.0219	−91 915	−91 846	−91 749
22.72	2.0133	2.0127	2.0118	0.000 60	0.000 60	0.000 60	0.0173	0.0179	0.0188	−92 197	−92 142	−92 058
28.16	2.0254	2.0253	2.0250	0.000 01	0.000 02	0.000 02	0.0029	0.0060	0.0062	−93 421	−93 273	−93 250
29	2.0263	2.0262	2.0260	0.000 01	0.000 01	0.000 01	0.0029	0.0029	0.0031	−93 463	−93 458	−93 440
32.88	2.0308	2.0308	2.0308	0.000 00	0.000 00	0.000 00	0.0000	0.0000	0.0000	−93 805	−93 805	−93 805

The y-axis of Figure 1.10 reads: $\dfrac{dE\,(mV)}{d/\,(°C)} \times 10^3$; the x-axis reads: Sulphuric acid concentration (% w/w)

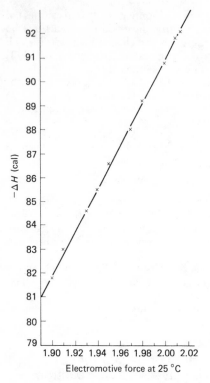

Figure 1.11 Lead–acid battery. Relationship between cell e.m.f. and heat content change (ΔH) at 25°C

is determined by calorimetry and the e.m.f. (E) and temperature coefficient of e.m.f. (dE/dT) are determined electrochemically, n can be found from Equation 1.20, as follows:

$$\Delta H = \frac{-nFE}{4.18} + \frac{FT}{4.18}\left(\frac{dE}{dT}\right)_p$$

$$= \frac{nF}{4.18}\left[\left(\frac{dE}{dT}\right)_p - E\right]$$

i.e.

$$n = \frac{4.18\,\Delta H}{F\left[\left(\frac{dE}{dT}\right)_p - E\right]}$$

$$= \frac{\Delta H}{23\,095\left[\left(\frac{dE}{dT}\right)_p - E\right]} \qquad (1.75)$$

1.15 Thermodynamic calculation of the capacity of a battery

In the reaction in which 1 mol of lead and lead dioxide react with 2 mol of sulphuric acid to produce 2 mol of lead sulphate and 2 mol of water (i.e. the discharge process in a lead–acid battery):

$$Pb + PbO_2 + 2H_2SO_4 + 2e = 2PbSO_4 + 2H_2O$$

the amount of electrical current produced is 2 F, i.e. 193 000 C and the heat content change in the reaction, ΔH, is approximately −90 500 cal.

If the initial concentration of sulphuric acid in the battery is 29% by weight and the final concentration at the end of discharge is 21% by weight, the cell e.m.f. values are (at 25°C) respectively 2.0262 and 2.0093 V, and the temperature coefficients at these e.m.f. values are 0.000010 and 0.000070 (Table 1.6).

In the above reaction 2 mol (196 g) of sulphuric acid are consumed, thus reducing the volume of electrolyte by 196/1.8305 = 107.07 ml (the relative density of 100% sulphuric acid is 1.8305) and 2 mol of water are produced increasing the volume of electrolyte by 36 ml. Thus the net volume reduction in the electrolyte accompanying the production of 93 000 C is 107.07 − 36 = 71.07 ml. This electrolyte volume is, of course, recovered during the next charge cycle when the sulphuric acid is regenerated.

By definition, the quoted values of ΔH and 193 000 C (2 F) refer to the above reaction in which 1 mol of lead and of lead dioxide react with 2 mol of sulphuric acid.

Knowing ΔH, E, dE/dT and the cell temperature (in kelvins) it is possible to calculate the ampere hour capacity (C) of this cell from Equation 1.18:

$$\Delta H = -\frac{nF}{4.18}\left[E - T\left(\frac{dE}{dT}\right)\right]$$

$$= -\frac{3600 \times C}{4.18}\left[E - T\left(\frac{dE}{dT}\right)\right]$$

i.e. $nF = 3600C$. Therefore

$$C = -\frac{4.18 \times \Delta H}{3600\left[E - T\left(\frac{dE}{dT}\right)\right]} \qquad (1.76)$$

as ΔH is negative, C will have a positive value.

At 25°C, in 29% by weight sulphuric acid,

$$E = 2.0262\,V$$

$$\frac{dE}{dT} = 0.000\,04 \quad \text{(mean value of 21–29\% by weight sulphuric acid)}$$

$$\Delta H = -93\,458$$

$$T = 298\,K$$

$$C = 53.9\,C$$

This is illustrated in Figure 1.12.

We thus have the capacity of this battery based on the molar reaction accompanied by the production of 95 458 cal.

On this basis it is possible to devise a calorimetric method for obtaining the capacity of any battery, as will be discussed later.

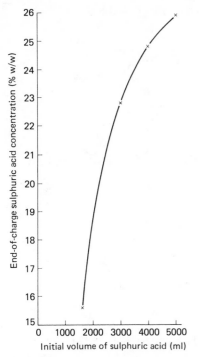

Figure 1.12 Effect of initial volume of sulphuric acid electrolyte (29% w/w or 35.10 g/100 ml) in a lead–acid battery on final end-of-discharge concentration

In the system described above it was assumed that the initial concentration of sulphuric acid at the start of the discharge was 29% by weight (i.e. the normal concentration recommended for a lead–acid battery) and the final concentration of sulphuric acid at the end of discharge was 21% by weight. During discharge, 196 g of sulphuric acid were consumed and 36 g of water produced. It remains to calculate the initial volume of 29% sulphuric acid that must be used in the battery such that, when it has supplied 196 g of sulphuric acid to form lead sulphate, the final acid concentration of the electrolyte will be 21%.

At the start of discharge there are:

100% by wt H_2SO_4 / 29% by wt H_2SO_4

29 g   100 g

i.e.

$$\frac{29}{1.8305} = 15.842 \text{ ml} \qquad \frac{100}{1.2104} = 82.617 \text{ ml}$$

(where 1.8305 and 1.2104, respectively, are the relative densities of 100% and 29% by weight sulphuric acid). Therefore

Volume or weight of water present
= 66.775 ml or g per 82.617 ml solution
= 80.824% w/v

and

29 g 100% H_2SO_4/100 ml solution

$$= \frac{29 \times 100}{82.617}$$

$$= 35.102\% \text{ w/v}$$

At the end of discharge:

100% by wt H_2SO_4 / 21% by wt H_2SO_4

21 g   100 g

i.e.

$$\frac{21}{1.8035} = 11.472 \text{ ml} \qquad \frac{100}{1.1471} = 87.176 \text{ ml}$$

(where 1.8035 and 1.1471, respectively, are the specific quantities of 100% and 21% by weight sulphuric acid). Therefore

Volume or weight of water present
= 75.704 ml or g per 87.176 ml solution
= 86.840% w/v

and

29 g 100% H_2SO_4/100 ml solution

$$= \frac{21 \times 100}{87.176} = 24.089\% \text{ w/v}$$

If V_I and V_F, respectively, are the volumes (in ml) of 29% by weight and 21% by weight sulphuric acid present in the battery at the start and end of the discharge,

At start of discharge

100% by wt H_2SO_4	H_2O/100 ml 29% by wt H_2SO_4
35.102 g	80.824 g

Therefore

$$\frac{V_I \times 35.102}{100} \text{ g} \qquad \frac{V_I \times 80.824}{100} \text{ g}$$

At end of discharge

	H_2O/100 ml 21% by wt H_2SO_4
24.089 g	86.840 g

Therefore

$$\frac{V_F \times 24.089}{100} \text{ g} \qquad \frac{V_F \times 86.840}{100} \text{ g}$$

It is known that

$$\frac{V_I \times 35.012}{100} - \frac{V_F \times 24.089}{100} = 96 \text{ g } 100\% \ H_2SO_4 \quad (2 \text{ mol})$$

and

$$\frac{V_F \times 86.840}{100} - \frac{V_I \times 80.824}{100} = 36 \text{ g water} \quad (2 \text{ mol})$$

Solving for V_I and V_F,

$$V_I = 1624.5 \text{ ml}$$
$$V_F = 1553.4 \text{ ml}$$
$$V_I - V_F = 71.1 \text{ ml volume reduction (theoretical)}$$

We can now completely define the battery. When 1 mol of each of lead and lead dioxide in contact with 1624.5 ml 29% by weight sulphuric acid is discharged, then when the lead and lead dioxide are completely reacted the final volume will have reduced to 1553.4 ml and the final acid concentration will be 21% by weight sulphuric acid. Two faradays (193 000 C) and ΔH (−93 458 cal) will have been evolved for the consumption of 2 mol (196 g) of sulphuric acid. If when a battery of unknown capacity, C, is discharged from V_1 ml of $A\%$ by weight to V_2 ml of $B\%$ by weight sulphuric acid, and it is observed calorimetrically that X calories are evolved, corrected for any ohmic (Joule heating) contribution (X has negative value as heat is evolved), then if E, T and dE/dT are known from Equation 1.76:

$$C = - \frac{4.18 \times X}{3600 \left[E - T \left(\dfrac{dE}{dT} \right) \right]} \text{ A h}$$

V_1 and A will be known, so it is possible to calculate V_2 and B and hence the weights of sulphuric acid, lead and lead dioxide consumed during the discharge.

A more generalized approach to the calculation of the volume of sulphuric acid to introduce into a lead–acid battery to optimize performance characteristics is given in Section 1.16.

1.16 Calculation of initial volume of sulphuric acid

The following calculation determines the initial volume of sulphuric acid of known weight concentration, required at the start of discharge, to provide a predetermined final acid weight concentration after the consumption in the cell reaction (i.e. at end of discharge) of a predetermined weight of sulphuric acid. If

C_1 = initial concentration (% w/w) sulphuric acid (at start of discharge)
C_2 = final concentration (% w/w) of sulphuric acid (at end of discharge)
S_1 = relative density of sulphuric acid of concentration C_1
S_2 = relative density of sulphuric acid of concentration C_2
S_3 = relative density of concentrated (100% w/w) sulphuric acid (= 1.8305)

at the start of discharge we have $C_1\%$ H_2SO_4, i.e.

C_1 g 100% H_2SO_4 per 100 g $C_1\%$ H_2SO_4

i.e.

$\dfrac{C_1}{1.8305}$ ml 100% H_2SO_4 per $\dfrac{100}{S_1}$ ml $C_1\%$ H_2SO_4

comprising

$\dfrac{C_1}{1.8305}$ ml 100% H_2SO_4

and

$\dfrac{100}{S_1} - \dfrac{C_1}{1.8305}$ ml water

i.e.

$\dfrac{100}{S_1}$ ml total solution

Therefore

g H_2SO_4 per 100 ml ($C_1\%$ w/w) acid $= C_1S_1$

g H_2O per 100 ml ($C_1\%$ w/w) acid $= \left(\dfrac{100}{S_1} - \dfrac{C_1}{1.8305} \right) S_1$

At the end of discharge, we have $C_2\%$ H_2SO_4,

i.e.

C_2 g 100% H_2SO_4 per 100 g $C_2\%$ H_2SO_4

i.e.

$\dfrac{C_2}{1.8305}$ ml 100% H_2SO_4 per $\dfrac{100}{S_2}$ ml $C_2\%$ H_2SO_4

comprising

$\dfrac{C_2}{1.8305}$ ml 100% H_2SO_4

and

$\dfrac{100}{S_2} - \dfrac{C_2}{1.8305}$ ml water

i.e. $100/S_2$ ml total solution.

g H_2SO_4 per 100 ml ($C_2\%$ w/w) acid $= C_2S_2$

g H_2O per 100 ml ($C_2\%$ w/w) acid $= \left(\dfrac{100}{S_2} - \dfrac{C_2}{1.8305} \right) S_2$

If V_I and V_F, respectively, denote the volumes of acid in the cell at the start and the end of the discharge process, the weight (in g) of 100% H_2SO_4 in the cell at the start of discharge is

$$W_1 = \frac{V_I \times C_1 \times S_1}{100} \tag{1.77}$$

the weight of 100% H_2SO_4 in the cell at the end of discharge is

$$W_2 = \frac{V_F \times C_2 \times S_2}{100} \tag{1.78}$$

the weight of 100% H_2SO_4 consumed during discharge is

$$W_A = W_1 - W_2$$

$$= \frac{V_I \times C_1 \times S_1}{100} - \frac{V_F \times C_2 \times S_2}{100} \qquad (1.79)$$

the weight of water in the cell at start of discharge is

$$W_3 = V_I \left(\frac{100}{S_1} - \frac{C_1}{1.8305}\right) S_1 \qquad (1.80)$$

the weight of water in the cell at the end of discharge is

$$W_4 = V_F \left(\frac{100}{S_2} - \frac{C_2}{1.8305}\right) S_2 \qquad (1.81)$$

and the weight of water consumed during discharge is

$$W_w = (W_3 - W_4)$$

$$= V_I \times S_1 \left(\frac{100}{S_1} - \frac{C_1}{1.8305}\right)$$

$$- V_F \times S_2 \left(\frac{100}{S_2} - \frac{C_2}{1.8305}\right) \qquad (1.82)$$

As it is known from theory that for every 196 g (2 mol) of sulphuric acid consumed 36 g (2 mol) of water are produced, if the weight of sulphuric acid consumed is W_A g then $(W_A \times 36)/196$ g of water are produced, i.e.

$$W_w = \frac{W_A \times 36}{196}$$

Thus, if

$$W_A = \frac{V_I \times C_1 \times S_1}{100} - \frac{V_F \times C_2 \times S_2}{100}$$

then

$$\frac{36 \times W_A}{196} = \frac{V_F \times S_2}{100} \left(\frac{100}{S_2} - \frac{C_2}{1.8305}\right)$$

$$- \frac{V_I \times S_1}{100} \left(\frac{100}{S_1} - \frac{C_1}{1.8305}\right)$$

These two equations can be solved for V_I and V_F:

$$V_F = \frac{W_A \left(\dfrac{36}{196} + \dfrac{100}{C_1 S_1} - \dfrac{I}{1.8305}\right)}{\left(1 - \dfrac{C_2 S_2}{C_1 S_2}\right)} \quad \text{cm}^3 \qquad (1.83)$$

$$V_I = \frac{100}{C_1 S_1} \left(W_A + \frac{C_2 S_2}{100} V_F\right)$$

$$= W_A \left[\frac{\dfrac{100}{C_1 S_1} + \dfrac{C_2 S_2}{C_1 S_1} \left(\dfrac{36}{196} + \dfrac{100}{C_1 S_1} - \dfrac{I}{1.8305}\right)}{\left(1 - \dfrac{C_2 S_2}{C_1 S_1}\right)} \right]$$

$$\qquad (1.84)$$

and the net volume decrease in electrolyte occurring on discharge is

$$V_I - V_F = W_A \left(\frac{I}{1.8305} - \frac{36}{196}\right)$$

$$= 0.3627 \times W_A \qquad (1.85)$$

1.17 Calculation of operating parameters for a lead–acid battery from calorimetric measurements

Suppose that when the electrochemical cell reaction

$$\frac{n}{2}Pb + \frac{n}{2}PbO_2 + nH_2SO_4 = nPbSO_4 + nH_2O$$

is carried out in a calorimeter the observed number of calories produced during a discharge is denoted by Q. As well as heat produced in the electrochemical reaction, additional heat will be produced due to Joule heating:

$$\frac{I^2 \times R \times t}{4.18} \quad \text{cal in } t \text{ s} \qquad (1.86)$$

(where I is current in amps, and R is the electrical resistance of cell and external contacts).

Every attempt should be made to minimize Joule heating by using low electrical resistance components in the experimental arrangement.

Suppose that the ohmic heating component is denoted by Q' calories during the discharge; then the heat production due to the above electrochemical reaction above is given by

$$Q'' = Q - Q'$$

If Q'' is known, it is then possible (using Equation 1.20) to calculate n; as heat is evolved Q'' has a negative value:

$$Q'' = \frac{n \times F}{4.18} \left[E - T\left(\frac{dE}{dT}\right)\right] \qquad (1.20)$$

i.e.

$$n = -\frac{Q'' \times 4.18}{F\left[E - T\left(\dfrac{dE}{dT}\right)\right]} \qquad (1.87)$$

where Q'' is the calories evolved during discharge due to electrochemical reaction, n the number of moles of sulphuric acid involved in the electrochemical reaction, $F = 1F$ (96 500 C), E is the cell e.m.f. (V), T the cell temperature (K) and dE/dT the temperature coefficient of e.m.f.

The weight of sulphuric acid (W_A) is given by

$$W_A = 98 \times n = -\frac{Q'' \times 4.18 \times 98}{F\left[E - T\left(\dfrac{dE}{dT}\right)\right]} \qquad (1.88)$$

Now $nF = 3600\,C$, where C is the A h capacity of the battery. Therefore

$$C = \frac{nF}{3600} = -\frac{Q'' \times 4.18}{3600\left[E - T\left(\dfrac{dE}{dT}\right)\right]} \qquad (1.89)$$

To take a particular example, if Q'' is $-150\,000\,cal$ per discharge,

$E = 2.010\,V$

$\dfrac{dE}{dT} = 0.000\,01$

$T = 293\,K\,(20°C)$

then

$$n = \frac{150\,000 \times 4.18}{96\,500\,(2.0100 - 293 \times 0.000\,01)}$$

$\quad = 3.237\,mol\,sulphuric\,acid$

$\quad = molecular\,weight\,of\,sulphuric\,acid \times n$

$W_A = 98 \times n$

$\quad = 317.226\,g\,sulphuric\,acid$

$$C = \frac{150\,000 \times 4.18}{3600\,(2.0100 - 293 \times 0.000\,01)}$$

$\quad = 86.76\,A\,h$

Suppose that the initial concentration of sulphuric acid at the start of discharge is $C_1\%$ by weight, its relative density is S_1 and its volume is V_I ml, all of which are known ($C_1 = 29$, $S_1 = 1.2104$, $V_I = 1624\,ml$). As $317.226\,g$ of sulphuric acid have been consumed in the electrochemical reaction we can, from Equation 1.84, find the final volume concentration (C_2) and relative density (S_2) of sulphuric acid at the end of the discharge:

$$V_I = W_A \left[\frac{100}{C_1 S_1} + \frac{\dfrac{C_2 S_2}{C_1 S_1}\left(\dfrac{36}{196} + \dfrac{100}{C_1 S_1} - \dfrac{1}{1.8305}\right)}{1 - \dfrac{C_2 S_2}{C_1 S_1}}\right]$$

$$(1.84)$$

i.e.

$$C_2 S_2 = \frac{C_1 S_1\left(\dfrac{V_I}{W_A} - \dfrac{100}{C_1 S_1}\right)}{\dfrac{36}{196} - \dfrac{1}{1.8305} + \dfrac{V_I}{W_A}} \qquad (1.90)$$

Since

$C_1 = 29\%\,by\,weight$

$S_1 = 1.2104$

$C_1 S_1 = 35.102\,g\,sulphuric\,acid/100\,ml$

$V_I = 1624\,ml$

$W_A = 317.22\,g$

Then

$C_2 S_2 = 16.75\,g\,sulphuric\,acid/100\,ml$

and, since S_2 (the relative density of 100% sulphuric acid) is 1.8305,

$$C_2 = \frac{16.75 \times 100}{100 + 16.75\left(1 - \dfrac{1}{1.8305}\right)}$$

$C_2 = 15.56\,g\,sulphuric\,acid/100\,g$

Therefore

$$S_2 = \frac{16.75}{15.56} = 1.076$$

The final volume (V_F ml) of electrolyte is then given by Equation 1.85:

$$V_F = V_I - \left(\frac{1}{1.8305} - \frac{36}{196}\right)$$

$$= 0.3627 W_A = 1509\,ml$$

The volume reduction in acid during discharge is

$$V_F - V_I = 1624 - 1509 = 115\,ml$$

1.18 Calculation of optimum acid volume for a cell

It is obvious in the battery discussed above that discharge has occurred to such an extent that the acid relative density has been reduced to the point that cell e.m.f. would reduce appreciably, i.e. we have entered the region of overdischarge, which is not recommended. To overcome this a further battery is studied in which the initial volume, V_I, of acid is increased from 1.624 litre by various amounts up to 5 litres.

The relative density (S_2), percentage weight concentration (C_2) and final volume (V_F) of electrolyte at end of discharge are calculated as above. The results are presented in Table 1.8. This table enables one to select, for the particular cell in question, a volume of acid to use in the cell which

Table 1.8 Effect of initial volume (V_I) of electrolyte on final acid gravity (S_2), weight concentration (C_2) and final electrolyte volume (V_F)

V_I (ml)	$C_2 S_2$ (g H_2SO_4/100 ml)	C_2 (g/ml³)	S_2 (% H_2SO_4 by wt)	V_F (ml)
1624	16.75	15.56	1.076	1509
3000	25.5	22.85	1.112	2885
4000	28.0	24.54	1.127	3885
5000	29.4	25.93	1.134	4885

$S_1 = 1.2104$, $C_1 = 29$, $C_1 S_1 = 35.102$, $W_A = 317.22$

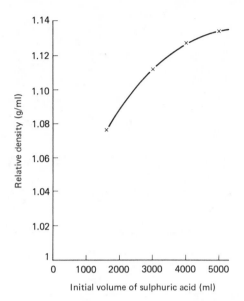

Figure 1.13 Effect of initial volume of sulphuric acid electrolyte (29% w/w or 35.10 g/100 ml) in a lead–acid battery on final acid concentration at end of discharge

will be adequate to maintain the necessary acid concentration in the electrolyte throughout the discharge, i.e. will ensure that the cell e.m.f. is satisfactory throughout discharge. From Figures 1.12 and 1.13 it would seem that an initial electrolyte volume of about 2500 ml would be adequate to ensure an acid concentration of 21% by weight at the end of charge. This corresponds to a relative density of about 1.10.

1.19 Effect of cell layout in batteries on battery characteristics

Consider a series of m 2-V cells connected in series, each cell having an e.m.f. of E_c volts, an internal electrical resistance of $R_c = 0.1\,\Omega$ and producing a current of I_c amps:

Total e.m.f., $E_T = mE_c$ (1.91)

Total internal resistance, $R_T = mR_c$ (1.92)

Total current, $I_T - \dfrac{E_T}{R_T} = \dfrac{mE_c}{mR_c} = \dfrac{E_c}{R_c}$ (1.93)

If $n = 3, 6$ and 9 cells in series, the corresponding characteristics are as shown in Table 1.9 and the total internal resistance and e.m.f. increase as the number of cells is increased but the current remains constant.

Suppose now that these cells are arranged in parallel. If the internal resistance of each cell is denoted by $R_c = 0.1\,\Omega$, its e.m.f. by $E_c = 2$ V and its

Table 1.9 Battery characteristics: cells in series

m	E_T (V)	R_T (Ω)	I_T (E_T/R_T) (A)	Heating (Joule) effect* (cal/t s)	Power (W)
3	6	0.3	20	28.7	120
6	12	0.6	20	57.4	240
9	18	0.9	20	86.1	360

* $\dfrac{I_T^2 R_T \times t}{4.18}$, where $t = 1$

current by I_c amps, then for n 2-V cells in parallel,

E_T (total e.m.f.) $= E_c = 2$ V

$\dfrac{1}{R_T} = n \times \dfrac{1}{R_c}\,\Omega$

(i.e. $R_T = R_c/n$) and

Current $= \dfrac{E_T}{R_T}$

If $n = 3, 6$ and 9 cells in parallel, the corresponding characteristics are as shown in Table 1.10. Thus, for an arrangement of nine 2-V cells each with a total internal electrical resistance of $0.1\,\Omega$, if all the cells are arranged in series a current of 20 A is delivered at 18 V, whereas if all the cells are arranged in parallel a current of 180 A is delivered at 2 V. With

Table 1.10 Battery characteristics: cells in parallel

n	E_T (V)	R_T (Ω)	I_T (E_T/R_T) (A)	Heating (Joule) effect* (cal/t s)	Power (W)
3	2	0.1/3 = 0.0330	60.0	28.4	120
6	2	0.1/6 = 0.0167	119.8	57.3	239.6
9	2	0.1/9 = 0.0111	180.2	86.2	360.4

* $\dfrac{I_T^2 \times R_T \times t}{4.18}$, where $t = 1$

either cell arrangements, the heat producing capacity is 9.56 cal/cell per second. In general, higher currents and lower internal resistances but lower potential differences result when cells are arranged in parallel to produce a battery.

In practical battery arrangements, in order to obtain a compromise between the various characteristics required in the battery, it is common practice to adopt a layout that combines batteries in parallel and in series. Thus, a 48-cell 2-V battery might be arranged in any of the layouts of n files in parallel of m cells in series shown in Table 1.11.

Table 1.11 Practical battery arrangements

	Number of files (n)	Number of batteries in each series (m)
A	1	48
B	2	24
C	3	16
D	4	12
E	6	8
F	8	6
G	12	4
H	16	3
I	24	2
J	48	1

The battery characteristics that would result from these various cell arrangements are discussed below.

Consider a single cell of e.m.f. E_c and internal electrical resistance R_s, with an ampere hour capacity of C_c A h and producing a current of I_c amps. For this cell,

$$I_c = \frac{E_c}{R_c} \qquad (1.93)$$

and

$$C_c = I_c h = \frac{E_c h}{R_c} \qquad (1.94)$$

where h is the duration of the discharge in hours.

Consider next a series of m such cells in series. The total e.m.f. (E_T), internal electrical resistance (R_T), current (I_T) and ampere hour capacity (C) of this arrangement are as follows:

$$E_T = mE_c \qquad (1.91)$$

$$R_T = mR_c \qquad (1.92)$$

$$I_T = \frac{mE_c}{mR_c} = \frac{E_c}{R_c} \qquad (1.93)$$

$$C_T = mI_c = \frac{mE_c}{R_c} = mI_T \, h \qquad (1.95)$$

$$\text{Duration of discharge} = \frac{C_T}{mI_T} \, h \qquad (1.96)$$

Consider now an arrangement consisting of n files in parallel of m cells in series, i.e. an $m \times n$ cell battery, in this example a 48-cell battery. The total e.m.f. (E_{TOT}), internal electrical resistance (R_{TOT}), current (I_{TOT}) and ampere hour capacity (C_{TOT}) of this arrangment are as follows:

$$E_{TOT} = mE_c \qquad (1.91)$$

$$\frac{I}{R_{TOT}} = \frac{n \times I}{mR_c}$$

i.e.

$$R_{TOT} = \frac{m}{n} R_c \qquad (1.97)$$

$$I_{TOT} = \frac{E_{TOT}}{R_{TOT}} = \frac{nE_c}{R_c} = \frac{mE_c}{R_{TOT}} \qquad (1.98)$$

$$C_{TOT} = m \times I_{TOT} \times h$$

$$= mnh \frac{E_c}{R_c}$$

$$= \frac{m^2 h E_c}{R_{TOT}}$$

$$= \frac{mh E_{TOT}}{R_{TOT}} \qquad (1.99)$$

The duration of discharge (in hours) is

$$\frac{C_{TOT} \times R_c}{m \times m \times E_c} = \frac{C_{TOT} \times R_{TOT}}{m^2 \times E_c}$$

$$= \frac{C_{TOT} \times R_{TOT}}{m E_{TOT}}$$

$$= \frac{C_{TOT}}{m I_{TOT}} \qquad (1.100)$$

Using these equations it is now possible to calculate for any arrangement of n files in parallel of m cells in series the values of E_{TOT}, R_{TOT}, I_{TOT}, C_{TOT}, the Joule heating effect and the wattage if values are assumed for the following characteristics of each individual cell in the arrangement: $E_c = 2$ V, internal electrical resistance $(R_c) = 0.1 \, \Omega$/cell, duration of discharge $= 4$ h. These results are given in Table 1.12 and Figures 1.14 and 1.15. Depending on the arrangement of forty-eight 80 ampere hour capacity 2-V cells we can obtain a range of amperages and voltages ranging from 20 A at 96 V (48 cells in series) to 960 A at 2 V (48 cells in parallel). It can be seen from Table 1.12 and Figures 1.14 and 1.15 that if, for example, 24 V at 80 A is required then four strings in parallel of 12 cells are required, i.e. $n = 4$, $m = 12$. Duration of discharge, Joule heating and wattage are unaffected by cell arrangement. Internal electrical resistance is highest when all the cells are in series and decreases dramatically as parallel arrangements of cells are introduced. The above calculations do not take into account the external electrical resistance of the battery. This factor is included in the methods of calculation discussed below. The choice of connection of cells in series or in parallel or a combination of both depends on the use to which the cells are to be put. For the best economy in electrical energy, as little as possible should be wasted as heat internally. This requires the internal resistance of the combination to be as small as possible compared to the external resistance, and the cells are therefore grouped in parallel. For the quickest action, such as

Table 1.12 Characteristics of series–parallel multi-cell batteries (48-cell)

Number of strings of cells in parallel (n)	Cells in series in each string (m)	$\dfrac{m}{n}$	Internal resistance, R_{TOT} (Ω)	E_{TOT} (V)	I_{TOT} (A)	C_{TOT} $\left(\dfrac{mhE_{TOT}}{R_{TOT}}\right)$* (A h)	Duration of discharge, h (h)	Joule heating $\left(\dfrac{I^2_{TOT} R_{TOT}}{4.18}\right)$ (cal/s)	Power ($E_{TOT} \times I_{TOT}$) (W)
1	48	48	4.8	96	20	3840	4	459	1920
2	24	12	1.2	48	40	3840	4	459	1920
3	16	5.33	0.533	32	60	3840	4	459	1920
4	12	3.00	0.300	24	80	3840	4	459	1920
6	8	1.33	0.133	16	120	3840	4	459	1920
8	6	0.75	0.075	12	160	3840	4	459	1920
12	4	0.33	0.033	8	240	3840	4	459	1920
16	3	0.19	0.019	6	320	3840	4	459	1920
24	2	0.083	0.0083	4	480	3840	4	459	1920
48	1	0.021	0.0021	2	960	3840	4	459	1920

* $h = 4\,\text{h}$ discharge per cell

may be required in a circuit of high induction, for example an electromagnet, the total resistance of the circuit should be as high as possible. The cells are therefore connected in series.

To obtain the maximum current in a given external circuit the grouping of cells varies according to the particular value of the external resistance. The calculations below take into account both the internal electrical resistances (R_c) of each cell and the external resistance of the whole battery (R_{ext}). To obtain maximum current, it is necessary to group the cells partly in series and partly in parallel, so that the combined internal electrical resistance of the combination is as nearly as possible equal to the external resistance.

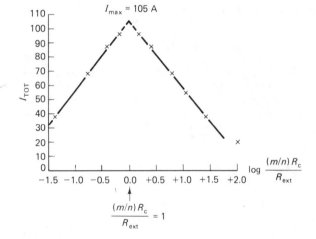

Number of cells in series, m

1	2	3	4	6	8	12	16	24	48
48	24	16	12	8	8	4	3	2	1

Number of strings of cells, n

$$I_{TOT} = \frac{mE_c}{(m/n)\,R_c + R_{ext}}$$

Assuming $E_c = 2$ V,

Internal resistance per cell, $R_c = 0.1\ \Omega$

External resistance, $R_{ext} = 0.05\ \Omega$

m = number of cells in series in each string
n = number of strings in parallel

i.e.

$$I_{TOT} = \frac{2m}{m/n \times 0.1 + 0.05}$$

I_{max} is achieved when m is 4 to 6 and n is 12 to 8

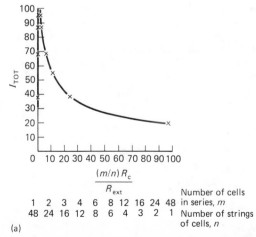

(a)

Figure 1.14 Various configurations of 48 two-volt cells. Effect of configuration on e.m.f., current and electrical resistance

(b)

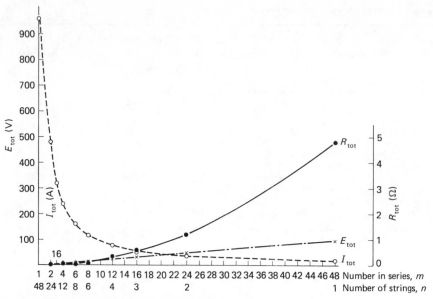

Figure 1.15 Various configurations of 48 two-volt cells. Effect of configuration on e.m.f., current and electrical resistance

Consider the case where there are n files in parallel of m cells in series. Each cell has an internal resistance of R_c and an e.m.f. of E_c, the external resistance of the whole battery being R_{ext}. The total number of cells, N, is given by:

$$N = nm$$

and

$$\text{Total e.m.f., } E_{TOT} = mE_c \qquad (1.91)$$

$$\text{Total current, } I_{TOT} = \frac{mE_c}{R_{TOT} + R_{ext}}$$

$$= \frac{mE_c}{m/nR_c + R_{ext}} \qquad (1.101)$$

where R_{TOT} is the total internal resistance, R_{ext} is the total external resistance, and R_c is the internal resistance of each of the mn cells in the battery. Eliminating m from this equation ($m = N/n$),

$$I_{TOT} = \frac{NnE_c}{NR_c + R_{ext}n^2} \qquad (1.102)$$

For a maximum value of I_{TOT} (i.e. I_{max}), forming the expression: dI_{TOT}/dn and equating to zero, we find that

$$\frac{dI_{TOT}}{dn} = \frac{d\left(\dfrac{NnE_c}{NR_c + R_{ext}n^2}\right)}{dn}$$

$$= NR_c - R_{ext}n^2 = 0 \qquad (1.103)$$

Hence

$$NR_c = R_{ext}n^2$$

and

$$\frac{mR_c}{n} = R_{ext}$$

but

$$\frac{mR_c}{n} = R_{TOT} \qquad (1.104)$$

where R_{TOT} is the total internal resistance of the group of $n \times m$ cells. Hence

$$R_{TOT} = R_{ext} \qquad (1.105)$$

Thus, to obtain the maximum current (I_{max}) from a group of $m \times n$ cells, the external resistance, R_{ext}, for the group of cells should be equal to the combined internal resistance ($R_{TOT} = (m/n)R_c$) of the group of N cells.

As

$$\frac{NnE_c}{NR_c + R_{ext}n^2} = I_{TOT} \qquad (1.102)$$

$$\frac{mR_c}{n} = R_{ext} \qquad (1.104)$$

$$I_{max} = \frac{NnE_c}{NR_c + R_{ext}n^2}$$

$$= \frac{nE_c}{2R_c}$$

$$= \frac{mE_c}{2R_{ext}} \qquad (1.106)$$

If we consider the example of a battery comprising 48 two-volt cells (i.e. $N = 48$), each cell having an e.m.f., E_c, of 2 V, it is required to calculate the number, n, of files in parallel of m cells in series to achieve the maximum current, I_{max}, from the battery. It is assumed that the battery has a total external electrical resistance, R_{ext}, of 0.05 Ω and that the internal resistance per cell, R_c, is 0.1 Ω, i.e. the total internal resistance of a 48-cell battery is R_{TOT} $m/n \times 0.1$. The total internal plus external resistance of the battery is $R = (m/n)R_c + R_{ext}$ Ω. Considering all the possibilities between $m = 48$, $n = 1$, i.e. 48 cells in series, and $m = 1$, $n = 48$, i.e. 48 cells in parallel, and all possible configurations of cells between these limits, it is possible to calculate the total current, I_{TOT}, produced by the battery:

$$I_{TOT} = \frac{mE_c}{\dfrac{m}{n}R_c + R_{ext}} \tag{1.101}$$

As the maximum current, I_{max}, is obtained when $(m/n)R_c = R_{ext}$ it is possible to obtain I_{max} graphically by plotting

$$\frac{\dfrac{m}{n}R_c}{R_{ext}}$$

or, better:

$$\log \frac{\dfrac{m}{n}R_c}{R_{ext}}$$

against the calculated value of I_{TOT} and to read off the maximum current, I_{max}, corresponding to

$$\frac{\dfrac{m}{n}R_c}{R_{ext}} = 1$$

or

$$\log \frac{\dfrac{m}{n}R_c}{R_{ext}} = 0$$

The data for the above battery are tabulated in Tables 1.13 and 1.14 and plotted in Figure 1.16. It can be seen clearly from the log plot in Figure 1.17 that a maximum achievable current of 96 A will be obtained from a battery comprising 12 strings of four cells in series (i.e. $n = 12$, $m = 4$), or eight strings of six cells in series (i.e. $n = 8$, $m = 6$). If currents other than the maximum are required, the values of

Table 1.13 Demonstrating that maximum current is obtained when total internal resistance of battery equals total external resistance

Internal resistance per cell, $R_c = 0.01$ Ω
External resistance of battery, $R_{ext} = 0.05$ Ω
Voltage per cell, $E_c = 2$ V
Duration of discharge of each cell, $h = 4$ h

Number of strings of cells, n	1	2	3	4	6	8	12	16	24	48
Number of cells in series per string, m	48	24	16	12	8	6	4	3	2	1
m/n	48	12	5.333	3.000	1.330	0.750	0.330	0.190	0.083	0.021
$(m/n)R_c$ (a)	4.800	1.200	0.533	0.300	0.133	0.075	0.033	0.019	0.0083	0.0021
$R_{TOT} = (m/n)R_c + R_{ext}$	4.850	1.250	0.583	0.350	0.183	0.125	0.083	0.069	0.0583	0.0521
$\dfrac{(m/n)R_c}{R_{ext}} = \dfrac{a}{0.05}$	96	24	16	6	2.66	1.50	0.660	0.380	0.167	0.042
$\log\left(\dfrac{(m/n)R_c}{R_{ext}}\right)$	1.982	1.380	1.028	0.778	0.425	0.176	−0.180	−0.420	−0.778	−1.377
$E_{TOT} = mE_c$	96	48	32	23	16	12	8	6	4	2
Current, I_{TOT} $= \dfrac{mE_c}{(m/n)R_c \times R_{ext}}$ (A)	19.79	38.40	54.89	68.57	87.43	96.00	96.03	86.95	68.61	38.39
Capacity, C_{TOT} $= \dfrac{mhE_{TOT}}{R_{TOT}}$ (A h)	3800	3686	3512	3291	2798	2304	1537	1043	549	153

Table 1.14 Effect of cell configuration on e.m.f. and current achieved in arrangements of 48 two-volt cells

Internal resistance per cell, $R_c = 0.1$
External resistance, $R_{ext} = 0.05\ \Omega$
Voltage per cell, 2 V

Number of strings of cells, n	1	2	3	4	6	8	12	16	24	48
Number of cells in series per string, m	48	24	16	12	8	6	4	3	2	1
m/n	48	12	5.333	3.000	1.330	0.750	0.330	0.190	0.083	0.021
$\log m/n$	1.681	1.072	0.727	0.477	0.124	−0.125	−0.523	−0.721	−1.081	−1.678
Current, $I_{TOT} = \dfrac{mE_c}{(m/r)R_c + R_{ext}}$ A	19.79	38.40	54.89	68.57	87.43	96.00	96.03	86.95	68.61	38.39
Total resistance, R, $= (m/n)R_c + R_{ext}$	4.850	1.250	0.583	0.350	0.183	0.125	0.083	0.069	0.583	0.0521
Voltage, $V = I_{TOT} \times R$	96	48	32	24	16	12	8	6	4	2
Wattage, $V \times I_{TOT}$	1899	1843	1756	1645	1400	1152	768	522	274	76.8
Joule heating $= \dfrac{I_{TOT}^2 \times R}{4.18}$ cal/s	454	440	420	393	335	275	183	124.8	64.9	18.4
Current* as % of I_{max} $= \dfrac{I_{TOT} \times 100}{I_{max}}$	18.8	36.6	52.2	65.3	83.3	91.4	91.4	82.8	65.3	36.6

* $I_{max} = 105$ A

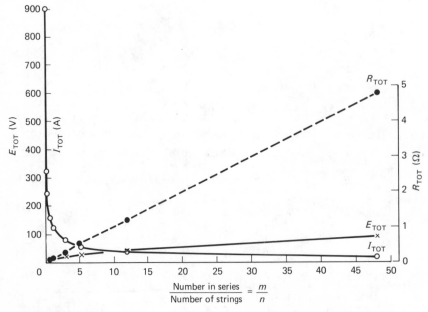

Figure 1.16 48-cell 2 V/cell lead–acid battery. Demonstration that maximum internal resistance of battery, $(m/n)R_c$, equals total external resistance (R_{ext}), i.e. $(m/n)R_c = R_{ext}$ or $m\,R_c/n\,R_{ext} = I$

Table 1.15 Characteristics of two battery configurations

	12 strings of 4 cells*	8 strings of 6 cells†
I_{max} (A) $= \dfrac{mE_c}{(m/n)R_c + R_{ext}}$	96	96
External resistance, R_{ext} (Ω)	0.05	0.05
Internal resistance, R_{TOT} $= (m/n)R_c$ (Ω)	0.0333	0.075
Total resistance, R_{TOT} $= R_{ext} + (m/n)R_c$ (Ω)	0.0833	0.125
Voltage, $E_{TOT} = I_{max}R_{TOT}$ (V)	8	12
Wattage, $V \times I_{max}$	768	1152
Joule heating, $\dfrac{I_{max}^2 R_{TOT}}{4.18}$ (cal/s)	183.6	275.6
A h capacity, $\dfrac{mhE_{TOT}}{R_{TOT}}$ (A h)	1536	2304

(where h = duration of discharge of each cell = 4 h)

* = 12, m = 4, N = 48
† n = 8, m = 6, N = 48

n and m required to achieve these currents can be deduced from Figure 1.17. The other characteristics of these two battery configurations are shown in Table 1.15.

It is seen that, with a set of 48 two-volt cells, to obtain near maximum current we have a choice of using 12 strings of four cells in series or eight strings of six cells in series. Both supply a maximum current of 96 A, the former at 8 V and the latter at 12 V. The higher e.m.f. battery has a higher available wattage and a greater Joule heating effect in parallel with its greater internal and total electrical resistance. There is more available capacity from the battery configuration comprising longer strings of cells in series, i.e. higher values of m (Table 1.13). If a voltage greater or less than 12 V is required, two choices are available:

1. Arrange the configuration of cells such that a higher (or lower) voltage is obtained at the cost of not operating at maximum current producing efficiency.
2. Increase (or decrease) the total number of cells in the battery.

1.19.1 Operation at lower current efficiency

As shown in Figure 1.17, the maximum current that could be expected for a battery consisting of 48 two-volt cells with an internal resistance, R_c, of 0.1 Ω/cell and a total external resistance, R_{ext}, of 0.10 Ω is 105 A. Because of the practical difficulties of arranging 48 cells in series–parallel, the maximum current achieved was 96 A (i.e. 91.4% of theoretical maximum) at 8 V (12 strings of four cells) or at 12 V (eight strings of six cells). Table 1.14 shows the currents and voltages that would be achieved with a wide range of other cell configurations for a 48-cell battery.

It is seen (Figure 1.17) that, depending on cell configuration, voltages between 96 V and 2 V can be obtained. When the voltage is 96 V, the current produced is only about 19% of the maximum value of 96 A that can be obtained as discussed above. Conversely, when the voltage is 2 V only 37% of the maximum current is produced. Figure 1.17 indicates that intermediate voltages, say 24 V, can be obtained with not too serious a loss of current from the maximum. There is a steady decrease in wattage available in going from an all-series battery (1899 W) to an all-parallel battery (76.8 W). Joule heating and wattage decrease in the same order (Figure 1.18).

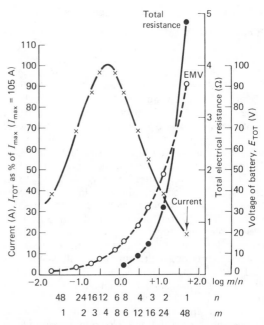

Figure 1.17 Various configurations of 48 two-volt lead–acid cells in series and parallel. Effect of configuration on current, voltage and total electrical resistance. Internal resistance, R_c = 0.1 Ω/cell. Total external electrical resistance (R_{ext}) = 0.05 Ω

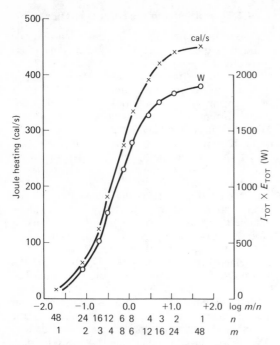

Figure 1.18 Various configurations of 48 two-volt lead–acid cells in series and parallel. Effect of configuration on Joule heating and wattage

1.19.2 Change in number of cells in a battery

It has been shown that, for a battery consisting of 48 two-volt cells (internal resistance, $R_c = 0.1\,\Omega$/cell and total external resistance of battery, $R_{ext} = 0.05\,\Omega$), the maximum achievable current (I_{max}) of 96 A (105 A theoretical) is obtained with 12 strings of four cells in series or eight strings of six cells in series. The voltages, respectively, of these two cell configurations are 8 V and 12 V. The following treatment applies if it is required to obtain the maximum current at a higher (or lower) voltage, or a higher maximum current at the same voltage.

If the required battery voltage is 96 V, then, clearly, a basic configuration of 48 two-volt cells (i.e. $m = 48$) in series is required to produce this voltage. The greater the number of strings of 48 in series connected together the greater will be the current. It remains to calculate the number of strings needed to produce the required current. The results in Table 1.16 and Figure 1.19 clearly illustrate the method of obtaining the number of strings in parallel of 48 cells in series to achieve any required current in the range 20–100 A at 96 V. It is interesting to note that the maximum current (I_{max}) of 98 A and also the maximum capacity of 18 797 A h (Table 1.16) would be obtained with 10 strings in parallel of 48 cells in series.

This is in agreement with the statement in Equation 1.104 that maximum current is achieved when total internal resistance of $m \times n$ cells equals total external resistance of battery, i.e.

$$\frac{m}{n} R_c = R_{ext}$$

If $R_c = 0.01\,\Omega$/cell, R_{ext} is $0.5\,\Omega$ per 48 cells, i.e.

$$\frac{0.05 \times m \times n}{48}\,\Omega \text{ per } m \times n \text{ cells}$$

Then

$$\frac{m}{0.1n} = \frac{0.05 \times m \times n}{48}$$

i.e.

$$n = \sqrt{\left(\frac{48}{0.1 \times 0.05}\right)} \approx 10$$

i.e. 10 files of 48 cells in series.

By these methods, it is possible to design the layout of a battery to produce any required combination of current and voltage.

In all the above calculations, for simplicity, standard values have been chosen for the internal resistance per cell, R_c, and the total external resistance of the battery, R_{ext}. In fact, of course, the particular values of the parameters that the battery designer is presented with will affect battery performance parameters such as current obtainable, wattage, joule heating and available A h capacity. The effect of variations of resistance parameters on battery performance will now be examined.

Tables 1.17 and 1.18, respectively, calculate the total internal plus external resistance (R_{TOT}) and currents and capacities obtained for an 8 V configuration of $n = 12$ strings, in parallel, of $m = 4$ cells in series, and a 24 V configuration of $n = 4$ strings of $m = 12$ cells in series. For any particular fixed value of the external resistance, R_{ext}, of the battery an increase of the internal resistance per cell, R_c, decreases the current available and capacity and also the Joule heating effect and wattage. The same comments apply when the internal resistance is fixed and the external resistance is increased.

In general, current available and capacity decrease as the total electrical resistance $(m/n)\,R_c + R_{ext}$ is increased. Figure 1.20 shows the expected linear relationship between current and the reciprocal of the electrical resistance and Figure 1.21 shows the resistance–capacity relationships. Naturally, these relationships change as the configuration of the 48 cells is changed.

It is obviously desirable from the point of view of maximizing current production when designing batteries to take every possible step to keep internal and external electrical resistances to a minimum as far as this is compatible with other aspects of battery design such as maintaining the mechanical strength of components.

Table 1.16 Selection of number (n) of strings of 48 two-volt cells in series (m = 48) to produce a required current (I_{TOT}) at 96 V

m	n	nm	m/n	R_{ext}* (Ω)	R_{int}† (Ω)	R_{TOT}‡ (Ω)	R_{int}/R_{ext}§	I_{TOT}** (A)	Capacity†† (A h)
48	1	48	48	0.05	4.8	4.85	96	19.79	3 800
48	2	96	24	0.1	2.4	2.5	24	38.4	7 373
48	3	144	16	0.15	1.6	1.75	10.66	54.8	10 522
48	4	192	12	0.2	1.2	1.4	6	68.6	13 171
48	5	240	9.6	0.25	0.96	1.21	3.84	79.3	15 225
48	6	288	8	0.30	0.8	1.10	2.67	87.3	16 761
48	7	336	6.86	0.35	0.686	1.036	1.96	92.7	17 798
48	8	384	6	0.4	0.6	1.00	1.5	96	18 432
48	9	432	5.33	0.45	0.533	0.983	1.184	97.9	18 797
48	10	480	4.8	0.50	0.48	0.98	0.96	97.9	18 797
48	20	960	2.4	1.0	0.24	1.24	0.24	77.4	14 861
48	30	1440	1.6	1.5	0.16	1.66	0.107	57.8	11 098
48	40	1920	1.2	2.0	0.12	2.12	0.06	45.3	8 698
48	50	2400	0.96	2.5	0.096	2.596	0.038 4	37.0	7 104
48	60	2880	0.8	3.0	0.08	3.08	0.026 7	31.1	5 971
48	70	3360	0.686	3.5	0.0686	3.568	0.019 6	26.9	5 165
48	80	3840	0.6	4.0	0.06	4.06	0.015	23.6	4 531
48	90	4320	0.533	4.5	0.0533	4.53	0.011 8	21.2	4 070
48	100	4800	0.48	5.0	0.048	5.048	0.009 6	19.0	3 648
48	110	5280	0.436	5.5	0.0436	5.544	0.007 93	17.3	3 322
48	120	5760	0.4	6.0	0.04	6.04	0.006 66	15.9	3 053
48	130	6240	0.369	6.5	0.0369	6.537	0.005 67	14.7	2 822
48	140	6720	0.343	7.0	0.0343	7.034	0.004 9	13.6	2 611
48	150	7200	0.32	7.5	0.032	7.532	0.004 26	12.7	2 438
48	160	7680	0.3	8.0	0.03	8.03	0.003 75	11.9	2 285

Internal resistance per cell, R_c = 0.1 Ω

* External resistance per mn cells, R_{ext} = 0.05 Ω/48 cells

$$= \frac{0.05\,nm}{48}\ \Omega/nm \text{ cells}$$

† Total internal resistance, $R_{int} = (m/n)R_c$

‡ Total resistance, $R_{TOT} = (m/n)R_c + R_{ext}$

$$= \frac{(m/n)R_c}{R_{ext}}$$

§ $\dfrac{R_{int}}{R_{ext}}$

** Total current, $I_{TOT} = \dfrac{mE_c}{(m/n)R_c + R_{ext}}$

†† Capacity $= \dfrac{mhE_{TOT}}{R_{TOT}} = \dfrac{48 \times 4 \times 96}{R_{TOT}}$

where h = duration of discharge of a single cell = 4 h

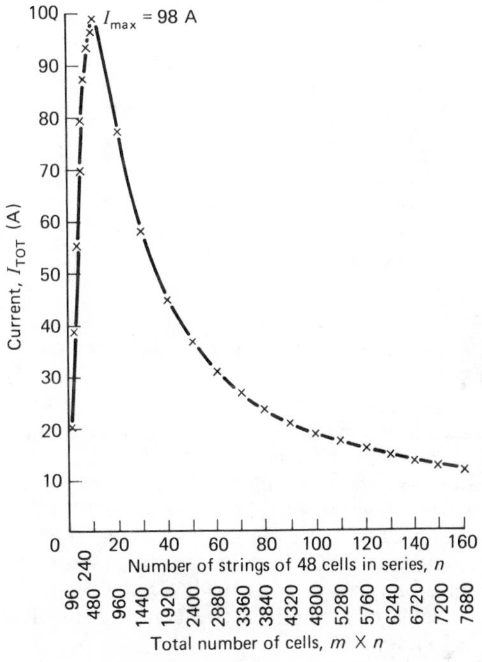

Figure 1.19 Selection of number of strings (n) of 48 two-volt cells in series to produce a required current (I_{TOT}) at 96 V. Internal resistance (R_c) per cell = 0.1 Ω. External resistance per cell = 0.05 Ω per 48 cells, i.e. 0.05 $mn/48$ Ω/cell

Table 1.17 Effect of internal, external and total electrical resistance on available current from a 48 two-volt (E_c) cell battery (8 V) comprising 12 strings of four cells in series

m/n	0.333		0.333			0.333	
R_{ext} (total for battery) (Ω)	0.1		0.5			2	
R_c, internal resistance/cell (Ω)	0.1	0.2	0.1	0.2	0.5	0.2	0.5
$R_{TOT} = (m/n)R_c + R_{ext}$	0.1333	0.1666	0.5333	0.5666	0.6665	2.0666	2.1665
$1/R_{TOT}$	7.50	6.00	1.87	1.76	1.50	0.484	0.461
$E_{TOT} = mE_c$	8	8	8	8	8	8	8
Current, $I_{TOT} = \dfrac{mE_c}{(m/n)R_c + R_{ext}}$ (A)	60	48	15	14.1	12	3.87	3.69
Joule effect, $\dfrac{I_{TOT}^2 \times R_{TOT}}{4.18}$ (cal/s)	114.8	91.8	28.7	26.9	22.9	7.40	6.87
Wattage, $I_{TOT} \times E_{TOT}$ (W)	480	384	120	112.8	96	31.0	29.5
Capacity, $\dfrac{m \times h \times E_{TOT}}{R_{TOT}}$ (A h)	960	768	240	225	192	61.9	59.0

Table 1.18 Effect of internal, external and total electrical resistance on available current from a 48 two-volt (E_c) battery (24 V) comprising four strings of twelve cells in series

m/n	3.000		3.000			3.000	
R_{ext} (total for battery) (Ω)	0.1		0.5			2	
R_c, internal resistance per cell (Ω)	0.1	0.2	0.1	0.2	0.5	0.2	0.5
$R_{TOT} = (m/n)R_c + R_{ext}$	0.400	0.700	0.800	1.100	2.000	3.600	4.500
$1/R_{TOT}$	2.50	1.42	1.25	0.909	0.500	0.277	0.222
$E_{TOT} = mE_c$ (V)	24	24	24	24	24	24	24
$I_{TOT} = \dfrac{mE_c}{(m/n)R_c + R_{ext}}$ (A)	60	34.3	30	21.8	12.0	6.7	5.3
Joule effect, $\dfrac{I_{TOT}^2 \times R_{TOT}}{4.18}$ (cal/s)	344.4	197.0	172.2	125.0	68.9	38.7	30.2
Wattage, $I_{TOT} \times E_{TOT}$ (V A)	1440	823	720	523	288	160.8	127.2
Capacity, $\dfrac{mhE_{TOT}}{R_{TOT}}$ (A h)	2880	1646	1440	1046	576	321	254

1.20 Calculation of energy density of cells

It has been shown (Section 1.17) that, in a lead–acid battery discharge of 1 mol of each of lead and lead dioxide from an initial sulphuric acid concentration of 29% by weight to a final concentration of 21% by weight, 53.9 A h are produced when the cell temperature is 25°C, i.e. capacity = 53.9 A h at a voltage of about 2 V. Thus 107.8 W h are produced. The weights in grams of active materials involved are as follows:

Lead	207.2
Lead dioxide	223.2
Sulphuric acid	196.0
Total active material	626.4 g

Thus the gravimetric energy density in W h/kg of active material is:

$$\frac{107.8 \times 1000}{626.4} = 172 \text{ (or 250.5 if electrolyte is excluded)}$$

This can be considered to be the maximum theoretical energy density of the lead–acid battery.

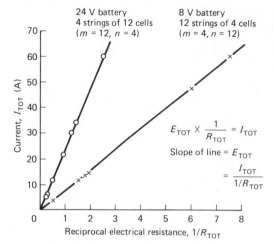

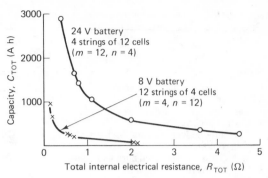

Figure 1.20 Plot of reciprocal of total electrical resistance $1/R_{TOT}$ versus current, I_{TOT}

Figure 1.21 Effect of total internal electrical resistance, R_{TOT}, of battery on available A h capacity, C_{TOT}

In practice, of course, the acid is dilute, not concentrated, and this adds to the cell weight, as does the weight of the lead supporting grids for the active material, the terminal posts, connecting bars, etc. The weight of the cell case and, if multi-cell batteries are being considered, the weight of the outer battery case must also be considered. The practical energy density of the battery is thus considerably less than the calculated theoretical value. Part of the technology of battery designing is to keep extraneous weight to a minimum, thereby increasing the energy density of the battery.

The alternative method of expressing energy density, which is applicable when the space occupied by a battery rather than its weight is the prime consideration, is in W h/dm³.

1.21 Effect of discharge rate on performance characteristics

Consider a 20 A h lead–acid battery being completely discharged at the 20 h rate. This battery will deliver 20 A for 1 h at a voltage of 1.7 V. If the battery is completely discharged more rapidly, say during 1 min, it will deliver a higher current for 1 min (1/60 h) at a lower voltage (1.2 V). Under these more rapid discharge conditions, its available capacity is 8 A h. The discharge current is 60 × 8 = 480 A.

The data in Table 1.19 give an analysis of the effect of discharge rate on energy density, capacity and heat evolution occurring during discharge.

It is seen that, in parallel with the reduced available capacity (3) at the higher discharge rate, there occurs a reduction in coulombs delivered (5) and energy density (10). The loss in available

capacity and energy density at higher discharge rates is due to the inability of the cell reaction to keep pace with the demand for current, principally caused by the inability of fresh electrolyte from the bulk of the cell to quickly replace the depleted electrolyte at the surface of and within the plates. Although the total heat evolution (ohmic plus chemical (14)) during the rapid discharge is lower, it is seen (15) that the number of calories produced per second during the rapid discharge is about 30 times greater than at the lower discharge rate.

1.21.1 Effect of discharge rate on capacity, energy density, Joule heating current, voltage and resistance

Figure 1.22 shows a family of voltage–time curves obtained by discharging a 2 V lead–acid battery at various discharge currents between 1 and 10 A. Figure 1.23 shows the fall-off in available capacity that occurs as the discharge current is increased. Table 1.20 tabulates all basic data for this set of discharges. Figures 1.24–1.29 show the effect of the duration of the discharge on discharge current and end-point voltage (Figure 1.24), discharge current and A h capacity (Figure 1.25), electrical resistance and discharge current (Figure 1.26), energy density (W h/kg) and discharge current (Figure 1.27), energy density (W h/kg) and A h capacity (Figure 1.28) and energy density (W h/kg) and end-point e.m.f. (Figure 1.29).

Figures 1.24 and 1.25 illustrate the optimum discharge time of 2.4 min for this cell which delivers 430 A at 1.44 V and has an available capacity of 17.2 A h. In general, curves of this type enable one to select the desired combination of the various parameters that would be required in battery service.

Figures 1.27–1.29 illustrate the interrelationship of discharge time with discharge current, end-point e.m.f., A h capacity and energy density. Thus, if 2 min and 30 min discharges are adopted, the effect

Table 1.19 Effect of discharge rate on energy density and other parameters

	1 h *discharge*	1 min *discharge*
1. Duration of discharge (min)	60	1
2. Voltage, E (V)	1.7	1.2
3. Capacity (A h)	20	8
4. Discharge current (A)	20	480
5. Coulombs delivered during discharge	72 000/h	28 800/min
6. Volt coulombs	122 400	34 560
7. Watts ($E \times A$) delivered during discharge	34/h	576/min
8. Watts delivered per second	0.009 44	9.6
9. Watt hours delivered during discharge	34	9.6
10. Practical energy density (W h/kg)	34	9.6
11. Ohmic (Joule) heating during discharge (cal) $J = \dfrac{E \times A \times \text{seconds}}{4.18}$ $\quad = \dfrac{A^2 \times R \times \text{seconds}}{4.18}$	29 282/h	8268/min
12. Internal resistance, E/A (Ω)	0.085	0.0025
13. Heat content change of cell reaction during discharge (negative sign = heat evolved) $\Delta H = \dfrac{-nF}{4.18}\left[E - T\left(\dfrac{\mathrm{d}E}{\mathrm{d}T}\right)\right]$ cal (see Equation 1.20) (assuming $\mathrm{d}E/\mathrm{d}T = 0.000\,25$, $n = 2$, $F = 96\,500$ C, $T = 25°C$ (298 K))	$-75\,053$/h	$-52\,890$/min
14. Total heat evolution during discharge, $J + \Delta H$ cal	104 335/h	60 390/min
15. Heat evolution (cal/s)	29	1006

Table 1.20 Discharges of 2 V lead–acid battery at different rates – basic data

t (min)	30	20	10	4	2	1
E (V)	1.6	1.59	1.56	1.51	1.41	1.26
Capacity (A h)	30	30	25	20	16	10
Current (A)	60	90.09	149.70	299.85	480.48	602.41
Coulombs	108 000	108 108	89 820	71 964	57 658	36 144
Volt coulombs	172 800	171 892	140 119	108 666	81 298	45 541
Watts	96.0	143.2	233.5	452.8	677.4	759.0
Watts/second	0.0533	0.1193	0.3892	1.8867	5.6450	1.2650
Watt hours	48.00	47.72	38.92	30.18	22.58	12.66
Energy density (W h/kg)	48.00	47.72	38.92	30.18	22.58	12.66
Joule (ohmic) heating	41 340	41 110	33 517	25 998	19 447	10 895
$R = E/A$ (Ω)	0.0267	0.0176	0.0104	0.005 03	0.002 93	0.002 09
$\Delta H = \dfrac{-3600\ \text{C}}{4.18}\left[E - T\left(\dfrac{\mathrm{d}E}{\mathrm{d}T}\right)_P\right]$	$-39\,415$	$-39\,157$	$-31\,982$	$-24\,722$	$-18\,400$	$-10\,208$
Joule $+ \Delta H$	80 755	80 267	65 499	50 720	37 847	21 103
Heat/second						
Total	44.86	66.90	109.16	211.33	315.39	351.71
Joule	22.97	34.25	55.86	108.32	162.06	181.58
ΔH	21.89	32.63	53.30	103.00	153.3	170.13
ΔH as % of total	48.8	48.8	48.8	48.7	48.6	48.4
Joule as % of total	51.2	51.2	51.2	51.3	51.4	51.6

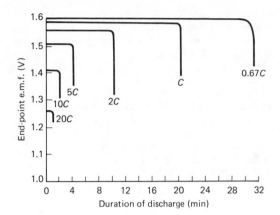

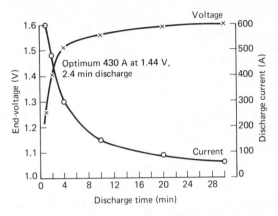

Figure 1.24 2 V lead–acid battery: relationships of discharge time with end-point e.m.f. and discharge current

Discharge time (min)	Discharge rate*	Discharge current (A)	End-point (V)	Capacity (A h)
30	0.67 C	1	1.6	30
20	C	1.5	1.59	30
10	2 C	2.5	1.56	25
4	5 C	5	1.51	20
2	10 C	8	1.41	16
1	20 C	10	1.26	10

* C = 20 min rate

Figure 1.22 Discharge curves of a 2 V lead–acid battery at various discharge rates

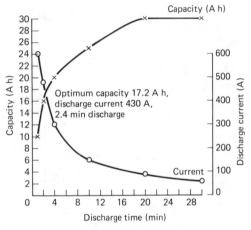

Figure 1.25 2 V lead–acid battery: relationship of discharge time with A h capacity and current

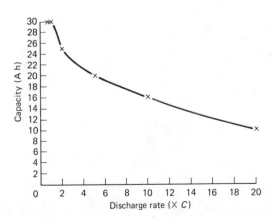

Figure 1.23 2 V lead–acid battery: influence of discharge rate on capacity

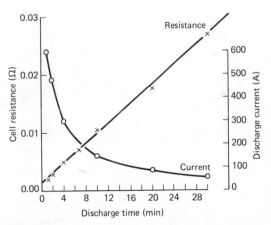

Figure 1.26 2 V lead–acid battery: relationship of discharge time with electrical resistance and discharge current

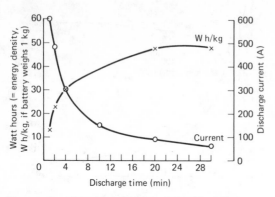

Figure 1.27 2 V lead–acid battery: relationship of discharge time with energy density and discharge current

of this on the above parameters would be as illustrated in Table 1.21. Information of this kind enables trade-offs to be made in the various aspects of battery performance.

Another aspect of battery performance can be studied by preparing plots of discharge current against end-point e.m.f. and electrical resistance (Figure 1.30) and A h capacity and energy density (Figure 1.31).

Again considering the effect of carrying out the discharge at two different currents 100 A and 600 A on battery operating parameters, we have the effects illustrated in Table 1.22.

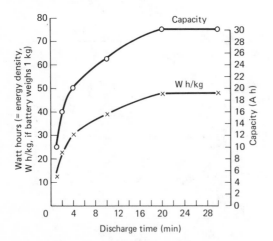

Figure 1.28 2 V lead–acid battery: relationship of discharge time with energy density and A h capacity

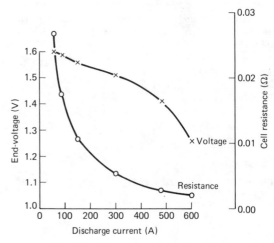

Figure 1.30 2 V lead–acid battery: relationship of discharge current with end-point e.m.f. and electrical resistance

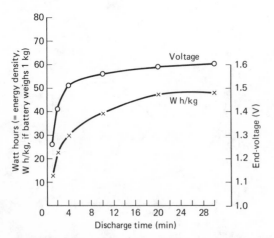

Figure 1.29 2 V lead–acid battery: relationship of discharge time with energy density and end-point e.m.f.

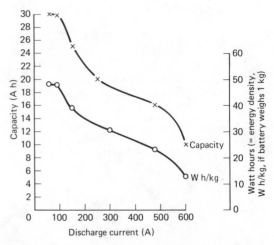

Figure 1.31 2 V lead–acid battery: relationship of discharge current with A h capacity and energy density

Table 1.21 Effect of discharge time on operation parameters of a 2 V lead–acid battery

Discharge time (min)	Discharge current (A)	End-point e.m.f. (V)	Capacity (A h)	Energy density (W h/kg)
2	480	1.41	16	22.5
30	60	1.60	30	48

Table 1.22 Effect of discharge current on operating parameters of a 2 V lead–acid battery

Discharge current (A)	End-point e.m.f. (V)	Electrical resistance (Ω)	Capacity (A h)	Energy density (W h/kg)
100	1.58	0.0155	29.7	46.5
600	1.26	0.002	10.0	12.5

1.22 Heating effects in batteries

Four types of heating effect will be discussed:

1. The chemical heating effects due to the heat of reaction of lead and lead dioxide with sulphuric acid during battery charging. This becomes a cooling effect when the battery is being discharged.

$$PbO_2 + Pb + 2H_2SO_4 = 2PbSO_4 + 2H_2O + 89\,400\,cal$$
$$2PbSO_4 + 2H_2O = PbO_2 + Pb + 2H_2SO_4 - 89\,400\,cal$$

This heating effect is identified with the heat content change of the reaction (ΔH) which, according to Equation 1.20, is given by the following expression:

$$\Delta H\,(cal) = \frac{-nF}{4.18}\left[E - T\left(\frac{dE}{dT}\right)_p\right]$$
$$= \frac{-3600C}{4.18}\left[E - T\left(\frac{dE}{dT}\right)_p\right]$$

where C = A h capacity of cell. Note that when heat is evolved, e.g. during the discharge of a lead–acid battery, ΔH has a negative value.

2. Chemical heating effect due to dissociation of water to hydrogen and oxygen towards the end of charge. This only applies to certain types of battery, e.g. lead–acid:

$$H_2 + \tfrac{1}{2}O_2 = H_2O + 57\,410\,cal$$

This type of heating occurs to an appreciable extent only when batteries are fully charged, i.e. on overcharge, and will not be discussed further.

3. Ohmic heating effects due to Joule heating effects related to operating current and the electrical resistance of the battery. From Equation 1.86,

$$\text{Joule heating} = \frac{I_{TOT}^2 \times R_{TOT} \times t}{4.18}\quad cal/t\,s$$

4. Localized heating effects due to tracking of current along low-conductivity paths on the top of the battery.

1.22.1 Consideration of chemical (ΔH) and ohmic (Joule) heating effects in a lead–acid battery

Information on the chemical and ohmic Joule heating effects during the discharge of a lead–acid battery is given in Table 1.20, which tabulates the extent of those effects during the discharge at various rates of a 2 V lead–acid battery.

This information is summarized in Figures 1.32 and 1.33 which, respectively, plot discharge time and current *versus* the chemical (ΔH) and ohmic (Joule)

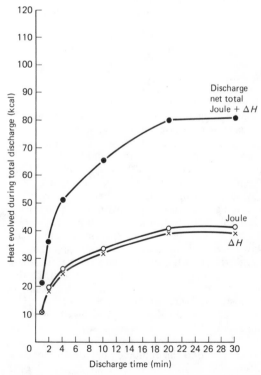

Figure 1.32 Discharge of a 2 V lead–acid battery: chemical (ΔH), ohmic (Joule) and net total (Joule + ΔH) heating effect

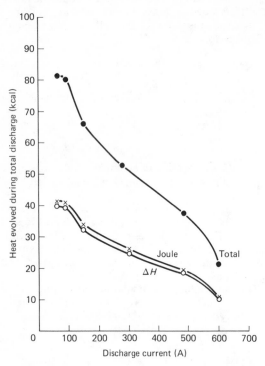

Figure 1.33 Discharge of a 2 V lead–acid battery: chemical (ΔH), ohmic (Joule) and net total (Joule + ΔH) heating effect

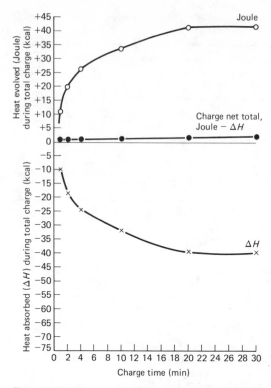

Figure 1.34 Discharge of a 2 V lead–acid battery: chemical (ΔH), ohmic (Joule) and net total (Joule + ΔH) cooling effect

contributions to the heat produced during a total discharge. It can be seen that, in this particular example, regardless of the discharge time, the chemical heating effect is very similar to the ohmic heating effect. It should be noted that the above comments apply only to the total discharge of a lead–acid battery when both the chemical and Joule effects contribute to the production of heat. Obviously, during charge, the chemical effect is one of cooling (i.e. ΔH is positive), whilst the Joule effect still leads to heating (Equation 1.86 – Joule heating is positive whether current is being withdrawn during discharge (positive) or supplied during charge (negative), as the square of the current is used in the calculation of heat production). During the total battery charge in this particular example, there is a small net heating effect of the battery as illustrated in Figure 1.34, which shows heat evolution due to ohmic heating, heat absorption (cooling) due to the chemical reaction, and a net heat evolution due to the combined heating and cooling effects.

It is assumed, in the above example of the charging of a lead–acid battery, that the charging

currents used are the same as the discharge currents used in the battery discharge as quoted in Table 1.20, i.e. 60–600 A during 30 s to 1 min charges or discharges. In fact, of course, during an actual battery charge, the charge currents and the duration of the charge may differ from these values and this will have an effect on chemical and ohmic heat evolution as discussed later.

The heat data quoted in the examples cited are the total heat evolution or absorption occurring during a total charge or a total discharge, i.e. during the whole discharge or charge process. Calculations of this type enable the number of kilocalories produced to be calculated, and provide data on the amount of cooling that would be required if there is a likelihood of the battery developing excessive temperatures in use. Cooling can be achieved by correct spacing of batteries, natural or forced air cooling, water cooling (internal or external), or the use of cooling fins.

An additional useful parameter which can be used when studying the heating effects of batteries in use is the average number of calories produced per second during a charge or a discharge. During battery discharge this parameter is obviously dependent on the discharge time and discharge current. Greater heating effects per unit time would be expected at higher discharge currents, i.e. lower discharge times. This is illustrated in Figures 1.35 and 1.36, which are prepared from data in Table 1.20 and plot average kilocalories per second for various discharge times between 1 and 30 min (Figure 1.35) corresponding to discharge currents between 602 and 60 A (Figure 1.36). It is seen that at the highest discharge rate quoted, i.e. 1 min discharge at 602 A between 0.3 and 0.4 kcal/s are produced on average, throughout the discharge and this may well indicate that cell cooling is required during such rapid battery discharges.

Discharge time/heat evolution graphs

In addition to heat evolved during a complete discharge and mean heat evolved per second, other battery parameters such as end-point volts, discharge current, A h capacity, energy density and electrical resistance are plotted against discharge time of a 2 V lead–acid battery (Figures 1.37–1.46). In Figures 1.47–1.54 these same parameters are plotted against discharge current.

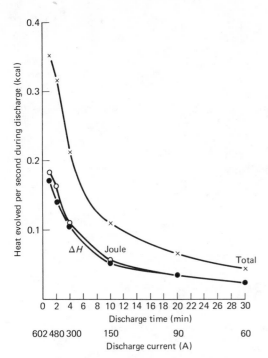

Figure 1.35 Discharge of a 2 V lead–acid battery: heat production per unit time as a function of discharge time

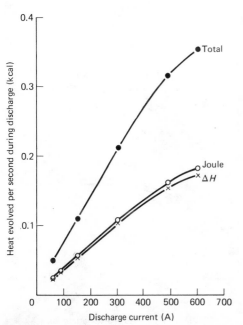

Figure 1.36 Discharge of a 2 V lead–acid battery: heat production per unit time as a function of discharge current

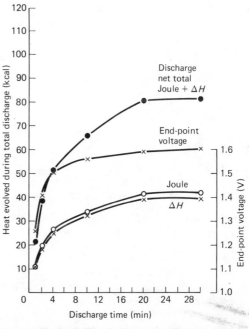

Figure 1.37 Discharge of a 2 V lead–acid battery: heat evolved during whole discharge, and end-point volts as a function of discharge time

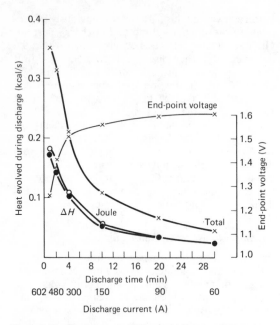

Figure 1.38 Discharge of a 2 V lead–acid battery: heat production per unit time, and end-point voltage as a function of discharge time

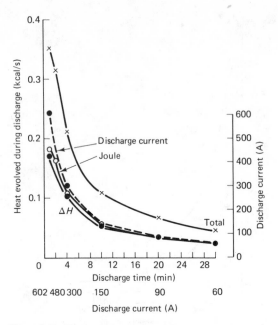

Figure 1.40 Discharge of a 2 V lead–acid battery: heat production per unit time, and discharge current as a function of discharge time

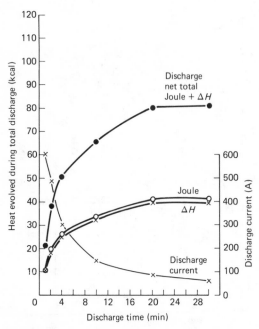

Figure 1.39 Discharge of a 2 V lead–acid battery: heat evolved during whole discharge, and discharge current as a function of discharge time

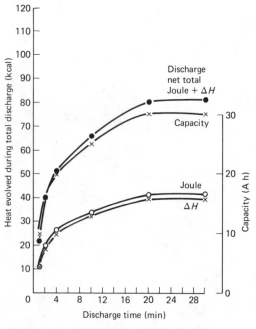

Figure 1.41 Discharge of a 2 V lead–acid battery: heat evolved during whole discharge and A h capacity as a function of discharge time

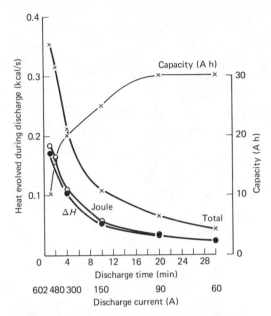

Figure 1.42 Discharge of a 2 V lead–acid battery: heat production per unit time and A h capacity as a function of discharge time

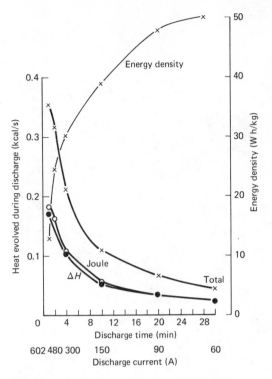

Figure 1.44 Discharge of a 2 V lead–acid battery: heat production per unit time and energy density as a function of discharge time

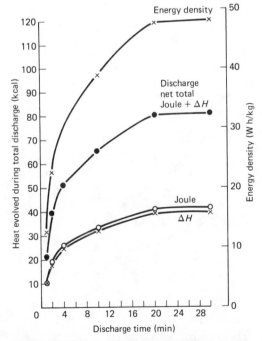

Figure 1.43 Discharge of a 2 V lead–acid battery: heat evolved during whole discharge and energy density as a function of discharge time

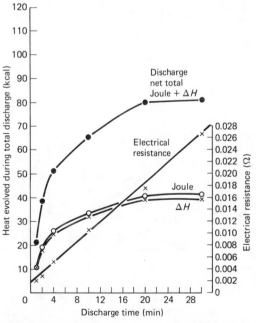

Figure 1.45 Discharge of a 2 V lead–acid battery: heat evolved during whole discharge and cell electrical resistance as a function of discharge time

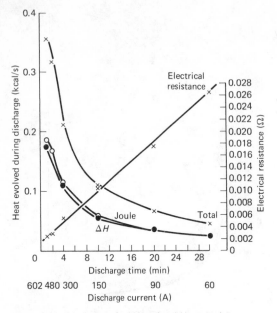

Figure 1.46 Discharge of a 2 V lead–acid battery: heat production per unit time and cell electrical resistance as a function of discharge time

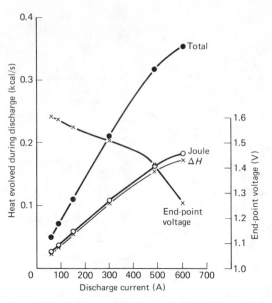

Figure 1.48 Discharge of a 2 V lead–acid battery: heat production per unit time and end-point voltage as a function of discharge current

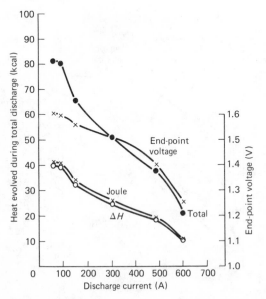

Figure 1.47 Discharge of a 2 V lead–acid battery: heat evolved during whole discharge and end-point voltage as a function of discharge current

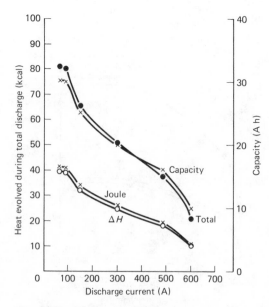

Figure 1.49 Discharge of a 2 V lead–acid battery: heat evolved during whole discharge and A h capacity as a function of discharge current

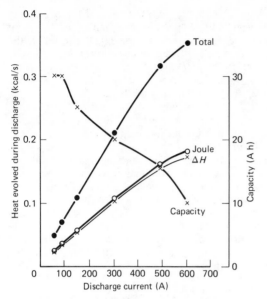

Figure 1.50 Discharge of a 2 V lead–acid battery: heat production per unit time and A h capacity as a function of discharge current

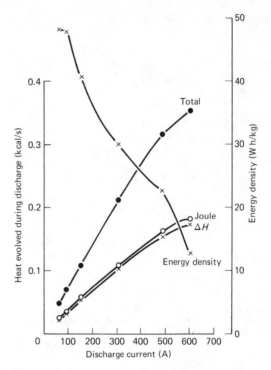

Figure 1.52 Discharge of a 2 V lead–acid battery: heat production per unit time and energy density as a function of discharge current

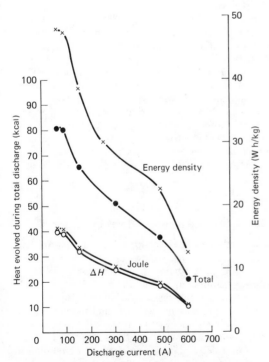

Figure 1.51 Discharge of a 2 V lead–acid battery: heat evolved during whole discharge and energy density as a function of discharge current

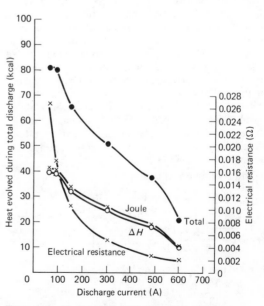

Figure 1.53 Discharge of a 2 V lead–acid battery: heat evolved during whole discharge, and cell electrical resistance as a function of discharge current

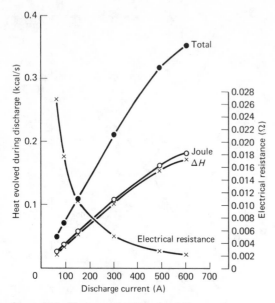

Figure 1.54 Discharge of a 2 V lead–acid battery: heat production per unit time, and cell electrical resistance as a function of discharge current

1.22.2 Effect of charging current and cell resistance on heat production in lead–acid batteries

During the charging of a lead–acid battery two processes occur which are associated with the production or loss of heat. The heat content ΔH of the reaction of lead sulphate to produce lead and lead dioxide has a positive value indicating that heat is absorbed from the surroundings to enable this process to occur, i.e. the electrolyte cools down. The other processes, ohmic (Joule) heating, depending on the charging current and the resistance of the cell always has a positive value, i.e. heat is evolved. Whether a battery cools down or heats up during charge depends on the relative magnitude of these two opposite cooling and heating effects which, in turn, depend on the operating conditions during battery charge.

The cooling effect is given by:

$$\Delta H = \frac{3600C}{4.18} \left[E - T \left(\frac{dE}{dT} \right)_p \right] \begin{array}{l} \text{cal during the charge} \\ \text{time } (t) \end{array}$$

i.e. $\Delta H = \dfrac{3600}{4.18 \times t} \left[E - T \left(\dfrac{dE}{dT} \right)_p \right]$ cal/s

The heating effect is given by:

$$J = \frac{I^2 \times R \times t}{4.18} \quad \text{cal during the charge time } (t)$$

i.e. $\quad \Delta H = \dfrac{I^2 R}{4.18}$ /s

where C is the capacity increase of battery during charge (A h), E the thermodynamic cell e.m.f. at end of charge (assumed to be 2.01 V), $(dE/dT)_p$ the temperature coefficient of cell e.m.f. (assumed to be 0.000 25 V/°C), T the cell temperature (K), I the charging current (A), R the cell internal resistance (Ω), and t the duration of charge (s).

If $J - \Delta H$ is negative, the cell will cool down during charge. Conversely, if $J - \Delta H$ is positive, the cell will heat up during charge.

In practical cell charging, there is an additional factor to be contended with, namely loss of calories by convection and conduction to the surroundings when $J - \Delta H$ is positive and a gain of calories from the surroundings when $J - \Delta H$ is negative. This factor is ignored in the calculations that follow but is easily allowed for in any such calculations.

Consider the case of a 2 V lead–acid battery that requires charging to 30 A h at 20°C during 2 h. The cooling effect is:

$$\Delta H = \frac{3600 \times 30}{4.18} \quad (2.01 - 293 \times 0.000 25)$$

$$= 50 041 \text{ cal/2 h}$$

$$= 6.950 \text{ cal/s}$$

The heating effect, J, will depend on the charging current and the cell resistance. If $J - 6.950$ is negative the cell will cool down; if $J - 6.950$ is positive the cell will heat up. Table 1.23 tabulates the heat losses or heat gains obtained with cells of four different assumed internal resistances, namely, 0.002, 0.005, 0.015 and 0.027 Ω, when charged at various currents between 20 and 1000 A. Cell cooling, during charge, occurs only at relatively low charging currents of the order of 20–50 A. Cell heating would not be a serious problem at charging currents under 200–300 A with cells in the 0.002–0.027 Ω electrical resistance range, especially when the natural heat losses from the cell due to convection and conduction are taken into account.

1.22.3 Localized heating effects in batteries

Additional heating effects of a localized nature can arise in batteries due to low-resistance short-circuit conducting paths being formed between cells in batteries arranged in series. Figure 1.55 shows such an effect where the resistance between batteries A and B in series is denoted by R_1 and the current by I_1, and the resistance between two cells A and D in parallel is denoted by R_2 and the current by I_2. Such conduction between cells can occur, for example, if free electrolyte has been spilt on the battery top, providing a low-resistance conducting path for the current. A particular example of this is batteries

Table 1.23 Heating and cooling effects during charge of 2 V lead–acid battery at 20°C. Influence of charge current and cell resistance

Charging current (A)	Assumed cell resistance (Ω)							
	0.002		0.005		0.015		0.027	
	Cell heating (+) or cooling (−) (cal/s)	e.m.f. change*	Cell heating (+) or cooling (−) (cal/s)	e.m.f. change*	Cell heating (+) or cooling (−) (cal/s)	e.m.f. change*	Cell heating (+) or cooling (−) (cal/s)	e.m.f. change*
20	−6.76	0.04	−6.47	0.10	−5.52	0.3	−4.37	0.54
50	−5.76	0.10	−3.96	0.25	+2.02	0.75	+9.20	1.35
100	−2.17	0.20	−5.01	0.5	+28.93	1.5	+57.64	2.7
200	+12.17	0.40	+40.89	1.0	+136.57	3.0		
300	+36.07	0.60	+100.69	1.5				
400	+69.53	0.80	+184.41	2.0				
500	+112.55	1.0	+292.05	2.5				
600	+163.13	1.2						
1000	+471.05	2.0						
1500	+717.0	3.0						

* Current × cell resistance

used in electric locomotives used in coal mines. Such batteries, in addition to free sulphuric weight, have coal dust on the battery tops. A slurry of coal dust in acid is a very good conductor and, as will be seen, such low-resistance conducting paths can conduct high currents and lead to localized high ohmic (Joule) heating effects on the battery top, which in certain circumstances can generate so much heat that battery ignition is a possibility.

From Equation 1.86, the calories generated per second during the passage of a current of I amps through a resistance of $R\,\Omega$ is given by

$$\frac{I^2 R}{4.18} \quad \text{cal/s}$$

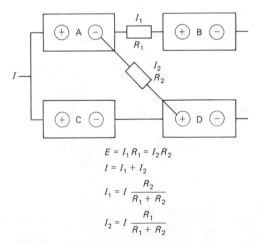

$$E = I_1 R_1 = I_2 R_2$$
$$I = I_1 + I_2$$
$$I_1 = I\,\frac{R_2}{R_1 + R_2}$$
$$I_2 = I\,\frac{R_1}{R_1 + R_2}$$

Figure 1.55 Distribution of current in a divided circuit on a battery

Therefore, calories generated per second along conducting path AB (Figure 1.55) is

$$\frac{I_1^2 R_1}{4.18} = \frac{I^2}{4.18}\left(\frac{R_2}{R_1 + R_2}\right)^2 R_1 \tag{1.107}$$

and calories generated per second along conducting path AD is

$$\frac{I_2^2 R_2}{4.18} = \frac{I^2}{4.18}\left(\frac{R_1}{R_1 + R_2}\right)^2 R_2 \tag{1.108}$$

Consequently:

$$\frac{\text{cal/s along path AB}}{\text{cal/s along path AD}} = \frac{R_2}{R_1} \tag{1.109}$$

Using Equations 1.107 and 1.108 it is possible to calculate the calories per second generated along the two paths for any particular values of R_1, R_2 and current (I).

Table 1.24 shows the heat productions along the two paths AB and AD when the current is 50 A, the internal resistance (R_1) is 0.1 Ω and the resistance between parallel strings of cells (R_2) is varied between 0.001 and 0.1 Ω. It can be seen that the lower resistance, R_2, of the conducting path AD is relative to the resistance, R_1, of the path AB (i.e. the lower the value of R_2/R_1), the greater is the proportion of the total heat evolution that is generated along path AD. This is demonstrated graphically in Figure 1.56.

Figure 1.57 illustrates the effect of increase of current flowing on heat production in a low-resistance ($R_2 = 0.005\,\Omega$) conducting path in a battery. Because of the exponential nature of this curve, low-resistance conducting paths between cells could lead to the development of a thermal runaway

Table 1.24 Localized ohmic heating effects on a battery top due to low-resistance intercell conducting paths (current, $I = 50$ A)

Intercell resistance (cells in series) ($R_1\ \Omega$)	Intercell resistance (cells in parallel) ($R_2\ \Omega$)	R_2/R_1	Heat evolved corresponding to resistance (cal/s)		Total heat evolved (cal/s) (path AB + path AD)	Heat generated along low-resistance path AD as % of total evolved (cal/s)
			$R_1\ \Omega$ (path AB)	$R_2\ \Omega$ (path AD)		
0.1	0.001	0.01	0.005 86	0.586	0.592	98.9
0.1	0.002	0.02	0.022 9	1.148	1.171	98.0
0.1	0.005	0.05	0.135	2.709	2.844	95.2
0.1	0.01	0.1	0.493	4.939	5.432	90.9
0.1	0.05	0.5	6.632	13.275	19.907	66.7
0.1	0.1	1.0	14.950	14.950	29.9	50.0

situation with consequent risk of battery fires. It is essential, therefore, in the design and maintenance of batteries, to avoid the development of low-resistance paths between cells.

1.23 Spontaneous reaction in electrochemical cells

Most electrochemical cells are based on the occurrence of a chemical reaction within the cell between the materials comprising the two electrodes. It may sometimes be necessary to establish whether a particular reaction would form the basis of an electrochemical cell. The following treatment enables calculations to be made to determine whether a given reaction will or will not occur spontaneously, i.e. whether it is theoretically possible.

Consider the general equation

$$aA + bB + \ldots \rightleftharpoons lL + mM + \ldots$$

and suppose that the reactants A, B, etc., and the products L, M, etc., are not necessarily present in their equilibrium amounts, but in any arbitrarily chosen state. The word 'state' here refers essentially to the partial pressure or concentration of any substance present in the system; the temperature may be regarded as constant. If the molar free energies of the various substances taking part in the reaction are F_A, F_B, ..., F_L, F_M, ..., in these arbitrary states, then

Free energy of products $= lF_L + mF_M + \ldots$
Free energy of reactants $= aF_A + bF_B + \ldots$

The increase of free energy, ΔF, accompanying the reaction with the reactants and products in the specified states is the difference between these two quantities; hence,

$$\Delta F = (lF_L + mF_M + \ldots) - (aF_A + bF_B + \ldots) \quad (1.110)$$

The molar free energy, F, of a substance in any state can be expressed in terms of its activity, a, in that state by means of Equation 1.110, which may be written as

$$F = F^0 + RT \ln a \quad (1.111)$$

where F^0 is the molar free energy in the standard state of unit activity.

If the values of F_A, F_B, ..., F_L, F_M, ..., in Equation 1.110 are replaced by the corresponding

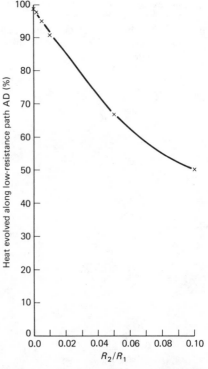

Figure 1.56 Distribution of heat production in a divided circuit on a battery (current 50 A)

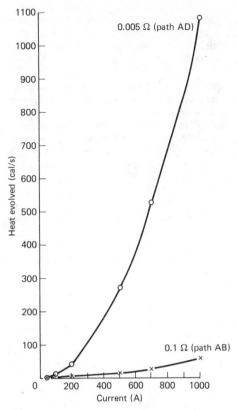

Figure 1.57 Effect of current on distribution of heat production in a divided circuit on a battery: 0.005 Ω lower resistance *versus* 0.1 Ω higher resistance

expressions derived from Equation 1.111, it is seen that

$$\Delta F = (l(F^0_L + RT \ln a_L) + m(F^0_M + RT \ln a_M) + \ldots) \\ - (a(F^0_A + RT \ln a_A) + b(F^0_B + RT \ln a_B) + \ldots)$$

(1.112)

where a_A, a_B, ..., a_L, a_M, are the activities of the various substances involved in the reaction in their arbitrary state. Upon rearranging Equation 1.112, the result is

$$\Delta F = \Delta F^0 + RT \ln \left(\frac{a^l_L \times a^m_M \times \ldots}{a^a_A \times a^b_B \times \ldots} \right)$$

(1.113)

where ΔF^0, the increase in free energy accompanying the reaction when all the reactants and products are in their respective standard states, is given by an equation analogous to Equation 1.110, i.e.

$$\Delta F = (lF^0_L + mF^0_M + \ldots) \\ - (aF^0_A + bF^0_B + \ldots)$$

(1.114)

It will be noted that the quantity whose logarithm is involved in Equation 1.113, which may be called the arbitrary reaction quotient, is exactly similar in form

to the expression for the equilibrium constant. In the latter, the activities of reactants and products are the values when the system as a whole is in equilibrium, whereas in the reaction quotient in Equation 1.113 .the activities correspond to the arbitrary specified states for the various substances, and these may or may not correspond to a condition of equilibrium.

For a system in equilibrium the free energy change, ΔF, is zero, provided the temperature and pressure at equilibrium are not allowed to alter. Consequently, when the arbitrarily chosen conditions represent those for equilibrium, ΔF in Equation 1.113 may be put equal to zero, so that

$$\Delta F^0 = - RT \ln \left(\frac{a^l_L \times a^m_M \times \ldots}{a^a_A \times a^b_B \times \ldots} \right)_e$$

(1.115)

where the subscript, e, indicates that the activities are those for the system at equilibrium. Since the standard free energy change ΔF^0 refers to the reactants and products in the definite states of unit activity, it is apparent that ΔF^0 must be constant; consequently, the right-hand side of Equation 1.115 must also be constant. The gas constant R has, of course, a definite value and so it follows that, if the temperature T is constant,

$$\frac{(a^l_A \times a^m_M \times \ldots)}{(a^a_A \times a^b_B \times \ldots)} = \text{constant} = K$$

(1.116)

This is the exact expression for the equilibrium constant K, as given in Equation 1.113; it has been derived from thermodynamic considerations alone, without the assumption of the law of mass action. It may be simplified for systems which do not depart appreciably from ideal behaviour, i.e. for reactions in solution mole fractions. In dilute solutions, concentrations may be employed. In general, however, with concentrated solutions such as cell electrolytes, it is best to use activities.

1.23.1 The reaction isotherm

If the symbol K for the equilibrium constant, as given by Equation 1.116, is substituted in Equation 1.115, the result is

$$\Delta F^0 = - RT \ln K$$

(1.117)

This is a very important equation, as it relates the standard free energy change of a reaction to the experimentally determinable equilibrium constant. If this value for ΔF^0 is now inserted in Equation 1.113, it follows that the free energy change for the reaction with reactants and products in any arbitrary state is given by

$$\Delta F = -RT \ln K + RT \ln \left(\frac{a^l_L \times a^m_M \times \ldots}{a^a_A \times a^b_B \times \ldots} \right)$$

(1.118)

where the a values refer to the activities in these arbitrary states. The arbitrary reaction quotient in terms of activities may be represented by the symbol Q_a, that is

$$Q_a = \frac{a_L^l \times a_M^m \times \ldots}{a_A^a \times a_B^b \times \ldots} \qquad (1.119)$$

so that Equation 1.118 becomes

$$\Delta F = -RT \ln K + RT \ln Q_a \qquad (1.120)$$

which is a form of what is known as the reaction isotherm. It is evident that, if the arbitrary states happen to correspond to those for the system at equilibrium, Q_a will become identical with the equilibrium constant K, since the expressions for both these quantities are of exactly the same form (see Equations 1.116 and 1.119). By Equation 1.120 the value of ΔF would then be zero, as indeed it should be for an equilibrium system.

The standard state of unit activity may be defined in any convenient manner and so it is obvious that the standard free energy change ΔF^0 must depend upon the particular choice of standard states.

Attention may be drawn to the fact that although ΔF^0 depends on the particular standard states that are chosen, the value of ΔF is as it should be, independent of their nature. This may be readily seen by writing Equation 1.120 in the simple form

$$\Delta F = RT \ln \frac{Q_a}{K} \qquad (1.121)$$

from which it is evident that ΔF is determined by the ratio of Q_a to K. Consequently, as long as both of these quantities are expressed in terms of the same standard states, that is, in terms of the same units, the result will be independent of the particular standard state employed.

For reactions in dilute solution, the standard state is chosen as the (hypothetical) ideal solution of unit concentration, i.e. 1 mol (or 1 gram-ion) per litre, or of unit molality, i.e. 1 mol/1000 g of solvent. Under these conditions the standard free energy is given by

$$\Delta F^0_c = -RT \ln K_c \qquad (1.122)$$

and the reaction isotherm becomes

$$\Delta F = -RT \ln K_c + RT \ln Q_c = RT \ln \frac{Q_c}{K_c} \qquad (1.123)$$

where Q_c is the arbitrary reaction quotient with the states of the reactants and products expressed in terms of concentrations in their ideal solutions. If the solutions are sufficiently dilute, the actual concentrations may be employed in place of the ideal values.

1.23.2 Criteria of spontaneous reaction

The essential importance of the reaction isotherm lies in the fact that it provides a means of determining whether a particular reaction is possible or not, under a given set of conditions. For a thermodynamically irreversible process taking place at constant temperature and pressure, ΔF must be negative; that is, the free energy of the system diminishes. If a particular physical or chemical change is to be theoretically possible it must be able to occur spontaneously; spontaneous processes are, however, irreversible in the thermodynamic sense and hence it follows that a reaction can take place spontaneously only if it is accompanied by a decrease of free energy, i.e. ΔF must be negative, at constant temperature and pressure. This result applies to any process, physical or chemical; it is immaterial whether the latter is reversible, in the chemical sense, or if it goes to virtual completion.

If the value of ΔF under a given set of conditions is positive, the reaction cannot possibly occur spontaneously under those conditions, although it may be able to do so if the conditions are altered. By writing the reaction isotherm in the general form

$$\Delta F = RT \ln \frac{Q}{K}$$

it can be seen that a particular process will be possible theoretically if the reaction quotient Q is less than the corresponding equilibrium constant K; ΔF will then involve the logarithm of a fraction and hence will be negative. If, however, the arbitrary concentrations are such as to make Q greater than A, the value of ΔF will be positive, and the reaction will not be possible under these conditions.

It should be borne in mind, of course, that a change in temperature may affect the value of the equilibrium constant to such an extent that a reaction that previously could not occur spontaneously, for a given value of the quotient Q, can now do so. At some temperatures K is less than Q, but as the temperature is altered K may become greater than Q; the sign of ΔF will thus reverse from positive to negative, and the reaction becomes possible, when the temperature is changed.

It is of interest to consider the physical significance of the conclusion that a reaction will be able to take place spontaneously if Q is less than A, but not if Q is greater than A. A comparison of Equations 1.116 and 1.119, which define K and Q respectively, shows that when Q is less than K, the arbitrary activities (or concentrations) of the products are relatively less than those in the equilibrium state. The occurrence of the reaction will mean that the amounts of the products are increased, while those of the reactants decrease correspondingly. In other words, the state of the system will tend to move spontaneously towards the state of equilibrium. On the other hand, if Q is greater than A, the products will be present in excess of the equilibrium activities and for the reaction to continue would mean a still greater departure from equilibrium; such a change would never occur spontaneously. In these circumstances, of course, the reaction would tend to take place in the reverse direction.

Although any reaction accompanied by a decrease of free energy is theoretically possible, this is no indication that the process will actually occur with measurable speed. In a series of analogous reactions, the rates at which the processes occur are roughly in the order of the free energy decrease, but in general, for different reactions, there is no connection between the magnitude of the decrease of free energy and the rate at which the reaction occurs. For example, at ordinary temperatures and pressures the free energy change for the combination of hydrogen and oxygen to form water has a very large negative value, yet the reaction, in the absence of a catalyst, is so slow that no detectable amount of water would be formed in years. The passage of an electric spark or the presence of a suitable catalyst, however, facilitates the occurrence of a reaction which the free energy change shows to be theoretically possible.

As shown above, a reaction can take place spontaneously if it is accompanied by a decrease of free energy, i.e. when ΔF is negative. If the reaction can be made to take place, for the passage of n faradays, in a reversible cell of e.m.f. equal to E, then by Equation 1.23

$$\Delta F = -nFE \qquad (1.23)$$

so that ΔF is negative if E is positive. Consequently, when the e.m.f. of a reversible cell is positive, the corresponding cell reaction will take place spontaneously.

The e.m.f. of a cell and even its sign depend on the activities, or concentrations, of the reactants and products of the reaction taking place in the cell; hence, the value of the free energy change will vary in an analogous manner. This is in agreement with the conclusions reached above. For many purposes it is convenient to consider the free energy change ΔF^0 associated with the reaction when all the substances are in their standard states of unit activity. The appropriate form of Equation 1.23 is

$$\Delta F^0 = -nFE^0_{cell} \qquad (1.124)$$

where E^0_{cell} is the standard e.m.f. of the reversible cell in which the given reaction occurs. The value of E^0_{cell} can be obtained by subtracting the standard oxidation potentials of the electrodes constituting the cell, i.e.

$$E^0_{cell} = E^0_{left} - E^0_{right}$$

where E^0_{left} and E^0_{right} are the standard oxidation potentials of the left-hand and right-hand electrodes, respectively, as given in Table 1.2. The standard free energy change of the cell reaction can then be derived from Equation 1.124.

Consider, for example, the Daniell cell:

$$Zn \mid ZnSO_4.(a_{Zn^{2+}} = 1) \parallel CuSO_4 (a_{Cu^{2+}} = 1) \mid Cu = (-0.340)$$

for which the reaction is

$$Zn + Cu^{2+} = Zn^{2+} + Cu$$

for the passage of 2 F. The standard oxidation potential of the left-hand (Zn, Zn^{2+}) electrode is $+0.761$ V (Table 1.2), while the standard oxidation potential of the right-hand (Cu, Cu^{2+}) electrode is -0.340 V; the standard e.m.f. of the complete cell, E^0_{cell} is thus $+0.761 - (-0.340) = 1.101$ V at 25°C, and by Equation 1.124:

$$\Delta F^0 = -2 \times 96\,500 \times 1.101 = -212\,500\,J$$
$$= -212\,500 \times 0.2390 = -50\,790\,cal$$

i.e. -50.79 kcal, at 25°C.

Since E^0_{cell}, the standard e.m.f. of the cell, is positive, the standard free energy change, ΔF^0, is negative, and the reaction as written is spontaneous; hence, metallic zinc can react spontaneously with cupric ions at unit activity to produce metallic copper and zinc ions also at unit activity.

Suppose it is required to determine whether the reaction

$$Cd + 2H^+ (a = 1) = Cd^{2+} (a = 1) + H_2 (1\,atm.)$$

that is, the displacement of hydrogen ions from solution by metallic cadmium, is possible theoretically when all the substances are in their standard states. This reaction would occur in the cell

$$Cd \mid Cd^{2+} (a = 1) \parallel H^+ (a = 1) \mid H_2(1\,atm.)$$
$$+0.402 \qquad\qquad\qquad 0$$

the standard e.m.f. of which is the same as the oxidation potential of the cadmium electrode, i.e. $+0.402$ V at 25°C. Since E^0_{cell} is positive, ΔF^0 is negative and the reaction should be capable of taking place spontaneously; the actual value of the standard free energy change could be determined, if required, from Equation 1.124, using $n = 2$ for the cell reaction as written above.

An illustration of another kind is provided by the cell

$$Ag \mid AgClO_4 (a = 1) \parallel Fe(ClO_4)_2, Fe(ClO_4)_3 (a = 1) \mid Pt$$
$$-0.799 \qquad\qquad\qquad\qquad -(-0.771)$$

in which the reaction is

$$Ag(s) + Fe^{3+} = Ag^+ + Fe^{2+}$$

for the passage of 1 F. The standard oxidation potential of the left-hand (Ag, Ag^+) electrode is -0.799 V, while the reduction potential of the right-hand (Fe^{2+}, Fe^{3+}, Pt) electrode is $-(-0.771)$, i.e. $+0.771$ V (Table 1.2). The standard e.m.f. of the cell depicted is thus $-0.799 + 0.771 - 0.028$ V; since E^0_{cell} is negative, ΔF^0 is positive, and the reaction as written will not occur spontaneously for the reactants and products in their standard states. For the reverse reaction, however, ΔF^0 will be negative, so that the process

$$Fe^{2+} + Ag^+ = Fe^{3+} + Ag(s)$$

can be spontaneous if all the substances taking part are at unit activity.

An examination of the foregoing results, or a general consideration of the situation, will reveal the fact that the standard e.m.f. of a cell is positive when the standard oxidation potential of the left-hand electrode is greater algebraically than the standard oxidation potential of the right-hand electrode; that is, when the former lies above the latter in Table 1.2. When this is the case, the cell reaction will be capable of occurring spontaneously, oxidation taking place at the left-hand side and reduction at the right-hand side. It follows, therefore, that any system in Table 1.2 should be able, theoretically, to reduce any system lying below it in the table, while it is itself oxidized, provided all the substances concerned are in their standard states of unit activity. Thus, as seen above, zinc (higher in the table) reduces cupric ions to copper (lower in the table), while it is itself oxidized to zinc ions; similarly, cadmium reduces hydrogen ions to hydrogen gas, and is itself oxidized to cadmium ions. In general, any metal higher in Table 1.2 will displace from solution, i.e. reduce, the ions of a metal (or of hydrogen) lower in the table of standard oxidation potentials.

These conclusions are strictly applicable only when the ions are all at unit activity. By changing the activity it is possible for a process to be reversed, particularly if the standard potentials of the systems involved are not far apart. For example, copper should be unable to displace hydrogen ions from solution, since the Cu, Cu^{2+} system has a lower oxidation potential than the H_2, H^+ system; this is true in so far as copper does not normally liberate hydrogen from acid solution. However, if the concentration of the cupric ions is decreased very greatly (for example, by the formation of complex ions), the oxidation potential is increased until it is greater than that of hydrogen against hydrogen ions in the same solution. In these circumstances, the displacement of hydrogen ions by metallic copper, with the evolution of hydrogen gas, becomes possible.

Similar observations have been made in connection with the Ag, Ag^+ and Fe^{2+}, Fe^{3+} systems; as seen above, if all the substances are in their standard states of unit activity, the spontaneous reaction should be the reduction of silver ions to metallic silver by ferrous ions, as is actually the case. The standard oxidation potentials of the two systems are not very different, although that of the Fe^{2+}, Fe^{3+} system is the higher.

Although the standard potentials provide some indication, therefore, of the direction in which a particular reaction may be expected to proceed spontaneously, especially if the potentials are appreciably different for the two systems involved, the results may sometimes be misleading. The real criterion, which is always satisfactory, is that the e.m.f. of the actual cell, i.e. E, with the substances at the given activities, and not necessarily E^0_{cell}, when the activities are all unity, should be positive for the reaction to be spontaneous. In other words, the actual oxidation potential of the left-hand electrode must be greater algebraically than that of the right-hand electrode if the reaction occurring in the cell is to proceed spontaneously.

1.23.3 Equilibrium constants

For many purposes it is more convenient to calculate the equilibrium constant of a reaction, instead of the free energy change; this constant provides the same information from a slightly different viewpoint. The equilibrium constant is related to the standard free energy change by Equation 1.117, namely

$$\Delta F^0 = -RT \ln K$$

and since ΔF^0 is equal to $-nFE^0_{cell}$, by Equation 1.124 it follows that:

$$E^0_{cell} = \frac{RT}{nF} \ln K \qquad (1.125)$$

or at 25°C, with E^0_{cell} in volts,

$$E^0_{cell} = \frac{0.059\ 15}{n} \log K \qquad (1.126)$$

By means of these equations the equilibrium constant of any reaction can be readily calculated from the standard e.m.f. of the reversible cell in which the reaction occurs.

The reaction in the Daniell cell, for example, is

$$Zn(s) + Cu^{2+} = Zn^{2+} + Cu(s)$$

for the passage of 2 F, i.e. $n = 2$, and the equilibrium constant is given by

$$K = \left(\frac{a_{Zn^{2+}}}{a_{Cu^{2+}}} \right)_e \qquad (1.127)$$

where the subscript, e, is used to show that the activities are the values when the reacting system attains equilibrium. The activities of the solid zinc and copper are, as usual, taken as unity. The standard e.m.f. of the cell, as seen above, is equal to $E^0_{Zn} - E^0_{Cu}$, i.e. 1.101 V at 25°C; hence, by Equations 1.126 and 1.127,

$$1.101 = \frac{0.059\ 15}{2} \log \left(\frac{a_{Zn^{2+}}}{a_{Cu^{2+}}} \right)$$

$$K = \left(\frac{a_{Zn^{2+}}}{a_{Cu^{2+}}} \right)_e = 1.7 \times 10^{37}$$

The ratio of the activities of the zinc and copper ions in the solution at equilibrium will be approximately equal to the ratio of the concentrations under the same conditions; hence, when a system consisting of metallic zinc and copper and their bivalent ions in aqueous solution attains equilibrium, the ratio of the zinc ion to the cupric ion concentration is extremely large. If zinc is placed in a solution of cupric ions, the latter will be displaced to form metallic copper until the $c_{Zn^{2+}}/c_{Cu^{2+}}$ ratio in the solution is about 10^{37}. In other words, the zinc will displace the copper from the solution until the quantity of cupric ions remaining is extremely small.

1.24 Pressure development in sealed batteries

This discussion is concerned with the development of gas pressures due to hydrogen and oxygen in sealed lead–acid batteries, although the comments made would apply, equally, to other types of batteries where hydrogen and oxygen are produced. As a lead–acid battery approaches the end of charge, i.e. its voltage exceeds 2 V, and also on overcharge, an increasing proportion of the charge current is used up not in charging the plates but in electrolysing the sulphuric acid to produce hydrogen and oxygen. Besides being wasteful in charging current and damaging to the battery, this gas production represents a loss of water from the electrolyte, which, unless the water is replenished, makes the acid more concentrated with consequent adverse effects on battery performance and, eventually, physical deterioration of the battery. The concept of a sealed battery is to reduce or eliminate gassing during charging and discharging of the battery so that water loss does not occur from the electrolyte. In this situation the battery can be fully sealed during manufacture (with the proviso that a pressure relief valve is supplied to relieve gas pressure if untypical conditions develop) and will not require topping up with water during its life.

Two main approaches have been made in the design of sealed batteries. In one approach, gassing is controlled at a very low level by attention to battery design such as using calcium–lead grid alloys and avoiding the presence of free electrolyte by attention to charging methods. In the other method, any hydrogen or oxygen produced is recombined back to water by means such as catalytic conversion or methods based on the third electrode principle. Reconversion of hydrogen and oxygen to water does, however, present problems. It is a fact that towards the end of charge, and on overcharge, the hydrogen and oxygen are not produced in exactly the stoichiometric amounts as indicated by the equation

$$H_2O = H_2 + \tfrac{1}{2}O_2$$

Were this the case the electrolysis gas would contain 66.7% v/v hydrogen and 33.3% v/v oxygen (i.e. the stoichiometric composition) and, provided the recombination device were efficient, complete recombination of these gases to water would occur:

$$H_2 + \tfrac{1}{2}O_2 = H_2O$$

In fact, as will be discussed later, in the later stages of charge and the earlier stages of discharge, the hydrogen–oxygen mixture in a sealed battery has a composition which is non-stoichiometric. An excess of either hydrogen or oxygen remains over the stoichiometric composition and this excess does not, of course, react. Therefore, a pressure build-up occurs in the cell. Whilst cells can be designed to tolerate a certain pressure build-up, there is a practical limit to this. During a complete charge/discharge cycle, the total amounts of hydrogen and oxygen produced are stoichiometric; and consequently, in the long term, complete recombination would occur and internal cell pressure would be relieved, i.e. cell pressure would not continually increase with continued cycling of the battery but a maximum pressure excursion would occur within each single charge/discharge cycle.

1.24.1 Overvoltage

At a platinized platinum cathode, hydrogen is liberated practically at the reversible hydrogen potential of the solution. With other electrodes, e.g. lead, a more negative potential is required to secure its liberation. The difference between the reversible hydrogen potential and the actual decomposition potential in the same solution is known as the hydrogen overvoltage of the metal. Approximate determinations of the hydrogen overvoltage can be made by observing the potential of the lead cathode when the current–voltage curve shows that appreciable electrolysis is taking place, or by making the cathode very small and observing its potential when the first visible bubbles of hydrogen occur. Hydrogen overvoltages in the range 0.36–0.64 V have been obtained by these methods at a lead cathode.

Similar considerations apply in the case of liberation of oxygen at the lead dioxide anode. The reversible anode oxygen potential for the liberation of oxygen at the lead dioxide anode is considerably more positive than the value calculated from free energy data and is, in fact, in the region 0.4–0.5 V.

Although the decomposition potential of an aqueous solution of sulphuric acid to produce hydrogen and oxygen is constant at about 1.7 V with smooth platinum electrodes, due to the overvoltage phenomenon, the value is different if other materials are employed as electrode materials. If the cathode is lead and the anode is platinum, for example, the decomposition potential increases to about 2.2 V.

The decomposition voltage for the electrolysis of sulphuric acid to hydrogen and oxygen is about 1.7 V and the hydrogen overvoltage at the cathode is

0.6 V; thus hydrogen does not start to be evolved in a lead–acid battery until the charging potential reaches 2.3 V. Similarly, the oxygen overvoltage at the anode is 0.5 V; thus oxygen does not start to be evolved until the charging potential reaches 2.2 V. In this sense, in a lead–acid battery, the anode and cathode behave independently of each other, each releasing oxygen and hydrogen, respectively, as dictated by the electrode e.m.f.

During the discharge of a lead–acid battery the following reactions occur:

$$H_2SO_4 = 2H^+ + SO_4^{2-}$$

1. At the positive electrode (anode):

$$PbO_2 + H_2SO_4 + 2H^+ + 2e = PbSO_4 + 2H_2O$$

above 2.2 V (oxygen evolution):

$$PbO_2 + 2H^+ + 2e + SO_4^{2-} = PbSO_4 + H_2O + \tfrac{1}{2}O_2$$

2. At the negative electrode (cathode):

$$Pb + SO_4^{2-} = PbSO_4 + 2e$$

above 2.3 V (hydrogen evolution):

$$Pb + SO_4^{2-} + 2H^+ = PbSO_4 + H_2$$

During charge the above reactions occur in reverse.

According to the Tafel relationship:

$$i = Ke^{-aE} \tag{1.128}$$

where i is the current (A), b a constant characteristic of the electrode, E the potential (V) of the cathode or the anode, and a is a constant identified as

$$\frac{F}{2RT}$$

where $F = 96\,500$ C, R is the gas constant (1.987) and T is the temperature (K).

Hence

$$i = k \exp\left(-FE/2RT\right) \tag{1.129}$$

Taking logarithms of Equation 1.129,

$$\ln i = \text{constant} - \frac{FE}{2 \times RT} \tag{1.130}$$

or

$$\log i = \text{constant} - \frac{FE}{2 \times 2.303 \times RT} \tag{1.131}$$

or

$$-\frac{dE}{d \log i} = \frac{2 \times 2.203 \times RT}{F} \tag{1.132}$$

Inserting values in Equation 1.132, at 18°C (291 K):

$$-\frac{dE}{d \log i} = \frac{2 \times 2.303 \times 1.987 \times 291}{96\,500} = 0.116\,\text{V}$$

i.e. the cathode potential becomes 0.116 V more negative for each ten-fold increase in the current

and the anode potential becomes 0.116 V more positive for each ten-fold increase in the current.

If E_c denotes the e.m.f. of the cathode and i_c the current flowing, and E_a denotes the e.m.f. of the anode and i_a the current flowing, then from Equation 1.131:

$$\log i_c = C - \frac{FE_c}{2 \times 2.303 \times RT}$$

and

$$\log i_a = C - \frac{FE_a}{2 \times 2.303 \times RT}$$

then

$$\log \frac{i_a}{i_c} = \frac{F(E_c - E_a)}{2 \times 2.203 \times RT} \tag{1.133}$$

At 18°C

$$\log \frac{i_a}{i_c} = 36.233\,(E_c - E_a)$$

Assume that E_c, the e.m.f. of the cathode, is 2.3 V (hydrogen liberation) and assume a range of values of 2.2–2.35 V for E_a, the e.m.f. of the anode (oxygen liberation), i.e. the anode e.m.f. starts off being less than and finishes up being greater than the cathode e.m.f.

Having ascertained i_a/i_c from the known values of E_c and E_a, assume that i_a has a value of 10 A, i.e.

$$i_c = \frac{10}{i_a/i_c}$$

(Table 1.21). From the anode reaction:

$$PbO_2 + 2H^+ + 2e + SO_4^{2-} = PbSO_4 + H_2O + \tfrac{1}{2}O_2$$

$2 \times 96\,500$ C liberate $11\,200$ cm^3 oxygen, i.e. a current of 10 A for 1 s liberates

$$\frac{10 \times 11\,200}{2 \times 96\,500} = 0.5803\,\text{cm}^3 \text{ oxygen/s}$$

From the cathode reaction:

$$PbSO_4 + 2H^+ = PbSO_4 + H_2$$

$2 \times 96\,500$ C liberate $22\,400$ cm^3 hydrogen, i.e. a current of $10/(i_a/i_c)$ A for 1 s liberates

$$\frac{10 \times 22\,400}{(i_a/i_c) \times 2 \times 96\,500} = \frac{10i_c}{i_a} \times 0.1160\,\text{cm}^3 \text{ hydrogen/s}$$

From these data, it is possible, as shown in Table 1.25, to calculate the volume of oxygen produced per second at the anode and the volume of hydrogen produced per second at the cathode, hence the total gas production and the gas composition. It is seen in this particular example that depending on the relative electrode e.m.f. the generated gas can contain between nil and 99.2% v/v hydrogen, compared with 66.7% v/v for the theoretical stoichiometric composition for a 2:1 v/v hydrogen/

Table 1.25 Electrolysis of sulphuric acid in the lead–acid battery

Cathode potential E_c (V)	Anode potential E_a (V)	$E_c - E_a$ (V)	$\log(i_a/i_c)$ (Equation 1.133)	i_a/i_c	i_c A (if i_a assumed 10 A)	Gas evolution per second			Composition of generated gas (% v/v)	
						Oxygen $\left(\dfrac{i_a \times 11\,200}{2 \times 96\,500}\right)$	Hydrogen $\left(\dfrac{10 i_c}{i_a} \times 0.1160\right)$	Total	Oxygen	Hydrogen
2.30	2.20	+0.10	3.6233	4200	0.002 238	0.5803	0.000 276	0.5806	99.9	0.1
2.30	2.22	+0.08	2.898	791	0.012 64	0.5803	0.001 466	0.5818	99.7	0.3
2.30	2.25	+0.05	1.811	64.70	0.154 5	0.5803	0.017 92	0.5982	97.0	3.0
2.30	2.28	+0.02	0.7247	5.305	1.885 0	0.5803	0.218 7	0.7990	72.6	27.4
2.30	2.29	+0.01	0.3623	2.304	4.340	0.5803	0.503 6	1.0839	53.5	46.5
2.30	2.295	+0.05	0.1811	1.517	6.591	0.5803	0.764 7	1.3450	43.1	56.9
2.30	2.32	−0.002	−0.7247	0.1884	53.08	0.5803	0.157 0	6.7373	8.61	91.39
2.30	2.35	−0.05	−1.811	0.0158	632.9	0.5803	73.417	73.997	0.78	99.22

oxygen mixture as would be produced by the direct electrolysis of sulphuric acid not in a lead–acid battery situation. The dependence of gas composition on anode and cathode potentials is clearly shown in Figure 1.58. This figure shows that the evolved gas has a stoichiometric composition (66.7% v/v hydrogen, 33.3% oxygen) when both the anode and cathode have a potential of 2.3 V. When the anode potential is less than 2.3 V the gas mixture is oxygen rich and when it is more than 2.3 V the mixture is hydrogen rich.

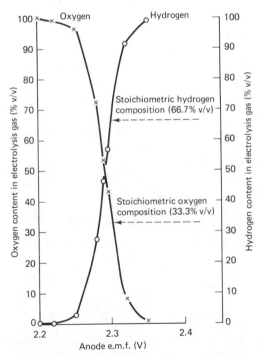

Figure 1.58 Dependence of electrolysis gas composition on e.m.f. Cathode e.m.f. is assumed fixed at 2.30 V and cathode current at 10 A

The above comments apply during battery charging, i.e. when the electrode e.m.f. values are increasing. During discharge the situation is reversed. Thus, over a complete charge/discharge cycle, the total yield of gas has a stoichiometric composition, i.e. 66.7% v/v hydrogen and 33.3% v/v oxygen. An effective catalyst recombination device would fully recombine this to water, thereby making up the electrolyte volume to its original value.

The difficulties are:

1. Devising a truly efficient recombination device that operates successfully in a battery environment.
2. Handling the pressure excursions that occur at certain parts of the charge/discharge cycle due to the presence of excess over stoichiometric amounts of unrecombined oxygen or hydrogen.

Even small departures from stoichiometric gas composition at certain periods of the charge/discharge cycle would cause excessive gas pressure build-up within the cell as shown below. Suppose that the dead space above the electrolyte in a sealed cell was 100 cm³ and that this space was filled with gas having a composition 70% v/v hydrogen and 30% oxygen, i.e. 70 cm³ hydrogen and 30 cm³ oxygen. With effective recombination the 30 cm³ of oxygen would consume 60 cm³ hydrogen leaving 10 cm³ of hydrogen unconsumed. If 100 ml of 70:30 v/v hydrogen/oxygen were being generated per minute the dead space would contain 10 ml of unconsumed hydrogen, i.e. the internal cell pressure would be

$$\frac{10}{100} = 0.1 \times 760 \text{ mmHg}$$

At the end of 10 min the internal pressure would be

$$10 \times 0.1 \times 760 = 760 \text{ mmHg}$$

and at the end of 30 min the pressure would be

$$30 \times 0.1 \times 760 = 2280 \, \text{mmHg}$$

i.e. three atmospheres. Clearly, the occurrence of internal pressures between one-tenth of an atmosphere and three atmospheres or higher has implications in the problems of designing cells capable of withstanding such pressure cycling.

It is very important, therefore, when designing sealed cells to avoid these problems as far as possible by reducing gas production to an absolute minimum during charge and discharge so that pressure excursions are kept to a minimum. The non-stoichiometry problem cannot be avoided, but, by keeping gas evolution to a minimum, maximum internal cell pressure can be kept low and gas loss by venting avoided so that in a subsequent stage of the charge/discharge cycle excess oxygen and hydrogen can be recombined back to water, thereby avoiding any long-term reduction in electrolyte volume.

2

Guidelines to battery selection

Contents

The first choice to be made is whether a primary or secondary battery is required. Usually, there is no doubt about the requirement in this respect. We shall therefore proceed to a discussion of the selection of the particular type of primary or secondary battery required for the application in mind.

2.1 Primary batteries

Until the 1970s, primary batteries were predominantly zinc-anode-based systems. Performance of these cells has undergone progressive improvements through development of the original Leclanché (carbon–zinc) system and introduction of new couples such as zinc–mercuric oxide, alkaline manganese dioxide, and zinc–silver oxide. Figure 2.1 shows these improvements for one aspect of cell behaviour – energy density. More recently, significant advances in energy density have been achieved, together with improvements in other areas, such as low-temperature performance and storage capability, through the development of lithium-anode-based systems and specialist couples using anode materials such as cadmium, magnesium and indium–bismuth.

Table 2.1 shows, in order of increasing gravimetric energy density, the gravimetric and volumetric energy densities, the open circuit and on-load cell e.m.f. and the minimum and maximum recommended operating temperatures of a wide range of primary batteries. These are some, but not all, of the factors that must be taken into account when selecting a type of battery.

Gravimetric energy density controls the weight of the battery required for a given energy output. Primary carbon–zinc and alkaline manganese dioxide batteries have relatively low gravimetric energy densities in the ranges, respectively, of 55–77 and 66–99 W h/kg or 120–152 and 122–268 W h/dm^3 on a volumetric basis. Only mercury–cadmium batteries and specialist batteries developed for particular applications, such as the thermally activated batteries and cuprous chloride and silver chloride type seawater-activated batteries, have lower gravimetric energy densities than the carbon–zinc and alkaline manganese dioxide types. More recently developed primary batteries, such as the manganese dioxide–magnesium perchlorate (90–110 W h/kg, 120–130 W h/dm^3), mercury–zinc (99–123 W h/kg, 300–500 W h/dm^3) and silver–zinc (110–267 W h/kg, 215–915 W h/dm^3) batteries, have appreciably higher gravimetric (and volumetric) energy densities (up to 330 W h/kg, 610 W h/dm^3), as do some of the batteries based on magnesium electrodes and organic electrolytes, as opposed to the normal aqueous electrolyte systems.

A further, even more recently developed type of primary battery, with appreciably higher energy

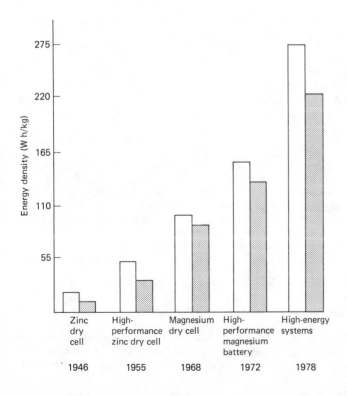

Figure 2.1 Improvement in primary battery performance at 21°C since 1946: □, initial; ■, after 2 years storage

Table 2.1 Energy densities of primary batteries in order of increasing W h/kg energy density

Battery system	Open circuit voltage per cell (V)	On-load voltage per cell (V)	Practical energy density		Operating temperature (°C)
			W h/kg	W h/dm³	
Thermal cells	*2.30–3.00	–	10–30	–	
	†2.70	2.20–2.60	1–9	2.4–18	−54 to 93
Water-activated cuprous chloride	1.50–1.60	1.20–1.40	11–110	18–213	−45 to 93
Mercury–cadmium	0.90	0.75	22	73	−40 to 70
Water-activated silver chloride	1.60–1.70	1.20–1.50	22–155	43–305	−45 to 93
Carbon–zinc	1.5	1.2	55–77	120–152	−7 to 54
Alkaline manganese	1.5	1.25	66–99	122–268	−30 to 54
Mercury–cadmium–bismuth	1.17	–	77	201	−20 to 90
Zinc chloride	1.5	1.2	88	183	−18 to 71
Manganese dioxide–magnesium perchlorate (reserve)	*1.90–2.00	1.50–1.55	90–110	120–130	–
	†1.85–2.70	1.35–1.60	44–110	60–180	−40 to 74
Mercury–zinc	1.35	1.25	99–123	300–500	−20 to 54
Magnesium–organic electrolyte	2.70–3.00	1.70–2.70	133–330	430–610	−54 to 74
Silver–zinc (monovalent)	*1.6	1.5	110–126	400–550	−40 to 54
	†1.85	1.30–1.55	110–267	215–915	0 to 54
Lithium–iodine	2.80	–	200	530	–
Zinc–air	*1.40	1.20	>220	180–900	–
	†1.40–1.50	1.2–1.30	155–330	180–490	−29 to 52
Lithium–sulphur dioxide	2.90	2.75	260–330	420	−40 to 60
Lithium–vanadium pentoxide	3.4	2.4	264	660	
Lithium–thionyl chloride	3.60	3.20–3.40	>660	1080	−54 to 60

*† Data from different sources

densities, is that based on lithium and organic electrolytes. These comprise, basically, three types of battery, the lithium–sulphur dioxide, lithium–vanadium pentoxide and lithium–thionyl chloride types, with energy densities, respectively, of 260–300 W h/kg (420 W h/dm³), 264 W h/kg (660 W h/dm³) and greater than 660 W h/kg (1080 W h/dm³). An example of a high energy density specialist battery is the lithium–iodine system with a gravimetric energy density of 200 W h/kg (530 W h/dm³).

Mechanically rechargeable zinc–air batteries can give twice the energy density of carbon–zinc types of cell currently in production. Energy densities up to 220 W h/kg are well within the capability of this system, compared with the 55–77 W h/kg obtainable from carbon–zinc types of primary cell.

Care is required in interpreting energy density data supplied to users by battery manufacturers. The figure of least value to the user is the theoretical energy density of the cell reaction involved. This figure takes into account only the weight of the active materials and ignores the weight of the cell construction materials (containers, terminals, separators). Others might publish energy densities for single cells, which is of more value. When cells are made up into batteries, further components contri-

bute weight but not electrical energy to the battery and this further complicates the interpretation of energy density data. These factors need to be taken into account when determining energy density data for the same type of cell supplied by different types of cell now available.

Frequently, the practical or commercial energy density of a battery is between one-half and one-third of the theoretical value, and this factor must be very carefully taken into account when selecting a type of battery.

As previously mentioned, energy density can be expressed in two ways; W h/kg and W h/dm³. It would be expected, therefore, that there would exist an approximate relationship between these two quantities, provided that the mean density did not vary too much from one battery type to another. That this is indeed so is shown by examination of Figures 2.2 and 2.3 in which W h/kg is plotted against W h/dm³. With some notable exceptions, e.g. lithium–thionyl chloride (Figure 2.2) a reasonably smooth plot is obtained which suggests that the mean densities (W h/kg)/(W h/dm³) of the various types of battery are in the range 2.0–3.2 g/cm³.

It will be appreciated, however, that high energy density is only one of the parameters which must be taken into consideration when selecting a primary

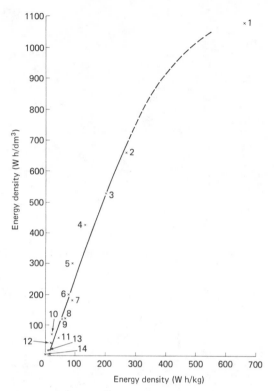

Figure 2.2 Gravimetric *versus* volumetric energy densities (commercial) of primary batteries: minimum quoted literature values

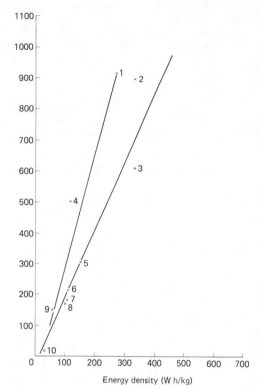

Figure 2.3 Gravimetric *versus* volumetric energy densities (commercial) of primary batteries: maximum quoted literature values

battery for a particular application. For example, great improvements in operation at low temperature have also been achieved during the 1970s. In this context manganese dioxide–magnesium perchlorate cells have excellent operating characteristics at −40°C with little loss of capacity, even though this type of cell has a lower energy density than alternative new types of cell now available.

Operating temperature ranges for various types of primary battery are listed in Table 2.1. The carbon–zinc battery, for example, ceases to operate at −7°C and the alkaline manganese at about −30°C, in contrast to manganese dioxide–magnesium perchlorate batteries which still successfully operate at −40°C and magnesium–organic electrolyte batteries at −54°C. If battery operating temperature in the application in mind is likely to deviate from 10 to 25°C, operating temperature will often have an influence on the type of battery selected for the application.

Another parameter that has greatly improved in recent years is shelf life and, in fact, the mercury–cadmium cell has a shelf life of 5–10 years

(and a wide operating temperature range) which, to the user, may outweigh the fact that its energy density is not as high, for example, as that of the mercury–zinc cell. The principal types of dry primary cell and their average characteristics are given in Table 2.2. This table shows some of the important parameters that must be taken into account when selecting cells for a particular application. These five types of cell will fulfil the needs of many applications. However, where particularly outstanding properties in one area or another are required it will be necessary to consider alternative newer types of cell, such as magnesium perchlorate–manganese dioxide types, lithium–organic electrolyte based systems or other types such as lithium–iodine or thermal batteries.

Another parameter which will have a very definite bearing on the choice of battery type is cell voltage, particularly on-load voltage, which are tabulated in Table 2.1 for various types of primary battery. It can be seen that these range from 0.75 V/cell (mercury–cadmium) to over 3 V with some of the lithium and magnesium based organic electrolyte systems.

Table 2.2 Principal types of dry battery and average characteristics

Common name	Carbon–zinc	Carbon–zinc (zinc chloride)	Alkaline manganese dioxide	Mercuric oxide	Silver oxide
Electrochemical system	Zinc–manganese dioxide (usually called Leclanché or carbon–zinc)	Zinc–manganese dioxide	Zinc–alkaline manganese dioxide	Zinc–mercuric oxide	Zinc–silver oxide
Voltage per cell (average on-load voltage in parentheses) (V)	1.5 (1.2)	1.5 (1.2)	1.5 (1.25)	1.35 (1.25)	1.6 monovalent (1.5)
Negative electrode	Zinc	Zinc	Zinc	Zinc	Zinc
Positive electrode	Manganese dioxide	Manganese dioxide	Manganese dioxide	Mercuric oxide	Monovalent silver oxide
Electrolyte	Aqueous solution of ammonium chloride and zinc chloride	Aqueous solution of zinc chloride	Aqueous solution of potassium hydroxide	Aqueous solution of potassium hydroxide or sodium hydroxide	Aqueous solution of potassium hydroxide or sodium hydroxide
Type	Primary	Primary	Primary and rechargeable	Primary	Primary
Number of cycles	10–20	10–20	50–60 rechargeable only		
Input if rechargeable			Approximately 100% of energy withdrawn rechargeable only		
Overall equation of reaction	$2MnO_2 + 2NH_4Cl + Zn \rightarrow ZnCl_2.2NH_3 + H_2O + Mn_2O_3$	$8MnO_2 + 4Zn + ZnCl_2 + 9H_2O \rightarrow 8MnOOH + ZnCl_2.4ZnO.5H_2O$	$2Zn + 3MnO_2 \rightarrow 2ZnO + Mn_3O_4$	$Zn + 2MnO_2 \rightleftharpoons ZnO + Mn_2O_3$	$Zn + Ag_2O \rightarrow ZnO + 2Ag$
Typical commercial service capacity	60 mA h to 30 A h	Several hundred mA h to 9 A h	Several hundred mA h to 23 A h	16 mA h to 28 A h	35 mA h to 210 mA h
Commercial energy density (W/kg)	55–77	88	Primary: 66–99 Rechargeable: 22	99–123	110–126
Commercial energy density (W/dm³)	120–152	183	Primary: 122–263 Rechargeable: 61–73	300–500	400–550
Practical current drain rates:					
Pulse	Yes	Yes	Yes	Yes	Yes
High (>50 mA)	15 mA/cm² of zinc area (D cell)	23 mA/cm² of zinc area (D cell)	31 mA/cm² of zinc area (D cell)	No	No
Low (<50 mA)	Yes	Yes	Yes	Yes	Yes
Discharge curve shape	Sloping	Sloping	Sloping	Flat	Flat
Temperature range (°C):					
Storage	−40 to +48	−40 to +71.1	−40 to +48.9	−40 to +60	−40 to +60
Operation	−6.7 to +54.4	−17.8 to +71.1	−28.9 to +54.4	0 to +54.4	0 to +54.4

Effect of temperature on service capacity	Poor low temperature	Good low temperature relative to carbon–zinc	Good low temperature	Good high temperature, poor low temperature – depends on construction	Poor low temperature – depends on construction
Impedance	Low	Low	Very low	Low	Low
Leakage	Medium under abusive conditions	Low	Rare	Some salting	Some salting
Gassing	Medium	Higher than carbon–zinc	Low	Very low	Very low
Reliability (lack of duds 95% confidence level)	99% at 2 years	99% at 2 years	99% at 2 years	99% at 2 years	99% at 2 years
Shock resistance	Fair to good	Good	Fair to good	Good	Good
Cost Initial	Low	Low to medium	Medium plus	High	High
Operating	Low	Low to medium	Medium to high at high power requirements	High	High
Features	Low cost; variety of shapes and sizes	Service capacity at moderate to high current drains greater than carbon–zinc; good leakage resistance; low-temperature performance better than carbon–zinc	High efficiency under moderate and high continuous conditions; good low-temperature performance; low impedance	High service capacity–volume ratio; flat voltage discharge characteristic; good high temperature	Moderately flat voltage discharge characteristics
Limitations	Efficiency decreases at high current drains; poor low-temperature performance		Primary type expensive for low drains. Rechargeable: limited cycle life; voltage-limited taper current charging	Poor low-temperature performance on some types	
Applications	Radios, barricade flashers, telephone amplifiers, marine depth finders, toys, lighting systems, signalling circuits, novelties, flashlights, photo-flashguns, paging, laboratory instruments	Cassette players and recorders, calculators, motor-driven toys, radios, clocks	Radios (particularly high current drain), bicycle lights and horns, shavers, electronic flash, lighting systems, movie cameras, radio-controlled model plane ignition, toys, tape recorders, television sets, walkie-talkies, cassette players and recorders, calculators, motor-driven toys, clocks, photo-flash, heavy duty lighting, any high-current drain, heavy discharge schedule use	Secondary voltage standard, television sets, radios, photoelectric exposure devices, walkie-talkies, paging, radiation detection, test equipment, hearing-aids, transistorized equipment, electronic watches	Hearing-aids, reference voltage source, photoelectric exposure devices, instruments, electronic watches

Table 2.3 Some applications of primary batteries

Application	Battery system	Load requirement	Life required (approx.)	Acceptable voltage regulation (%)	Reason for battery choice
Roadside hazard lamp (flashing)	Carbon–zinc	6 V 0.3 A bulb approx. 60 pulses/min	1200 h	40	Economy; availability
Roadside hazard lamp	Air depolarized	3.8 V 250 mA bulb	120 h continuous	20	Economy
Personal locator beacon	Mercury–zinc	13.4 V 30 mA continuous at 0°C	30 h	20	Operation at 0°C; good storage
Location beacon	Mercury–indium–bismuth	14.7 V 60 mA pulse train	Operation after storage at 70°C; 300 h	5	Operation after storage at 70°C
Sealed metering and indicator	Mercury–cadmium	15 mA pulse sporadically	48 months	10	Special termination; good storage
Portable warning lamp	Zinc–air	250 mA	10 h continuous	5	Low weight requirement and voltage regulation; economy
Roadside hazard lamp continuously burning	Carbon–zinc	12 V 0.1 A bulb	63 h	30	Economy; availability
Alarm system	Air depolarized	3.0 mA continuously	18 months	20	Economy
Laser beam detector	Mercury–zinc	13.4 V 20 mA intermittently	100 h	5	Special termination; good storage
Memory core standby supply	Mercury–indium–bismuth	20 µA	24 months	5	Special termination; good storage
Intruder alarm system	Carbon–zinc	12 V 300 mA	8 h after 12 months storage	30	Economy; availability
Radio location buoy	Air depolarized	750 mA pulse 2 mA quiescent	12 months	20	Weight; economy
Solid-state relay	Mercury–zinc	13.4 V 10 µA continuously	18 months	1	Long life and voltage regulation
Emergency and portable fluorescent handlamp	Carbon–zinc	12 V 600 mA	3 h continuously	25	Economy; availability
Detonator	Mercury–zinc	1.34 V 500 mA pulse	One shot after 18 months	10	Good storage and special termination
Direct reading frequency meter	Carbon–zinc	9 V 25 mA	200 h at 4 h/day	25	Size; economy; availability
Radio microphone	Mercury–zinc	13.4 V 15 mA 2 h/day	50 h	10	Voltage regulation; special shape
Wide-range oscillator	Carbon–zinc	18 V 25 mA	200 h at 4 h/day	25	Size; economy; availability
Communication transceiver	Carbon–zinc	12 V T_x 1.2 A R_x 0.6 A	24 h continuous on $T_x/R_x = 1/9$	30	Economy; weight
Tape recorder	Carbon–zinc	6 V 200 mA	20 days at 30 min/day	35	Availability; economy
Miniature ignition device	Carbon–zinc	400 mA pulses	6–9 months	30	Miniaturization

2.1.1 Selection of primary batteries

Selecting a battery can be as simple as buying a cell for a pen light or as complicated as specifying a source of stored energy for a satellite transmitter.

Primary batteries are used extensively in a wide range of applications, some of which are shown in Table 2.3 for the more conventional types of battery. Applications range from the sophisticated to the mundane, covering the use of the smallest 15 V Leclanché layer stack battery made (10 mm diameter and 27 mm height), which has an application in a CMOS liquid crystal display, and every-day applications such as road hazard lamps and intruder alarm circuits.

To date, there is no single battery system that has every advantage over all the other systems, and therefore a procedure for selecting the most suitable system is necessary. The factors involved in the selection of primary batteries suitable to meet a particular requirement are extremely complex. It is essential at an early stage in the design to liaise with the technical department of the battery manufacturer to ensure that the characteristics of the battery and the equipment are matched so that the user of the equipment obtains the best possible performance from both.

There are three basic applications for which primary batteries are used:

1. Miniature equipment (worn or carried unobtrusively in use).
2. Equipment that is portable in use (often moved during or between discharge periods).
3. Transportable equipment (not frequently carried) and standby systems.

Each of these can be based on one of several electrochemical systems. Batteries in miniature equipment are discussed in Section 2.1.2, and in portable equipment in Section 2.1.3. Transportable batteries are usually of the rechargeable type and are dealt with in Chapter 2.2.

One of the problems facing a designer is selecting the correct system for the application from a choice of many. Some of the available systems are listed in Table 2.1, which gives open-circuit voltage, average on-load voltage, energy densities and operating temperatures for various systems, including all those commonly available for commercial use.

Although the many types of battery available may seem to make a proper choice difficult, the problem can be somewhat simplified by first outlining the application requirements and then matching a battery to the job. The preliminary information that must be available before a battery can be specified is set out below. The discussion of the basic characteristics and features of various batteries in the following sections will indicate which one(s) are most suitable for the application. Unfortunately, the ideal characteristics cannot be found in any one battery design; nor can the characteristics of one battery always be compared directly with those of another. Therefore optimum performance of a battery in an application can usually be achieved best by meeting the critical needs of the application and subordinating the others.

The selection of a battery is best achieved by setting out a list of minimum requirements, conditions and limitations, as follows:

1. Maximum permissible voltage at the beginning of discharge.
2. Normal voltage during discharge (voltage stability on load).
3. End-voltage, that is, voltage at which equipment ceases to function properly.
4. Current–voltage relationships: constant current (amps), constant resistance (ohms), constant power (watts).
5. Type of discharge and current drain (duty cycle): continuous, intermittent, continuous with pulses.
6. Storage and service life.
7. Environmental conditions in storage and in service.
8. Physical restrictions such as dimensions and weight.
9. Special requirements.

It is important that all these points be considered because they are interdependent. This information will allow the battery manufacturer to recommend to the equipment designer a battery suitable for the proposed application.

Table 2.4 gives some idea of the relative running costs between three different primary systems: carbon–zinc, alkaline manganese and mercury–zinc of the R20 size cell covering heavy, medium and light duties. In this comparison of running costs no account is taken of voltage regulations. Although the voltage tolerances differ widely between systems (10% for mercury–zinc, 20% for alkaline manganese, 25% for air depolarized and up to 40% for carbon–zinc), it should be noted that good voltage regulation may be obtained from any battery system using devices such as Zener diodes. Thus, it can be seen that by careful design of a given circuit the lower-cost system could be used, or, because of the more stable voltage of another system, the number of cells in a given battery could be reduced, which could reduce the volume and weight. The operating voltage range of the equipment determines the extent to which the available capacity in a battery is realized. A cut-off voltage (that is, complete useful discharge) of about 0.9 V/cell is desirable but not always possible for many types of cell.

Current is directly linked with voltage tolerance, regulation and load and, by taking into account the period of time for which the current is required, the

Table 2.4 Running cost comparison, relative to carbon–zinc cell size R20

System	Retail price	Cost per hour		
		Heavy-duty* (30 mA, 30 min/day)	Medium-duty* (37.5 mA, 4 h/day)	Light-duty* 10 mA, 12 h/day)
Carbon–zinc (Vidor SP2)	1.0	1.0	1.0	1.0
Alkaline manganese	5.5	2.6	2.8	5.0
Mercury–zinc	26.0	1.0	10.0	18.0

* End-voltage 0.8 V/cell

required capacity in ampere hours can be determined. Ampere hour capacity of the zinc–carbon system can be obtained provided that the period of time for which current is required is known and the end-voltage to which the equipment can work efficiently is given. Electrical loading and required service life will determine the required capacity of the battery and this is the basic parameter in deciding the type of cell to be used. Duty cycle may also be a factor in the choice of battery. Applications requiring heavy pulses of short duration will be met by a battery with a different internal construction from that used for a very low continuous current duty.

Battery volume is dependent on the required voltage, ampere hour capacity, temperature and termination and is dictated in many cases by the commercially available types such as the round cell range and layer stack batteries. Primary batteries are made up of individual cells; in the case of carbon–zinc batteries these can be round or flat cells. A primary battery made from flat cells, known as a 'layer stack', generally gives a higher voltage and consequently lower capacity than a round cell battery of the same volume. The trend is to fit all round cells into equipment, but by careful selection of round cells and a layer stack battery the volume required by the batteries can be reduced by more than 40% in some cases. Such combinations have been very successful in reducing the size of military portable communication transceivers. In special applications for industrial and military use, shape is generally as important as volume.

If a standard commercial battery or cell is chosen, the size and type of terminals are already specified, but care is none the less required; for example, on round cells the tolerance allowable (BS 397: 1976, *Primary Cells and Batteries*, covering R20, D size (American Standards Association) and Vidor SP2) is 2.3 mm in height and 1.6 mm in diameter. The difference in size between cells from different manufacturers is particularly apparent when several cells are connected end to end in series. The

construction of a battery container to allow for the difference can be difficult and expensive. The use of layer stack batteries having connectors, in which there is no such build-up of tolerances, can therefore greatly simplify the construction of the battery container. In applications for industrial and military use, termination is of major importance and much attention is paid to terminal design so that it can meet the environmental testing demanded.

Allowable size and weight will sometimes determine which battery is selected in spite of other requirements. A premium is usually paid for small size with high output capacity. Bulky and heavy batteries can be reliable as well as economical if their size can be accommodated.

All electrochemical systems are temperature sensitive, some giving better performance than others at the extremes; service temperature affects both capacity and life. The suggested operational temperature ranges of some battery systems are given in Table 2.5.

The majority of battery manufacturers publish discharge figures taken at 20°C. By the correct matching of size and current drain, the carbon–zinc and other primary systems can successfully be operated at lower temperatures than those indicated, but liaison with the battery manufacturer is essential.

Table 2.5

Suggested operational temperature range (°C)	Battery system
−7 to +54	Carbon–zinc
−29 to +50	Zinc–air
−30 to +50	Alkaline manganese
−20 to +50	Mercury–zinc
−40 to +70	Mercury–cadmium
−20 to +90	Mercury–indium–bismuth
−40 to +60	Lithium-based system,
−54 to +60	depending on type

To select a system that can be stored and operated at the extreme temperatures of −40 to +90°C – a requirement with which battery manufacturers are sometimes presented – is difficult and expensive; therefore it is important to specify temperature requirements, particularly the dwell times at the extremes. This is essential information which the designer of the equipment should be able to give to the battery manufacturer.

Table 2.6 gives a guide to the period, at 20°C, for which some battery systems can be stored, at the end of which the cells would have retained between 80 and 90% of their capacity.

Table 2.6

System	Shelf life at 20°C before use* (months)
Carbon–zinc	18–30
Alkaline manganese	18–30
Mercury–zinc	24–30
Mercury–indium–bismuth	36
Mercury–cadmium	48
Air depolarized	48
Silver oxide–zinc	18
Lithium–sulphur dioxide	60

* Varies with cell size

Shock or vibration may indicate the need for a rugged battery construction. Unusual rates of acceleration or high-altitude operation are also vital environmental considerations. Storage time and temperature under any of these conditions should be noted.

It should be noted from Table 2.3 that batteries with different types of electrochemical system are chosen for different types of duty application. Obviously, some batteries based on a particular type of electrochemical system are more suited to particular applications than are others, and here one is confronted with the dilemma that any particular primary battery manufacturer will have only some of these electrochemical systems included in their range and that the one recommended may not be absolutely the best from the total range available. It is advisable, therefore, to discuss battery requirements at an early stage of design with several battery suppliers covering the whole range of types of battery.

2.1.2 Batteries in miniature equipment

Miniature applications have become more important in recent years with the general acceptance of the behind-the-ear hearing-aid and the advent of the electronic watch. High energy density per unit volume is the prime requirement for a battery in these products. The mercuric oxide–zinc, silver oxide–zinc, zinc–air and lithium-based systems appear to be likely contenders for this market. Although the last two types of battery have been produced in sizes suitable for miniature applications, they are not widely available in this format. These systems will therefore be discussed later in their usual cylindrical form, and the conclusions drawn then may explain the difficulties that have prevented their wide acceptance.

The mercuric oxide–zinc cell for miniature applications is usually based on the familiar 'button' construction using a compressed cathode of mercuric oxide and graphite (added for conductivity) in a plated steel can. The cell seal is supported by a cathode sleeve on top of which is placed a synthetic separator and an electrolyte absorbing pad; the electrolyte is a solution of potassium hydroxide. The amalgamated zinc anode is added and the cell sealed with a polymeric gasket and a metal top cap.

The mercury cell has a low internal resistance and high cathode efficiency. Discharge characteristics are substantially flat, an obvious advantage for hearing-aid use. Capacity retention of the system on storage is good. Multi-cell batteries using the mercury system are available for applications requiring higher voltages, and some cylindrical sizes are produced. In general, the high cost of the system restricts it to those uses where space is at a premium or where voltage regulation is critical.

In hearing-aids, the current drain may be of the order of 1 mA for a total discharge of several days. The low-resistivity separator is able to retain the droplets of mercury that form as the cathode is discharged. In a watch, the batteries must perform adequately over a period of months of years, and the discharge pattern may be one of short periods of a drain of tens of milliamperes superimposed on a microampere continuous drain. Under these conditions a separator with carefully controlled properties is required to avoid possible mercury penetration and short-circuiting of the cell.

Silver oxide–zinc cells are often specified for electronic watch applications. Sodium hydroxide electrolyte, which has a lower conductivity than potassium hydroxide, is often used because it has a lower tendency to 'creep' at the seal. The separator in the silver oxide system must retain soluble silver species produced by chemical dissolution of the oxide, and a multiple layer separator of low-porosity film achieves this.

There are two types of silver oxide cell: one has a cathode of monovalent silver oxide (Ag_2O) and the other type uses divalent silver oxide (AgO). The latter type has a higher theoretical potential (1.8 V) and, because there is an additional chemical reduction from AgO to Ag_2O it has a higher capacity. (The theoretical energy density is 424 W h/kg.) The two-stage reduction process would normally result in a discharge curve with two plateaux at 1.7

and 1.5 V and the voltage drop in the middle of discharge may necessitate a voltage regulator in the equipment. If the surface layer of the electrode is of Ag_2O, however, discharge takes place at the lower potential throughout. In order to achieve voltage stability, the surface may be treated to reduce AgO to Ag_2O or, in various patented arrangements, a 'dual-oxide' system may be adopted. Higher raw material costs mean that silver oxide cells are more expensive than their mercury equivalents.

2.1.3 Portable-in-use batteries

This category includes test equipment, portable radio apparatus, lighting and calculators, and is a much larger market than the miniature or transportable sectors.

Lithium has several advantages as a possible anode material for an electrochemical power source. It has a low equivalent weight and density and is the most electronegative element that is solid at normal temperatures. However, lithium reacts with water, the common electrolyte solvent, and with most non-metallic elements and compounds; it is therefore normally essential to use a non-aqueous electrolyte.

Solid electrolyte lithium batteries have been produced for low-drain applications. The Catalyst Research Corporation produce a lithium–iodine solid electrolyte system, rated at $20\,\mu A$ for heart pacemaker use.

For a battery that is to sustain high current densities at ambient temperatures, a solid electrolyte is unlikely to be acceptable. A suitable electrolyte solvent must provide stable solutions of the electrolyte over a wide temperature range and be stable towards lithium and the cathode material.

In order to exploit the value of a lithium-based system to the maximum, the positive electrode (cathode) material should also be of high energy density. The search for the ideal combination of cathode material and electrolyte has attracted a great deal of effort in the 1970s. The lithium–polycarbon–monofluoride system is one of two lithium systems that have been commercially promoted. Developed by the Matsushita Electric Industrial Co. in Japan, the cells are available in several cylindrical sizes. The patented cathode material is of the form $(CF_x)_n$ where x has a value between 0.5 and 1.0 and is formed by reacting carbon with fluorine under various conditions of temperature and pressure, depending on the type of carbon used as the starting material. Except where batteries are intended for low-rate applications, acetylene black or graphite is added to the electrode to improve conductivity. The electrolyte is lithium tetrafluoroborate dissolved in α-butyrolactone. Honeywell Inc. and the Mallory Battery Co. in the United States have introduced lithium batteries based on the lithium–sulphur

dioxide electrochemical couple. The positive active material in these batteries, liquid sulphur dioxide, is dissolved in an electrolyte of lithium bromide, acetonitrile and propylene carbonate, and is reduced at a porous carbon electrode.

Both types of lithium battery have a spiral-wound electrode pack, made up from rectangular foil electrodes. Lithium foil is rolled on to an expanded metal mesh current collector as the negative electrode, and is separated from the similarly supported cathode by a polypropylene separator.

Practical open-circuit voltages of the lithium–polycarbon–monofluoride and lithium–sulphur dioxide systems are approximately 2.8 V and 2.9 V respectively at 20°C. The high voltage means that these batteries are not interchangeable with other electrochemical systems in existing equipment, unless a 'dummy' cell is also included.

The high volumetric energy densities reflect the high voltages of the lithium-based systems. One reason for some lack of acceptance in miniature applications is that although one lithium cell could be specified where it is necessary to use two mercury cells in series, a lithium button cell would have a capacity approximately one-half that of the equivalent mercury cell, and the frequency of battery changing would in extreme cases be correspondingly increased.

The volumetric ampere hour capacity of mercuric oxide–zinc cells is higher than that of lithium-based systems. However, in many cases using two lithium cells in parallel or one larger lithium cell will give the same ampere hour capacity that can be achieved in an equal or even smaller volume than an equivalent two-cell series mercury–zinc battery of similar voltage. This is illustrated in Table 2.7, which gives a comparison of lithium–sulphur dioxide and mercuric oxide–zinc cells.

Thus one lithium–sulphur dioxide cell (voltage 2.75 V) occupies about 30% more space than two series mercuric oxide–zinc cells (voltage 2.5 V). Admittedly, this compares the worst cited case for lithium against the best for mercuric oxide–zinc.

Table 2.7

	Lithium–sulphur dioxide	Mercuric oxide–zinc
Volumetric energy density (W h/dm^3)	420	500 (best)
Normal working voltage (V)	2.75	1.25
Volumetric capacity (A h/dm)	153	400
Relative cell volume per A h	2.6	1.0

Higher energy density systems such as lithium–vanadium pentoxide and lithium–sulphur dioxide would show significant volume savings over an equivalent ampere hour mercuric oxide–zinc system. In fact lithium–sulphur dioxide systems are being increasingly considered for high-rate miniature power source applications including military applications where it is found that a two-cell mercuric oxide–zinc battery can occupy considerably more space than the equivalent lithium–sulphur dioxide cell; there are the added advantages inherent in the lithium–sulphur dioxide system of excellent storage life (5–10 years, that is, very low self-discharge), wide operating temperature range (−50 to +60°C) and stable voltage characteristics on load.

The reactivity of lithium necessitates controlled-atmosphere assembly – in some cases the use of expensive materials in the cell construction to avoid corrosion, and in some cases (for example, lithium–sulphur dioxide) the provision of a sophisticated seal design.

Although lithium batteries have high-rate discharge capability, their use at very high rates or accidental shorting could result in temperatures leading to seal failure or explosion. Manufacturers have incorporated vents and/or fuses to minimize these risks.

The zinc–air system, which attracted a great deal of investment in the late 1960s and early 1970s to make a consumer product in standard cylindrical sizes, suffered initially from four problems. It was difficult to produce air-breathing cathodes of consistent quality; the need to allow air into the cell led to electrolyte leakage; carbonation of the electrolyte occurred on long-term discharge; during intermittent discharge oxygen ingress products caused wasteful corrosion of the active material.

Despite these initial difficulties, commercially available D and N cells were available in the UK in the early 1970s. The ECL D-size cell outperformed D-size nickel–cadmium cells under most conditions, except at sub-zero temperatures, in a military 'manpack' radio application. The cost of this zinc–air cell was about 25% of that of an equivalent nickel–cadmium cell and proved to be more economical. However, the nickel–cadmium cell had the advantage of being rechargeable. (Primary zinc–air batteries were used extensively during the Vietnam War.) The N-size cell has been used successfully for the small pocket paging equipment market. The demise of zinc–air D and N cells rests on economic factors rather than on technical grounds. The picture for zinc–air button cells is quite different; Gould have manufactured them in the USA for several years for applications such as hearing-aids, watches, etc., and Gould and Berec have both been marketing them in the UK since 1980.

Leclanché and alkaline manganese batteries retain the great majority of the primary battery market for portable-in-use applications. They are widely available in a variety of equivalent cylindrical sizes.

The Leclanché system has the property of 'recovering' during rest periods in the discharge regimen. This property and the low cost resulting from high-volume automated manufacture make the Leclanché ideal for the intermittent usage pattern found in most consumer appliances. Figure 2.4 shows the effect of increased current drain on the service life of the SP2 (R20 size) battery. The discharge period is 5 h/day to an end-voltage of 0.9 V.

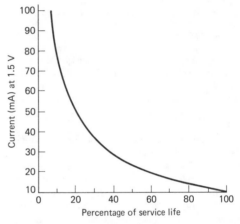

Figure 2.4 SP2 (R20) Leclanché cells. Effect of increased current drawn on service life

For higher current drains, for example in motorized equipment, a high-power (HP) battery should be used. HP batteries contain electrolyte or chemically prepared manganese dioxides, which have enhanced discharge properties. Figure 2.5

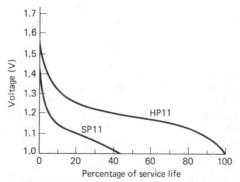

Figure 2.5 Heavy duty effect. SP11 and HP11 (R14) alkaline manganese dioxide batteries. Beneficial effect of using a high-power battery (HP11) on service life; high current drain (10 Ω, 2 h/day discharge)

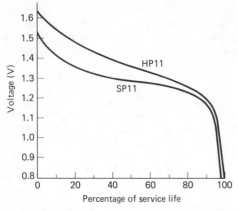

Figure 2.6 Low discharge effect. SP11 and HP11 (R14) alkaline manganese dioxide batteries. Effect of using a high-power battery (HP11) on service life; high current drain (300 Ω, 2 h/day discharge)

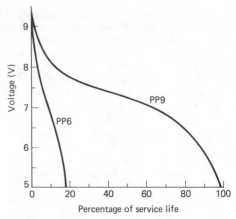

Figure 2.7 Size effect. Leclanché batteries: comparative service lives of two 9 V power packs, PP9 and PP6, on a 450 Ω, 4 h/day discharge

compares SP11 and HP11 (R14 size) types on a 10 Ω, 2 h/day discharge – the advantage of using the high-power type is obvious. For low-rate discharges, the additional cost of the HP version is not warranted; Figure 2.6 compares the same cells on a 300 Ω, 2 h/day discharge.

The storage properties of modern Leclanché batteries are much better than those of early dry batteries because of technological improvements. For example, average discharge levels of SP11 and HP11 batteries, on radio and tape recorder discharge respectively, after 2 years' storage at 20°C, are over 95% of those obtained for similar batteries discharged within a few weeks of production.

Transistor applications require multiples of the unit cell voltage; the layer stack ('power pack') format provides a convenient battery in a smaller volume than the equivalent combination of round cells. Only one unit needs to be replaced, and connection is by simple non-reversible press studs ensuring correct polarity. Selection of the most economical layer stack battery should be made in conjunction with the battery manufacuer. Specification of a battery which is too small for the current drain can lead to unnecessarily frequent battery replacement. Figure 2.7 shows the comparative service lives of two 9 V power packs, PP9 and PP6, on a 450 Ω, 4 h/day discharge.

The additional cost of the alkaline manganese construction is usually not warranted for intermittent discharges, despite its improved performance. For example, comparing an HP2 Leclanché cell with an Mn1300 (IECLR 20) alkaline manganese cell on a 5 Ω, 2 h/day discharge to an end-voltage of 1.1 V, the Mn1300 gives twice the discharge life but costs three times as much as the HP2. Only on continuous high-rate discharges does the alkaline manganese system have an undoubted economic advantage. It is also recommended for low-temperature discharges.

Magnesium cells incorporate a manganese dioxide cathode, magnesium anode and magnesium perchlorate or bromide electrolyte. They have excellent shelf storage properties, and a higher energy density than the Leclanché battery. However, they have a non-standard voltage and a characteristic voltage drop at the beginning of discharge. On intermittent use, gassing at the magnesium electrode means that the batteries cannot be properly sealed.

Where a rechargeable power source is required for portable-in-use apparatus, the nickel–cadmium system is generally specified. Nickel–cadmium batteries and cells are available in a wide range of cylindrical and button sizes, and they are ideally suited to high-rate applications. Both sintered electrode cylindrical batteries and mass plate electrode button cell constructions are available. The following discussion refers largely to the former type, since these occupy the major part of the market.

Because of the low internal resistance of nickel–cadmium batteries, constant-current charging is recommended to avoid elevated battery temperatures and thermal runaway. Batteries that have been discharged to 1.0 V may be brought to the fully charged state in 12 h by charging at the C_8 rate. The nickel–cadmium system will accept overcharge more readily than the lead–acid battery, but for permanent trickle charging a lower charge current than the above is recommended.

2.2 Secondary batteries

Table 2.8 shows, in order of increasing gravimetric density, the gravimetric and volumetric densities, the open circuit and on-load cell e.m.f. values and the minimum and maximum recommended operating temperatures of a range of secondary batteries.

Table 2.8 Energy densities of secondary batteries

Battery system	Open circuit voltage per cell (V)	On-load voltage per cell (V)	Practical energy density		Operating temperature (°C)
			W h/kg	W h/dm^3	
Sealed lead–acid					
Eagle Picher	2.12	1.50–2.00	18–33	43–85	−40 to 60
Sonnenschein	–	–	23–34	51–83	–
Other sources	2.10	–	22–33	49–83	−51 to 70
Unsealed lead–acid	2.10	–	15–26	31–122	−54 to 54
Silver–cadmium – as cells					
Eagle Picher	1.40	0.80–1.10	24–73	40–180	−40 to 43
Yardney	–	–	40–75	91–165	–
Silver–cadmium – as batteries					
Eagle Picher	1.40	0.80–1.10	18–44	24–122	−40 to 43
Yardney	–	–	18–51	–	–
SAFT	–	–	30–75	40–150	−47 to 75
Other sources	–	–	77	171	–
	1.40	–	24–120	37–250	−48 to 71
Sealed nickel–cadmium					
Eagle Picher	1.30–1.35	1.00–1.25	18–44	37–120	−40 to 70
Nife	–	–	24–35	60–91	–
button mass	–	–	21–22	61–62	–
round mass	–	–	22	69	–
round sintered	–	–	22	78	–
rectangular mass	–	–	18	37	–
rectangular sintered	–	–	21	52	–
other sources	1.35	–	26–37	61–85	−40 to 60
Vented nickel–cadmium					
Nife	1.35		26–44	61–99	−51 to 60
Nickel–zinc					
Eagle Picher	1.85	1.5–1.65	29–55	60–110	−40 to 93
Yardney	–	–	44–77	79–134	–
Other sources	1.71	–	up to 66	134	–
	–	–	33–77	67–134	−48 to 71
Alkaline manganese rechargeable*	–	–	50	–	–
Silver–zinc (theoretical 440 W h/kg)					
Eagle Picher:					
as cells	1.85	1.30–1.55	55–209	80–415	−40 to 74
as batteries	1.85	1.30–1.55	37–114	55–262	−40 to 74
Yardney	–	–	70–120	120–250	
SAFT	–	–	50–150	80–300	−40 to 75
Other sources	–	–	55–220	80–610	−48 to 71
	–	–	123	220	–
	–	–	130	–	–
Silver–zinc (remotely activated)	1.85	1.20–1.55	11–110	24–240	−35 to 49
Nickel–hydrogen					
Eagle Picher	1.40	1.20–1.30	56–67	55	−18 to 27
Silver–hydrogen					
Eagle Picher	1.50	1.00–1.10	90–100	60	4 to 27
Zinc–chlorine*	–	–	130	–	–
Zinc–air*	–	–	150	–	–
Sodium–sulphur*	–	–	240	–	–
Lithium–chlorine*	–	–	330	–	–
Lithium–sulphur*	–	–	370	–	–

* At 5 h rate

Table 2.9 Mean secondary battery data

	On-load voltage per cell (V)	Operating temperature (°C)		Mean energy density		Attractive features
		Minimum	Maximum	W h/kg	W h/dm³	
Unsealed lead–acid	1.50–2.00	−54	54	20	76	Inexpensive, reliable life, cycling ability
Sealed nickel–cadmium	1.00–1.25	−40	65	24	65	Reliable cycling ability over a period measured in years
Sealed lead–acid	1.50–2.00	−45	65	27	65	
Unsealed nickel–cadmium	1.00–1.25	−51	60	35	80	
Silver–cadmium	0.80–1.10	−44	58	53	120	High energy density, ability for voltage regulation throughout cycle life of 0.5–3 years
Nickel–zinc	1.50–1.65	−44	82	56	106	Low and high rate capabilities over wide temperature range
Silver–zinc (remotely activated)	1.20–1.55	−35	49	60	132	Unlimited shelf life, instantaneous activation at temperatures of −54 to 93°C
Nickel–hydrogen	1.20–1.30	−18	27	62	55	Long life, rechargeable system, capable of operating over years
Silver–hydrogen	1.00–1.10	4	27	95	60	High energy density, rechargeable capability for 1–3 years' operation
Silver–zinc	1.30–1.55	−42	−73	113	223	Where maximum energy density and voltage regulation is required throughout a cycle life period of 0.5–1 year
Gravimetric energy density only						
Zinc–chlorine	–	–	–	130	–	Very high density applications, e.g. electric automobiles
Zinc–air	–	–	–	150	–	
Sodium–sulphur	–	–	–	240	–	
Lithium–chlorine	–	–	–	330	–	
Lithium–sulphur	–	–	–	370	–	

Table 2.10 Ranking of energy density and on-load voltage of secondary batteries

Ranking in terms of gravimetric energy density (W h/kg)		Ranking in terms of on-load voltage (V)	
Lead–acid	~20	Silver–cadmium	0.81–1.10
Nickel–cadmium	~35	Silver–hydrogen	1.00–1.10
Silver–cadmium	53	Nickel–cadmium	1.00–1.25
Nickel–zinc	56	Nickel–hydrogen	1.20–1.30
Nickel–hydrogen	62	Silver–zinc	1.30–1.55
Silver–hydrogen	95	Nickel–zinc	1.50–1.65
Silver–zinc	113	Lead–acid	1.50–2.00
Zinc–chlorine	130		
Zinc–air	150		
Sodium–sulphur	240		
Lithium–chlorine	330		
Lithium–sulphur	370		

For simplicity, Table 2.8 tabulates the mean data and some features that make these batteries attractive for particular applications. It is now possible from the data in Table 2.9 to approximately rank secondary batteries in order of increasing gravimetric energy density and on-load voltage, to take but two of the important battery parameters. Table 2.10 shows, for example, that lead–acid, nickel–cadmium, silver–cadmium, nickel–zinc and nickel–hydrogen are in the lower energy density groups; silver–hydrogen, silver–zinc, zinc–chlorine and zinc–air are in the medium energy density group; and sodium–sulphur, lithium–chlorine and lithium–sulphur are in the high energy density group. This general impression is confirmed by examination of Figure 2.8 which presents gravimetric energy density data at the 5 h rate of discharge, obtained from another source.

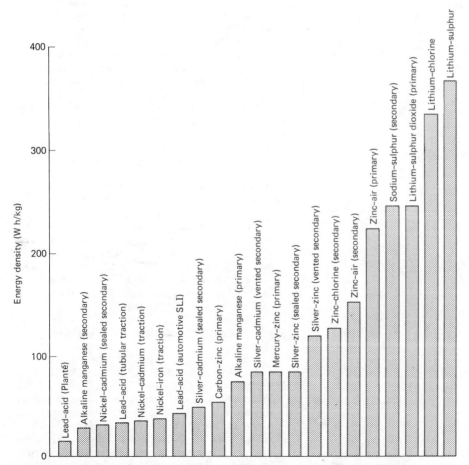

Figure 2.8 Typical energy densities at 5 h rate of discharge for known electrochemical cells

With data of this type, it is possible to make a start in the process of selecting a suitable type of secondary battery to meet a particular application. However, as discussed in Section 2.1 on primary batteries, many other considerations would apply in making the final selection.

As in the case of primary batteries (see Figures 2.2 and 2.3) it is possible to plot commercial volumetric *versus* gravimetric energy densities of secondary batteries as shown in Figure 2.9, which indicates a reasonably good correlation between the two parameters. Average densities are between 1.8 and $2.3\,g/cm^3$ compared to 2.0–$3.2\,g/cm^3$ for primary batteries.

The number of applications calling for an integrated energy system has been increasing impressively. To meet the varied power needs of these applications, manufacturers have created a number of variations within each family of available systems.

Battery designs that provide such advantages as recyclability, extended shelf life prior to use, remote activation, adjustment in load current with minimum voltage variations, and operation under extreme temperature conditions, have been developed over the years. All these performance characteristics, however, cannot be found in any one battery design; a few are incompatible and some features must be subordinated to others. To achieve optimum performance in each application, therefore, it is necessary to direct the design of the battery systems towards meeting the critical needs of the application.

Five major types of secondary rechargeable batteries are lead–acid, nickel–cadmium, silver–zinc, silver–cadmium and nickel–zinc, and these are discussed in more detail below.

2.2.1 Lead–acid batteries

The lead–acid battery is the most widely used of the five mentioned, its main application being in the automotive field. Its advantages are low cost, high voltage per cell and good capacity life. Its disadvantages are that it is relatively heavy, it has poor low-temperature characteristics and it cannot be left in the discharged state for too long without being damaged.

Since the late 1950s sealed rechargeable batteries based on the lead–lead dioxide couple have become available and while these are not produced in the same quantities as the non-sealed type their uses are increasing. For transportable applications weight is not of major importance, while the power requirements of the apparatus may be considerable. Here the lead–acid battery has the advantage of low cost. Transportable batteries are usually of the rechargeable type and are discussed later. The disadvantages associated with this system in its usual form, the SLI (starting, lighting and ignition) battery for vehicles, have been largely overcome by the low-maintenance types of lead–acid battery now available. One type, the Varley system, uses thin plates interleaved with highly absorbent separators and compressed to form a block which absorbs the electrolyte. This type of battery is vibration and shock resistant. It has a low internal resistance and for a given size can give a greater high-rate capacity and short-circuit current than its free-acid counterpart.

Charging may be carried out with a simple charger under controlled-time conditions. The rated capacity is given at the C_{20} rate, that is, the capacity obtained when a fully charged battery is discharged to bring it to an end-voltage of 1.75 V/cell in 20 h.

Sealed nickel–cadmium batteries have a higher initial cost, but these batteries may be economical for some transportable applications, where their long cycle life and indefinite shelf life in any state of charge without electrode deterioration are advantageous. They are particularly suitable for trickle charge uses such as emergency lighting.

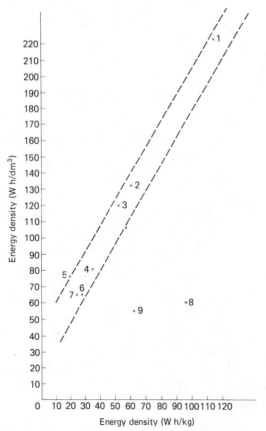

Figure 2.9 Gravimetric *versus* volumetric energy densities (commercial) of secondary batteries. Mean literature values

2.2.2 Nickel–cadmium batteries

The nickel–cadmium battery is mechanically rugged and long lived. In addition it has excellent low-temperature characteristics and can be hermetically sealed. Cost, however, is higher than for either the lead–acid or the nickel–zinc battery and, by comparison, its capacity on light drain in terms of watt hours per kilogram is also poorer than for nickel–zinc.

For many everyday batteries, the choice is still between the lead–acid and the nickel–cadmium systems, rather than the more recently developed systems discussed below.

Each of these two main types of battery has its own advantages and the choice between nickel–cadmium and lead–acid batteries depends very much on the particular application and on the performance characteristics required.

In addition to these two types of battery, there are several others which, because of their higher cost, are not used as extensively, but are nevertheless of direct interest in particular fields. This is because of their outstanding performance characteristics in certain specialized applications, e.g. silver–zinc and silver–cadmium batteries.

2.2.3 Silver–zinc batteries

Rechargeable silver–zinc batteries can provide higher currents, more level voltage and up to six times greater watt hour capacity per unit weight and volume than the lead–acid, nickel–zinc and nickel–cadmium storage batteries. Because it is capable of delivering high watt hour capacities at discharge rates less than 30 min, the silver–zinc battery is used extensively for missile and torpedo applications. Its high energy density makes it attractive in electronics applications, satellites and portable equipment where low weight and high performance are prime considerations. It is highly efficient and mechanically rugged, operates over a wide temperature range and offers good shelf life; quick readiness for use and the ability to operate at $-40°C$ without heating are two of the features of this battery. It is available in both high-rate and low-rate cells. Until now, the fact that it is more expensive, sensitive to overcharge and has a shorter cycle life than ordinary storage batteries has limited the silver–zinc battery to applications where space and weight are prime considerations. However, long-life silver–zinc batteries have been developed in which some 400 cycles over a period of 30 months' application have been achieved.

2.2.4 Silver–cadmium batteries

The silver–cadmium battery combines the high energy and excellent space and weight characteristics of the silver–zinc battery with the long-life,

low-rate characteristics and some resistance to overcharge of the nickel–cadmium battery. The battery also provides high efficiency on extended shelf life in charged or uncharged conditions, level voltage and mechanical ruggedness. Watt hour capacity per unit of weight and volume are two to three times greater than those of a comparable nickel–cadmium battery and it has superior charge retention. The silver–cadmium battery promises great weight and space savings and superior life characteristics to those of the nickel–cadmium battery currently used as storage batteries in most satellite programmes.

Today a silver–zinc system offers the greatest available energy density in terms of watt hours per kilogram. There are newer so-called high energy density couples which have been under development for many years; the effective energy density of many of these systems tends to decline as they are developed close to the point of practical utilization. In addition, chronic safety problems have already caused serious difficulties with the lithium systems and are potentially dangerous in others, most of which are high-temperature systems based on volatile materials. The use of silver as a couple obviously increases initial costs (although silver costs are recoverable) when compared to other existing systems such as lead–acid, nickel–cadmium, etc. When space and weight are limiting factors, the silver–zinc system is a very attractive proposition.

Other metal couples that are considered at present to be of great potential are the nickel–hydrogen and nickel–zinc systems. These may be batteries of the future in applications such as utilities load levelling and electric vehicles; the latter type is, in fact, now in commercial production.

2.2.5 Nickel–zinc batteries

With the development of new separators and improved zinc electrodes, the nickel–zinc battery has now become competitive with the more familiar battery systems. It has a good cycle life and has load–voltage characteristics higher than those of the silver–zinc system. The energy per unit of weight and volume are slightly lower than those of the silver–cadmium system. Good capacity retention (up to 6 months) has made the nickel–zinc battery a more direct competitor of the silver–zinc and silver–cadmium systems. Nickel–zinc batteries are not yet available in a sealed form.

2.2.6 Cadmium–air batteries

Rechargeable cadmium–air and zinc–air batteries are currently still only at the development stage and may not be commercially available for several years. Certainly, sealed versions of these batteries can only

Table 2.11 Rechargeable battery systems – basic parameters

	Nickel–cadmium		Lead–acid		Nickel–zinc	Silver–zinc	Silver–cadmium
	Sealed	Vented	Storage	Sealed			
Cathode	nickel hydroxide	nickel hydroxide	lead dioxide	lead dioxide	nickel oxyhydroxide	silver oxide	silver oxide
Anode	cadmium	cadmium	lead–antimony	lead–calcium	zinc	zinc	cadmium
Electrolyte	potassium hydroxide	potassium hydroxide	sulphuric acid	sulphuric acid	potassium hydroxide	potassium hydroxide	potassium hydroxide
Open-circuit voltage (V)	1.35	1.35	2.10	2.10	1.71	1.86	1.40
Energy density (W h/kg)	26–37	26–44	15–26	22–33	33–77	55–220	24–120
Energy density (W h/dm³)	61–85	61–91	31–122	49–83	67–134	80–610	37–250
Max. discharge rate	15C	50C	20C	20C	10C	50C	10C
Average discharge voltage (V)	1.20 (1 h rate)	1.22 (1 h rate)	1.80 (1 h rate)	1.90 (2 h rate)	1.60 (1 h rate)	1.50 (1 h rate)	1.10 (1 h rate)
Normal operating temperature (°C)	−40 to +60	−51 to +60	−54 to +54	−51 to +71	−48 to +71	−48 to +71	−48 to +71
Normal storage temperature (°C)	−51 to +71	−51 to +74	0 to 21	0 to 21	−48 to +38	−48 to +38	−48 to +38
Charging method	constant current	const. current const. voltage	taper current const. voltage	taper current	const. current const. voltage	const. current const. voltage	const. current const. voltage
Charge and retention (50% cap. 27°C)	60 days	1 year	90 days	1.5 years	1 year	1 year	2 years
Cycle life	100–1500	100–3000	10–600	100–150	100–200	10–200	100–500
Initial cost	high	high	low	low	high	high	high

be considered to be a prospect for the future. The cadmium–air system has a theoretical energy density of 445 W h/kg. The cadmium anode used is the type that has demonstrated good stability and low self-discharge in other alkaline systems. The air cathode is similar to that used in the mechanically rechargeable zinc–air battery. The battery consists of two air cathodes in parallel, positioned on the two sides of a plastics frame, and one anode. The bifunctional air electrode can be used for both discharge and charge. Over 300 cycles have been obtained with the best combination of materials.

There are several problems associated with the operation of the cadmium–air battery, such as the loss of cadmium on cycling, cadmium penetration and poisoning of the air electrode by a soluble cadmium species migrating to the air electrode. Water must be added to this cell periodically because water loss occurs by transpiration of water vapour through the air electrode – a characteristic of batteries using an air electrode. Operating cell voltage varies between 0.70 and 0.85 V at the $C/2$ to $C/10$ rate (that is, the discharge current is numerically equal to one-half to one-tenth of capacity, C). At these rates practical batteries produce 80–90 W h/kg and 14–24 W h/dm^3.

2.2.7 Zinc–air batteries

Development of the zinc–air electrically rechargeable battery is under way. Experimental cells have given 155–175 W h/kg at the $C/5$ rate of discharge. These energy densities are approximately twice those of the best existing rechargeable systems.

However, there is the problem of internal shorting after several cycles as a result of zinc dendrite growth. This problem can be overcome to a great extent by proper selection of separator materials. It was also found that the air electrodes, which contain platinum as the catalyst and have been used successfully in the mechanically rechargeable zinc–air battery and the cadmium–air electrically rechargeable battery, do not function in the charging as well as the discharging mode. This is because platinum on the anode surface acts as a low hydrogen overvoltage site, thereby enhancing zinc self-discharge and reducing the available capacity of the zinc electrode. Thus, until an adequate air electrode is developed, a third electrode will be required for charging purposes.

Table 2.11 compares some basic parameters of the five main types of rechargeable battery system. Some of these data must, of course, be interpreted with caution – particularly energy density data. Discharge characteristics for these five types of battery are compared in Figure 2.10.

As is seen in Table 2.12, the energy density data quoted by different battery manufacturers for any particular type of battery vary over a wide range – more so in the case of the newer types of battery such as nickel–zinc, silver–zinc and silver–cadmium where battery designs have not yet been fully optimized. In addition to battery design, several factors such as battery size and whether the data quoted refer to a cell or a battery (where the additional weight or volume of the outer case affects the calculated energy density) have to be taken into account. For silver–zinc cells Yardney quote

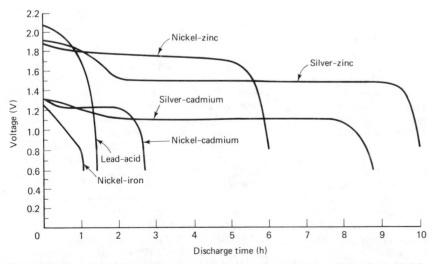

Figure 2.10 Typical discharge characteristics of various secondary battery systems of equal weight discharging under the same conditions

Table 2.12 Energy density data quoted by different manufacturers for rechargeable batteries

Type	Source	W h/kg	W h/dm^3
Nickel–zinc	Theoretical	374	–
	Yardney	44–77	79–134
	Eagle Picher	29–75	52–126
	Other sources	up to 66	134
	Other sources	33–77	67–134
Nickel–cadmium sealed	Nife	24–33	61–90
	Button mass	21–22	61–62
	Round mass	22	69
	Round sintered	27	78
	Rectangular mass	18	37
	Rectangular sintered	21	52
	Other sources	26–37	61–85
	(Vented)	26–44	61–99
Silver–zinc	Theoretical	440	–
	Yardney	70–120	150–250
	Eagle Picher	55–209 (as cells)	80–415
	Eagle Picher	37–114 (as batteries)	55–262
	Other sources	123	220
	Other sources	130	–
	Other sources	55–220	80–610
Silver–cadmium	Theoretical	310	–
	Yardney	48–75 (as cells)	91–165
	Yardney	18–51 (as batteries)	–
	Eagle Picher	24–73 (as cells)	40–171
	Eagle Picher	18 to 44 (as batteries)	24–122
	Other sources	77	171
	Other sources	24–120	37–250
Lead–acid	Sonnenschein (sealed)	23–34	51–83
	Other sources	22–33	49–83

70–120 W h/kg depending on type, while Eagle Picher quote 55–209 W h/kg for cells and, as would be expected, a lower value of 37–114 W h/kg for batteries (where extra non-capacity-producing weight is required). The theoretical energy density of the silver–zinc cell is 440 W h/kg (that is, only weight of active material is taken into account, while weight of battery case, separators and terminals is ignored). On a volumetric basis Yardney quote 150–250 W h/dm^3 and Eagle Picher 80–415 W h/dm^3 as cells and 55–262 W h/dm^3 as batteries. These data highlight the differences between the products and the effect on energy density of whether the data are calculated for a cell or for a battery. Comparable data for the silver–cadmium system are as above.

It is thus seen that considerable caution must be exercised when interpreting claims for energy density in manufacturers' literature and, indeed, this comment would apply to many of the quoted performance data of cells and batteries.

2.3 Conclusion

In conclusion, it can be said regarding the battery selection process, whether primary or secondary batteries are being considered, that the design engineer is faced with many commercial portable power systems in a wide variety of models. The battery selection process cannot therefore be reduced to an exact science. Seldom does any one battery system meet all the requirements for a given application. The selection of a battery is further complicated by the fact that the performance characteristics of battery systems vary with temperature, current drain, service schedules, etc. Consequently the selection process usually involves a trade-off or compromise between battery requirements and battery system characteristics.

The battery or portable power source is an integral part of the electrical system and should be considered as early as possible in the design process.

In selecting a battery the following four basic steps should be followed:

1. Determine the battery requirements, including:
   (a) *physical* – size and weight limitations, shape, shock and vibration resistance, operating position, acceleration, high-altitude use;
   (b) *electrical* – voltage, current drain (initial and operating), constant or interrupted demand, discharge schedule;
   (c) *environmental* – storage and operating temperatures, moisture and humidity factors;
   (d) *special considerations* – cost, replace or recharge, service life, shelf life, operating schedule, activation, type of terminals, end-point voltage (if equipment will not operate below a certain critical voltage).
2. Establish relative importance of requirements – determine those that are mandatory and those that are desirable. List the requirements in order of importance.
3. Compare the characteristics of each battery system with the battery requirements. For each requirement list those systems that can meet the requirement.
4. Determine necessary compromises. The selected system must meet the mandatory requirements. Trade off on the desirable requirements, beginning with those of least importance.

Through proper battery system design following the demands of the user, one of the following battery systems can be established:

1. An automatically activated primary.
2. A manually activated primary.
3. A rechargeable secondary.

When designing for any one of the above batteries, the following special considerations for each of the systems should be kept in mind:

1. Automatically activated primary:
   (a) Method of activation – mechanical or electrical.
   (b) Activation time required to distribute electrolyte.
   (c) Wet stand time after activation prior to application of load.
   (d) Orientation during activation – upright, upside-down, etc.
   (e) Orientation during operation.
   (f) Temperature in storage prior to use.
2. Manually activated primary:
   (a) Soak time after activation prior to use.
   (b) Cycle life (if any) – number of charge/discharge cycles.
   (c) Wet shelf life (if any) – wet stand time after activation.
   (d) Temperature in storage prior to use.
3. Rechargeable secondary:
   (a) Cycle life – number of charge/discharge cycles.
   (b) Wet shelf life – wet stand time after activation.
   (c) Capacity as a function of cycling – capacity at first cycle compared with capacity at last cycle.
   (d) Charge retention – ability to stand in charged condition.
   (e) Method of recharging.
   (f) Maintenance.

Because of the varying performance characteristics of battery systems under different conditions, it is virtually impossible to select the battery best suited for a particular application. In selecting the proper battery for a specific application, the designer should consult battery specialists for assistance early in the design process. Since the battery is an integral part of the electrical system, unwarranted and costly compromises may be avoided by coordinating battery selection and product design.

Part 1

Battery Characteristics

3

Lead–acid secondary batteries

Contents

3.1 Open-type lead–acid batteries

The lead–acid battery is still the most widely used, its main application being in the automotive field, although it has a growing number of other applications. Its advantages are low cost, high voltage per cell and good capacity life. Its disadvantages are that it is relatively heavy, it has poor low-temperature characteristics, and it cannot be left in the discharged state for too long without being damaged. Since the late 1950s sealed rechargeable batteries based on the lead–lead dioxide couple have become available and while these are not produced in the same quantities as the non-sealed type their uses are increasing.

3.2 Non-spill lead–acid batteries

Several manufacturers supply small lead–acid batteries which, although not sealed, are specifically designed to be spillproof. Thus, although these batteries will require topping up with electrolyte during their life, they have specific advantages in the field of portable battery-operated equipment. Sonnenschein manufacture a range of spillproof accumulators with capacities between 2 and 9.5 A h designed for use with photoflash equipment and other portable units (Table 50.23). These batteries are housed in a tough transparent plastics case and feature a labyrinth-like construction below the vent opening through which no acid will pass even with extreme changes of position and vibration. The vent does, however, permit the passage of gases during charging. Each individual cell has an indicator based on floats, which accurately indicates the state of charge. The batteries at −40°C retain 38% of the capacity they possess at 20°C. These batteries are available in two versions; standard and permanent.

The standard batteries are traditional lead accumulators. When not in use, depending on the ambient temperature, recharging is required every 4–6 weeks to compensate for the self-discharge. They are extraordinarily resistant to frequent cycling operation and are outstanding for requirements where regular use without long idle periods is likely to occur. Even in case of intensive use they have an excellent lifetime. Permanent batteries, in contrast, may be stored at an average ambient temperature of 20°C for some months without maintenance and should be recharged after 10 months (older batteries after 6 months). Discharged permanent batteries should be recharged within 2 weeks. Their low self-discharge is achieved through the use of a special lead alloy. These batteries are intended especially for use in applications where they are not used regularly. Their expected lifetime is higher than that of the standard batteries.

Varley Gates Dry Accumulators supply a range of non-spill lead–acid batteries in the capacity range 9–50 A h (Table 50.32).

3.3 Recombining sealed lead–acid batteries

There are two categories of sealed lead–acid cell. These are the non-recombining or partially recombining type, such as those manufactured by Sonnenschein and by Crompton-Parkinson Ltd, and the fully recombining types, as manufactured by the General Electric Company and by the Gates Rubber Company. The fully recombining types are also produced in the UK under licence by Chloride Energy Ltd under the trade name Cyclon.

Particularly towards the end of charge and when being overcharged, the sulphuric acid electrolyte in lead–acid batteries undergoes electrolysis to produce hydrogen and oxygen. Consequently, in service, the electrolyte level drops and the concentration of sulphuric acid increases, both of which are deleterious to battery performance and, unless attended to by periodic topping up with distilled water, will lead to the eventual destruction of the battery. Aware of this danger manufacturers recommend a periodic topping up of the electrolyte to the prescribed mark with distilled water.

The need for regular topping up has in the past limited the applications in which lead–acid batteries can be used. Manufacturers have adopted two methods of avoiding the need to top up lead–acid batteries:

1. The development of non-recombining or partially recombining batteries in which, by attention to battery design (new lead alloys for grids, etc.) and by using more sophisticated battery charging methods, gassing is reduced to a minimum and topping up is avoided.
2. The development of fully recombining types of batteries in which any hydrogen and oxygen produced by gassing is fully recombined to water, thereby avoiding loss of electrolyte volume.

Both methods have been used to produce a range of non-spill either partially or fully recombining sealed lead–acid batteries which are now finding an ever-increasing range of applications for this type of battery. The theory and design of both of these types of batteries are discussed in Part 2.

3.3.1 Partially recombining sealed lead–acid batteries

Various manufacturers supply these batteries, including Sonnenschein (West Germany), Dryfit batteries (available up to 36 A h capacity), Crompton-Parkinson (UK; up to 45 A h capacity), Eagle

Picher (USA), Carefree (up to 44 A h capacity) and Yuasa (Japan; up to 8 A h capacity) – see Part 5 for details; 6 and 12 V versions are also available. These batteries have several design features, which contribute to their non-maintenance characteristic and their ability not to require overcharging to maintain full capacity. In fact, no water needs to be added to these batteries during their entire life. If, because of improper charging or large variations in temperature, a gas pressure does build up, the safety valves ensure that this gas can escape immediately. Afterwards, the safety valves automatically shut off the electrolyte space from the outside atmosphere. Assuming that the operating and environmental conditions are satisfactory, this type of battery is sealed and does not gas. Sonnenschein Dryfit batteries, for example, meet the requirements of the German Physikalische-Technische Bundesanstalt for use in hazardous areas containing combustible substances of all the explosive classes within the range of inflammability G1 to G5. Their method of construction also meets the requirements of VDE D171/1–69.

Usually, a gelled electrolyte is used so that no emissions of acid or electrolysis gas occur during charging or use. These batteries have a low internal resistance and consequently a high current delivery capability. The batteries can be used in any position and they function at full efficiency in a completely inverted position, although this should not be made a deliberate feature in equipment design.

Many types of recombining sealed lead–acid battery feature calcium–lead grids rather than the conventional antimonial lead grids in order to reduce self-discharge rates and gassing on stand and towards the end of charge. The Gates product uses pure lead grids.

Partially recombining sealed lead–acid batteries may be charged and discharged over an ambient temperature ranged from -20 to $+50°C$. If continuous operation at one or more of the extreme values is required, a temperature sensor should be used to optimize charging. Compared to this, nickel–cadmium batteries, depending on their type, are severely limited by an allowable temperature range of $0–45°C$ for charging. To operate properly the lead–calcium acid battery must be matched with a constant potential charging system, that is, one in which the current acceptance becomes a function of the back e.m.f. of the battery. It is necessary that the battery sustain a constant on-charge voltage initially and throughout its life. Constant current charging may also be used in trickle charge applications.

Under constant potential charging the maintenance-free lead–calcium acid battery will accept relatively high currents initially when the system is most efficient. This stage is followed by a second period when the back e.m.f. of the battery begins progressively to increase and to control the amount of current accepted. At full charge and under proper voltage regulation, the current accepted will be reduced to a few milliamps input, thus restricting the degree of overcharge which protects the unit from excessive electrolyte loss.

This type of battery further supports this function by the fact that it develops a very high hydrogen and oxygen overvoltage which enhances the efficiency of conversion and ensures a sharp and reproducible rise in on-charge voltage to regulate current acceptance best.

Partially recombining lead–acid batteries of the same capacity can be operated in series or in parallel. To charge batteries in series, a high-voltage low-current source is needed, while units in parallel require a low-voltage high-current source. The constant potential setting for series charging is equal to the sum of that calculated for each battery while the voltage setting for units in parallel is that required for one unit. Charging batteries in parallel is the recommended procedure.

Partially recombining sealed lead–acid batteries have a gravimetric energy density in the range 30–35 W h/kg at the 20 h capacity rate.

Comparisons of the energy density of such batteries with that of nickel–cadmium batteries are interesting as in many applications these two types of battery are competing with each other. Rectangular mass and sintered plate nickel–cadmium batteries exhibit lower energy densities (17 and 25 W h/kg respectively) than sealed lead–acid and cylindrical sintered plate nickel–cadmium batteries (35–37 W h/kg). Yuasa quote 21–30 kg/W h energy density for their range of partially recombining sealed lead–acid batteries (Table 50.26).

3.3.2 Fully recombining sealed lead–acid batteries

General Electric are developing a range of sealed cylindrical 2.00 V lead–acid rechargeable cells and also batteries. The first cell in this range to be offered was the D cell, which is rated at 2.5 A h at the 10 h rate. It is 61 mm in height and 34 mm in diameter.

The construction of this cell is similar to that of standard cylindrical nickel–cadmium cells. It is sealed, has a safety vent and makes use of a cylindrical spiral-wound plate design.

Stable capacity/voltage in long-term overcharge is well suited to standby applications. Special alloy plate grids and efficient oxygen recombination enhance long life during overcharge. Drying out is not a problem. Sealed lead–acid cells are available as individual cells, standard batteries of convenient sizes and voltages, and also as assemblies of custom battery packs.

Life is a function of use and environment, but the spiral-wound sealed lead–acid cell combines all the features – oxygen recombination, low impedance, seal design, resealing vent, low alloy plate grid – to extend the life of the cell beyond what other lead–acid designs offer. The General Electric sealed lead–acid system offers a number of desirable discharge characteristics. The spiral plate design enhances long life in both float and cycle applications. Since the spiral configuration results in lower impedance, the cell may be discharged at higher rates. State of charge is conveniently determined by measuring open-circuit cell voltage after a short stabilizing period. A voltmeter is all that is required to measure state of charge. The Chloride-Gates Cyclon range of sound electrode combination batteries and the General Electric range discussed above are equivalent and hence the Chloride product, available in the UK, will not be discussed in great detail here.

These batteries are contined in a polypropylene case with a metal outer canister and are available as single cells and in battery packs including a lantern battery. Batteries will be specially designed and produced to meet customers' requirements.

The standard types of cell now available are the D, X and BC cells (Table 3.1).

Temperature ranges are −65 to +65°C (storage and discharge) and −40 to +65°C (charge). Storage lives from full charge are 18 years at 0°C, 3 years at 25°C and 1 year at 40°C. The expected cycle life varies from 300 to 2000 cycles and the float life is 8–10 years. Charging is by most conventional methods including constant voltage, constant current, taper charging, pulse charging, etc., according to the application. Self-discharge is typically 6–8% per month at room temperature.

The manufacturers claim that the D cell will give instantaneous discharges of up to 100 A and sustained discharges of up to 30 A. Double these figures can be obtained with the X cell, and up to 800 A with the BC cell.

Batteries (6 and 12 V) are available with capacities in the range 2.7–52 A (20 h rate), 2.5–50 A (10 h rate).

Table 3.1 Characteristics of standard cells

	D	X	BC
Nominal voltage (V)	2.0	2.0	2.0
Capacity rating (A h)			
10 h rate	2.5	5.0	25
	(250 mA)	(500 mA)	(2.5 A)
1 h rate	2.8	3.2	20
	(2.5 A)	(5 A)	(25 A)
Peak power rating (W)	135	200	600
	(at 135 A)	(at 200 A)	(at 600 A)
Internal resistance ($\times 10^{-3}\,\Omega$)	10	6	2.2
Diameter (mm)	34.2	44.4	65.7
Height (mm)	67.6	80.6	176.5
Weight (kg)	0.181	0.369	1.67

4

Nickel batteries

Contents

4.1 Nickel–cadmium secondary batteries

The nickel–cadmium battery is mechanically rugged and long lived. In addition, it has excellent low-temperature characteristics and can be hermetically sealed. Cost, however, is higher than for either the lead–acid or the nickel–zinc battery and, by comparison, its capacity on light drain in terms of watt hours per kilogram is also poorer than for nickel–zinc.

Cells using the nickel–cadmium couple can be divided into two main groups: cells having thick plates in which the active material is compressed into finely perforated metal sheets in the form of tubes or pockets, and cells with thin sintered plates in which the active material is deposited in porous metal supports. The latter type has a very low internal resistance and a high load capacity which result in improved power-to-volume and power-to-weight ratios. The batteries are manufactured as open or semi-open type cells in which the electrolysis gas escapes through a vent (which require periodic topping up with water to replace the electrolyte) and as fully sealed batteries which are designed in such a way that gas evolution does not occur and consequently topping up of the electrolyte with water is not required. In general, the open or semi-open types of battery are in the higher capacity range, used for applications such as traction and large emergency power installations. Sealed nickel–cadmium batteries that are usually in the lower capacity range (up to about 30 A h) are the type that find extensive application in the electronics, small appliances, domestic products, defence equipment and space research markets.

4.1.1 Open and semi-open, low-maintenance, unsealed nickel–cadmium batteries

Various manufacturers supply small low-maintenance cells and batteries. Varta supply a non-sealed double cell (DTN) for portable lamps.

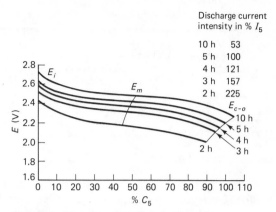

Figure 4.1 Discharge curves showing dependence on capacity available at 20°C of a Varta double cell DTN battery (Courtesy of Varta)

This battery is available in the capacity range 4.5–11 A h (Table 4.1); it is recommended for applications such as portable lamps where its light weight, even voltage level, good capacity, long service life, lack of maintenance and low self-discharge rate are of advantage. Discharge curves for a DTN battery are shown in Figure 4.1.

Chloride Alkad supply their Unibloc range of alkaline batteries, which are virtually maintenance free but are not sealed. These range from 7.5 to 27 A h capacity (Table 4.2). Unlike traditional alkaline batteries, which are built up from individual separately packaged cells, Unibloc starts off as a five 1.2 V cell battery in a lidded container (Figure 4.2). The Unibloc range is available in four basic capacities, each providing low-rate discharge performance:

7.5 A h	Type 5LP7
12.5 A h	Type 5LP12
15 A h	Type 5LP15
27 A h	Type 5LP27

The five cells are incorporated into a resilient plastics case containing sufficient extra potassium

---

Table 4.1 Varta DTN range of low-maintenance nickel–cadmium batteries

Charging Battery type	Type No. 333	Nominal capacity in 5 h (A h)	Nominal discharge current in 5 h (A)	Median discharge voltage (V)	Nominal charging current in 7 h (A)	Charging voltage (V)	Filled weight (kg)	Quantity of electrolyte (dm³)
DTN4.5K	2452 010	4.5	0.9	2.4	0.9		0.42	0.08
DTN6.5K	2652 010	6.5	1.3	2.4	1.3	increasing	0.61	0.11
DTN7K	2702 010	7.0	1.4	2.4	1.4	from 2.7	0.75	0.14
DTN12K	3122 020	11.0	2.2	2.4	2.2	to 3.6	0.90	0.20

Table 4.2 Unibloc batteries: dependence of maintenance-free intervals on capacity and charging methods

Type	*Capacity at* 10 h *rate* (A h)	*Maintenance interval* (years)	
		Fully automatic operation without boost (float voltage 1.47 V/cell)	*Close voltage application with 6-monthly boost of* 1.65–1.70 V/cell *(float voltage* 1.42 V/cell)
5LP7	7.5	8	14
5LP7	12.5	5	9
5LP12	15.0	4	7.5
5LP27	27.0	2.5	5

hydroxide electrolyte to enable periodic topping up of the battery to be extended up to a maximum of 20 years on a constant 1.42 V/cell float charge without any boost charging, although 7–10 years will probably be the average.

Unibloc batteries are recommended for applications such as switch tripping, emergency lighting and fire alarm operation.

For most applications, constant-voltage charging is recommended to minimize electrolyte loss which is dependent on the float charge voltage, the capacity of the battery and the frequency of boost-charging. Table 4.2 shows how the maintenance interval across the Unibloc range of batteries is related to battery capacity and charging method.

The discharge performance of Unibloc batteries is shown in Figure 4.3.

Simple snap-on or bolt-on terminals
5-in-1 composite cell construction
Easy-to-clean surfaces provide maximum protection against corrosion and dirt
Extra-generous electrolyte reserve means negligible maintenance

Figure 4.2 Chloride Alkad Unibloc nickel–cadmium battery construction (Courtesy of Chloride Batteries)

Discharge (V/cell)
1.4
1.3
1.2
1.1
1.0
0.9
0.8
0.7
C/10
C/5
C/3
C/2
C
1 2 3 4 5 10 15 30 45 1 2 3 4 5 10 15 20
Time (min)
(h)

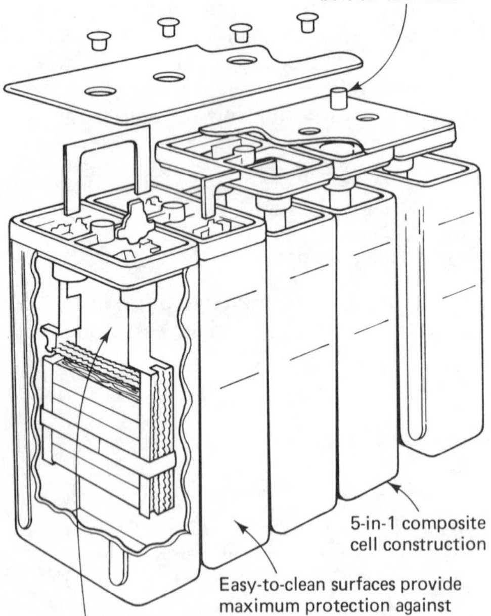

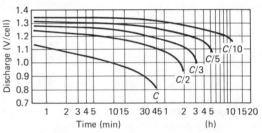

Figure 4.3 Discharge curves at 25°C: Chloride Alkad Unibloc nickel–cadmium battery (Courtesy of Chloride Batteries)

4.1.2 Sealed nickel–cadmium batteries

Characteristics of sealed nickel–cadmium batteries

Sealed nickel–cadmium batteries were first developed in the 1950s from vented nickel–cadmium batteries and therefore originally used the same active material and similar intercell components that had proved their reliability in production and service. Because the active material in the vented batteries was a powdered mass, it was logical for the earlier ranges of sealed batteries to use the same material. Since then materials and production techniques have improved, enabling sealed battery performance to be improved by optimizing the active material. These changes in active material, while not significantly improving the performance at low discharge rates (10 or 5 h rate), have changed the voltage characteristics at the higher rates of discharge. An example of this improvement for a Varta 0.225 A h cell (10 h rated, 22 mA discharge current at 10 h rate) is shown in Figure 4.4 by comparison of the characteristics in (a) original plate materials and (b) improved plate materials. This improvement in discharge performance has been

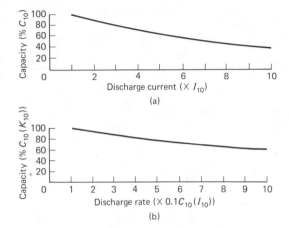

Figure 4.4 Effects of improvement in plate active materials since the 1950s on the performance of the Varta 0.225 A h nickel–cadmium sealed battery: available capacity at 20°C (a) before and (b) after improved materials were incorporated (Courtesy of Varta)

achieved without adversely affecting the cell's good charge retention or external dimensions.

Sealed nickel–cadmium cells are normally rated at either the 5 or 10 h rate, that is, the discharge current that can be taken from a cell to discharge it fully in 5 or 10 h. Sintered cells are rated at the 5 h rate (I_5) and, occasionally, particularly with US manufacturers, at the 1 h rate (I_1), and mass plate cells at the 10 h rate (I_{10}). The nominal on-load voltage of a nickel–cadmium cell is 1.2 V and this voltage is said to be the average value for the total discharge period. The cell's fully charged open-circuit voltage, if taken within 24 h of terminating a charge, can vary between 1.35 and 1.4 V. The voltage characteristic is flat for most of the discharge curve, even at comparatively high discharge rates, therefore the discharge voltage is not indicative of the cell's state of charge. At rates of discharge up to ten times the I_{10} rate, a cell is said to be fully discharged at 0.9 V. The voltage of a cell when discharged below 0.9 V falls rapidly to zero and it is therefore possible for some cells in a battery to reverse their polarity. Provided the discharge rate is at the I_{10} rate or less, this reversal will not harm the nickel–cadmium cell, but with maximum life in view it should be avoided.

Sealed nickel–cadmium cells and batteries exhibit relatively constant discharge voltages. They can be recharged many times. They are small convenient packages of high energy output, hermetically sealed in leak-resistant steel cases, and will operate in any position. The cells have very low internal resistance and impedance, and are rugged and highly resistant to shock and vibration. Some general comments on sealed nickel–cadmium batteries are made below.

Temperature of operation

Cells and batteries may be charged, discharged and stored over a very wide temperature range. The stated maximum permissible temperature ranges for these batteries are as follows:

Charge	0 to +45°C
Discharge	−20 to +45°C
Storage	−40 to +50°C

Discharge and storage at +60°C are permissible for a maximum of 24 h (see also Table 4.3).

Use at high temperatures, however, or charging at higher than recommended rates, or repeated discharge beyond the normal cut-offs, may be harmful. In the case of button cells which do not contain a safety vent, charging at temperatures lower than those recommended may cause swelling or cell rupture.

For low-temperature operation (0°C or below), sintered cells are recommended. At these temperatures, the cell's internal resistance increases and this can greatly affect the voltage characteristic at high discharge rates. The low temperature charge problems of sealed nickel–cadmium cells are caused partly by the poor charge efficiency of the negative electrode at these temperatures. Also, at low temperatures, the cell's ability to recombine oxygen during overcharge is inhibited and therefore, under normal circumstances, charging at low temperatures is not recommended. In addition, at sub-zero temperatures hydrogen evolution may be promoted (cathode overcharge), causing venting of gas and consequent electrolyte loss. The cells have a negative temperature coefficient of approximately 4 mV/cell per °C. When selecting cells for low-temperature operation, both their reduced capacity and changed cell voltage should be taken into consideration.

Efficiency

Sealed nickel–cadmium cells with mass plate electrodes have an ampere hour efficiency of approximately 72%, while those with sintered electrodes have approximately 84% efficiency at about 20°C. (The effect of temperature will be discussed later.) For this reason the charging factor for mass plate cells is 1.4 and for sintered plates 1.2; therefore to fully charge a fully discharged cell, either 40% (1.4) or 20% (1.2) more capacity than has been withdrawn must be put back into the cell.

The watt hour efficiency is the quotient of the discharged energy and the necessary energy to fully charge a cell. Comparing the mean discharge voltage with the mean charge voltage confirms a figure of approximately 61% for a cell with mass electrodes and 73% for a cell with sintered electrodes when discharged at the I_{10} rate (both at approximately 20°C).

Table 4.3 Effect of high and low temperatures on storage, discharging and charging of Eveready nickel–cadmium cells and batteries

	Low temperature	*High temperature*
Storage (all types)	At −40°C	At 60°C
	No detrimental effect. However, cells or batteries should be allowed to return to room temperature prior to charging	No detrimental effect. However, self-discharge is more rapid starting at 32°C and increases as temperature is further elevated
Discharge (all types)	At −20°C	At 45°C
	No detrimental effect, but capacity will be reduced as shown by curves	No detrimental effect
Charge CF and CH types (10 h rate)	At 0°C	At 45°C
	Cells or batteries should not be charged below 0°C at the 10 h rate	Cells or batteries show charge acceptance of approximately 50%
CF types (1–3 h rate)	At 15.6°C	At 45°C
	Cells or batteries should not be charged below 15.6°C at the 1 h rate or below 10°C at the 3 h rate	Cells or batteries show charge acceptance of approximately 90%
Button cells	At 0°C	At 45°C
	Cells or batteries should not be charged below 0°C at the 10 h rate	Cells or batteries show charge acceptance of approximately 60%. Also possible detrimental effect on cycle life

Storage

Sealed nickel–cadmium batteries can be stored indefinitely, ideally in a clean, dry atmosphere, in a discharged state, unlike lead–acid batteries which will sulphate if stored in a discharged state. Nickel–cadmium batteries can be stored in any state of charge, without a significant loss of life.

Maintenance charging is not required. However, after prolonged storage, up to three cycles of charge and discharge may be required to achieve the battery's rated capacity. It is recommended that the first charge following storage should be for 24 h at the I_{10} rate.

During operation and storage, it is possible that crystals may form in the area of the sealing between the positive and the negative poles. This is due to minute pores in the sealing ring allowing electrolyte to combine with carbon dioxide in the air to form potassium carbonate crystals. This crystallization has no detrimental effect on the electrical properties or life expectancy of the battery but, if it is thought to be aesthetically unacceptable, removal of the crystals with a dry cloth and then a smear of silicone grease will inhibit further growth.

Internal resistance

Unlike open-cell batteries, where the internal resistance decreases linearly with an increase of capacity, the internal resistance of a sealed nickel–cadmium battery is dependent mainly on plate surface area, distance between the plates, separator resistance, amount and density of electrolyte, and temperature. The internal resistance will also change with the state of charge. As an example, the d.c. internal resistance of typically fully charged 1 A h sealed cells is as follows:

Standard button cell	110 mΩ/cell
Heavy-duty button cell	50 mΩ/cell
Rolled sintered cell	19 mΩ/cell

In some applications it is an advantage to know the a.c. internal resistance. As a comparison, the a.c. resistance at 1000 Hz is:

Standard button cell	42 mΩ/cell
Heavy-duty button cell	14 mΩ/cell
Rolled sintered cell	17 mΩ/cell

Cyclic life

The life of a sealed battery depends on the operating conditions. It is principally affected by the depth of discharge (capacity discharged/rated capacity) in a charge/discharge application, or in ampere hours overcharged in a permanent charge application. Life is also affected by temperature, though less so at high temperatures, and by end-point requirements regarding rate and capacity (increased cycle life will ordinarily be the result of a shallow discharge regimen).

Any treatment that causes a cell to vent itself is harmful. Frequent or extended venting of even properly valved cells eventually destroys them. In rating cycle life, the end of life for a sealed nickel–cadmium cell is considered to be when it no longer provides 80% of its rated capacity. The discharge currents used in determining the cycle lives listed below are the 10 h rate for button cells and the 1 h rate for cylindrical types. The charge current is terminated after return of approximately 140% of the capacity previously removed. If a cell can be considered to be satisfactory while delivering less than the arbitrary 80% end-point figure, the reliability characteristic of the sealed nickel–cadmium battery corresponds to the 'bathtub' curve of electronics components. In cycling applications, as in the supply to a portable transceiver, the service life of a battery is most conveniently expressed as the number of cycles of charge and discharge, and the failure rate in failures per cycle.

In applications with permanent maintenance charge, the life of a battery is expressed as the multiple of C_5 A h of overcharge or as hours of operation, and the failure rate as failures per operating hour. As an example, Figure 4.5 shows the estimated cycling life at 20°C as a function of discharge for SAFT sealed nickel–cadmium batteries.

The following results have been obtained at 20°C: burn-in period lasts a few cycles or a few tens of C_5 A h of overcharge; failure rate of the order of 10^{-5} per cycle with rapid charge at $2C_5$ A and discharge over several hours; failure rate also of the order of 10^{-5} per cycle with permanent maintenance charge at rates below $0.05C_5$ A.

During the service life of sealed nickel–cadmium batteries obtained from a reputable manufacturer, there are practically no failures through loss of performance, and the mortality law is very probably of an exponential nature, under the following types of service condition:

1. 500 cycles with charge for 14 h at $0.1C_5$ A and complete discharge over several hours.
2. 1000 cycles with the same type of charge but discharge 50% of capacity over several hours.
3. At least 1500 cycles with rapid charge at $2C_5$ A and discharge over several hours; the failure rate in this case is typically of the order of 10^{-7} failures per cycle.
4. At $2000C_5$ A h with overcharge at a rate below $0.05C_5$ A, the maximum service life is 8 years (at a rate of $0.02C_5$ A): $2000C_5$ A h corresponds to 100 000 h or a theoretical service life of 11 years.

As a further example of the life expectancy of sealed nickel–cadmium batteries Varta claim that their products are not affected by prolonged storage provided the recommended storage temperatures (−40 to +50°C) are not exceeded. Thus less than 20% capacity loss occurs in 2 years storage at −20°C, approximately 40% in 10 or 12 months at 20°C, and proportionally higher rated capacity loss at higher temperatures. In applications of these cells where they are operated in a purely cyclic mode (that is, charge/discharge cycles), mass plate batteries will give 300 cycles at the 10 h rate to approximately 80% of nominal capacity. Partial discharging will significantly increase the number of cycles achieved. Similarly, discharging at high rates will reduce the expected life of the battery.

Sintered electrode batteries supplied by Varta, when continuously trickle charged at rates between one-fifth and one-half of the 10 h rate, will give, typically, a life of approximately 5 years to 60–70% of nominal capacity.

When a fast charge is applied to a sintered cell battery at a maximum of $20 \times I_{10}$, provided the charge is terminated at 80% capacity, a life can be achieved similar to that of a sintered cell battery at normal rates of charge.

Energy density

The energy contents of the various types of sealed nickel–cadmium battery supplied by one particular manufacturer are as shown in Table 4.4.

Cylindrical, rectangular and button designs of sealed nickel–cadmium battery are available from most suppliers. These three types are available with compressed and with sintered electrodes. Batteries with sintered electrodes have a very low internal

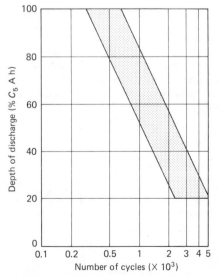

Figure 4.5 Estimated life cycle at 20°C as a function of discharge (SAFT sealed nickel–cadmium batteries) (Courtesy of SAFT)

Table 4.4 Energy contents of various sealed nickel–cadmium batteries

	Volumetric energy density (W h/dm³)	Gravimetric energy density (W h/kg)
Button cell (mass plate)	61–62	21–22
Round cell (mass plate)	69	22
Round cell (sintered plate)	78	27
Rectangular (mass plate)	37	18
Rectangular (sintered plate)	52	21

resistance and a high load capacity which results in an improved power to size and weight ratio. These three types of nickel–cadmium cell are discussed in detail in Part 2.

Generally speaking, nickel–cadmium button cells are available in a lower capacity range than cylindrical cells. Thus one manufacturer supplies button cells in the 20–1000 mA h range and cylindrical cells in the 150–4000 mA h range. Voltages of between 1.2 V and 14.4 V (12 cell) are available.

4.1.3 Nickel–cadmium button cells

Nickel–cadmium button cells have a nominal discharge voltage of 1.2 V per cell. The rated capacity C_5 (Table 4.5) at 20°C is defined as the capacity in ampere hours obtained at the 5 h rate of discharge ($0.2C_5$ A) to an end-voltage of 1.10 V/cell after a full charge at $0.1C_5$ A for 14 h. Rates of charge and discharge are expressed in multiples of the rated capacity C_5, thus $0.2C_5$ A represents 20 mA for a cell having a rated capacity C_5 of 100 mA h. $5C_5$ A represents 500 mA for the same cell. The range of button cells available from one leading manufacturer is tabulated in Table 4.5.

The life of button cells (SAFT, VB and VBE types, Table 4.5) can extend over 10 years, but this is dependent on operating conditions and is principally influenced by the depth of discharge (discharged capacity/rated capacity) in a cycling application and by the number of ampere hours

overcharged in a continuous charge application. The great reliability of button cells makes their life comparable with that of most electronics components. During cycling, approximately 500 cycles are obtained with 100% depth of discharge and 2000 cycles with 50% depth of discharge. After 5 years' continuous charge at $0.05C_5$ mA, the failure rate for SAFT VB series button cells is not greater than 5 per 1000. Under the same conditions their VBE series button cells offer ten times better reliability; a failure rate of 5 per 10 000.

Button cells with mass plate electrodes are produced by Varta in two versions. DK type cells are assembled using one positive and one negative electrode separated by porous insulating material. These cells are suitable for operation at normal temperatures and discharge rates of up to 10 times the nominal discharge current. DKZ cells have two positive and two negative electrodes which lower the internal resistance and therefore make the cell suitable for high rates of discharge (20 times the nominal discharge rate) and also for low-temperature operation. Button cells are suitable for applications where low cost and good packing density are required. The charge retention of mass plate button cells is superior to that of all other forms of sealed nickel–cadmium cell. Many types of button cell can be made into stack batteries by connecting them end on end, in sleeves, thereby ensuring a minimum internal resistance.

4.1.4 Nickel–cadmium cylindrical cells

Nickel–cadmium cylindrical cells are designed for a wide range of applications requiring compact rechargeable batteries with high output capacity, heavy loads at steady voltages, the ability for rapid charge and discharge, and exceptionally long life. Designed to be dimensionally similar to the principal cylindrical primary cells, of internationally standardized dimensions, these cells usually incorporate thin sintered plate construction. Complete spirally wound plate assemblies are tightly fitted into a steel container to form cells. The container is closed by a cover having an automatic self-releasing

Table 4.5 Sealed button cells (nominal discharge voltage 1.20 V) available from SAFT

Serial No.	Type	Rated capacity (C_5 mA h)	Maximum dimensions (mm)		Maximum weight (g)	Maximum discharge rates (A)	
			Height	Diameter		Continuous	Pulse
VBE	4	40	6	15.7	3.5	0.2	2.3
VB and VBE	10	100	5.3	23.0	7	1	11
	22	220	7.8	25.1	12	2.2	18.5
	30	300	5.5	34.7	18	3	23
	60	600	9.8	34.7	31	6	29

safety valve for high pressures should excess gas pressure be caused by maltreatment. Contact between the plates and the cover on the case is usually through a connection electrically welded to the edges of the plates. This method of assembly improves battery performance on charge and discharge. A typical range of cylindrical cells available from one supplier (SAFT) is listed in Table 4.6.

Cylindrical cells are characterized by a nominal voltage of 1.2 V/cell and their remarkable power (1.2 A h/cell) is, for example, capable of supplying a current of 100 A at 0.6 V for 0.3 s. Up to 170 A peak discharge can be tolerated with a SAFT VR cell (Table 4.6) and the ultra-rapid charge time on a discharged battery can be as low as 60 s when using the appropriate charging equipment involving monitoring cell temperature, voltage or time of charge. For uncontrolled rapid charge SAFT have developed a special range of batteries. These batteries can be discharged continuously at 20 times the normal rate and 40 times the normal rate for short periods. Apart from ensuring the correct charging method, maintaining end-voltages and temperature limits, no special care is needed. The rugged construction results in a battery that can withstand shocks and vibration. The cylindrical cell is available in high reliability versions (SAFT VRE type – Table 4.6) and versions for space applications.

Cylindrical cells when discharged at 1 V/cell at normal temperatures supply currents of the following order:

1.2 A h	cell 12A
2 A h	cell 14A
4 A h	cell 28A
7 A h	cell 35A
10 A h	cell 80A

Short-duration peaks (maximum power) can be obtained (300 ms at 0.65 V/cell), typically:

1.2 A h	cell 93A
2 A h	cell 93A
4 A h	cell 145A
7 A h	cell 170A
10 A h	cell 250A

Cylindrical cells and batteries are designed for use in applications that require good high rates of discharge performance (continuous and peak). Due to their low internal resistance, higher discharge currents can be drawn and also substantially better voltages than apply in the case of cells available a few years ago.

SAFT also produce the VX series of cylindrical cells (Table 4.6). These have the same characteristics as those of the standard VR range. The VX cells have the novel conception of a plate which improves

Table 4.6 Sealed cells supplied by SAFT

Serial No.	Type	Equivalent primary battery	Rated capacity C_5 (A h)	Maximum dimensions (mm)		Maximum weight (g)	Maximum discharge rates (A)		Maximum power
				Height	Diameter		Continuous	Pulse	
VR	0.1 I/3AA	R6	0.100	17.4	14.5	11	1.5	15	10
1.2 V/cell	0.45 I/2A		0.450	28.1	17.3	21	4.5	31	20
	0.5 AA	R6	0.500	50.2	14.5	24	5.0	24	16
	0.7 I/2C		0.700	26.2	26.4	39	7.0	46	7
	1.2 RR		1.2	42.1	22.7	51	12	93	1.2
	1.8 C	R14	1.8	49.7	25.9	77	14	93	–
	2.5 2/3D		2.5	43.5	32.9	105	20	120	78
	4D	R20	4	60.5	32.9	150	28	145	95
	7F		7	91.3	32.9	240	35	170	110
	10 SF		10	89.2	41.7	400	75	280	170
VX	AA	R6	0.200	50.2	14.5	21	5	24	
	RR		0.500	42.1	22.7	45	12	93	
	C	R14	0.700	49.7	25.9	70	14	93	
	D	R20	1.6	60.5	32.9	135	28	145	
VRE	1 RR		1.0	43.2	23.2	49	9	70	
	1.6C		1.6	48.8	26.4	74	11	70	
	3.5 D		3.5	59.2	33.4	150	18	92	
	6 F		6.0	90.0	33.4	240	24	123	
VRS	D		3.5	64.8	33.1	174	11	–	
	F		6.0	93.8	33.1	270	18	–	

the recombination of oxygen produced on overcharge. They may thus be charged or overcharged at higher rates than those for standard VR cells. VX cells have a smaller capacity than the corresponding standard cells (about 50%) but the power characteristics are the same.

Cylindrical cells with mass electrodes (Varta D series)

These cells are available from Varta. They are physically equivalent to primary cells and are suitable for either one- or two-cell operations. These cells are suitable for comparatively low rates of discharge where good charge retention is important. They are suitable for a continuous discharge up to 10 times nominal discharge rate.

Cylindrical cells with sintered electrodes (Varta RS series)

These cells are suitable for applications where a high rate of discharge, or operation at extremes of temperature, or permanent trickle charge or fast charging is required. The cells are constructed from rolled sintered positive and negative plates, separated by a highly porous separator which absorbs all the free electrolyte within the cell. A safety vent is incorporated within each cell to enable gas, which may build up under fault conditions, to be released. Batteries containing RS cells are available either in standard formats or designed for specific applications.

4.1.5 Nickel–cadmium rectangular cells

Rectangular or prismatic cells offer the highest capacities within the sealed nickel–cadmium range, and in fact differ very little in construction from the heavy-duty nickel–cadmium type. Capacities are up

to 23 A h at 1.22 V. This battery weights 1.390 kg, its dimensions being $51 \times 91 \times 125$ mm. Permissible operating temperatures are the same as for the button type. These cells comprise an assembly of two groups of rectangular thin sintered plates and the separator in rectangular nickel-plated steel cases. Current take-off is through terminal studs. They have medium-pressure (4×10^5 N/m^2) self-reclosing safety valves. SAFT, for example, supply three sizes of rectangular cell (VD series). There are three sizes of cell with capacities below 10 A h: VO4C (3.8 A h), VO4US (4.4 A h) and VO9 (9.9 A h) (Table 4.7). There are also high-reliability sealed VO cells for space applications (VOS cells). VO cells and batteries may be continuously discharged at currents up to $5C_5$ A. They can supply short-duration peaks of up to $10C_5$ A.

Sealed rectangular cells with mass electrodes (Varta D23 type) are suitable for continuous discharge up to 10 times the nominal discharge current. They are fitted with a safety vent which prevents excessive internal pressure damaging the cell case in fault conditions.

Rectangular cells with sintered electrodes, SD series (Varta CD type), are suitable for applications in which a high rate of discharge, or operations at extremes of temperature, or permanent trickle is required. The cells are constructed from cut sintered positive and negative plates, separated by a highly porous separator which absorbs all the free electrolyte within the cell. A safety vent is incorporated within each cell to enable gas which may build up under fault conditions to be released. All cells in this range have the cases connected to the positive electrode. All cases are not insulated but intercell separators and nickel-plated connective links are available. The CD series is available made up in metal boxes.

A range of rectangular cells available from SAFT is listed in Table 4.7.

Table 4.7 Sealed rectangular cells available from SAFT

Serial No.		Rated capacity (C_5 mA h)	Maximum dimensions (mm)			Maximum weight (g)	Maximum discharge rates (A)	
			Height	Width	Depth		Continuous	Pulse
VO	4C	3.8	74	64	19	180	20	38
	4US	4.4	74	64	19	220	22	44
	9	9.9	107	98	19	400	50	100
VOS	4	4	78.5	53	20.5	230	20	–
	7	7	112.5	53	20.5	360	35	–
	10	10	91	76.2	28.9	500	40	–
	23	23	165	76.2	28.9	1060	60	–
	26	26	186	76.2	28.9	1200	80	–

4.1.6 Comparison of characteristics of lead–acid and nickel–cadmium sealed batteries

For many everyday battery applications, the choice is still between the lead–acid and the nickel–cadmium systems, rather than the more recently developed systems. Each of these two main types of sealed battery has its own advantages and the choice between nickel–cadmium and lead–acid batteries depends very much on the particular application and on the performance characteristics required.

The characteristics of the two types of battery are compared in Table 4.8. Although this information will be of assistance to the designer in helping to select whether a lead–acid or a nickel–cadmium battery is most appropriate to the application in mind, it is strongly recommended that a final decision should not be made until the problem has been discussed with the suppliers of both types of battery. The lead–acid batteries referred to in Table 4.8 are the Dryfit type supplied by Accummulator-enfabrik Sonnenschein (West Germany), although no doubt in many respects the comparisons will hold good for sealed lead–acid batteries produced by other suppliers.

Figure 4.6 plots the energy density (W h/kg) as a function of the 20 h capacity for Dryfit batteries and nickel–cadmium batteries supplied by Varta and SAFT. It can be seen that the rectangular mass and sintered plate nickel–cadmium batteries exhibit lower energy densities than Dryfit batteries and cylindrical sintered plate nickel–cadmium batteries. The increase in capacity which occurs in Dryfit batteries after several charging cycles together with the associated increase in energy density has not been taken into consideration in the data presented in Figure 4.6. Also not considered is the increase in weight when five nickel–cadmium cells are connected in a supplementary housing to form a 6 V battery. The usual cylindrical sinter cell batteries, the large button cells and Dryfit batteries are thus seen to occupy approximately the same W h/kg range with the bandwidth of the range being in general defined by the individual type variants.

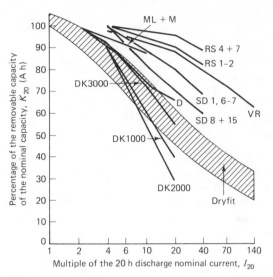

Figure 4.7 Removable capacity dependence on charge current: Dryfit and nickel–cadmium batteries. Sinter round cells: RS 1–7, 1–7 A h (Varta); VR (SAFT). Prismatic sinter round cells: SD 1–15, 1–15 A h (Varta). Prismatic mass cells: D (Varta). Button cells with mass electrodes: DK 1000–3000 (Varta). Prismatic sinter cell batteries: ML, M (Varta) (Courtesy of Dryfit)

The relationship between multiples of the 20 h nominal discharge current and the available percentage of the nominal capacity for Dryfit and nickel–cadmium batteries is shown in Figure 4.7. The shaded area represents the variation of capacity of Dryfit batteries from low discharge currents to very high currents. The lower edge of the band is for a new battery after recharging while the upper edge represents the characteristic after 30–50 charging cycles or several months of trickle charging. From 1 to $40 \times I_{20}$, the various types of sintered nickel–cadmium cell perform better than the Dryfit batteries, and similar or worse performance is obtained up to $20 \times I_{20}$ for the various button cells with mass electrodes.

Looking at the high current end of the curves in Figure 4.7, it is clear that definite limits are placed on nickel–cadmium batteries. These limits are imposed by the maximum allowable discharge current as shown in Figure 4.8. The upper curve emphasizes the exceptionally high maximum load capacity of Dryfit batteries, especially at the lower end of the capacity range where it is a number of times greater than that of nickel–cadmium batteries.

The self-discharge properties of sealed nickel–cadmium batteries differ greatly for mass plate and sinter plate cells, but compared to the Dryfit battery they are substantially inferior. As Figure 4.9 indicates, mass plate nickel–cadmium cells discharge at 20°C to half their original value after 13 months. The decrease in the first months is exceptionally steep and not as flat as that for the

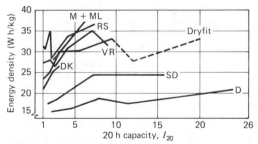

Figure 4.6 Energy density comparisons: Dryfit and different types of sealed nickel–cadmium battery (6 V). M + ML, prismatic sintered cells in plastic cases; RS, sintered round cells (Varta); VR, sintered round cells (SAFT); SD, prismatic sintered cells (Varta); D, prismatic mass cells (Varta); DK (Courtesy of Dryfit, SAFT and Varta)

Table 4.8 Comparison of sealed lead–acid and nickel–cadmium batteries

Property	Lead–acid (Dryfit)	Nickel–cadmium
Capacity range	A300 series, standby operation 1–9.5 A h; A200 series, high cyclic applications 1–36 A h	Mass cells, 10 mA h to 23 A h; sinter cells 100 mA h to 15 A h
Cell voltage	2.1 V	1.25 V
Number of cells/12 V battery	6	10
Case material	Plastics	Nickel-plated steel
Stability of case dimensions during charging	Do not change	Button cells increase by up to 3%
Stabilization of electrolyte	With gelled acid	–
Gassing	Practically no escape of gas	Practically no escape of gas
Cell closure	Self-closing safety valve	Plastics seal or self-closing or non-resealing safety vent
Energy density	The same order of magnitude for Dryfit, nickel–cadmium button cells, sintered cylindrical cells and sinter cells in plastics cases. Rectangular nickel–cadmium mass and sinter cells in steel cases exhibit lower energy densities	
Current-dependent capacities	Average	Button cells, average; sintered cylindrical cells, above average; rectangular sinter cells, below average
Allowable continuous current-carrying capacity	Very good	Sinter cells, good/very good; mass cells, poor
Self-discharge	Very low	Mass cells, slightly higher than Dryfit; sinter cells, exceptionally high
Series connections	Permitted	Not allowed for small capacities, not recommended for high capacities
Parallel connections	Permitted	Permitted
Storage life when discharged	Weeks to over 1 year depending on temperature	Weeks to over 1 year depending on temperature
Charging temperature range	−30 to +50°C	Mass cells, 0 to 45°C; sinter cells, −30 to +50°C
Discharge temperature range	−45 to +50°C	Mass cells, −20 to +45°C; sinter cells, −40 to +50°C
Rapid charging temperature range	−30 to +50°C	Not recommended for mass cells; sinter cells, −30 to +50°C
Charging methods	Voltage limited	Current limited
Charging time	No time limit	Above I_{20} dependent on current
Shortest charging time	1 h approx.	Mass cells, 3 h with supplement voltage control; sinter cells, 15 min for part charge with voltage and temperature control
Chargers	Require voltage stabilization	Mass and sinter cells, no stabilization required for cyclic operation
Quick charge units – trickle charge	Normal, but with higher performance. Constant	Mass cells, permitted only up to 0.1 I_{20}; sinter cells constant but adequate current limit
Cyclic service life	Up to 500 deep discharge cycles and 1000 shallow discharge cycles	Mass cells, approx. 300 shallow discharge cycles; sinter cells, 500–1000 deep discharge cycles or up to 15 000 shallow discharge cycles
Service life under trickle charge	5 years approx.	Mass cells, 4–5 years at small current up to 0.1 I_{10}; sinter cells, approximately 10 years

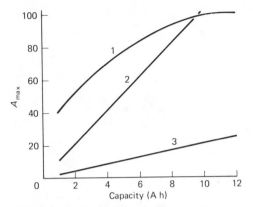

Figure 4.8 Maximum allowable permanent discharge currents: Dryfit and nickel–cadmium batteries with mass electrodes and sintered electrodes: 1, Dryfit; 2, nickel–cadmium with sintered electrodes; 3, nickel–cadmium with mass electrodes (Varta) (Courtesy of Dryfit)

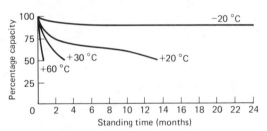

Figure 4.9 Self-discharge of sealed nickel–cadmium cells at different temperatures (Courtesy of Dryfit)

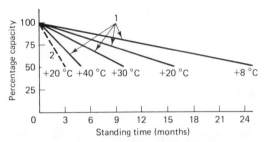

Figure 4.10 Self-discharge of Dryfit batteries at different temperatures compared with standard battery in lead–antimony alloy. 1, Dryfit battery (calcium–lead alloy); 2, conventional antimonial lead–acid battery (Courtesy of Dryfit)

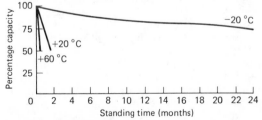

Figure 4.11 Self-discharge of sealed sintered electrode nickel–cadmium cells at different temperatures (Courtesy of Dryfit)

sealed lead–acid battery (16 months to lose half original capacity) – see Figure 4.10. The construction of the mass plate cell, whether button cells with single or double electrodes, or rectangular cells, makes practically no difference. The self-discharge of sinter cells, as shown in Figure 4.11, is many times greater. After 6 weeks at 20°C, half the capacity has already been lost. Compared to lead–acid batteries, longer storage of a deeply discharged (due to self-discharge) nickel–cadmium battery does not damage it, whereas such treatment of discharged lead–acid batteries results in an irreversible loss in capacity. However, the nickel–cadmium battery cannot be used for applications demanding immediate availability without continuous recharging and therefore increased maintenance costs.

Dryfit batteries may be charged and discharged over an ambient temperature range of −20 to +50°C. If continuous operation at one or more of the extreme values is required, a temperature sensor should be used to optimize charging. This can, however, be omitted if the battery is operated infrequently at extremes of temperature during the course of its life. The overcharging at high temperatures and the charging, which cannot be completed in its usual time at low temperatures because of the lack of temperature compensation, affect the quality of the battery to only a minor degree. Short-term temperatures up to 75°C can be tolerated. No precautions need be taken against the lower temperature of about −50°C, since the gelled electrolyte in the charged state cannot freeze solid. The nickel–cadmium battery also works at its best only if charging takes place at 20°C. Many types are severely limited by an allowable temperature range of 0 to 45°C for charging. Charging to a certain achievable percentage of full charge occurs in the freezing range and charging above 45°C is possible only for special types using sintered plate constructions. Compared to lead–acid batteries, charging at temperatures between 40 and 50°C is possible only at the expense of considerable loss of capacity. Mass electrode cells exhibit even more changes in voltage at extreme temperatures, especially low temperatures. Sinter electrode cells, despite their increased self-discharge, are preferable at extreme temperatures. Charging of sinter cells at low temperatures requires a reduction in the usual charging current, which in turn leads to increased charging times.

4.2 Nickel–iron secondary batteries

Nickel–iron batteries commercially available today are virtually the same as those developed by Thomas Edison approximately 70 years ago. The system offers very long cycle life but, primarily as the result of the iron electrode, a low energy density at high discharge rates has made this system unattractive for electric vehicles. Other problems associated with the iron electrode are related to a low hydrogen overvoltage potential resulting in poor charge efficiency, substantial self-discharge and poor utilization of the active material (approximately 25% of the theoretical capacity is achieved). However, as a result of abundant material, inexpensive design and potential high capacity, the iron electrode is experiencing a renaissance.

Recent reports have been very encouraging. A pilot production facility is reported which is producing nickel–iron batteries with energy densities greater than 44 W h/kg. The iron electrode is receiving particular interest in Europe and significant improvements in the performance of nickel–iron batteries have been reported. One group reports electrodes that yield capacities 2 to 3 times greater than that of commercially available iron electrodes, provide higher numbers of cycles and require an overcharge of only 15%.

Eagle Picher's experience with the nickel–iron system has been equally encouraging. They have 40 A h nickel–iron cells on cycle test with energy densities between 40 and 50 W h/kg.

Eagle Picher are active in the development of the nickel–iron battery system. Their development programme is primarily targeted for developing the system for use in electric vehicle propulsion. However, other applications are lift trucks, deep mining vehicles, off-peak utility storage, standby storage and solar. Eagle Picher, through a cooperative agreement with the Swedish National Development Company, has brought into the programme advanced iron electrode technology previously not available in the United States. The presently developed technology offers to the user a battery with high specific energy, low maintenance, deep discharge capability and long cycle life. Full-scale electrodes have exceeded 1300 deep discharge cycles. Full-scale cells as well as 6 V modules have been fabricated and show no capacity fade under 100% depth-of-discharge cycling. Energy efficiencies are typically 70% or greater (see Table 4.9).

Table 4.9 Nickel–iron battery performance

	1980 to present state of the art	Goals
Specific energy at battery level (W h/kg)	50	60
Energy density at battery level (W h/l)	102	120
Specific peak power at battery level (W/kg)	110	125
Deep discharge (100%) cycle life	1000	2000

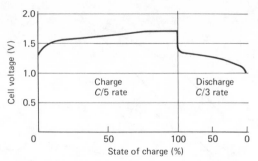

Figure 4.12 Nickel–iron cell charge/discharge profile (Courtesy of Eagle Picher)

Figure 4.12 shows the voltage profile for a single nickel–iron cell tested to the constant current modes as indicated.

The set of discharge curves shown in Figure 4.13 was taken from actual tests of a 280 A h cell. Note the retention of capacity at the higher discharge rates, indicating the good power characteristics of this type of battery. The 6 V modules shown in

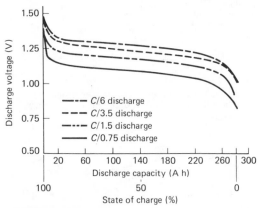

Figure 4.13 Nickel–iron cell constant current discharge profiles (Courtesy of Eagle Picher)

Figure 4.14 have been fabricated by Eagle Picher, and placed under test. Nominal capacities of these modules are 280 A h when discharged at the 3 h rate. Nominal operating voltage is 6.15 V. Capacities of these modules have been stable under 100% depth of discharge (DOD) cycling. Cycle life has presently reached 300 cycles as testing continues. Energy densities are 50 W h/kg and 102 W h/l respectively, with an energy efficiency of 70% or greater. A return factor of 110% has been established as the most optimum for the nickel–iron system with respect to required maintenance. Tests to date show no thermal problem to exist with the currently established design when the module is subjected to the expected vehicle driving profile.

The nickel–iron system may suffer the disadvantage of being a vented system, which will require

Figure 4.14 280 A h nickel–iron electric vehicle battery (Courtesy of Eagle Picher)

periodic maintenance. However, the cost advantage is significant in this system's favour. Combining a projected cost of $50–100/kW h with extremely long cycle life, this system could offer an economical type of battery for use in electric vehicles.

4.3 Nickel–zinc secondary batteries

With the development of new types of separator and zinc electrodes, the nickel–zinc battery has now become competitive with the more familiar battery systems. It has a good cycle life and has load–voltage characteristics higher than that of the silver–zinc system. The energy available per unit weight and volume are slightly lower than those of the silver–cadmium system. Good capacity retention (up to 6 months) has made the nickel–zinc battery a more direct competitor of the silver–zinc and silver–cadmium systems. Nickel–zinc batteries are available from only a limited number of suppliers (e.g. Medicharge, UK). Yardney have pioneered the development of the nickel–zinc couple for electric traction, and a number of road vehicles that are now operating are powered by Yardney nickel–zinc experimental batteries. If development work is successful, the nickel–zinc battery will be a useful contender for electric vehicle propulsion in that it will offer a significant performance improvement over lead–acid at a modest increase in price. Of particular importance in the consolidation of the design of this battery is the effect of charging on cell performance, on frequency of reconditioning and on battery cycle life. Certain techniques have been identified by Yardney that have a favourable effect on these parameters. The main characteristics of the nickel–hydrogen battery are listed in Table 4.10.

Eagle Picher offer a range of vented nickel–zinc batteries in the capacity range 2–35 A h. These are commercially available and are particularly recommended for field communication power supplies.

They are recommended for use whenever one or more of the following characteristics are required: high voltage (1.5–1.6 V working voltage per cell, the highest available in alkaline batteries), wide temperature tolerances (−39 to +81°C operating temperature, −53 to +73°C storage temperature), moderate cycle life (150 cycles maximum), and recycle flexibility. Eagle Picher claim that, with further improvements in design, coupled with a lowering of unit costs through mass production, the nickel–zinc battery holds promise of the lowest potential production cost per unit of energy supplied. The batteries have high-rate low-temperature capabilities, a flat discharge voltage for a large portion of the discharge, which is maintained over a wide current range, and can be left fully discharged for long periods without deterioration. These batteries are claimed to have an excellent shelf life even when wet and a charge retention of 60% for 30 days. Charging is non-critical and is carried out for 4 h at constant current until the voltage rises to 2.1 V, when charging is stopped. The batteries can be fully recharged with 10–15% overcharge.

4.4 Nickel–hydrogen secondary batteries

At the time of writing, no commercial design of these batteries exists. However, Eagle Picher, and no doubt other companies, are conducting long-term development on such designs, which it is hoped will become competitors to lead–acid and nickel–cadmium batteries available for non-space application, in addition to their already established space applications where cost has not been an important consideration. The basic nickel–hydrogen system consists of a catalytic gas electrode (negative electrode) coupled with the nickel electrode from the nickel–cadmium system (positive electrode). Electrochemically, the reactions at the positive electrode are the same as those occurring in the parent systems. At the negative electrode, hydrogen is displaced from water by electrolysis during charge. During discharge the reverse process occurs, namely the electrochemical discharge of hydrogen ions in solution.

Using current hermetically sealed aerospace nickel–cadmium systems as a cost comparison, it is projected by Eagle Picher that their present design concepts will lead to a reduction in nickel–hydrogen system cost from approximately 25% greater to approximately 50% less than current sealed nickel–cadmium system cost.

Development activities conducted over several years have demonstrated the potential superiority of the nickel–hydrogen system for long-life secondary battery applications. The attractiveness of the

Table 4.10 Characteristics of Yardney nickel–zinc batteries

Theoretical energy density	374 W h/kg
Practical energy density	44–77 W h/kg and 79–134 W h/dm^3 depending on cell model and conditions of use
Open-circuit cell voltage	1.75–1.8 V (i.e. the highest operating voltage of any of the alkaline batteries)
Nominal voltage under load	1.6 V
Voltage regulation	±5% at a fixed load and temperature limits within ±5°C
Recommended discharge cut-off voltage	1.0 V
Cycle life	Dependent on model and conditions of use; typical number of complete charge/discharge cycles obtainable:

Depth of discharge (%)	Approx. cycles
95–75	100–200
75–50	200–400
50–25	over 400

Wet life	Up to 3 years when manufacturer's recommended charge and discharge rates are used and operating and storage temperatures do not exceed 37°C
Dry storage life	Up to 5 years
Gassing on discharge or stand	Slight
Operational temperature range	+73 to −21°C; down to −53°C with heaters; for optimum cell performance, from +56 to −45°C
Storage temperature range	Wet, +37 to −47°C; dry +73 to −64°C
Operating attitude	In any position although, for optimum service, upright is recommended
Internal resistance	Very low; varies with cell model, temperature and rate of discharge
Resistance to mechanical stress	Excellent, extremely rugged, leakproof and spillproof. Construction similar to design of Yardney Silvercel batteries which have met the stringent requirements of Spec. MIL E-5272A. Can be packaged to meet the most severe requirements
Charging time	Can be fully recharged within 10–20 h, depending on requirements and type of cell
Charge retention	Up to 70% of nominal capacity after 6 months charged stand at room temperature

proposed nickel–hydrogen system stems from the following factors:

1. System design does not impose the acceptance of new technologies on the battery user. The basic nickel–hydrogen cell consists of the familiar alkaline electrolyte system using electrode and separator materials adapted from other established battery designs.
2. By the substitution of a troublesome negative electrode with an ultra-lightweight catalytic gas electrode, a system is created which offers higher energy density (55–70 W h/kg), longer cycle life, deep depth of discharge capability, and an inherent insensitivity to overcharge and overdischarge modes.

Generally, current 'long-life' rechargeable battery applications using either non-sealed or semi-sealed lead–acid and nickel–cadmium systems impose a significant weight penalty, require equipment accessibility design for periodic maintenance, and endanger surrounding equipment because of potential electrolyte leakage. True hermetically sealed systems are generally not practical for non-space applications because of difficulty in determining the system state of charge, particularly full state of charge. For safe operation, the hermetically sealed designs require a sophisticated charge-control system and, because of temperature sensitivity, must be operated under close environmental control.

The nickel–hydrogen system offers for the first time a true hermetically sealed design capable of thousands of maintenance-free cycles without the need for either complex charge control circuitry or close environmental control. In addition to these attributes, which render the system very attractive from an operational and energy density standpoint, the internal cell pressure offers a direct measure of the system state of charge. Using a pressure transducer technique, a simple and reliable charge control system can be designed.

Since the nickel–hydrogen cell exhibits a relatively high internal pressure during operation (3.5–4.0 kN/m^2, full state of charge), unusual cell container configurations are necessary to maintain lightweight designs.

5

Silver batteries

Contents

5.1 Silver oxide–zinc primary batteries

The weight and size of a silver–zinc battery or cell is less than half that of a lead–acid or nickel–cadmium type of similar capacity. The silver oxide–alkaline zinc (Ag_2O KOH Zn) primary battery is a major contribution to miniature power sources, and is well suited for hearing-aids, instruments, photoelectric exposure devices, electronic watches and as reference voltage sources.

Silver oxide batteries provide a higher voltage than mercuric oxide batteries. They offer a flat voltage characteristic and have good low-temperature characteristics. Their impedance is low and uniform. The silver oxide cell operates at 1.5 V (open circuit voltage 1.6 V) while mercury cells operate at about 1.35 or 1.4 V. Two major suppliers, Union Carbide and Mallory, supply silver–zinc button cells in capacity ranges between 35 and 210 mA h and 36 and 250 mA h respectively. Cylindrical and button cells are available from various suppliers. The impedance of silver oxide batteries for hearing-aid use is low and consistent. It does not rise appreciably until after the voltage of the battery has fallen below a useful operating level.

Silver oxide cells have excellent service maintenance; generally 90% after 1 year's storage at 21°C, and for up to 2 years without serious detriment. Cells will last up to 5 years in inactive sealed storage conditions.

There are numerous applications in the aerospace field, particularly where a high energy density battery is required which can offer both fast activation and a long wet-stand charged life. It was to meet this demand that Yardney Electric launched the primary PM Silvercel battery. This manually filled silver–zinc battery, which has a higher specific energy output than alkaline manganese, carbon–zinc or mercury types of battery, provides rapid activation and an activated stand time far longer than that of any other high-energy primary. Additionally, it offers a limited recyclability which allows performance testing before actual use. Silver–zinc batteries have surpassed the stringent mechanical requirements of missile applications and proved their ruggedness and reliability in the field. The following advantages are offered by this battery system over alternative systems: up to 75% higher capacity, or higher capacity with optimum life, or higher voltages at peak pulses. By carefully choosing the critical application requirement that must be met, whether it be capacity, life or voltage, and then optimizing these in relationship to each other, it is possible to satisfy the single most stringent demand. The characteristics of the Yardney PM cell are identical to those of the Yardney HR and LR Silvercel rechargeable secondary battery system with the exception that cycle life and wet life are somewhat limited.

Details of the types of silver–zinc batteries available from some suppliers are tabulated in Table 5.1.

5.1.1 Remotely activated silver oxide–zinc batteries

Silver–zinc primary batteries can be remotely activated for single use within seconds or fractions of a second, even after long-term storage, by inserting the electrolyte under pressure. In addition to a unit containing dry charged plates they also contain an

Table 5.1 Equivalents in silver–zinc cells using the silver oxide system

Nominal voltage (V)	Nominal capacity (mA h)	Height (mm)	Diameter (mm)	Duracell (Mallory)	IEC	National	Ray-o-Vac	Ucar	Varta
1.5	115	4.19	11.56	WS11	SR43	WS11	RW14, RW34	301	528
1.5	83	3.43	11.56	WS12					529
1.5	–	6.33	11.56	IOL14	SR44	WL14	RW22, RW42	357	541
1.5	85	3.48	11.56	IOL120					549
1.5	70	2.79	11.56	IOL122					
1.5	75	5.33	7.75	IOL123	SR48	WS6	RW18, RW28	309	546
							RW28, RW48	393	
1.5	120	4.19	11.56	IOL124	SR43	WL11			
1.5	38	3.58	7.75	IOL125	SR41	WS1/	RW24, RW44	386	548
						WL1	RW47	384	527
								392	547
1.5	250	4.83	15.49	IOL129			RW15, RW25	355	
							RW35		
1.5	200	5.59	11.56	WS14		WS14	RW12, RW32	303	521
1.5	180			MS76H				EPX77	7301
1.5	70			MS13H					7306
1.5	120			MS41H					7308

electrolyte vessel from which the cells are filled by electrical or mechanical means. Pile batteries are a new development of remotely activated primary batteries which confer a very high density on the battery.

The battery has no moving parts. It has an extended shelf life. It can be activated directly into a load or in parallel with a mains system. The open-circuit voltage of this battery is 1.6–1.87 V with a working voltage of 1.20–1.55 V. The battery operating temperature is −29 to +71°C and, with water assistance, −48 to +71°C. The energy density is 11–66 W h/kg or 24–213 W h/dm^3, depending on battery size. Units are available in the weight range 0.14–135 kg.

5.2 Silver–zinc secondary batteries

Work on the development of silver–zinc batteries began in 1942. This battery contains silver oxide as the positive electrode and zinc as the negative electrode and has an alkaline electrolyte. This combination results in what is, for alkaline batteries, a very high constant discharge voltage of approximately 1.8 or 1.5 V/cell respective to the two-step voltage discharge characteristic of silver–zinc batteries.

Rechargeable silver–zinc batteries can provide higher currents, more level voltage and up to six times greater ampere hour capacity per unit weight of battery. Thus, depending on the type of silver–zinc battery and by whom it was manufactured, one can obtain 70–120 W h/kg (as batteries) or 55–120 W h/kg (as cells) (theoretical value 440 W h/kg) for a silver–zinc battery compared to 30 W h/kg for a lead–acid battery and 25–30 W h/kg for a nickel–zinc or a nickel–cadmium battery (Figure 5.1). Similarly, the silver–zinc battery gives appreciably greater volumetric outputs (W h/dm^3)

than the other types of battery. Eagle Picher, for example, produce units with a practical volumetric energy density of 50–415 W h/dm^3 as cells and 55–262 W h/dm^3 as batteries. These are thus an attractive type of battery in circumstances where space availability is low or battery weight has to be kept low, and where the power output requirement is high. Because it can generate the same amount of electrical energy as other battery systems, in packages as little as one-fifth of their weight, the silver–zinc battery provides a long-sought answer to the design engineer's critical need for space and weight savings. Because it is capable of delivering high watt hour capacities at discharge rates less than 30 min, the silver–zinc battery is used extensively, for example, in missile and torpedo applications. Its high energy density makes it attractive in electronics applications, satellite and portable equipment where low weight, high performance and low cost are prime considerations. It is highly efficient and mechanically rugged, operates over a wide temperature range, and offers good shelf life; quick readiness for use and the ability to operate at temperatures down to −40°C without heating, and up to 73°C, are two of the features of this battery. The corresponding battery storage temperatures are −53°C and 73°C. Eagle Picher claim a dry shelf life of 2.5 years and a wet shelf life of up to 1 year at 21°C for their silver–zinc batteries. In excess of 500 cycles are claimed with a charge retention of up to 1 year at 26°C. Either constant-current or constant-potential (1.96–1.98 V) charging methods can be used and 16 h is a typical recharge time.

The extremely low internal resistance of silver–zinc batteries permits discharges at rates as high as 30 times the ampere hour capacity rating, and its flat voltage characteristic enables highest operational efficiency and dependability. Cells have been built with capacity ranging from 0.1 A h to 20 000 A h.

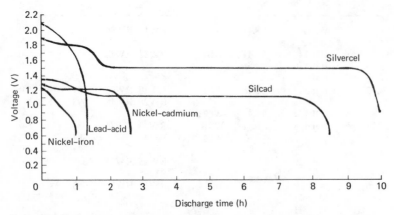

Figure 5.1 Comparison of discharge characteristics of a silver–zinc battery with other types (all the same weight) under the same conditions: Silvercel, Union Carbide silver–zinc; Silcad, Union Carbide silver–cadmium (Courtesy of Union Carbide)

The cells are leakproof and spillproof. The silver–zinc battery is available in high-rate and low-rate cells, i.e. it can be subject to high or low rates of discharge.

Rechargeable silver–zinc batteries are available as cells for discharges at high discharge rates (high-rate batteries) where the total energy must be delivered within 1 h or as low-rate versions with discharge times of 2 h or more. The latter type can be discharged in from 2 h down to a few minutes. The high-rate and low-rate systems differ principally in their electrode surface areas and the thickness of the separators between the plates, both of which have an effect on the internal resistance of the cell.

Silver–zinc batteries are also available in modular packs; one supplier (Yardney) supplies packs in normal capacities between 3 and 80 A h. These batteries have voltages between 2.8 and 15.8 V and equivalent energy densities of about 100 W h/kg, which exceeds the energy of a nickel–cadmium battery by four times and a lead–acid battery by five times. They are, of course, more expensive, but in many applications their characteristics make them an attractive alternative.

Until recently, the fact that the rechargeable silver–zinc battery has been more expensive and more sensitive to overcharge and has a shorter cyclic life than conventional storage batteries has limited it

Table 5.2 Characteristics of Yardney HR and LR Silvercell silver–zinc batteries

Chemical reaction	$Ag + Zn(OH)_2 = AgO + Zn + H_2O$ discharged charged
Energy output	20–120 W h/kg; as much as 220 W h/kg for certain types
Open-circuit voltage	1.86 ± 0.04 V (70–100% charged) 1.60 ± 0.04 V (30–70% charged)
Nominal voltage under load	1.50 V
Plateau voltage regulations	$\pm 2\%$ at a fixed load and temperature limits within $\pm 5°C$
Recommended discharge cut-off voltage	1.00 V
Recommended charge cut-off voltage	2.05 V

Cycle life (before output drops to 80% of nominal capacity – depends on model and conditions of use)	HR cells: 5–20 cycles (1 h rate)
	              10–15 cycles (30 min rate)
	               5–10 cycles (time-limited discharge)
	               1 cycle (6 min rate or higher, full-capacity discharge)
	LR cells: 80–100 cycles

Dry storage life	Up to 5 years
Operating (wet) life	HR cells: 6–12 months LR cells: 1–2 years
Gassing on discharge or stand	Very slight
Operational temperature range	+7 to $-23°C$ down to $-50°C$ with heaters; optimum cell performances at 51–10°C
Storage temperature range	Wet, 37 to $-48°C$; dry, 73 to $-65°C$
Operating attitude	In any position, although for optimum service upright is recommended
Internal resistance	Very low, varies with cell model, temperature and rate of discharge
Resistance to mechanical stress	Excellent; Silvercell batteries have met the stringent requirements of Spec. MIL-E-5272A and are currently used in missiles, rockets, torpedoes, etc.; they are extremely rugged, leakproof and spillproof, and can be packaged to meet the most severe requirements
Charging time	Can be fully recharged in 10–20 h, depending on requirements and type of cell
Charge completion	Indicated by sharp rise in cell voltage at 2.0 V/cell during charge
Charge retention	Up to 85% of nominal capacity after 3 months charged stand at room temperature

These data apply mainly to individual cells; when the cells are assembled to form batteries, performance will be determined by specific application and packaging requirements. HR = high-rate batteries; LR = low-rate batteries

to applications where space and weight are prime considerations. However, long-life silver–zinc batteries have now been developed in which up to 400 cycles over a period of 30 months can be achieved with ease.

The data below indicate, in brief, the performance capabilities achieved with silver–zinc cells supplied by Yardney showing the requirements which can be met:

1. *Vibration*: 30–40 sinusoidal vibration.
2. *Random Gaussian vibration*: 5–2000 Hz band, with as high as $60\,g$ r.m.s. equivalent in both non-operational and operational conditions; hard mounted unit.
3. *Shock*: tested up to $200\,g$ in all directions.
4. *Acceleration*: $100\,g$ in all directions of sustained acceleration, except in the direction of the vent for vented cell systems.
5. *Thermal shock*: from $+70$ to $-50°C$ and repeated, with stabilization occurring at each temperature.
6. *Altitude*: successful activation, and operation at a simulated altitude of 60 km, for a period of 1 h without pressurization; with pressurization there is no limit.

Table 5.2 gives the basic characteristics of the high-rate and low-rate versions of the Yardney Silvercel.

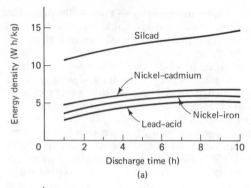

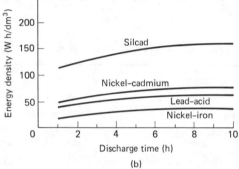

Figure 5.2 Comparison of energy density at various discharge rates: Silcad silver–cadmium batteries (Courtesy of Union Carbide)

5.3 Silver–cadmium secondary batteries

The silver–cadmium battery has a voltage of 1.4–1.1 V/cell (open circuit), 0.9–1.14 V/cell (closed circuit) and combines the high energy and excellent space and weight characteristics of the silver–zinc battery with the long life, low-rate characteristics and resistance to overcharge of the nickel–cadmium battery. Operating temperatures are between $-43°C$ and $75°C$. The silver–cadmium battery also provides high efficiency on extended shelf life in the charged or the uncharged condition, level voltage and mechanical ruggedness. Watt hour capacity per unit of weight and volume is two to three times greater than that of a comparable nickel–cadmium battery, and the silver–cadmium battery has superior charge retention. The principal characteristics of silver–zinc and silver–cadmium cells and batteries are compared and their important differences shown in Table 5.3. The silver–cadmium battery promises great weight and space savings and superior life characteristics to those of nickel–cadmium batteries currently in use as storage batteries in most satellite programmes. Yardney have reported 6000 cycles with a 50% depth of discharge in satellite applications.

Prior to the introduction of silver–cadmium batteries most available types of battery fell into the categories of low-energy long-life types or high-energy short-life types. Silver–cadmium batteries go a long way to combining the best features of both types giving two to three times more energy density per unit of weight and volume than the lead–acid, nickel–cadmium and nickel–iron systems, and at the same time offering ruggedness and long life (Figure 5.2).

Although the theoretical energy density of the silver–cadmium battery is 310 W h/kg, the actual energy density of practical batteries is in the range 24–73 W h/kg as cells and 18–44 W h/kg as batteries. The practical volumetric energy density is 36–171 W h/dm^3 as cells and 24–122 W h/dm^3 as batteries.

The low internal resistance of silver–cadmium batteries permits heavy drains and wide variations in load at virtually constant voltage. Leakproof, spillproof packaging with high impact strength helps ensure dependable operation under severe environmental and mechanical stress conditions. The extremely low internal resistance of silver–zinc batteries permits discharges at rates as high as 30 times the ampere hour capacity rating, and its flat voltage characteristic allows the highest operational efficiency and dependability. Cells have been built

Table 5.3 Comparison of characteristics of Yardney Silvercell and Silcad batteries

	Silvercel (silver–zinc)	Silcad (silver–cadmium)

Energy output
  W h/kg
  W h/dm³
	88–110
	150–190
		48–75
		90–160

Open-circuit voltage	1.82–1.86 V	1.40 ± 0.04 V (70–100% charged)
Nominal voltage under load	1.5 V	1.1 V
Plateau voltage regulation at fixed load and ±5°C	2%	5%

Cycle life (dependent on model and usage conditions); typical life to 80% of nominal	LR series:
	  80–100 cycles
	HR series:
	  1 h rate, 15–20 cycles
	  30 min rate, 10–15 cycles
	  10 min rate, 5–10 cycles
	  6 min rate, single cycles
		150–300 cycles (3000 cycles under ideal conditions)

Operating life; wet shelf life (varies with temperature and state of charge)	LR series: 1–2 years HR series: 6–12 months	2–3 years
Dry storage life	Up to 5 years	Up to 5 years
Operating temperature range	+73 to −21°C with heaters; optimum performance 10–51°C	+73 to −21°C with heaters; optimum performance 10–51°C
Storage temperature range	Wet: +37 to −47°C Dry: +73 to −64°C	Wet: +37 to −47°C Dry: +73 to −64°C
Operating attitude	Any position except inverted; for optimum service, use upright	Any position except inverted; for optimum service, use upright
Internal resistance	Very low, varies with cell model design, temperature and discharge rate	Very low, varies with cell model design, temperature and discharge rate
Resistance to mechanical sress	Excellent	Excellent
	Both types have met the stringent requirements of Spec. MIL-E-5272A and are currently used in missiles, rockets, torpedoes and Moon/space applications; extremely rugged, leakproof and spillproof, they can be packaged to meet the most severe requirements	
Charging times (dependent on cell design)	10–20 h	10–20 h
Charge completion	Sharp voltage rise at 2.0 V/cell	Sharp voltage rise at 1.55–1.6 V/cell
Charge retention	Up to 85% of nominal capacity after 3 months charged stand at room temperature	Up to 85% of nominal capacity after 3 months' charged stand at room temperature

with capacity ranging from 0.1 A h to 20 000 A h. The use of silver in silver–zinc and silver–cadmium cells obviously increases initial costs (although silver costs are recoverable) and introduces security problems when compared to other battery systems such as lead–acid, nickel–cadmium, etc. However, when space and weight are limiting factors, the silver–zinc system is a very attractive proposition.

The full range of characteristics of silver–cadmium batteries is given in Table 5.4.

These batteries are available in high and low rate versions (0.1–300 A h) and in modular packs (2–10 A h) from Yardney and other suppliers.

The major application for the high-rate series is for service conditions where maximum energy density and voltage regulation at high rates of discharge are required throughout a cycle life of 6–18 months. This system also has an exceptional ability to retain its charge during long periods of wet storage.

Table 5.4 Characteristics of Yardney Silcad silver–cadmium batteries

Chemical reaction	$Ag + Cd(OH)_2 = AgO + Cd + H_2O$ discharged $\qquad$ charged
Energy output	Theoretical 310 W h/kg; 48–75 W h/kg; 90–160 W h/dm^3
Open-circuit voltage	1.40 ± 0.04 V (70–100% charged) 1.16 ± 0.04 V (30–70% charged)
Nominal voltage under load	1.1 V
Plateau voltage regulations	$\pm 5\%$ at a fixed load and temperature limits within $\pm 5°C$
Recommended discharge cut-off voltage	0.60 V
Recommended charge cut-off voltage	1.65 V (indicated by sharp rise in cell voltage at 1.55–1.60 V/cell during charge)
Cycle life	Dependent on model and conditions of use; typical number of complete charge/discharge cycles obtainable: *Depth of discharge* (%) $\qquad$ *Approx. cycles* 95–75 $\qquad\qquad\qquad$ 200–300 75–50 $\qquad\qquad\qquad$ 300–600 50–25 $\qquad\qquad\qquad$ over 600
Wet life	Up to 3 years when manufacturer's recommended charge and discharge rates are used and operating and storage temperatures do not exceed 37°C
Dry storage life	Up to 5 years
Gassing on discharge or stand	Very slight
Operational temperature range	$+73$ to $-21°C$; down to $-53°C$ with heaters; for optimum cell performance, from 10 to 51°C
Storage temperature range	Wet, $+37$ to $-47°C$; dry $+73$ to $-64°C$
Operating attitude	In any position although, for optimum service, upright is recommended
Internal resistance	Very low; varies with cell model, temperature and rate of discharge
Resistance to mechanical stress	Excellent, extremely rugged, leakproof, spillproof. Construction similar to design of Yardney Silvercel batteries which meet requirements of Spec. MIL-E-5272A, and are currently used in missiles, rockets, torpedoes, etc. Can be packaged to meet the most severe requirements
Charging time	Can be fully recharged within 10–20 h, depending on requirements and type of cell (in some cases rapid charging possible)
Charge retention	Up to 85% of nominal capacity after 3 months charged stand at room temperature

The major application for the low-rate series is for service conditions where maximum energy density and voltage regulation at low rates of discharge are required throughout a cycle life of 12–36 months. The system also has an exceptional ability to retain its charge during long periods of wet storage.

Constant-current charging is recommended for silver–cadmium batteries, with a charge rate of one-tenth nominal capacity to 1.65 V/cell or 1.55 V/cell minimum. Alternatively, constant-potential charging can be adopted with an input of 1.65 ± 0.02 V. With constant-potential charging the current is limited to one-fifth of nominal capacity charging to the trickle charge rate.

5.4 Silver–hydrogen secondary batteries

Eagle Picher have also explored the possibilities of developing a silver–hydrogen battery using similar design concepts to those of their nickel–hydrogen system. The silver–hydrogen couple is believed to be capable of achieving energy densities in the range 90–110 W h/kg with maximum cell operating pressure of 2.75 kN/m^2 and an energy volume relationship of between 61 and 79 W h depending on cell design. The silver–hydrogen system is believed to be capable of 600 deep discharge cycles (80–90% depth of discharge).

6

Alkaline manganese batteries

Contents

6.1 Alkaline manganese primary batteries

Alkaline manganese batteries are a direct competitor to the standard carbon–zinc (Leclanché) battery over which, in certain applications, they enjoy advantages in performance. The primary alkaline manganese battery, which first became available in 1960, represents a major advance over the standard carbon–zinc battery and meets the growing need for a high-rate source of electrical energy.

Alkaline manganese batteries have a high depolarizing efficiency. On continuous or heavy drains they perform very well and have distinct advantages over the relatively cheaper carbon–zinc batteries, on a performance per unit cost basis. Both alkaline manganese and mercury batteries have a superior operating life to carbon–zinc batteries. For applications where voltage stability and extremely small size are not of prime importance, the alkaline manganese battery may be considered preferable to the more expensive mercury battery. As with the mercury battery, the alkaline manganese battery has a long shelf life, is self-venting and is of corrosion-free construction. Alkaline manganese batteries are capable of providing heavy currents for long periods, which makes them very suitable for applications requiring high surge currents, and can operate successfully at temperatures between −20 and +70°C; in certain cases even higher temperatures can be tolerated.

The voltage of an alkaline manganese cell is 1.5 V in standard N, AAA, AA, C and D cell sizes. Batteries are available with voltage up to 9 V and in a number of different service capacities. The closed-circuit voltage of an alkaline manganese battery falls gradually as the battery is discharged (Figure 6.1). This stability is superior to that obtained with a carbon–zinc battery but distinctly inferior to that obtained with a mercury–zinc battery.

Alkaline manganese batteries perform similarly to carbon–zinc batteries in that the service hours delivered are greater when the cut-off voltage is lower. The cut-off voltage should be made as low as possible so that the high energy density of the cell can be used. Service capacity remains relatively constant as the discharge schedule is varied. Capacity does not vary as much with current drain as for the carbon–zinc battery. Service capacity ranges from several hundred milliampere hours to up to tens of ampere hours depending on current drain and cut-off voltage.

The alkaline manganese battery is intended for applications requiring more power or longer life than can be obtained from carbon–zinc batteries. Alkaline manganese batteries contain 50–100% more total energy than a conventional carbon–zinc battery of the same size. Energy densities of alkaline manganese types are superior to those of carbon–zinc batteries of the same size. In a conventional carbon–zinc battery, heavy current drains and continuous or heavy-duty usage impair the efficiency to the extent that only a small fraction of the built-in energy can be removed. The chief advantage of the alkaline manganese battery lies in its ability to work with high efficiency under continuous or heavy-duty high-drain conditions where the carbon–zinc battery is unsatisfactory. Under some conditions, alkaline manganese batteries will provide as much as seven times the service of standard carbon–zinc batteries. Discharge characteristics of alkaline manganese and carbon–zinc batteries are compared in Figure 6.2

Although alkaline manganese batteries can outperform carbon–zinc batteries in any type of service, as indicated earlier, they may now show economic advantage over them at light drains, or under intermittent-duty conditions, or both. For example, with intermittent use at current drains below about 300 mA the D cell size alkaline type, while performing very well, will begin to lose its economic advantage over carbon–zinc batteries.

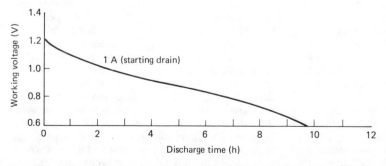

Figure 6.1 Voltage discharge characteristics of an alkaline manganese battery (D cell) discharged continuously (Courtesy of Union Carbide)

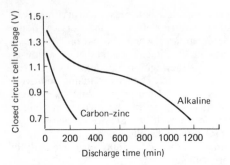

Figure 6.2 Comparison of discharge characteristics of alkaline manganese and carbon–zinc D-size cells for 500 mA starting, discharged continuously at 21°C (Courtesy of Union Carbide)

6.2 Alkaline manganese secondary batteries

The alkaline manganese battery is an example of a type of battery that, until quite recently, was available only in the non-rechargeable (primary) form but has now become available in a rechargeable (secondary) form. Certain manufacturers supply rechargeable alkaline manganese cells and batteries; for example, Union Carbide offer the Eveready rechargeable alkaline manganese battery range. These batteries use a unique electrochemical system, are maintenance free, hermetically sealed, and will operate in any position. They have been designed for electronic and electrical applications where low initial costs and low operating costs, compared with the cost of nickel–cadmium cells, are of paramount importance. Rechargeable alkaline manganese cells can also be used as an alternative to dry cells, since, although they have a higher initial cost, they are cheaper overall owing to their recycling property. They therefore bridge the gap between low initial cost batteries and the more expensive but recyclable secondaries. The alkaline manganese dioxide battery cannot be recharged as many times as the nickel–cadmium type, but the initial cost is only a fraction of the cost of an equivalent nickel–cadmium battery. Its operating cost is also normally lower.

Present types of Eveready battery available from Union Carbide include 4.5 and 7.5 V batteries using D size cells, and 6, 13.5 and 15 V batteries composed of G size cells. Specifications are listed in Table 6.1 and average performance characteristics are for the 4.5 V cell shown in Figure 6.3.

The discharge characteristics of the rechargeable alkaline manganese dioxide battery are similar to those exhibited by primary batteries; the battery voltage decreases slowly as energy is withdrawn from the battery. The shape of this discharge curve changes slightly as the battery is repeatedly dis-

Table 6.1 Characteristics of cells in Eveready rechargeable alkaline manganese dioxide batteries

Cell size	Voltage (V)	Average operating voltage (V)	Rates capacity (A h)	Max. recommended discharge current (A)
D	1.5	1.0–1.2	2.5	0.625
G	1.5	1.0–1.2	5.0	1.250

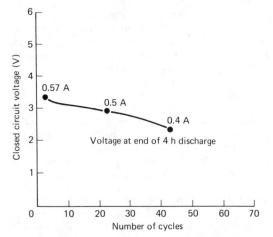

Figure 6.3 Average performance characteristics: voltage *versus* cycles at 21.1°C, Eveready rechargeable alkaline manganese dioxide batteries, 4.5–15 V. (Discharge cycle: 4 h through 5.76 Ω. Charge cycle: 16 h voltage taper current charging – voltage limit 5.25 V, filtered regulated power supply 1% regulation, with 2.4 Ω limiting resistor.) (Courtesy of Union Carbide)

charged and charged. The total voltage drop for a given energy withdrawal increases as the number of discharge/charge cycles increases. Coupled with this is the fact that the available energy per cell lessens with each discharge/charge cycle even though the open-circuit voltage remains quite constant; this is shown graphically in Figure 6.4. When a rechargeable battery of the alkaline manganese dioxide type is discharged at the maximum rate for a period of time to remove the rated ampere hour capacity and then recharged for the recommended period of time, the complete discharge/charge cycle can be repeated many times before the battery voltage will drop below 0.9 V/cell in any discharge period, depending on cell size. Decreasing either the discharge current or the total ampere hour withdrawal, or both, will increase the cycle life of the battery by a significant percentage. Conversely, if the power demands are increased to a point where they exceed the rated battery capacity, the cycle life will decrease more quickly than the increase in

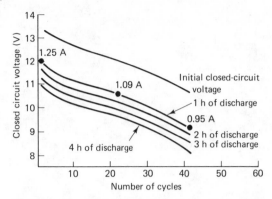

Figure 6.4 Changes in battery performance with every fifth cycle: Eveready 561 alkaline manganese dioxide battery on 9.6 Ω load for 4 h. (Discharge cycle: 4 h through 9.6 Ω. Charge cycle: 16 h voltage-limited taper current charging – voltage limit 17.5 V, filtered regulated power supply 1% regulation, with 4.0 Ω limiting resistor.) (Courtesy of Union Carbide)

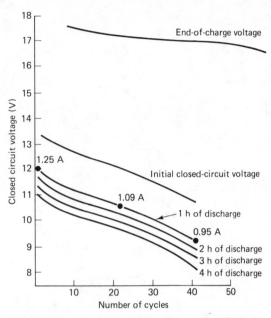

Figure 6.5 Typical voltage performance on voltage-limited taper current charge: Eveready 561 rechargeable 15 V alkaline manganese dioxide battery. (Charge and discharge conditions as for Figure 6.4.) (Courtesy of Union Carbide)

power demand. During the early part of its cycle life there is a very large power reserve in the Eveready rechargeable alkaline manganese dioxide battery. In these early cycles, the battery terminal voltages may measure 1.0–1.2 V/cell after the battery has delivered its rated ampere hour capacity. This reserve power can be used in situations where maximum total battery life can be sacrificed for immediate power. During the latter part of cycle life, there is little or no reserve power and the terminal voltage of the battery will fall to between 1.0 and 0.9 V/cell at the end of discharge. Figure 6.5 shows the maximum and minimum discharge voltages that the battery will reach on evenly spaced cycles throughout its cycle life; the curves show 561 battery performance on a 9.6 Ω load for 4 h.

Eveready rechargeable alkaline manganese dioxide batteries have excellent charge-retention properties. The new battery is fully charged and has the charge-retention characteristics of a primary battery. These characteristics also apply during the charge/discharge cycling of the battery. With this type of battery it is necessary to discharge it to its rated capacity before it will be capable of standing any overcharge.

7

Carbon–zinc and carbon–zinc chloride primary batteries

Contents

7.1 Carbon–zinc batteries

The standard carbon–zinc Leclanché battery is expected to remain one of the most widely used dry primary batteries in the future because of its low cost and reliable performance. This type of battery is manufactured in many styles by many companies – one company alone manufactures over 100 versions with voltages varying from 1.5 V to approximately 500 V and with various shapes, sizes and terminal arrangements.

The chemical efficiency of a carbon–zinc battery improves as current density decreases. This reveals an important application principle: consistent with physical limitations, use as large a battery as possible. Over a certain range of current density, service life may be tripled by halving the current drain. This is equivalent to using a larger battery for a given application and so reducing current density within the cells. This is true to a certain point beyond which shelf deterioration becomes an important factor.

The service capacity depends on the relative time of discharge and recuperation periods. The performance is normally better when the service is intermittent. Continuous use is not necessarily inefficient if the current drain is very light.

Figure 7.1 illustrates the service advantage to be obtained by proper selection of a battery for an application. The figure indicates how the rate of discharge and frequency of discharge affect the service efficiency of a battery. The energy–volume ratio of a battery using round cells is inherently poor because of voids occurring between the cells. This factor accounts for an improvement in energy–volume ratio of nearly 100% for flat cells compared with round cell assemblies. Two basic

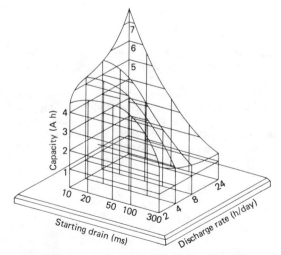

Figure 7.1 Battery service life as a function of initial current drain and duty cycle: D size carbon–zinc battery (Courtesy of Union Carbide)

types of carbon–zinc Leclanché cell are available; the SP type and the high-power HP type. The HP type is intended for applications where high current drains are involved; the SP type is recommended for low current drain applications.

There is no simple or rapid method for determining the service capacity of a dry battery. Tests must be run which closely duplicate the class of service for which the battery is intended. The schedule of operation is very important, except for very light drains. The service capacity of a battery used 2 h/day on a given drain will be considerably different from that of the same battery used 12 h/day. There is no relationship between continuous-duty service and intermittent service. It is therefore impossible to rate the merits of different batteries on intermittent service by comparing results of continuous-duty tests.

Another fallacy concerning dry batteries is that relative 'quality' or service capacity of a battery can be determined by amperage readings. This is not true, and in most instances gives results that are totally misleading. The D size photoflash round cell and the flashlight cell are identical in size and shape. However, the photoflash cell, which will show more than twice the amperage of the flashlight cells, has less service capacity in typical flashlight uses.

The short-circuit amperage of a zinc–carbon cell may be adjusted over a wide range by varying the carbon and electrolyte content of the depolarizing mix. Carbon contributes nothing to the service capacity of the zinc–carbon cell and is used primarily to control cell resistance. It is obvious that, as carbon is added to a cell, depolarizer must be removed; this means that service capacity is reduced.

Dry batteries can be tested with a loaded voltmeter to check their present condition. A meter test, however, will give no indication of remaining service capacity unless the exact history of the battery is known and can be compared on a capacity against meter-reading basis with other batteries tested in similar service. A loaded voltmeter is considered the best spot-check device, since open-circuit voltmeter readings give no indication of internal resistance, and a short-circuit amperage reading is damaging.

The internal resistance of cells becomes important when battery-operated devices require a high current for short periods of time. If the internal resistance of a cell is too high to provide the current, a larger cell may be used. The internal resistance of unused carbon–zinc cells is low and is usually negligible in most applications. Internal resistance may be measured by flash current or short-circuit amperage, which is defined as the maximum current, observed on a dead-beat (damped) ammeter, which a cell or battery can deliver through a circuit of 0.01 Ω resistance, including the ammeter.

Table 7.1 Approximate internal resistance of standard carbon–zinc and carbon–zinc chloride cylindrical cells

ANSI cell size	Average flash current (A)		Approx. internal resistance (Ω)	
	Standard carbon–zinc	Zinc chloride	Standard carbon–zinc	Zinc chloride
N	2.50	–	0.60	–
AAA	3.80	–	0.40	–
AA	5.30	4.50	0.28	0.33
C	3.90	6.50	0.39	0.23
D	5.60	8.50	0.27	0.18
F	9.00	11.30	0.17	0.13
G	12.00	–	0.13	–
6	30.00	–	0.05	–

Amperage readings are not necessarily related to service capacity. Amperage is usually higher in large cells.

The internal resistance increases with storage time, use and decreasing temperature. The cell dries out with age. During discharge some of the ingredients are converted to different chemical forms which have higher resistance. Table 7.1 lists the approximate internal resistance, as determined by the flash current, of several round cells of the carbon–zinc and carbon–zinc chloride types.

Carbon–zinc cells (1.5 V) come in a variety of standard sizes covered by the ANSI designation; the type numbers of 1.5 V cells available from one particular supplier (Union Carbide) under each ANSI designation are listed in Table 7.2.

7.2 Carbon–zinc chloride batteries

The zinc chloride cell is a special modification of the carbon–zinc cell. It differs principally in the electrolyte systems used. The electrolyte in a zinc chloride cell contains only zinc chloride, while in a carbon–zinc cell the electrolyte contains a saturated solution of ammonium chloride in addition to zinc chloride. The omission of ammonium chloride in the zinc chloride cell improves the electrochemistry of the cells but places greater demands on the cell seal.

Zinc chloride cells are able to operate at higher electrode efficiencies than carbon–zinc cells and hence have a higher useful current output and will operate at higher current drains than carbon–zinc cells of the same size. In addition, the voltage lever

Table 7.2 Type numbers of Union Carbide carbon–zinc Leclanché dry cells available under various ANSI designations

Union Carbide type number	ANSI designation	Voltage (V)	Suggested current range (A)
HS6, IF6, IS6, IS6T, EA6F, EA6, EA6FT, EA6ST	6	1.5	0–1.50
HS150, HS50, 11T0, 8 500 001, T50, 1050, 950, 1250	D	1.5	0–1.15
HS35, 935, T35, 835, 1035, 1235	C	1.5	0–0.08
HS15, 815, 915, 1015, 1215	AA	1.5	0–0.025
904	N	1.5	0–0.02
201	WO	1.5	0–0.06
812, 912	AAA	1.5	0–0.02
E340E	N	1.5	0–0.02

under load holds up longer into the discharge. Zinc chloride cells also perform better at low temperatures.

The internal resistance of zinc chloride (and carbon–zinc) cells increases with storage time, use and decreasing temperature. Table 7.1 compares the approximate internal resistance, as determined by flash currents of several types of round cell of the carbon–zinc and zinc chloride types.

Mercury batteries

Contents

8.1 Mercury–zinc primary batteries

Mercury batteries have an appreciably higher energy-to-weight ratio than carbon–zinc batteries, resulting from the high energy density of the materials used in their construction. Thus mercury batteries are only one-third the size of conventional dry batteries of the same capacity.

Characteristics and features of the mercuric oxide electrochemical system include the following:

1. Long life.
2. High capacity-to-volume ratio, resulting in several times the capacity of alkaline manganese and carbon–zinc cells in the same volume, or proportionally reduced volume for the same capacity.
3. Flat discharge characteristic.
4. Higher sustained voltage under load.
5. Relatively constant ampere hour capacity.
6. Low and substantially constant internal impedance.
7. No recuperation required, therefore the same capacity is obtained in either intermittent or continuous usage.
8. Good high-temperature characteristics.
9. High resistance to shock, vibration and acceleration.
10. Electrically welded or pressure intercell connections.
11. Single or double steel case encapsulation.
12. Chemical balance – all the zinc is converted at the end of the battery's life.
13. Automatic vent.
14. Resistance to vacuum and pressure.
15. Resistance to corrosive atmospheres and high relative humidity.

The ampere hour capacity of mercury cells and batteries is relatively unchanged with variation of discharge schedule and, to some extent, with variation of discharge current. They have a relatively flat discharge characteristic (see Figure 8.1; compare this with Figure 6.1 for alkaline manganese types). The mercury system withstands both continuous and intermittent discharge with relatively constant ampere hour output and this allows the capacity rating to be specified. Rest periods are not required as in the case of carbon–zinc batteries.

Mercuric oxide batteries are available with two formulations designed for different field usage. In general, the 1.35 V cells (that is, 100% mercuric oxide depolarization) or batteries using these cells are recommended for voltage reference sources (that is, high degree of voltage stability) and for use in applications where higher than normal temperatures may be encountered (and for instrumentation and scientific applications). The 1.4 V cells (mixture of manganese dioxide and mercuric oxide depolarization) or batteries using these cells are used for all other commercial applications. The 1.4 V cells or batteries should be used for long-term continuous low-drain applications if a very flat voltage characteristic is not needed.

Although Union Carbide and Mallory are two of the major producers of mercury–zinc cells and batteries, there are, of course, other important producers such as Ever Ready (Berec) (UK), Crompton-Parkinson (Hawker Siddeley) (UK) and Varta (West Germany). Table 8.1 gives the type numbers of equivalent cells that meet a given International Electrochemical Commission (IEC) designation as produced by Union Carbide, Mallory, Ever Ready (Berec) and Varta. In many instances equivalent cells are available from more than one manufacturer. In such instances comparative cost quotations would be of interest to the intending purchaser.

8.2 Mercury–indium–bismuth and mercury–cadmium primary batteries

Mercury–indium–bismuth and mercury–cadmium batteries are available from some manufacturers (e.g. Crompton-Parkinson). These are alkaline systems recommended for applications where high reliability in particularly onerous long-term storage and use conditions is a prime requirement.

For example, the particular advantage of mercury–cadmium is that it will show good performance

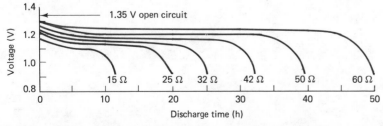

Figure 8.1 Typical voltage discharge characteristics of mercury–zinc cells under continuous load conditions at 21°C. At 1.25 V, equivalent current drains for the resistances are: 15 Ω, 83 mA; 25 Ω, 50 mA; 32 Ω, 40 mA; 42 Ω, 30 mA; 50 Ω, 25 mA; 60 Ω, 20 mA (Courtesy of Union Carbide)

Table 8.1 Some equivalent mercury–zinc cells from different manufacturers

IEC designation	Union Carbide Eveready No.	Voltage (V)	Capacity (mAh)	Mallory Duracell No.	Voltage (V)	Capacity (mAh)	Varta No.	Voltage (V)	Ever Ready (Berec) No.	Voltage (V)
MR41	325	1.35	50							
MR08	E400N	1.35	80							
	E400	1.40	80							
NR48	E138, E13E	1.40	85	RM13H	1.40	85	7106	1.40	RM13H	1.40
MR48	323	1.35	95	WH6	1.35	95				
NR08	343	1.35	120	RM575H	1.40	100	7109	1.40	RM575H	1.40
MR42				WH12NM	1.35	110				
NR43				RM41	1.40	150				
MR43	354	1.35	150	PX675	1.35	210	7001	1.35	MP6754	1.40
MR07	EPX675	1.35	220	MP675H	1.40	210	7103			
NR07				SP675	1.40	230				
				WH3	1.35	220				
MR44	313	1.35	220							
MR9	E625	1.40	350	PX625	1.35	250	7002	1.35	PX625	1.35
	E630	1.40	350	RM625N	1.35	350			RM625N	1.35
	E625N	1.35	250	PX640	1.35	500				
	EPX625	1.35	250	RM640H	1.40	500				
	EPX13	1.35	250							
MR52	E401N	1.35	800							
NR52	E401, E401N	1.40	800							
MR1										
MR50	EPX1	1.35	1000	RMIN	1.35	1000			RMIN	1.35
MR7	E1	1.40	1000	PX1	1.35	1000			PX1	1.35
NR50				RM1	1.40	1000				
NR1				MP401	1.40	1000				
MR17	E3	1.35	2200							
NR6	E502E	1.40	2400	ZM9C	1.40	2400			ZM9C	1.40
MR6	E9	1.40	2500							
	E502	1.35	2400							
MR51				RM12R	1.35	3600	7101	1.40	RM1H	1.40
				RM675H	1.40		7102	1.40	RM401H	1.40
				RM625			7107	1.40		
				RM312H			7108	1.40		
				RM41H			7201	1.50		
				PX825						

at low temperatures even below −20°C, where most other primary systems just would not function. However, as the temperature drops, the operating voltage also drops as expected. Away from these very low temperatures, mercury–indium shows a better power-to-volume ratio than mercury–cadmium, and still gives exceptional performance in difficult environments. Both these systems also have the advantage of level voltage discharge characteristics. Typical applications include reactor and high-temperature processing, telemetry systems and military uses.

In mercury–cadmium batteries, the battery contains a cadmium anode (similar to that used in a nickel–cadmium cell), a mercuric oxide cathode, and a potassium hydroxide electrolyte. Batteries are fabricated in prismatic or flat button designs. They have an open-circuit voltage of approximately 0.9 V and an operating voltage between 0.75 and 0.9 V depending on the discharge load. Energy densities for mercury–cadmium systems range between 44 and 66 W h/kg depending on discharge rate, which is lower than that of a mercury–zinc cell. The cells is characterized by good stability because both electrodes exhibit a high degree of insolubility in the electrolyte. Resulting excellent storage characteristics (5–10 years) are its main advantage. These cells also have inherently better high-temperature storage than other primary batteries. Operating characteristics of mercury–cadmium cells are shown in Figure 8.2. The low voltage and high cost of materials are the main drawbacks of the system as a primary battery.

Mallory also supply a mercury–cadmium battery in their Duracell range. This battery (No. 304116) is of 12.6 V with a capacity of 630 mA h to an end-point of 10.3 V, or 900 MA h to an end-point of 7.0 V, both on 13 000 Ω at 20°C. This cylindrical battery weighs 127 g and occupies a volume of

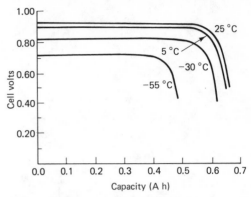

Figure 8.2 Operating characteristics for 0.6 A h mercury–cadmium cell at various temperatures. Hourly discharge rate is 2% of capacity, i.e. *C*/50 (Courtesy of Crompton-Parkinson)

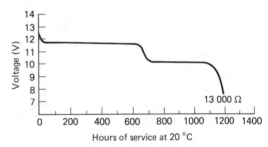

Figure 8.3 Mallory No. 304116, 12.6 V mercury–cadmium battery; two-stage discharge curve. Current drain at 11.5 V: 13 kΩ, 0.884 mA (Courtesy of Mallory)

40.5 cm^3 and is recommended for use in applications such as alarm systems using voltage-sensitive low-battery indicators. Typical two-step discharge characteristics for this battery are shown in Figure 8.3.

9

Lithium batteries

Contents

9.1 Introduction

There are many types of lithium cell either under commercial consideration or being commercially produced. A list of representative types of lithium batteries with their characteristics is given in Table 9.1. A look at these characteristics should help engineers evaluating batteries for equipment use.

Lithium-based batteries are quickly entering mainstream electronic designs, particularly in consumer, portable equipment and non-volatile memory backup applications where small size, long life and low cost are the primary requirements. Specifically, watches, pacemakers, Kodak's disc camera and Mostek's dual-in-line packaged 16K CMOS RAM (with on-board battery backup) are good examples of devices making use of the newer lithium battery technologies.

The practicality of lithium is being greatly enhanced by the current progress in CMOS technology. Indeed, lithium batteries can easily provide standby power for the current crop of 'pure' CMOS devices (i.e. those with complete p- and n-channel pairing throughout), which operate at nanoampere current levels. As low-power CMOS device technology advances – gaining a greater share of semiconductor applications along the way – lithium battery use will undoubtedly expand to serve these new areas.

The electrochemical potential of lithium-based power sources has been known for a long time. Advances in materials processing, materials handling and non-aqueous electrolytes in the 1960s have enabled manufacturers to reduce this potential to practical cell hardware and demonstrate the benefits that were theoretically possible.

During the past few years, it has become clear that the need for greater power from hermetically sealed cells could not be met with the mercury, silver or alkaline manganese dioxide systems. To fill the need for a higher energy density system at a reasonable cost, various manufacturers have developed practical organic electrolyte lithium systems. The lithium battery offers gravimetric energy densities of up to 330 W h/kg, nearly three times that of mercury and silver, and four times that of alkaline manganese. The volumetric energy density is 50% greater than that of mercury batteries and 100% greater than that of alkaline manganese batteries. Lithium cells offer the facility of reducing size and weight in existing applications, and making new lighter weight designs possible. In addition, the excellent shelf life offers new possibilities for designers.

Three principal types of lithium organic electrolyte battery are currently available; the lithium–thionyl chloride system, the lithium–vanadium pentoxide system and the lithium–sulphur dioxide system. These batteries all have high-rate capabilities. The approximate open-circuit equilibrium cell voltages for these various cathode systems and for some other systems that have been considered are shown in Table 9.2.

The energy density superiority of lithium systems is shown graphically in Figure 9.1. The three lithium systems shown represent actual performance achieved by battery manufacturers. The gains in energy density seen in the data shown in Figures 9.2 and 9.3 can be attributed in part to the light weight

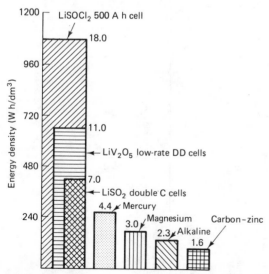

Figure 9.1 Comparison of energy density of lithium cells and other types of cell (Courtesy of Honeywell)

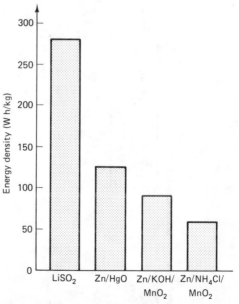

Figure 9.2 Comparison of energy density of lithium–sulphur and other types of cell (Courtesy of Honeywell)

Table 9.1 Representative types of lithium battery

Chemistry	Electrolyte	Initial voltage (V)	Discharge curve	Supplier	Part no.	Capacity (mA h)	Current	Operating temperature (°C)	Shape and size (mm)*	Price (100 qty) ($)‡
Lithium–iodine	Solid	2.8	Sloping	Catalyst Research Catalyst Research Medtronics	1935 2736 Custom†	200 350 NA	Low μA Low μA NA	−55 to +70 −55 to +125 NA	Coin 19 × 3.5 Coin 25 × 3.6 Custom	6.88 5.62 NA
Lithium–carbon monofluoride	Organic	2.7	Flat	Panasonic Ray-o-Vac Ray-o-Vac	BR2/3A BR-1225 BR-2325	1200 35 160	To 1 A Low μA Low μA	−40 to +85 −40 to +85 −40 to +85	2/3A Coin 12 × 2.5 Coin 23 × 2.5	1.00 1.50 1.50
Lithium–manganese dioxide	Organic	3.0	Sloping	SAFT SAFT Duracell Duracell	LM 2020 LM 2425 DL 2430 DL 2/3A	90 200 200 1	To 10 mA pulse To 15 mA pulse Low μA 100 mA	−20 to +50 −20 to +50 0 to +40 0 to +40	Cylinder 20 × 2.0 Coin 24 × 2.5 NA 2/3A	1.30 1.30 1.00 3.50
Lithium–sulphur dioxide	Organic	2.8	Flat	SAFT Duracell	LX 16/34 LO 26SX	1000 8500	1 to 100 mA To 1 A	−40 to +55 −40 to +52	Cylinder 16 × 34 D	5.60 10.00
Lithium–thionyl chloride	Inorganic liquid	3.5	Flat	SAFT Union Carbide Tadiran Tadiran	LS6 L31 TL 2150 TL 2100	1700 1000 850 1750	50 mA 40 mA 100 mA 100 mA	−40 to +71 −60 to +75 −55 to +75 −55 to +75	Cylinder 50 × 13 Cylinder 46 × 13 1/2AA AA	8.05 1.46 NA NA
Lithium–copper oxide	Organic	2.0	Flat (at 1.5 V)	SAFT	LC6	3400	1 mA	−55 to +125	AA	1.30

NA, not available
* Size of batteries can be decoded from part numbering
† Medtronics specializes in custom shapes
‡ Most batteries available off the shelf

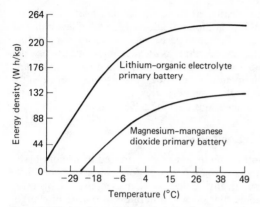

Figure 9.3 Comparison of energy density of organic *versus* magnesium batteries at various temperatures at nominal 30 h rate (Courtesy of Honeywell)

Table 9.2 Open-circuit equilibrium cell voltages for various cathode systems

System	Formula	Cell voltage (V)
Thionyl chloride	$SOCl_2$	3.6
Vanadium pentoxide	V_2O_5	3.4
Sulphur dioxide	SO_2	2.9
Molybdenum trioxide	MoO_3	2.9
Copper fluoride	CuF_2	3.4
Silver chromate	Ag_2CrO_4	3.0
Copper sulphide	CuS	2.2

Table 9.3 Weights of various standard size D-cells

System	Weight for D-cells (g)
Lithium–sulphur dioxide	94
Carbon–zinc	94
Magnesium perchlorate–manganese dioxide	99
Lithium–thionyl chloride	100
Lithium–vanadium pentoxide	128
Alkaline manganese dioxide	128

of active components used in lithium cells. The volumetric energy density of lithium systems is, however, not always so greatly different from that of aqueous systems. The higher gravimetric energy density of the lithium system is demonstrated by the comparison of weights of standard size D-cells of various systems given in Table 9.3.

Typical voltage characteristics on medium load for a lithium type (lithium–sulphur dioxide) battery is compared with those for other types of primary battery in Figure 9.4. The outstandingly higher voltage of the lithium system is apparent.

The construction of organic lithium cells is somewhat different from that of mercury and alkaline manganese cells. A lithium foil anode, a separator and a carbonaceous cathode are spirally wound together. This assembly is placed in a steel case, and the anode and cathode are connected with welded tabs to the case and top assembly. Since the electrolyte is non-aqueous, there is no hydrogen gas evolved during discharge. The assembly contains a vent to prevent the build-up of high internal gas pressure resulting from improper use or disposal.

Lithium organic cells offer higher energy density, superior cold temperature performance, longer active life and greater cost effectiveness.

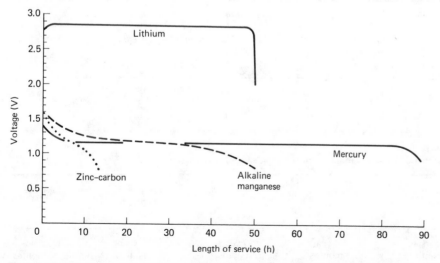

Figure 9.4 Typical voltage characteristics on medium load of lithium and other types of battery (Courtesy of Honeywell)

9.1.1 Higher energy density

The higher energy density is attributable, to a large extent, to the operating voltage of most lithium-based electrochemical couples. Lithium, having the highest potential of metals in the electromotive series, provides an operating voltage of about twice that of traditional systems. The voltage discharge profiles of the systems mentioned above are shown in Figure 9.5.

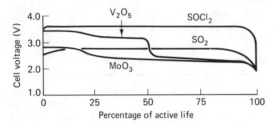

Figure 9.5 Voltage discharge profile of lithium–vanadium pentoxide (V_2O_5), lithium–thionyl chloride ($SOCl_2$) and lithium–sulphur dioxide (SO_2) and lithium–molybdenum trioxide (MoO_2) cells (Courtesy of Honeywell)

9.1.2 Superior cold temperature performance

Because of the non-aqueous nature of the electrolytes used in lithium systems, the conductivity of these systems at cold temperatures is far superior to that of previously available systems. Table 9.4 compares the relative performance of two lithium systems with the cold temperature performance of other systems. The lithium–vanadium pentoxide and lithium–sulphur dioxide data shown are from

tests conducted on practical hardware configurations and at moderate discharge rates. The numbers given reflect the percentage of room temperature performance that is achievable at the colder temperatures. Lithium–vanadium pentoxide and lithium–sulphur dioxide systems will operate satisfactorily at temperatures as low as −55°C with efficiencies approaching 50%.

9.1.3 Long active shelf life

Another significant benefit offered by lithium-based systems is their potential for long active shelf life. Hermetically sealed cells, made possible by using systems that do not generate gas during discharge, protect the cell from impurities from external environments and prevent leakage of electrolyte from the cell. This, in addition to the absence of self-discharge reactions, or the low rate at which they occur, gives lithium systems in active primary batteries the potential for 5–10 years' shelf life without providing special storage environments. Also, the discharge products of most lithium systems are such that they do not contribute to, or increase the rate of, self-discharge. Therefore, lithium cells can also be reliably used intermittently over several years in applications where it is advantageous or required. A comparison of the projected shelf life for various cell systems is shown in Table 9.5. In this comparison, acceptable shelf life is defined as the time after which a cell will still deliver 75% of its original capacity. Hermetically sealed lithium–sulphur dioxide double-C cells after storage for 180 days at 71°C deliver 88% of their fresh cell capacity when discharged at room temperature into an 8 Ω

Table 9.4 Cold temperature performance of lithium–organic electrolyte systems

Temperature (°C)	Percentage of room temperature performance					
	Lithium–vanadium pentoxide	Lithium–sulphur dioxide	Mercury	Magnesium	Alkaline	Carbon–zinc
−7	88	96	0	58	15	5
−29	78	85	0	23	3	0
−40	73	60	0	0	0	0

Table 9.5 Comparison of projected shelf life for various lithium–organic electrolyte cell systems

Storage temperature (°C)	Lithium	Mercury	Magnesium	Alkaline manganese dioxide	Carbon–zinc
21	10 years +	3–4 years	5–7 years	2–3 years	1–2 years
54	12 months +	4 months	7 months	2 months	1.5 months

load. By some standards 180 days' storage at 71°C is approximately equivalent to over 12 years at room temperature. This would imply that these cells would exhibit about 1% annual degradation at room temperature. This has been confirmed by discharge of non-hermetically sealed Honeywell lithium–sulphur dioxide double-C cells which were stored for approximately 1 year at 27°C. After storage these cells exhibited no losses in performance under identical test conditions, based on lot acceptance test data generated with fresh cells.

9.1.4 Cost effectiveness

In high-volume production the lithium-based systems will compete directly with mercury and silver systems on a cost per watt hour basis. Cost comparisons with alkaline systems will be marginal on a per watt hour basis, but cost advantages can be realized by considering the cost advantages of other systems with that achievable with a lithium power source. The increases in overall system cost effectiveness can be achieved as follows:

1. More payload may be possible when using lithium batteries due to their smaller size and/or lighter weight.
2. More operating life will be achievable with the same size but lighter battery.
3. Performance at colder temperatures will be possible.
4. Longer active life will result in reduced maintenance costs associated with battery replacements, especially in remote locations, or in other applications where replacing the battery is a labour-intensive function.
5. Replacement inventory costs will be reduced because periodic replacement of batteries will be significantly decreased or eliminated.
6. No special storage provisions are required.

Even though cell costs may be high initially, people responsible for maximizing a system's overall cost effectiveness must look at the total life-cycle tests and take these points into consideration. As well as lithium–vanadium pentoxide, lithium–sulphur dioxide and lithium–thionyl chloride systems, there exist other types of cell containing organic electrolytes and lithium. For example, the lithium–copper fluoride couple has a theoretical voltage of 3.5 V and a theoretical energy-to-weight ratio of 1575 W h/kg. Approximately 330 W h/kg have been obtained in experimental cells at the 100 h rate using this couple in a propylene carbonate electrolyte. Cells consisting of lithium–cupric chloride have operated down to −40°C and give 100 W h/kg at high discharge rates (approximately 10 h rate). However, both of these systems have poor activated shelf life and are currently practical only in reserve structures.

There are in fact many other types of lithium cell at present under commercial consideration, and a range of lithium cells that have been considered by the Catalyst Research Corporation for use for watch and calculator applications is discussed below.

Because the introduction of lithium power sources to the electronics industry is so recent, many potential users are not aware that lithium batteries are not all alike. Lithium is only the first name of any lithium power source. Just as there are many zinc batteries available (zinc–carbon, zinc–silver oxide, zinc–mercuric oxide) there are many varieties of lithium system, each with its own peculiar internal chemistry and construction. Several of these systems are briefly described below.

Table 9.6 shows the lithium systems available for commercial use from Catalyst Research Corporation. Most are button cells approximately 20–30 mm in diameter and 1.5–3 mm thick. All are high-voltage cells with moderate or high internal resistances. The first two have seen use in watches

Table 9.6 Types of lithium cell available from Catalyst Research Corporation

System	Energy density (W h/dm^3)	Seal	Voltage (V)	Electrolyte	Volume change	Self-discharge	Internal resistance	Separator added
Lithium–manganese dioxide	440	Crimp	2.9	Liquid organic	?	?	Moderate	Yes
Lithium–carbon fluoride	330	Crimp	2.8	Liquid organic	?	Moderate	Moderate	Yes
Lithium–thionyl chloride	500	Hermetic	3.6	Liquid inorganic	Small shrinkage	?	Moderate	Yes
Lithium–lead iodide	470	Hermetic	1.9	Solid salt	?	Low	High	Yes
Lithium–iodine P2VP	600	Hermetic	2.8	Solid salt	None	<5% in 10 years	High	No; forms *in situ*

or calculators only; the remaining three have also been used in the pacemaker field.

Like any common batteries, lithium batteries will rupture if exposed to fire. The low-rate lithium batteries, intended for watches, should be safe if used witthin manufacturers' specified temperatures. Thick separators in these low-rate cells prevent shorting and their small size permits easy heat dissipation if any local internal reactions should occur. In fact, a good case can be made that most low-rate lithium cells are safer than zinc–mercury cells, which can introduce poisonous mercury into the atmosphere when incinerated. SAFT supply lithium–copper oxide and lithium–silver chromate cells. Matsushita supply lithium–carbon fluoride cells. Venture Technology (formerly Ever Ready (Berec)), Ray-o-Vac and Sayo (Japan) supply lithium–manganese dioxide cells. A further type of cell which has reached commercial realization is the Venture Technology lithium–ferrous sulphide cell (1.5 V/cell nominal). It has a volumetric energy density of 500–900 W h/dm^3 and has an excellent shelf and working life and very good low-temperature performance, with similar applications to the lithium–manganese dioxide cell.

9.2 Lithium–sulphur dioxide primary batteries

Of all the lithium battery systems developed over the last decade, the system based on sulphur dioxide and an organic solvent is now acknowledged to have emerged as the most successful both commercially and technically. Other lithium batteries are capable of delivering as great or greater energy densities, in particular the lithium–thionyl chloride systems. However, the latter system may give rise to spontaneous decomposition of explosive violence in high-rate batteries, while the other lithium systems do not have the high-rate capabilities of the lithium–sulphur dioxide battery.

The reactivity of lithium necessitates controlled-atmosphere assembly during manufacture, and in some cases the use of expensive materials in the cell construction to avoid corrosion and the provision of a sophisticated seal design.

Although lithium batteries have high-rate discharge capability, their use at very high rates or accidental shorting could result in temperatures leading to seal failure or explosion. Manufacturers incorporate vents and/or fuses to minimize these risks.

Honeywell Inc. and the Mallory Battery Company in the USA have introduced lithium batteries based on the lithium–sulphur dioxide electrochemical couple. The positive active material in these batteries, liquid sulphur dioxide, is dissolved in an electrolyte of lithium bromide, acetonitrile and propylene carbonate, and is reduced at a porous carbon electrode.

This type of battery has a spiral-wound electrode pack, made from rectangular foil electrodes. Lithium foil is rolled on to an expanded metal mesh current collector as the negative electrode, and is separated from the similarly supported cathode by a polypropylene separator. Two types of cell construction are used: jelly-roll electrodes in crimp-sealed or hermetically sealed cylindrical cells, and large 20–100 A h 12 V flat-plate electrodes in large reserve batteries. It is a relatively high-pressure system and cells must have safety vents to avoid explosion in the event of accidental incineration (see Part 2 for further details of construction).

The lithium–sulphur dioxide cell has an open circuit voltage of 2.92 V at 20°C and a typical voltage under rated load of 2.7 V compared with 1.5 V for most conventional types of battery. The specific voltage on discharge is dependent on the discharge rate, discharge temperature and state of charge. The end or cut-off voltage is 2 V. The lithium system is capable of maintaining more stable voltage plateaux at higher currents than any other cell of comparable size.

It is also claimed that the lithium–sulphur dioxide system operates very efficiently over a wide range of temperatures (typically from −40 to +70°C), achieving higher discharge rates at lower temperatures than are possible with other types of cell, which provide little service below −18°C. The cells can be operated with success at elevated temperatures. When operated at very high currents outside the recommended limits, the cells may produce heat and high internal gas pressures, and these conditions should be avoided. The cells are, however, an excellent source for high pulse currents. Multicell batteries contain additional insulation as well as a fuse for protection from mishandling such as short-circuits.

A typical energy density of a lithium–sulphur dioxide cell is 420 W h/dm^3 or 260 W h/kg according to one manufacturer and 330 W h/kg and 525 W h/dm^3 according to another. These are nearly three times the values expected for mercury–zinc cells, four times that of an alkaline manganese dioxide cell and two to four times higher than that of conventional zinc and magnesium type batteries.

The high volumetric energy densities reflect the high voltages of the lithium-based systems. One reason for some lack of acceptance in miniature applications is that although one lithium cell could be specified where it is necessary to use two mercury cells in series, a lithium button cell would have a capacity approximately one-half that of the equivalent mercury cell, and the frequency of battery charging would in extreme cases be correspondingly increased.

The volumetric ampere hour capacity of mercuric oxide–zinc cells is higher than that of lithium-based systems. However, in many cases, using two lithium cells in parallel or one larger lithium cell will give the same ampere hour capacity as can be achieved in an equal or even smaller volume by an equivalent two-cell series mercury–zinc battery of similar voltage. This is illustrated in Table 9.7. One

Table 9.7 Comparison of lithium–sulphur dioxide and mercuric oxide–zinc cells

	Lithium–sulphur dioxide	Mercuric oxide–zinc
Volumetric energy density (W h/dm^3)	420	500 (at best)
Normal working voltage (V)	2.75	1.25
Volumetric capacity (A h/dm)	153	400
Relative cell volume (per A h)	2.6	1.0

lithium–sulphur dioxide cell (voltage 2.75 V) occupies about 30% more space than two series mercuric oxide–zinc cells (voltage 2.5 V). Admittedly, this compares the worst cited case for lithium against the best for mercuric oxide–zinc. Higher energy density systems such as lithium–vanadium pentoxide and lithium–sulphur dioxide would show significant volume savings over an equivalent ampere hour mercuric oxide–zinc system. In fact lithium–sulphur dioxide systems are being increasingly considered for high-rate miniature power source applications including military applications where it is found that a two-cell mercuric oxide–zinc battery can occupy considerably more space than the equivalent lithium–sulphur dioxide cell.

The lithium–sulphur dioxide system is versatile and relatively inexpensive. This battery has excellent storage characteristics. Honeywell claim that the batteries should store for 12 years at 20°C. The battery can be supplied either as reserve batteries with capacities between 20 and 100 A h or as active batteries in the 0.7–20 A h range (see Table 56.2).

The lithium–sulphur dioxide battery has a high power density and is capable of delivering its energy at high current or power levels, well beyond the capability of conventional primary batteries. It also has a flat discharge characteristic.

Lithium–sulphur dioxide batteries are subject to the phenomenon known as voltage delay, a characteristic shared with lithium–vanadium pentoxide and lithium–thionyl chloride batteries. After extended long-term storage at elevated temperatures, the cell may exhibit a delay in reaching its operating

voltage (above 2.0 V) when placed on discharge, especially at high current loads and at low temperatures. This start-up delay is basically caused by film formed on the anode, the characteristic responsible for the excellent shelf life of the cell. The voltage delay is minimal for discharge temperatures above −30°C. No delay is measurable for discharge at 21°C even after storage at 71°C for 1 year. On −30°C discharges, the delay (time to 2 V) was less than 200 ms after 8 weeks' storage at 71°C on discharges below the 40 h rate. At higher rates, the voltage delay increased with an increase in storage temperature and time. However, even at the 2 h discharge rate, the maximum start-up time is 80 s after 8 weeks' storage at 71°C. After 2 weeks' storage, the start-up time is only 7 s. The start-up voltage delay can be eliminated by preconditioning with a short discharge at a higher rate until the opening voltage is reached, since the delay will return only after another extended storage period.

As mentioned above, the lithium–sulphur dioxide system has emerged as the leading candidate among the high energy density batteries for high-rate applications. Other lithium batteries are capable of delivering as great or greater energy densities, in particular the lithium–thionyl chloride systems (see below). However, the latter system may give rise to spontaneous decomposition of explosive violence in high-rate batteries while the other lithium systems do not have the high-rate capabilities of those using soluble depolarizers.

Lithium–sulphur dioxide cells are available in a variety of cylindrical cell sizes from companies such as Honeywell and Duracell International, capacities available ranging from 0.45 to 21 A h. Larger cells are under development. A number of the cells are manufactured in standard ANSI (American National Standards Institute) cell sizes in dimensions of popular conventional zinc primary cells. While these single cells may be physically interchangeable, they are not electrically interchangeable because of the high cell voltage of the lithium cell.

Standard and high-rate cells are available. The standard cell is optimized to deliver high-energy output over a wide range of discharge loads and temperatures. The high-rate cell is designed with longer and thinner electrodes than the standard cell and delivers more service at a high discharge rate (higher than the 10 h rate) and at low temperatures. At lower discharge rates, the service life of the high-rate cell is less than that delivered by the standard cell.

In addition, Duracell manufacture a lithium limited (or balanced) cell (designated SX). The cell is designed with a stoichiometric ratio of lithium to sulphur dioxide in the order of 1:1 rather than the excess of lithium used in the other designs. The lithium-limited feature ensures the presence of sulphur dioxide throughout the life of the cell to protect the lithium from chemically reacting with the other cell components. This design has been found successfully to withstand extended reverse discharge below 0 V at rated loads. In addition, these cells do not produce the toxic chemicals that form when standard cells are fully discharged, thus simplifying disposal procedures. The lithium-limited cell does, however, deliver lower capacity at low discharge rates, compared with the standard cell (below the 5 h rate).

9.3 Lithium–thionyl chloride primary batteries

This system uses a lithium anode and a gaseous cathode dissolved in an inorganic electrolyte. It has a 3.63 V open-circuit voltage and a typical voltage under rated load of 3.2–3.4 V. Like the lithium–sulphur dioxide system, it has a very flat discharge profile through 90% of its life. Cell construction is similar to that of the lithium–sulphur dioxide cell, except that hermetically sealed cells are mandatory at present. The lithium–thionyl chloride system is a very low-pressure system and, because of that, it is potentially superior to lithium–sulphur dioxide systems in high-temperature and/or unusual form factor applications. The cells are manufactured without any initial internal gas pressure and, because the discharge reaction generates only a limited amount of gas, the need for venting is eliminated. The system appears to be safe in low-rate cell designs, and may be safe, if properly vented, in high-rate cell designs; however, there is an insufficient database on the system (particularly in the high-rate configuration) to make that claim with a high degree of confidence. One manufacturer claims to supply cells that have an energy density in excess of $1100 \, W \, h/dm^3$ and $660 \, W \, h/kg$. Another manufacturer claims an energy density of $800 \, W \, h/dm^3$, compared to $400 \, W \, h/dm^3$ for zinc–mercury, $200 \, W \, h/dm^3$ for zinc–carbon and $300 \, W \, h/dm^3$ for alkaline manganese dioxide (the corresponding $W \, h/kg$ data for the four types of cell are 420, 100, 80 and 100).

Cells operate between −55 and +71°C, and a shelf life of up to 10 years can be expected due to the negligibly low self-discharge rate of these cells. Earlier versions of this battery exhibited severe passivation of the lithium anode, which severely limited shelf life. For example, at discharge current density as low as $0.6 \, mA/cm^2$, significant initial voltage drop and voltage delay were observed at 25°C after storage periods as short as 1 week at 72°C. The primary cause of the voltage drop is the formation of a film, which results in excessive anode passivation. This becomes evident on closed circuit as a sharp initial voltage drop and a long recovery before the voltage stabilizes (that is, voltage delay).

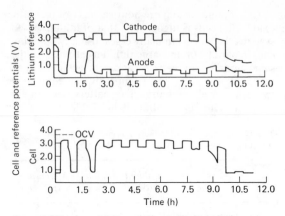

Figure 9.6 Honeywell lithium–thionyl chloride cell: polarization and voltage delay of a lithium–thionyl chloride cell at 24°C after 3 weeks' storage at 24°C; loads 120 mA (30 min) to 45 mA (30 min); electrolyte 1.5M LiAlCl₄ (commercial grade) SOCl₂ (Courtesy of Honeywell)

This is clearly illustrated in Figure 9.6, which shows the strong anode polarization and initial cell voltage drop followed by a slow recovery to a useful cell voltage. Investigations by workers at Honeywell have shown that the passivation film on the anode produced in the lithium–aluminium chloride electrolyte consists of lithium chloride. They have also shown that the formation of this film can be prevented by the inclusion of 5% sulphur dioxide in the electrolyte. It is significant that effective control of lithium passivation appears to be critically dependent on the sulphur dioxide concentration. As shown in Figure 9.7, discharge performance after storage can be adversely affected at greater than 5 weight % level. At these higher sulphur dioxide levels after 2 weeks at 74°C, cells were found to be anode limited and showed severe polarization under a 120 mA load when they were discharged at −29°C.

Reserve and active type cells are available in the capacity range of less than 1 to 17 000 A h.

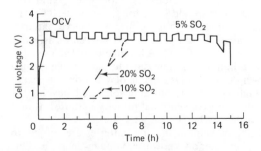

Figure 9.7 Discharge performance of a Honeywell lithium–thionyl chloride cell at 24°C after 4 months' storage at 74°C; loads 120 mA (30 min) to 45 mA (30 min); nominal fresh cell discharge time 16 h; electrolyte 1.5M LiAlCl₄.SOCl₂ 5, 10 and 20 wt% SO₂ (Courtesy of Honeywell)

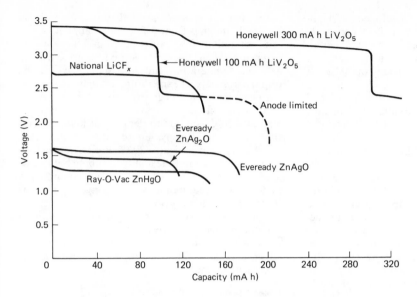

Figure 9.8 Button cell comparison: Honeywell lithium *versus* other systems (Courtesy of Honeywell)

System	Diameter (mm)	Height (mm)	Rate	Nominal voltage (V)	Capacity	
					mA h	mW h
ZnAgO	12	4	65K	1.5	170	255
ZnHgO	12	4	65K	1.35	150	203
ZnAg$_2$O	12	4	65K	1.5	120	180
LiV$_2$O$_5$	28	3	5K	3.3	100	330
LiV$_2$O$_5$	28	5	5K	3.3	300	990
LiCF$_x$	22	2.5	5K	2.65	140	370

9.4 Lithium–vanadium pentoxide primary batteries

This system utilizes the lithium anode, a carbon–vanadium pentoxide cathode and a double-salt metal fluoride electrolyte (lithium hexafluoroarsenate dissolved in methyl formate). It has a unique two-plateau discharge profile (see Figure 9.8) of approximately 3.4 V for the first 50% of life and 2.4 V for the last 50%. When both plateaux are used it offers an energy density (660 W h/dm^3, 264 W h/kg) intermediate between that of the lithium–sulphur dioxide and the lithium–thionyl chloride systems. It is a relatively low-pressure system and low-rate cells using it need not be vented. It is ideal for those applications in which safety is of the utmost importance. This battery has excellent storage characteristics (Figure 9.9).

The voltage discharge profile of the lithium–vanadium pentoxide battery is compared with that of lithium–thionyl chloride and lithium–sulphur dioxide systems in Figure 9.5. Lithium–vanadium pentoxide systems operate satisfactorily at tempera-

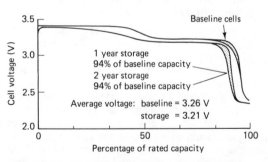

Figure 9.9 Honeywell lithium–vanadium pentoxide cell: discharge performance after 1 and 2 years' active storage at 24°C, current density ≈0.4 mA/cm^2 (Courtesy of Honeywell)

tures as low as −55°C with efficiencies approaching 50% (Table 9.3).

Honeywell can supply small lithium–vanadium pentoxide cells with capacities up to 100 mA h, glass ampoule reserve cells with capacities up to 500 mA h, cylindrical, button, prismatic and flexible cell configurations, and a range of active cells.

Honeywell have recently developed a high-integrity crimp-sealed lithium–vanadium pentoxide button cell. This has an open-circuit voltage of 3.42 V and a voltage under load of more than 3 V. The G3093 model has a rated capacity of 300 mA h and the G3094 of 100 mA h, both under the rated load. The energy density is 180–300 W h/dm^3 and 132 W h/kg. The cells are 29.08 mm in diameter, 25–50 mm thick and weigh 6–8 g. The operational temperature range is −29 to +49°C and they have a projected storage capability at 23°C of more than 20 years.

Development work carried out by Honeywell has shown that although a solution of lithium hexafluoroarsenate dissolved in methyl formate at a concentration of 2 mol/l is one of the most conductive organic electrolyte solutions known, at high temperatures it decomposes to produce gases including carbon monoxide and dimethyl ether. This instability becomes apparent in active cells through case swelling caused by a rise in internal pressure, which is often accompanied by self-discharge due to cell distortion. In reserve cells instability becomes evident by premature cell activation caused by rupture of the glass electrolyte ampoule.

As a result of their investigation work, Honeywell concluded that electrolyte decomposition at elevated operating temperatures in lithium–vanadium pentoxide cells could be considerably reduced if the electrolyte were made basic. Thus an electrolyte with the composition 2M lithium hexafluoroarsenate (LiAsF$_6$) plus 0.4M lithium borofluoride (LiBF$_4$) is now used in these cells.

9.5 Lithium–manganese dioxide primary batteries

This is claimed to be a reliable high-density miniature power source with a long shelf life and good low-temperature performance, which is safe, leakproof and non-corrosive. The lithium–manganese dioxide battery is a 3 V system combining a lithium anode and a manganese dioxide cathode in a lithium perchlorate electrolyte. The electrolyte is dissolved in an organic solvent (a mixture of propylene carbonate and dimethoxyethane), and the system is completely non-aqueous. The problem of gas evolution due to dissociation of water has now been solved and lithium–manganese dioxide cells will not bulge during storage or under normal operating conditions. The system offers a stable voltage, starting at approximately 3.3 V, and may be considered fully discharged at a cut-off voltage of 2 V. The high voltage is supported by the high energy density associated with lithium, making the system attractive as a substitute for high-energy silver oxide in 3 V and 6 V photographic applications. The energy output of the lithium cell is up to ten times that of a zinc alkaline cell.

Lithium–manganese dioxide batteries are suitable for loads ranging from a few microamps to a few tens of milliamps, with potential for upward extension. The cells may be stored for up to 6 years at room temperature and still retain 85% of the original capacity. Temperature excursions to 70°C are permissible, although extensive storage beyond 55°C is discouraged. The system will work between 50 and −20°C, subject to derating at low temperature. In some designs, up to 70% of the rated capacity is delivered at −20°C. Properties include good storage life and the ability to supply both pulse loads and very low currents. This combination matches the requirements for many applications incorporating microprocessors.

Table 9.8 compares the energy density of Duracell lithium–manganese dioxide button and cylindrical cells with those of conventional mercury–zinc, silver–zinc and zinc–alkaline manganese dioxide and carbon–zinc cells.

Table 9.8 Comparison of energy density of lithium–manganese dioxide cells with conventional types

Cell type	Energy density	
	W h/dm^3	W h/kg
Button cells (low-rate, C/200 rate)		
LiMnO$_2$	610	225
ZnHgO	425	92
ZnAgO	535	135
Cylindrical cells, N size (moderate-rate, 100 mA discharge)		
LiMnO$_2$	400	215
Zinc–alkaline manganese dioxide	180	63
Carbon–zinc	60	32

Lithium–manganese dioxide cells are manufactured in a variety of button cell and cylindrical cell forms ranging in capacity from 30 to 1400 mA. Larger capacities are under development by Duracell. Ratings are C/200 h rate for low-rate cells, and C/30 h rate for high-rate and cylindrical cells. In some instances, interchangeability with other battery systems is provided by doubling the size of the cell to accommodate the 3 V output of the lithium–manganese dioxide cell compared to the 1.5 V of the conventional primary cell.

9.6 Lithium–copper oxide primary batteries

SAFT supply this type of battery. The particular advantages claimed for lithium–copper oxide batteries are long operating life, long shelf life (up to 10 years projected) and high operating temperature (tested between −20 and +50°C). Volumetric capacity (A h/dm^3) is 750 compared with 300 for alkaline manganese dioxide, 400 for mercury–zinc and 500 for lithium–sulphur dioxide. (Corresponding W h/kg values are 300, 100, 100 and 300.) The annual self-discharge rate at 20°C is 2–3%. Battery packs containing groups of cells connected in series/parallel are available for specific applications.

Cylindrical 1.5 V batteries are available in 3.6 A h (type LC01), 1.6 A h (type LC02) and 0.5 A h (type LC07). Applications of this type of cell include oil well logging equipment, microprocessors, telephone systems, public address systems, high-temperature heat counters, CMOS memory protection, telex systems, high-temperature devices and specialized industrial and military applications.

9.7 Lithium–silver chromate primary batteries

SAFT supply a range of these 3.1 V cells with capacities in the range 130–3100 mA h (cylindrical cells) and 2090–2450 mA h (rectangular cells), see Part 6.

The special features of these cells are high reliability (better than 0.7×10^{-8}), indication of end of life by a second plateau voltage, a discharge capability up to 100 µA, and low internal impedance. Applications of lithium–silver chromate cells include cardiac stimulators and other human implantable devices, high reliability memory protection for inertial navigation systems, and RAM backup power sources.

9.8 Lithium–lead bismuthate primary cells

SAFT supply this type of cell. It has a nominal voltage of 1.5 V, exhibits a high energy density (400–500 W h/dm^3, 90–150 W h/kg), has long life expectancy and is interchangeable with normal IEC button cells with a nominal voltage of 1.5 V. The cells possess a low internal resistance which makes them particularly suitable for analogue quartz watches and pulsing electronic devices. The characteristics of these cells make them interchangeable alternatives in the many applications currently using the traditional silver oxide and mercury oxide button cells. The annual self-discharge rate of lithium–lead bismuthate button cells is 2–3%. The operating temperature range is −10 to +45°C. The cells have a discharge pulse capability up to 500 µA. Applications include, in addition to quartz analogues and LCD watches, cameras, calculators and scientific instruments and any applications in which they replace silver oxide button cells. SAFT supply these 1.5 V button cells at three capacities: 185 mA h (type LP1154), 110 mA h (type LP1136) and 50 mA h (type LP1121).

9.9 Lithium–polycarbon monofluoride primary batteries

In order to exploit the value of a lithium-based system to the maximum the positive electrode (cathode) material should also be of high energy density. The search for the ideal combination of cathode material and electrolyte attracted a great deal of effort in the 1970s. The lithium–polycarbon monofluoride system has been commercially developed by the Matsushita Electric Industrial Co. in Japan. The cells are available in several cylindrical sizes. They have an open circuit voltage of 2.8 V. The patented cathode material is of the form $(CF_x)_n$, where x has a value between 0.5 and 1.0, and is formed by reacting carbon with fluorine under various conditions of temperature and pressure, depending on the type of carbon used as the starting material. Except where batteries are intended for low-rate applications, acetylene black or graphite is added to the electrode to improve conductivity. The electrolyte is lithium tetrafluoroborate dissolved in α-butyrolactone. These cells have a spiral wound electrode pack made from rectangular foil electrodes. Lithium foil is rolled on to an expanded metal mesh current collector as the negative electrode and is separated from the similarly supported cathode by a polypropylene separator.

9.10 Lithium solid electrolyte primary batteries

Duracell supply these batteries, which consist basically of a 2 V system capable of supplying currents up to a few hundred microamps at room temperature. The electrolyte is a dry mixture of lithium iodide, activated alumina and lithium hydroxide sandwiched between the lithium anode and the lead/lead iodide/lead sulphide cathode. The absence of liquid means that the solid electrolyte cell is intrinsically leakproof. The chemistry is exceptionally stable, and it is possible to accommodate discharge temperatures up to 125°C without degradation in performance. Similarly, the cell may be stored at temperatures up to 200°C without serious losses. Perhaps the most interesting quality is the 15–20 years' projected storage life at room temperature. This is compatible with the design life of professional equipment containing microprocessors, making it possible to fit the cell as a permanent component.

The solid electrolyte cell relies on ionic conduction in the solid state. This is a low-rate process where temperature is an important variable. Discharge efficiency on maximum load is particularly affected and a 350 mA h cell is typically rated at 1 µA at room temperature – well below its maximum capability. This reduces its dependence on temperature and guarantees a high discharge efficiency over a wide range of temperatures. The closely matched temperature characteristics of solid electrolyte batteries and CMOS logic circuits is another property which makes the solid electrolyte battery ideally suited for memory-retention applications. The major advantages of the solid electrolyte battery are as follows:

1. *Virtually unlimited shelf life* The Duracell solid electrolyte battery has a projected shelf life in excess of 20 years under normal storage conditions and is capable of extended storage at temperatures as high as 120°C.
2. *Wide operating temperature range* Solid electrolyte cells can operate from 40 to over 120°C. Operation at higher temperatures is possible with modified designs. The current capability is a function of temperature. At 95°C the current capability is 10–20 times the room temperature performance; however, at −40°C it is only 2–3% of that at room temperature.
3. *High energy density* A volumetric energy density of 300–500 W h/dm^3 is superior to most conventional battery systems.
4. *High voltage density* The thin cell structure and high cell voltage (1.9 V) gives a high voltage-to-height ratio.
5. *No gassing, corrosion or leakage* The use of solid cell components and the absence of chemical reactions eliminates gassing and leakage.
6. *Hermetic, leakproof design* Only one 'seal' is required per battery.
7. *Safety* Neither short-circuit nor voltage reversal causes pressure build-up or chemical reaction.

Solid electrolyte batteries are currently available in a button or circular disc configuration, with a nominal 25.4 mm diameter and rated at 350 mA h. Table 9.9 summarizes the major physical and electrical characteristics of these batteries.

Table 9.9 Duracell solid electrolyte batteries

	Duracell type no.	
	305127	*305159*
Nominal voltage (V)	2.0	4.0
No. of cells	1	2
Rated capacity* (mA h)	350	350
Dimensions		
Diameter (mm)	28.9 ± 0.13	29.7 ± 0.13
Height (mm)	2.54 ± 0.25	5.8 ± 1.8
Volume (cm^3)	1.44	4.04
Weight (g)	7.25	15.85

* Rated at 1 μA discharge at 21°C

Duracell solid electrolyte batteries are designed primarily for low-power, long-service-life applications, and should be used in accordance with the manufacturers' specifications. Although such conditions should be avoided, these cells can withstand short-circuit or voltage reversal. No explosion due to pressure build-up or chemical reaction can occur. Prolonged short-circuiting will in fact result in a separation between the electrode and the electrolyte, rendering the cell inoperative.

The following handling procedures and precautions are recommended:

1. The solid electrolyte battery can be discharged at temperatures up to 125°C and stored, with little loss in capacity, at temperatures up to 200°C. However, for best performance, excessively high temperatures should be avoided. Temperatures above 200°C may cause bulging and failure of the hermetic seal. The cell should not be incinerated unless suitable procedures are followed and appropriate precautions taken at the disposal site.
2. High-impedance voltmeters must be used for measurement, preferably with $10^9 \Omega$ or higher input impedances. Batteries with open circuit voltages of 10 V or less may be checked rapidly using meters with $10^7 \Omega$ input impedance. Prolonged measurements using such instruments will result in gradual voltage decrease with time due to battery polarization.
3. Care must be exercised to prevent shunting of battery terminals by the human body (fingers, hands, etc.). Such shunting can correspond to a load resistance of the order of $10^5 \Omega$, which will load down the battery accordingly. The higher the battery voltage, the greater the transient effect of such handling and the longer the time subsequently required for recovery to full open-circuit voltage level. No permanent damage occurs in such accidental short-duration shunting.
4. Batteries must be protected against ambient high-humidity environments, which might pass through dew point resulting in moisture condensation across the battery terminals. Such condensation would electrically shunt, shorting out the battery.
5. The battery is a relatively high-impedance device, so short-circuit currents are of relatively low magnitude. However, discharge of this system depends on slow solid-state diffusion phenomena. Too rapid discharge, such as shunting, involves such rapid lithium solution at the anode to electrolyte interface that vacancies may be produced irreversibly. Extended shorting, for example periods of the order of 30 min, can render a cell completely inoperable.
6. Care must be exercised in handling, lead attachment, etc., to prevent mechanical damage

to hermetic glass seal feedthroughs. Loss of hermeticity will ultimately make the battery completely inoperable.

7. Cells should not be opened, crushed, punctured or otherwise mutilated.

9.11 Lithium–iodine primary batteries

9.11.1 Pacemaker batteries

Since 1972, the Catalyst Research Corporation (USA) has been the supplier of the lithium–iodine batteries intended initially for cardiac pacemakers and other implantable devices. They supply 70% of pacemaker batteries used in the world. They claim that the capacity of this battery is four times that of the Mallory RM-I mercury–zinc pacemaker battery and that the lithium–iodine battery operates more efficiently at body temperature than at room temperature. Since the lithium–iodine reaction generates no gas, the cell can be hermetically sealed. A feature of this type of battery is its extremely high reliability. At the time of writing, Catalyst Research had supplied 150 000 batteries for use in cardiac pacemakers without a single failure or premature rundown. Catalyst Research Corporation lithium–iodine batteries are marketed in the UK by Mine Safety Appliances Ltd.

The lithium–iodine battery differs from other lithium batteries in the following respects:

1. Lithium–iodine batteries contain no liquid electrolyte.
2. Lithium–iodine batteries do not require a separator material to keep the anode from contacting the cathode.
3. The total volume of the cell remains the same during discharge; there is neither swelling nor shrinking of the cell during discharge.
4. The cell is hermetically sealed with a metal-to-metal fusion weld and glass-to-metal seals.
5. No gas is produced by the reactions and no gas is given off by the cell.

A full specification of the Catalyst Research Corporation current pacemaker battery model 802/23 is given in Table 9.10. The projected electrical performance over an 18–20 year period is shown in Figure 9.10.

9.11.2 Non-medical lithium–iodine batteries

Figure 56.26 gives performance characteristics of the Catalyst Research Corporation S23P-1.5 cell. The largest cell available in this range, the D27P-20, with a capacity of 0.260 A h, would be useful for long-life liquid crystal display applications or in analogue watches with stopping motors. It should be ideal for

Table 9.10 Characteristics of Catalyst Research Corporation model 802/23 solid lithium–iodine pacemaker battery

Nominal size	$23 \times 45 \times 13.5$ mm
Volume	11.2 cm^3
Weight	30 g
Density	2.7 g/cm^3
Lithium area	17.1 cm^2
Voltage	2.80 V under no load
Nominal capacity	2.3 A h
Energy	6.0 W h
Energy density	530 W h/dm^3, 200 W h/kg
Self-discharge	<10% in 10 years
Seal	Heliarc welded with glass–metal hermetic seals; less than 4.6×10^{-8} maximum helium leak, by helium-backfill method
Insulation resistance	$>10^{10}$ Ω from pin to case
Storage temperature	40 or 50°C, brief excursions to 60°C

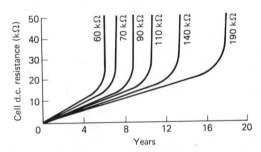

Figure 9.10 Catalyst Research Corporation lithium–iodine Pacemaker Model 802/23; projected performance over 20 years (Courtesy of Catalyst Research Corporation)

calculators where, at a 50 μA drain, it could deliver more than 5000 h of use over as much as a 10-year period. Its performance is shown in Figure 56.27.

Catalyst Research also supply 460 mA h (type 2736) and 870 mA h (type 3740) batteries.

Lithium–iodine batteries offer energy densities on a volume basis of 1.5–2 times that of conventional batteries and on a weight basis of up to 3 times that of conventional button cells such as the zinc–mercuric oxide types. More important is the fact that these batteries offer 10 years of storage, rather than the 2–3 years expected of conventional batteries, which has considerable implications for user–designer applications; for example, the 10-year calculator or digital watch becomes possible. These cells will accept a recharge provided charging currents are restricted to the same maximum as imposed on the discharge current. This might be useful in solar cell, rechargeable watch or calculator applications.

10

Manganese dioxide–magnesium perchlorate primary batteries

Contents

In recent years many companies have attempted, successfully, to replace the zinc can anode of the carbon–zinc primary cell with magnesium to take advantage of magnesium's greater electrochemical potential. The magnesium cell has an open-circuit voltage of around 2.0 V, compared to 1.5 V for the zinc dry cell. In addition to this higher voltage, the magnesium dry cell has several other advantages over the conventional dry cell. It has the capability of giving twice the gravimetric energy density (W h/kg) of the dry cell at medium discharge rates (10–50 h).

Magnesium dry cells also exhibit excellent storage characteristics. These batteries can be stored at high temperature (75% of initial capacity after 12 weeks at 71°C). Figure 10.1 shows the initial capacity and storage characteristic of a magnesium and an equivalent zinc dry cell. The storage ability of the

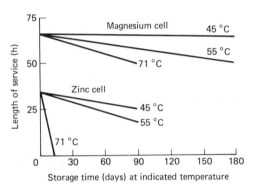

Figure 10.1 Comparison of capacity and storage for magnesium and zinc dry cells (Courtesy of Eagle Picher)

magnesium dry cell means that refrigerated storage need not be provided, as is often the case with conventional zinc dry batteries in military depots where large quantities of batteries are used. This cell has excellent operating characteristics at −40°C with little loss of capacity.

An undesirable characteristic of the magnesium dry cell is its 'delayed action', which involves a transient voltage drop that occurs at the instant the cell is subjected to load. Voltage then gradually recovers to normal. Delayed action is measured by the time a magnesium cell takes to recover to the lowest usable voltage (end-voltage) after it has been placed under load. This interval is attributed to the time required to break down the corrosion-inhibiting oxide or hydroxide film that forms at the magnesium surface. There is a direct relationship between the duration of delay and the load. Higher current drain results in a greater delay. For the battery to function when a load is applied, it is necessary to break down the protective film. This

creates a delay characteristic; that is, the time required for the cell to achieve its normal on-load voltage after the load is applied. Experience has indicated that this delay time depends on the ambient temperature, length of storage and state of discharge. However, the delay time would be less than 3 s in the case of a fresh battery.

The selection of proper magnesium alloys, electrolytes and inhibitors (for example, low concentrations of chromates) has reduced the delay to below 0.3 s for most applications. Nevertheless, the existence of a delay time must be given due consideration in certain applications.

These batteries are characterized by a high energy density in terms of both weight and volume. This results from the fact that the magnesium cell operates at a high cell voltage and that magnesium is a lightweight metal. These advantages become most obvious in a multicell battery in which the higher cell voltage permits the use of fewer cells than other primary systems. Thus magnesium batteries are lighter, smaller and have a higher reliability factor, which is inherent in the use of fewer cells.

Manganese dioxide–magnesium perchlorate batteries are available in reserve and non-reserve forms.

10.1 Reserve type

The reserve battery is activated by electrolyte addition, either manually or automatically, by electrical or mechanical means. The heat of corrosion of the magnesium anode, at a controlled rate, enhances operation over a wide range of temperature and discharge rates. The batteries are non-hazardous, being vented to atmosphere, and non-explosive; the electrolyte is far less corrosive than that used in conventional alkaline batteries.

The positive plates are high-conductivity expanded metal grids and the negative plates are fabricated from magnesium sheet. The electrolyte consists of a concentrated solution of magnesium perchlorate. Battery cases are fabricated in fibreglass, acrylonitrile butadiene styrene polymer, paper, rubber or nylon. The theoretical energy density of this battery is 242 W h/kg against a practical energy density of 110 W h/kg (130 W h/dm^3) as cells and 88 W h/kg (120 W h/dm^3) as batteries. The open-circuit voltage is 2.0 V and the nominal working voltage 1.55 V/cell. The output range available is 2–300 W with rated capacities between 3 and 120 A h.

The operating and storage temperatures, respectively, are −54 to +18°C and −68 to +18°C. Batteries are available with volumes of 4 g cm^3 upwards and 28 g cm^3 upwards. A shelf life of 3–5 years is claimed at 20°C, reducing to 3 months at 71°C.

11

Magnesium–organic electrolyte primary batteries

Theoretical considerations predict that organic compounds, such as *meta*-dinitrobenzene (*m*-DNB), are especially attractive as cathode materials in primary cells. For example, if one assumes the complete reduction of *meta*-dinitrobenzene, a yield of 1915 W h/kg is possible.

Considerable effort has been spent by various battery manufacturers, including Marathon, on the development of magnesium–magnesium perchlorate–*meta*-dinitrobenzene cells. They have been made with a cylindrical structure similar to that used with the manganese dioxide–magnesium perchlorate dry cells. The organic cell requires the use of a higher surface area electrolyte absorbent (carbon black) to be mixed with the *meta*-dinitrobenzene that is used with manganese dioxide. Also, the ratio of the *meta*-dinitrobenzene to carbon black is lower (of the order of 2:1) and the water content is higher than for the manganese dioxide cell. Water participates in the electrochemical reaction of the *meta*-dinitrobenzene cell by supplying hydrogen and hydroxyl ions. The operating voltage of this cell is lower than for the other dry cells; however, it has a flat discharge curve (Figure 11.1). Up to 121 W h/kg have been obtained in practice for an A size cell at

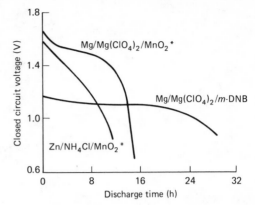

Figure 11.1 Discharge curves for different types of dry cell at 21°C (A-cells, 16.66 Ω drain). *Synthetic manganese dioxide (Courtesy of Chloride Batteries)

21°C at the 25 h rate (time for 'total' discharge), compared to 100 W h/kg for the equivalent size manganese dioxide–magnesium perchlorate cell. Improvements are required in the high-rate characteristics of this system, however.

12

Metal–air cells

Contents

12.1 Zinc–air primary batteries

Metal–air and fuel cells are not discussed in detail in this book. A brief discussion follows in which the characteristics of zinc–air primary cells are compared with those of other types of primary cell. Details are also given of zinc–air button cells.

In relation to their physical size, zinc–air batteries store more energy per unit of weight (in excess of 220 W h/kg) than any other primary type. Other superior features include the ability to supply high currents for longer periods and the maintenance of an almost constant voltage throughout the discharge period. Up to ten times the rated current can be supplied for pulse loads. Batteries are supplied in standard international sizes and the largest has a capacity of 18 A h at 2 A within a size of 33 mm diameter × 61 mm long and a weight of 85 g.

One type of zinc–air primary battery is that in which the active electrodes are inserted into the electrolyte, the so-called mechanically rechargeable system. The zinc–air system is capable of operating down to −18°C and giving 80% of the 21°C performance at this lower temperature. Some start-up time is necessary for the battery to warm itself and to achieve proper air convection. When this is achieved, proper operation can be sustained. At the higher temperature, there is a drying-out problem, which can be alleviated by designing the cells with a reservoir of water. At 54°C approximately 60% of 21°C capacity is obtained. Of course, these figures, for operation at the temperature extremes, depend on rate of discharge. These zinc–air batteries are capable of 100 cycles of discharge (replacement of anode–electrolyte composites). This is highly desirable because of the cost of the air cathodes, which contain an expensive noble metal catalyst. A projected development is a one-shot throwaway version of these batteries using low-cost catalysts for the air electrodes such as spinels and mixed oxides, and activated carbons such as Darco G 60. Two versions of the throwaway are possible; a reserve type in which water will be added to the cells just before use, and a non-reserve type, which will just require the removal of plastics strips that cover the air ports.

The demise of the zinc–air cell has been brought about by economic rather than technical factors. The picture for zinc–air button cells is quite different and Gould have manufactured these for several years for applications such as hearing-aids and watches. Gould and Berec have been marketing them in the UK since 1980.

12.2 Zinc–air secondary batteries

Development of the rechargeable zinc–air battery is under way. Experimental cells have given 155–175 W h/kg at the $C/5$ rate of discharge. These energy densities are approximately twice those of many of the best existing rechargeable systems. However, there is the problem of internal shorting, after several charge/discharge cycles, which occurs as a result of zinc dendrite growth through the separator. This problem can be overcome to a great extent by proper selection of separator materials. It was also found that the air electrodes, which contain platinum as a catalyst and have been used successfully in the mechanically rechargeable zinc–air battery (Section 12.1 and Chapter 26) and the electrically rechargeable cadmium battery, do not function in the charging as well as in the discharging mode. This is because platinum on the anode surface acts as a low hydrogen overvoltage site, thereby enhancing zinc self-discharge and reducing the available capacity of the zinc electrode. Thus, until an adequate air electrode is developed, a third electrode will be required for charging purposes.

12.3 Cadmium–air secondary batteries

Rechargeable cadmium–air and zinc–air batteries are currently still only at the development stage and may not be commercially available for several years. Certainly sealed versions of these batteries can only be considered to be a project for the future. On the other hand, non-rechargeable one-shot types are available and are discussed in Chapter 26.

The cadmium–air system has a theoretical energy density of 445 W h/kg. The cadmium anode used is the type that has demonstrated good stability and low self-discharge in other alkaline systems, such as the nickel–cadmium battery. It has been found that optimum cadmium anode performance is achieved with a sponge-type anode prepared from cadmium oxide, a small amount of nickel carbonyl powder and an extender such as iron or titanium oxide, which is pressed on to the nickel grid. This negative electrode had a porosity of about 60% when fully charged. Perlon, cellophane, polyethylene and inorganic materials are used as separators. A solution of about 30% by weight of potassium hydroxide is usually used as electrolyte. The air cathode is similar to that used in the mechanically rechargeable zinc–air battery. The battery consists of two air cathodes in parallel, positioned on the two sides of a plastic frame, and one anode. The functional air electrode can be used for both discharge and charge. Over 300 cycles have been obtained with the best combination of materials.

There are several problems associated with the operation of a cadmium–air battery. One is the loss of cadmium on cycling the batteries; use of an extender and of air scrubbed free of carbon dioxide significantly reduces this problem. Cadmium penetration of the separator is another problem, which can be minimized by eliminating carbonate from the cell (i.e. by using carbonate-free electrolyte and air scrubbed free from carbon dioxide), by using proper separation between anode and cathode and by limiting overcharging. A third problem is that of poisoning of the air electrode by a soluble cadmium species, believed to be cadmium hydroxide ($Cd(OH)_3$), migrating to the air electrode; the addition of zincate or aluminate ions is claimed to reduce this effect. Water must be added to the cell periodically because water loss occurs by transpiration of water vapour through the air electrode, a characteristic of batteries using air electrodes.

Operating cell voltage varies between 0.2 and 0.85 V at the $C/2$ to $C/10$ rate (i.e. the discharge current is numerically equal to one-half to one-tenth of capacity C). At these rates practical batteries produce 80–90 W h/kg (as opposed to a theoretical 445 W h/kg) and, on a volumetric basis, 14–24 W h/dm^3.

It is seen that many problems remain to be solved if cadmium–air batteries are to become accepted as serious competitors to other available types of battery. The same comment applies to zinc–air batteries.

13

High-temperature thermally activated primary reserve batteries

Contents

Thermal electrochemical batteries have been manufactured since 1946 to meet the needs of applications requiring instant power, fast activation, short operating life, high energy density, high amperage and a shelf life claimed by various manufacturers to be between 12 and over 20 years. Battery voltages from 2.3 to 500 V are available. Thermal batteries contain an electrochemical system that is inert until activated by application of heat. They may be stored on load or open circuit. The cell electrolyte used is a mixture of anhydrous salts which conduct current only when molten. A heat source, which can be ignited either by an electric match or a mechanical primer (percussion cap), is an integral part of a thermal battery. When the battery is ignited, this heat source develops sufficient heat energy to melt the electrolyte and thus permit the battery to deliver a considerable amount of electrical power. Activation times are rapid (as low as 0.3–2.5 s). The batteries have an open-circuit voltage of 2.3–3.0 V/cell. Voltage regulation is within 10% of nominal under constant resistive load. The battery operating range covers the interval 1–300 s. There are two principal types of high-temperature thermally activated battery; those based on calcium anodes and those based on lithium anodes. Within each of these types there are many possible configurations available.

13.1 Performance characteristics of calcium anode thermal batteries

Catalyst Research Corporation have designed and developed a high-efficiency electrochemical system for use in high current density applications. This 'Power Cell' system has the flexibility for use in both short- and long-life battery requirements. Available power output is approximately 25% greater than that of conventional thermal battery systems. Performance data on this battery are given in Table 60.3.

Catalyst Research Corporation also supply fast-activation thermal batteries. These are to meet the needs of applications requiring battery power to fire squibs, electric matches, dimple motors, explosive bolts, etc., after battery activation.

Table 60.2 presents size and performance for some typical units which have been supplied in the three categories of fast-start batteries; high current pulse, medium life and high current. Catalyst Research Corporation also produce a high-voltage thermal battery designed to meet the need for high-voltage low-current power supplies. Typical specifications for such units are shown in Table 60.4.

Calcium anode thermal batteries are capable of operation at temperatures between −54 and +73°C at accelerations up to 2500g lateral, under a shock of up to 200g and vibration of 30g and also at spin rates of 400 rev/s.

Table **13.1** Firing energies of Catalyst Research Corporation calcium anode thermal batteries

	No fire	*All fire*
Match	0.25 A for 5 s	0.50 A for 50 ms
Igniter	1 A for 5 min	3.5 A for 20 ms

Activation of the battery by ignition or an internal heat source can be achieved by several methods:

1. *Electrical* High- and low-energy matches and squibs with wire bridges are available. Firing energies of two of the units are shown in Table 13.1.
2. *Mechanical* An M42G primer similar to that in a .38 calibre bullet, with a modified charge, has been used extensively in thermal batteries. Spring or explosive-driven firing pins are used to deliver the 1800 cm$g$ of energy required for primer firing. A second method of mechanical initiation uses a self-contained inertial starter which is activated by forces of 300–600g imposed on the battery.

The corresponding parameters for calcium anode thermal batteries supplied by Eagle Picher are similar (see Table 13.2).

Table **13.2** Eagle Picher calcium anode thermal batteries

High-rate capacity	10 A for 60 s
Low-rate capacity	250 mA for 300 s
Activation time	100 ms to 3.5 s
Operating temperature	−54 to +93°C
Storage temperature	−54 to +74°C
Storage life	Up to 10 years at storage temperature
Vibration	Withstand 300g all areas
Shock	Withstand 3000g longitudinal
Acceleration	Withstand 450g longitudinal

13.2 Performance characteristics of lithium anode thermal batteries

Quite large thermal batteries of this type are being produced; one lithium–iron sulphide battery produced by Mine Safety Appliances weighs 3.8 kg, consisting of 30 series cells housed in a container 89 mm in diameter by 235 mm long. This battery shows smooth discharges and freedom from electrical disturbance for average current densities varying from 0.3 to 1.86 A/cm^3. To 80% of the peak voltage, durations range from 3 to 21 min. It is also noteworthy that the cell capacity is very constant at 275 A minimum over this range of discharge rates, although there are indications of some reduction at the highest current density. To compare the performance with typical results obtained for the Ca/LiCl–KCl, $CaCrO_4$/Fe system for this capacity of lithium–iron sulphide cell, the current density performance is at least 15 times better than with Ca/$CaCrO_4$, for discharge densities of over 2–3 min, and analysis of the discharge curves also shows that the energy density achieved by the cell (including pyrotechnic) is 113 W h/kg, with 63 W h/kg for the battery. The volumetric energy density of the battery is 113 W h/dm^3.

A further battery design gives a duration of 7 min, a short activation time and a low internal resistance for good voltage regulation for a wide range of current drains. It uses two parallel stacks of 75 mm nominal diameter cells, housed in a container 89 mm in diameter and 190 mm long. The battery weight is 2.9 kg. At an average discharge current of 20 A (0.23 A/cm^2 electrode current density), 8 min discharge duration is obtained to 80% of the maximum voltage for both conditioning temperatures. The battery internal resistance remains constant for the discharge duration to 80% of mean voltage, with a value of 0.09–0.11 Ω. The batteries tested after initial conditioning at +50°C activate in approximately 1.6 s giving 12.8 A at 25.5 V, and those conditioned at −30°C in approximately 3.0 s. The battery design uses anodes with excess coulombic capacity (380 A minimum) and considerable improvement in activation time is possible if these are matched to the rated cathode capacity of 200 A minimum.

Various other battery designs to meet particular requirements are available from Mine Safety Appliances, who claim that these batteries have a wide range of possible military uses. Designs optimized for maximum energy can achieve energy densities up to 100 W h/kg and power densities of up to 1500 W h/kg.

14

Zinc–chlorine secondary batteries

The rechargeable zinc–chlorine battery currently under development by Energy Development Associates, Michigan is proposed as a new approach to energy-saving power systems for vehicles. This is an interesting new concept in battery design, which may prove to have applications in the small-battery field. The object that Energy Development Associates have in mind is to produce a battery capable of propelling a 1 tonne vehicle 322 km at a steady speed of 80.5 km/h with 120 km/h excursions. Comparative claimed performance characteristics for the zinc–chlorine battery with those of the conventional lead–acid electric vehicle battery are given in Table 14.1.

The performance figures become important when measured against the needs of a 1 tonne car. Experimentally, it has been found that, with a 50 kW h, 363 kg battery system, the vehicle could accelerate from 0 to 64 km/h in 10 s. This requires a minimum power density of 10 W/kg. To maintain an average speed of 80.5 km/h on level ground requires 12 kW, equivalent to an energy density of 23 W h/kg. At a 4 h discharge rate, this would yield a 322 km range between charges.

As can be seen from Table 14.1 the lead–acid battery will accelerate the vehicle, but its low energy density seriously limits the operating range. The zinc–chlorine battery, on the other hand, yields an energy density of 154 W h/kg (well above the 23 W h/kg required for a viable vehicle). Note also that the zinc–chlorine battery offers better performance in terms of rechargeability, cycle life and volume, in addition to a significant cost advantage.

Table 14.1 Comparative performance data on zinc–chlorine and lead–acid electric vehicle batteries

	Zinc–chlorine	Lead–acid
Delivered energy (4 h rate) (kW h)	50	50
Weight (kg)	32	137
Power output (1 min peak) (kW)	40	150
Minimum charge time (h)	4	2
Cycle life	500–5000	300–1000
Volume (dm^3)	285	570
Overall energy efficiency (%)	70	75
Energy density (W h/kg)	154	35
Cost (estimated) ($)	500–750	1000–1500
Average power to load (kW)	12.5	
Peak power for 30 s (kW)	40	
Peak power for 2 s (kW)	50	
Energy per unit weight at 4 h rate (W h/kg)	154	
Energy per unit volume at 4 h rate (W h/dm^3)	177	
Peak power density for 30 s (W h/kg)	125	
Peak power density for 2 s (W h/kg)	154	

15

Sodium–sulphur secondary batteries

The sodium–sulphur battery is a relatively high ampere hour rechargeable battery, envisaged as a power source on electric vehicles and trains. It was patented by the Ford Motor Company of Dearborn, USA in 1967. Chloride UK and British Rail have also taken out patents on the sodium–sulphur battery. Although developments have been in progress for over 10 years, batteries are not expected to be commercially available before the end of the 1980s. Its operation relies on the recently discovered property of β-alumina, namely that it combines very low electronic conductivity with an unusually high ionic conductivity, which is specific to sodium ions as charge carriers.

The cell voltage, 2.08 V, is derived from the chemical reaction between sodium and sulphur to produce sodium polysulphide; and the theoretical energy density, about 750 W h/kg, is high compared with that of the lead–acid battery (about 170 W h/kg). The novelty of the properties discovered in sodium β-alumina may be estimated from the fact that, although its melting point is around 2000°C, several amps per square centimetre of current may be passed across the electrolyte at 300°C. The cell operates at about 350°C. The key to the rapid sodium ion mobility lies in the crystal structure and the low potential energy path for sodium migration that this produces.

This battery has attracted worldwide commercial interest because of the low cost and ready availability of its chemical components and its very high energy density, some five times greater than that of conventional lead–acid batteries. In the USA, beta batteries are being developed mainly for load levelling in the electricity supply industry; 100 MW h units with 10 MW capacity are being designed and tested. In Europe the main thrust of research is directed towards automotive traction applications, especially for urban transport and delivery vehicles and also for rail cars. The engineering and materials problems associated with this battery are severe, and research involving industrial, government-funded and university laboratories has been in progress for many years in order to improve battery lifetimes and to increase the commercial attractiveness of the system.

16 Other fast-ion conducting solid systems

Many research centres in the USA and Europe have started programmes to discover new fast-ion conducting solids. To date, this search has not been very successful. No sodium or other alkali metal ion conducting materials have yet been found that show significantly improved electrical properties over β-alumina, and none has as low an activation energy for ionic mobility (0.13 eV).

One or two fast-iron conductors have been known for a considerable time. Thus, β-silver iodide transforms to α-silver iodide at 147°C, and the first-order phase charge is accompanied by a 1000-fold increase in ionic conductivity. The structure of the phase was studied in 1934 by Strock, who postulated that the silver ions were distributed over a large number of nearly equivalent sites, whereas the anions formed a regular well ordered lattice. The entropy change accompanying this transition can be thought of as due to the 'melting' of the ordered Ag^+ sublattice of the β-phase.

A similar phenomenon was found with lithium sulphate and extensively studied in the 1960s. The aim of most modern research into fast-ion conductors is to produce materials that show highly conductive behaviour at relatively low temperatures for convenient incorporation into electrochemical devices. Thus, in 1967, $RbAg_4I_5$ and KAg_4I_5, which have conductances at ambient temperatures of over 10 s/m, were described – the same value as that of molar aqueous solutions of potassium chloride. These are optimized Ag^+ conductors having features in common with α-silver iodide.

One new and important concept which has emerged, involving fast-ion conductivity, is the insertion compound or solid-solution electrode. In a solid-solution electrode there is both rapid ion transport and electronic conductivity, but a third vital ingredient is the presence of a range of stoichiometry involving the mobile ion. Fast-ion conducting materials of this type, therefore, can be used not as electrolytes, as is the case for β-alumina, but as battery electrodes with either a liquid electrolyte or a solid electrolyte. In the latter case fast-ion conducting materials are used to make an all-solid-state battery. In the former case, although the arrangement of solid and liquid phases is formally similar to that found in conventional lead–acid or alkaline batteries, the energy densities obtainable using modern materials are much higher than was previously thought possible.

In contrast to solid electrolytes, where sodium is generally found to be the most mobile of the alkali ions, most good solid-solution electrode materials so far discovered act best as conductors of lithium ions:

$$x Li^+ + TiS_2 + x e \rightleftharpoons Li_x TiS_2$$

although this distinction may not survive further research.

One of the most attractive solid-solution electrodes so far reported is based on the semiconducting layered transition metal dichalcogenide titanium disulphide, TiS_2. This is a layer material with adjacent layers of sulphur atoms held together only by van der Waals bonding. Titanium disulphide, and other similar chalcogenides of the transition metals, can absorb alkaline cations between the sulphur layers. There is a slight expansion of the crystal axis perpendicular to the layers but otherwise no change in the crystal structure. For lithium passing into titanium disulphide there is a continuous region of non-stoichiometry from TiS_2 to $LiTiS_2$. Moreover, the partial free energy of intercalation is relatively constant across the whole range. Titanium disulphide is quite light, cheap and readily available and the lithium–titanium sulphide battery developed by Exxon, which has an organic electrolyte and works at room temperature, has a theoretical energy density of 480 W h/kg. This is already quite close to that of the sodium–sulphur couple, and there is the advantage of room temperature operation, the absence of liquid alkaline metals and, possibly, considerably reduced problems in fabrication and corrosion.

A wide range of solid-solution electrode mixed conductors has been investigated, including:

1. Tungsten and vanadium bronzes which have channels that can incorporate metal atoms.
2. Non-stoichiometric silver sulphide, which has a remarkably high diffusion coefficient for silver, but a very limited range of composition.
3. Alloys such as Li_xAl and Li_xSi.
4. Alkali metal β-ferrites, which are isomorphic with β-alumina.
5. Graphite and modified graphites such as $(CF_x)_n$.

Recently a great deal of interest has been shown in polyacetylene films that can incorporate alkali metal ions reversibly at a cathode to form electronically conducting compounds of the form $(CHNa_y)_x$. An 'all-polymer' solid-state battery has been developed in which the electrolyte is a sodium iodide–polyethylene oxide and the electrodes are doped polyacetylenes:

$$(CHNa_y)_x(s) \quad \text{Note. } PEO.NaI(s) | (CHI_z)_x(s)$$

Such a power source has a relatively high energy density but the overall internal resistance of the cell is high.

Solid-solution electrodes therefore provide a field of complementary scientific interest to solid electrolytes and have at least equal potential application in battery systems. Apart from the presence of rapid ion transport, the extra keys to the utility of solid-solution electrodes lie in the interfacial aspects of ions being able to pass directly between the electrolyte phase and the electrode without any change in crystal structure of the electrode or the necessity to form a new compound by electrocrystallization (compare the electrode reactions in lead–acid and alkaline batteries), and the ability of the electrode to absorb a sufficient quantity of the active ion to provide systems with very attractive energy densities.

A third, more traditional, concept being applied in advance battery design is the application of fused-salt electrolytes. The use of fused salts, of course, implies elevated temperatures, but fused-salt electrolytes can allow the drawing of higher current densities than are currently possible with solid-electrolyte-based batteries. The most notable current development using a fused-salt electrolyte is the lithium–iron sulphide battery, developed at the Argonne National Laboratory (USA). This battery operates at around 400°C and uses a fused-salt eutectic electrolyte mixture of lithium chloride and potassium chloride. Using a lithium–aluminium anode, the theoretical energy density is 650 W h/kg, similar to that of sodium–sulphur, and the expected actual energy densities of the two systems are comparable.

17

Water-activated primary batteries

Contents

There are three main types of water-activated battery: magnesium–silver chloride, zinc–silver chloride and magnesium–cuprous chloride. These batteries are all activated by water or seawater and have obvious applications in marine and aircraft safety lights, buoys and flashing beacons, military power systems, communications and portable electronic devices, sonobuoys, torpedoes, flares, mines, pingers, balloons and life-jackets. These batteries have a long storage life provided they are kept in a sealed condition. Once unsealed and immersed in water they must be considered expended. They cannot be recharged.

17.1 Magnesium–silver chloride batteries

Durations are between several seconds and 20 days for magnesium–silver chloride batteries. An example of this type of cell is the L18A diver's underwater searchlight battery supplied by McMurdo Instruments Ltd (Aquacells). This cell is 103 × 35 × 25 mm, weighs 95 g and has an output of 4 V for 1.5 h at a 1 A loading. The L37 submarine escape hatch lighting battery is a 24 V emergency supply system which is designed to operate at pressures up to 2756 kPa. The battery is 305 × 254 × 72 mm, weighs 6.6 kg and will operate two 24-V 36-W lamps for 8 h. The L43H distress beacon (13.5 V) battery for ships and aircraft is 175 × 51 × 125 mm, weighs 110 g and has a 90 h discharge time. McMurdo also supply a range of Aquacells for life-rafts and life-jackets (Table 17.1).

Magnesium–silver chloride batteries have an energy density of 30–120 W h/kg, 40–250 W h/dm^3 and a power density of 1200 W/kg between −20 and +60°C, and cells up to 150 kW are obtainable. Designs in the range 1–250 V are available.

17.2 Zinc–silver chloride batteries

These cells have similar applications to those of magnesium–silver chloride batteries. They have an energy density of 15–60 W h/kg (power density 100 W/kg) and 20–150 W h/dm^3 (volumetric power density 2400 W/dm^3). Operating temperatures are between −30 and +60°C, and cells up to 2500 W are available from SAFT. These cells are water or seawater activated, and have a long dry shelf life, high performance and a long discharge time capability at low rates (up to 300 days). Designs are available in the 1–50 V range.

17.3 Magnesium–cuprous chloride batteries

These cells have a similar energy density (20–90 W h/kg and 10–150 W h/dm^3) and operate in the temperature range −20 to +60°C. They are available with up to 50 W output and in the 1–100 V range.

Table 17.1 Magnesium–silver chloride cells for life-jackets and life-rafts available from McMurdo Instruments

Designation	Application	Dimensions (mm)	Weight (g)	Power output	Duration (h)
L8.I	Life-jacket	93 × 26 × 12	34	1.5 V, 0.3 W	8
M8M	Life-jacket	93 × 26 × 12	34	1.5 V, 0.165 W or 1.5 V, 0.25 W	20 14
L12B	Life-jacket and life-raft	110 × 30 × 22	80	3 V, 1 W or 3 V, 0.25 W	12 24
L50	Life-raft	117 × 66 × 36	120	3 V, 0.5 W	12

Part 2

Battery theory and design

18

Lead–acid secondary batteries

Contents

18.1 Chemical reactions during battery cycling

The basic cell reactions in the traditional lead–acid battery are as follows:

$$PbO_2 + Pb + 2H_2SO_4 \underset{charge}{\overset{discharge}{\rightleftharpoons}} 2PbSO_4 + 2H_2O \quad (18.1)$$

The reaction at the positive electrode:

$$PbO_2 + 3H^+ + HSO_4^- + 2e^- \underset{charge}{\overset{discharge}{\rightleftharpoons}} 2H_2O + PbSO_4 \quad (18.2)$$

At the negative electrode:

$$Pb + HSO_4^- \underset{charge}{\overset{discharge}{\rightleftharpoons}} PbSO_4 + H^+ + 2e^- \quad (18.3)$$

When the cell is recharged, the primary reaction taking place is as shown in Equation 18.1. Finely divided particles of lead sulphate are being electrochemically converted to sponge lead at the negative electrode and lead dioxide at the positive by the charging source driving current through the battery. As the cell approaches complete recharge, where the majority of the lead sulphate has been converted to lead and lead oxide, the overcharge reactions begin. For typical lead–acid cells, the result of these reactions is the production of hydrogen and oxygen gas and subsequent loss of water.

Further details of the influence of theory on the design of lead–acid batteries is given below. This discussion refers particularly to a motive power lead–acid battery. The same general principles apply to any conventional open type of lead–acid battery.

The negative plates have a spongy lead as their active material, while the positive plates of the lead–acid cell have an active material of brown lead dioxide. The plates are immersed in an electrolyte of dilute sulphuric acid.

Through a quirk of history in the development of motive power batteries throughout the world, a tubular positive plate is used in Europe while in the USA a flat plate is preferred. The basic differences are as follows.

The tubular positive consists of rods of antimonial lead, which are surrounded by sleeves of an inert porous material such as terylene. The annular spaces in the tubes of the sleeves and around the rods are filled with the active material lead dioxide. In the case of the flat plate design, the plate is made from lead alloy grids with lattices containing lead dioxide. The negative plate of the tubular cell, which must match the electrical capacity of the positive plate to enable efficient chemical reaction to take place, is of a similar design to that of the flat plate positive, but the lattices in this case are filled with a sponge of pure lead. The positive and negative plates are placed into a container, each positive being placed next to a negative and so on. In the lead–acid cell, there are always an odd number of plates, the extra one being a negative. Therefore, in a 13-plate cell, there will be seven negative and six positive plates.

Contact between plates of opposite polarity must be avoided, to prevent short-circuiting. This is normally achieved by inserting the negative plate into a microporous envelope. Microporous envelopes have good insulating properties but low electrical resistance, thus allowing a free flow of ions and diffusion of electrolyte.

The element is manufactured by assembling positive and negative plates alternately, at the same time placing the negative plate into the separator sleeve. The positive plates are interconnected by welding in position a precast terminal post; the negative plates are similarly connected. The cell is completed by heat-sealing the lid in position on the box.

Cells are said to be in series when the positive pole of one cell is connected to the negative pole of the adjacent one, and this arrangement is continued for any desired number of cells. The voltage of cells in series is additive. The capacity in ampere hours of the battery will, however, still be that of a single cell. Cells are connected in parallel when all the positive poles are joined together and all the negative poles are similarly connected. The voltage of cells so connected is that of a single cell, but the capacity of the combination is the sum of the individual cell capacities.

Cell boxes were traditionally made from hard rubber, but there is now an increasing use of plastic materials. Polypropylene is one such material. It has the advantage of being more robust, and it allows a polypropylene heat-sealed joint to be made between cell box and lid, creating a significant improvement over the box-to-lid seals previously used.

Batteries do not work on alternating current (a.c.), so to recharge a discharged battery direct current (d.c.) has to flow into the battery in the opposite direction to that of the discharge. The graph in Figure 18.1 shows the variation in voltage of a motive power cell during a 10–12 h recharge.

The size of a motive power cell is defined by its dimensions and capacity. In other words, the amount of active surface available in the plates of a given cell. This can be obtained, for example, by 7 plates of height A or 14 plates of height $\frac{1}{2}A$.

Both cells will produce approximately the same electrical storage capacity. The required height of a cell is confined by the space available within the design of the vehicle it is to be fitted to.

There follows a more detailed explanation of the chemical reaction which takes place while a battery is 'cycling'; that is, changing from fully charged to discharged and back again.

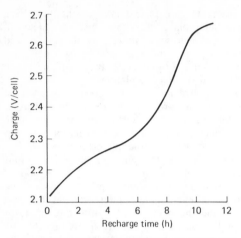

Figure 18.1 Variation of voltage of lead–acid cell during recharge

18.1.1 Discharging

When a battery is delivering energy it is said to be discharging. The energy is produced by the acid in the electrolyte gradually combining with the active material of the plates. This combination produces lead sulphate in both negative and positive plates. A cell is completely discharged when both plates are entirely sulphated, and as they are now composed of identical material the terminal voltage collapses. In practice, of course, discharging would be stopped long before the plates reached this condition.

18.1.2 Charging

The object of charging is to drive all the acid out of the plates and return it to the electrolyte. A direct current is passed through the cell in the opposite direction to that during discharge, liberating the acid from the plates, i.e. the concentration of acid in the electrolyte increases. This reverses the action of the discharge and restores the battery to its original charged condition. When the cell is fully charged, the active material of the positive plates is lead dioxide, and that of the negative plates is metallic lead in spongy form. The concentration of acid in the electrolyte at this stage is at its maximum.

18.1.3 Characteristic voltage

The nominal voltage of a lead–acid cell is 2 V, which remains unaltered by the number of plates or their capacity. In practice, the voltage of a cell does vary slightly according to the state of the charge, the cell temperature, the charge or discharge current, and the age of the cell.

18.1.4 Voltage on charge

When placed on charge there is an immediate rise in the battery voltage, mainly due to the sudden increase in density of electrolyte in the pores of the active material. The subsequent rise of voltage is governed by the rate at which acid is produced in the plates and the rate of diffusion into the free electrolyte of the cell.

When the voltage on charge reaches approximately 2.40 V, there is a fairly sharp rise in voltage. At this stage there is almost complete conversion of lead sulphate. Most of the charge is now being used in dissociating the water of the sulphuric acid solution into hydrogen and oxygen and the cell begins to gas freely. When this happens the cell voltage rises, levels off and finally shows no further increase. Recharge is considered to be complete when the voltage and relative density of the electrolyte remain constant for about 3 h. Figure 18.1 shows the variation of voltage of a motive power cell during a recharge period of 10–12 h.

18.1.5 Voltage on discharge

The effect of discharge on voltage is the reverse of that on charge. The internal resistance of the cell creates a voltage drop when a current is passing, causing the voltage during discharge to be less than it is on open circuit. This can be expressed as:

Voltage on discharge =
Open circuit voltage − (Current × Internal resistance)

Figure 8.2 shows the effect on voltage of a discharge at the 5 h rate, in a typical lead–acid cell.

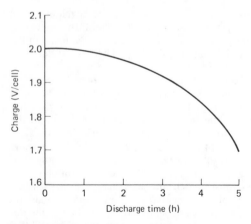

Figure 18.2 Variation of voltage of lead–acid battery on discharge (Courtesy of Chloride Batteries)

18.1.6 Capacity

The capacity of the battery will vary according to the current at which it is discharged. The higher the current being taken out of the battery, the lower the available capacity. For example, if a battery of 500 A h capacity is discharged at the 5 h rate, it will give 100 A for 5 h. The same battery discharged at 200 A, however, will give current for only 2 h, thereby providing a capacity of 400 A h at the 2 h rate of discharge. This is because, at higher rates, the voltage drop is more rapid and the final voltage is reached more quickly.

Motive power battery capacities are normally given at the 5 or 6 h rate of discharge. This hypothetical, continuous rate of discharge relates most closely to the actual performance of an industrial truck; stop–start operation over an 8 h shift. Ambient temperatures also affect the capacity of a battery. At low temperatures the capacity is considerably reduced.

18.2 Maintenance-free lead–acid batteries

Any battery, when overcharged, will liberate hydrogen and oxygen gases as water is decomposed. For a battery to be maintenance free, it is necessary to retard gas liberation, otherwise the electrolyte would be depleted prematurely and catastrophic failure would result.

There are two principal types of maintenance-free lead–acid battery:

1. The type featuring calcium–lead alloys and immobilized sulphuric acid electrolyte, which reduces but does not completely eliminate gassing, i.e. there is electrolyte volume reduction. Such cells are usually manufactured containing a reserve of electrolyte so that topping up is not required during battery life.
2. The type in which complete recombination of electrolysis gases occurs, i.e. virtually no electrolyte loss occurs.

18.2.1 Calcium–lead alloy batteries

This type of lead–acid battery is typified by batteries in the Sonnenschein Dryfit range. Under constant-potential charging, the maintenance-free lead–calcium–acid battery will accept relatively high currents initially when the system is most efficient. This stage is followed by a second period when the back e.m.f. of the battery begins progressively to increase to control the amount of current accepted. At full charge and under proper voltage regulation, the current accepted will be reduced to a few milliamps input, thus restricting the degree of

overcharge which protects the unit from excessive electrolyte loss.

Various designs of these batteries further support this function by developing a very high hydrogen and oxygen overvoltage, which enhances the efficiency of conversion and ensures a sharp and reproducible rise in on-charge voltage to regulate current acceptance best.

Absolute recombination in a battery of hydrogen and oxygen produced towards the end of charge or on overcharge is achieved only in a closed container in which the gases are held under relatively high pressure. In such a system equilibrium occurs where the amount of oxygen electrochemically generated at the positive plates is chemically reacted with spongy lead to form lead oxide, which then reacts with sulphuric acid to form lead sulphate and water. The oxygen is returned to the electrolyte in the form of water and the constant applied current is used up electrochemically to convert the lead sulphate back to spongy lead. This cycle can be repeated many times without losses to the system.

The act of absolute recombination can present a problem. It has been demonstrated that a sealed lead–acid battery can develop thermal runaway in a manner similar to sealed nickel–cadmium cells. This is a condition whereby a battery on charge under voltage regulation begins to accept progressively higher currents resulting in the build-up of excessive internal heat. Both recombination and temperature decrease the back e.m.f. of the battery, allowing higher and higher currents to be accepted. The result is an uncontrollable system which can become hazardous.

Low-pressure vented maintenance-free batteries have the ability to recombine these gases. However, the degree of recombination is not complete since, from the hazard point of view, it is not considered advisable to operate batteries under excessively high gas pressures.

In practical designs of low-pressure maintenance-free batteries, reduction or elimination of gassing towards the end of charge and self-discharge is handled by battery design features such as special grid alloys and immobilized electrolytes, while gassing on overcharge is avoided by attention to charging method and charging control.

18.2.2 Use of calcium alloys to control gassing on overcharge or during self-discharge

With lead–acid batteries, electrolyte decomposition into hydrogen and oxygen is accompanied by reduction of electrolyte volume, thereby increasing the concentration of the electrolyte. Damage is then caused by increased corrosion of the positive plate due to the highly oxidizing effect of the atomic oxygen being liberated. A further reason for the evolution of oxygen and for corrosion is anodic oxidation due to anode potential. At the same time there is a risk of explosion since the oxygen–hydrogen mixture evolved can be ignited at relatively low concentrations and temperatures. In partially recombining sealed lead–acid batteries, liberation of gas is reduced to minimum values by using a lead–calcium alloy containing only 0.08% calcium rather than the more usual lead–antimony alloy which contains 4–6% antimony. It is thus possible to reduce the self-discharge and the associated liberation of gas by a factor of about 5. In addition, other factors affect the self-discharge rate of conventional lead–acid batteries when not in use. One example is the virtually unhindered passage of atmospheric oxygen into the cell interior and then to the negative plate. This effect is hindered in many battery designs by the provision of one-way self-sealing vents. It is for these reasons that although the capacity of a conventional 5% antimonial lead–acid battery loses about half its initial capacity during 3 months at 20°C, the same decrease would take about 16 months in the case of a 0.08% calcium–lead battery (and more than 24 months at 8°C).

An additional advantage of using a lead–calcium alloy is that it allows the charging voltage to be limited to values of between 2.25 and 2.3 V/cell while still being able to achieve complete charging at 20°C. These voltages lie below the values at which the evolution of gas begins. At 40°C the required charging voltages lie at 2.2–2.25 V, which again is below the gassing voltage. During neither regular operation nor when continuously connected to a charger is there creation of such a high gas pressure that it could lead to opening of the relief valves installed in each cell. When stored under open-circuit conditions a lowering of pressure is brought about by the 'cold' combustion of small quantities of gas present, and occurs to a certain extent at the negative plate.

18.2.3 Immobilization of electrolyte

One of the design features of most of the available types of partially recombining lead–acid batteries is that the sulphuric acid electrolyte is immobilized with thixotropic silicic acid. A constant regeneration of the thixotropic gel structure takes place through changes in volumes and the formation of minute bubbles of gas in the pores of the electrode. This gives rise to a gentle mechanical movement of the gel, thereby avoiding loss of contact between the electrolyte and the plates. The thixotropically immobilized 'dry' electrolyte allows the electrolysis gases to contact the negative plate at which oxygen recombination occurs, thereby producing a reduction in gas pressure. The electrolyte used in batteries obtainable from certain suppliers, e.g. Sonnenschein Dryfit batteries, also contains phosphoric

acid additives to improve the constancy of recoverable charge during battery life. Lead–acid batteries are characterized by the presence of the relatively large volume of free electrolyte or, in the case of partially recombining batteries, of the gel. This is in contrast to nickel–cadmium batteries, in which there is a much smaller amount of potassium hydroxide electrolyte, which does not change greatly in volume during charge and discharge. In the alkaline battery the porosity of the electrodes and the fleecy plastics separator (commonly Nylon) absorb and consequently immobilize most of the electrolyte. In alkaline button cells, the formation of damaging interstices is prevented by a spring mechanism, which compresses the group of plates. This spring simultaneously provides contact and a space for gas at the negative electrode.

Calcium–lead alloy batteries, because of the gelled electrolyte, are insensitive to orientation, as are sealed nickel–cadmium cells. They can therefore be stored, charged or discharged in any position, even upside-down. This characteristic has obvious advantages in cyclic or float applications such as power tools, portable radios, lamps, electronic, communications and medical equipment, alarms and emergency lighting equipment.

A design of this type of battery is shown in Figure 18.3.

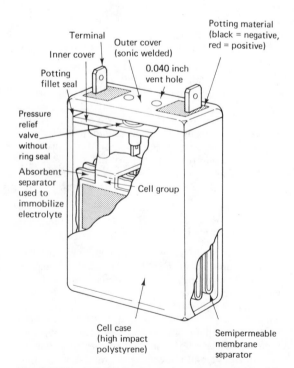

Figure 18.3 Eagle Picher Carefree maintenance-free lead–acid battery construction (Courtesy of Eagle Picher)

18.2.4 Fully recombining lead–acid battery

This is typified by the Gates D-cell lead electrode combination battery, supplied by General Electric Co. (USA) and also available under the trade name of Cyclon from Chloride Batteries (UK).

A unique aspect of the Gates cell is that the majority of the oxygen generated within the cell on overcharge (up to the $C/3$ rate) is recombined within the cell. The pure lead grids used in the construction minimize the evolution of hydrogen on overcharge. Although most of the hydrogen is recombined within the cell, some is released to the atmosphere.

Oxygen will react with lead at the negative plate in the presence of sulphuric acid as quickly as it can diffuse to the lead surface. Hydrogen will be oxidized at the lead dioxide surface of the positive plate at a somewhat lower rate, as shown by the following overcharge recombination reactions:

$$Pb + HSO_4^- + H^+ + \tfrac{1}{2}O_2 \rightleftharpoons PbSO_4 + H_2O$$
$$PbO_2 + HSO_4^- + H^+ + H_2 \rightleftharpoons PbSO_4 + 2H_2O$$

In a flooded lead–acid cell, this diffusion of gases is a slow process and virtually all of the hydrogen and oxygen escapes from the cell rather than recombines.

In the Gates cell, the closely spaced plates are separated by a glass mat separator, which is composed of fine glass strands in a porous structure. The cell is filled with only enough electrolyte to coat the surfaces of the plates and the individual glass strands in the separator, thus creating the 'starved electrolyte' condition. This condition allows for homogeneous gas transfer between the plates, which is necessary to promote the recombination reactions.

The pressure release valve maintains an internal pressure of 40–60 psi. This condition aids recombination by keeping the gases within the cell long enough for diffusion to take place. The net result is that water, rather than being released from the cell, is electrochemically cycled to take up the excess overcharge current beyond what is used for conversion of active material. Thus the cell can be overcharged sufficiently to convert virtually all of the active material without loss of water, particularly at recommended recharge rates.

At continuous high overcharge rates (e.g. $C/3$ and above), gas build-up becomes so rapid that the recombination process is not as highly efficient and oxygen as well as hydrogen gas is released from the cell.

The D-cell was the first cell available within this range. It is rated at 2.5 A h at the 10 h rate. It is 61 mm in height and 34 mm in diameter. The construction of this cell is similar to that of standard cylindrical nickel–cadmium cells. It is sealed, has a safety vent, makes use of a cylindrical spiral-wound plate design for high energy density and low internal impedance, and can be charged and discharged in any orientation.

The starved system, porous separator and oxygen recombination allow efficient space utilization for active material in the spiral-wound sealed lead–acid cell, resulting in a 15–50% increase in volumetric energy density over gelled-electrolyte systems.

Low impedance is derived from the tightly wound design with multiple interconnecting tabs. In addition, the spiral-wound design maintains place spacing better than flat-plate designs; therefore the impedance remains more constant over life (as low as 10 mΩ compared with 23.5 mΩ for the closest gelled-electrolyte battery).

The spiral-wound plate roll is mechanically structured to prevent interplate movement found in flat-plate design.

The sealed design, accomplished with a welded cell container, self-resealing pressure safety vent and oxygen recombination, obviates the need to add water. Oxygen recombination prevents electrolysis of the water in the electrolyte during overcharge. This design does not vent hydrogen and oxygen from the electrolyte during normal operation.

18.2.5 Gas recombination technology in automotive and commercial vehicle lead–acid batteries

Several principles have been used by various manufacturers in the production of low-maintenance batteries, i.e. batteries which do not require topping up of the electrolyte during service. Catalytic recombination of electrolysis gases is not used for this type of battery.

In general, for automotive, vehicle and traction batteries the antimony-free grid technology such as used, for example, by Sonnenschein in their sealed batteries for power tools is not now used. A few years ago automotive SLI battery production fell into two categories based on grid alloy composition. It was either an antimonial lead or a calcium–lead battery. Within the last few years, however, this distinction has become less clear as many manufacturers have gone into what could be termed hybrid construction, which uses a low-antimony positive grid and a lead–calcium alloy negative grid. In taking this approach, many battery manufacturers have been able to minimize some of the potential shortcomings associated with the purely lead–calcium alloy batteries. At the same time, they have also minimized many of the undesirable traits of the conventional antimonial alloy battery.

The progression from a lead–calcium alloy battery to a hybrid construction employing a low-antimony positive grid and a lead–calcium alloy negative grid has reduced the necessity for a microporous separator in all battery designs. Enveloped polyethylene microporous separators are still widely used in batteries of the lead–calcium type, and are recommended at virtually all plate spaces. However, as one moves to the hybrid construction, the necessity for a microporous separator in all battery designs has been reduced. When plate spacings are less than about 1 mm, enveloped microporous separators are still recommended. This separator construction precludes side

and bottom shorts due to mossing or dendrite growth. At the closer plate spacing, shorting through the separator can be a problem and the use of a microporous polyethylene separator virtually eliminates this failure mode. Low electrical resistance leaf-type separators can be substituted for enveloped microporous separators when plate spacings are greater than 1 mm since the propensity for shorting at the greater plate spacing is reduced.

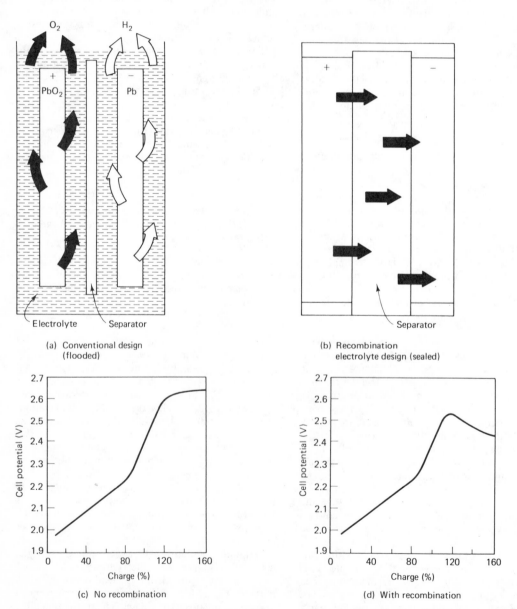

(a) Conventional design (flooded)

(b) Recombination electrolyte design (sealed)

(c) No recombination

(d) With recombination

Figure 18.4 Recombination technology in lead–acid batteries. In a conventional cell, oxygen from the positive plate rises to the top of the electrolyte and is lost. In a cell utilizing recombination electrolyte (RE) technology, oxygen passes through the special separator to the negative plate and, ultimately, back into the electrolyte (Courtesy of Yuasa)

18.2.6 Highly porous separator, low free acid volume technology

In a lead–acid battery, when the positive and negative plates become fully charged they start gassing. The positive plates give off oxygen and the negative plates give off hydrogen (oxygen is usually given off from the positive a little before hydrogen evolution starts at the negative), reducing the water content of the electrolyte and therefore the water level in the battery. Unless this water is replaced by topping up, the battery will eventually cease to operate.

The problem has been partly solved with the introduction of maintenance-free batteries, with their improved grid alloys and reduced gassing and water loss. The drawback with this type of battery is the extra head space required to provide a reservoir of acid, enabling the battery to last 3–4 years. In addition, there is still the possibility of corrosion from acid spray.

In a conventional battery the gasses 'bubble' to the surface through the electrolyte – the easiest path to take. In a recombination electrolyte battery the electrolyte is completely absorbed in the plates and the highly absorbent and very porous separators consisting, usually, of glass microfilms, which are closely wrapped around the plates and take up within its pores all the liquid electrolyte. There is no electrolyte above or below the cell assembly as in a conventional battery.

When the positive plates become charged and give off oxygen, the easiest path for the gas to take is through the highly porous separators to the negative plates. Here the oxygen combines with the material of the plates to form lead sulphate, the same material produced during discharging. Further charging breaks this down to lead, so the oxygen is returned to the electrolyte. There is therefore a closed loop, since the continual arrival of oxygen at the negative plate means that it never reaches a state of gassing, so hydrogen is never evolved. The principles of the closed loop system are outlined in Figure 18.4. With no water loss and no gases permanently evolved, the battery can be completely sealed and never needs topping up.

In the event of severe charging abuse, hydrogen can be produced, plus oxygen at a rate above the capacity of the battery to recombine it. In this situation, a valve on top of the battery will release the pressure safely and then reseal itself.

A good example of the application of recombination electrolyte technology to automotive batteries is the Chloride Automotive Ltd Torque Starter battery, introduced in 1983. Chloride claim that the battery is a much safer product to handle, being totally sealed with no free electrolyte to spill. However, under conditions of abuse, the generation of gas could be greater than the recombination

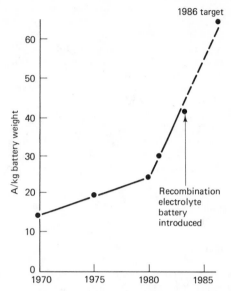

Figure 18.5 Cold cranking amps/kg battery weight (current at −18°C 30 s duration to 7.2 V) (Courtesy of Chloride Batteries)

process could cope with, and for this reason a small pressure valve has been built into the lid.

Excessive gassing occurs for only one reason – the voltage across the terminals is too high, because of either faulty alternator regulation or mains charging at too high a setting.

Figure 18.5 shows the improvement in energy density from 30 A/kg in 1981 (non-RE) to 40 A/kg in 1983 (RE).

Since a recombination electrolyte battery is still a lead–acid battery, one would not expect to find large differences between it and conventional flooded lead–acid batteries. However, there are some differences:

1. *Charging* This is best done by a charger which has control over the voltage such that it does not exceed 15 V. With this type of control the commencing charge current can be as high as the conventional electrolyte battery will accept at 15 V or below.

   Most mains chargers, if any control exists, can control the current only, in which case the value of the current should not exceed 3 A, as this is the maximum value the recombination will accommodate with a 12 V battery when it is approaching a fully charged condition. If the charger has no control at all then 3 A (average) or 4 A (RMS) is the largest charger rating that should be used in order to avoid raising the battery voltage too high. Even from the completely flat condition the charge is complete within 15 h.

2. *Boost charging* From what has been said so far it can be seen that fast chargers or boost chargers which do not control the voltage cannot be used with recombination electrolyte batteries.

3. *Jump starting* This is carried out in exactly the same way as with any conventional battery. If the two vehicles are allowed to run with the engines on medium throttle and the jump leads connected, a flat battery will normally accept charge at a very high rate. Provided the jump starting technique is the safe approved method, this form of boost charging is safer than using a mains boost charger no matter what the types of battery involved.

4. *Temperature* A recombination electrolyte battery conducts heat from its plates to the outside very badly since there is only gaseous contact between the container wall and the inner components. In a flooded product the acid acts as the main heat transfer medium and is much more efficient at equalizing temperature differences. The net result is that the recombination electrolyte battery requires a longer time for the inner components to fall to the outside temperature; overnight, for example. This means that the battery appears to perform better at low temperatures, because the inner components may not actually be that cold.

5. *Internal resistance* The internal resistance of the battery is lower mainly because of the properties of the unique separator. This means fewer internal losses when delivering high current, of course.

6. *Voltage* The voltage is inherently higher because of the increased gravity of the acid, and this in conjunction with the lower internal resistance means that the load voltage is also higher. Fewer internal losses and a higher voltage mean that the starter motor can utilize more of the energy available in the battery.

The combined effects of 4, 5 and 6 means that a properly matched starter motor can crank the engine faster at low temperatures while in most cases drawing a smaller current from the battery. This is a very important feature, which enables a further stage of system optimization to take place.

The most common presentation of starter motor/battery characteristics is shown in Figure 18.6 and it can be seen that current is on the horizontal axis and the vertical scales are battery terminal volts, starter terminal volts, pinion speed, output power and electrical efficiency.

The test rig actually measures the first four items, and a computer calculates internal resistance and voltage and automatically plots the composite curves. To produce these curves, the brake resistance is changed from zero to the stalling torque over a period of about 10 s, so the test is appropriate to short time cranking. Tests can be done on batteries chilled to −12, −18 or −30°C. The rig can also be operated at a specific brake resistance to simulate the engine resistance at a specified temperature.

The system design should, of course, ensure that the most difficult cranking condition does not take the operating point too far down the slope of the

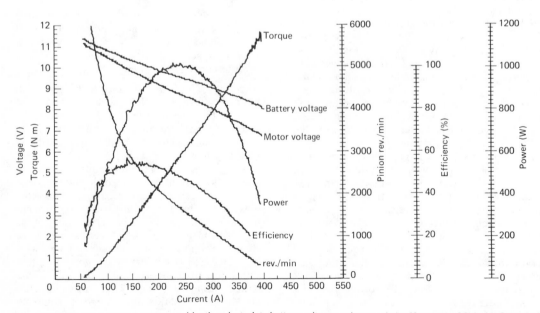

Figure 18.6 Chloride torque starter recombination electrolyte battery volt-amps characteristics (Courtesy of Chloride Batteries)

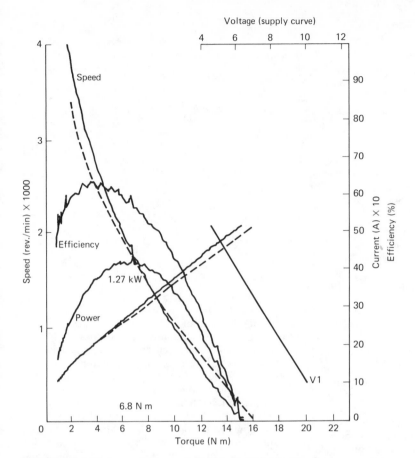

Figure 18.7 Chloride torque starter recombination electrolyte battery. Torque – rev./min curves. Starter motor weight 5.7 kg, free spin at 11.71 V, 9633 rev./min at 8 s, current 73 A at 8 s, coast time 1.85 s (Courtesy of Chloride Batteries)

output power and efficiency curve otherwise very large currents are required. The curves may be plotted with torque on the horizontal axis, which in many ways is more logical since in this context it is the independent variable. An example is shown in Figure 18.7.

Figure 18.8 shows a curve automatically plotted using a test rig during a discharge at −18°C of a 10-plate recombination electrolyte battery under constant-current conditions complying with SAE specifications. For comparison purposes the characteristics of a 12-plate flooded battery are superimposed.

The test requires that after 30 s of discharge at, say, 440 A the voltage must be 7.2 V or more. It can be clearly seen that, for batteries having the same rating on this basis, the recombination electrolyte battery has a higher voltage up to 30 s and then classically falls below that of the flooded product thereafter (owing mainly to acid starvation). However, in starting a petrol engine, probably only the first 15 s of test have any real relevance (if indeed

this rather artificial constant-current discharge test has any relevance at all to cranking).

18.3 Important physical characteristics of antimonial lead battery grid alloys

18.3.1 Mechanical strength

Typical mechanical properties required for battery grids are given in Table 18.1. The necessary mechanical strength of cast lead–antimony grids can be achieved by means of homogeneous or heterogeneous alloying (natural hardness) or age-hardening.

Table 18.1 Required mechanical properties for industrial battery grids

Brinell hardness, H_B	12–14 kg/mm^2
Tensile strength, σ_B	4.5–6.5 kg/mm^2
Elongation, δB	>3%

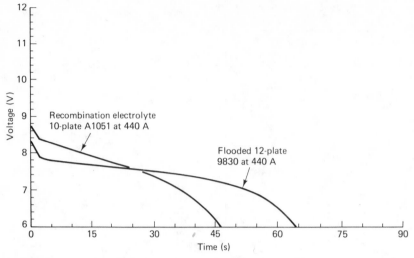

Figure 18.8 Chloride torque starter recombination electrolyte battery. Discharge at −18°C under constant current conditions (SAE) (Courtesy of Chloride Batteries)

Hardening by alloying (natural hardness)

Figure 18.9 illustrates the increase of hardness by using antimony as an alloying element. The curve shows the hardness of as-cast grids *versus* antimony content of the alloys. In the case of low antimony contents (<3% antimony), the antimony forms a solid solution with lead. An alloying element converted into a solid solution always hardens the solvent metal. When the antimony content is increased further, the solubility of antimony in lead is exceeded and two solid phases appear, forming an intimate mixture (the *eutectic* mixture). The percentage of the eutectic mixture increases with rising antimony content above 4% and produces a frame structure of the alloy. This hard eutectic frame structure brings about the high natural hardness of lead–antimony alloys containing more than 4% antimony.

Age-hardening

The above-mentioned solid solution of antimony in lead causes supersaturation when the usual techni-

que of grid casting is applied, because at room temperature the solubility of antimony in lead is extremely low compared to the 3.5% solubility at 252°C. The reduction of supersaturation by precipitation of finely dispersed antimony within the lead–antimony solid solution grains causes the ever-present age-hardening.

The supersaturation, and therewith the age-hardening of the grid, can be increased by fast cooling immediately after casting or by applying additional heat treatment (solution treatment and sudden quenching from about 250°C). Figure 18.10 shows the effect of different treatments on age-hardening. The precipitation of very fine planes (about 10^5–10^{10} planes/cm^2) of almost pure antimony, which are arranged parallel to the (111) planes of lead, causes the age-hardening.

Figure 18.10 also shows that the age-hardening effect decreases with increasing antimony content. This is reasonable, because the hardness of the casted alloy is determined more and more by the

Figure 18.9 Natural hardness of lead–antimony alloys (Courtesy of the Swiss Post Office, Bern)

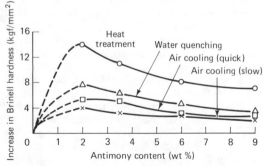

Figure 18.10 Age-hardening of lead–antimony alloys (Courtesy of the Swiss Post Office, Bern)

framework of the eutectic mixture, which also reduces nucleation difficulties for antimony. Therefore, with increasing antimony content, increasing amounts of the originally supersaturated antimony (which produces age-hardening by fine dispersion) are lost due to precipitation of antimony within the eutectic mixture.

Total hardness

The total hardness finally achieved is the sum of natural hardness and hardness due to age-hardening (depending upon age-hardening conditions). Figure 18.11 shows the total hardness attained after 3 weeks. As can be seen, a definite pretreatment can

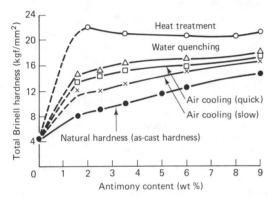

Figure 18.11 Total hardness (natural hardness plus age-hardening) (Courtesy of the Swiss Post Office, Bern)

raise the hardness of grids made from 2% antimony alloys to values which are known for alloys with higher antimony contents. However, it must be mentioned that the amount of hardness due to age-hardening decreases again when the grids are exposed to higher temperature (between 50 and 200°C). This is caused by the dissolving or coarsening of finely dispersed antimony at higher temperature, so that the effectiveness of the antimony dispersion in producing age-hardening is decreased.

Increase of age-hardening by further additives

The age-hardening is also increased by further additives, e.g. arsenic. The influence of arsenic is more effective when the supersaturation of antimony in lead is relatively low (e.g. for grids not subjected to special heat treatment). Figure 18.12 shows this effect; arsenic not only increases, but also accelerates the age-hardening process.

18.3.2 Castability

In addition to the mechanical strength required for further treatment of the grids, castability is an

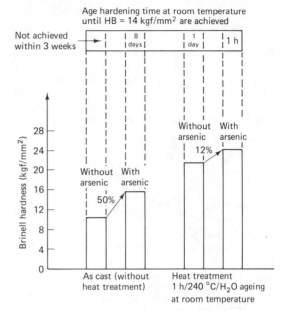

Figure 18.12 Influence of arsenic on the age-hardening of lead–2% antimony alloy (Courtesy of the Swiss Post Office, Bern)

important factor for producing sound grids at tolerable cost. The main difficulty with casting low-antimony alloys is that the decrease of antimony content is accompanied by the appearance of hot cracks if no special precautions are taken. For example, Figure 18.13 shows the cast structure of grids containing 2% antimony. The solidification

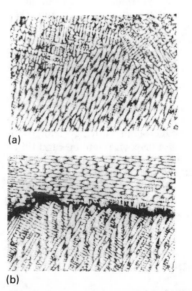

Figure 18.13 Coarse dendritic cast structure with cracks (lead–2% antimony alloy). (a, b) Industrial grid (×100) (Courtesy of the Swiss Post Office, Bern)

takes place in a coarse dendritic structure containing cracks along grain boundaries. Inadequate grid quality (caused by these cracks) was one of the main reasons why low antimony alloys were formerly not used on a large scale in the battery industry.

The poor grid quality of low antimony alloys can be overcome by the addition of selenium. This addition forms a fine globulitic solidification which results in fewer casting faults (Figure 18.14). The

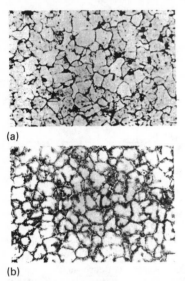

(a)

(b)

Figure 18.14 Fine globulitic cast structure (lead–2% antimony–0.02% selenium alloy). (a) Automotive grid (×200), (b) industrial grid (×200) (Courtesy of the Swiss Post Office, Bern)

grain size of an alloy containing 2% antimony decreases from about 100 μm to about 60 μm when 0.02% selenium is added. With respect to grain refinement, selenium proved to be superior to all other additives. The formation of dendrites, which disturbs the feeding capacity of the mould during casting and leads to casting faults, is almost fully suppressed. With fine globulitic solidification, uniform mechanical properties are achieved in all directions, and hence ductility is increased.

Using the combination of alloying elements of antimony (1.5–3.5%) arsenic, selenium (and tin to increase the castability) with lead, it is feasible to manufacture battery grids with very low antimony contents, which achieve the necessary mechanical strength within suitable ageing times, and show the necessary grid quality.

18.4 Lead alloy development in standby (stationary) batteries

Until 10 years ago, the grids of lead–acid batteries were usually made of lead–antimony alloys containing 5–11 wt% antimony. The necessary mechanical strength and castability are easily achieved with this content of antimony. However, the unavoidable corrosion of the positive grid liberates antimony from the grid, which proves to be both favourable as well as unfavourable with respect to battery performance. On the one hand, antimony stabilizes the active material of the positive electrode, the cycle life of the battery is improved and passivation effects disturbing mainly the discharge mechanism are not observed when alloys with a high antimony content are used for positive grids. On the other hand, antimony migrates to the negative plate where it is precipitated and reduces the hydrogen overvoltage. This leads to lower charge voltage, increased self-discharge and therefore increased water loss of the battery.

As a consequence, most battery manufacturers tried to minimize or even eliminate the antimony addition, especially in batteries for stationary applications where smaller demands are made in respect to cycle service.

It has been found recently that, when selenium is used as an alloying constituent in grid metal, the antimony content can be reduced to the point that the dxrawbacks of antimony are almost eliminated, while the positive effects of antimony are largely retained.

Lead alloys characterized by the addition of selenium exhibit a fine grain structure even at very low antimony contents (<2%). This structure yields acceptable quality of the castings as indicated by mechanical strength, natural hardness, age-hardening and total hardness, and also yields acceptable castability characteristics. The hardness necessary for handling the battery plates is achieved by age-hardening due to finely dispersed antimony.

Corrosion stability at constant potential or constant current proves to be slightly superior, but of the same order of magnitude as conventional antimony alloys (≥4%).

However, at constant cell voltage, the amount of antimony released from the positive grid is too small to affect the potential of the negative electrode markedly. Therefore the very low initial gassing rate, which is of the same order as experienced with antimony-free batteries, remains almost constant.

Of special importance is that the well known stabilizing effect of antimony with respect to the positive active material is virtually maintained with these selenium alloys. This results in capacity stability as well as good cycling performance.

Because of the extremely high potential of the positive electrode, only lead can be used as grid material in stationary and, indeed, all types of lead–acid battery. This lead is unavoidably subject to gradual erosion from corrosion. For battery applications this corrosion rate must be reasonably low, otherwise the battery life is limited by the corrosion rate of the positive grid.

The potential of the electrode (the most important parameter) determines the electrochemical corrosion of lead. Therefore, potentiostatic corrosion tests are very suited to comparing corrosion rates for different grid alloys. Figure 18.15 shows the results obtained for different lead–antimony alloys. The samples used in these experiments were metal rods prepared carefully to avoid any sources of error caused, for example, by poor casting quality. The applied corrosion potential corresponds to about 2.6 V cell voltage.

As can be seen in Figure 18.15, the potentiostatic corrosion rate is of a comparable order of magnitude for all tested antimonial lead alloys. On the basis of

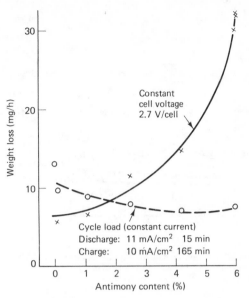

Figure 18.16 Corrosion test on battery grids (tubular type, 70 A h, plate) (Courtesy of the Swiss Post Office, Bern)

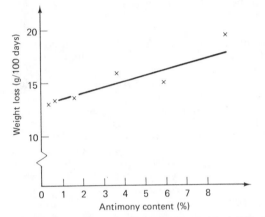

Figure 18.15 Lead corrosion at constant potential −1.45 V with reference to the Hg/Hg$_2$SO$_4$ electrode (Courtesy of the Swiss Post Office, Bern)

these experiments the selenium alloys can be expected to perform slightly better with regard to corrosion attack than conventional lead–antimony alloys.

Corrosion tests on battery grids are usually carried out in cells using a negative battery plate as the counter electrode. Figure 18.16 shows results for battery grids with reference to antimony content. As can be seen, the result is fundamentally different, depending on the testing procedure. When a cycling regimen is applied, which means constant current is impressed on the electrode, the differences in corrosion attack are small. Actually the performance should be expected to be very similar to the result shown in Figure 18.15 because constant current more or less means constant potential of the positive electrode.

However, when constant cell voltage is applied in the experiment, the antimony content of the corroding grid largely determines the result, as can be seen in Figure 18.16. The reason is that antimony, released at the corroding positive grid, is precipitated at the negative electrode, which results in a reduced hydrogen overvoltage. This means that the rate of hydrogen evolution is the same at a less negative electrode potential. Therefore, the negative electrode potential is shifted in a positive direction and consequently the positive electrode potential is raised because the cell voltage is kept constant. The rise in positive electrode potential leads to an increase in corrosion, which again produces antimony at a faster rate and therefore gives rise to increased contamination of the negative electrode. In this way a self-accelerating process originates, causing the strong effect of antimony content shown in Figure 18.16. The corrosion test at constant cell voltage therefore does not compare pure corrosion rates, but rather is a mixture of corrosion and potential shift by antimony contamination of the negative electrode. However, this test reflects the situation in many battery applications, because lead–acid batteries are to a large degree charged at constant voltage, especially stationary batteries which are kept on a constant floating voltage.

Therefore, battery performance in the field to a large extent corresponds to the constant voltage corrosion test. Compared with the bare grid, the rate of antimony release is somewhat reduced at a complete positive plate, because antimony is absorbed to a marked degree by lead dioxide.

This antimony effect on battery performance is shown by the results of an accelerated floating test shown in Figure 18.17. Industrial batteries of the same type were used, with the only difference that the positive grids consisted of various alloys. The floating voltage of 2.23 V/cell was applied as usual for stationary batteries. The acceleration of the test

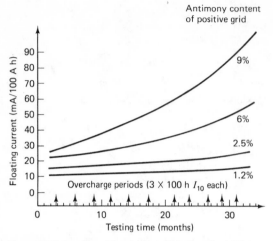

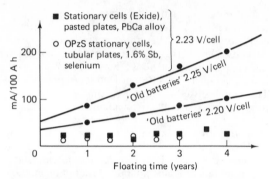

Figure 18.18 Floating test on stationary batteries: I. T. Gerber, Technische Mitteilunger PTT, Nos 6/7 and 11, 1976, Bern, Switzerland (8–11% antimony alloys) (Courtesy of the Swiss Post Office, Bern)

Figure 18.17 Accelerated floating test. Tubular cells, 280 A h, floating voltage 2.23 V/cell (Courtesy of the Swiss Post Office, Bern)

was achieved by periods of heavy overcharge of 3 times 100 h at a 3 month time interval.

Figure 18.17 shows the situation of constant voltage floating described above in relation to Figure 18.16. At an antimony content of less than 2% in the positive grid, the amount of antimony reaching the negative electrodes is obviously too small to cause a marked effect on the electrode potential, and the floating current is therefore stabilized at the low initial level.

In Figure 18.18, results of floating tests on stationary batteries with selenium alloys are compared with tests published by the Swiss Post Office in 1976. For his well founded experiments, Gerber used industrial batteries supplied by several manufacturers. The grid material of these batteries contained about 10% antimony. Some results on stationary batteries equipped with lead–calcium alloys are also given in Figure 18.18.

This comparison again shows the advantage of the extreme reduction of antimony content that is feasible in selenium alloys. The floating current at 2.23 V/cell is even smaller than the floating current

for 'old batteries' at 2.20 V/cell. Furthermore, for selenium batteries the increase of floating current is much smaller, which means that the self-accelerating process of corrosion and potential shift has nearly ceased. Topping up intervals are long and battery shelf life is high.

As already mentioned, the disadvantages of antimony apart, the antimony content in the positive grid material contributes beneficial effects for the lead–acid battery. Antimony stabilizes the positive active material, which results in good cycling performance and less sensitivity to deep discharges. This beneficial effect of antimony is illustrated in Figure 18.19, which shows the results of a cycling test, usually applied for motive power batteries. Each cycle in this test includes a discharge of 75% of nominal capacity; two cycles are performed per day. At intervals of about 50 cycles, the capacity is determined by complete discharge. For the single cell, the test is finished as soon as the capacity is less than 80% of nominal capacity.

The cells used for the test shown in Figure 18.19 differed only in antimony content of the positive grid. The stabilizing effect of the antimony is obvious. While the cells without any antimony failed after about 200 cycles, the cells equipped with selenium alloy containing only 1.6% antimony

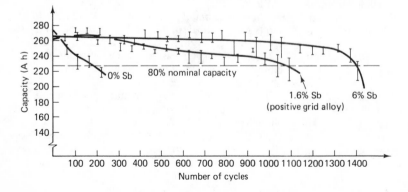

Figure 18.19 Cycling test according to IEC Publication 254 (Courtesy of the Swiss Post Office, Bern)

performed about 1000 cycles. This stabilizing effect is obviously an important advantage of selenium alloys. Even for batteries that are not cycled, this stabilizing effect means less sensitivity to occasional deep discharges.

18.5 Separators for lead–acid automotive batteries

Typical properties for separators for antimonial lead and non-antimonial lead automotive batteries are shown in Tables 18.2 and 18.3.

Table 18.2 Typical properties of separators for antimonial automotive batteries

	Cellulose	PVC
Electrical resistance $(m\Omega/cm^2)$	3.0–4.9	3.0–5.0
Backweb thickness (mm)	40–91	25–36
Pore size (μm)	35	24
Porosity (%)	60	35–40
Mechanical strength	Good	Fair
Corrosion resistance	Fair	Good
Enveloping capability	No	With difficulty
Selling price	Lowest	Moderate

Table 18.3 Typical properties of separators for non-antimonial automotive batteries

	Microporous polyethylenes	Sintered PVC
Electrical resistance ($m\Omega/cm^2$)	1.4–1.8	2.5–3.5
Backweb thickness (mm)	25	25–35
Acid displacement (cm^3/m^2)	146	250
Porosity (%)	55	35
Oxidation resistance	Good	Good
Maximum pore size (μm)	<0.1	24
Flexibility	Good	Poor
Sealability	Good	Satisfactory
Chemical purity	Good	Good
Cold voltage $-30°C/30\,s$ (V)	8.76	7.98

The following are considered to be the key properties for separators for non-antimonial lead maintenance-free automotive batteries:

1. Low electrical resistance to maximize cold cranking voltage.
2. Thin backweb to minimize acid displacement.
3. Good oxidation resistance to give separator integrity at the end of battery life.
4. Small pore size to minimize dendritic shorting.
5. Flexibility and capability of being enveloped.
6. Good chemical purity, i.e. low concentrations of chloride and certain metals (iron, manganese, copper).

Recent work has shown that two parameters, namely low electrical resistance and low separator thickness, have a profound effect on cold cranking performance of low-maintenance calcium–lead alloy automotive batteries. Figures 18.20 and 18.21 show the beneficial effect of decreasing the electrical resistance of the separator on the cell voltage in tests run at $-18°C$ and $-30°C$ on low-maintenance lead–acid batteries.

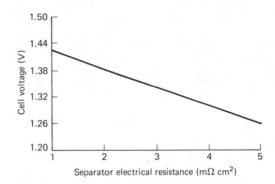

Figure 18.20 18°C cold cranking performance: effect of separator electrical resistance. 30 s cell voltage at 280 A, nine-plate calcium alloy single cells SAE J-537; test procedure (Courtesy of W. R. Groce)

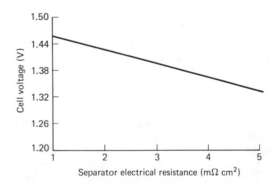

Figure 18.21 $-30°C$ cold cranking performance: effect of separator electrical resistance. 30 s cell voltage at 210 A, 9-plate calcium alloy single cells, SAE J-537; test procedure (Courtesy of W. R. Groce)

Figure 18.22 shows that reducing the separator thickness improves cell voltage under high-rate discharge conditions. Although decreasing the separator thickness has a beneficial effect on cold cranking performance of the battery, it does have a deleterious effect on reserve capacity. This is because, with thinner separators, the volume of sulphuric acid adjacent to the plates is diminished, thereby decreasing the utilization of the lead and lead dioxide active material in the plates. This is illustrated in Figure 18.23. Thus there has to be a trade-off in selection of separator thickness to achieve a desired compromise between acceptable

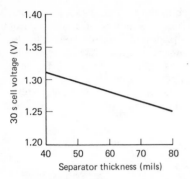

Figure 18.22 Effect of separator thickness on cold performance. –18°C cold cranking test at 280 A, SAE J-537; test procedure. Nine-plate antimonial alloy cells/ cellulosic separators; 29 observations (Courtesy of W. R. Groce)

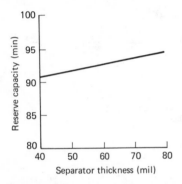

Figure 18.23 Effect of separator thickness on reserve capacity, SAE J-537; test procedure, 9-plate antimonial alloy cells/ cellulosic separators. Reserve capacity = 87.1 + 0.096 × separator thickness (Courtesy of W. R. Groce)

cold cranking performance on the one hand and acceptable reserve capacity on the other.

The shift away from a lead–calcium alloy battery to a hybrid construction employing a low-antimony positive grid and a lead–calcium alloy negative grid has reduced the necessity for having a microporous separator in all battery designs. Enveloped polyethylene microporous separators are still widely used in batteries of the lead–calcium type and are recommended at virtually all plate spaces. However, as one moves to the hybrid construction, the necessity for a microporous separator in all battery designs has been reduced. When plate spacings are less than about 1 mm, enveloped microporous separators are still recommended. This separator construction precludes side and bottom shorts due to mossing or dendrite growth. At the closer plate spacing, shorting through the separator can be a problem and the use of a microporous polyethylene separator virtually eliminates this failure mode. Low electrical resistance leaf-type separators can be substituted for enveloped microporous separators when plate spacings are greater than 1 mm, since the

the propensity for shorting at the greater plate spacing is reduced. The need for a low electrical resistance leaf-form separator can currently be satisfied by what is generically called a glass separator. The glass separator, because of its low electrical resistance, allows a significant improvement in the cold cranking performance of the battery, compared with similar constructions using either cellulosic or sintered PVC separators.

The separator needs for cycling batteries are somewhat different from those for automotive batteries. In the case of cycling batteries, low electrical resistance is not as important a parameter, but good oxidative stability, low acid displacement, improved voltage control and increased resistance to shorting are some important properties. The conventional glass matted paper separator, the cellulosic separator and the glass matted and a glass matted glass separator are suitable for this application.

18.6 Further reading

Beckmann, H. (1930) The microporous rubber diaphragm in storage batteries. *Elektrotechnische Zeitschrift*, **51**, 1605–1607 (in German)

Booth, F. (1970) *Comparative Study on Battery Separators for Lead–Acid Starter Batteries* (printed as manuscript, 12pp.)

Dafler, J. R. (1978) Resin-bonded cellulose separators: an overview with prognoses. *Journal of the Electrochemical Society*, **125**, 833–842

Dafler, J. F. *et al.* (1976) An X-ray diffraction, scanning electron microscopy study of separator failure. *Journal of the Electrochemical Society*, **123**, 780–789

Goldberg, B. S. *et al.* (1983) Accelerated cycle life testing of lead–acid golf car batteries and the influence of separator type on battery life, energy consumption and operating cost. *Journal of Power Sources*, **10**, 137–148

John, P. J. (1965) Battery separators. *Batterien*, **19**, 827–833

Karr, C. (1980) The effect of separator upon cranking capacity. *IBMA 43rd Convention*, Chicago, 1980, 49–53

Landers, J. J. (1970) Requirements and characteristics of secondary battery separators. In *Proceedings of the Meeting of the Electrochemical Society*, February 1970, pp. 4–24

Lundquist, J. T. Jr (1983) Separators for nickel–zinc batteries. *Journal of Membrane Science*, **13**, 337–347

Orsino, J. A. *et al.* (1962) *Structure and Properties of Storage Battery Separators*. Publication No. 253-62, National Lead Co., Brooklyn, NY, USA

Palanichamy, S. *et al.* (1968) Importance of physical characteristics of separators on the performance of lead–acid battery. In *Proceedings of the Symposium on Lead–Acid Batteries*, Indian Lead–Acid Information Center, Calcutta, 1968, pp. 36–40

Robinson, R. G. *et al.* (1963) Separators and their effect on lead–acid battery performance. *Batteries*, **1**, 15–41

Sundberg, E. G. (1970) A new microporous polymeric separator. *Proceedings of the Meeting of the Electrochemical Society*, February 1970, pp. 32–56

Vijayavalli, R. *et al.* (1976) *Separators for Lead–Acid Batteries*, Indian Lead Zinc Information Center, New Delhi, 10 pp.

Yamasaki, K. *et al.* (1983) A new separator for industrial lead–acid batteries. *New Materials + New Processes*, **2**, 119–123

Yankow, L. *et al.* (1983) Studies of phenolic resin-based microporous separator materials. *Journal of Applied Electrochemistry*, **13**, 619–622

Zehender, E. *et al.* (1964) The influence of separators on antimony migration in lead–acid batteries. *Electrochimica Acta*, **9**, 55–62 (German with English summary)

19

Nickel batteries

Contents

19.1 Nickel–cadmium secondary batteries

Both sealed and open nickel–cadmium batteries are based on similar chemical reactions. The sealed type is designed to be maintenance free and under normal conditions will not release gas, whereas an open battery has been designed to release gases when necessary, and requires topping up and, in some cases, a complete electrolyte change. A battery cell consists of two plates containing nickel hydroxide (positive active material) and cadmium hydroxide (negative active material) kept apart by a separator to maintain electronic insulation. An approximation of the chemical equation is given below (see Figure 19.1).

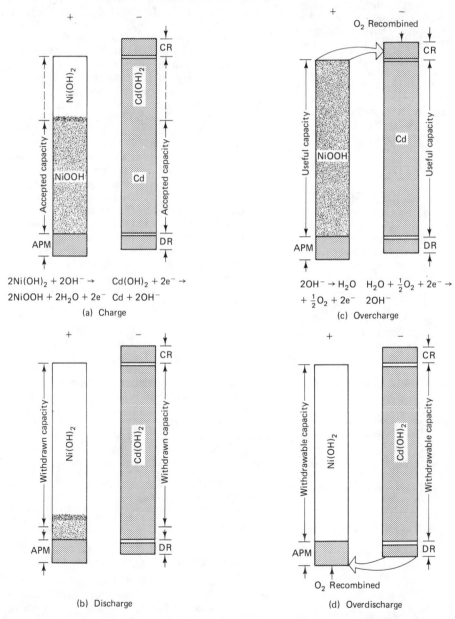

$$2Ni(OH)_2 + 2OH^- \rightarrow$$
$$2NiOOH + 2H_2O + 2e^-$$

$$Cd(OH)_2 + 2e^- \rightarrow$$
$$Cd + 2OH^-$$

(a) Charge

$$2OH^- \rightarrow H_2O$$
$$+ \tfrac{1}{2}O_2 + 2e^-$$

$$H_2O + \tfrac{1}{2}O_2 + 2e^- \rightarrow$$
$$2OH^-$$

(c) Overcharge

(b) Discharge

(d) Overdischarge

Figure 19.1 The electrochemical process in sealed nickel–cadmium batteries. (a) Charge on the positive electrode produces nickel(III) hydroxide, NiO(OH); charge on the negative electrode produces metallized cadmium, Cd. (b) Discharge on the positive electrode produces nickel(II) hydroxide, Ni(OH)$_2$; discharge on the negative electrode produces cadmium hydroxide, Cd(OH)$_2$. See Equations 19.1 and 19.2. Electrolyte, dilute potassium hydroxide; APM, antipolar mass; CR, charge reserve; DR, discharge reserve (Courtesy of Varta)

On discharge (Figure 19.1),

$$2NiO(OH) + Cd + 2H_2O \underset{charge}{\overset{discharge}{\rightleftharpoons}} 2Ni(OH)_2 + Cd(OH)_2$$

| Positive electrode | Negative electrode | Positive electrode | Negative electrode |

$$(19.1)$$

Two parts of the charged positive mass of nickel(III) hydroxide change to nickel(II) hydroxide, while the negative mass of cadmium oxidizes to cadmium hydroxide, and two molecules of water are taken from the electrolyte. The above equation holds good on charge or discharge only so long as convertible material is still available. If, for instance, on charge the positive mass of nickel(II) hydroxide in the positive electrode is completely converted to nickel (III) hydroxide, then the reaction is modified and oxygen is evolved; that is, the plate is in an overcharged state:

$$Cd(OH)_2 \underset{overdischarge}{\overset{overcharge}{\rightleftharpoons}} \tfrac{1}{2}O_2 + Cd + H_2O \qquad (19.2)$$

If the same condition is applied to the negative plate, hydrogen gas will be evolved. With an open battery, both conditions will occur at the end of every charge and water will be decomposed.

Note that the electrolyte potassium hydroxide is not mentioned in Equations 19.1 and 19.2.

Water is produced when the battery is charging and taken up when it is discharging; when the electrolyte is in the correct proportion, this water has very little effect on the operation of the cell. In comparison, on discharge, if one of the electrodes becomes exhausted and the discharge current is maintained by the voltage of the other cells, then hydrogen is given off by the positive plate and oxygen by the negative plate. Also, in a discharged condition, the battery polarity can be reversed.

19.1.1 Sealed nickel–cadmium batteries

In the modern sealed nickel–cadmium battery the active mass is modified so that the battery is 'safe from overcharge' and protected against 'overdischarge' due to oxygen only being evolved. In the case of overcharge this is achieved by the negative plate having a surplus amount of cadmium hydroxide incorporated in its structure, which ensures that the negative plate is partially charged when the positive plate is fully charged; this surplus mass is called the 'charge reserve' (Figure 19.2). The process is shown in Figure 19.1(c), which also shows that the oxygen produced on the positive plate is recombined on the negative plate. This electrochemical reaction causes heat to be generated within the battery, unlike an open battery where the gases would be allowed to vent or escape from the plate area. It should be noted that the resultant temperature rise is not caused by the internal resistance of the battery. The capacity of a sealed battery increases in proportion to its volume, but the surface area, which determines the temperature rise, does not increase by so great a proportion. Therefore, it can be seen that it is necessary to pay more attention to heat generation on overcharge in the larger-capacity sealed cells than in smaller-capacity batteries; generally 2 A h can be used as a dividing line.

When the overcharge current is terminated the free oxygen in the battery continues the reaction at the negative electrode. The negative electrode itself supplies the necessary electrons and some of the metallic cadmium is oxidized to cadmium hydroxide. The oxygen pressure is therefore decreased, and after a short time a vacuum may be formed.

The ability of the negative plate to consume the oxygen produced by the positive plate on overcharge is due to the following design features being incorporated into sealed batteries:

1. The capacity of the positive plate is less than that of the negative plate.
2. The plate spacing is small (usually about 0.2 mm).
3. The separators are very porous.
4. The quantity of electrolyte is just on the saturation limit of the plates and separator.

To protect the sealed battery against overdischarge, at least one manufacturer (Varta) adds a controlled quantity of negative mass to the positive electrode and this is called the *antipolar mass* (see Figure 19.3). The antipolar mass does not disturb the normal function of the positive electrode during charge and discharge since it is electrochemically

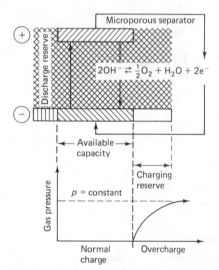

Figure 19.2 Schematic diagram of the charging process in sealed nickel–cadmium batteries with negative charging reserve (Courtesy of Varta)

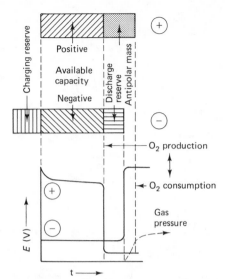

Figure 19.3 Schematic diagram of the discharge process in sealed nickel–cadmium batteries with antipolar mass in the positive electrode (Courtesy of Varta)

ineffective cadmium hydroxide; this reaction is shown in Figure 19.1(d).

Thus, with the help of charge reserve and antipolar mass, sealed batteries can be made comparatively safe in normal operating conditions.

There are three main configurations of nickel–cadmium batteries – button cells, cylindrical cells and rectangular cells. These are discussed below.

Button cells

These cells have the form of a button in various thicknesses. They are composed of a stack of disc-shaped sintered plates and separators held in two nickel-plated steel cups, one fitting into the other and pressed together with an insulating gasket (Figure 19.4). There is also available a high-

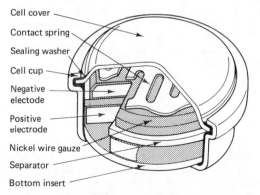

Figure 19.4 Eveready (Berec) sealed nickel–cadmium button cell construction (Courtesy of Eveready (Berec))

reliability version of the button cell with improved performance on high-rate discharge at very low temperatures, e.g. SAFT VBE services button cells (Table 51.1). These have the same dimensions and capacities as the standard button cells and are designed to withstand more severe operating conditions. They have an improved voltage characteristic at low temperatures, high discharge rates, and higher mechanical strength and reliability.

Button-type cells consist of a cup (positive pole) and a battery cover (negative pole); the electrodes are tablets wrapped in a microporous nickel or nickel-plated wire gauze and separated by a fine gauze separator, which is frequently made of non-woven polyamide. The capacity of this type is up to about 3 A h at a voltage of 1.24 V. Typically the size of the battery for this capacity is 50 mm diameter × 25 mm thick, with a weight of 135 g, and permissible operating temperatures of −20 to +45°C. These cells are also available in welded and plastics-covered packs of up to ten cells. SAFT, for example, supply five sizes of button cell from 40 to 600 mA h in their standard (VB) and improved performance (VBE) series. Button cells can be assembled in a rigid case to make packs with discharge voltages between 2.4 V (2 cells) and 28.8 V (24 cells). SAFT supply packs made up from VM 4, 10, 22, 30 and 60 cells (capacity 40, 100, 220, 300 and $600C_5$ mA, Table 51.1). The final assemblies have, of course, the same rated C_5 capacity as the individual cells but vary in discharge voltage between 2.4 V (2 cells) and 28.8 V (24 cells).

Varta and other suppliers produce assemblies of two to ten button cells. Assemblies are available in stock formats, including insulating sleeve, plastic cassettes, plastic boxes and sheet steel cases. Also available are environment-proof assemblies in which the interconnected cells are resin potted in a plastics container. An advantage of button cell packs is that these can operate as very low compact low-voltage capacitors giving much greater energy storage in a much smaller space. VB button cell packs can deliver very high currents, the maximum discharge rate being $10C_5$ mA. It should be ensured that the packs are not discharged to below 1 V/cell. Short-duration peak discharges may be as high as $50–100C_5$ mA at 0.8 V/cell.

Cylindrical cells

A cutaway diagram of a typical cylindrical cell is shown in Figure 19.5. Cylindrical cells can be obtained in standard battery packs (6–12 V) and extended range modular packs (12 and 24 V, Table 51.1), both in a range of sizes. Standard packs provide for applications requiring high performance on discharge at large currents either continuously or in short-duration peaks. The extended range modular packs are supplied in a plastics case in the range

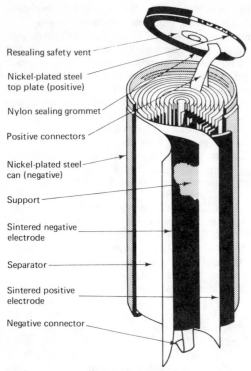

Resealing safety vent

Nickel-plated steel top plate (positive)

Nylon sealing grommet

Positive connectors

Nickel-plated steel can (negative)

Support

Sintered negative electrode

Separator

Sintered positive electrode

Negative connector

Figure 19.5 Ever Ready (Berec) sealed cylindrical nickel–cadmium cell construction (Courtesy of Ever Ready (Berec))

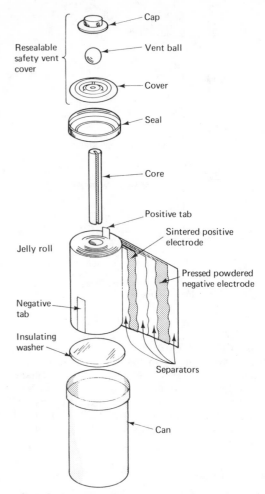

Resealable safety vent cover

Cap

Vent ball

Cover

Seal

Core

Positive tab

Sintered positive electrode

Pressed powdered negative electrode

Jelly roll

Negative tab

Insulating washer

Separators

Can

Figure 19.6 Eveready 'Sub C' sealed high-rate nickel–cadmium battery with jelly roll construction (Courtesy of Union Carbide)

0.5–10 A h at 12 V and 24 V. (Rated capacity is for discharge at the 5 h rate at 20°C after nominal charge and for a terminal voltage of 1.0 V/cell.) The packs can be discharged steadily at up to the rated capacity and can be given a nominal charge at a constant current of $0.1C_5$ for 14 h irrespective of their initial state of charge. Alternatively, the packs can be given a fast charge. They will withstand a continuous charge of up to $0.05C_5$ and will operate in the temperature range −40 to +50°C. They may be stored indefinitely in any state of charge without deterioration in the above temperature range.

The design of Union Carbide Eveready high-rate cylindrical nickel–cadmium cells features a pressed powder negative electrode, which consists of dry blended active materials pressed into an expanded metal carrier. This is claimed to give the cell outstanding cycle life and long-term overcharge capability with essentially no fade and with little or no memory effect. A number of the Eveready cylindrical cells feature the 'jelly roll' construction illustrated in Figure 19.6.

Most of the Eveready high-rate cylindrical cells have a resealing pressure vent; the others have a puncture-type failsafe venting mechanism. The resealing pressure vent is shown, for the 'Sub C' cell, in Figure 19.6; this vent permits the cell to release

excess gas evolved if the cell is abused, for example. When the internal pressure has dropped to an acceptable level, the vent will reseal, permitting the cell to be recycled in the normal manner with little or no further loss of electrolyte or capacity.

Manufacturers usually provide a complete specification for each of their range of batteries. A sample specification for a battery in the Union Carbide range is shown in Table 19.1 and Figure 19.7.

Large vented nickel–cadmium rectangular batteries

A typical cell design of a large vented nickel–cadmium battery is shown in Figure 19.8. The schematic cross-section through a pocket plate shows that the active mass is supported by pockets of perforated, corrosion-protected steel tape (Figure 19.9). The pockets are sealed by folding the tape and then thoroughly connected together mechanically. The active materials used are:

1. Positive plate: nickel hydroxide.
2. Negative plate: cadmium hydroxide.

The plate edges are supported by strips of nickel-coated steel and then pressed together. This method gives a good electrical contact and conductivity to the plate terminals. Plates of the same polarity are welded together to the connector, thus forming a plate group.

Positive and negative plate sets are assembled to form the plate group of one cell. Separators form

Table 19.1 Example specification for Union Carbide sealed nickel–cadmium cells: the Eveready B20 and B20T cells (1.2 V, 20 mA h)

Voltage taps	$-$, $+1.2$
Average service capacity (to 1.1 V) rated at 2 mA	20 mA h
Terminals	Flat contacts or two solder tabs
Average weight	1.13 g
Volume (by displacement)	490 mm^3
Cell	One button (standard moulded electrode)

Average performance characteristics at 21.1°C. B20 and B20T should be discharged only with currents up to 2 mA. With a 2 mA discharge current the cell voltages are:

Initial voltage (voltage under load after 2 mA h have been removed from a fully charged cell)	1.32 V
Average voltage (voltage under load at mid-life, after 10 mA h have been removed from a fully charged cell)	1.22 V
Final voltage (voltage under load after 20 mA h have been removed from a fully charged cell)	1.10 V

Temperature characteristics. Ranges of temperature applicable to operation of the B20 and B20T cell are:

Charge	0 to 45°C
Discharge	-20 to $+45$°C
Storage	-40 to $+60$°C
Charging rate for fully discharged cells	14 h at 2 mA
Charging voltage (for partially discharged cells, reduce time or current proportionally)	1.35–1.50 V
Trickle charge	Not recommended

Internal resistance (B20 or B20T cell) varies with state of charge (tolerance of $\pm20\%$ applies):

Charged	1000 mΩ
Half-discharged	5340 mΩ
Discharged	7800 mΩ

The impedance (B20 or B20T cell) varies with state of charge and frequency as follows (tolerance of $\pm20\%$ applies):

Frequency (Hz)	*Impedance, no load* (mΩ)		
	Cell charged	*Cell half charged*	*Cell discharged*
50	2100	2600	4900
100	1900	2400	4300
1 000	1500	1800	2500
10 000	1250	1450	1700

Soldering	Do not make soldered connections to the B20 cell, use the B20T cell for this purpose

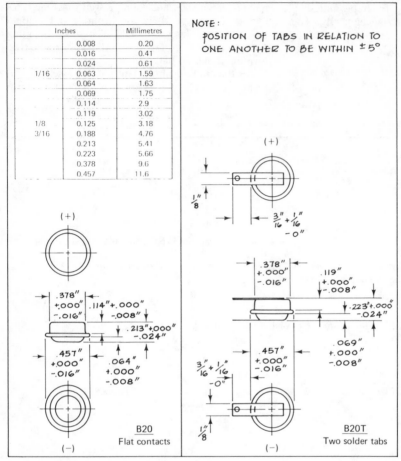

Inches		Millimetres
	0.008	0.20
	0.016	0.41
	0.024	0.61
1/16	0.063	1.59
	0.064	1.63
	0.069	1.75
	0.114	2.9
	0.119	3.02
1/8	0.125	3.18
3/16	0.188	4.76
	0.213	5.41
	0.223	5.66
	0.378	9.6
	0.457	11.6

NOTE:
POSITION OF TABS IN RELATION TO
ONE ANOTHER TO BE WITHIN ±5°

B20
Flat contacts

B20T
Two solder tabs

Figure 19.7 Sample specification: Eveready B20T sealed nickel–cadmium cell, 1.2 V, 20 A h capacity (Courtesy of Union Carbide)

the intermediate layer. The cells for normal discharge purposes have perforated corrugated separators.

The cells for high-current discharge purposes have pin separators.

The terminal bolts are made of nickel-coated steel, giving corrosion resistance, and these are welded to a similar group bar which in turn is welded to the positive and negative plates. The pillars are then located and sealed into the lid with lock nuts. The large diameter of these lock nuts ensures a low voltage drop. Cells of high capacity have a maximum of four terminals.

The filling and venting hole has a captive plug. As an option, a screw plug or explosion-inhibiting vents are available.

Until a few years ago, cell containers and cell lids were made of nickel-coated steel. Today, more and more, they are made of corrosion-proof plastic, which is highly stable against shock and requires less internal insulation within the cell. Its transparency allows an 'at-a-glance' check of the electrolyte level. The cell lids are welded to the container when they are made of steel. If they are of plastic, they are welded or bonded to the container with a solvent adhesive.

The number of cells necessary to form a nickel–cadmium battery is calculated from the required nominal voltage. For instance, a 24 V battery consists of 20 1.2-V cells.

Depending on the size of the cells, from 3 to 10 cells form a unit within a block or a wooden crate. Large cells with a capacity of more than 400 A h are installed as single units.

Cells in steel containers are provided in wooden crates only and are located as a totally insulated unit on lugs welded to the container. Cells in plastic containers are installed in a crate or put together to form a block. These blocks are preferred on economic low-volume considerations. Cells are

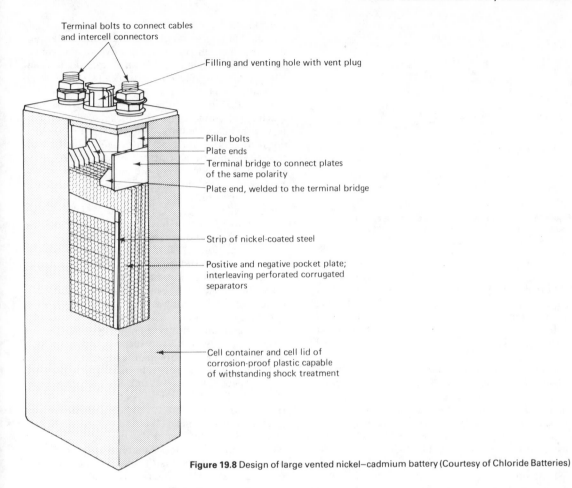

Terminal bolts to connect cables and intercell connectors

Filling and venting hole with vent plug

Pillar bolts

Plate ends

Terminal bridge to connect plates of the same polarity

Plate end, welded to the terminal bridge

Strip of nickel-coated steel

Positive and negative pocket plate; interleaving perforated corrugated separators

Cell container and cell lid of corrosion-proof plastic capable of withstanding shock treatment

Figure 19.8 Design of large vented nickel—cadmium battery (Courtesy of Chloride Batteries)

available on single-tier or double-tier stands or as an insulated unit in a special battery cubicle. By using inter-tier or inter-block connectors, the cells form a complete battery.

If discharged with a constant current, there is, at the beginning of the discharge, as shown in Figure 19.10, a relatively fast voltage drop. Thereafter, at approximately the nominal voltage of 1.2 V, the cell voltage follows a flat line. At the end of the discharge, there is again a fast drop to the final discharge voltage level.

If charged, the chemical reactions are reversed, i.e. there is a reduction of oxygen in the active material of the negative plate and an oxidation of the active material of the positive plate. As shown in Figure 19.10, the charging voltage follows a relatively flat line within the range of 1.4–1.5 V. At a voltage of 1.55 V/cell, gassing occurs and the

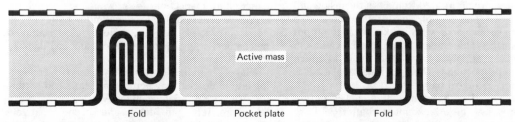

Active mass

Fold   Pocket plate   Fold

Figure 19.9 Design of the pocket plate in large vented nickel—cadmium battery. Pocket plate with perforated upper and lower tap (Courtesy of Chloride Batteries)

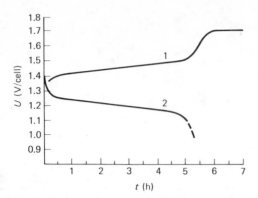

Figure 19.10 Voltage characteristics of nickel–cadmium battery: 1, charged; 2, discharged (Courtesy of Chloride Batteries)

battery has then reached approximately 80–90% of the energy taken out. After a relatively steep ascent up to approximately 1.7 V, the charging curve follows a horizontal line again and the battery is now fully charged.

Comparison of vented lead–acid and nickel–cadmium batteries for standby and motive power applications A nickel–cadmium battery contains plenty of electrolyte. Positive and negative electrodes are kept apart by separators whose only purpose is to keep the electrodes electrically insulated from each other. Lead–acid batteries, on the other hand, have separators which also have the task of preventing sedimentation from the active material and the formation of lead excrescences on the plates. These excrescences can otherwise cause short-circuiting. Since the separators are to a large extent responsible for the internal electrical resistance in the cells, the nickel–cadmium battery has low internal resistance due to the almost entirely open separators in the form of widely spaced grids or rods, and can therefore give excellent load performance.

The active material of most modern types of nickel–cadmium battery is enclosed in pockets of perforated steel strips, which are joined to the plate materials. The steel strips, which form the walls of the pockets, are perforated from both the inside and the outside. This double perforation creates a maximum surface area which makes for high output performance. This means that nickel–cadmium batteries can supply up to three times as much current in proportion to their nominal capacity as is normal for lead–acid batteries.

As well as their extremely high power output characteristics, nickel–cadmium batteries have many advantages from the electrical point of view. In contrast to lead–acid batteries, the electrolyte of the alkaline battery does not change during charging and discharging. It retains its ability to transfer ions between the cell plates irrespective of the charge level. Nickel–cadmium batteries can be left uncharged or partially charged because there is no sulphating process, unlike lead–acid batteries.

Like the plates, the current collectors and mechanical connections of a nickel–cadmium battery are made entirely of steel. The alkaline electrolyte does not react with steel, which means that the supporting structure of the battery stays intact and unchanged for the entire lifetime of the battery. This should be compared with a lead–acid battery, in which the supporting parts and the current collectors are made of lead, which corrodes in sulphuric acid. For this reason the strength of a lead–acid battery decreases progressively.

The corrosion in a lead–acid battery can lead to what is known as sudden death. A supporting part of a plate or some important component is eaten through, breaks and causes a short-circuit. The battery suddenly stops working, or else the performance drops off rapidly. The steel construction of a nickel–cadmium battery guarantees that this can never happen, which is why nickel–cadmium batteries give better reliability than lead–acid batteries.

The alkaline electrolyte in nickel–cadmium batteries keeps its density and low freezing point constant when discharging. As a lead–acid battery runs down, the density of the sulphuric acid decreases, which means that the electrolyte freezes at low temperatures and the battery is damaged. Nickel–cadmium batteries therefore perform better at low temperatures than lead–acid batteries. The risk of damage to the batteries from cold is therefore practically non-existent.

The life of nickel–cadmium batteries is also less affected by high temperatures than that of lead–acid batteries. These batteries work in both tropical heat and arctic cold. The battery range covers temperatures from $-50°C$ to $+55°C$.

All open batteries emit hydrogen and oxygen during the final stage of charging. In nickel–cadmium batteries these are the only gases given off, whereas lead–acid batteries also give off corrosive sulphuric acid fumes.

Modern lead batteries are generally fitted with acid-separating valves, which need regular maintenance and inspection to prevent the escape of corrosive acid fumes. Nickel–cadmium batteries can be safely kept beside sensitive electrical and mechanical equipment without causing any damage to it.

Nickel–cadmium batteries withstand overcharging and undercharging better than lead–acid batteries. They are not deformed by overcharging and can be completely run down and even negatively charged without damage, and can withstand short-circuiting.

Table 19.2 Maximum allowable impurity levels in potassium hydroxide electrolyte (relative density 1.28) used in nickel–cadmium batteries

Impurities	*Maximum concentration (mg/l)*
1. Metals of the hydrochloric acid and hydrogen sulphide group, e.g. arsenic, antimony, tin, bismuth, cadmium, mercury, silver, copper, lead	5
2. Metals of the ammonium sulphide group, e.g. iron, aluminium, chromium, manganese, zinc, cobalt, nickel	20
3. Sodium	5000
4. Halogen and cyanide	150
5. Nitrogen	100
6. Sulphur	5
7. Carbon dioxide	1000
8. Oxidizable carbon (in organic compounds)	50
9. Silicic acid (SiO_2)	50
10. Other non-metals, e.g. phosphorus, boron	10

Table 19.2 presents a list of maximum levels of impurities that can be tolerated in the potassium hydroxide electrolyte used for nickel–cadmium batteries.

19.1.2 The memory effect in nickel–cadmium batteries and cells

The memory effect in a battery is the tendency of the battery to adjust its electrical properties to a certain duty cycle to which it has been subjected for an extended period of time. For instance, if a battery has been cycled to a certain depth of discharge for a large number of cycles, then on a subsequent normal discharge the battery will not give more capacity than that corresponding to the applied cycling regimen.

The memory effect was first noticed by workers at Nife Jungner with sealed sintered plate satellite cells. In the satellite programmes the cells were subjected to well defined discharge/recharge cycles. When after such tests at the 25% discharge level the cells were tested for capacity in a normal manner, it was found that the cells performed in accordance with Figure 19.11. The capacity was only about 25% to the normal cut-off voltage of 1.0 V. However, full capacity was obtained to a cut-off voltage of 0.6 V.

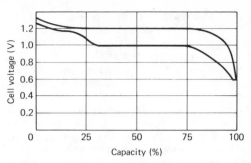

Figure 19.11 Memory effect in sealed nickel–cadmium batteries: tests at 25% discharge levels (Courtesy of Union Carbide)

Thus the capacity was available but at a much lower voltage than normal. Accordingly, the effect can be described as an apparent reduction in cell capacity to a predetermined cut-off voltage resulting from highly repetitive use patterns.

The memory effect becomes progressively pronounced as the number of charge/discharge cycles increases. The effect is more likely to occur when the amount of overcharge in each cycle is small and the rate of discharge is high. Also, an elevated temperature will accelerate the development of the memory effect. The memory effect can normally be erased by a full discharge followed by a full charge; thus it is a temporary effect. The memory effect does not manifest itself when the cells are subjected to random charge/discharge regimens, which is typical for most applications. It is important to note that it is only the sintered type of nickel–cadmium cell that exhibits the memory effect. The effect has been noticed in both sealed and vented sintered plate cells. Pocket plate nickel–cadmium cells and batteries do not develop the memory effect under any circumstances.

Although it is not fully clear what causes the memory effect, indications are that memory is connected mainly with physical changes in the negative electrode involving the formation of intermetallic compounds of nickel and cadmium. These are produced by interactions between charged, negative active material and the large surface area of the sintered nickel support. Scanning electron micrographs have shown that negative electrodes with memory contain a greater number of large cadmium hydroxide particles than do normal negative electrodes. During discharge the reactions proceed uniformly when the cadmium active material consists of very small particles as is the case in a cell without memory effect. On the other hand, if the cadmium material contains both smaller and larger crystals, as in a cell with memory, the smaller particles will react first and become discharged at a normal voltage, and then the larger particles will discharge. Because the current density will be higher during the discharge of the larger particles, the polarization will increase and there will be a loss of voltage in the cell. Accordingly, the discharge curve will exhibit two voltage plateaux.

The reasons why the cadmium active material changes its particle size distribution during memory cycling are not fully understood. It is well known that on cycling sintered-type cadmium electrodes do not exhibit crystal growth concurrent with a redistribution of the cadmium to the outer regions of the electrodes. However, this mechanism does not seem to be applicable in the present case. Instead, the present theory is that there is a loss of contact between the active particles and the electron conducting material of the electrode occurring during the memory cycling, resulting in the special kind of crystal growth previously discussed.

The difference between sintered plate cells and pocket-type cells with regard to memory may be connected with the fact that the pocket cadmium active material contains an addition of finely divided iron compounds. This addition is made to prolong life by preventing recrystallization and agglomeration of cadmium particles. It seems probable that the iron addition will not only prevent the normal tendency for crystal growth of the cadmium material, but will also eliminate the particle size redistribution that causes the memory effect.

19.1.3 Recent developments in rechargeable nickel–cadmium battery design

As an indication of possible future developments in nickel–cadmium sealed battery design, there follows a brief discussion of one such design under development by Eagle Picher in the USA. This cell incorporates the superior gas recombination design provisions of the nickel–hydrogen system. The evolution of this hybrid metal-to-gas couple has resulted in the development of certain interesting and very successful internal cell design innovations, and this new technology could easily be applied to the sealed nickel–cadmium system.

By incorporating the gas recombination design provisions of the nickel–hydrogen cell, the requirement for a high porosity separator material in the sealed nickel–cadmium cell is eliminated. High electrolyte absorption and retention, and non-temperature and non-time (cycle life) degradable inorganic separator materials such as potassium titanate or asbestos may be used. This same gas recombination also renders the sealed nickel–cadmium system less sensitive to electrolyte level increases of up to 50%, eliminating the need to operate the cell in a virtually starved electrolyte condition. Finally, by incorporating a small section of the nickel–hydrogen gas electrode material in unique electrical contact with the cell positive terminal, a hydrogen gas recombination mechanism

(during charge and overcharge) is introduced. If, during operation, the cell is stressed to the point at which the hydrogen gas overvoltage potential of the negative (cadmium) electrode is achieved, the evolved gas is rapidly recombined at a low equilibrium pressure and one of the more serious problems associated with the sealed nickel–cadmium system is reduced.

The modified Eagle Picher nickel–cadmium cell incorporates the 'split negative' design in which a single negative or cadmium electrode is replaced by two cadmium electrodes each of which is approximately one-half the thickness of the original. The two back-to-back negative electrodes are then separated by a gas spacer component to facilitate gas access, as in the nickel–hydrogen cell. To further enhance oxygen gas recombination, a thin film of Teflon may be applied to the back of each negative electrode (Figure 19.12).

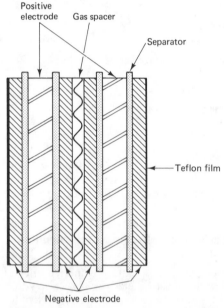

Figure 19.12 'Split negative' electrode stack design: Eagle Picher sealed nickel–cadmium cell (Courtesy of Eagle Picher)

To introduce a hydrogen gas recombination ability into the design, a section of the catalytic gas electrode from the nickel–hydrogen system is incorporated. A strip of this material, approximately the same width as the narrow edge of the cell, is wrapped in a U-shaped fashion around the entire edge of the electrode stack and connected electrically to the positive terminal of the cell.

Eagle Picher fabricated test cells incorporating various design features as follows:

Design A: Non-woven nylon (Pellon 2505) separator material.

Design B: Non-woven nylon (Pellon 2505) separator material with gas electrode.

Design C: Asbestos mat (fuel cell grade) separator material with gas electrode.

Design D: Asbestos mat (fuel cell grade) separator material.

Design E: Non-woven (Pellon 2505) separator material with Teflon film.

Design F: Asbestos mat (fuel cell grade) separator material with Teflon film.

The stainless steel container, incorporating two ceramic-to-metal seals, was approximately 114.3 mm tall (excluding terminals) by 76.2 mm wide, by 22.9 mm thick. To simulate a thin negative electrode design using available materials, the positive electrodes used consisted of two 0.0064 mm electrodes in a back-to-back configuration offering in effect a 0.0127 mm electrode (positive). Each test cell incorporated five 0.0127 mm positive electrodes and ten 0.0064 mm negative electrodes (Figure 19.12). The capacity of a cell was approximately 20 A h and the negative-to-positive capacity ratio was approximately 1.3:1. All cells were activated with electrolyte to a level of approximately 3.5 cm^3/A h.

The influence of cell design on overcharge pressure characteristics, that is, potential improvement in oxygen recombination characteristics, is illustrated in Figure 19.13.

With the exception of design D, all cells using the asbestos designs appear to recombine oxygen gas satisfactorily at this rate. The high charge voltage associated with design D may have resulted in the hydrogen gas overvoltage potential being exceeded and hence the generation of high pressure. The

hydrogen gas overvoltage potential may have also been exceeded with design C but, in this case, the design incorporated a gas electrode which facilitated hydrogen gas recombination. The relatively lower charge voltage associated with design F may have resulted from the application of the Teflon film which enhanced oxygen gas recombination and maintained the negative electrodes at a lower state of charge. The general higher charge voltage associated with these asbestos separator designs is attributed to a very loose electrode stack assembly required by the use of available components. The nylon separator stacks did not fit snugly in the cell container and the asbestos material was only approximately one-half the thickness of the nylon material. To ensure hydrogen gas evolution for an evaluation of the gas electrode design, the nylon separator cells, designs A, B and E, were subjected to an overcharge at a C/10 rate and a temperature of approximately 0°C. The results of this test are presented in Figure 19.14.

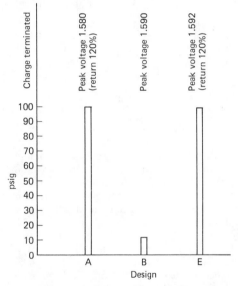

Figure 19.14 Gas electrode *versus* non-gas electrode design: Eagle Picher split negative electrode stack design sealed nickel–cadmium cell. Charge 200%, rate C/10, temperature 0°C (Courtesy of Eagle Picher)

The high charge voltages associated with all three cells indicated that the hydrogen gas overvoltage potential was exceeded; however, design B, incorporating the gas electrode, was capable of recombining the hydrogen gas and was stabilized at a low pressure. The high pressures associated with designs A and E were not appreciably reduced after several days of open-circuit stand.

Electrolyte quantity sensitivity of the advanced design cells with nylon separators (designs A, B and E) was evaluated by increasing the electrolyte

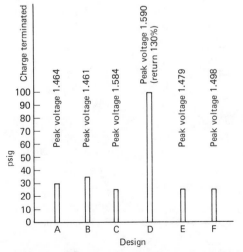

Figure 19.13 Overcharge pressure characteristics: Eagle Picher split negative electrode stack design, sealed nickel–cadmium cell. Charge 200%, rate C/5, temperature 20°C (Courtesy of Eagle Picher)

quantity in each cell by approximately $0.5\,cm^3/A\,h$ and then subjecting the cells to an overcharge at a $C/10$ rate and a temperature of approximately 20°C. The preceding steps were then repeated until all cell pressures exceeded $6.9 \times 10^5\,N/m^2$ during overcharge. The results of this testing are presented in Figure 19.15.

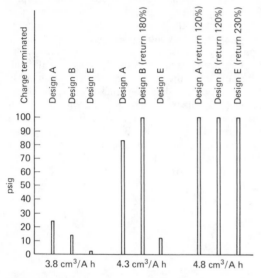

Figure 19.15 Overcharge pressure against electrolyte quantity: Eagle Picher split negative electrode stack design sealed nickel–cadmium cell. Charge 250%, rate $C/10$, temperature 20°C (Courtesy of Eagle Picher)

As may be seen, all cells were able to accommodate the first increment of electrolyte without significant gas recombination degradation. However, after the second increment, the cell designs without the Teflon backing on the negative electrode (designs A and B) exhibited a notable pressure increase, probably as a result of negative electrode flooding. After the third increment of electrolyte, the Teflon film cell design (design E) finally exceeded the $6.9 \times 10^5\,N/m^2$ limit, but even in this case the recombination rate was still sufficient to require an extended period of overcharge to achieve this limit.

Although the test results on the advanced design cell were not conclusive, the projected improved performance characteristics of the advanced design were realized. The split negative design successfully permitted the incorporation of a low-porosity separator material which would significantly increase the list of potential candidate materials. In addition, the split negative design resulted in an improved oxygen gas recombination rate (in this case up to a $C/5$ overcharge rate). The incorporation of a gas electrode was also successful in the recombination of hydrogen gas (again at least up to a $C/10$ overcharge rate). The application of a thin Teflon film to the back of the split negative electrodes was beneficial to all designs, particularly with respect to increases in electrolyte quantities. Although all cells accommodated greater electrolyte quantities than would normally be tolerated in a conventional cell, an even greater quantity could be accommodated in the Teflon film design.

Developments of this type in nickel–cadmium cell design promise even better cell characteristics in the future. In the particular example of battery design improvement discussed above, only minimal further development should evolve a system which offers both greater reliability and a significant improvement in cycle-life capacity.

The teflonated negative plate referred to above was studied as part of an investigation into the elimination of potential failure modes and the enhancement of reliability of the nickel–cadmium system intended for use in satellites. The presence of a thin layer of porous Teflon film wrapped around the surface of the negative electrode was believed to offer the advantage of reducing or eliminating cadmium migration, which is known to occur in present systems over time. The results obtained from cell testing have been very encouraging. The Teflon film does not affect the operating voltage of

Table 19.3 Cycle test sequence of porous Teflon membrane in sealed nickel–cadmium batteries: Eagle Picher cell RSN-50 (50 A h capacity)

Cycle	Cell potential on charge (V)	Cell pressure on charge (i.w.g.)	End-of-discharge voltage (V)
Near Earth orbit cycle			
Cycle: 56 min, 14 A charge 23 min, 14 A discharge 22 min			
Temperature: 4.4°C			
1	1.458	−27	1.232
640	1.463	−24	1.242
1270	1.460	−25	1.238
1920	1.459	−25	1.236
2650	1.462	−25	1.235
Synchronous orbit cycle			
Cycle: 24 h, 5 A charge 22.8 h, 25 A discharge 1.2 h			
Temperature: 26.6°C			
3	1.405	−14	1.217
14	1.412	−2	1.194
27	1.413	2 psi	1.199
44	1.383	10 psi	1.197
110	1.396	10 psi	1.190
Accelerated synchronous orbit cycle			
Cycle: 12 h, 5 A charge 10.8 h, 25 A discharge 1.2 h			
Temperature: 4.4°C			
156	1.474	−5	1.206
162	1.480	−10	1.200
185	1.480	−10	1.202

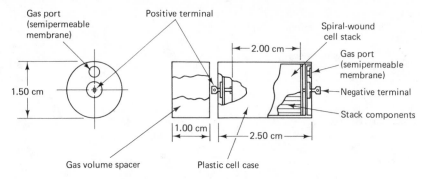

Figure 19.16 Test cell of Eagle Picher nickel–hydrogen battery (Courtesy of Eagle Picher)

the cell on charge or discharge, but does significantly improve the oxygen gas recombination rate.

After completion of approximately 3000 low Earth orbit cycles followed by approximately 300 synchronous orbit cycles (Table 19.3) the Teflon film design batteries were opened up and it was shown that the non-woven, nylon separator (Pellon 2505) was remarkably free of any indication of cadmium migration.

19.2 Nickel–hydrogen and silver–hydrogen secondary batteries

These batteries are still at or slightly beyond the experimental stage. The basic nickel–hydrogen system consists of a catalytic gas electrode (negative) coupled with a nickel electrode (positive). Electrochemically the reactions occurring at the positive electrode are the same as those occurring in the parent systems. At the negative electrode, hydrogen is displaced from water by electrolysis during charge. During discharge the reverse process occurs, namely, the electrochemical discharge of hydrogen ions in solution.

Figure 19.16 shows a laboratory test cell system designed for evaluation of various nickel–hydrogen electrode and electrode stack designs. The system offers a very heavy-duty pressure vessel capable of withstanding virtually any pressure developed during system tests, and offers a convenient means of instrumentation to obtain various data of interest. In this case 'pineapple slice' electrode designs are stacked together to form a 50 A h capacity electrode stack. All support mechanisms and electrical connections are accommodated through the centre of the stack, permitting placement of electrode edges very close to the cell container wall, which enhances thermal properties. In addition, this design significantly reduces the length of the path that the reactant gas must travel to reach any point on the electrode surfaces. Figure 19.17 shows prototype nickel–hydrogen cell designs manufactured by

Eagle Picher. The cells exhibit the spherical geometry characteristics of pressure vessels. The cell in Figure 19.17 yields 20 A h capacity at approximately 55 W h/kg and is referred to as the 'prismatic spheroid' design. The electrode stack is contained within the prismatic area defined by the broken lines, and the spherical portions of the container serve as free volume for the storage of hydrogen gas. The ratio of free volume to ampere hour capacity is selected so that the cell will operate at a maximum pressure of $3.5 \, kN/m^2$.

It is observed that the prismatic spheroid design is not a self-contained pressure vessel – the two flat surfaces must be supported. The design theory considers that two parallel flat surfaces will greatly facilitate packaging of cells into a battery configuration. From an energy density standpoint, the transfer of strength requirements from individual cells to a common battery frame supporting a number of cells offers a significant weight advantage. The flat surfaces are also in intimate contact with the electrode stack, providing effective areas for the application of thermal control measures.

This cell also features two specially designed high-pressure ceramic-to-metal seals incorporating caustic-resistant zirconium and nickel–gold braze

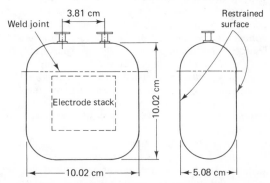

Figure 19.17 Prismatic spheroid cell design of Eagle Picher nickel–hydrogen battery, 20 A h capacity (Courtesy of Eagle Picher)

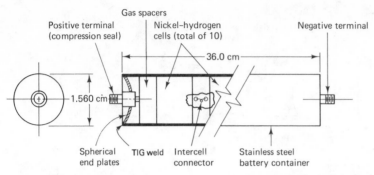

Figure 19.18 Eagle Picher nickel–hydrogen battery design (Courtesy of Eagle Picher)

materials. The seal body is internal to the cell and the complete assembly, in the configuration shown, is rated in excess of 13.8 MN/m². Prototype units have demonstrated an energy density of 60 W h/kg at a maximum operating pressure of 3.5 kN/m². This corresponds to a volumetric energy density of 67 W h/dm³ and a volumetric capacity of 55 A h/dm³.

The cylindrical cell design shown in Figure 19.18 also yields 50 A h capacity, but at a slightly lower energy density of approximately 59 W h/kg. The cylindrical cell design is a self-contained pressure vessel whereas the prismatic spheroid design requires partial support from the battery structure.

The proposed Eagle Picher concept for assembling nickel–hydrogen cells into batteries is presented in Figure 19.18. Although the cell represented is the cylindrical half-cell design, the same basic concept would apply to the prismatic spheroid design. The ten cells connected in series produce a 12 V, 3 A h system. The arrangement of these cells in the battery produces the necessary individual cell support and results in a strength requirement for the end-plate and retainer rod components sufficient only to restrain the force exerted by one cell.

In addition to the cells in a nickel–hydrogen battery operating as isolated pressure vessels each with their own hydrogen supply, Eagle Picher have developed the novel concept of a battery with a common gas manifold arrangement. This concept connects all cells within the battery to a common gas source, significantly improving cell performance uniformity, and prevents pressure differentials from occurring between cells during operation. The successful application of this concept permitted the consideration of multiple cells within a single pressure vessel battery in nickel–hydrogen system design.

This leads to an interesting design modification in the metal–hydrogen system. If each cell is designed with an excess of positive material, the cell will be gas limited on discharge and all cells in the series string will be exhausted at exactly the same time. This feature would allow batteries to be discharged

to a very low voltage without concern for one or more cells dropping out before the rest of the group. A metal–hydrogen cell is capable, however, of operating in reversal without degradation and prevention of this type of failure mode does not carry the same impact as it would in current secondary systems.

Other advantages of this design include the reliability and economics of the use of a single pressure vessel for a battery. In addition, the internal cells, which are not required to be sealed or to withstand pressure, may be of simple design and fabricated with cheaper materials. Individual cells will not require the relatively expensive ceramic-to-metal seals.

The energy density of the common pressure vessel design does not appear to be quite as attractive as the concept might indicate. Although individual cells may be fabricated of lighter materials, the increased size of the pressure vessel requires greater strength and increased material thickness, which more than compensate for the preceding weight reduction. For operation at a maximum pressure of 3.5 kN/m², the achievable energy density for the nickel–hydrogen system is 60 W h/kg.

A third approach to the design of nickel–hydrogen systems considers operation at very high pressures. By increasing the maximum operating pressure from 3.5 kN/m² to approximately 14 MN/m², the volume of a modular metal–hyrogen cell may be reduced to a value comparable to that of a sealed nickel–cadmium cell of the same capacity. This design would have obvious advantages with respect to volume-critical applications.

However, the energy density of a modular cell operating at a maximum pressure of 14 MN/m² is not as attractive as that of a cell operating at 3.5 kN/m². The cell container material thickness would have to be increased to withstand the higher pressure, although it would not be a directly proportional increase because the reduction in the size of the pressure vessel increases strength with respect to internal pressure. The problem results from the fact

Table 19.4 Comparison of Eagle Picher metal–hydrogen systems

	A Modular cell design (3.5 kN/m^2)	B Common pressure vessel (3.5 kN/m^2)	C High-pressure cell design (14 MN/m^2)	D Low-pressure cell design (hydride)
Nickel electrode				
Energy density (W h/kg)	70	60	60	60
Volumetric energy density (W h/dm^3)	73	73	183	213
Volumetric capacity (A h/dm^3)	0.055	0.055	0.146	0.170
Silver electrode				
Energy density (W h/kg)	110	90	90	90
Volumetric energy density (W h/dm^3)	79	79	305	366

that a decrease in the volume of the vessel does not produce an equivalent reduction in the surface area of the vessel. The projected energy density for the nickel–hydrogen system is again 60 W h/kg, with 90 W h/kg for the silver–hydrogen system.

The advantages of the high-pressure design are demonstrated in the volume parameters. A volumetric energy density of 183 W h/dm^3 and a volumetric capacity of 146 A h/dm^3 are achievable with the nickel–hydrogen system.

Another approach to the design of nickel–hydrogen systems involves designs operating at relatively low pressure, eliminating the need for a pressure vessel. Hydrogen readily reacts with certain alloys to form a hydride and may be chemically stored in this fashion rather than as a gas. Many metals and intermetallic compounds are known to form hydrides. Compounds such as lanthanum nitride adsorb and desorb large quantities of hydrogen gas under relatively low pressures at room temperature. This hydride limits the maximum operating pressure for a metal–hydrogen system to approximately 0.7 kN/m^2. Operation of modular cells at this pressure is not only attractive from a safety standpoint, but permits the use of conventional, sealed nickel–cadmium, prismatic cell and battery assembly technology.

Again, the projected energy density of the low-pressure cell design does not appear to be as high as that of the modular cell design operating at 3.5 N/m^2. The lower operating pressure permits the use of lightweight cell container materials; however, the weight of current hydride compounds more than compensates for this advantage. The projected energy density for the nickel system is 60 W h/kg. The hydride systems offer an advantage with respect to minimizing cell volume and an improvement in the volume parameters. The volumetric energy density of the nickel–hydrogen system is 213 W h/dm^3 with a volumetric capacity of 170 A h/dm^3.

Table 19.4 presents comparative data on energy density for the four designs of nickel–hydrogen system discussed above; data are also given for silver–hydrogen systems.

Of the various designs, Eagle Picher appear to have made most progress with the modular cylindrical design (A in Table 19.4). They report that the performance recorded for this type of nickel–hydrogen cells and batteries is very encouraging. Such systems have successfully completed thousands of cycles under both static and dynamic environmental conditions, under both simulated low Earth orbit and synchronous orbit cycle regimens. In addition, to demonstrate the deep depth-of-discharge (DOD) capabilities of the nickel–hydrogen system 'cylindrical' 50 A h cell designs have been placed on 100% DOD cycles and have to date successfully passed approximately 600 cycles. The nickel–hydrogen battery has a projected capability of more than 600 deep discharge cycles (in excess of 70% DOD) at an energy density of up to 70 W h/kg. Representative results of this test are shown in Figure 19.19.

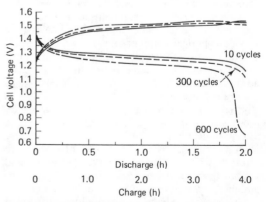

Figure 19.19 Cycling of 50 A h Eagle Picher nickel–hydrogen RNH-50-3 battery, cylindrical modular design. Charge: 15 A, 4 h. Discharge: 25 A, 2 h. Ambient temperature 21°C. 100% DOD cycles (Courtesy of Eagle Picher)

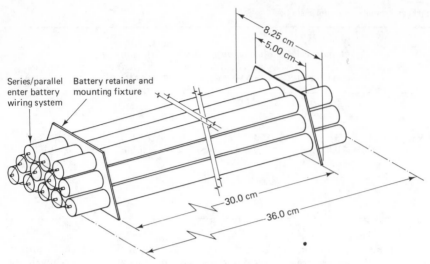

Figure 19.20 Eagle Picher 36 V nickel–hydrogen multiple battery unit design (Courtesy of Eagle Picher)

A low-cost nickel–hydrogen system proposed by Eagle Picher is based on the multiple cell per single battery pressure vessel concept. Additional system cost reduction is proposed through use of standardized components, ease of manufacture and an inherent design versatility able to meet various voltage and capacity requirements with only minor modifications.

Figure 19.18 shows the proposed nickel–hydrogen battery design. Essentially, the battery container consists of a long cylinder or tube of a length selected to accommodate a desired number of cells. The system depicted is a 10-cell 12 V battery which will yield 3 A h capacity at an energy density projected to be in excess of 55 W h/kg. During assembly, the long tube is loaded with pre-manufactured cells spaced apart and positioned by the gas volume spacer component. The spacer component length is selected to render sufficient free volume to store the reactant gas at a desirable maximum pressure. The battery end-plate is of a recessed spherical configuration to facilitate tungsten inert gas (TIG) welding of the interface joints and to ensure a reliable hermetic seal. Each end-plate accommodates a single low-cost high-pressure compression seal.

System versatility could be greatly enhanced by grouping the proposed battery designs into multiple battery units. Figure 19.20 shows a 12-battery nickel–hydrogen system. In this diagram, the battery has been wired to produce 36 V and yield 12 A h capacity. Because the batteries are self-contained pressure vessels, the retaining and mounting components may be of ultra-lightweight design

resulting in a battery system that exhibits a very high energy density. Eagle Picher project that the proposed design concept will reduce nickel–hydrogen system cost from approximately 25% greater to approximately 50% less than current sealed nickel–cadmium system cost. With demonstrated superior performance characteristics, the proposed cost reduction design should render the nickel–hydrogen system an attractive alternative to lead–acid and nickel–cadmium systems in many non-space applications.

19.3 Nickel–zinc secondary batteries

These batteries are not yet available in a sealed form. This type of battery comprises a positive nickel electrode of the sintered or pasted type similar to that used in the nickel–cadmium battery. The zinc electrode is usually prepared from powdered zinc oxide, often mixed with a Teflon binder, and pressed on to an expanded silver grid. The zinc electrode is made somewhat oversized compared to the nickel electrode and is contoured to minimize zinc anode shape changes during cycling of the cell. Typical separator materials are fibrous sausage skin or cellophane and Perlon wrapping. Potassium hydroxide solutions between 30 and 40% by weight are the usual electrolytes for the cell. Cell constructions, prismatic and flat, are similar to those used for nickel–cadmium cells.

The basic electrochemical reaction occurring in the cell is as follows:

$$2NiO(OH) + Zn + 2H_2O \rightleftharpoons 2Ni(OH)_2 + Zn(OH)_2$$

20

Silver batteries

Contents

20.1 Silver oxide–zinc primary batteries

The are two types of silver oxide cell: one has a cathode of monovalent silver oxide (Ag_2O) and the other type uses divalent silver oxide (AgO). The latter type has a higher theoretical potential (1.8 V) and, because there is an additional chemical reduction from AgO to Ag_2O, it has a higher capacity (the theoretical energy density is 424 W h/kg).

The cell reactions are as follows:

$$AgO + Zn + H_2O \rightleftharpoons Ag + Zn(OH)_2$$

Negative electrode:

$$Zn + 2OH^- \rightarrow Zn(OH)_2 + 2e^- \qquad E_0 = -1.245 \text{ V}$$
$$Zn(OH)_2 + 2KOH \rightleftharpoons K_2Zn(OH)_4$$

Positive electrode:

$$2AgO + H_2O + 2e^- \rightarrow Ag_2O + 2OH^- \qquad E_0 = +0.57 \text{ V}$$
$$Ag_2O + H_2O + 2e^- \rightarrow 2Ag + 2OH^- \qquad E_0 = +0.344 \text{ V}$$

The two-stage reduction process would normally result in a discharge curve with two plateaux at 1.7 and 1.5 V, and the voltage drop in the middle of discharge may necessitate a voltage regulator in the equipment. If the surface layer of the electrode is of Ag_2O, however, discharge takes place at the lower potential throughout. In order to achieve voltage stability, the surface may be treated to reduce AgO to Ag_2O or, in various patented arrangements, a 'dual oxide' system may be adopted. Higher raw material costs mean that silver oxide cells are more expensive than their mercury equivalents.

The silver oxide battery consists of a depolarizing silver oxide cathode, a zinc anode of high surface area and a highly alkaline electrolyte. The electrolyte is potassium hydroxide in hearing-aid batteries. This is used to obtain maximum power density at hearing-aid current drains. The electrolyte in watch batteries may be either sodium hydroxide or potassium hydroxide. Sodium hydroxide electrolyte, which has a lower conductivity than potassium hydroxide, is often used because it has a lower tendency to 'creep' at the seal. Mixtures of silver oxide and manganese dioxide may be tailored to provide a flat discharge curve or increased service hours. The separator in the silver oxide system must retain silver soluble species produced by chemical dissolution of the oxide, and a multiple layer separator of low-porosity film achieves this.

A cutaway of a silver oxide button cell is shown in Figure 20.1. The plated steel cell cap functions as the negative terminal of the cell. The zinc anode is a high-purity amalgamated zinc powder and the cathode is a compressed pellet of silver oxide plus graphite for conductivity. An absorbent pad of a non-woven natural material holds the alkaline electrolyte, which is a strongly alkaline potassium hydroxide solution. The separator is a synthetic

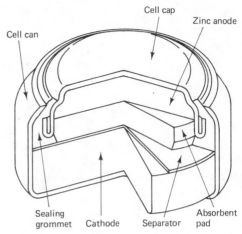

Figure 20.1 Cutaway view of a silver oxide cell (Courtesy of Eagle Picher)

ion-permeable membrane. A sealing grommet both seals the cell and insulates the positive and negative terminals. The plated steel can act as a cell container and as the positive terminal of the cell. The inner surface of the cell is of a metal electrochemically compatible with zinc, to minimize wasteful corrosion and harmful gas evolution, and the cell can is nickel-plated steel, which is highly resistant to the electrolyte.

The manner in which the cell is designed results in high volumetric efficiency. An effective radial seal is a unique feature of the construction of many types of silver–zinc cell (for example, Union Carbide cells). Briefly, the radial sealing system, developed in the late 1950s, incorporates the use of a nylon gasket and a top which is a gold-plated bi-clad stainless steel anode cup which serves as the negative terminal. The cathode cup is usually a nickel-plated steel can, which serves as the positive terminal. The radial seal is formed during the final stages of cell manufacture. The cell can is subjected to an operation that actually reduces the diameter of the can. This process tightly squeezes the nylon gasket against the bi-clad stainless steel top, creating the initial radial seal. This operation is possible because a stainless steel anode top can withstand the extreme pressure during the diameter reducing operation. Secondly, the selection of nylon is important because nylon, after being squeezed, tries to regain its original position, i.e. it has a memory. Once the can diameter is reduced, a secondary seal is effected by crimping the edge of the can over the gasket. Again the use of a nylon gasket is significant since nylon will continue to exert pressure as a result of the second sealing operation. The radial seal technique is highly effective in providing excellent protection against the incidence of salting.

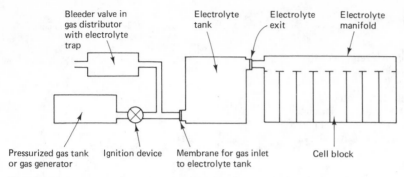

Bleeder valve in gas distributor with electrolyte trap

Electrolyte tank

Electrolyte exit

Electrolyte manifold

Pressurized gas tank or gas generator

Ignition device

Membrane for gas inlet to electrolyte tank

Cell block

Figure 20.2 Schematic layout of remotely activated Silberkraft silver–zinc battery (Courtesy of Dryfit)

20.1.1 Remotely activated silver oxide–zinc batteries

Silver–zinc primary batteries can be remotely activated for single use within seconds or fractions of a second, even after long-term storage, by inserting the electrolyte under pressure. In addition to a unit containing dry charged plates they also contain an electrolyte vessel from which the cells are filled by electrical or mechanical means. Pile batteries are a new development of remotely activated primary batteries, which consist of bipolar electrodes that confer a very high density on the battery. A schematic representation of a remotely activated silver–zinc battery is shown in Figure 20.2. Such batteries are recommended for use in aerospace, torpedo and pyrotechnic ignition applications.

In an alternative design of remotely activated primary silver–zinc batteries the electrolyte reservoir is sealed at both ends with high-pressure (6.89 MPa burst strength) metal diaphragms. The gas generator to pump in the electrolyte is activated by a squib or a primer, which is electrically, mechanically or inertially activated. The expanding gas bursts the diaphragms and forces the electrolyte into the manifold and cells. In this cell design a vented system enhances the rapid clearing of the manifold and eliminates excessive intercell leakage paths. The battery has no moving parts and has an extended shelf life. It can be activated directly into a load or in parallel with a mains system. The open-circuit voltage of this battery is 1.6–1.87 V with a working voltage of 1.20–1.55 V. The battery operating temperature is −29 to +71°C and, with water assistance, −48 to +71°C. The energy density is 11–66 W h/kg or 24–213 W h/dm^3, depending on battery size. Units are available in the weight range 0.14–135 kg. The dry storage charge retention characteristics of this battery are excellent at temperatures up to 52°C.

Eagle Picher supply gas-generating pyrotechnic devices, which can be used for the activation of

remotely activated batteries by transferring electrolyte from a reservoir to the battery cells. Eagle Picher have developed over 70 gas-generator configurations with capabilities ranging from 20 to

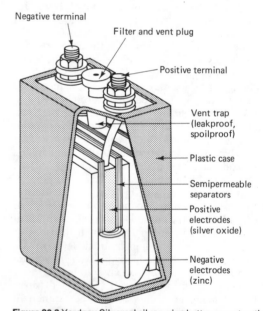

Negative terminal

Filter and vent plug

Positive terminal

Vent trap (leakproof, spoilproof)

Plastic case

Semipermeable separators

Positive electrodes (silver oxide)

Negative electrodes (zinc)

Figure 20.3 Yardney Silvercel silver–zinc battery construction. Positive electrode manufactured from special grades of silver with or without silver collectors. Negative electrode manufactured from zinc with or without collectors. Semipermeable and ion-exchange separators designed and patented by Yardney. Strong electrolyte solution of potassium hydroxide. Positive and negative terminals of silver-plated steel for maximum corrosion resistance and minimum contact resistance, or special non-magnetic alloys. The filling vent allows activation of the cell with electrolyte, while the vent plug or screw valve, when placed in the cell, will allow gassing and prevent electrolyte spillage. Vent trap, designed to be moulded as part of the cover, prevents leakage and spillage, and so allows the cell to operate in any position. Case and cover moulded in plastics such as ABS, SAN, nylon, modif polypropylene oxide, polyarylether or polysulphone (Courtesy of Yardney)

$200\,000\,cm^3$ (standard condition gas output), many of which have been used in a majority of missile/spacecraft systems.

20.2 Silver–zinc secondary batteries

These cells are leakproof, spillproof and have high impact strength. The assembled cells can be enclosed in sturdy lightweight cases for special mechanical and environmental conditions. Batteries can also be designed and built to meet the most stringent requirements.

A typical battery design is shown in Figure 20.3. This battery contains silver oxide as the positive electrode and zinc as the negative electrode, and has an alkaline electrolyte.

20.3 Silver–cadmium secondary batteries

The design of this battery is very similar to that of the silver–zinc battery, with the exception that the zinc negative electrode is replaced by one fabricated from cadmium.

21

Alkaline manganese batteries

Contents

21.1 Alkaline manganese primary batteries

Cylindrical alkaline manganese batteries have an inverted construction compared with the familiar Leclanché carbon–zinc battery construction. A typical construction of this type of battery is shown in Figure 21.1.

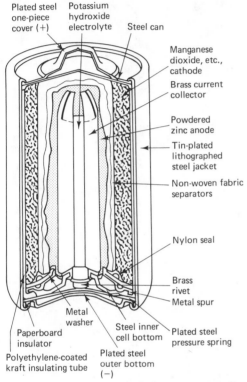

Figure 21.1 Construction of an alkaline manganese cell (Courtesy of Union Carbide)

Plated steel one-piece cover (+)
Potassium hydroxide electrolyte
Steel can
Manganese dioxide, etc., cathode
Brass current collector
Powdered zinc anode
Tin-plated lithographed steel jacket
Non-woven fabric separators
Nylon seal
Brass rivet
Metal spur
Metal washer
Steel inner cell bottom
Plated steel pressure spring
Paperboard insulator
Plated steel outer bottom (−)
Polyethylene-coated kraft insulating tube

In the cylindrical alkaline manganese battery the positive terminal is formed by a stud, in contact with the cathode at the top of the battery, fixed to a steel can which surrounds the positive electrode. The hermetically sealed can and steel jacket take no part in the electrochemical reactions of the cell. Where acetylene black is generally used as an additive to the manganese dioxide in a Leclanché battery (because of its adsorptive properties), the higher conductivity of the potassium hydroxide electrolyte used in the alkaline manganese battery allows the use of highly conductive graphite as a cathode (depolarizer) additive. The highly compressed cylindrical manganese dioxide cathode is lined with an absorbent separator and the gelled anode of amalgamated high-surface-area zinc powder inserted. Contact to the negative terminal at the base of the battery is by an internal metal contact.

This design gives a high current density per unit volume of active materials, low resistance and impedance, high service capacity, and a relatively constant capacity over a range of current drawing and discharge schedules. Cylindrical alkaline manganese cells are fitted with an outer steel jacket which is electrically isolated from the cell can. This type of cell differs from the carbon–zinc type primarily in that a strongly alkaline electrolyte is used instead of the acidic ammonium chloride or zinc chloride medium used in the carbon–zinc design.

21.2 Alkaline manganese secondary batteries

The individual cells that comprise the rechargeable alkaline dioxide battery use electrodes of zinc and manganese dioxide with an electrolyte of potassium hydroxide. These are put together in a special inside-out cell construction and each cell is then hermetically sealed. The voltage per cell is 1.5 V. They are made in two sizes, D and G. A cutaway view of a Union Carbide Eveready cell is given in Figure 21.2.

Finished batteries of 3 V and above are constructed by connecting the required number of the appropriate cell size in series and sealing them in a metal case.

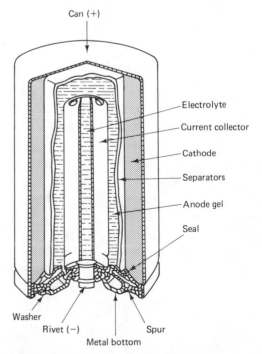

Can (+)
Electrolyte
Current collector
Cathode
Separators
Anode gel
Seal
Washer
Rivet (−)
Spur
Metal bottom

Figure 21.2 Eveready rechargeable alkaline manganese dioxide cell construction (Courtesy of Union Carbide)

22

Carbon–zinc and carbon–zinc chloride batteries

Contents

22.1 Carbon–zinc primary batteries

The electrochemical system of the carbon–zinc (Leclanché) battery uses a zinc anode and a manganese dioxide cathode and an electrolyte of ammonium chloride and zinc chloride dissolved in water. Various grades of manganese dioxide are used for Leclanché cells. These range from the natural ores used for inexpensive low-rate cells to the synthetically produced or electrolytically prepared grades used in more expensive high-rate cells. Powdered carbon is used in the cathode depolarizing mix, usually in the form of acetylene black, to improve conductivity of the mix and to retain moisture.

The standard carbon–zinc dry battery is not designed to be recharged. The basic cell is made in many shapes and sizes, but two general categories exist for round cells, which are available as unit cells or in assembled batteries and flat cells.

The difference between the two is mainly physical. The chemical ingredients are the same for both: carbon, depolarizing mix, separator, electrolyte and zinc. A flat cell, such as the Union Carbide 'Mini-Max', uses these materials in a laminated structure, while in a round cell they are arranged concentrically (see Figures 22.1 and 22.2).

In flat cells a zinc plate is coated with carbon to form a duplex electrode – a combination of the zinc of one cell and the carbon of the adjacent one. The Union Carbide Mini-Max cell contains no expansion chamber or carbon rod as does the round cell. This increases the amount of depolarizing mix available per unit cell volume and therefore increases the energy content. In addition, the flat cell, because of its rectangular form, reduces waste space in assembled batteries.

22.2 Carbon–zinc chloride primary batteries

The zinc chloride Leclanché cell is a special modification of the carbon–zinc cell. They differ principally in the electrolyte systems used. The electrolyte in a zinc chloride cell contains only zinc chloride, while in a carbon–zinc cell the electrolyte contains a saturated solution of ammonium chloride in addition to zinc chloride. The omission of ammonium chloride improves the electrochemistry of the cells but places greater demands on the cell seal. Zinc chloride cells therefore have either a new type of seal not previously used in carbon–zinc cells or an improved conventional seal so that their shelf life is equivalent to that of carbon–zinc cells. Electrode blocking by reaction products and electrode polarization at high current densities are minimized by the more uniform and higher diffusion rates that exist in an electrolyte containing only zinc

chloride. Because of their ability to operate at high electrode efficiencies, the useful current output of zinc chloride cells is usually higher than that of carbon–zinc, and zinc chloride cells will operate at

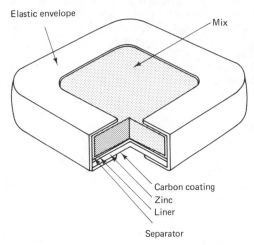

Figure 22.1 Cross-section of a Union Carbide Mini-Max carbon–zinc flat cell (Courtesy of Union Carbide)

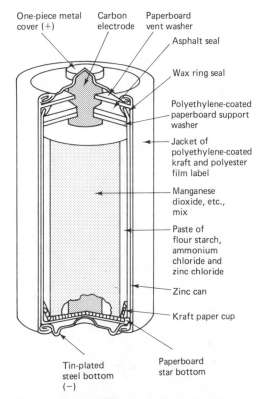

Figure 22.2 Cross-section of a standard carbon–zinc round cell (D-size). The rod in the centre of the cell is carbon and functions as a current collector (Courtesy of Union Carbide)

high current drains for a considerably longer time than carbon–zinc cells of the same size. In addition, the voltage level under load holds up longer.

Because of the electrochemical reactions that occur in the zinc chloride cell during use, water in the cell is consumed (by a reaction product which is an oxide compound) along with the electrochemically active materials, so that the cell is almost dry at the end of its useful life.

A cross-section view of a D-size zinc chloride cell is shown in Figure 22.3.

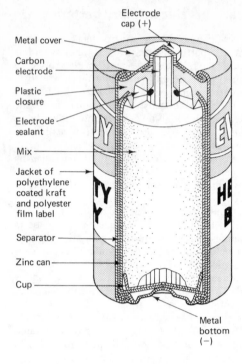

Figure 22.3 Cross-section of a zinc chloride cell (D-size) (Courtesy of Union Carbide)

23

Mercury–zinc batteries

Contents

23.1 Mercury–zinc primary batteries

Mercury batteries and cells are produced in cylindrical and button types. Electrochemically, both are identical and they differ only in can design and internal arrangement. The anode is formed from cylinders or pellets or powdered high-purity amalgamated zinc, or a gelled mixture of electrolyte and zinc. The depolarizing cathode is compressed mercuric oxide–manganese dioxide in sleeve or pellet form (cell voltage 1.4 V) or pure mercuric oxide (cell voltage 1.35 V), and the electrolyte, which does not participate in the reaction, is concentrated aqueous sodium or potassium hydroxide. The cathode is separated from the anode by an ion-permeable barrier. In operation this combination produces metallic mercury, which does not inhibit the current flow within the cell (Figure 23.1). Zinc is thermodynamically unstable with respect to water and in strongly alkaline solutions will tend to self-discharge with the evolution of hydrogen. Zinc anode corrosion is reduced to acceptable levels by amalgamating the zinc powder with mercury, which increases the hydrogen overpotential of the zinc and by ensuring that the surface area of the porous

electrodes is not too high. Also it is necessary to use only materials of high hydrogen overpotential (silver, lead, tin and copper, not iron or nickel) in contact with both the zinc electrode and the electrolyte. It is for this reason that the cell has a centre case (nickel-plated steel) to provide mechanical strength and an inner case of suitable high hydrogen overpotential to avoid corrosion of the anode. The inside of the cell top is also electrochemically compatible with the zinc anode. The cell components used provide precise mechanical assemblies having maximum dimensional stability and a marked improvement in performance over batteries of the carbon–zinc type.

Most major producers of mercury–zinc cells and batteries use a vented construction, which renders their products leakproof and free from bulging under all normal working conditions. Self-venting occurs automatically if, for example, excessive gas is produced within the cell under sustained short-circuit conditions. This type of cell is manufactured in a wide variety of structures to provide batteries of varying size, voltage and capacity. A wide range of purpose-built batteries can also be obtained. For maximum reliability intercell connection is by spot

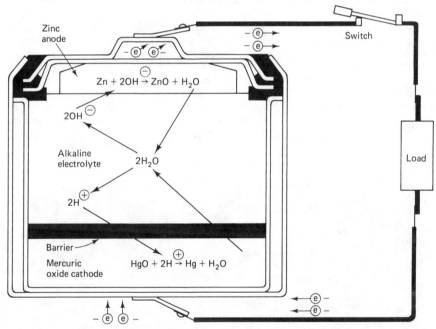

Figure 23.1 Electrochemical processes occurring in the mercury–zinc cell. (1) With the switch open, the zinc anode and the mercuric oxide cathode are both charged with respect to the alkaline electrolyte: Zn ($+1.32$ V) and HgO$^-$ (-0.03 V) + cell voltage = 1.35 V, against normal hydrogen electrode in the same electrolyte. (2) With the switch closed, the cell voltage forces electrons to flow through the external circuit in accordance with Ohm's law. (3) At the anode, loss of electrons through the external circuit disturbs the anode equilibrium. Hydroxyl ions from the electrolyte discharge at the anode to restore the anode charge, forming zinc oxide and water. (4) At the cathode, gain of electrons upsets the cathode equilibrium. Hydrogen ions from the electrolyte discharge at the cathode, forming mercury and water. There is no net change of water content in the electrolyte. (5) Overall chemical reaction may be expressed as Zn + HgO → ZnO + Hg with the liberation of 820 mA h of electricity for each gram of zinc in the cell (Courtesy of Union Carbide)

welded strips of nickel-plated steel. Outer containers are made from materials ranging from cardboard to metal.

Some design features used by various manufacturers in the construction of mercury–zinc cells are as follows:

1. Depolarizing cathodes of mercuric oxide, to which a small percentage of graphite is added, are shaped as shown in Figure 23.1 and either consolidated to the cell case (for flat electrode types) or pressed into the cases (of the cylindrical types).
2. Anodes are formed of amalgamated zinc powder of high purity, in either flat or cylindrical shapes.
3. A permeable barrier of specially selected material prevents migration of any solid particles in the cell, thereby contributing effectively to long shelf and service life and also inhibiting the diffusion and migration of soluble mercuric hydroxide species from the cathode, which causes self-discharge of the zinc anode.
4. Insulating and sealing gaskets are moulded of nylon, polyethylene or Neoprene, depending on the application for which the cell or battery will be used.
5. Inner cell tops are plated with materials that provide an internal surface to which zinc will form a zinc amalgam bond.
6. Cell cases and outer tops of nickel- or gold-plated steel are used to resist corrosion, to provide greatest passivity to internal cell materials and to ensure good electrical contact.
7. An outer nickel-plated steel jacket is generally used for single cells. This outer jacket is a necessary component for the 'self-venting construction' used on some cells, which provides a means of releasing excess gas in the cell. Venting

occurs if operating abnormalities such as reverse currents or short-circuits produce excessive gas in the cell. At moderately high pressures, the cell top is displaced upwards against the external crimped edge of the outer jacket, tightening that portion of the seal and relieving the portion between the top and the inner steel cell case. Venting will then occur in the space between the internal cell container and the outer steel jacket. Should any cell electrolyte be carried into this space, it will be retained by the safety absorbent ring. Corrosive materials are therefore not carried with the escaping gas through the vent hole at the bottom of the outer steel jacket. After venting excess gas and reducing the internal pressure, the cell stabilizes and reseals the top seal, continuing normal operation in the circuit.

23.2 Mercury–zinc cardiac pacemaker batteries

The first pacemaker batteries were produced in about 1973, and since then many improvements have been made in their design. The layout of a Mallory mercury–zinc pacemaker battery is shown in Figure 23.2. The anode consists of amalgamated zinc, which is in contact with the welded top assembly, the inner face of which is also amalgamated to achieve electrochemical compatibility with the anode. It is surrounded by the absorbent barrier, which is impregnated with a solution of sodium hydroxide saturated with zinc oxide. This type of electrolyte minimizes hydrolysis of the zinc, which causes production of hydrogen and results in capacity loss. The annular cathode consists of mercuric oxide mixed with silver powder, which forms a solid amalgam with the mercury appearing

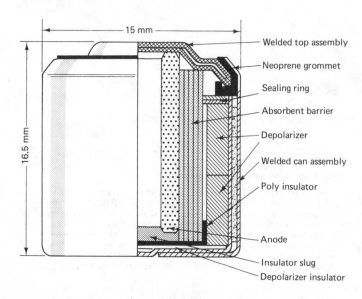

Figure 23.2 Mallory mercury–zinc pacemaker battery type 1 (Courtesy of Mallory)

during discharge. It is in contact with the welded can assembly, which forms the positive terminal of the battery. The sealing ring and insulators, which divide the battery into two compartments, reduce the risk of short-circuit and electrical leakage. The gasket is made of Neoprene; this material keeps its elasticity for a long period of time even under mechanical stress, thus ensuring the best possible seal.

The battery shown in Figure 23.2 has a nominal voltage of 1.35 V and a nominal capacity of 1000 mA h defined for a load of 1250 Ω, with a cut-off voltage of 0.9 V. For lighter loads, tests on early production batteries have given the results shown in Table 23.1.

The figures in Table 23.1 are to be considered as minima. One sees immediately that reducing the drain does not result in a proportional increase of life. To meet the demand for longer service life, Mallory have developed a new cell (RM2 type) (Figure 23.3). Its construction differs from the earlier one in a number of respects, as follows:

1. A collector in the form of a spring ensures a better contact with the anode material and increases its efficiency.
2. The multi-turn microporous barrier is heat sealed.

Table 23.1 Mallory cell 317653 discharge at 73°C

Load (kΩ)		Minimum	Average	Maximum
11	mA h	954	1066	1080
	months	10.5	12.0	12.25
22	mA h	690	941	1035
	months	15.2	20.7	23.2
33	mA h	622	797	1000
	months	20.8	26.0	33.0

3. Another microporous barrier is clamped over the absorbent-separator assembly to close the anodic compartment.
4. A secondary barrier has been added to minimize penetration of zinc dendrites.
5. The single can is made of pure nickel.
6. The gasket is moulded around a single biplate top.

Its nominal capacity is 1800 mA h and a capacity realization of 93% over 28 months' discharge is claimed. The service life of this battery is likely to be less than 5 years.

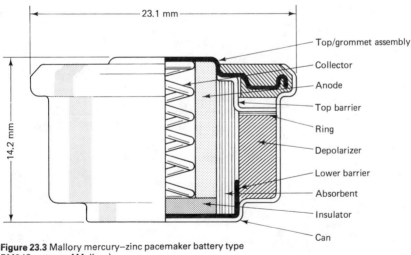

Figure 23.3 Mallory mercury–zinc pacemaker battery type RM2 (Courtesy of Mallory)

23.1 mm

14.2 mm

Top/grommet assembly
Collector
Anode
Top barrier
Ring
Depolarizer
Lower barrier
Absorbent
Insulator
Can

24

Lithium batteries

Contents

24.1 Lithium–sulphur dioxide primary batteries

The system uses a lithium anode, a gaseous sulphur dioxide cathode (about 70% of the weight of the electrolyte depolarizer) and an electrolyte comprising lithium bromide dissolved in acetonitrile.

The discharge at the anode may be represented by the half-cell reaction:

$$Li = Li^+ + e^-$$

The cathode reaction may be written as:

$$2SO_2 + 2e^- = S_2O_4^{2-}$$

and the overall cell reaction as

$$2Li + 2SO_2 = Li_2S_2O_4 \quad \text{(lithium dithionite)}$$

The discharge of lithium–sulphur dioxide cells is accompanied by the precipitation of lithium dithionite in the cathodes, but the precipitate appears not to lessen the rate capability of the cells noticeably. An optimal electrolyte composition can be specified that permits the cathodes to operate along a maximal conductivity path during discharge.

The high-rate capability of the cells is favoured by the high conductivity of the electrolyte and the small effect of temperature on the conductivity. Heat dissipation structures are recommended for lithium–sulphur dioxide cells and batteries that operate at high power levels. Lithium–sulphur dioxide cells can operate at 460 W/kg while delivering 100 W/kg in suitable high-rate configurations.

On the basis of postulated cell reactions, the discharge process of the lithium–sulphur dioxide cell may be illustrated schematically as shown in Figure 24.1. Lithium ions formed at the anode are transported to the cathode where charge neutralization occurs and insoluble lithium dithionite deposits at and inside the porous carbon electrode where the sulphur dioxide is reduced. Depending on the current density, the precipitate may deposit throughout the porous cathode (at low current densities) or predominantly at the surface of the cathode facing the anode (at high current densities).

The nature of this discharge process shows that the conductivity of the electrolyte plays an important role in determining the rate capability of the system, both from the point of view of the rate at which lithium ions can be transported to the cathode and from the point of view of the depth to which the discharge reaction can penetrate into the cathode.

Figure 24.2 shows the composition–conductivity contour of the lithium bromide–sulphur dioxide–acetonitrile electrolyte over a wide temperature range. A maximum conductivity of about $6 \times 10^{-2}/(\Omega\,cm)$ is obtained at 25°C. It may be seen that the conductivity contours form a ridge which follows closely the sulphur dioxide depletion line. The latter is the extended line connecting the sulphur dioxide apex in the diagram and the point representing the initial composition of the electrolyte used in the cells. To enhance the rate capability of the cells, an electrolyte composition should be selected on the sulphur dioxide-rich side of the conductivity maximum and represented by the start of the arrow in Figure 24.2 (71–75% sulphur dioxide). This choice

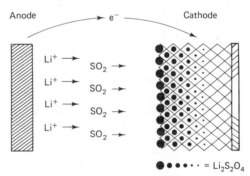

Figure 24.2 The change in lithium bromide–sulphur dioxide electrolyte conductivity during discharge at 25°C (Courtesy of Honeywell)

is arbitrary to some extent, except for its location along the conductivity ridge. It was dictated by the recognition that, under the dynamic conditions of discharge, sulphur dioxide depletion will occur inside the porous cathode. Thus, by choosing a sulphur dioxide-rich electrolyte, depletion of the sulphur dioxide in the cathode will in fact lead to an increased conductivity of the electrolyte in the cathode and an enhanced participation of the interior of the cathode in the discharge process. At low rates of discharge, the composition point moves towards the end of the arrow. At high rates of discharge, the sulphur dioxide utilization approaches 40–50% and the overall electrolyte composition at the end of discharge may be

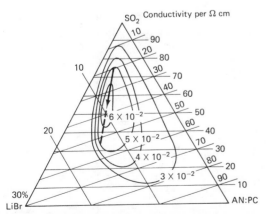

Figure 24.1 Schematic representation of the discharge process of a lithium–sulphur dioxide cell at 25°C. AN:PC = 3:1 by volume (Courtesy of Honeywell)

Table 24.1 Effect of electrolytes and storage conditions on discharge performance of Honeywell lithium–sulphur dioxide reserve batteries at 560 mA h

Electrolyte	No. of cells	Electrolyte storage	Life (h) to		Average voltage (V)		Capacity (A h) to		Cathode A h/g to 2.71 V
			2.71 V	2.00 V	2.71 V	2.00 V	2.71 V	2.00 V	
At −2°C									
1M LiBr:acetonitrile:SO_2	2	No	231	251	2.840	2.818	0.655	0.706	1.64
0.25M LiAsF₆:acetonitrile:SO_2	1	No	164	165	2.853	2.851	0.468	0.471	1.17
0.50M LiAsF₆:acetonitrile:SO_2	1	No	208	215	2.851	2.843	0.593	0.610	1.48
1M LiBr:acetonitrile:SO_2	2	4 weeks at +71°C	226	247	2.830	2.793	0.640	0.690	1.60
0.25M LiAsF₆:acetonitrile:SO_2	2	4 weeks at +71°C	174	180	2.808	2.795	0.487	0.501	1.22
0.50M LiAsF₆:acetonitrile:SO_2	2	4 weeks at +71°C	256	296	2.802	2.775	0.716	0.820	1.80
1M LiBr:acetonitrile:SO_2	2	4 weeks at −54°C	226	248	2.832	2.797	0.640	0.694	1.60
0.25M LiAsF₆:acetonitrile:SO_2	3	4 weeks at −54°C	209	237	2.824	2.783	0.590	0.660	1.48
0.50M LiAsF₆:acetonitrile:SO_2	2	4 weeks at −54°C	223	238	2.823	2.800	0.630	0.665	1.58
At +32°C									
1M LiBr:acetonitrile:SO_2	2	No	148	154	2.848	2.833	0.420	0.435	1.05
0.25M LiAsF₆:acetonitrile:SO_2	2	No	143	144	2.845	2.845	0.409	0.410	1.02
0.50M LiAsF₆:acetonitrile:SO_2	1	No	175	179	2.844	2.836	0.498	0.508	1.25
1M LiBr:acetonitrile:SO_2	2	4 weeks at +71°C	147	152	2.848	2.836	0.419	0.431	1.05
0.25M LiAsF₆:acetonitrile:SO_2	2	4 weeks at +71°C	151	153	2.843	2.838	0.428	0.435	1.07
0.50M LiAsF₆:acetonitrile:SO_2	2	4 weeks at +71°C	169	181	2.846	2.825	0.481	0.510	1.21
1M LiBr:acetonitrile:SO_2	2	4 weeks at −54°C	155	161	2.822	2.807	0.437	0.452	1.09
0.25M LiAsF₆:acetonitrile:SO_2	3	4 weeks at −54°C	145	146	2.853	2.850	0.413	0.416	1.03
0.50M LiAsF₆:acetonitrile:SO_2	2	4 weeks at −54°C	176	184	2.850	2.836	0.502	0.520	1.26

Cells were discharged through a constant resistive load of 1000 Ω, which corresponds to a current density of 0.24 mA/cm³ of 560 mA h reserve batteries at −2°C

represented by a composition point close to the maximum point on the conductivity ridge.

The effect of temperature on the conductivity is shown in Figure 24.3, for an electrolyte composition of 74% sulphur dioxide. It may be seen that a decrease in temperature from 25 to −50°C decreases the conductivity by a factor of only about one-half.

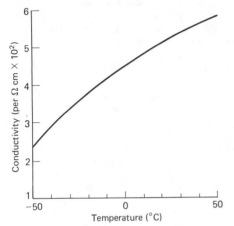

Figure 24.3 The effect of temperature on lithium bromide–sulphur dioxide electrolyte conductivity, 74% sulphur dioxide (Courtesy of Honeywell)

This small effect of the temperature on the conductivity explains the excellent high-rate capability of the lithium–sulphur dioxide system at low temperatures, coupled with the high exchange currents for the lithium oxidation and the sulphur dioxide reduction reactions. The small decrease of the conductivity with a lowering of the temperature can be attributed to the beneficial effects of the sulphur dioxide and solvent on the viscosity of the electrolyte.

Honeywell have described their work on the development of an alternative electrolyte for a multi-cell lithium–sulphur dioxide reserve battery. In developing a multi-cell lithium reserve battery, the lithium bromide–sulphur dioxide acetonitrile electrolyte system used in their primary batteries was found to be unstable when stored by itself at high temperature – a functional capability required for all reserve applications. In addition to consumption of the oxidant sulphur dioxide in reactions causing instability, some of the products of electrolyte degradation are solid, which would cause major problems in activation. Primary active cells after storage do not undergo such degradation reactions.

The following reactions are believed to be responsible for electrolyte decomposition in reserve batteries:

1. Reaction of sulphur dioxide with alkali metal halides:

$$4LiBr + 4SO_2 \rightarrow 2Li_2SO_4 + S_2Br_2 + Br_2$$

2. Polymerization of acetonitrile:

$$nCH_3C \equiv N \rightarrow (-C=N-)_n$$
$$| $$
$$CH_3$$

3. Reaction(s) of acetonitrile or its polymer with intermediates and/or products of the LiBr–SO$_2$ reaction.

As lithium bromide appeared to initiate the reactions causing electrolyte instabilities, Honeywell investigated other lithium salts for use in reserve battery electrolytes and concluded that lithium hexafluoroarsenate (LiAsF$_6$) combined with acetonitrile and sulphur dioxide was a suitable electrolyte, which did not exhibit discoloration or deposition of solids during storage. Table 24.1 compares the performance of batteries made up using the lithium bromide- and lithium hexafluoroarsenate-based electrolytes. Clearly, the 0.5 molal lithium hexafluoroarsenate electrolyte is functionally equivalent or superior to the lithium bromide electrolyte.

No adverse effects after 4 weeks' storage at −54 or +71°C are observed. On the contrary, the 0.5 molal LiAsF$_6$–acetonitrile–SO$_2$ solution significantly improved in performance after 4 weeks' storage at +71°C.

Two typical cell constructions are used: jelly-roll electrodes in crimp sealed on hermetically sealed cylindrical cells, and large 20–100 A h, 12 V flat-plate electrodes in large reserve batteries.

It is a relatively high-pressure system and cells must have safety vents to avoid explosion in the event of accidental incineration. Diagrammatic representations of Honeywell reserve and non-reserve (active) lithium–sulphur dioxide systems are shown, respectively, in Figures 24.4 and 24.5.

The external configuration of the reserve battery (Figure 24.4) is that of a right cylinder, which in cross-section reveals three main portions of the internal design:

1. The electrolyte storage reservoir sections.
2. The electrolyte manifold and activation system.
3. The reserve cell compartment.

Referring to the 20 A h battery cross-section, approximately half of the internal battery volume contains the electrolyte reservoir. The reservoir section consists primarily of a collapsible bellows in which the electrolyte solution is stored during the reserve phase of the battery life-cycle. The reservoir section contains a sufficient quantity of electrolyte solution (a mixture of sulphur dioxide, acetonitrile and the electrolyte salt) to provide the capacity rating of the battery. Surrounding the bellows, between it and the outer battery case, is space that holds a specific amount of Freon. The Freon gas is selected such that its vapour pressure always exceeds that of the electrolyte, thereby providing the driving force for

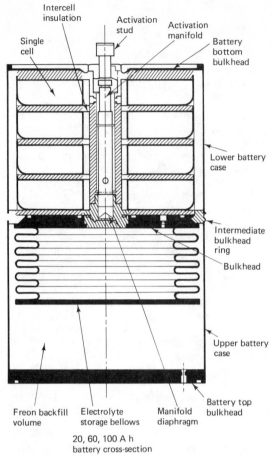

Figure 24.4 Honeywell reserve lithium–sulphur dioxide cell (Courtesy of Honeywell)

eventual liquid transfer into the cell chamber section once the battery has been activated.

In the remaining half of the battery volume, shown at the top of the figure, is the centrally located electrolyte manifold and activation system housed in a 15.88 mm diameter tubular structure plus the series stack of four toroidally shaped cells, which surround the manifold/activation system.

The manifold and cells are separated from the reservoir by an intermediate bulkhead. In the bulkhead there is a centrally positioned diaphragm of sufficiently thin section that it can be pierced by the cutter contained within the manifold. In fabrication, the diaphragm is assembled as part of the tubular manifold, which in turn is welded as a subassembly to the immediate bulkhead.

Figure 24.6 is a more detailed cross-sectional view of the electrolyte manifold and activation system with the major components identified. The activating mechanism consists of a cutter (5) which is manually moved into the diaphragm, cutting it and thereby allowing electrolyte to flow. The movement

of the cutter is accomplished by the turning of an external screw (7), which is accessible in the base of the battery. The cutter section and the screw mechanism are isolated from each other by a small collapsible metal cup (9), which is hermetically sealed between the two sections. This prevents external electrolyte leakage. The manifold section is a series of small non-conductive plastics tubes (10) connected at one end to the central cylinder (4) and to each of the individual cells at the other end. The long length and small cross-sectional area of the tubes minimize intercell leakage losses during the time that electrolyte is present in the manifold structure.

Using the lithium–sulphur dioxide system, four individual cells are required to achieve the voltage range desired for the total battery. The number of cells is, of course, adjustable with minor modification to meet a wide range of voltage needs. Each cell contains flat circular anodes and cathodes, which are separately wired in parallel to achieve the individual cell capacity and plate area needed for a given set of requirements. The components, with intervening separators, are alternately stacked around the cell centre tube, after which the parallel connections are made. The cells are individually welded about the inner tube and outer perimeter to form hermetic units ready for series stacking within the battery. Connections from the cells are made to external terminals, which are located in the bottom bulkhead of the battery.

Figure 24.7 is a photograph of the major battery components before assembly. The components shown are fabricated primarily from 321 stainless steel, and the construction is accomplished with a series of TIG welds.

When the battery is activated by cutting the diaphragm in the intermediate bulkhead, all the liquid stored in the reservoir does not enter the cell compartments. The amount of liquid initially placed in the cells is approximately 70% of the total contained in the reservoir. The balance remains in the manifold section and the uncollapsed portion of the bellows. The vapour pressure of the Freon gas behind the bellows maintains it in a collapsed position and therefore in the liquid state. As discharge proceeds in the cells, generating free volume, the bellows continue to collapse, forcing additional oxidant into the cells and permitting continued discharge. When the remaining 30% of the oxidant enters the cells, the bellows have bottomed on the intermediate bulkhead, which relieves the Freon pressure from the cell section. This permits vaporization of the sulphur dioxide, which consequently isolates the cells from one another by vapour locks created in the small tubes. The activation system, therefore, is dynamic in nature for the initial phases of the discharge life. For the concept to be practical, the hardware design

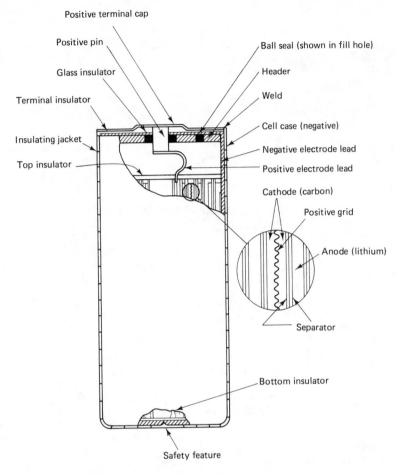

Figure 24.5 Honeywell active lithium–sulphur dioxide cell showing vent concepts (Courtesy of Honeywell)

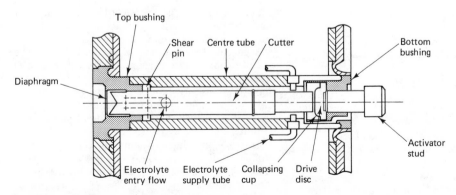

Figure 24.6 Honeywell lithium–sulphur dioxide battery electrolyte manifold and activation system (Courtesy of Honeywell)

Figure 24.7 Honeywell lithium–sulphur dioxide battery 20 A h reserve battery components (Courtesy of Honeywell)

must minimize intercell leakage, maximize activator efficiency in terms of liquid delivered, and ensure the maintenance of liquid flow during the early phase of discharge.

Figure 24.8 presents data recorded during a test at 10°C with a 20 A h activator. A constant Freon pressure of 3.72 kg/cm² absolute was noted throughout the electrolyte delivery cycle. This is as expected since, at this temperature, sufficient Freon is packaged to remain in a saturated condition within the activator. On activation, the electrolyte was allowed to flow into an evacuated volume, which simulated the internal volume environment of a cell stack. The tubes used for transfer of the liquid were identical to those used in a finished battery.

In this test, 305 cm³ of electrolyte were packaged, 270 cm³ of which were delivered to the simulated battery volume within 75 s after activation. This represented a delivery efficiency of approximately 88%, which provides a sufficient quantity of electrolyte/oxidant to the cell stack to achieve the 20 A h output required.

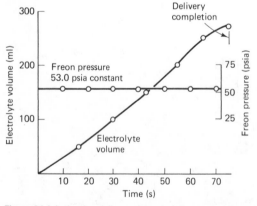

Figure 24.8 Activation delivery characteristics against time at −10°C. 20 A h activator hardware, 305 ml electrolyte packaged, 270 ml (88.5%) delivered, four-capillary tube with 10°C delivery (Courtesy of Honeywell)

24.2 Lithium–thionyl chloride primary batteries

The cell reactions in the lithium–thionyl chloride cell are postulated as follows:

$$4Li + SOCl_2 \rightarrow 4LiCl + SO_2 + S$$

On a partially discharged cell stored at 60°C and above, a possible consistent reaction could be:

$$2SOCl_2 + 3S \rightarrow 2S_2Cl_2 + SO_2$$

The electrolyte used in lithium–thionyl chloride batteries consists of a solution of 1.5M lithium aluminium chloride ($LiAlCl_4$) dissolved in thionyl chloride. This electrolyte is usually doped with 5% sulphur dioxide to prevent passivation of the anode by a lithium chloride film. The anode is pure lithium foil mounted on a stainless steel collector plate, and the cathode is a cold-pressed sheet of 80% Shawinigan carbon and 20% Teflon mounted on stainless steel. The active lithium–thionyl chloride battery is available in capacities of 1.3, 160, 380 and 517 A h. The battery is constructed in a jelly-roll configuration with the electrodes separated by a glass mat separator.

Lithium–thionyl chloride is often cited as having the highest performance potential of any of the lithium cells. It is particularly promising for high-current applications, though it has been used for low currents as well.

According to some experts in the field, this type of cell might replace carbon and alkaline C and D cells for general consumer and industrial applications. It is currently limited to military applications. Lithium–thionyl chloride cells being produced now have five to eight times the capacity of an ordinary carbon cell, so the higher cost ($4–9) is within reason. The growing inclination of retailers to stock products with batteries on the shelf is another good argument for using long-life lithium cells as an alkaline replacement.

Like the sulphur dioxide combination, thionyl chloride has been haunted by safety questions. One

(a) The Eagle Picher GAP 9059 used in strategic defence systems

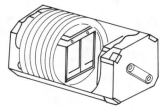

(b) The Eagle Picher GAP 9047 used in air-to-air missiles

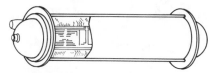

(c) The Eagle Picher GAP 9049 used in re-entry vehicles

(d)

Figure 24.9 Eagle Picher lithium–thionyl chloride reserve batteries (Courtesy of Eagle Picher)

safety-related problem exists when a gas forms during the cell's operation, placing the cell under pressure. If heating occurs, there could be an explosion, or at least an output of toxic gases through the safety vents.

At the same time, one applications disadvantage of thionyl chloride is that it has the same reaction delays and position sensitivity of sulphur dioxide. In fact, the response delays can be several seconds long, which could be a problem in memory-backup use.

Eagle Picher also manufacture lithium–thionyl chloride reserve batteries. Three types of electrolyte reservoir configurations are available, as shown in Figure 24.9, with the following characteristics:

1. *The spherical reservoir:*
   (a) Smallest absolute volume.
   (b) Lightest weight.
   (c) Isolation between electrolyte and pressure medium.
2. *The cylinder-piston reservoir:*
   (a) Smallest packaged volume.
   (b) Flexibility in geometric shape.
   (c) Isolation between electrolyte and pressure medium.
3. *The coil tubing reservoir:*
   (a) Unlimited flexibility in geometric shape.

The reserve battery operating concept developed by Eagle Picher is illustrated in Figure 24.10.

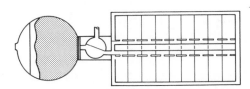

(a) Unactivated condition

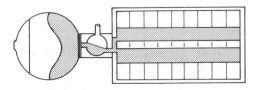

(b) Partially activated condition

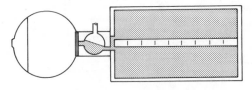

(c) Completely activated condition

Figure 24.10 Eagle Picher lithium–thionyl chloride reserve battery operating concept (Courtesy of Eagle Picher)

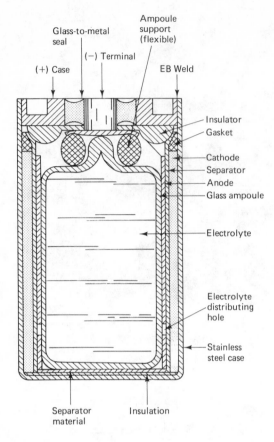

(a) G2666

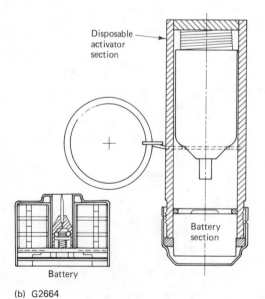

(b) G2664

Figure 24.11 Honeywell lithium–vanadium pentoxide reserve cells: typical construction (Courtesy of Honeywell)

24.3 Lithium–vanadium pentoxide primary batteries

Diagrammatic representations of reserve and non-reserve (active) lithium–vanadium pentoxide batteries supplied by Honeywell are shown, respectively, in Figures 24.11 and 24.12.

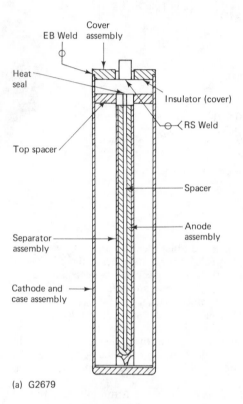

(a) G2679

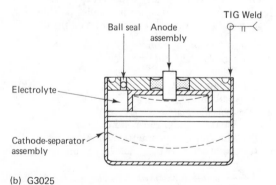

(b) G3025

Figure 24.12 Honeywell lithium–vanadium pentoxide cells: typical construction of active non-reserve cells (Courtesy of Honeywell)

24.4 Lithium solid electrolyte primary batteries

The Duracell solid electrolyte cell is made of the following materials:

Anode: high-purity lithium sheet.
Cathode: mixture of lead iodide, lead sulphide and lead.
Electrolyte: blend of lithium iodide and activated alumina.

At the anode, the lithium loses electrons forming lithium ions (Li^+). The ions travel though the solid electrolyte layer and the electrons travel through the external load to reach the cathode. At the cathode, the lithium ions react with the composite cathode material and the incoming electrons to form the discharge products. The discharge reactions can be expressed by the following equations:

$$2Li + PbI_2 \rightarrow 2LiI + Pb$$
$$2Li + PbS \rightarrow Li_2S + Pb$$

Lithium iodide is virtually a pure ionic conductor. The ionic conductivity is $10^{-7}/(\Omega\,cm)$ at room temperature. The conductivity can be enhanced by incorporating high-surface-area alumina in the solid lithium iodide. The solid electrolyte used in these cells has an ionic conductivity of about $10^{-5}/(\Omega\,cm)$ at room temperature, which enables the cell to deliver currents of $10\,\mu A/cm^2$ at 20°C with high utilization of the active materials.

These batteries are available in button and circular disc designs.

24.5 Lithium–iodine primary batteries

Lithium iodide provides a good example of a solid which has a moderately high conductivity owing to a large number of mobile cation vacancies. In Figure 24.13 the movement of such a defect from left to right is seen to result in the net transfer of a positive charge from right to left, i.e. the Li^+ vacancy behaves in effect as a singly charged anion.

Despite its rather modest ambient conductivity (5×10^{-5} S/m), lithium iodide forms the electrolyte of a battery which is now one of the commonest power sources for implantable cardiac pacemakers. The battery is successful in this application since only small currents are drawn (typically $25\,\mu A$), thus keeping the iR drop in the cell low. It also meets the requirement of long service-life (8–10 years), and the high open circuit voltage and low equivalent mass of lithium guarantee a high energy density, so minimizing the weight of the implanted device. Finally, in this all-solid-state cell, it is possible to avoid many of the problems associated with conventional batteries, such as leakage, self-discharge and separator failure, and so produce a power source of exceptional reliability.

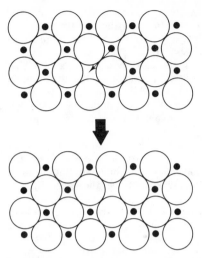

Figure 24.13 Migration of a cation vacancy within a lithium iodide lattice, where ● = Li^+ and ○ = I^-

A typical cell is constructed by placing in contact a lithium anode and an electronically conducting charge-transfer complex of iodine and poly-2-vinylpyridine (P2VP) containing an excess of iodine. A thin layer of lithium iodide is immediately formed, which becomes the electrolyte of the cell and prevents any further direct reaction between the active cell components.

Catalyst Research Corporation (USA) is a major manufacturers of lithium–iodine batteries used for pacemaker and other applications. The Catalyst Research Series 800 cells use the 'lithium envelope' concept. The cell is constructed with a centrally located cathode current collector and a lithium envelope which surrounds and contains the iodine depolarizer material (Figure 24.14). This depolariz-

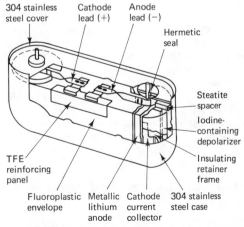

Figure 24.14 Catalyst Research Corporation model 802/803 solid lithium–iodine cell (Courtesy of Catalyst Research Corporation)

er material is corrosive to the stainless steel case and must be kept from contacting the case for maximum cell life. A second barrier, formed from fluorocarbon plastics, surrounds the lithium envelope, insulates it from the case, and provides a second envelope for the containment of the depolarizer. The corrosive effect of iodine-containing depolarizer on stainless steel is lower than that of depolarizer made with more active halogens, such as bromine or chlorine.

The modes of failure of this battery are slow. It cannot suffer from separator rupture since the lithium iodide layer is self-healing. Nor does it suffer from electrolyte leakage since the electrolyte is solid.

A cell is formed when a lithium anode is brought into contact with a cathode composed of iodine and P2VP depolarizer. Between these components forms the third component necessary for any cell – the electrolyte (in this case solid lithium iodide). As current is drawn from the cell, lithium is oxidized and iodine reduced forming additional lithium iodide. At the anode:

$$2Li = 2Li^+ + 2e^- \tag{24.1}$$

At the cathode ($n = 6$):

$$2Li^+ + 2e^- + P2VPnI_2 = P2VP(n-1)I_2 + LiI \tag{24.2}$$

Net reaction:

$$2Li + P2VPnI_2 = P2VP(n-1)I_2 + 2LiI \tag{24.3}$$

The lithium iodide, which serves as both electrolyte and separator, accumulates as the cell is discharged increasing the internal resistance. The result is an initial linear decline in voltage as can be seen in Figure 24.15. Later, when the P2VP loses most of its iodine, the cathode itself begins to rise in resistance. This results in the 'knee' seen in the voltage–time curves of pacemaker cells shown in Figure 24.15. The sharpness of this 'knee' is a worst case approximation. Under a 90 kΩ load, at least 5 months should elapse as cell impedance increases from 15 to 40 kΩ.

The open circuit voltage of the cell is 2.8 V and the overall energy density is typically 150–250 mW h/g. As the cell is discharged, lithium ions are transported through the electrolyte to the cathode where iodide ions are being formed. Thus, as the discharge continues, the thickness of the electrolyte increases, until ultimately its impedance becomes the current limiting factor.

24.6 Lithium–manganese dioxide primary batteries

Some details on the construction of this type of cell are shown in Figure 24.16. The current collector is a sheet of perforated stainless steel. The stainless steel

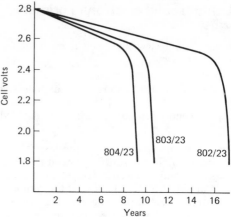

(a) Voltage under 190 kΩ load

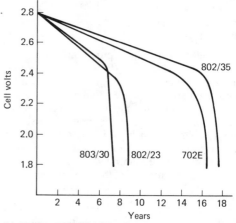

(b) Voltage under 90 kΩ load

Figure 24.15 Projected performance of Catalyst Research Corporation Lithiode TM lithium–iodine cells based on 5 years' test data (Courtesy of Catalyst Research Corporation)

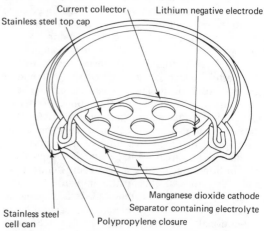

Figure 24.16 Construction of a lithium–manganese dioxide cell (Courtesy of Eagle Picher)

top cap functions as the negative terminal and the stainless steel cell can as the positive terminal of the cell. The polypropylene closure is highly impermeable to water vapour and prevents moisture entering the cell after it has been sealed. The separator is of non-woven polypropylene cloth and contains the electrolyte, a solution of lithium perchlorate in a mixture of propylene carbonate and dimethoxyethane. The lithium negative electrode is punched from sheet lithium and the manganese dioxide cathode is made from a highly active electrolytic oxide.

25

Manganese dioxide–magnesium perchlorate primary batteries

The magnesium dry cell consists of a magnesium anode, a manganese dioxide–carbon cathode and an electrolyte, which is usually magnesium perchlorate or magnesium bromide. The basic cell reaction is:

$$Mg + 2MnO_2 + H_2O \rightarrow Mg(OH)_2 + Mn_2O_3$$

The cell reactions for this battery system are as follows. Anode reaction:

$$Li \rightarrow Li^+ + e^-$$

Cathode reaction:

$$Mn^{IV}O_2 + Li^+ + e^- \rightarrow Mn^{III}O_2 \, (Li^+)$$

Total cell reaction:

$$Li + Mn^{IV}O_2 \rightarrow Mn^{III}O_2 \, (Li^+)$$

Manganese dioxide is reduced from the tetravalent to the trivalent state by lithium; $Mn^{III}O_2 \, (Li^+)$ signifies that the lithium ion enters into the manganese dioxide crystal lattice.

A Kraft paper separator keeps the anode and cathode apart but permits ionic conduction. The magnesium anode is made up of AZ21A alloy. The construction of the cell is similar to that of the conventional cylindrical cell, with the magnesium anode also serving as the cell container. The cross-section of one such cell is shown in Figure 25.1. Present cells have a mechanical seal with a vent to permit the escape of hydrogen gas formed by the following parasite reaction:

$$Mg + 2H_2O \rightarrow Mg(OH)_2 + H_2$$

The construction of a typical manganese dioxide–magnesium perchlorate non-reserve button cell is shown in Figure 25.2. Constructed in a manner similar to that of the paper-lined cylindrical carbon–zinc cell, the can, extruded from a special magnesium alloy, serves as the anode. The cathode consists of manganese dioxide and carbon black, with magnesium perchlorate or bromide in water as the electrolyte. An absorbent paper separator electrically insulates the two electrodes while permitting ionic flow (current flow) when a load is placed across the cell on discharge. The open end of the can is sealed with a plastics cover, which is constrained by means of a plated steel ring. The cover uses a safety vent to guard against abuse such as accidental shorting of the battery, which might cause gassing and excessive internal pressure.

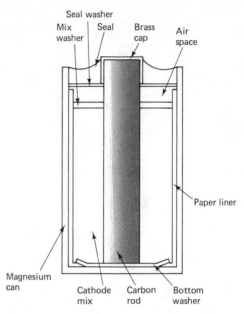

Figure 25.1 Cross-section of magnesium perchlorate–manganese dioxide dry cell (Courtesy of Eagle Picher)

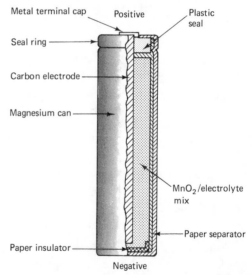

Figure 25.2 Marathon CD-size magnesium cell (Courtesy of Marathon)

26

Metal–air batteries

Contents

26.1 Zinc–air primary batteries

Although zinc–air batteries have been in use for many years, their use has been restricted to very low-rate applications. As a result of the work performed on fuel cells and advancements in technology for making thin gas electrodes that operate at high current densities, zinc–air cells are now being examined as a power source. Cells have been designed with a structure that contains two cathodes (air electrodes) in parallel and one anode (Figure 26.1(a)). The two cathodes are cemented to a frame to form the cell container. With air cells, it is not necessary to incorporate a depolarizer, such as manganese dioxide, into the cell. Instead, oxygen from the air is used as the depolarizer. The replaceable element in the cell consists of a highly porous zinc anode, containing the proper amount of potassium hydroxide in dry form and wrapped with appropriate separator materials. When a cell is ready to be activated, the cell container is filled with water. The zinc anode with the dry potassium hydroxide and separator is then inserted into the cell. The cell is now ready for use. Multi-cell batteries can be constructed. Spacers are placed between cells to permit air flow by convection. For heavy loads and large batteries, convection may not suffice and blowers are required. When spent, the zinc anodes are replaced with new ones for another discharge cycle. A clamping mechanism on one side of the battery is desirable to maintain proper pressure on the cells once the anodes are inserted into the cells. Such zinc–air batteries are capable of giving energy densities up to 220 W h/kg, which is much higher than for the magnesium–manganese dioxide systems.

A constructional diagram of the Berec zinc–air button cell is given in Figure 26.1(b).

McGraw Edison supply a range of Carbonaire zinc–air cells and batteries. These batteries use a carbon–zinc couple and a caustic potash and lime electrolyte, and are activated by the addition of water. The lime combines with, and thereby removes, the zinc reaction products, thus extending the useful life of the electrolyte.

McGraw Edison supply a range of batteries for use in navigational aids, railway lamps, etc. The batteries as supplied are contained in plastic bags to prevent deterioration of the zinc anode during storage by ingress of air and moisture. The batteries are activated simply by topping up with water. Individual cells can be connected in series or in

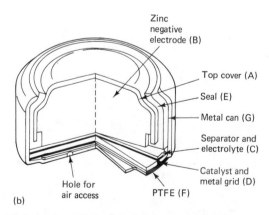

Figure 26.1 (a) Zinc–air cell. (b) Construction of the Berec zinc–air cell: A, the plated steel cell cap functions as the negative terminal of the cell; B, high-purity amalgamated zinc-powder anode, which also retains the alkaline electrolyte; C, synthetic ion-permeable membrane separator; D, cathode of carbon/catalyst mixture with a wet-proofing agent coated on to a plated steel mesh support, and with an outer layer of gas-permeable hydroplastic PTFE; E, sealing grommet to seal and insulate the positive and negative terminals, and seal the cathode to the stepped base; F, permeable diffusion membrane distributes air from the access holes uniformly across the cathode surface; G, plated steel can forms a support for the cathode and acts as a cell container and the positive terminal of the cell – the holes in the cathode can permit air access once a sealing strip is removed (Courtesy of Ever Ready (Berec))

parallel with external connections. Normally these batteries require no routine maintenance or inspection for the first year. Should the service period run well over a year, an annual visual inspection of battery solution levels is desirable. The transparent case permits the user to see when each battery is ready for replacement.

McGraw Edison supply 1100–3300 A h versions of these batteries with voltages between 6 and 22.5 V.

26.2 Metal–air secondary batteries

See Chapter 12, Sections 12.1–12.3.

27

High-temperature thermally activated primary batteries

Contents

27.1 Calcium anode-based thermal batteries

Thermal batteries produced by various suppliers (Catalyst Research Corporation, Eagle Picher, Mine Safety Appliances) have an electrochemical system that consists of an active calcium anode, a fused salt electrolyte (lithium chloride–potassium chloride eutectic mixture) and a cathode consisting of a relatively inert metal current iron or nickel collector containing calcium chromate (or tungstic oxide, ferric oxide, vanadium pentoxide cathode active materials). In typical thermal battery designs the nickel or iron cathode cup, possibly with an added grid, contains a cathode active (on a depolarized) layer, one or two electrolyte layers, a central calcium anode layer followed by further electrolyte and depolarizer layers, and a cathode cell cover. The edge of the closed cell is crimped over and sealed with a gasket. This cell is thus a double cell having a common anode. The cathode active material (depolarized) is fabricated as a chemical layer on inorganic fibre paper, and the electrolyte is dip coated on woven glass tape. The open-cell system is a simpler construction in which the electrolyte and depolarizer chemicals are fabricated into a two-layer pellet disc using an inorganic binder. Simple disc cathodes and anodes are used, connected together as a dumb-bell to link adjacent cells in a series configuration.

Batteries are made up from both the closed and open systems in any series or parallel configuration of stacked cells interleaved with the thermite layers required to activate the battery. Complete batteries are thus usually cylindrical, in the form of hermetically sealed metal canisters. Terminals providing connections to cells are located in glass or ceramic insulating seals.

Before activation, the electrolyte is an inert solid. Since the case is hermetically sealed there is no deterioration of the electrochemical system. Battery storage and operation over a wide range of environmental conditions can be achieved. Because of heat losses and ultimate resolidification of the electrolyte, the active life of thermally activated reserve batteries is necessarily short, generally less than 5 min. Typical battery designs are shown in Figure 27.1. In this type of cell the cathode and electrolyte are combined in a single homogeneous 'DEB' pellet, consisting of Depolarizer, Electrolyte and Binder. An electrically conductive heat pellet functions as the cell heat source and intercell connector.

Significant advantages come from the use of the molten salt electrolytes, usually lithium chloride–potassium chloride eutectic (61 mol% lithium chloride), which has a melting point of 352°C. First, a wider span of operating temperatures can be used than in other types of system without

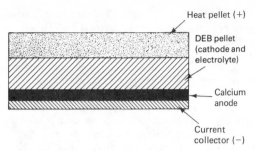

Figure 27.1 Typical thermal battery design (Courtesy of Catalyst Research Corporation)

serious degradation of performance. Using the internal heat source to raise the temperature of the electrolyte above its melting point, a range of ambient temperatures from -40 to $+70°C$ is common and wider ranges of -55 to $+100°C$ can also be accommodated. Secondly, a high cell voltage, up to 3.0 V/cell, can be obtained. A wide voltage band can be used in this battery system since neither oxidation nor reduction of the eutectic alkali-metal chloride electrolyte occurs over the range of 3.0 V, unlike aqueous electrolyte systems or even molten alkali metal sulphates. Furthermore, conductivity of molten alkali electrolytes is considerably greater than that of aqueous electrolytes. Lithium chloride–potassium chloride eutectic has a specific conductance of $2.1/(\Omega\,cm)$ at 900°C as against $0.2/(\Omega\,cm)$ for saturated aqueous sodium chloride and $0.039/(\Omega\,cm)$ for 0.1N hydrochloric acid at 15°C. Thus significantly higher currents for limited periods can be carried in molten salt batteries compared with other systems.

A typical design of a calcium chromate thermal cell produced by Catalyst Research Corporation (Figure 27.2) consists of a folded bimetal dumb-bell and pad of pyrotechnic paper. The catholyte disc consists of glass fibre tape impregnated with a molten mixture of lithium chloride, potassium chloride and potassium dichromate. The tape acts as a binding matrix for the molten salt when the battery is active.

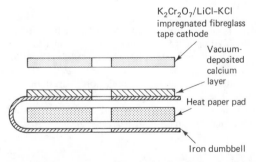

Figure 27.2 Cell cross-section of a Catalyst Research calcium–calcium chromate thermal cell (Courtesy of Catalyst Research Corporation)

The pyrotechnic paper or 'heat paper' supplies the thermal energy to elevate the thermal battery cell to operating temperatures. The pad consists of a ceramic fibre paper, which acts as a binder/carrier for a slurry of zirconium metal fuel and barium chromate oxidant. The paper is extremely fast burning, making it ideal for a fast activation cell. Because the remaining ash of the heat pad is electrically non-conductive, it is necessary to provide a means for series intercell connection.

The intercell connection is provided, along with the calcium anode, by a dumb-bell shaped piece of steel strip, which is folded around the heat pad. The anode is a vacuum-deposited calcium layer on the steel base, which forms a bimetal assembly.

27.1.1 Configurations

Thermal batteries are manufactured in a wide range of configurations to suit various types of application. Some of the configurations available from Catalyst Research Corporation are illustrated in Figures 27.3–27.8.

Figure 27.3 shows the cell configuration of the cup cell design. This was the first thermal battery design used in production, and it is still in use in many applications in the USA and the UK. The cup cell features a common anode and two depolarizer pads, thereby doubling the cell area. The eutectic lithium chloride–potassium chloride electrolyte is impregnated on to glass tape to serve as a separator. These

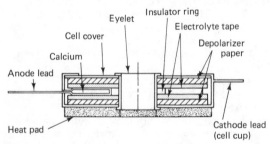

Figure 27.3 Construction of a tungstic oxide closed cell (Courtesy of Catalyst Research Corporation)

are solidified and die-cut to fit the cell. The heat source is zirconium–barium chromate, which forms an insulating ceramic mass on combustion. A cathode lead on the periphery of the cell serves as an intercell connector. Tungstic oxide and calcium chromate depolarizers are commonly used in this design. The active positive material is often mixed with materials such as glass and quartz fibres and formed paper-thin to fit the cell.

Open cell/tape electrolyte

An extension of the cup cell technology involved elimination of the cup itself. The cell configuration

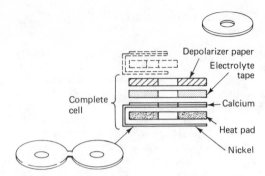

Figure 27.4 Construction of a tungstic oxide open cell (Courtesy of Catalyst Research Corporation)

shown in Figure 27.4 has much in common with the cup cell design, in that the electrolyte is impregnated on to glass tape and fibre pad depolarizers such as tungstic oxide are used. The dumb-bell shaped part, coated with calcium on one side, is folded around heat paper discs and serves as the connector between cells.

Elimination of the cup is made possible by reducing the quantity of electrolyte and by restricting the battery to low or moderate dynamics. This design is useful for light to moderate loads ($<50 \, mA/cm$) for durations of 1–1.5 min. A significant feature of this design is the thin cell, approximately 1.5 mm thick, including heat pad. The coulombic capacity per unit cell area of the design is, however, limited to levels below that of cup designs, particularly since the latter have two depolarizer pads and twice the cell area within the cell.

Potassium dichromate depolarizer is used in an open cell, with lithium chloride–potassium chloride impregnated on glass tape. This depolarizer allows very high current density for fractions of seconds to about 5 s.

Pellet-type construction

This construction involves mixing finely divided inorganic binder, which acts as a separator, with finely ground electrolyte eutectic salts and compacting them into pellets. Finely ground positive active material (for example, calcium chromate) may be either mixed with the original binder or the basis for another pellet. The pellet cell may be a single or double layer.

Multi-layer pellets/heat paper designs

With the development of pellet electrolyte cells, sufficient electrolyte and depolarizer could be included in a cell to provide a coulombic capacity equivalent to the cup design without the necessity of a complex cell configuration. Binders added to the

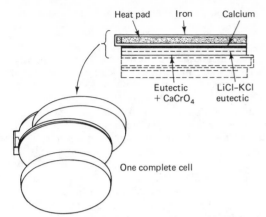

Heat pad   Iron   Calcium

Eutectic
+ CaCrO₄   LiCl–KCl
                  eutectic

One complete cell

Figure 27.5 Construction of an open pellet cell (Courtesy of Catalyst Research Corporation)

electrolyte tend to retain the shape of the pellet during thermal battery operations.

One such configuration is shown in Figure 27.5. The two-layer pellet consists of an electrolyte layer (lithium chloride–potassium chloride) with a binder such as kaolin or micro-size silica, and a second layer composed of calcium chromate, plus some electrolyte. Calcium metal is used as the anode and zirconium–barium chromate heat paper is again the heat source, with dumb-bell shaped parts serving as intercell connections.

Glazed depolarizers

In the thermal battery configuration shown in Figure 27.6, one side of one face of nickel strip is coated with a glaze of vanadium pentoxide using boric oxide as a flux. The other side of the nickel is coated with calcium and the composite folded around heat

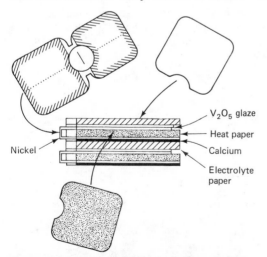

V_2O_5 glaze

Heat paper

Calcium

Nickel

Electrolyte
paper

Figure 27.6 Construction of a glazed vanadium pentoxide depolarizer (Courtesy of Catalyst Research Corporation)

paper. The resulting open-circuit voltage is approximately 3.2 V. An asbestos mat serves as a matrix for electrolyte. The cell has a square configuration and a thickness of only 1.0 mm. The thin cell characteristics and high voltage allow a significant voltage per unit height of cell stack.

Figure 27.7 shows the $Mg/LiCl–KCl/V_2O_5–B_2O_3$ three-layer pellet system.

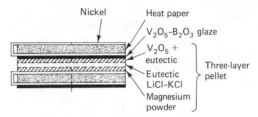

Nickel   Heat paper

V_2O_5–B_2O_3 glaze

V_2O_5 +
eutectic

Eutectic
LiCl–KCl   Three-layer
                  pellet

Magnesium
powder

Figure 27.7 Construction of a magnesium–vanadium pentoxide pellet cell (Courtesy of Catalyst Research Corporation)

Heat pellet/DEB pellet

In 1959, Catalyst Research Corporation conceived a cell design in which no distinct electrolyte layer was necessary; instead lithium chloride–potassium chloride–calcium chromate and kaolin binder were formed into a homogeneous pellet. Since it was homogeneous, it could not be misassembled during construction as is the case with multi-layer pellets. Shortly thereafter a similar pellet was conceived by Sandia, where the binder was Cab-O-Sil, a microporous silica. Using a new heat source, iron-rich iron–potassium chlorate pellets, and the homogeneous DEB pellet, an advanced cell design was created (Figure 27.8) with the advantages of ease of assembly, low cost, high dynamic stability and suitability for longer life than is possible with heat paper/two-layer pellet systems. After combustion, the heat pellet is an electronic conductor, supplying intercell connection. The heat pellet burns at a lower rate than heat paper so that heat pellet batteries generally start under load about 0.2 s slower than heat paper batteries. On combustion, however, the heat pellet retains its shape and, since

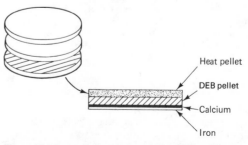

Heat pellet

DEB pellet

Calcium

Iron

Figure 27.8 Construction of a heat pellet–DEB pellet cell (Courtesy of Catalyst Research Corporation)

it has a higher enthalpy than heat paper ash, it serves as a heat reservoir, retaining considerable heat, reaching lower peak temperatures and re-emitting its heat back to the cell as the electrolyte starts to cool.

The characteristics of some of the range of thermal batteries available from Eagle Picher are shown in Table 27.1.

Table 27.1 Characteristics of thermal batteries available from Eagle Picher

---

EAO-6266: 101.6 mm diameter five-cell stacks

Voltage	27 ± 3 V
Life	at 71°C, 221 s
	at 24°C, 366 s
	at −54°C, 317 s
Load	33 A nominal
Size	121 mm diameter × 193.7 mm long
Weight	5.2 kg
Watt hours	89.1 = 17.2 W h/kg

EAP-6230: 101.6 mm diameter five-cell stacks, DEB type

Voltage	28.5 ± 3.5 V
Life	at 38°C, 24 min
	at 5°C, 16 min
Load	4.3 A nominal
Size	122 mm diameter × 104 mm long
Weight	2.86 kg
Watt hours	48.3 = 16.9 W h/kg

EAP-6251: 76 mm diameter parallel stacks

Voltage	27 ± 3 V
Life	at 71°C, 110 s
	at 24°C, 200 s
	at −40°C, 140 s
Load	23 A nominal
Size	94 mm diameter × 145 mm long
Weight	3 kg
Watt hours	31 = 10.4 W h/kg

EAT-6261: 76 mm diameter, parallel stacks

Voltage	$+33 \pm 5.5$ V
	-33 ± 5.5 V
Life	at 71°C, 170 s
	at 24°C, 200 s
	at −56°C, 135 s
Load	4.2 A each section
Size	89 mm diameter × 117 mm long
Weight	1.9 kg
Watt hours	20 = 8.1 W h/kg
	(138.6 × 2 sections)

EAP-6273

Voltage	12 ± 3 V
Life	at 24°C, 350 s
Load	500 A nominal
Size	177.8 mm diameter × 406 mm long
Weight	20.4 kg
Watt hours	583.3 = 38.6 W h/kg

27.1.2 Emergency starting power thermal batteries

The EAP-6273 12-V battery (see Table 27.1) by Eagle Picher was specifically designed to provide emergency starting power for military vehicles and aircraft, and provides a good example of an application of thermal batteries.

The basic design of this battery is eight parallel cell stacks of 152.4 mm diameter. The cell area of 181 cm^2 per cell for the eight sections results in 1449 cm^2 of active area. At a discharge rate of 0.7 A/cm^2, the EAP-6273 could produce 1000 A. This rate is well within the normal operating limits of the standard calcium–calcium chromate system. The cell is fabricated in one piece using the standard two-layer (DEB) configuration and is 152.4 mm in diameter, 2.2 mm thick, and contains a 12.7 mm centre hole. Each cell was pressed at 18.1×10^4 kg or about 1000 kg/cm^2.

The header is of standard 304 stainless steel 9.5 mm thick. Because of the high current, four terminals for each negative and positive lead are used. At the 800 A level, each terminal would carry 200 A. The battery is connected externally by paralleling the four positive and the negative terminals at the load connections.

The battery case is assembled from 321 stainless steel welded tubing with a 9.5 mm steel plate welded in the bottom. The burst strength of this container was calculated to be in excess of 454 kg.

Because of the total height of the cell stacks (approximately 356 mm) the stack was assembled in two separate sections, each containing four parallel stacks of cells. These are wrapped and inserted into the case separately. The case is lined with 12 mm of thermoflex and 12 mm of asbestos in the bottom. Each battery section is wrapped with 3.8 mm of glass cloth. Asbestos for the top of the battery and the header insulation are added to obtain a closing pressure of 10–22 kg. The total battery weighs 20.4 kg and is 406.4 mm tall and 177.8 mm in diameter.

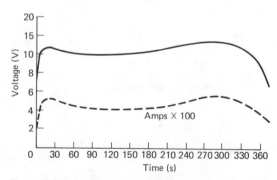

Figure 27.9 Performance of an Eagle Picher EAP-6273 thermal battery under 0.024 Ω load (Courtesy of Eagle Picher)

The load desired is $0.024\,\Omega$, capable of dissipating $25\,kW$. A $15.8\,mm$ outside diameter stainless steel pipe is cut to length and welded to two copper bars. Connections are made to allow water to flow through the pipe. A $1000\,A$ current shunt is bolted into series with one copper bar to provide current monitoring capabilities.

The performance of the EAP-6273 is shown in Figure 27.9. Essentially the battery performs as a $12\,V$ battery delivering $550\,A$. Additional current could be drawn on if necessary, as would be required for an electric motor starter. However, the power developed would be capable of starting any $12\,V$ vehicle requiring up to $800\,A$ of peak current loads.

27.2 Lithium anode thermal batteries

One of the constraints in designs of calcium anode-based thermal batteries has been the limited duration at high current densities due to a low electrochemical efficiency of the calcium anode, arising from the existence of side-reactions involving the electrolyte and the electrode active materials. It has been demonstrated by Mine Safety Appliances Ltd that significant improvements in thermal battery performance can be achieved by the use of lithium as the active anode material in the cell, coupled with new cathodes. These batteries are in production and are capable of operating at electrode current densities of up to $2.0\,A/cm^2$, with long battery durations. A capability has been demonstrated of achieving $100\,W\,h/kg$ compared with approximately $20\,W\,h/kg$ in designs of conventional calcium batteries, such as the Ca/LiCl–KCl and $CaCrO_4$/Fe systems. These latter designs, although they give very satisfactory performance for applications with durations of up to approximately 5 min, show a very significant drop in electrode current density, which results in a reduction in the power and energy density as discharge durations are extended.

Mine Safety Appliances have designed a system with lithium as the active anode material and ferrous sulphide as the cathode, which shows significant improvements in performance over the conventional systems. The Li–FeS_2 system is based on a pile-type construction, which allows the battery voltage to be easily adjusted by varying the number of series cells to meet different application requirements. The components making up the cells are the lithium, which is held in the anode assembly by a porous support material, the cathode (depolarizer layer), which is a thin disc of iron disulphide with a proportion of electrolyte, and an electrolyte layer, which is a disc of the electrolyte with a binder added to immobilize the electrolyte when it is in the active molten state. The cells are interleaved with the heat source, also a disc of a pyrotechnic mixture of iron powder and potassium perchlorate. This component is electrically conducting and acts as the series connection between the cells. The battery is activated by the igniter flame being transmitted via pyrotechnic fuse-strips to ignite the individual pyrotechnic discs. The battery cell stack is thermally insulated and contained in a mild steel case, hermetically closed by argon arc welding. The electrical output is through terminals in glass-to-metal compression seals.

A typical layout of such a battery is shown in Figure 27.10. The electrolyte–catholyte cathode pellet is pelletized powder made by the compaction of two distinct powders in separate layers. The electrolyte layer is required to isolate the anode electrically from the iron pyrite cathode layer, which is electrically conductive. The electrolyte is a mixture of lithium–potassium chloride eutectic electrolyte and ceramic binder. The cathode is a mixture of iron pyrite and electrolyte.

The iron-based pyrotechnic is a pelletized wafer or pill of iron powder and potassium perchlorate.

The anode assembly consists of an iron cup, which is crimped around an elemental lithium carrier matrix. Here, a binding agent is used which binds the lithium by surface tension in a manner similar to the 'gelling' of electrolyte by ceramic binders. This anode has numerous advantages when compared with lithium–aluminium or lithium–silicon alloys for

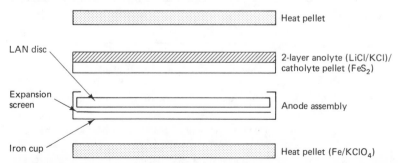

Figure 27.10 Cell cross-section of Catalyst Research Corporation's lithium–iron sulphide battery (Courtesy of Catalyst Research Corporation)

Table 27.2 Comparison of properties of calcium–calcium chromate and lithium–iron sulphide systems

	Calcium–calcium chromate	Lithium–iron sulphide
Energy density (W/s)	600	1600
Activation time (s)	0.03 possible, 0.3 typical	0.3 possible
Voltage range	2.5–500	2.5–500
Multiple voltage taps	Up to 5 voltages/unit are common	4 voltages/unit currently employed
Dependability	0.995	0.995
Maintenance	Nil	Nil
Shelf life (years)	15	$\geqslant 15$ expected
Severe environments		
Operating temperature (°C)	−55 to +93	−54 to +74
Units of shock during activated life	10 000	10 000
g acceleration during activated life	2500	2500
Humidity/pressure	Unaffected	Unaffected
Activation	Electrical percussion mechanical shock	Electrical percussion mechanical shock

the bulk of thermal battery applications. These include greater design versatility, higher power and energy densities, lower cost, excellent storage characteristics, higher current densities, very broad operational temperature range and excellent safety characteristics.

27.3 Lithium alloy thermal batteries

A further improvement on the lithium anode–iron sulphide (FeS_2) thermal cell is one in which the anode consists of a lithium alloy (usually lithium–aluminium alloy, although lithium–silicon and lithium–boron alloys have also been used) and an iron sulphide cathode. The load-carrying capacity of this $LiAl–FeS_2$ (abbreviated $LAN–FeS_2$) system is two to three times better than that of the calcium–calcium chromate system. The $LAN–FeS_2$ system has a low and constant internal impedance which makes it ideal for both long discharge lives and pulse applications.

This type of battery can now replace many remotely activated silver oxide–zinc applications, resulting in an equal or smaller size, lightweight battery at lower cost.

A comparison of some of the important characteristics of the calcium–calcium chromate and $LAN–FeS_2$ systems is given in Table 27.2.

Table 27.3 summarizes the characteristics of the main types of thermal battery.

27.4 Theory of thermal batteries

27.4.1 The pyrotechnic heat sources

The two principal pyrotechnic heat sources used in thermal batteries are a paper-type composition of zirconium and barium chromate powders supported on inorganic fibres and a pressed tablet made up of iron powder and potassium perchlorate, commonly known as a heat pellet.

The zirconium–barium chromate heat paper is manufactured from pyrotechnic-grade zirconium

Table 27.3 Types of thermal battery

Electrochemical system: anode/electrolyte/cathode	Operating cell voltage (V)	Characteristics and/or applications
Ca/LiCl–KCl/WO$_3$	2.4–2.6	Used principally for fuse applications where a low level of electrical noise is essential and where dynamic environments are not severe
Ca/LiCl–KCl/CaCrO$_4$	2.2–2.6	Used in applications requiring short-term operation in severe dynamic environments
Mg/LiCl–KCl/V$_2$O$_5$	2.2–2.7	Used in applications requiring short-term operation in severe dynamic environments
Li(M)/LiCl–KCl/FeS$_2$	1.6–2.1	Overall advantages: low electrical noise, can operate in severe dynamic environments, long service life (up to 1 h)

and barium chromate, with a particle size of about 1–10 μm. Inorganics such as ceramic and asbestos fibres are added. The mix is formed into paper as individual sheets by use of a mould or continuously by use of a paper-making process. The resultant sheets are cut into parts and dried. Once dry, they must be handled very carefully since they are very susceptible to ignition by static discharge. Heat paper has a burning rate of about 10–15 cm/s and a heat content of about 1675 J/g. Heat paper combusts to an inorganic ash with high electrical resistivity. Consequently, its use necessitates addition of nickel or iron electrode collectors and intercell connectors to conduct current around each cell's heat paper pad.

Fine iron powder (1–10 μm) and potassium perchlorate are blended dry and pressed to form heat pellets. The iron content ranges from 80 to 88% by weight and is considerably in excess of stoichiometry. Excess iron provides the combustible pellet with sufficient electronic conductivity, eliminating the need for intercell connectors. The heat content of iron–potassium perchlorate pellets ranges from 920 J/g for 88% iron to 1420 J/g for 80% iron. Burning rates of pellets are generally slower than those of heat paper, and the energy required to ignite them is greater. The heat pellet has higher activation energy and is therefore less susceptible to inadvertent ignition during battery manufacture. However, the battery must be designed so that there is good contact of the heat pellet with the first fire or ignition source.

After combustion, the heat pellet is an electronic conductor, simplifying intercell connection and battery design. The heat pellet burns at a lower rate than heat paper, and so heat pellet batteries generally start about 0.2 s slower than heat paper batteries under load. Upon combustion, however, the heat pellet ash retains its original shape and, since it has a higher enthalpy than heat paper ash, it serves as a heat reservoir, retaining considerable heat, reaching lower peak temperatures, and releasing its heat to the cell as the electrolyte starts to cool. The heat paper combusts to shapeless refractory oxides causing slumping of the battery stack during ignition and less resistance to environmental effects.

27.4.2 Methods of activation

Thermal batteries are initiated by either mechanical action using a percussion-type primer or an electrical pulse to an integral electric match (squib). For most military applications, safety considerations require the squib to be non-ignitable under a load of 1 W, 1 A. In heat pellet batteries, an intermediate heat paper firing train is often added to carry the ignition from the primer or squib to each pellet.

27.4.3 Insulation materials

Thermal batteries are designed to maintain hermeticity throughout service life, even though internal temperatures approach 600°C. Thermal insulation used to minimize peak surface temperatures must be anhydrous and, if organic, must have high thermal stability. Dehydrated asbestos, ceramic fibres, and Kapton (polyimide) have been used. Insulative layers are tightly bound around the periphery of the battery stack, and several asbestos or ceramic fibre discs are placed on each end to ensure a tight pack in the metal can. Special end reservoir pellets are often used. Long-life (>10 min) thermal batteries often use a very low thermally conductive product called 'min-K' (Johns-Manville Co.), which is manufactured from titania and silica.

27.4.4 Chemistry

Calcium–calcium chromate thermal batteries

The overall reaction for the calcium–calcium chromate system is dependent on the discharge parameters. A postulated cell reaction is:

$$3Ca + 2CaCrO_4 + 6LiCl \rightarrow 3CaCl_2 + Cr_2O_3.2CaO + 3Li_2O$$

although the exact composition of the mixed oxide has not been determined. It is possible that lithium formed by reaction of calcium with lithium chloride electrolyte enters into the electrochemical reaction.

Side-reactions occur between calcium and LiCl–KCl, which can limit the full utilization of coulombic capacity between calcium and suitable depolarizers. The reaction:

$$Ca + 2LiCl \rightleftharpoons 2Li + CaCl_2$$

occurs spontaneously in a thermal battery with the result that an alloy (Li_2Ca) is formed which is liquid at thermal battery operational temperatures. This melt, known as an 'alloy', can be responsible for internal shorts, resulting in intermittent cell shorts (noise) and cell misfunction during the discharge. This alloy permits high anode current density but limits the use of the couple on open circuit or light loads because an excess causes cell shorting. The rate of alloy formation can be controlled by deactivating or passivating the cell to slow down the Ca + LiCl reaction. Techniques include controlling by current density, acetic acid treatment of calcium, addition of passivating agents and excess binder to the electrolyte, and reduction of electrolyte pellet density.

Calcium chloride, formed by the reaction of calcium with lithium chloride, reacts further with potassium chloride to form the double salt $CaCl_2.KCl$. This salt has a melting point of 575°C and has been identified as a precipitate in calcium anode thermal cells. It has been suggested that it can coexist with molten chloride electrolytes up to about 485°C.

The self-discharge reaction of calcium with calcium chromate is highly exothermic, forming complex chromium(III) oxides. Above about 600°C the self-discharge reaction accelerates, probably due to the markedly increasing solubility of the chromate in the chloride electrolyte. This acceleration increases the rate of formation of calcium–lithium alloy. The resulting thermal runaway is characterized by short battery lives, overheating and cell step-outs, shorts, and noise characteristics of excess alloy.

The calcium–calcium chromate couple can be separated by a discrete electrolyte (plus binder) layer or can be a homogeneous pelletizing mix of depolarizer, electrolyte and binder. The calcium/DEB battery developed in the 1960s has been used considerably in military fusing.

Much of the effort associated with thermal battery design involves determining the amount of pyrotechnic heat to give acceptable performance over the necessary temperature range. Many of the required temperature extremes range from −55 to +75°C for initial ambient. This 130°C difference is well within the operational limits of 352°C, the melting point of lithium chloride–potassium chloride eutectic, and 600°C, the approximate temperature of thermal runaway. However, as discussed earlier for calcium–calcium chromate, a temperature of at least 485°C is necessary for minimal performance at moderate drain ($>50\,\mathrm{mA/cm^2}$) due to freeze-out of double salt ($KCl.CaCl_2$) below that temperature. Thus, it is seen that the working temperature range is reduced to 115°C whereas military extremes exceed this by as much as 15°C.

Figure 27.11 shows the performance and physical characteristics of a $28 \pm 4\,\mathrm{V}$ Ca–DEB pellet–heat pellet battery over the temperature range −54 to +71°C. The life to a 24 V limit is shown as a function of temperature at a 1.5 A drain. The heat balance is adjusted for optimum performance at about 15°C, with the cold performance limited to freeze out on the one hand, and hot performance limited by self-discharge on the other. It is apparent that the thermal balance can be shifted to give optimum performance at something other than room temperature.

Lithium anode thermal batteries

Since lithium metal is molten at thermal battery discharge temperatures, it is retained on high surface area metals by immersion of the metal matrix in molten lithium to form anodes. Often this structure is contained within a metal cup to prevent leakage during cell operation. Another method is the fabrication of lithium alloy anodes, such as lithium–boron, lithium–aluminium and lithium–silicon, which are solid at battery discharge temperatures and thus offer the possibility of simpler

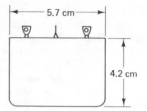

Nominal voltage: 28±4 V
Nominal current: 1.9 A
Operating temperature range: −54 to +71°C
Nominal life: 60 s
Start time: 0.6 s
Volume: 105 cm³
Weight: 316 g

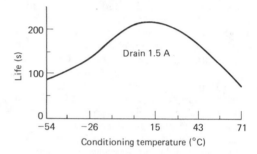

Figure 27.11 Thermal battery performance as a function of temperature

construction. However, the lithium alloys are more difficult to fabricate than the metal matrix anodes and do not achieve this same peak current density. Most of the lithium anode batteries currently use the lithium chloride–potassium chloride electrolyte and an iron disulphide (FeS_2) cathode.

The lithium–iron disulphide electrochemistry is more straightforward than that of the calcium chromate cell. The overall reaction is

$$2Li + FeS_2 \rightarrow Li_2S + FeS$$

Iron disulphide begins to decompose thermally at about 550°C, into sulphur and iron sulphide, but good cathode efficiencies have been obtained up to 600°C. Above 600°C, the rate of decomposition increases, but experience with this system shows that thermal runaway is not as much a problem as with calcium–calcium chromate.

Another advantage to the lithium–iron disulphide system is the absence of high-melting salt phases such as $CaCl_2.KCl$. The cell can thus operate close to 352°C. While calcium–calcium chromate may be used with homogeneous electrolyte–depolarizer blends, iron disulphide must be separated from the anode by a distinct electrolyte layer. Otherwise, the iron disulphide, which is a fairly good conductor, will be electronically shorted to the anode.

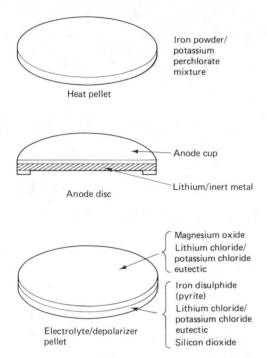

Heat pellet — Iron powder/ potassium perchlorate mixture

Anode cup

Lithium/inert metal

Anode disc

Magnesium oxide
Lithium chloride/ potassium chloride eutectic

Iron disulphide (pyrite)
Lithium chloride/ potassium chloride eutectic
Silicon dioxide

Electrolyte/depolarizer pellet

Figure 27.12 Typical LAN (lithium–iron disulphide) cell construction

Most experience has been with molten lithium anode batteries where the lithium is mechanically retained. Figure 27.12 shows one version of a lithium–iron disulphide thermal battery cell. The design uses a heat pellet, but heat paper versions are also feasible.

Figure 17.13 shows the average service life of a 14-cell lithium–iron disulphide thermal battery designed for power applications discharged at various constant-current loads to 24 V over a temperature range of −40 to +71°C. This battery has 5.1 cm external diameter and 4.4 cm height and uses a modification of the cell shown in Figure 27.12. For comparison, the performance of a similar-sized calcium–calcium chromate thermal battery is also plotted. The lithium–iron disulphide battery has a significantly higher capacity at the high power levels; at the lighter loads, the performance of both cells is limited by the cooling of the cell below the operating temperature.

Figure 27.14 shows the discharge curves comparing an LiAl/FeS$_2$ cell with a magnesium–iron disulphide cell as well as with a typical calcium–calcium chromate cell. Both the magnesium and lithium anode cells give longer performance than the calcium cell; the main difference between magnesium and lithium is the higher voltage of the lithium

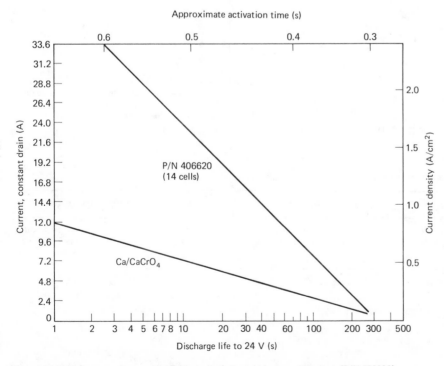

Figure 27.13 Average performance of lithium–iron disulphide thermal battery (P/N 406620) (Courtesy of Catalyst Research Corporation)

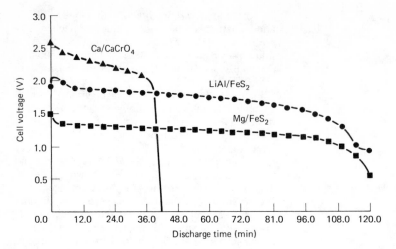

Figure 27.14 Discharge curves at 520°C for Mg/FeS$_2$, LiAl/FeS$_2$, Ca/CaCrO$_4$ thermal battery cells (Courtesy of Catalyst Research Corporation)

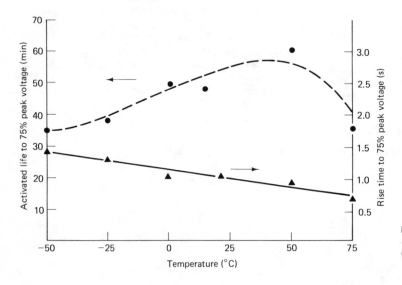

Figure 27.15 Activated life and rise time of a Li(Si)/FeS$_2$ thermal battery (Courtesy of Catalyst Research Corporation)

Table 27.4 Comparison of LiFeS$_2$ thermal batteries

Duration of discharge (min)	Li$_{(liq)}$/LiF–LiCl–LiBr/FeS$_2$		Li(Si)/LiCl–KCl/FeS$_2$	
	Power density (W/cm^3)	Energy density of cell (W h/l)	Power density (W/cm^3)	Energy density of cell (W h/l)
3.5	5.9	340		
5.0			4.3	360
5.5	5.2	480		
10.0			2.6	430
12	2.6	520		
21	1.9	520		
24	–	–	0.8	390
41	–	–	0.5	370

cell. Figure 27.15 shows the activated lifetime and rise time for a 60 min, 28 V, 0.5 A thermal battery with a volume of 400 cm^3 in the Li(Si)/LiCl.KCl/ FeS$_2$ electrochemical system. The rise time decreases with increasing temperature; the activated life, however, maximizes in the range of 25–50°C.

Table 27.4 presents data on two types of lithium–iron disulphide thermal battery and illustrates the advantage of the lithium anode systems. While the liquid lithium anode battery shows better performance than the lithium alloy battery, particularly in its rate capability, some of the advantage is due to size and design differences, which caused the smaller LiSi battery to cool more rapidly.

The performance data for the lithium–iron disulphide thermal battery cover a range from high power (½ min rate) to long life (40–60 min rate). Although the data are still based on prototype batteries, they already show a magnitude of improvement over the conventional thermal battery systems. The advantages are so pronounced that the lithium–iron disulphide thermal battery, it can be confidently predicted, will be the dominant system in the 1990s.

28

Zinc–chlorine secondary batteries

The chemical reactions occurring in the zinc–chlorine battery are very straightforward (Figure 28.1). During charge, an electrolyte solution of zinc chloride in water is passed through the battery with the aid of a circulator. As electrical direct current is passed through the battery from an external source of energy, zinc metal is deposited on the negative plates of the battery, and chlorine gas is generated at the positive plates. The chlorine gas is carried away with the circulating electrolyte stream, but the zinc metal remains at the negative electrodes as a uniform non-porous solid plate. Outside the plate area (the stack), the mixture of chlorine gas and excess electrolyte is cooled to about 0°C. At this point, a pale yellow solid, chlorine hydrate, forms by reaction between the chlorine gas and water from the electrolyte. Chlorine hydrate has the chemical composition, $Cl_2.6H_2O$. The solid chlorine hydrate is retained separate from the battery plates and the excess electrolyte is returned to the stack.

When the volume in the system assigned to the hydrate (the store) is full, the battery is deemed fully charged. At this time, the external source of energy is disconnected and the circulator and cooler are switched off. When the battery is discharged, circulation of the aqueous zinc chloride electrolyte is initiated. Since chlorine is slightly soluble in the electrolyte, this circulation carries chlorine to the positive plates of the stack and current may be withdrawn from the battery. As current passes to the external load, zinc metal reacts at the negative electrode to form zinc ions, and chlorine reacts at the positive electrode to give chloride ions. The net result is the production of zinc chloride, which is dissolved in the circulating electrolyte, and chlorine is removed from the electrolyte. The electrolyte passes from the stack to the store, where more chlorine dissolves in the electrolyte by decomposition of chlorine hydrate. This means that water is also added to the electrolyte from the store during discharge. This water provides for the zinc chloride added to the electrolyte in the stack. Thus the electrolyte concentration remains relatively constant during discharge. Electrolyte recirculated to the stack carries chlorine with it to sustain the current being withdrawn. These processes continue until the store contains no chlorine hydrate, at which point the battery is deemed fully discharged.

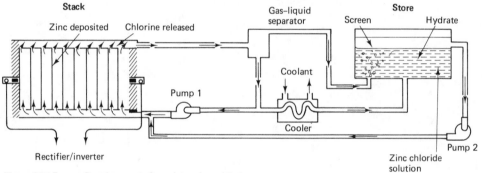

Figure 28.1 Energy Developments Associates zinc–chlorine battery (Courtesy of Energy Developments Associates)

29

Sodium–sulphur secondary batteries

Contents

The operation of this battery relies on the recently discovered properties of β-alumina; namely, that it combines very low electronic conductivity with an unusually high ionic conductivity which is specific to sodium ions as charge carriers.

In the sodium–sulphur battery, patented by Ford, for example, instead of solid electrodes separated by a liquid electrolyte (as in the conventional lead–acid car battery, for example), sodium β-alumina is used as a solid electrolyte, specifically conducting sodium ions between liquid electrodes of sodium metal and sulphur (Figure 29.1).

An alternative design has been developed by Chloride Silent Power (UK). This cell is unusual because it embodies liquid electrodes (sodium and sulphur) and a solid electrolyte (β-alumina). For simplicity, the cell reaction can be written:

$$2Na + 5S = Na_2S_5$$

During cell discharge sodium ions migrate through the solid electrolyte phase into the positive compartment, to combine with the S^{2-} ions produced there. The sodium ions are pumped through the β-alumina to the negative compartment during recharge.

The role of β-alumina (a ceramic material of approximate composition $1.22\,Na_2O\!:\!11\,Al_2O_3$) is critical. β-Alumina is a member of a class of solid materials with high ionic conductivities that have variously been called optimized ionic conductors, superionic conductors and, more usually, fast-ion conductors. The study of this type of substance constitutes the new field of solid-state ionics.

The structure of β-alumina itself is illustrated in Figure 29.2. This is a two-dimensional conductor; the Na^+ ions migrate fairly freely in conduction

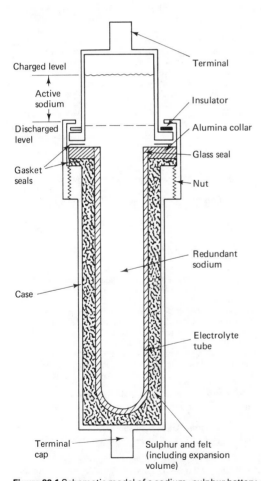

Figure 29.1 Schematic model of a sodium–sulphur battery which uses a sodium–β-alumina solid electrolyte as the separator between liquid electrodes (sodium anode and sulphur cathode); operating temperature 300–400°C (Courtesy of Ford Motors)

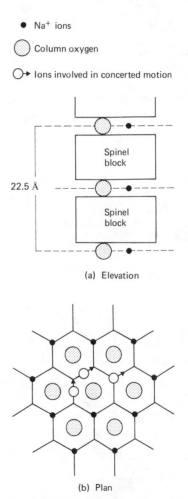

Figure 29.2 Structure of β-alumina showing (a) the alternation of spinel blocks and conduction planes; (b) the 'honeycomb pathway' along which Na^+ ions migrate and the kind of concerted mechanism that is envisaged (Courtesy of Chloride Silent Power)

plates which are located between the spinel blocks and contain many more possible sodium sites than there are ions to occupy them. The conductivity of β-alumina at 350°C (100 S/m) is similar to that of molten sodium nitrate, and reference is sometimes made to the quasi-liquid behaviour of the sodium ions in order to distinguish this type of electrolyte from, say, lithium iodide, where a point defect model is applied, such as lattice vacancies (Schottky defects) or interstitial ions (Frenkel defects).

Further, more recent information on the sodium–sulphur battery is given in the following references.

29.1 References on sodium–sulphur batteries*

Ford Aerospace

Bridges, D. W., Harlow, R. A., Haskins, H. J. and Minck, R. W. (1983) Operational experiences with sodium–sulfur batteries for utility and electric vehicle applications. *One Hundred Sixty-Fourth Meeting of the Electrochemical Society*, Washington, DC, USA, 9–14 October 1983

Crosbie, G. M., Tennenhouse, G. J., Tischer, R. P. and Wroblowa, H. S. (1984) Electronically conducting doped chromium oxides. *Journal of the American Ceramics Society*, **67**, 498

Gates, A. B. (1983) Status of the Ford Aerospace sodium–sulfur battery program. *Eighteenth Intersociety Energy Conversion Engineering Conference*, Orlando, Florida, USA, 21–25 August 1983

Halbach, C. R., McClanahan, M. L. and Minck, R. W. (1983) The sodium–sulfur battery. *Fall Meeting of the American Chemical Society*, Washington, DC, USA, 28–31 August 1983

Harlow, R. A. (1983) An overview of the DOE/Ford Aerospace/ Ford sodium–sulfur battery program. *Beta Battery Workshop V*, Washington, DC, USA, 18–20 October 1983

Harlow, R. A. and McClanahan, M. L. (1982) Sodium–sulfur battery development for electric vehicle and stationary energy storage applications. *Thirtieth Power Sources Conference*, Atlantic City, NJ, USA, 7–10 June 1982

Harlow, R. A. and McClanahan, M. L. (1984) The status of the Ford Aerospace sodium–sulfur battery development program. *Power Sources Conference*, Cherry Hill, NJ, USA, 11–14 June 1984

Harlow, R. A., McClanahan, M. L., Minck, R. W. and Gordon, R. S. (1982) Status of the Ford/Ford Aerospace/DOE sodium–sulfur battery program. *Fifth DOE Battery and Electrochemical Contractors' Conference*, Arlington, Virginia, USA, 7–9 December 1982

Harlow, R. A., McClanahan, M. L. and Minck, R. W. (1984) Sodium–sulfur battery development. *Extended Abstracts: Sixth DOE Electrochemical Contractors Review*, Washington, DC, USA, 25–28 June 1984

Haskins, H. J. and Bridges, D. W. (1982) Operation of a 100-kilowatt-hour sodium–sulfur battery. *Seventeenth Intersociety Energy Conversion Engineering Conference*, Los Angeles, California, USA, 8–12 August 1982, Paper No. 829 097

Haskins, H. J. and Minck, R. W. (1983) Design considerations for sodium–sulfur batteries for electric vehicles. *Electric Vehicle Council EXPO*, Dearborn, Michigan, USA, 4–6 October 1983

Haskins, H. J., Bridges, D. W. and Minck, R. W. (1983) Evaluation of a sodium–sulfur battery module for electric vehicles. *SAE International Congress and Exposition*, Detroit, Michigan, USA, 28 February–4 March 1983

Haskins, H. J., McClanahan, M. L. and Minck, R. W. (1983) Sodium–sulfur cells for high power spacecraft batteries. *18th Intersociety Energy Conversion Engineering Conference*, Orlando, Florida, USA, 21–26 August 1983, Paper No. 839 239

McClanahan, M. L. (1983) Shipping of Na/S cells and batteries. *Beta Battery Workshop V*, Washington, DC, USA, 18–20 October 1983

McClanahan, M. L. (1983) Electrolyte durability from a user's viewpoint. *Beta Battery Workshop V*, Washington, DC, USA, 18–20 October 1983

McClanahan, M. L. and Minck, R. W. (1982) Durability of beta-alumina from a user's viewpoint. *One Hundred Sixty-First Meeting of the Electrochemical Society*, Montreal, Quebec, Canada, 9–14 May 1982

Minck, R. W. (1982) Corrosion problems of the sodium–sulfur battery. *DOE Meeting on High Temperature Corrosion Problems in Battery Systems*, London, 13–14 July 1982

Minck, R. W. (1983) Practical experiences with beta-alumina. *Research Assistance Task Force Meeting on Beta-Alumina for Sodium–Sulfur Cells*, Oak Ridge, Tennessee, USA, 8–10 March 1983

Minck, R. W. (1983) SES and EV battery performance. *Beta Battery Workshop V*, Washington, DC, USA, 18–20 October 1983

Minck, R. W. (1983) SES and EV cell performance. *Beta Battery Workshop V*, Washington, DC, USA, 18–20 October 1983

Minck, R. W. (1983) The sodium–sulfur battery – applicability to EV's and AV's. *Electric and Hybrid Vehicle Systems Assessment Seminar*, Gainsville, Florida, USA, 15–16 December 1983

Minck, R. W. and Halbach, C. R. (1982) Characteristics of sodium–sulfur cells for diverse applications. *Seventeenth Intersociety Energy Conversion Engineering Conference*, Los Angeles, California, USA, 8–12 August 1982, Paper No. 829 096

Sernka, R. P. (1983) Reduction of sodium–sulfur battery development costs. *Beta Battery Workshop V*, Washington, DC, USA, 18–20 October 1983

Sodium–Sulfur Battery Development (1982) Interim Report, 1 March 1980 to 30 September 1981, DOE Contract No. DE-AMO2-79Ch10012

Wroblowa, H. S., Markovac, V., Tischer, R. P. and Crosbie, G. M. (1984) Corrosion of materials in high-temperature sodium–sulfur cells. *35th Meeting of the International Society of Electrochemistry*, 6 August 1984 (Extended Abstract No. A5-2)

Ford Motor Company

Crosbie, G. M., Tennenhouse, G. J., Tischer, R. P. and Wroblowa, H. (1983) Electronically conducting doped chromium oxides. *Eighty-Fifth Annual Meeting of the American Ceramic Society*, Chicago, Illinois, USA, 24–27 April 1983

Gupta, N. K. and Tischer, R. P. (1972) Thermodynamic and physical properties of molten sodium polysulphides and open circuit voltage measurements. *Journal of the Electrochemical Society*, **119**, 1033

Mikkor, M. (1982) Current distribution within sulfur electrodes of cylindrical sodium–sulfur cells. *Seventeenth Intersociety Energy Conversion Engineering Conference*, Los Angeles, California, USA, 8–12 August 1982, Paper No. 829 098

Tischer, R. P. (1983) Review of the chemistry of the sodium–sulfur battery. *Invited Symposium Presentation at the One Hundred Sixty-Third Meeting of the Electrochemical Society*, San Francisco, California, USA, 8–13 May 1983

Tischer, R. P., Ed. (1983) *The Sulfur Electrode*. Academic Press, New York

Tischer, R. P., Wroblowa, H. S., Crosbie, G. M. and Tennenhouse, G. J. (1983) Ceramic oxides as container/current collector materials for the sulfur electrode. *Proceedings of the Fourth International Symposium on Molten Salts*, held in conjunction with the *One Hundred Sixty-Third Meeting of the Electrochemical Society*, San Francisco, California, USA, 8–13 May 1983

*Supplied by Ford Motor Co., Wixom, Michigan, and reproduced by their kind permission.

Wroblowa, H. S. and Tischer, R. P. (1983) Carbon in sodium–sulfur cells. *DOE Workshop on Carbon Electrochemistry*, Case Western Reserve University, Cleveland, Ohio, USA, August 1983

Wroblowa, H. S., Tischer, R. P., Crosbie, G. M., Tennenhouse, G. J. and Markovac, V. (1983) Stability of materials for the current collector of the sulfur electrode. *Beta Battery Workshop V*, Washington, DC, USA, 18–20 October 1983

Ceramatec Inc.

Gordon, R. S. and Miller, G. R. (1983) Overview of Ceramatec's research and development activities related to the development of beta batteries. *Beta Battery Workshop V*, Washington, DC, USA, 18–20 October 1983

Miller, G. R. (1983) Corrosion of oxides. *Beta Battery Workshop V*, Washington, DC, USA, 18–20 October 1983

Miller, G. R. and Gordon, R. S. (1983) Manufacturing cost reduction of β-alumina. *Beta Battery Workshop V*, Washington, DC, USA, 18–20 October 1983

Rasmussen, J. R., Miller, G. R. and Gordon, R. S. (1982) Degradation and lifetime of β-alumina electrolytes. *One Hundred Sixty-First Meeting of the Electrochemical Society*, Montreal, Quebec, Canada, 9–14 May 1982

Chloride Silent Power Ltd (UK)

Lomax, G. R. (1971) *Some UK Progress in Sodium Sulphur Technology*. Chloride Batteries Ltd, UK

Part 3

Battery performance evaluation

30

Primary batteries

Contents

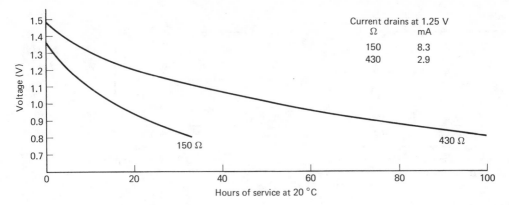

Figure 30.1 Typical discharge curve of a Mallory Duracell alkaline manganese 1.5 V dry cell (PX825) 300 mA h (Courtesy of Mallory)

30.1 Service time–voltage data

This is typified by the curves shown in Figure 30.1 for alkaline manganese dioxide 1.5 V dry cells in which cell voltage is plotted against hours of service obtained from the cell when it is operated at a particular standard temperature (usually 20°C). Obviously such curves must be prepared when the cell is under a defined standard ohmic load. The alkaline manganese dioxide cell has a voltage of 1.5 V/cell. If, as occurs in this example, it is used in an application requiring 1.25 V and the ohmic load is

150 Ω then the current drain is

$$\frac{1.25 \times 100}{150} = 0.83 \, \text{mA}$$

(see Figure 30.1). Cell voltage, capacity, current drain, ohmic load and temperature data should all be stated on such curves.

Voltage *versus* hours of service curves can be prepared at different temperatures to show the effect of cell temperature on hours of service obtainable.

30.1.1 Silver–zinc primary cells

Typical discharge characteristics at stated current drains and resistances for a full range of silver–zinc cells are reproduced in Figure 30.2.

Figure 30.3 shows typical voltage–discharge curves for silver–zinc cells, relating estimated hours of service obtained at 35°C with closed-circuit voltage at various starting drains and ohmic loads on a 24 h/day test schedule. Figure 30.4 shows discharge curves obtained for silver–zinc cells on a 16 h/day schedule at 21°C.

A typical performance profile at 27°C for an Eagle Picher remotely activated silver–zinc battery is shown in Figure 30.5.

30.1.2 Alkaline manganese dioxide primary cells

Service time–voltage data for a series of these cells with capacities between 700 and 10 000 mA h are shown in Figure 30.6.

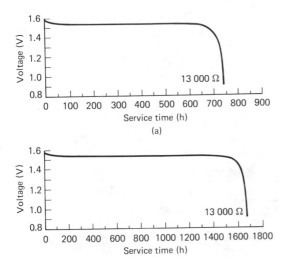

Figure 30.2 Service time–voltage data for silver–zinc cells: current drain at 1.55 V, 13 kΩ, 0.12 mA. (a) WS12, 83 mA h to 9 V on 13 000 Ω at 20°C, (b) WS14, 200 mA h, to 0.9 V on 10 000 Ω at 20°C (Courtesy of Union Carbide)

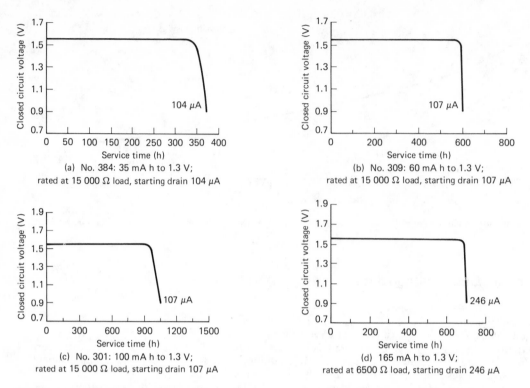

(a) No. 384: 35 mA h to 1.3 V;
rated at 15 000 Ω load, starting drain 104 μA

(b) No. 309: 60 mA h to 1.3 V;
rated at 15 000 Ω load, starting drain 107 μA

(c) No. 301: 100 mA h to 1.3 V;
rated at 15 000 Ω load, starting drain 107 μA

(d) 165 mA h to 1.3 V;
rated at 6500 Ω load, starting drain 246 μA

Figure 30.3 Union Carbide Eveready silver–zinc cell voltage–discharge curves: estimated hours of service at 35°C against closed-circuit voltage at stated starting drains and loads and on a 24 h/day schedule (Courtesy of Union Carbide)

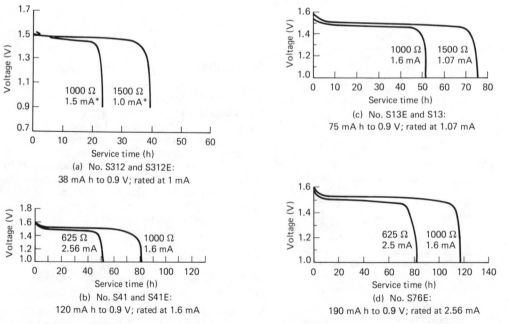

(a) No. S312 and S312E:
38 mA h to 0.9 V; rated at 1 mA

(c) No. S13E and S13:
75 mA h to 0.9 V; rated at 1.07 mA

(b) No. S41 and S41E:
120 mA h to 0.9 V; rated at 1.6 mA

(d) No. S76E:
190 mA h to 0.9 V; rated at 2.56 mA

Figure 30.4 Union Carbide silver–zinc cell voltage–discharge curves: estimated hours of service at 21°C against voltage at stated starting drains (* average drains) and loads and on a 16 h/day schedule (Courtesy of Union Carbide)

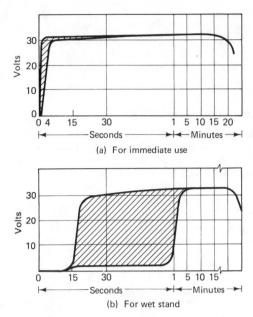

(a) For immediate use

(b) For wet stand

Figure 30.5 Eagle Picher remotely activated primary silver–zinc battery typical performance profile at 27°C (Courtesy of Eagle Picher)

30.1.3 Carbon–zinc primary cells

The closed-circuit or working voltage of a carbon–zinc cell falls gradually as it is discharged (Figure 30.7). The service hours delivered are greater as the cut-off or end-voltage is lower.

Typical cut-off voltages range from 0.65 to 1.1 V per 1.5 V cell, depending on the application. The cut-off voltage should be made as low as possible to use the available energy in the battery. This is sometimes done, if the equipment can tolerate it without causing failure and depending upon the character of the load, by using a slightly higher voltage battery than the application normally

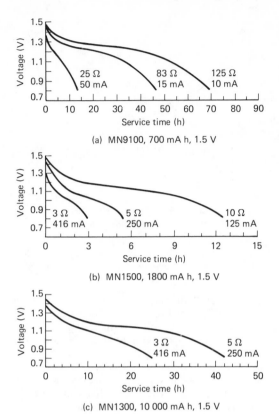

(a) MN9100, 700 mA h, 1.5 V

(b) MN1500, 1800 mA h, 1.5 V

(c) MN1300, 10 000 mA h, 1.5 V

Figure 30.6 Service time–voltage data for primary alkaline manganese dioxide cells, current drains at 1.25 V (Courtesy of Union Carbide)

requires. The cut-off voltage per cell is then lower and more efficient use of the battery can be achieved.

Voltage–hours of service curves under particular stipulated test conditions are given in Figure 30.8 for a 1.5 V carbon–zinc cell. Further details are given in Tables 54.1 and 54.2 (Part 6).

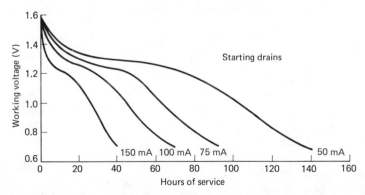

Figure 30.7 Voltage–discharge characteristics of a carbon–zinc D cell discharged 2 h/day at 21°C (fixed-resistance load) (Courtesy of Union Carbide)

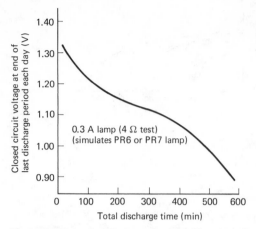

Figure 30.8 Union Carbide dry carbon–zinc 'Hercules' cells, 1.5 V: total discharge times at 21°C for different closed circuit voltages (Courtesy of Union Carbide)

Carbon–zinc batteries are normally designed to operate at 21°C. The higher the battery temperature during discharge, the greater the energy output. However, high temperatures reduce shelf life, and prolonged exposure to temperatures above 52°C causes very rapid deterioration of the battery.

Service life at low temperatures is reduced because of decreased chemical activity in the cell. The effects are more pronounced for heavy current drains than for light drains. When a standard carbon–zinc battery has reached a temperature of −18°C, little service is obtained except at light drains. At −23°C, the battery is usually inoperative unless special low-temperature electrolytes are used. Since a battery does not reach the temperature of its surroundings immediately, insulation is helpful.

The open-circuit voltage of a carbon–zinc cell decreases on average 0.0004 V/°C when the temperature is decreased from 25°C to −20°C. For practical purposes the actual working voltage at the terminals of the cell or battery is of more significance than the open-circuit voltage. Figure 30.9 shows the different voltage characteristics of a D-size carbon–zinc cell discharged at three different temperatures. In each case, the cells are discharged with a continuous current drain which has an initial value, when a 2.25 load is first applied, of 667 mA. (This simulates a 0.5 A lamp.)

30.1.4 Mercury–zinc primary cells

Typical discharge characteristics at stated current drains and resistances are given in Figure 30.10.

Figure 30.11 shows the discharge characteristics of a Vidor K13-4-1499 mercury–zinc battery. This battery measures 59 × 40 × 59 mm and weighs 285 g. The battery terminal open voltage is 13.56 V at 0°C, 13.57 V at 20°C and 13.58 V at 30°C. At a constant discharge of 10 mA the battery has a life of approximately 2 years at temperatures between −10 and +20°C, decreasing towards 1 year at temperatures outside this range. Typical discharge characteristics for this cell at temperatures between −10 and +20°C and on constant resistance loads are also shown in Figure 30.11. The discharge characteristics are little affected by whether the duty is continuous or intermittent.

Stabilized battery characteristics are unaffected by high temperatures. The mercuric oxide cell has good high-temperature characteristics: it can be used up to 65°C and operation at 140°C is possible for a few hours. In general, mercuric oxide batteries do not perform well at low temperatures. However, recent developments have produced several cell sizes that have good low-temperature characteristics. For the mercuric oxide batteries not in this group, there is a severe loss in capacity at about 4.4°C, and near 0°C the mercuric oxide cell gives very little service except at light current drains.

The life curve (Figure 30.12) shows the typical performance of an E12-type cylindrical mercuric oxide battery over the temperature range −23 to +71°C for current drains encountered in many applications. Successful operation at temperatures above 120°C for short periods has been reported, but it is recommended that 70°C is not exceeded.

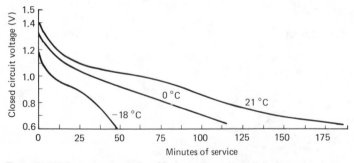

Figure 30.9 Voltage characteristics of D-size carbon–zinc cell discharged at three different temperatures with a continuous current drain starting at 667 mA (Courtesy of Union Carbide)

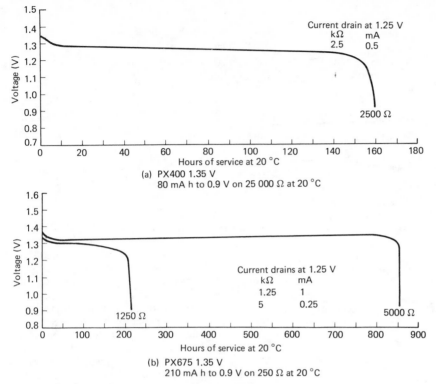

(a) PX400 1.35 V
80 mA h to 0.9 V on 25 000 Ω at 20 °C

(b) PX675 1.35 V
210 mA h to 0.9 V on 250 Ω at 20 °C

Figure 30.10 Mallory Duracell mercury–zinc cells and batteries, 1.35 V: voltage–discharge curves – estimated average service at 21°C against voltage (Courtesy of Mallory)

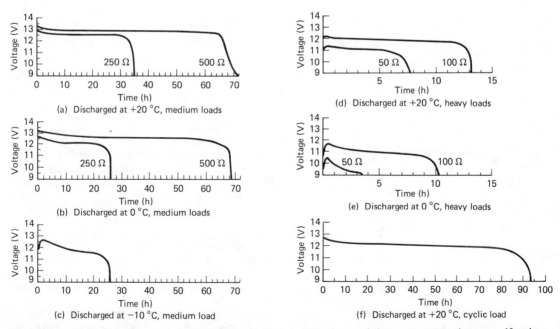

(a) Discharged at +20 °C, medium loads

(b) Discharged at 0 °C, medium loads

(c) Discharged at −10 °C, medium load

(d) Discharged at +20 °C, heavy loads

(e) Discharged at 0 °C, heavy loads

(f) Discharged at +20 °C, cyclic load

Figure 30.11 Vidor K13-4-1499 mercury–zinc dry battery: typical discharge characteristics at temperatures between −10 and +20°C (constant resistance load). (f) Typical cyclic discharge characteristics at 100 mA followed by 9 min at 10 mA; typical voltages at the end of each 100 mA cycle are plotted (Courtesy of Vidor)

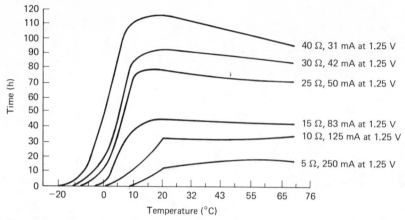

Figure 30.12 Mercury–zinc cell: life against temperature (E12 type, 3.6 A h capacity) (Courtesy of Vidor)

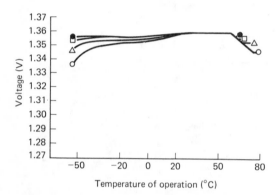

Figure 30.13 Effect of temperature on operating voltage of mercury–zinc cells at different current drains: △, 50 μA; □, 11 μA; ●, no current drain; ○, 100 μA. Initial reading of an E12 cell (Courtesy of Vidor)

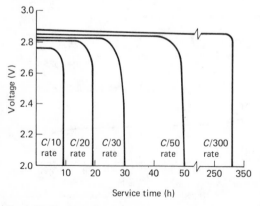

Figure 30.14 Typical discharge characteristics of a lithium–sulphur dioxide cell at various loads at 21°C (Courtesy of Honeywell)

Figure 30.13 indicates that voltage depression is slight at low temperatures when drains are 100 μA or less in the large cell types or when intermittent drains of this order are used with smaller cells.

Mallory produce special mercury cells, based on the wound-anode construction principle, which are particularly suitable for use at lower temperatures than those recommended for normal mercury cells. Thus Mallory types RM1438, RM1450R and RM2550R operate efficiently at a temperature 15°C lower than the standard mercury cells.

30.1.5 Lithium types

Lithium–sulphur dioxide primary batteries

Typical discharge curves for the cell at 21°C are given in Figure 30.14. The high cell voltages and flat discharge curves, until capacity is almost fully utilized, are characteristic for this type of cell. Another feature of the cell is its ability to be efficiently discharged over a wide range of current or power levels, from as high as the 2 h rate to low-drain continuous discharges for periods up to as long as 2 years, with good voltage regulation even at the extremes of discharge load. This cell is capable of delivering its energy at high current or power levels, well beyond the capacity of conventional batteries.

Mallory supply hermetically sealed lithium–sulphur dioxide organic electrolyte cells in the capacity range 1.1–10 A h with a nominal voltage of 3.00 V. Further details are given in Tables 56.2 and 56.3. Discharge curves for two of these batteries are given in Figure 30.15.

The lithium–sulphur dioxide cell is noted for its ability to perform over a wide temperature range from −54 to +71°C, giving a higher discharge rate at low temperatures than is possible with other types of

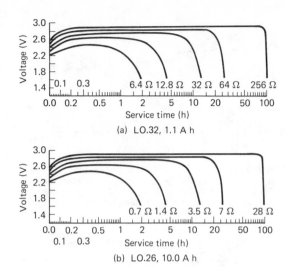

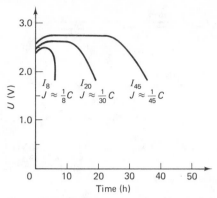

Figure 30.15 Discharge performance of Mallory lithium–sulphur dioxide batteries at 21°C (Courtesy of Mallory)

Figure 30.18 Discharge curve of a lithium–sulphur dioxide cell at −30°C (Courtesy of Mallory)

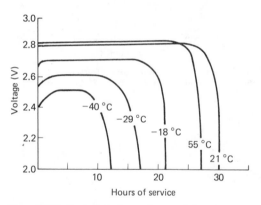

Figure 30.16 Typical discharge characteristics of a lithium–sulphur dioxide cell at various temperatures at $C/30$ rate (Courtesy of Mallory)

cells. Discharge curves for a standard 0.26 or D cell at various temperatures are shown in Figure 30.16. Of particular significance are the flat discharge curve over the wide temperature range, the good voltage regulation, and the high percentage performance available at the temperature extremes. Naturally the relative performance of the cell depends on discharge rate and improves at low discharge rates. Even at high discharge rates, however, a high percentage of the 21°C performance is available at temperature extremes.

The cells can be operated with success at elevated temperatures. When operated at very high currents outside the recommended limits, the cells may produce heat and high internal gas pressures and these conditions should be avoided. The cells are, however, an excellent source for high pulse currents. Multi-cell batteries contain additional insulation as well as a fuse for protection from mishandling such as short-circuits.

Figures 30.17 and 30.18 compare typical discharge performances of a lithium–sulphur dioxide battery at +20°C and −30°C, showing that very low temperatures do not have any particularly adverse effect on discharge characteristics.

Lithium–thionyl chloride primary batteries

Discharge performance data for an active lithium–thionyl chloride cell are shown in Figure 30.19. Figures 30.20 and 30.21 compare the discharge performance of lithium–thionyl chloride batteries at −29 and 24°C.

Lithium–manganese dioxide primary batteries

Typical discharge curves for lithium–manganese dioxide button cells at 20°C are given in Figure 30.22, which presents the typical discharge curve for the 2N-size cell. It may be seen that only the heaviest discharge rates give a sloping discharge profile, thus indicating good rate capability.

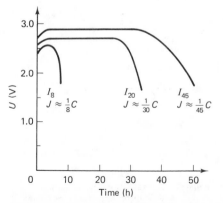

Figure 30.17 Discharge curve of a lithium–sulphur dioxide cell at 20°C (Courtesy of Mallory)

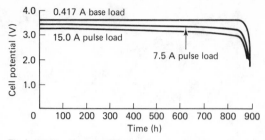

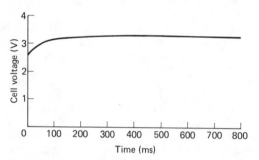

Figure 30.19 Honeywell lithium–thionyl chloride cell (G3037) Spacecraft B cell: discharge profile at 30°C; storage at 29 ± 4.5°C; capacity 381 A h; energy density 973 W h/dm³, 563 W h/kg (Courtesy of Honeywell)

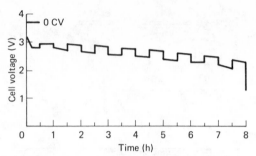

Figure 30.20 Honeywell lithium–thionyl chloride cell: discharge performance at −29°C after 2 weeks' storage at 74°C; loads 120 mA (30 min) to 45 mA (30 min): electrolyte 1.5 M LiAlCl$_4$.SOCl$_2$ plus 5 wt% sulphur dioxide (Courtesy of Honeywell)

Figure 30.21 Honeywell lithium–thionyl chloride cell: voltage delay at 24°C after 1 month's storage at 74°C; load 120 mA (3.3 mA/cm²); electrolyte 1.5 M LiAlCl$_4$.SOCl$_2$ + 5 wt% sulphur dioxide (Courtesy of Honeywell)

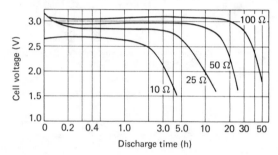

Figure 30.22 Typical discharge curves: Duracell lithium–manganese dioxide 2N size cell (20°C) (Courtesy of Duracell)

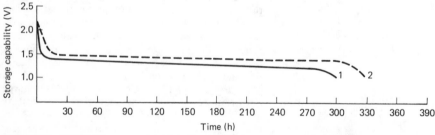

Figure 30.23 SAFT lithium–copper oxide cells: voltage–time shelf-life curves (Courtesy of SAFT)

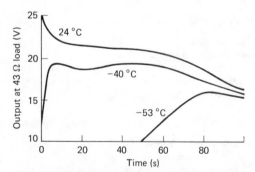

Figure 30.24 Eagle Picher reserve primary manganese dioxide–magnesium perchlorate battery: voltage–time curves on activation at various temperatures (Courtesy of Eagle Picher)

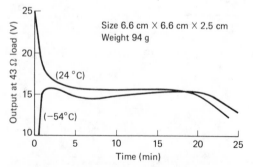

Figure 30.25 Eagle Picher reserve primary manganese dioxide–magnesium perchlorate battery: discharge curves after activation at −54 and +24°C (Courtesy of Eagle Picher)

Lithium–manganese dioxide cylindrical and button cells are eminently suited for applications requiring short-duration heavy-pulse discharge superimposed on a low background current. The cell can withstand 25 000 such discharges (1.25 s on, 15 min off) at 20°C before the cell voltage drops.

Lithium–copper oxide primary cells

Figure 30.23 shows a performance comparison of a lithium–copper oxide cell which has been stored at 'desert' temperatures for 5 years (solid line) against that of a new cell (dashed line).

30.1.6 Manganese dioxide–magnesium perchlorate primary batteries

Figure 30.24 shows the voltage service curves (43 Ω load) when a 92 g manganese dioxide–magnesium perchlorate reserve battery is activated by addition of electrolyte at −54, −40 and +24°C. Even at −54°C, a near maximum voltage is achieved within 80 s. The subsequent discharge curves at −54 and +24°C are shown in Figure 30.25.

30.1.7 Thermally activated primary batteries

A typical discharge curve for a calcium anode type thermal battery is shown in Figure 30.26. This curve illustrates the short activation time (0.26 s to peak voltage) and the rapid discharge of these batteries. A discharge curve for a larger thermal battery is shown in Figure 30.27. This battery has a 10 s activation time and maintains its voltage and current characteristics for 330 s.

30.2 Service life–ohmic load curves

Another way of expressing the performance of batteries is plots showing the effect of ohmic load on the battery *versus* hours of service obtained to a particular end-point voltage when the battery is at a particular standard temperature. This type of curve is typified in Figure 30.28, which shows the influence of ohmic load on hours of service obtained for 1.5 V alkaline manganese cells at 20°C to an end-point voltage of 0.8 V/cell. Such curves may also be prepared for a series of battery temperatures to show the effect of temperature on service life at a particular ohmic load.

Figure 30.29 shows the effect of ohmic load on the service life of 1.5 V alkaline manganese dioxide dry cells at 20°C.

Figure 30.30 shows the service life data at 20°C on constant-resistance load for various Duracell 1.35 and 1.4 V mercury–zinc cells for cells with nominal capacities of 85 and 3600 mA h.

Performance on constant resistance load curves for lithium–sulphur dioxide batteries are given in Figure 30.31. The effect of battery temperature between −40 and +52°C on these curves is also shown.

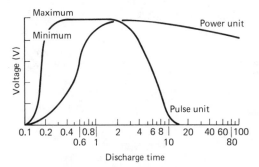

Figure 30.26 Typical discharge curves of Eagle Picher thermal batteries (Courtesy of Eagle Picher)

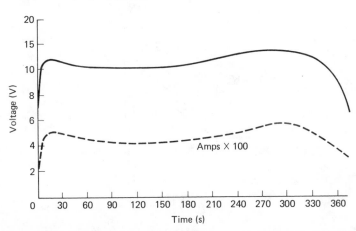

Figure 30.27 Performance of Eagle Picher EAP-6273 thermal battery under 0.024 Ω load (Courtesy of Eagle Picher)

30.3 Effect of operating temperature on service life

A very effective way of showing the effect of battery temperature on service life is to plot temperature *versus* the battery capacity obtained to the end-point voltage at that temperature expressed as a percentage of the battery capacity obtained at a standard battery temperature, usually 20 or 21°C. Alternatively, plots of service time achieved as a percentage of service time obtained at 20°C *versus* temperature provide similar information.

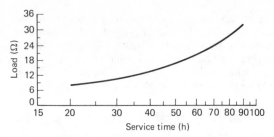

Figure 30.30 Effect of ohmic load on service life of Duracell Rmi2R mercury–zinc 1.35 V primary batteries at 20°C on constant resistance load, 3600 mA to 0.9 V on 25 Ω (Courtesy of Duracell)

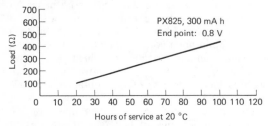

Figure 30.28 Service life–ohmic load curve, 300 mA h alkaline manganese dioxide cell (Courtesy of Mallory)

Silver–zinc primary batteries

Figure 30.32 shows the effect of temperature on available capacity for silver–zinc cells and batteries.

Lithium–sulphur dioxide batteries

The available capacity of this type of battery as a function of temperature and ohmic load is shown in Figure 30.33. Applications in which this system has been particularly valuable are those demanding high-rate operation at temperatures down to −40°C. Results covering the current range from 10 mA to 1 A and the temperature range +60 to −23°C are quoted in Table 30.1.

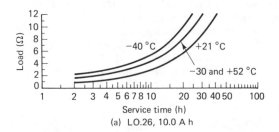

(a) LO.26, 10.0 A h

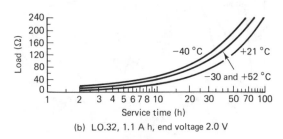

(b) LO.32, 1.1 A h, end voltage 2.0 V

Figure 30.31 Typical ohmic load–service life performance of Mallory primary lithium–sulphur dioxide cells on constant resistance load (Courtesy of Mallory)

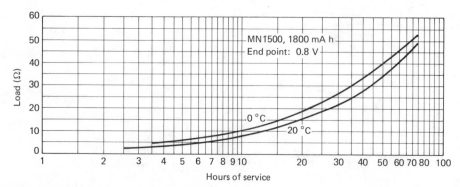

Figure 30.29 Effect of ohmic load on service life of Mallory alkaline manganese dioxide 1.5 V primary cells at 20°C on constant resistance load (Courtesy of Mallory)

It is interesting to note from the results quoted in the table that full nominal capacity is realized at 24°C regardless of load and that at 10 mA load full nominal capacity is realized regardless of temperature. The estimated current and temperature limits for the hermetic D-size cell for 80, 90 and 100% capacity realization are shown in Figure 30.34. It is interesting that the high-rate properties have been optimized to the 20–30°C temperature range without great sacrifice at other temperatures.

Lithium–thionyl chloride primary batteries

Figure 30.35 shows the effect of battery temperatures on available capacity for a lithium–thionyl chloride cell.

Table 30.1 High-rate discharge performance of Mallory lithium–sulphur dioxide hermetic D-size cell: test performed on resistive loads after 6 months' storage at 20°C

Nominal discharge current (mA)	Capacity (A h) at various discharge temperatures				
	−23°C	0°C	24°C	49°C	60°C
1000	3.7	5.8	9.7	6.9	6.8
250	7.1	8.3	10.0	8.2	8.8
50	9.7	9.7	10.1	8.9	9.5
10	10.1	10.3	10.2	10.5	10.5

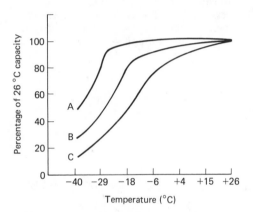

Figure 30.32 Eagle Picher remotely activated primary silver–zinc battery: effect of temperature on capacity at constant discharge (Courtesy of Eagle Picher)

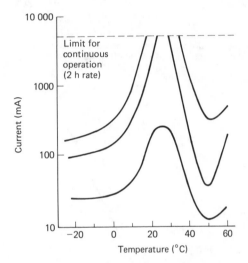

Figure 30.34 Mallory hermetic D-size lithium–sulphur dioxide cell: current and temperature limits for various capacity realizations (Courtesy of Mallory)

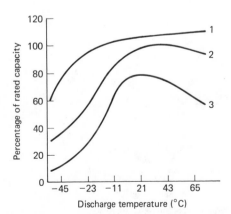

Figure 30.33 Honeywell lithium–sulphur dioxide cell: discharge as a function of load and temperature. 1, one-seventh rated load; 2, rated load; 3, seven times rated load (Courtesy of Honeywell)

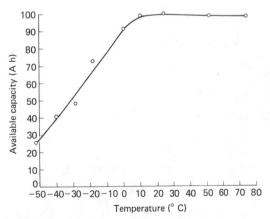

Figure 30.35 Effect of battery temperature on available capacity of a lithium–thionyl chloride cell (Courtesy of Mallory)

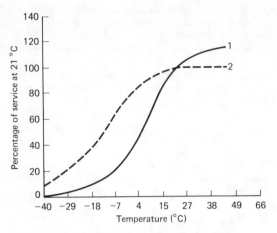

Figure 30.36 Union Carbide, Eveready E91 alkaline manganese 1.5 V dry cell: effect of temperature on service life at 21°C, 24 h/day discharge, 1.0 V cut-off. 1, starting drain 200 mA, load 7.5 Ω; 2, starting drain 20 mA, load 75 Ω (Courtesy of Union Carbide)

Alkaline manganese dioxide primary batteries

Figure 30.36 shows a curve obtained for a 1.5 V dry cell. It can be seen that a reduction in battery temperatures from 21°C to −2/+4°C approximately halves the service life attainable.

Carbon–zinc and carbon–zinc chloride primary batteries

Figure 30.37 shows the variation of service capacity with temperature for a general-purpose D-size carbon–zinc cell discharged continuously on a 2.25 Ω load to a 0.9 V cut-off. The initial curve drain (instantaneous value) is 667 mA. The load simulates a 0.5 A lamp.

Low temperatures, or even freezing, are not harmful to carbon–zinc cells as long as there is not repeated cycling from low or higher temperatures. Low-temperature storage is very beneficial to shelf life; a storage temperature of 4–10°C is effective.

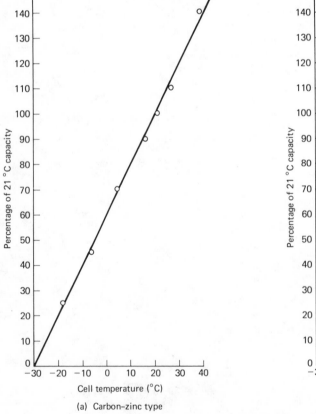

(a) Carbon-zinc type

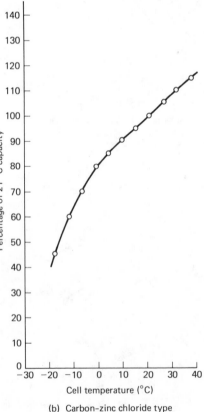

(b) Carbon-zinc chloride type

Figure 30.37 Effect of temperature on the capacity of D-sized cells: (a) carbon–zinc dry cells discharged continuously through a resistance of 2.25 Ω to an end-voltage of 0.9 V; (b) zinc chloride cells when discharged 4 min, 1 h, 8 h, 1 day with a 16 h rest through a resistance of 2.25 Ω to a cut-off voltage of 0.9 V (Courtesy of Union Carbide)

The zinc chloride cell performs better at low temperatures than the standard carbon–zinc cell. Figure 30.37(b) indicates the variation of service capacity with temperature for a D-size zinc chloride cell discharged on the 2.25 Ω light-industrial flashlight test. The load simulates a 0.5 A lamp.

30.4 Voltage–capacity curves

If a battery or a cell has a nominal capacity of say 10 A h at the standard rate of discharge, the available capacity to a particular end-voltage becomes lower when the battery is discharged at a high current. This is equivalent to saying that the higher the current at which a given battery is discharged, the lower is the available capacity of the battery and the shorter is the time during which that capacity is available (assuming that temperature is constant). Consider the example of a 6 A h nominal capacity manganese dioxide–magnesium perchlorate cell. This cell delivers its maximum capacity when it is continuously discharged for a period of about 100 h. However, when the continuous time of discharge exceeds 200 h, a decrease in available capacity becomes apparent. This is due to the self-discharge process at the magnesium anode. When discharged completely in relatively short, continuous periods of time, the magnesium battery, like other primary batteries, shows a reduction in available capacity.

Figure 30.38 shows voltage capacity discharge curves for such a 6 A h cell at 23°C when discharged to 1.0/1.1 V/cell at a range of discharge current rates between 0.06 A and 0.80 A. It can be seen that, when discharged at the standard 0.06 A discharge rate, a full available capacity of 6 A h is obtained, i.e. a life of 6/0.06 = 100 h is obtained. However, when the discharge rate is increased from 0.06 to 0.8 A the ampere hour capacity reduces from 6.0 A h

to 1.5 A h. That is, at 0.8 A h discharge rate the available capacity is only 25% of the nominal capacity. Correspondingly, the 0.8 A is available over 1.9 h (Figure 30.39) instead of 100 h.

Under intermittent conditions of use the realizable capacity of the magnesium battery depends on factors such as the magnitude of off-time period, the rate of discharge, the frequency of discharge intervals and the ambient temperature. The effect of each of these and their interrelationship makes it difficult to predict the capacity of the battery unless the mode of application is completely defined. In general, it may be stated that intermittent usage at heavy discharge drains is a favourable condition for optimum performance.

The magnesium battery performs better at higher operating temperatures than at 21°C, especially at the higher current drains. This can readily be seen by comparing the family of curves in Figure 30.40 (discharged at 71°C) with those in Figure 30.38 (discharged at 23°C). Like all other battery systems,

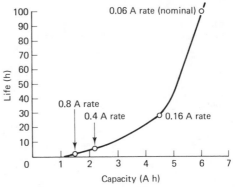

Figure 30.39 Marathon Magnesium CD cell, 23°C. Nominal capacity 6 A h at 0.06 A discharge rate (Courtesy of Marathon)

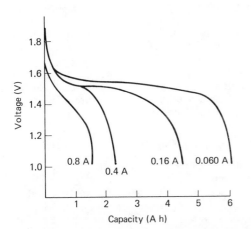

Figure 30.38 Marathon CD-size magnesium cell: constant-current discharge curves at 23°C (Courtesy of Marathon)

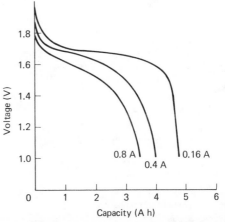

Figure 30.40 Marathon CD-size magnesium cell: constant-current discharge curve at 71°C (Courtesy of Marathon)

its performance diminishes when discharged at lower temperatures (see Figure 30.41). However, the magnesium battery is still able to operate at temperatures as low as −18°C especially at the light drain rates.

Naturally, the shape of voltage–capacity curves is dependent on battery temperature. Figure 30.42 shows discharge curves at −40°C and +52°C

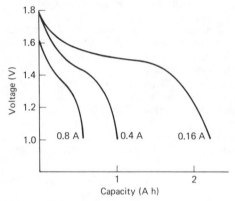

Figure 30.41 Marathon CD-size magnesium cell: constant-current discharge curves at 10°C (Courtesy of Marathon)

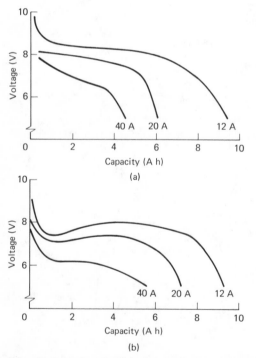

Figure 30.42 Rate–voltage characteristics of a 60 g, 656 cm³ Eagle Picher reserve primary manganese dioxide–magnesium perchlorate battery (10 A h) of dimensions 10.7 × 10 × 5.6 cm, (a) at 52°C, (b) at −40°C (Courtesy of Eagle Picher)

(rate–voltage characteristics) for a reserve primary 10 A h five-cell manganese dioxide–magnesium perchlorate battery. The batteries were, at each of these temperatures, discharged to 5.5 V at three different rates; 40, 20 and 12 A. If the 12 A discharge rate is considered to be nominal, i.e. capacity 9.2 A h to 5.5 V at −40°C and at +52°C, then the percentage capacity retentions at the three discharge rates, at each of the two battery temperatures, are as shown in Table 30.2.

Table 30.2 Percentage capacity retention

Temperature (°C)	Discharge rate		
	40 (A)	20 (A)	12 (A) (*nominal*)
+52	46 (6.0)	65 (18)	100 (46)
−40	56 (7.8)	78 (21.6)	100 (46)

Figures in parentheses are life in minutes to 5.5 V

Percentage capacity retention improves with decrease of discharge rate and, at higher discharge rates, with reduction in battery temperature. However, as can be seen from examination of Figure 30.42, at −40°C battery voltages (at all discharge rates) over the duration of life are considerably lower when discharges are conducted at a lower temperature (−40°C), and in many applications this factor outweighs small improvements in percentage capacity retention and life at lower temperatures. The best voltage characteristic is obtained at the higher temperature (52°C) and the lowest discharge rate (12 A).

As an example of a voltage–ampere hour capacity curve, for another type of battery, Figure 30.43 shows such a curve for an 1100 A h Edison Carbonaire zinc–air cell. This figure shows the effect of cell temperature on cell voltage at a constant continuous moderate discharge rate of 0.5 A, while Figure 30.44 shows the effect of reducing the discharge rate from 0.5 to 0.15 A. Again, as with the primary manganese dioxide–magnesium perchlorate cell, best voltage stability is shown at higher temperatures (21°C) and lowest discharge rates (0.15 A).

Voltage – percentage rated capacity curves

Plotting volts obtained on discharge to the cut-off voltage *versus* percentage of rated capacity at various discharge rates is another useful method of presenting data on battery performance. Such a curve for a 20-cell zinc–air battery is presented in Figure 30.45. Clearly, discharge at the 50 h rate gives the most stable voltage characteristic over the whole discharge process for this particular type of battery. The slower the discharge, the better the voltage characteristic. Figure 30.46 shows a similar discharge profile curve for a Honeywell lithium–sulphur dioxide cell at 21°C. Discharge was continued to a 2.0 V/cell cut-off. In this case, the cell was discharged at the rated load, 7× rated load and one-seventh rated load; again slow discharge, i.e. low load, leads to the best voltage characteristic.

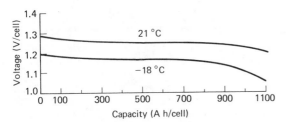

Figure 30.43 Voltage–capacity curves for discharge of one 1100 A h cell of a McGraw Edison Carbonaire zinc–air battery type ST.22 – discharge rate 0.5 A (Courtesy of McGraw Edison)

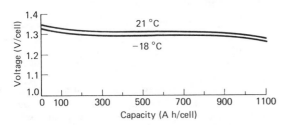

Figure 30.44 Voltage–capacity curves for discharge of one 1100 A h cell of a McGraw Edison Carbonaire zinc–air battery type ST.23 – discharge rate 0.15 A (Courtesy of McGraw Edison)

30.5 Shelf life–percentage capacity retained

Modern batteries retain their capacities well, some types more so than others, during storage before use. This applies particularly during storage at ambient temperatures near to 20°C. Obviously, capacity fall-off during storage is greater at elevated temperatures such as might occur in the tropics. Plots of percentage retained capacity *versus* shelf life at stated temperatures are a useful way of comparing this characteristic for various types of cell or battery. Figure 30.47, for example, shows a plot of percentage retained capacity *versus* shelf life in months at 20°C for a 1.5 V alkaline manganese dry cell when the battery is under an ohmic load of 25 Ω, i.e. a current drain of 60 mA. The preparation of such plots at a range of temperatures enables the temperature coefficient of the percentage capacity retention of the cell to be calculated.

30.5.1 Silver–zinc primary batteries

The dry storage charge retention characteristics of this battery are excellent at temperatures up to 52°C.

30.5.2 Alkaline manganese primary batteries

Alkaline manganese batteries retain their capacity for 30 months without significant loss. For example, after 12 months' storage at 20°C batteries are capable of realizing over 92% of their initial capacity. With the best type of cells, approximately 80% capacity is retained after 4 years' storage at 20°C. A typical performance curve is shown in Figure 30.47.

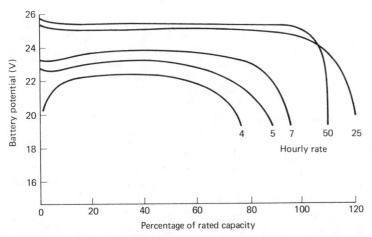

Figure 30.45 Discharge characteristics under continuous duty of a 20-cell zinc–air battery (Courtesy of McGraw Edison)

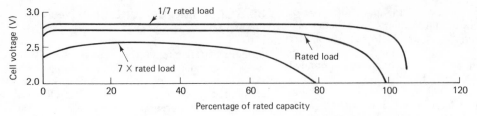

Figure 30.46 Honeywell lithium–sulphur dioxide cell room temperature discharge profile as a function of load, with 2.0 V cut-off (Courtesy of Honeywell)

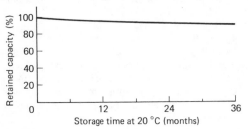

Figure 30.47 Typical capacity retention of an MN 1500 alkaline manganese cell, 25 Ω/20°C (Courtesy of Union Carbide)

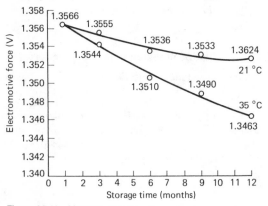

Figure 30.48 Mercury–zinc E1 cell: e.m.f. *versus* storage time

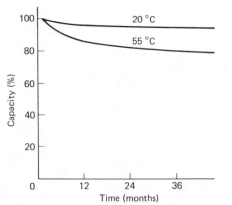

Figure 30.49 Available capacity of a Mallory lithium–sulphur dioxide cell after storage for up to 4 years at 20 and 55°C (Courtesy of Mallory)

30.5.3 Mercury–zinc primary batteries

The service capacity of mercuric oxide batteries after 1 year of storage at 21°C is generally more than 90% of the capacity of a fresh battery. Figure 30.48 shows the effect of storage time on e.m.f. at 21 and 35°C. Batteries can be stored for periods of up to 3 years; it is not, however, good practice to store them for long periods.

30.5.4 Lithium primary batteries

Lithium–sulphur dioxide batteries

Lithium–sulphur dioxide batteries have an excellent shelf life even at 71°C. Energy loss (W h/kg) is less than 10% during 2 years' storage at 71°C.

After extended storage, a reactivation time of the order of a few seconds is required for the system to achieve operating voltage. This is not a feature exclusive to lithium–sulphur dioxide cells, but one which is also exhibited by lithium–vanadium pentoxide and lithium–thionyl chloride cells.

On standing unloaded, lithium–sulphur dioxide cells develop a passive inhibiting film of thionite on the anode, formed by the initial reaction of lithium and sulphur dioxide, which prevents any further reactions or loss of capacity during storage. By virtue of this trait, shelf life is excellent over a wide range of temperature. Short-term accelerated tests indicate retention of 50% of capacity after 10 years. The passive film is quickly removed when a load is applied to the cell.

Lithium–sulphur dioxide cells now being supplied by leading manufacturers are claimed to retain 75% of initial capacity after 5 years' storage at 21°C. Corrected for sulphur dioxide leakage, true capacity loss is estimated at 6% in 5 years, or 1% per year. Capacity loss curves for lithium–sulphur dioxide batteries produced by Mallory are shown in Figure 30.49. For the Mallory cell approximately 80% of the nominal capacity of the battery is available after storage for 4 years at 55°C, compared with 95% when stored at 20°C (Figure 30.49). Hermetic cells retain 65% of initial capacity after 6 months at 72°C and about 50% after 12 months at 72°C or 6 months at 87°C.

Although permissible for shorter periods, prolonged storage at such high temperatures is detrimental to the capacity as well as to the high-rate properties of the system and should be avoided in applications in which such properties are important. Discharge tests at the 42 day rate (10 mA) at 49 and 60°C have indicated that no capacity was lost at these temperatures over the 42-day discharge period. The lithium system is capable of maintaining more stable voltage plateaux at higher currents than any other cell of comparable size. In 1975 Mallory introduced their new glass-to-metal hermetic seal to reduce sulphur dioxide losses from the cell and consequent capacity loss and corrosion effects during storage. Such cells with hermetic seals have been stored for 1.75 years at 21°C and 4 months at 72°C without leakage.

Lithium–manganese dioxide primary cells

The capacity or service life of lithium–manganese dioxide cells, normalized for a 1 g and a 1 cm³ cell, at various temperatures and loads, is summarized in Figure 30.50. These data can be used to calculate the

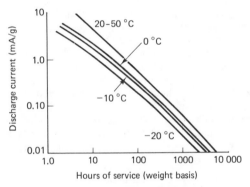

Figure 30.50 Service life of Duracell lithium–manganese dioxide cells (to 2 V/cell) (Courtesy of Duracell)

performance of a given cell or to select a cell of suitable size or weight for a particular application. It is to be noted that, since the actual cell performance is dependent on cell size and construction and other such factors, these data are only approximate. These cells are a relatively new type and consequently only limited storage life data are available. Projections sugggest that cells will provide 85% of initial capacity after 6 years storage at 20°C.

30.5.5 Manganese dioxide–magnesium perchlorate primary batteries

Field experience has shown that shelf life is an outstanding characteristic of manganese dioxide–magnesium perchlorate batteries. After 18 months' storage at room temperature, magnesium

batteries have yielded 90% of their initial capacity. Storage for 90 days at 55°C has shown that the battery still retains 85% of its initial capacity, and after storage at 71°C for 30 days the capacity retention is 85%. The shelf life of these batteries is demonstrated in Figure 30.51 for conditions in which the batteries were discharged under transit–receive conditions after storage at temperatures between 21 and 71°C.

Figure 30.52 shows the capacity retention *versus* time at two different temperatures for a reserve primary manganese dioxide–magnesium perchlorate battery.

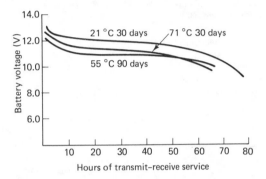

Figure 30.51 High-temperature storage capability of Marathon CD-size magnesium BA4836 batteries (Courtesy of Marathon)

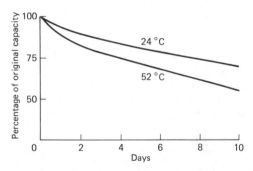

Figure 30.52 Eagle Picher reserve primary manganese dioxide–magnesium perchlorate battery: capacity retention in activated state, shown to 1.0 V/cell end-voltage (Courtesy of Eagle Picher)

30.6 Other characteristic curves

30.6.1 Internal resistance (impedance)–percentage discharge curves

Figure 30.53 shows a plot for a 1.5 V alkaline manganese dry cell at 21°C of internal resistance, i.e. impedance, *versus* percentage discharge. The change in resistance is more noticeable at lower temperatures, reaching a maximum when the cell is about 30% discharged (Table 30.3).

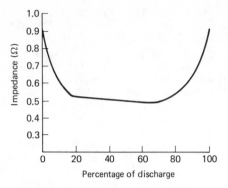

Figure 30.53 Cell impedance of a Duracell O-26 cell at 21°C (Courtesy of Duracell)

Table 30.3 Effect of temperature on internal resistance of a Union Carbide Eveready 1.5 V (E93) alkaline manganese cell

Temperature (°C)	Internal resistance (Ω)
21	0.18
0	0.24
−6.7	0.31
−17.8	0.46
−28.9	0.74
−40	1.14

30.6.2 Service time–wattage curves

Such plots are a useful means of evaluating the high-rate performance of cells and batteries. Figure 30.54 compares the high-rate capability of various primary batteries. The lithium–sulphur dioxide cell maintains a high capacity almost to 1 h on 10 W rate, whereas the zinc and magnesium batteries begin to drop off significantly at the 20–50 h rate.

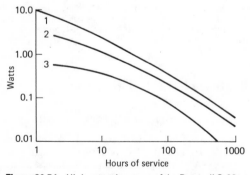

Figure 30.54 High-rate advantage of the Duracell O-26 lithium–sulphur dioxide cell at 21°C (1), compared with (2) a magnesium cell and (3) a zinc–carbon cell (Courtesy of Duracell)

30.6.3 Service time–discharge rate curves

The capacity or service life of a lithium–sulphur dioxide cell at various discharge rates and temperatures is shown in Figure 30.55. The data are normalized for one cell and presented in terms of hours of service at various discharge rates. The

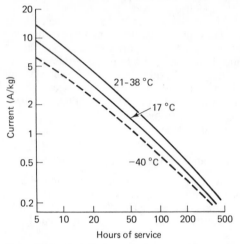

Figure 30.55 Projected service life of a Duracell O-26 lithium–sulphur dioxide cell (to 2 V/cell) (Courtesy of Duracell)

linear shape of these curves, except for the fall-off at high current levels and low temperatures, is again indicative of the capability of the lithium–sulphur dioxide battery to be efficiently discharged at these extreme conditions. These data are applicable to the standard cells and can be used in several ways to calculate the performance of a given cell or to select a cell of a suitable size for a particular application.

30.6.4 Capacity–current drain curves

Some types of battery exhibit exceptional drain carrying capabilities. With these types of battery, battery impedance (resistance) is low and essentially constant, even to low voltage cut-offs. The ampere hour capacity of most types of battery is not a fixed value. It varies with current drain, rate of discharge, operating schedule, cut-off voltage, cell temperature, recent history in terms of charge/discharge cycles, and the storage period of the battery before use. Table 30.4 shows approximate service capacities for carbon–zinc batteries for three different current drains. The values given are for fresh batteries at 21°C; the operating schedule is 2 h/day; and the cut-off voltage is 0.8 V per 1.5 V cell for all of the cells, which is consistent with the normal cut-off voltages of the cells in service. The data in Table 30.4 are based on starting drains and fixed resistance tests. From the voltage of the battery and the total number of cells, the number of 1.5 V cells in series and in parallel may be determined. Service

Table 30.4 Service capacity of Union Carbide carbon–zinc Leclanché cells

Cell	Starting drain (mA)	Service capacity (h)	Cell	Starting drain (mA)	Service capacity (h)
N	1.5	320	114	0.7	300
	7.5	60		3.5	57
	15.0	20		7.0	25
AAA	2.0	350	117	0.8	475
	10.0	54		4.0	98
	20.0	21		8.0	49
AA	3.0	450	118	0.8	525
	15.0	80		4.0	110
	30.0	32		8.0	54
B	5.0	420	127	1.0	475
	25.0	85		5.0	150
	50.0	32		10.0	72
C	5.0	520	132	1.3	275
	25.0	115		6.5	40
	50.0	53		13.0	16
D	10.0	525	135	1.3	550
	50.0	125		6.5	108
	100.0	57		13.0	52
F	15.0	630	148	2.0	612
	75.0	135		10.0	150
	150.0	60		20.0	60
G	15.0	950	155	5.0	620
	75.0	190		15.0	170
	150.0	78		30.0	74
6	50.0	750	161	3.0	500
	250.0	210		15.0	120
	500.0	95		30.0	55
55	15.0	635	165	3.0	770
	75.0	138		15.0	200
	150.0	61		30.0	90
105	0.4	210	172	5.0	780
	2.0	30		25.0	200
	4.0	8		50.0	90
109	0.6	710	175	5.0	1000
	3.0	155		25.0	260
	6.0	75		50.0	110
112	0.7	210			
	3.5	35			
	7.0	12			

capacity is given for single 1.5 V cells. If a battery uses cells in parallel, dividing the current drain the battery is supplying by the number of parallel strings of cells and looking up this value of current will give the service life.

Figure 30.56 shows actual capacities expressed as a function of discharge current of an LC01 lithium–copper oxide cell at three temperatures compared with a zinc–manganese battery of the same size.

30.6.5 Energy dependence on temperature curves

The useful energy content of a battery is designated in watt hours and is the product of ampere hour capacity and the average discharge voltage. To enable comparisons to be made between one battery and another this is often expressed as Wh/dm^3 on a volumetric basis or Wh/kg on a weight basis.

Another useful method of presenting the overall temperature performance of batteries is a plot

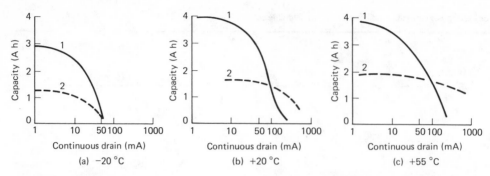

Figure 30.56 1, SAFT LC01 1.5 V (3.6 A h) lithium–copper oxide cell; 2, 'alkaline' K6 zinc–manganese dioxide cell. Capacity *versus* discharge current at various operating temperatures. The superiority of the lithium–copper oxide couple at low drain is evident (Courtesy of SAFT)

showing the energy dependence of W h/dm^3 on temperature at different continuous discharge rates. Figure 30.57 shows such curves for the Marathon CD-size magnesium cell for discharge rates between 0.16 and 1.6 A at a temperature of 24°C. The highest watt hour output (about 190 W h/dm^3) is obtained at the lowest discharge current and the highest temperature (0.16 A and 71°C).

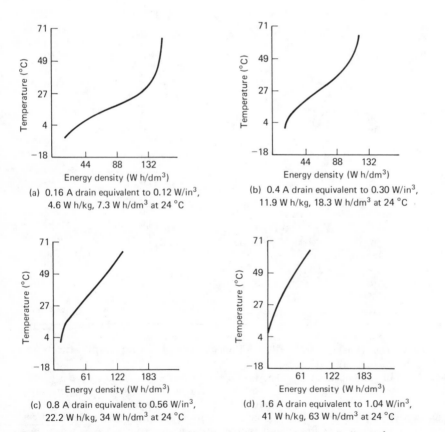

(a) 0.16 A drain equivalent to 0.12 W/in^3, 4.6 W h/kg, 7.3 W h/dm^3 at 24 °C

(b) 0.4 A drain equivalent to 0.30 W/in^3, 11.9 W h/kg, 18.3 W h/dm^3 at 24 °C

(c) 0.8 A drain equivalent to 0.56 W/in^3, 22.2 W h/kg, 34 W h/dm^3 at 24 °C

(d) 1.6 A drain equivalent to 1.04 W/in^3, 41 W h/kg, 63 W h/dm^3 at 24 °C

Figure 30.57 Effect of current density and temperature on the continuous discharge of a Marathon CD-size magnesium cell (Courtesy of Marathon)

31

Secondary batteries

Contents

31.1 Discharge curves

A classical method of representing a discharge curve for a secondary battery is to plot terminal voltage against time into the discharge at a stipulated battery temperature and constant current discharge rate. Such curves can be prepared at several different battery temperatures in order to establish the effect of temperature on the voltage–time relationship.

When stating the capacity obtained for a battery during continuous discharge, it is common practice to state the capacity available at a particular discharge rate and battery temperature when the battery is discharged to a particular end-point voltage per cell. The 20 h rate at 20°C is commonly used. Thus, if a battery is discharged continously for 20 h at 20°C, the nominal capacity is available, referred to as $1 \times C_{20}$ capacity. If the same battery were continuously discharged at a higher discharge current to the end-point voltage over, say, 10 h instead of 20 h, the capacity is referred to as $2C_{20}$; i.e.

$$\frac{20\,h\ discharge}{10\,h\ discharge} \times C_{20} = 2C$$

Similarly, when complete discharges are performed in 40, 50, 100, 200 and 400 h, the capacity obtained is referred to as $0.5C$, $0.4C$, $0.2C$, $0.1C$ and $0.05C$.

If the same battery is continuously discharged during a shorter period of time than 20 h at 20°C, the end-point voltage per cell will reduce and the full nominal capacity ($1 \times C_{20}$) will not be obtained, i.e. $2C$ (10 h discharge) is less than $1C_{20}$. Conversely, if the battery is discharged during a longer period of time than 20 h, the end-point voltage per cell will increase and the full nominal capacity ($1 \times C_{20}$) will be obtained. In fact, with some batteries, slightly more than the nominal capacity will be obtained in these circumstances, i.e. $0.05C$ (400 h discharge) $\geqslant$ $1C_{20}$.

In a typical case, if a battery with a nominal capacity of 20 A h at the 20 h rate ($1 \times C_{20}$) is continuously discharged at a higher discharge current for 5 h, i.e. at the 5 h discharge rate, $4C$ might be 80% of $1C_{20}$. That is, the capacity obtained would be 16 A h, i.e. $4C = 16$ A h.

A higher discharge current will, of course, be obtained during this discharge in return for this loss of capacity. Conversely, the voltages obtained during the 5 h discharge will be lower than those obtained during the 20 h discharge.

At the 20 h ($1C_{20}$) rate:

Current (I A) $\times$ Time (t h) = C_{20}
I A $\times$ 20 h = 20 A h

i.e. current of I amps for 20 h at V volts.

At the 5 h ($4C$) rate:

Capacity realized = (say) $0.8C_{20} = 16$ A h
x A $\times$ 5 h = 16 A h
$x = 3.2$ A

i.e. current of 3.2 A for 5 h at V_1 volts ($V_1 < V$).

Similarly at the 1 h rate the capacity returned ($10C_{20}$) might be only 50% of the nominal capacity obtained at the 20 h rate, i.e.

$$\frac{10}{1} \times C_{20} = 50\% \text{ of } 1 \times C_{20}$$
$$= 50\% \text{ of } 20 \text{ A h} = 10 \text{ A h}$$
$$y = A \times 1\,h = 10 \text{ A h}$$

i.e. current of 10 A for 1 h at V_2 volts ($V_2 < V_1$).

To obtain the capacity of a battery at the nominal (20 h) rate, i.e. C_{20} at 20°C, it will be necessary to condition the battery to 20°C then prepare constant current discharge voltage–time plots at a range of discharge currents such that the time taken for discharge to the end-point voltage ranges from appreciably below 20 h to appreciably above 20 h. Select the end-point voltages, i.e. the voltage at which the voltage starts to decrease rapidly, appropriate to each discharge current, and read off for each discharge current the corresponding end-point voltage and discharge time. From the products of discharge current and discharge time, calculate the corresponding capacities. Plot capacity against discharge time and read off the nominal capacity obtained during a 20 h discharge, i.e. C_{20}.

Knowing the nominal capacity (C_{20}) and the nominal discharge time (20 h), it is possible to calculate the nominal discharge current (I_{20}), the discharge current that will deliver the nominal battery capacity (C_{20} A h) in 20 h continuous discharge at 20°C. It is then possible to plot end-point voltages against discharge times (to each end-point voltage) and read off the voltage obtained during the 20 h discharge, and then to say whether a discharge at 20°C at I_{20} A for 20 h to a stated end-point voltage will deliver the nominal capacity of C_{20} A h. It is also possible from such curves to obtain these parameters for any other stated rate of constant discharge of the battery. If a battery is continuously discharged to the end-point voltage at 20°C (or any other stipulated temperature) for 20 h at a discharge current of I_{20} A it will deliver its nominal capacity of C_{20} A h. If the continuous discharge current is increased say to $2I_{20}$ or $10I_{20}$ the discharge time and end-point voltage both decrease and the capacity returned will be only a percentage of the nominal capacity, C_{20}.

This whole sequence of measurements is designed to determine the nominal capacity (C_{20}) of a battery at the nominal discharge rate (I_{20}) (or the capacity of a battery of any other discharge rate) at 20°C. It can be repeated at a variety of battery temperatures to obtain capacity data over a range of temperatures.

31.2 Terminal voltage–discharge time curves

The voltage characteristics for a cell can be portrayed as a plot of terminal voltage *versus* discharge time at a stipulated discharge rate and battery temperature.

31.2.1 Lead–acid batteries

Typically, sealed lead–acid batteries will deliver several hundred complete charge/discharge cycles. When subjected to less than 100% depth of discharge, cycle life will improve considerably. Depth of discharge is defined as the percentage of rated capacity removed when discharged under a fixed set of conditions.

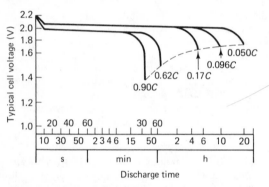

Figure 31.1 Typical voltage characteristics at various rates of discharge at 21°C: Eagle Picher, 1.5 V Carefree rechargeable lead–acid battery (Courtesy of Eagle Picher)

Table 31.1

Discharge rate (h)	Approximate percentage of rated capacity	End-voltage (V/cell)
20	100	1.75
10	97	1.70
5	88	1.65
1	62	1.50
0.5	52	1.00

The nominal capacity of a typical battery when discharged to various voltages as a percentage of its capacity when discharged to 1.75 V/cell is given in Table 31.1.

Figure 31.1 illustrates typical voltage curves of this type of battery at various rates of discharge.

Figure 31.2 shows the discharge performance of the Crompton-Parkinson range of partially recombining sealed batteries. As an example of the use of these curves, consider an SLA6-6.4 battery giving 1.5 A for nominal C_{20} capacity of 6.4 A h. Then

$$I_{20} = \frac{6.4}{20} = 0.32 \, \text{A}$$

hence

$$\frac{1.5}{0.32} \times I_{20} = 5 \times I_{20}$$

The $5 \times I_{20}$ curve in Figure 31.2 gives an end-voltage of about 5.1 V after about 3.5 h.

Figure 31.3 shows the relationship between terminal voltage and discharge time at various 20 h

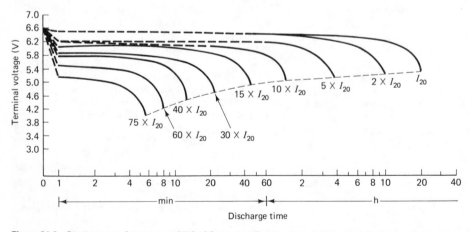

Figure 31.2 Discharge performance at 20°C of Crompton-Parkinson sealed lead–acid batteries. Curves show a change in voltage at the battery terminals with time during constant-current discharges at a number of different current rates. The current is given in multiples of I_{20} so that a single family of curves can be used for all the sizes of battery in the range. I_{20} is the current that will deliver the nominal battery capacity, C_{20}, in 20 h. Its value (A) is one-twentieth of the battery capacity (C_{20}/20). The lower end of each discharge curve indicates the recommended minimum voltage at the battery terminals for that particular rate of discharge; that is, the voltage at which the discharge should be discontinued (Courtesy of Crompton-Parkinson)

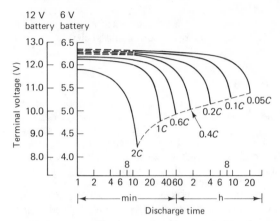

Figure 31.3 Discharge characteristic curves at 20°C for a Yuasa sealed lead–acid battery. C = A h rating at 20 h discharge rate to 1.75 V/cell (Courtesy of Yuasa)

discharge rates obtained for 6 V and 12 V sealed lead–acid batteries with calcium–lead alloy grids and Figure 31.2 illustrates the same curve for a 1.5 V battery. In these curves the nominal capacity of the battery is expressed when discharged at a 20 h rate to the final voltage of 1.75 V/cell. The curves show currents drawn at different discharge capacity rates at 20°C after a few charge/discharge cycles. It is clear from Figure 31.3 that the ampere hour capacities of the battery at a 20 h rate expressed as standard values are smaller when the battery is subjected to discharge at greater than the 20 h rate.

31.2.2 Nickel–cadmium batteries

During discharge, the average voltage of a sealed nickel–cadmium battery is approximately 1.2 V/cell. At normal discharge rates the characteristic is very nearly flat until the cell approaches complete

discharge. The battery provides most of its energy above 1 V/cell. If the cell is discharged with currents exceeding the rated value, however, the voltage characteristic will have more of a slope, a lower end-voltage will be necessary, and the ampere hours obtained per cycle will be reduced.

High-rate nickel–cadmium cells will deliver exceedingly high currents if they are discharged continuously under short-circuit conditions. Self-heating may do irreparable damage. If the output is withdrawn in pulses spaced to limit to a safe figure the temperatures of a few critical areas in the cell, high currents can be utilized.

The heat problems vary somewhat from one cell type to another, but in most cases internal metal strip tab connectors overheat and/or the electrolyte boils. General overheating is normally easy to prevent because the outside temperature of the battery can be used to indicate when rest, for cooling, is required. In terms of cut-off temperature during discharge, it is acceptable practice to keep the battery always below 65°C. The overheated internal connectors are difficult to detect. This form of overheating takes place in a few seconds or less, and overall cell temperature may hardly be affected. It is thus advisable to withdraw no more ampere seconds per pulse, and to withdraw it at no greater average current per complete discharge, than recommended for the particular cell in question. In special cases, where cooling of the cell or battery is likely to be poor (or unusually good), special tests should be run to check the important temperatures before any duty cycle adjustment is made.

Cylindrical cells

Figure 31.4 shows discharge curves for several cylindrical nickel–cadmium batteries of voltages between 6 and 14 V and capacities between 0.5 and

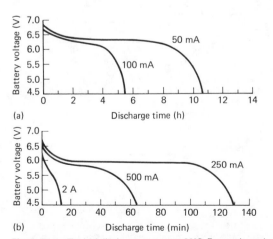

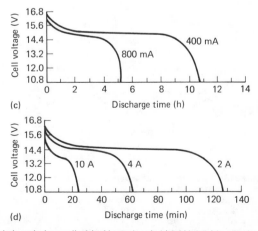

Figure 31.4 Typical discharge curves at 20°C. Eveready sealed nickel–cadmium cylindrical batteries: (a,b) 6.0 V, 0.5 A h capacity, (c,d) 14.4 V, 4.0 A h capacity (Courtesy of Union Carbide)

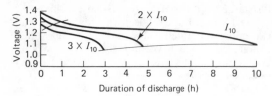

Figure 31.5 Discharge curves for Varta 151D sealed nickel–cadmium cells with mass electrodes (Courtesy of Varta)

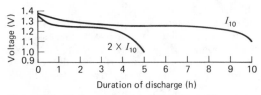

Figure 31.6 Discharge curves at 20°C for Varta RS sealed nickel–cadmium cells with sintered electrodes (Courtesy of Varta)

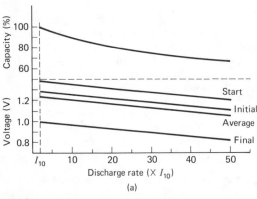

(a)

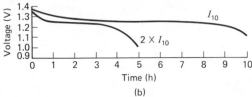

(b)

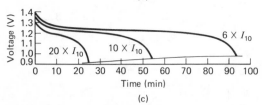

(c)

Figure 31.7 Discharge curves at various rates: Varta sealed nickel–cadmium rolled sintered cells up to 2 A h. (a) Starting voltage approx. 30 min after end of charge. Initial voltage after 10% of discharge capacity. Average voltage approx. 1.22 V. Final voltage approx. 0.99 V. (b,c) Voltage during discharge of a fully charged cell when discharged to below 1 V (Courtesy of Varta)

4 A h. Figures 31.5 and 31.6, respectively, compare discharge curves for typical sealed nickel–cadmium cylindrical cells made using mass electrodes and sintered electrodes.

Button cells

The available capacity from a cell is dependent on the discharge rate and is proportionally greater with a sintered plate cell. The maximum discharge current for Varta D and DK type mass plate cells is $10 \times I_{10}$ and for their DKZ type mass plate and RS and SD types of cell with sintered electrodes it is $20 \times I_{10}$. In pulse load applications these rates can be doubled. Figure 31.7 shows the discharge at various rates between I_{10} and $20I_{10}$ of fully charged rolled sintered cells of up to 2 A h capacity. The voltage of a cell when discharged below 1 V/cell rapidly falls to zero (Figure 31.7(b) and (c)) and it is then possible under these circumstances for some cells in a battery to reverse their polarity. Provided that the discharge rate is at the I_{10} rate or less, this reversal will not harm a nickel–cadmium sealed cell, but for maximum life it should be avoided. Discharge curves for a button cell with mass plate electrodes are shown in Figure 31.8.

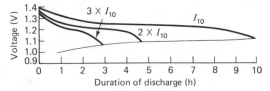

Figure 31.8 Discharge curves for a Varta 150 DK sealed nickel–cadmium button cell with mass electrodes (Courtesy of Varta)

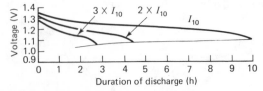

Figure 31.9 Discharge curves for Varta D23 type sealed nickel–cadmium rectangular cells with mass electrodes (Courtesy of Varta)

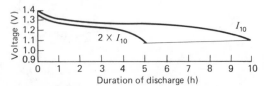

Figure 31.10 Discharge curves of a Varta SD 1.6 sealed nickel–cadmium rectangular cell with sintered electrodes (Courtesy of Varta)

Rectangular cells

Figures 31.9 and 31.10 compare discharge curves for sealed rectangular nickel–cadmium cells with mass electrodes and sintered electrodes, respectively.

31.2.3 Nickel–zinc batteries

Figure 31.11 shows charge/discharge characteristics for a nominal 5 A h nickel–zinc cell at 2 A and 75°C. This curve illustrates that following an input of 6 A h during a 3 h charge the cell has an output of 5 A h during the following 160 min discharge period.

31.2.4 Silver–zinc batteries

Figure 31.12 shows typical voltage characteristics for high-rate and low-rate Eagle Picher silver–zinc cells.

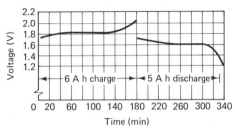

Figure 31.11 Typical charge/discharge characteristics of an Eagle Picher 5 A h nickel–zinc cell at 2 A and 24°C (Courtesy of Eagle Picher)

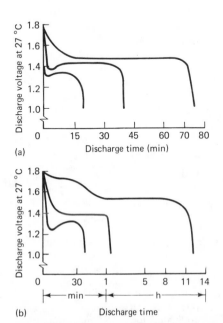

Figure 31.12 Voltage characteristics of Eagle Picher high-rate (a) and low-rate (b) rechargeable silver–zinc batteries at three different rates of discharge (Courtesy of Eagle Picher)

31.2.5 Silver–cadmium batteries

Figure 31.13 shows typical voltage characteristics for high-rate and low-rate silver–cadmium batteries at three different discharge rates.

Figure 31.14 shows a plot of plateau voltage *versus* discharge time at high and low rates of discharge for a silver–cadmium battery.

31.3 Plateau voltage–battery temperature curves

A useful method of studying the effect of temperature on the performance of lead–acid, silver–cadmium and silver–zinc batteries is to plot plateau

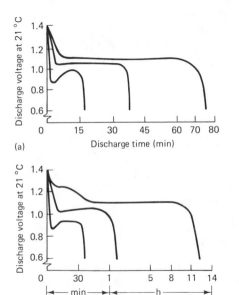

Figure 31.13 Characteristics of Eagle Picher high-rate (a) and low-rate (b) rechargeable silver–cadmium batteries at three different rates of discharge (Courtesy of Eagle Picher)

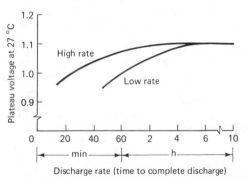

Figure 31.14 Characteristics of Eagle Picher high-rate and low-rate rechargeable batteries (Courtesy of Eagle Picher)

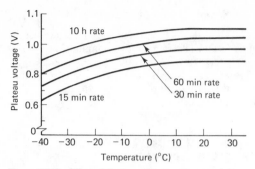

Figure 31.15 Effect of discharge temperature on plateau voltage for high-rate and low-rate rechargeable silver–cadmium batteries

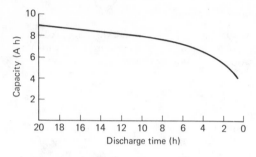

(a) VPT 6.9/8 (6 V, 9 A h)

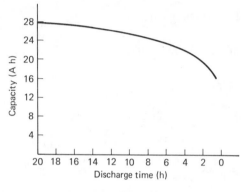

(b) VPT 6.15/30 (6 V, 28 A h)

Figure 31.16 Relationship between capacity returned and discharge rate at 20°C at 20 h rate for Varley 6 V non-spill lead–acid batteries (Courtesy of Varley)

voltage obtained at various rates of discharge against battery temperature at stipulated rates of discharge. Figure 31.15 shows such curves for high-rate and low-rate silver–cadmium batteries.

A small change in plateau voltage occurs for the silver–zinc system when battery temperature is lowered.

31.4 Capacity returned (discharged capacity)–discharge rate curves

Plots of the variation of the returned capacity of battery *versus* the duration of the discharge in hours at a stipulated discharge rate and battery temperature are a useful means of assessing battery performance. The slower the discharge the greater the capacity returned. Figure 31.16 illustrates such curves for a 6 V non-spill lead–acid battery (9 and 28 A h) discharged at 20°C at the 20 h rate. The relationship between capacity returned and discharge rate is discussed in more detail in Section 31.9.

31.5 Capacity returned (discharged capacity)–discharge temperature curves and percentage withdrawable capacity returned–temperature curves

As mentioned earlier, the nominal capacity of a battery delivered at the nominal 20 h discharge rate, when discharged at the nominal I_{20} discharge current, depends on the battery temperature. Generally speaking the higher the temperature the higher the capacity returned and vice versa. However, for many types of batteries, the capacity starts to decrease above a certain battery temperature.

Such data are invaluable for predicting the performance of batteries at extremes of service temperature conditions from arctic to tropical. A refinement of this method of presentation is to plot the available capacity obtained from the battery at the test temperature as a percentage of the nominal capacity at 20°C (i.e. percentage withdrawable capacity) against battery temperature for various discharge currents down to the end-point voltage.

Such data are discussed further below for several types of battery.

31.5.1 Lead–acid batteries

The curves in Figure 31.17 give information regarding the variation of available capacity of low-maintenance sealed lead–acid batteries, as a percentage of nominal capacity (withdrawable capacity), as a function of temperature over the range −30 to +50°C for three different loads with continuous discharge down to the relevant discharge cut-off voltage. For the values of the upper edges of the curves, the batteries were charged at an ambient temperature of 20°C voltage limited to 2.3 V/cell. For the lower edge of the curves, the batteries were charged at the indicated low ambient temperature and hence under somewhat less favourable conditions. The curves show the behaviour of the

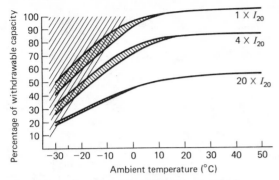

Figure 31.17 Percentage of withdrawable capacity at different temperatures for three load examples. Upper edge of curve: charging at 20°C, discharging at quoted temperature. Lower edge of curve: charging and discharging at quoted temperature (Dryfit low-maintenance lead–acid batteries) (Courtesy of Dryfit)

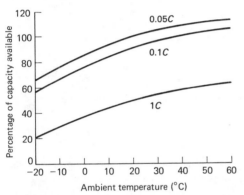

Figure 31.18 Temperature effects in relation to battery capacity of a Yuasa sealed lead–acid battery (Courtesy of Yuasa)

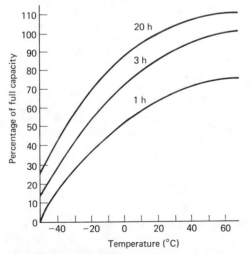

Figure 31.19 Capacity *versus* temperature at various rates of discharge. Eagle Picher Carefree rechargeable lead–acid battery (Courtesy of Eagle Picher)

batteries after a few charging cycles. When assessing both the upper and lower edges of the curves for the higher discharge currents, the dependence on the load must also be taken into account. This is already included in Figure 31.17. To prevent permanent damage to the capacity behaviour of the batteries at very low temperatures, the shaded areas in Figure 31.17 should be avoided. Permanent use at extremely high temperatures adversely affects the life of all types of battery, although for a few hours there is no objection to an ambient temperature up to a maximum of 80°C. For longer periods, allowance must be made for the cumulative permanent loss of capacity. Permanent use at temperatures above 50°C should be avoided.

Figure 31.18 shows a curve of percentage available capacity *versus* temperature for a low-maintenance partially recombining lead–acid battery. This battery functions satisfactorily over a wide range of service temperatures from sub-zero to high ambient temperatures. At higher temperatures an increased capacity is obtainable with a corresponding increase in self-discharge rate. Conversely, at lower temperatures available capacity and self-discharge rate are reduced. Outside the range 5–40°C for constant-voltage current-limited type chargers, a temperature compensation of −4 mV/°C per cell is applicable. Figure 31.19 illustrates the effect of discharge rate on the plot of temperature *versus* percentage withdrawable capacity for low-maintenance lead–acid batteries (see also Figure 31.18).

31.5.2 Nickel–cadmium batteries

Figure 31.20 illustrates the effect of battery temperature on capacity returned (discharged capacity) for 1.2 V, 20 μA h and 1 A h capacity nickel–cadmium button cells. In both cases capacities start decreasing below 10°C and above 38°C. Similar plots comparing the performance of sealed nickel–cadmium button

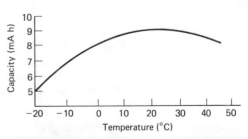

Figure 31.20 Sample capacity *versus* temperature curve for Eveready OB90 and OB90T sealed nickel–cadmium cells, 1.2 V, 20 A h capacity. Ranges of temperature applicable to operation are: charge, 0–45°C; discharge, −20 to 45°C; storage, −40 to 60°C. Discharge current 9 mA (Courtesy of Union Carbide)

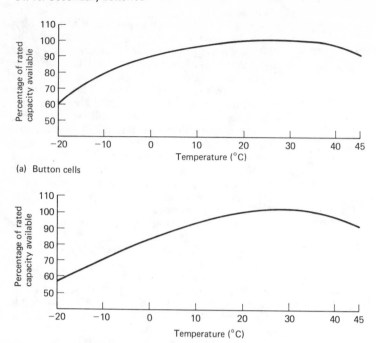

(a) Button cells

(b) Cylindrical cells

Figure 31.21 Effect of discharge temperature on capacity at temperatures between −20 and +45°C for Eveready sealed nickel–cadmium cells. Discharge after initial charge at room temperature for 14 h at the 10 h rate (Courtesy of Union Carbide)

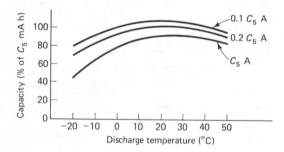

Figure 31.22 Influence of temperature on capacity of SAFT sealed nickel–cadmium sintered plate button cells (Courtesy of SAFT)

and cylindrical cells are shown in Figure 31.21. Figure 31.22 shows a typical plot of percentage withdrawable capacity *versus* temperature for a sealed nickel–cadmium sintered button cell.

31.6 Capacity returned (discharged capacity)–terminal voltage curves

As mentioned in Section 31.1, if the current drawn from a battery upon discharge is increased, the duration of the discharge to the point where the

end-point voltage is reached is decreased and, also, the voltage profile decreases, i.e. lower voltages are obtained, during a rapid discharge. When considering a battery for use in a particular application there will usually be a particular voltage requirement during discharge and this will dictate the discharge rate used and predetermine the capacity available. To obtain a desired combination of capacity and voltage at a particular temperature, compromise may be necessary and a decision to use a larger (or smaller) battery may have to be taken. Discharge current, discharge voltage capacity available, discharge duration and ambient temperature are all intimately connected. Plots of capacity returned *versus* terminal voltage (or percentage withdrawable capacity *versus* terminal voltage) and of capacity discharged *versus* current at various temperatures, if necessary, provide useful information when such decisions have to be taken.

31.6.1 Lead–acid batteries

Figure 31.23 illustrates the charge of terminal voltage as a function of capacity returned (discharged capacity) for a General Electric fully recombining D cell. Typical curves are shown for constant-current discharge rates at the standard conditions noted.

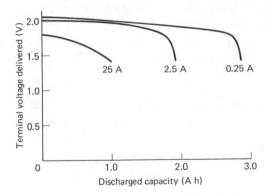

Figure 31.23 Typical discharge curves at various discharge rates (constant load) for a General Electric D cell. Standard conditions: charge 2.45 V at 23°C for 16 h, discharge at 23°C to 1.4 V (Courtesy of General Electric)

Both capacity and delivered voltage vary as inverse functions of discharge rate. At the lowest rates, voltage and capacity are the highest. A cut-off limit of the working voltage is arbitrarily set at 1.4 V. The voltage cut-off at that point is rapid, and little useful additional energy would be obtained from the cell by attempting to use it below this voltage level. As the curves in Figure 31.23 indicate, higher drain rates aggravate the condition of concentration polarization (a decrease in electrolyte concentration near the active materials in the plate caused by lack of sufficient ion diffusion) resulting in the reduction of available voltage and capacity.

The effect of temperature on the discharge curves of D cells is shown in Figure 31.24. All curves were obtained at 250 mA discharge. As the temperature is lowered, the rate of ion diffusion in the electrolyte is reduced, and consequently the delivered voltage and available capacity are reduced at lower cell temperatures.

31.6.2 Nickel–cadmium batteries

Figure 31.25 shows the change in terminal voltage as a function of discharged capacity for a Marathon 1.2 V sealed nickel–cadmium cell. Figure 31.26 compares the performance of a 12 V battery at various rates and at temperatures of +20 and −20°C. As would be expected, the highest discharge rate combined with the lower temperature gives the poorest performance.

Figures 31.27 and 31.28 show discharge curves obtained by a continuous discharge and pulse discharge, respectively, obtained with nickel–cadmium button cells.

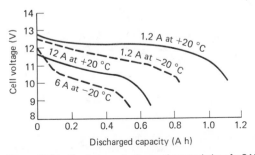

Figure 31.25 Discharge characteristics of a Marathon sealed nickel–cadmium cell (Courtesy of Marathon)

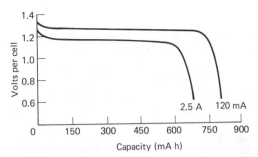

Figure 31.26 Continuous discharge characteristics of a SAFT 10 VR 1.2RR ten 1.2 A h cell nickel–cadmium battery (12 A h) at +20 and −20°C (Courtesy of SAFT)

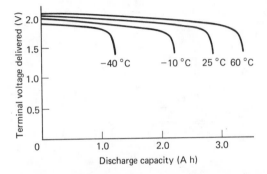

Figure 31.24 Typical discharge curves at various discharge temperatures (constant-current load) for a General Electric D cell. Standard conditions: charge 2.45 V at 23°C for 16 h, discharge 0.25 A to 1.4 V at 23°C (Courtesy of General Electric)

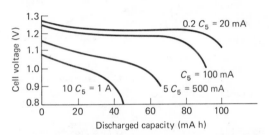

Figure 31.27 Discharge curves for a SAFT VB 10, 100C_5 (mA h) nickel–cadmium button cell (continuous discharge at 20 ± 5°C, maximum permissible continuous rate 1 A) (Courtesy of SAFT)

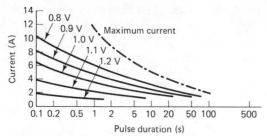

Figure 31.28 Discharge curves for a SAFT VB 10, 100C_5 (mA h) nickel–cadmium button cell (pulse discharge at 20 ± 5°C, 100% charged, maximum power for a 0.3 s pulse = 7 W, maximum permissible pulse rate = 11 A). The voltages shown are the maximum cell voltage for a pulse discharge of a given current and duration (Courtesy of SAFT)

31.6.3 Nickel–zinc and nickel–cadmium batteries

Typical discharge characteristics of a 5 A h nickel–zinc battery are shown in Figure 31.29.

Figure 31.30 shows initial and subsequent discharge curves after 2 years' continuous overcharge, without periodic discharges, of a nickel–cadmium battery. The first discharge after the 2 year charge period yields a slightly reduced voltage curve and 65% capacity. The second cycle after 2 years' continuous overcharge provides essentially the same discharge curve as the initial one.

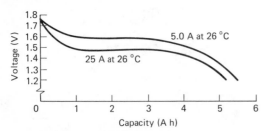

Figure 31.29 Typical discharge characteristics of an Eagle Picher NZS-5.0 (5 A h) nickel–zinc battery (Courtesy of Eagle Picher)

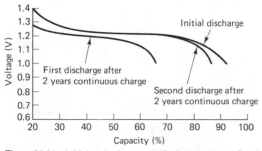

Figure 31.30 Initial and subsequent discharge curves after 2 years continuous overcharge without periodic discharges: Eveready sealed nickel–cadmium cylindrical cells (Overcharging at $C/20$ rate, 60 mA, CH 1.2 cell) (Courtesy of Union Carbide)

31.6.4 Silver–zinc and silver–cadmium batteries

Figure 31.31 shows typical capacity characteristics for high-rate and low-rate silver–zinc cells. Discharge characteristics at various rates (Figure 31.31(a)) and the effect of temperature (Figure 31.31(b)) on the discharge are also shown.

Similar curves for a silver–cadmium battery at various discharge rates and temperatures is shown in Figure 31.32.

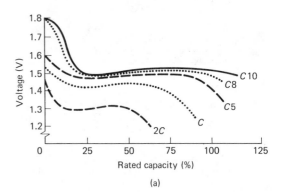

(a)

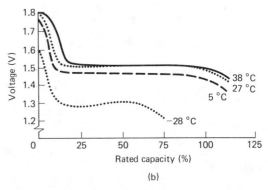

(b)

Figure 31.31 Characteristics of Eagle Picher rechargeable silver–zinc batteries. (a) Discharge at different rates at 27°C. (b) Discharge at different temperatures (Courtesy of Eagle Picher)

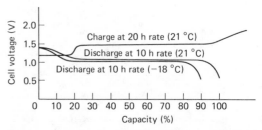

Figure 31.32 Typical charge and discharge characteristics of a Silcad silver–cadmium cell (Courtesy of Eagle Picher)

31.7 Withdrawable capacity–terminal voltage curves

An alternative method of presenting the dependence of voltage on discharged capacity is to plot percentage withdrawable capacity (i.e. available capacity as a percentage of nominal capacity) as a function of battery voltage. This type of plot is a useful means of assessing the overcharge capability of a battery.

31.8 Capacity returned (discharged capacity)–discharge current curves

This type of curve is illustrated in Figure 31.33, in which the discharge current is plotted against capacity returned (discharged capacity) to various end-voltages.

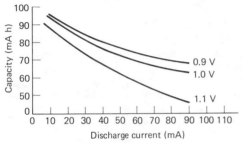

Figure 31.33 Typical capacity *versus* discharge current curves of Eveready sealed nickel–cadmium button cells, 1.2 V, 0.09 A h capacity (Courtesy of Union Carbide)

31.9 Discharge rate–capacity returned (discharged capacity) curves

A battery that is manufactured to have a capacity of, say, 20 A h when continuously discharged at the 20 h rate to a fixed end-voltage will, under these conditions, realize its maximum capacity of 20 A h and deliver a current of 1 A for 20 h ($1C_{20}$), or, within reason, any multiples of 20 A h, e.g. 0.5 A for 40 h ($0.5C_{20}$) or 2 A ($2 \times C_{20}$) for 10 h. Whereas the full capacity of 20 A h is always delivered (and sometimes slightly exceeded) when the continuous discharge period is increased above 20 h, no matter how long the discharge time, this is not so when the time for complete discharge is dramatically decreased to, say, 2 h or to 10 min. Although, in these circumstances, higher currents will be produced, the durations for which these currents are available will be lower than those calculated from the simple relationship above. This is due to certain chemical and physicochemical rate-determining processes occurring in the battery, which limit the achievement of the full 20 A h capacity. For this reason, if

rapid discharge is broken down into several separate discharges, with rest periods in between, the total capacity achieved will be nearer to the theoretical value of 20 A h due to the occurrence of time-dependent recovery processes within the battery. To illustrate the effect of discharge rate on capacity delivered, Figure 31.34 shows the relationship between discharge current and discharge time for a series of non-spill lead–acid batteries. For the 6 V, 12 A h VPT 6.13/12 battery, for example, 12 A h capacity is obtained when the discharge is performed relatively slowly at a discharge current of approximately 0.6 A over 20 h. When, however, the battery is discharged rapidly, say during 1 min at 10 A, a capacity of only 10 A h is obtained.

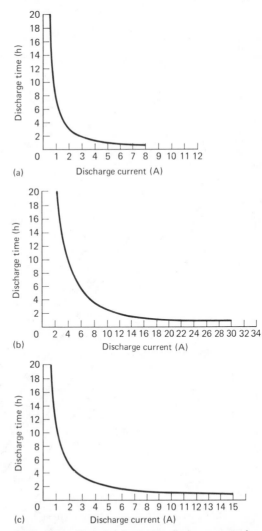

Figure 31.34 Discharge time *versus* discharge current for Varley non-spill lead–acid batteries: (a) VPT 6.9/8 (6 V, 9 A h), (b) VPT 12.7/50 (12 V, 50 A h), (c) VPT 6.13/12 (6 V, 12 A h) (Courtesy of Varley)

The decrease in cell voltage that accompanies the increase in discharge current for a nickel–cadmium cell is shown in Figure 31.35. This decrease in voltage is even more dramatic at lower temperatures.

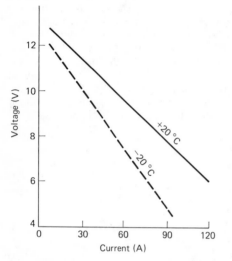

Figure 31.35 0.3 s peak discharge characteristics of a SAFT 10 VR 1.2 RR ten 1.2 A h cell nickel–cadmium battery (12 A h) at +20 and −20°C (Courtesy of SAFT)

Various manufacturers supply curves for determining the minimum battery size required for an intended application. Thus, for sealed lead–acid batteries, at a selected discharge current and discharge time, the required 20 h rate capacity of the battery can be read off (Figure 31.36); for example, 1 A for 40 min would require a 1.2 A h battery. In general, there is a relationship between depth of discharge of a battery and the number of charge/discharge cycles that can be expected of it during its life (Figure 31.37). Consequently, in many cases it is preferable to choose a higher capacity battery than the graph indicates if long life is a paramount consideration.

Relationships between discharge rate and capacity together with other relevant relationships (discharge rate–voltage, and discharge rate–energy density) are illustrated below for several types of battery.

31.9.1 Lead–acid batteries

Table 31.2 illustrates the effect of discharge rates between the 20 h rate and the 1 h rate on discharge capacity for a range of maintenance-free lead–acid batteries. Lower ampere hour capacities are obtained as the time for a complete discharge is decreased (i.e. as the discharge rate is increased) over the whole range of batteries.

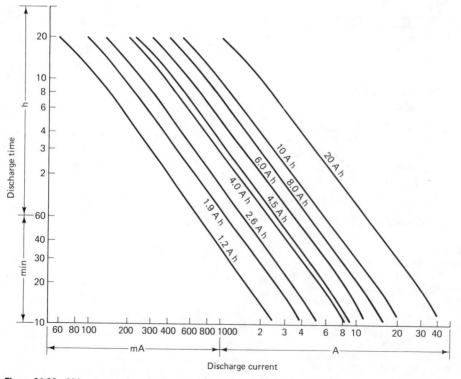

Figure 31.36 20 h rate capacity selection chart for a Yuasa sealed lead–acid battery at 20°C (Courtesy of Yuasa)

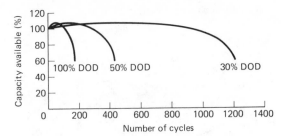

Figure 31.37 Cycle service life *versus* depth of discharge (DOD) of a Yuasa sealed lead–acid battery at 20–25°C. Discharge current 0.17CA (FV 1.7 V/cell), charging current 0.09CA, charging volume 125% of discharged capacity (Courtesy of Yuasa)

Table 31.2 Discharge capacity of Yuasa maintenance-free lead–acid batteries at various discharge rates

Type	Discharge capacity (A h)				
	20 h rate	10 h rate	5 h rate	3 h rate	1 h rate
NP1.2–6	1.2	1.1	1.0	0.9	0.7
NP2.6–6	2.6	2.4	2.2	2.0	1.6
NP4–6	4.0	3.7	3.4	3.1	2.4
NP4.5–6	4.5	4.2	3.8	3.5	2.7
NP6–6	6.0	5.6	5.1	4.6	3.6
NP8–6	8.0	7.4	6.8	6.2	4.8
NP10–6	10.0	9.3	8.5	7.7	6.0
NP1.9–12	1.9	1.8	1.6	1.5	1.1
NP2.6–12	2.6	2.4	2.2	2.0	1.6
NP6–12	6.0	5.6	5.1	4.6	3.6
NP20–12	20.0	18.6	17.0	15.4	12.0

31.9.2 Nickel–cadmium batteries

Figure 31.38 shows the decrease in capacity with increasing discharge rate for rolled sintered nickel–cadmium batteries when discharged at 0 and −20°C after an initial charge at room temperature. Extremes of temperature adversely affect the performance of sealed nickel–cadmium batteries, especially below 0°C. This reduction in performance at lower temperatures is mainly due to an increase in

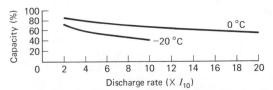

Figure 31.38 Extractable capacity *versus* discharge rate of a Varta sealed nickel–cadmium rolled sintered cell fully charged at room temperature to 2 A h (Courtesy of Varta)

internal resistance, and therefore, because of their inherently lower internal resistance, batteries incorporating sintered cells are normally recommended when lower operating temperatures are to be encountered.

31.10 Discharge rate–terminal voltage curves

Figure 31.39 shows the relationship between the discharge rate and voltage of rolled sintered nickel–cadmium batteries when discharged at 0 and −20°C after an initial charge at room temperature.

Figure 31.40 shows plots of voltage and capacity *versus* discharge current for nickel–cadmium batteries (0.4 and 5.6 A h sealed). It can be seen that both capacity and voltage decrease as the discharge current is increased.

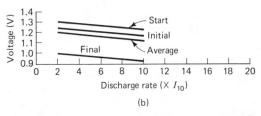

(a)

(b)

Figure 31.39 Voltage levels on discharge of a fully charged Varta sealed nickel–cadmium rolled sintered cell in relation to discharge current (up to 2 A h) at an ambient temperature of (a) 0°C, (b) −20°C (Courtesy of Varta)

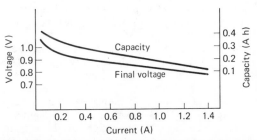

Figure 31.40 Capacity and final voltage *versus* discharge current of a Nife Jungner sealed nickel–cadmium cell type KR15/51, 0.4 A h (Courtesy of Nife Jungner)

31.11 Discharge rate–mid-point voltage curves

Discharge voltage varies with the state of charge, discharge rate and cell temperature. Figure 31.41 shows for a D-type General Electric sealed lead–acid battery that the mid-point voltage (that is, the voltage when the cell is half discharged in time) is linearly related to both rate and temperature and that the slope of the curves increases as cell temperature is reduced.

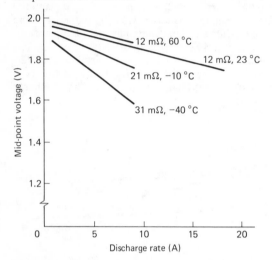

Figure 31.41 Mid-point voltage delivered as a function of discharge rate and temperature for a General Electric D cell charged at 2.45 V at 23°C for 16 h, discharged to 1.4 V at 23°C (Courtesy of General Electric)

The discharge capacity of the D cell ranges from 2.5 A h at approximately 0.4 A discharge rate (10 h rate) to 1.75 A h at approximately 3 A discharge rate (1 h rate), assuming that charging is carried out to 2.45 V at 23°C for 16 h and discharging is carried out at 23°C to 1.4 V.

31.12 Discharge rate–energy density curves

Figure 31.42 shows the effect of discharge rate in terms of C (nominal 1 h discharge rate) on energy per unit weight (Figure 31.42(a)) and energy per unit volume (Figure 31.42(b)) for silver–zinc batteries in the 0.1 and 300 A h capacity range. The distinct improvement in the energy per unit weight or volume of silver–zinc over nickel–cadmium, nickel–iron and lead–acid battery systems is very apparent. Figure 31.43 shows the effect of temperature on the energy per unit weight available from silver–cadmium batteries when discharged at various rates. The energy available starts to decrease as the cell temperature approaches −18°C.

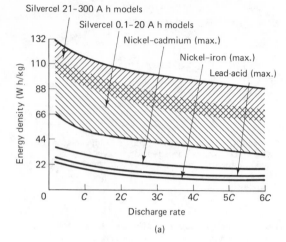

(a)

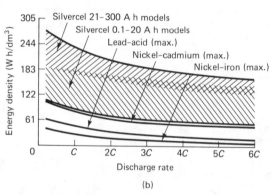

(b)

Figure 31.42 Effect of discharge rate (in terms of C, the nominal hour discharge rate) on energy output: Yardney Silvercel rechargeable silver–zinc battery compared with nickel–cadmium, nickel–iron and lead–acid batteries (Courtesy of Yardney)

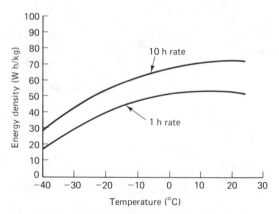

Figure 31.43 Effect of temperature on energy per unit weight of Silcad YS-18 rechargeable silver–cadmium cells (discharged without heaters) (Courtesy of Yardney)

31.13 Self-discharge characteristics and shelf life

When fully charged batteries are stored by suppliers or in the stores of the user then, depending on the type and characteristics of the battery and other factors such as storage time and temperature, the batteries will undergo a gradual process of self-discharge. The result is that when these batteries are called into service they will not have the design ampere hour capacity that would be expected of a newly manufactured battery. Usually, the full capacity can be restored to the battery by subjecting it to a process of acclimatization whereby it is given a few charge/discharge cycles under controlled conditions. After a few such cycles the full capacity of the battery is usually restored. In some circumstances, batteries in store are given a periodic topping up charge to restore their full capacity, which is then available when the batteries are brought into service.

Different types of batteries vary appreciably in their self-discharge rates; indeed, some types of batteries are designed to have inherently very low self-discharge rates. The self-discharge characteristics of various types of battery are discussed below.

31.13.1 Lead–acid batteries

Low-maintenance lead–acid batteries usually feature a special calcium–lead grid instead of the lead–antimony alloy used in conventional lead–acid batteries such as the traction battery. The elimination of antimony and other metallic impurities from the alloy increases the ability of the battery to retain its charge when not in use and gives the battery a shelf life of about five times as long as when the conventional lead–antimony alloy is used. Even at elevated temperatures the shelf life with calcium–lead alloys exceeds that obtained with conventional alloy.

Figure 31.44 shows a curve of percentage of nominal capacity at various storage times at four temperatures between 40 and 100°C for a low-maintenance lead–acid battery. Most manufacturers of this type of battery claim that during storage at normal temperatures around 15–20°C these batteries lose about 5% of their initial capacity during 12–16 months and an annual recharge is all that is necessary to reinstate the full charges (Figure 31.45). Another manufacturer of such batteries is more conservative, recommending storage at temperatures between 0 and 10°C with a 6-monthly recharge (Figure 31.46).

At 8°C the storage time to reach half the initial capacity exceeds 2 years, while a conventional lead–antimony alloy battery would reach this value in about 3 months at 20°C.

Figure 31.47 shows the effect of time and temperature on the self-discharge rate of a fully recombining lead–acid battery (the General Electric D cell). This shows, for example, that at 25°C the available capacity is reduced to 60% of fully charged capacity after approximately 300 days' storage.

31.13.2 Nickel–cadmium batteries

These cells have an acceptably low rate of self-discharge (Figure 31.48(a)). Self-discharge is very much dependent on temperature and increases rapidly with rising temperature. Self-discharge is

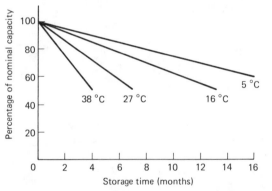

Figure 31.44 Typical self-discharge characteristics between 40 and 100°C of Eagle Picher Carefree rechargeable lead–acid batteries (Courtesy of Eagle Picher)

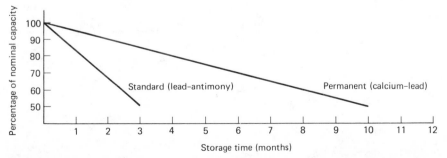

Figure 31.45 Self-discharge of Sonnenschein spillproof lead–acid standard and permanent batteries at 20°C (Courtesy of Dryfit)

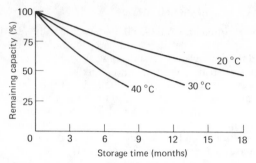

Figure 31.46 Self-discharge characteristics of a Yuasa sealed lead–acid battery (Courtesy of Yuasa)

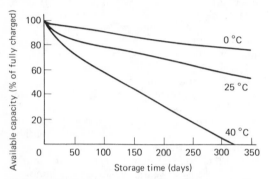

Figure 31.47 Capacity retained as a function of storage time at various temperatures: General Electric D cell (Charge, 2.45 V at 23°C for 16 h; discharge, 0.25 A at 23°C to 1.4 V) (Courtesy of General Electric)

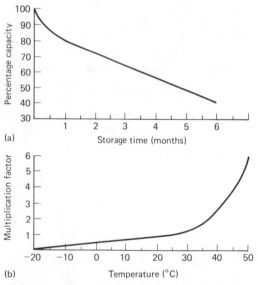

Figure 31.48 Nife Jungner NICA sealed nickel–cadmium cells: (a) self-discharge at 25°C (stored and charged at 25°C), (b) multiplication factor for calculating self-discharge at other temperatures between −20 and +50°C (Courtesy of Nife Jungner)

small at very low temperatures and is almost zero at −20°C (Figure 31.48(b)). These cells and batteries can be stored in the charged or discharged condition. When charged they should be handled with care to avoid short-circuits because the short-circuit current is very high. When storing for more than 6 months a storage temperature less than 15°C is recommended by the manufacturers. As in the case of lead–acid batteries, self-discharge performance varies appreciably from one manufacturer to another and this point should be carefully investigated before a battery selection is made.

Button cells These cells have a moderate self-discharge characteristic at normal temperatures, as shown in Figure 31.49. For applications requiring good charge retention, mass plate electrodes are preferred. When button cells have been stored for a long period (over 6 months), regardless of storage temperature, they should not immediately be charged, but should first be fully discharged and then charged once at half the normal rate, that is, 28 h at the 20 h rate.

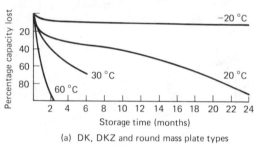

(a) DK, DKZ and round mass plate types

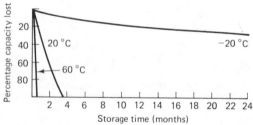

(b) RS and SD sintered plate types

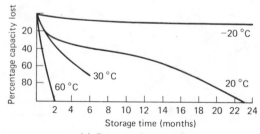

(c) D rectangular mass plate types

Figure 31.49 Self-discharge charge retention characteristics of Varta sealed nickel–cadmium batteries (Courtesy of Varta)

31.14 Float life characteristics

Batteries used in standby applications such as emergency power and lighting, where the demand for them to supply power is spasmodic, would often be returned to full charge by the float charging regimen in which the battery is kept on continuous charge. Under these conditions the battery is usually charged by the constant potential method at a potential sufficient to maintain it at full charge. In addition to making up any power loss due to use of the battery this also compensates for any long-term self-discharge processes that occur when the battery is not under load.

In the case of lead–acid batteries the float constant potential is maintained continually at 2.28–2.30 V/cell. The limiting current is between 1 and 20% of the rated capacity.

If float charging is carried out within the recommended limits, the batteries are likely to withstand overcharge for several years without significant deterioration occurring. The number of charge/discharge cycles obtained from a battery during its life (i.e. the cyclic life) is dependent on the depth of discharge (i.e. percentage of nominal capacity discharged) to which it is subjected on discharge. Thus one supplier of lead–acid batteries for standby applications (Crompton-Parkinson) claims a cyclic life of 200 cycles when the batteries are on deep discharge duty (i.e. 80% of nominal capacity still delivered after 200 cycles) improving to a cyclic life of 1500 cycles when the battery is on 30% partial discharge duty (i.e. 30% of nominal capacity removed on each discharge).

A good way of representing the cyclic life of batteries on such duties is to prepare a plot of percentage of nominal capacity available *versus* number of cycles for various percentage depths of discharge. Figure 31.50 presents such a curve for a sealed lead–acid battery supplied by Yuasa. It can be seen that about 200, 400 and 1200 cycles, respectively, are obtained for 80, 50 and 30% discharge per cycle.

Yuasa claim that, due to the use of heavy-duty grids, 1000 cycles are obtained for their batteries and a normal life of 4–5 years is expected in float charge on standby applications (based on normal charge voltage of 2.28–2.30 V/cell at 20°C).

Yuasa claim that the capacity of their batteries increases to a maximum value during the initial 50 cycles in service (Figure 31.50). If a battery with a higher rated capacity and one with a lower rated capacity are discharged to the same ampere hours, the battery with the higher rated capacity will have a longer life (more cycles) than the other because the depth of discharge is in inverse proportion to the capacity, resulting in a shallower discharge from, and a lighter burden on, the higher capacity battery.

Figure 31.51 shows the behaviour of batteries when they are discharged once every 3 months. The float service life is affected by the number of discharge cycles, depth of discharge, temperature in float charge and float voltage.

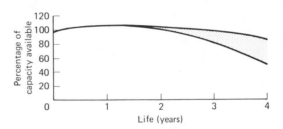

Figure 31.51 Float service life of a Yuasa sealed lead–acid battery at an ambient temperature of 20–22°C. Floating voltage 2.25–2.30 V/cell (Courtesy of Yuasa)

Another method of presenting the performance of this type of battery is to record the nominal capacity obtained when these batteries are discharged to various voltages as a percentage of their capacity when discharged to 1.75 V/cell (in the case of lead–acid batteries).

Typical data for a lead–acid battery are given in Table 31.3.

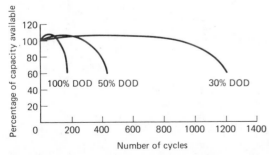

Figure 31.50 Cycle service life *versus* available capacity for various depths of discharge (DOD) of a Yuasa sealed lead–acid battery (Courtesy of Yuasa)

Table 31.3

Rate (h)	Approximate rated capacity (%)	End-voltage (V/cell)
20	100	1.75
10	97	1.70
5	88	1.65
1	62	1.50
0.5	52	1.00

Part 4

Battery applications

32

Lead–acid secondary batteries

Contents

32.1 Stationary type or standby power batteries

Large battery installations which are called upon only occasionally to supply power, usually in emergency or auxiliary power supply circumstances, are referred to as stationary or standby batteries. Such installations having total capacities in hundreds of ampere hours are commonly in use. Figure 32.1 illustrates a large secondary lead–acid battery installation.

Figure 32.1 Large stationary lead–acid battery installation (Courtesy of Chloride Batteries)

Standby power can be defined as a source of electricity that becomes available when the mains source of electricity ceases to be available. The choice of a standby system lies between batteries, usually lead–acid or nickel–cadmium, and generators, or a combination of the two. For many applications, the battery's ability to provide the power required instantly makes it more suitable. Indeed, even in large installations for which a generator is essential, batteries are often used both to start the generator and to provide power for the initial period until the generator is running to speed.

Standby power is required in many situations. For department stores, offices, factories, cinemas and other public places, emergency power to light people to safety without panic may be enough. In industry, the need varies from keeping fire and burglar alarms and other communications working to supplying emergency power to vital production processes or even to supply the power necessary to

effect a controlled safe shutdown of the process. Classic cases of the use of standby power are in an operating theatre and an airport where, in the event of mains failure, power is required for lighting, communications and the operation of vital equipment. Power stations, telephone exchanges, lighthouses and computer installations are other examples of the application of stationary or standby batteries.

32.1.1 Types of stationary or standby battery

Several types of standby lead–acid batteries are manufactured to meet the needs of different types of application.

Planté cells

For the long life with very high reliability needed in places like power stations and telephone exchanges, batteries are made up of cells of a kind named after Planté. They are kept on continuous trickle charge and are ready to spring into life immediately there is a mains failure. The capacity of a cell (the amount of electricity it can store) and its life depend largely on the design of its positive plate. In the Planté cell, this is cast from pure lead in a form that gives it a surface 12 times its apparent area. Figure 32.2 shows representations of a modern type of Planté grid. The negative plate is of the pasted grid type made by forcing lead oxide paste into a cast lead alloy grid. Positive and negative plates are interleaved and insulated from each other to prevent short-circuits.

The all-important surface of the positive plate is continuously regenerated during the life of the battery. Trickle charging compensates for natural open-circuit losses. The average life of a typical Planté battery is 20 years and some designs extend to 35 years and sometimes even longer with no fall in capacity. Although the initial cost is high, the average annual cost over the life of the battery can be lower than that of other kinds used for standby power purposes.

Cell capacities are rated by relating current flow to time in ampere hours. High-performance Planté cells up to 2000 A h have transparent plastic containers to allow visual checking of the acid level and general condition. Large Planté cells up to 15 000 A h, often found in large telephone exchanges, are most commonly housed in lead-lined wooden boxes.

Flat plate cells

Because of the high initial cost of Planté cells, specially designed flat plate cells have been developed to provide a cheaper but shorter-lived alternative source of standby power. The layout of a typical modern flat plate cell is shown in Figure 32.2.

The flat (or pasted) plate cell, pioneered in the 1880s, has a positive plate made from a lead alloy lattice grid into which a paste of lead oxide and sulphuric acid is pressed. The negative plate is also of the pasted grid type, but is thinner.

Although this is the basis of the modern car battery, car batteries are wholly unsuitable as an emergency power source because they are designed to give high current for a short time, as when starting a car engine.

Like the Planté battery, pasted plate batteries manufactured for standby use have transparent containers. They are available in capacities of up to 500 A h and have a life expectancy of 10–12 years to meet emergency lighting regulations.

Tubular cells

Cells with tubular positive plates are normally used to power electric trucks, on which daily recharging is needed. They are also suitable for certain standby applications. They deliver high power at low and medium rates of discharge and work well in adverse conditions.

The positive plate consists of lead alloy spines surrounded by synthetic fibre tubes filled with a mixture of lead oxides (Figure 32.3). The tubes keep the active material in contact with the conducting spines during expansion and contraction resulting from the charge and discharge cycle, and so contain the stresses that would break up other types of plate.

Tubular cells are ideal for any kind of application that requires frequent charge/discharge cycles. When on standby duty, they have a life expectancy of 10–12 years, compared to 5 years when powering an electric truck. They are less costly than Planté cells, but fall in capacity as time goes on.

Table 32.1 shows the types of application for each of these types of battery as recommended by a leading manufacturer of standby power batteries, Chloride Storage Batteries Ltd, UK.

32.1.2 Selection of suitable stationary or standby power batteries for the application

When someone who needs standby power has to select a battery, four questions arise:

1. What is the probability of power failure?
2. What is the likely duration of the failure?
3. What damage could power failure do to people, equipment and production?
4. What regulations have to be complied with?

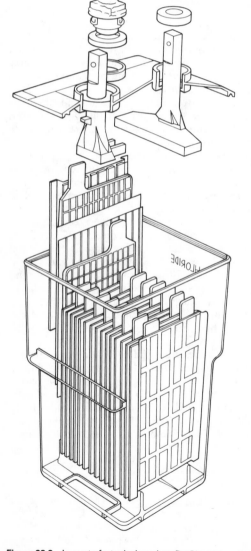

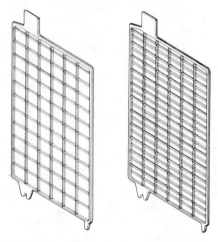

Figure 32.2 Layout of a typical modern flat Planté-type lead–acid standby battery (Courtesy of Chloride Batteries)

Table 32.1 Application of various types of lead–acid battery

	Planté YA/YC/YH	Flat plate FA/FC	Flat plate R range	Tubular	Marine
Telephone exchanges	●				
Mobile telephone exchanges				●	
Generating stations and substations	●				
Obsolescent generating stations (short-term)				●	
Emergency lighting	●	●			
Alarms	●	●			
Computer emergency power	●		●	●	
Engine starting	●		●		
Oil rigs			●	●	
Marine					●

Figure 32.3 Layout of a modern design of lead–acid tubular cell standby battery (Courtesy of Chloride Batteries)

It is generally accepted that lead–acid batteries are ideal for standby power systems, because they are reliable and relatively low-cost. When it comes to selection, the choice is between the kinds of battery referred to earlier. Batteries are tailor-made for specific standby applications, and Table 32.2 gives a rough guide to the suitability of different batteries for different jobs.

There is an all-too-common idea that 'a battery is a battery' and this sometimes leads to car batteries

Table 32.2

	Life expectancy (years)	Publication No.	Cell type range
Planté	25	SPUK$_{1/2}$	YA/YC/YH
Flat plate	10–12	SPUK$_3$	FA/FC
Flat plate	5–6	SPUK$_5$	R range
Tubular	10–12	SPUK$_4$	–
Marine	10–12	M$_4$	–

(Courtesy of Chloride Power Storage Ltd, Swinton, UK)

being used to supply standby power. Although this usually results in low initial cost, it has serious drawbacks. Car batteries are designed to supply high currents for short periods and are not intended to carry industrial type loads for any significant periods. Their life expectancy in standby applications is very low compared to that of Planté and other industrial batteries. More important, the use of car batteries may also invalidate the specifications of equipment makers and insurance companies.

Initial savings from incorrectly chosen batteries can evaporate rapidly in the face of short life, especially if the wrong size of battery is chosen as well as the wrong type. And the failure of an emergency lighting or fire alarm system can result in danger to human life or legal action. Expert opinion on whether the system is the right one for the job will be an important factor in deciding where the responsibility lies. The battery must therefore be selected from a range specifically designed for the purpose. A very important aspect involved in the selection of standby power batteries is to ascertain what is known as the load profile for the proposed application.

To be able to advise on this the battery manufacturer needs to know what is required in terms of current output, voltage and frequency of operation. If only one continuous load is to be supplied, sizing the battery can be simple. For example, an emergency lighting system drawing 20 A for 3 h on battery power would require a 60 A h battery at the 3 h rate. Since standby batteries are conventionally rated at the 10 h rate, this is equivalent to a 75 A h battery at the 10 h rate.

Load profile 1 (Figure 32.4(a))

A battery is like an athlete who can be just as exhausted after running 100 m as after 10 000; the difference is in the rate at which energy is used up.

Most industrial standby power systems now tend to be more complex because their total load is the product of many varied and intermittent loads. One battery may serve different purposes within the same building.

Load profile 2 (Figure 32.4(b))

If a battery's duty cycle can be arranged so that any high-current loads occur early, a battery of a smaller capacity can be used than if heavy currents are required at the end. This is because a piece of electrical equipment demands not only a certain current, but also a minimum voltage. If loads are likely to occur at any time in the work cycle a special approach will be needed in deciding the battery size. It is safest to assume that the loads will occur at the end of the work cycle so that sufficient battery capacity and voltage are available for the work. Otherwise the voltage may drop so low that the battery fails to do its job and this can be costly, if not dangerous.

Choosing the most suitable and economical battery can be bewildering, but most manufacturers now have computer programs to help select the most effective type and battery size. Their advice should always be sought.

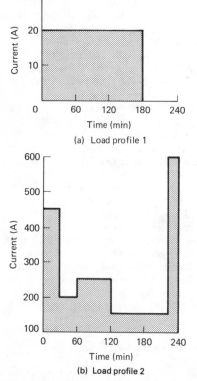

(a) Load profile 1

(b) Load profile 2

Figure 32.4 Load example profiles for lead–acid standby power battery type selection (Courtesy of Chloride Batteries)

In any event specifying the correct size and type of battery is of paramount importance. It is certainly prudent to allow a margin of safety, as one day lives may depend on the correct battery standing behind the system.

32.1.3 Charging and operation of standby power batteries

For completeness, this is included here. Battery charging is discussed more fully in Part 5.

The essential feature of any standby power system is immediate availability of the battery power if the normal mains supply fails. The battery must, therefore, always be in a fully charged condition. Charging equipment is essential to a reliable system, and the recommendations of the battery manufacturer should always be followed.

The basic charging equipment is a means of transforming the normal alternating current (a.c.) mains supply to a direct current (d.c.) matching the battery requirements. Two rates of charge are normally provided, quick boost and trickle. The quick boost rate must be capable of an output between 2.0 and 2.70 V/cell for recharging of the battery. The trickle charge output should be capable of compensating for the internal losses of the battery, at a voltage level of 2.25 V/cell. The outputs required, in terms of current, will depend on the ampere hour capacity of the battery.

During charging, 'gassing' will start when the cell voltage has reached approximately 2.30–2.40 V/cell. The cell voltage will continue to rise and the charge may be considered to be completed when the voltage and relative density have remained the same for approximately 3 h.

The voltage could be as high as 2.70 V and charging should not be allowed to continue beyond this level.

A recharge voltage table for a Planté cell is illustrated in Table 32.3.

Table 32.3 Recharge voltage table for a typical Planté cell at 15°C

Cell e.m.f.	
2.70	Top of charge volts at 7% of A h/10
2.40	Minimum top of charge volts to recharge with good gassing
2.25	Ideal for trickle charge float and minimum voltage for recharge, but subject to problems such as stratification
2.20	Floating/trickle charge, but with need for freshening charges
2.05	Open-circuit voltage for 1.210 electrolyte relative density

Trickle charging

The essential requirement of present applications of standby batteries is that the battery shall be ready instantly to perform its duty. One method of meeting this requirement is by trickle charging of the battery. After a battery has been discharged, it must be recharged at the recommended quick rate as soon as possible. Once restored to the fully charged condition, trickle charging will then keep it fully charged until the next demand. No amount of trickle charging will ever recharge a substantially discharged battery, but when it is part of the duty of a standby system to meet occasional brief demands the energy thus used can be gradually replaced by trickle charging at a suitable rate. It is satisfactory only if the amount of discharge is a very small proportion of the ampere hour capacity of the battery.

Float charging

Float charging is used where the connected load will not tolerate even a momentary interruption. A simple example of such a scheme would be a continuous process plant where even a millisecond failure of the supply would result in the complete breakdown of a manufacturing run, which could be extremely costly.

To guard against such interruptions, the battery, load and charger are connected in parallel, the charger being of the constant-voltage type, capable of delivering an output equivalent to the normal load demand plus the battery charging requirements. With the float system, should a failure occur, the battery takes over the load with no interruption whatsoever.

A constant voltage charger is a statically controlled rectifier, normally employing transductor or thyristor devices to correct for changes in output. The transductor can be regarded as an inductive choke, the a.c. impedance of which is varied by the degree of magnetic saturation produced by direct current in an auxiliary winding. Thyristor control is achieved by variation in the firing angle of the a.c. waveform.

The normal float level of a floating battery system is equivalent to 2.25 V/cell. This preset voltage is held within usual limits of $\pm 2\%$ irrespective of mains voltage variations of $\pm 6\%$, with load variations of 0–100% of the charger rating. Facilities are usually incorporated for quick boost facilities for recharging of the battery following a deep emergency discharge.

The selection of the correct charger for a battery-operated standby power system is of vital importance, and the advice of both battery and charger manufacturers should be sought to ensure the most efficient and properly matched combination.

32.2 Traction or motive power type

Traction or motive power lead–acid batteries are the power source used for every kind of electric vehicle from road transport to fork lift trucks in industry.

In many applications electrically operated vehicles are replacing vehicles with internal combustion engines. Electric vehicles are cheaper to run because of the lower fuel cost, and because of their relative simplicity they are easier to maintain than vehicles equipped with internal combustion engines. They also present fewer environmental pollution problems. In an electric vehicle the motor and controller are virtually maintenance-free and this is equally true of the battery and charger. With the introduction of automatic battery topping-up devices and self-compensating battery chargers, these once onerous battery and charger maintenance routines have now become almost non-existent. Present indications are that petrol and diesel fuel costs will probably escalate more steeply than electricity prices, and that the price differential will continue to remain in favour of electricity. Some data, supplied by Chloride Motive Power, on actual costs of electric *versus* diesel vehicles, are given in Tables 32.4 and 32.5. Both on vehicle maintenance and fuel cost, therefore, electric vehicles are displacing conventional vehicles in many applications.

The interior design of lead–acid batteries intended for traction purposes, such as milk floats and electric locomotives and fork life trucks, differs from that of stationary batteries because of their different service requirements.

32.2.1 Selection of motive power batteries

The duties the battery will be expected to perform can be broken down into a series of components, e.g. vehicle range, hills to be negotiated or loads to be lifted in the case of a fork lift truck, each of which will demand a certain amount of discharge current from the battery.

The amount of capacity available from a battery is dependent largely on the number and size of plates in the cells. Clearly, battery manufacturers cannot dictate to the vehicle designer how much room must be allowed for the battery, as this is only one of very many design considerations that must be taken into account.

Normally, the battery maker is presented with a given space into which the battery must fit, and this space is a compromise between overall design requirements and the need for a battery of sufficient size to make the vehicle a viable, economic proposition to the user.

In practice, therefore, the exact battery capacity is not usually calculated in advance for a particular duty. Instead, the type of truck required is

Table 32.4 Yardstick operating costs

30 cwt payload urban delivery vehicles	*Case 1*		*Case 2*	
	Electric	*Diesel*	*Electric*	*Diesel*
Capital costs (£)				
Vehicle	4800	7690	6 330	7300
Battery	3076	–	3 076	–
Charger	650	–	650	–
	8526	7690	10 056	7300
Standing charges/year (£)				
Vehicle depreciation (years)	320 (15)	1098 (7)	422 (15)	730 (10)
Battery depreciation (years)	615 (5)	–	615 (5)	–
Charger depreciation (years)	43 (15)	–	43 (15)	–
Interest on capital	639	577	754	548
Licence	–	101	–	101
Insurance	112	152	112	152
	1729	1928	1946	1531
Running costs/mile (p)				
Electricity 1.5 kW h/mile @ 2p/kW h	3.0	–	3.0	–
Derv 12/15 mpg @ £1.15	–	9.6	–	7.7
Maintenance materials, labour and overheads				
including tyres and lubricants	4.7	9.1	4.3	8.2
Total running costs/mile	7.7	18.7	7.3	15.9
Add standing charge reduced to cost/mile				
(9125/7500 miles/year) (p)	18.9	21.1	25.9	20.4
Total cost/mile	26.6	39.8	33.2	36.3
Index	100	150	100	109

Case 1: vehicle with open milk-float body used 365 days/year. Top speed 15 mph. Range capability is approximately 50 miles laden. This analysis is done on 25 mile range with 250 stops, which is comfortably within daily range of the electric with reducing load

Case 2: vehicle similar to case 1 but with aluminium box-van body with hinged rear doors. Driven less intensively than a milk-float, the vehicle averages 30 miles/day on general urban deliveries

Source: Electric Vehicle Association, 1980

determined by the maximum loads, and one of the standard batteries for the truck is used. There is usually some limited choice between the minimum capacity recommended and the maximum that can be fitted.

Experience shows that, over their lifetime, vehicles are required to fulfil a number of duties differing from their original application. So if a vehicle is to be flexible enough to be switched to a more demanding duty it makes sound economics to fit the largest possible battery initially. This will also help to avoid overdischarging the battery, which can cause it irreparable damage.

Where intensive and continuous operation of a vehicle is involved, and the largest battery that can be accommodated is not capable of completing the vehicle's duties between recharges, it is actually necessary to calculate the total capacity required.

Every electric vehicle, whether on the road or in industry, has a predetermined pattern of use. A milk

float is needed to cover a certain route, with a known number of stops, distance to be travelled, and the terrain it is required to negotiate. (It is important to know of any steep hills the vehicle must climb, for instance, as this will demand extra energy from the battery.) The needs of a fork lift truck can similarly, although less exactly, be calculated. How heavy are the loads to be lifted? What speed is it required to do? What travelling time must be allowed from the truck's base to where it will work? Are there any steep gradients involved?

Each of these facets will demand a known amount of discharge current from the battery and this is the basis of the calculation. The discharge current obtained from the battery is multiplied by the time (in minutes) taken to perform each movement, giving a series of figures in ampere minutes. These figures can be added together to give the total value of ampere minutes required to complete a delivery round or working shift.

Table 32.5 Yardstick operating costs

2-tonne lift capacity counterbalanced forklift trucks	Case 1			Case 2		
	Electric	*Diesel*	*Propane*	*Electric*	*Diesel*	*Propane*
Capital costs (£)						
Truck	8 595	9486	9486	8 595	9486	9486
Battery (number used)	2 043 (1)	–	–	4 086 (2)	–	–
Charger (8 or 12 h)	657 (12)	–	–	873 (8)	–	–
	11 295	9486	9486	13 554	9486	9486
Standing charges/year (£)						
Truck depreciation	716	1186	1186	1432	2371	2371
Battery depreciation	340	–	–	680	–	–
Charger depreciation	55	–	–	73	–	–
Interest on capital	847	711	711	1016	711	711
Rental of fuel facility	–	–	126	–	–	126
Total standing charge/year	1958	1897	2023	3201	3082	3208
Running cost/h (£)						
Electricity 4 units @ 2/3p	0.08	–	–	0.12	–	–
Gas oil 1.75 litres @ 25p	–	0.44	–	–	0.44	–
Propane 4 litres @ 13p	–	–	0.52	–	–	0.52
Maintenance labour + overheads						
Replacement parts and tyres						
including mid-life overhaul	0.57	0.89	0.89	0.57	0.89	0.89
	0.65	1.33	1.41	0.69	1.33	1.41
Add standing charge/year reduced						
to cost/h (£)	0.98	0.95	1.01	0.80	0.77	0.80
Total cost/h in service (£)	1.63	2.28	2.42	1.49	2.10	2.21
Index	100	140	148	100	141	148

Case 1: one truck with one battery operated on single-shift use

Case 2: one similar truck with two batteries operated with battery exchange on two-shift use

Source: Electric Vehicle Association, 1980

For example, a fork life truck lifting 2 tonnes might require a discharge current of 200 A for 0.25 min or 50 A min. This truck travels 6 m at 1 m/s, during which the discharge current is 80 A, or 480 A s (8 A min). These calculations are continued for the whole series of operations. The ampere minutes are added together and the total converted into ampere hours.

This gives the total number of ampere hours consumed by the truck's motors. However, other factors must be considered before the ampere hour capacity of the battery or batteries required can be decided. To ascertain current values and duration of discharge for a motive power battery several factors must be considered:

1. The actual value of the discharge currents. The higher the average discharge current, the lower the available capacity.
2. The total time over which the battery's discharge is spread. The shorter the time, the lower the available capacity.

Figure 32.5 shows a typical curve of capacity available at various discharge currents, and for discharges spread over various times. For example, when a 100 A h battery is discharged continuously at 100 A, the available capacity is 54% of nominal capacity and it will last 0.54 h, but when the same capacity is spread over 1 h, the available capacity is 61%.

Finally, when spread over 8 h, it is almost 88%. Because the required battery capacity depends on the percentage current, and the percentage current itself is dependent on the chosen battery capacity, one must choose the battery first and then check the currents to determine whether the chosen battery allows sufficient margin.

However, it is generally satisfactory to assume that the total number of ampere hours consumed by

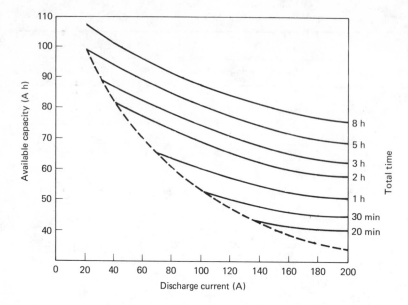

Figure 32.5 Capacity on intermittent discharge at 30°C. (Nominal cell capacity 100 A h in 5 h) (Courtesy of Chloride Batteries)

the motor should be divided by 0.7 to take account of the factors discussed above.

The result of this calculation will give a total of ampere hours that would completely discharge the battery. At this stage, it must still be considered a theoretical calculation, as in practice this is an unacceptable premise in view of the following:

3. The vehicle's eventual duty may, on occasion, be heavier than at first assumed.
4. As a battery becomes older, its ability to give full rated capacity reduces slightly.

In either of these events, to assume full discharge from a battery in the original calculation would leave the vehicle unable to perform satisfactorily.

To cover these factors, it is accepted that a battery should not be regularly discharged by more than about 80% of its normal capacity. This is also good battery operating practice to protect it against overdischarge and general abuse.

Therefore, the ampere hour value should now be divided by a further factor of 0.80.

Figure 32.6 shows how a battery's life and therefore the cost per kilowatt hour of output is affected by the depth to which the battery is discharged each day; thus the necessity for the above calculation.

To summarize so far, we have reached a calculation of:

Nominal 5-h capacity required

$$= \text{ampere hours consumed by the motor} \times \frac{I}{0.70} \times \frac{I}{0.80}$$

$$= \text{ampere hours consumed by the motor} \times \frac{I}{0.56}$$

This final figure represents the total nominal battery capacity required for the duty of the vehicle.

It will now be apparent whether one, two or three batteries will be needed, and, at this stage, it is important to lay down a suitable working schedule, which will ensure that the work of each battery is shared equally between recharges.

On no account should a battery be worked until the vehicle's performance begins to deteriorate. It is far better either to ensure that the duty does not make this demand, or to change the battery before recharging. In terms of efficient performance, and increased battery life, it is far more economic to organize battery charging routines that avoid this situation.

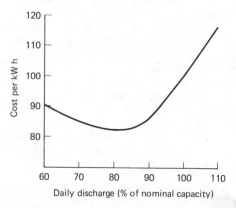

Figure 32.6 Battery cost for different extents of daily discharge (Courtesy of Chloride Batteries)

32.3 Starting, lighting and ignition (SLI) or automotive batteries

These are discussed in Part 6.

32.4 Partially recombining sealed lead–acid batteries

These batteries may be used in either cyclic or float service in a wide range of applications which include: point of sales terminals of centrally controlled cash register systems, electronic cash registers, portable test equipment, fire and security systems, communications equipment, photographic equipment, portable power tools, computer memory systems, portable televisions, hand lamps, propelled toys, power packs, invalid carriages, intruder and fire alarm systems and medical instrumentation. The batteries feature lead–calcium alloy grids for achieving long and reliable service and also a suspended electrolyte system. These batteries are available in capacities between 5 and 50 A h.

33

Nickel batteries

Contents

33.1 Nickel–cadmium secondary batteries

33.1.1 Open and semi-open smaller types

The wide applicability and range of sealed nickel–cadmium batteries is a result of their remarkable performance characteristics, some features of which are:

1. Total absence of maintenance.
2. Very long operational life.
3. Ability to accept permanent overcharge.
4. Very favourable prolonged storage time characteristics in stock, whatever the state of charge; batteries produced by some manufacturers are better in this respect than those produced by others. If charge acceptance deteriorates markedly after a few months' storage the capacity can be restored by cycling.
5. Ability to be overcharged at high rates.
6. Rapid and ultra-rapid chargeability.
7. Constancy of discharge voltage characteristics even at very high discharge rate.
8. Mechanically robust and resistant to electrical damage (overcharge reversal).
9. Operation over a wide temperature range: -40 to $+50°C$.

Some applications of vented large nickel–cadmium batteries are discussed below.

Large nickel–cadmium batteries are made in a variety of designs to suit different types of application. The major applications are discussed below:

1. Central batteries for emergency, safety and auxiliary lighting are used for public meeting places (theatres, warehouses, sports halls, etc.) and factories. They permit the safe evacuation of buildings in the case of mains failure, by providing secondary illumination.
2. Nickel–cadmium batteries are used for special emergency power supplies, as required in medical centres, intensive care units and operating theatres and similar applications where lighting interruption can be a matter of life or death.
3. Engine starting batteries are used to start motor generator sets as required for standby power systems. The batteries have to start the engines in bad weather conditions where low temperatures may prevail. These engine starting batteries must be highly reliable, have a high performance capability and be free from the effects of high stress and vibration, such as occurs in vehicles and the engine rooms of ships.
4. Backup standby batteries provide power in the event of a mains failure, for example in the operation of a crane to prevent the load being released. Another application is in computers and processing systems.

5. Switch tripping batteries in high-voltage distribution systems are well suited for use with distribution ring mains. They offer reliability of operation and provide the high current necessary to close the circuit breakers and switches associated with power stations and substation networks.
6. Small hand lamps for powerful light output.
7. Train lighting batteries are installed in containers below the carriages and supply the coach lights during steps at stations or in the case of a power supply failure. Nickel–cadmium cells are also used for the uninterrupted emergency power supply to railway main stations and signalling systems.
8. Marine applications require an uninterruptible power supply to the emergency systems for radar, communications, steering and navigation, and for the emergency lighting on board.
9. Air transport batteries provide the ground power supplies and also the support battery systems on board aircraft. Also, at the airport, all the essential lighting and communication relies on the batteries. In this method of working, the system will operate in the standby mode, giving uninterrupted power supply on mains failure.

33.1.2 Sealed nickel–cadmium batteries

The application of sealed nickel–cadmium batteries may be subdivided into four main categories, which are discussed below: self-contained supply, emergency supply, auxiliary supply, starting and high-power supplies.

Self-contained supply

Sealed nickel–cadmium batteries are mainly used in portable devices as the only source of energy. Examples of this type of application are: transceivers, razors, portable equipment and tools, remote radio and ultrasonic controls, flashguns, toys, portable spray equipment, toothbrushes, hair curlers, hearing aids, cine cameras, calculators, computers, model ships, etc., test and control instruments, survey equipment, rangefinders, sonar, timers, digital counters, domestic electricity and telephone meters, clocks, time and security check clocks, ticket, money and petrol dispensers, identity check machines, electronically controlled machine tools, lamps, tape recorders, television dictation equipment, measuring and test instruments, medical equipment, miners' lamps, transceivers, telephones, incubators, heated clothing, underwater lamps, portable refrigerators, telephone answering machines, metal detectors, invalid chairs and control instruments.

In these types of application, sealed nickel–cadmium batteries are ideal sources of energy, meeting the requirements of medium energy at low rate and freedom from a.c. supply loads. Thus, in the operation of an electronic flashgun, the battery is capable of recharging the capacitor in a few seconds at a current of $10-20C_5$ A. Battery recharge takes 30 min (rapid charge) or 3 h (accelerated charge).

SkyLab, the manned space vehicle developed by NASA, was powered by eight secondary nickel–cadmium batteries (as well as 16 primary or secondary silver–zinc batteries). Nickel–cadmium batteries are used to power the NATOIII communication satellite.

Emergency supply

In the last few years, the need for emergency supplies during a.c. supply failure has been considerably extended in electronics and security equipment. Sealed nickel–cadmium batteries are kept on constant charge when the a.c. supply is present and incorporated in utilization circuits, which must continue to function during a.c. supply failure. Sealed nickel–cadmium batteries are particularly suited for these applications by reason of their ability to accept overcharge, their freedom from maintenance and their reliability.

The principal applications in emergency supplies are in electronics, emergency lighting, fire and burglar detection, signalling and alarms, medical apparatus and surveillance systems (including defibrillators, resuscitation units, portable electrocardiographs and blood transfusion pumps), switch tripping, logic memories, control of switchgear, circuit breakers and signalling.

As an example of these applications consider the case of a self-contained emergency lighting unit for a building. The unit has a charge circuit which applies a permanent charge to the battery, at a rate of $0.05C_5$ A. Complete charging of the battery is required in 24 h. When the a.c. supply fails the battery is connected to the lamps by a relay; when the a.c. supply is re-established the lamps go out and the battery is recharged.

Consider as a further example an electrical supply for the memory in a data system. In order not to lose data in case of a.c. supply failure, certain data systems are fitted with self-contained emergency power supplies for relatively low powers. Data systems are generally supplied from the a.c. supply and are particularly sensitive to very short breaks and transient voltage variations. To overcome the weaknesses of the a.c. supply, there are various types of equipment, in particular inverters, supplied by battery–charger combinations. The inverter derives its supply with a.c. supply present from the rectifier and with a.c. supply absent from the battery.

In these systems the battery serves also as a filter for commutating currents injected by the inverter when switching on or removing the load. For example, a 300 ms hold for a 5 kV A inverter with 110 V input can be obtained with a 4 A h battery (110 VR 4D SAFT). The peak current is 65 A ($16C_5$ A) and the battery is charged at a permanent rate of $0.02C_5$ A with excellent reliability. The theoretical service life of this unit is 9 years.

Examples of the application of nickel–cadmium batteries in emergency power and lighting applications include multistorey buildings, aviation safety, railway vehicles, ships, tunnels, sports areas, shops and offices. Nickel–cadmium batteries are also used for starting up diesel generators to supply electrical power in the event of a mains failure (Figure 33.1).

Figure 33.1 Nickel–cadmium battery for starting up diesel generators for emergency lighting (Courtesy of Nife Jungner)

A good example of a situation in which there is a critical requirement for the very rapid availability of emergency power is an operating theatre. Here, in the event of a mains failure, the operating staff must have a system that restores light and power for instruments within a fraction of a second. Such systems are usually based on the nickel–cadmium battery with associated electronic equipment. Another, lesser, example is a bell-ringing tower. Losing control of all the bells at the same instant of time could have unwanted effects both on the music produced and the well-being of the ringers.

An example in which the proper supply of emergency lighting and power is mandatory is an airport. Traffic interruptions due to bad weather at airports is an increasing problem in today's intensive air traffic. Therefore, airports are equipped with different landing aids such as runway lights, instrument landing systems and radar. An interruption in the supply of electrical power to this equipment means that an approaching aircraft will lose guidance to the runway and a critical situation might develop. For this reason, emergency power must be available to use in case of interruption of the normal power supply. The International Civil Aviation Organization (ICAO) has made standards for equipment of airports to allow safe landing at different visibilities.

These standards have different categories. The lowest category is 'Non-Instrument Runway', where 1.5 km visibility is required. The highest category is 'Precision Approach Runway Category III', intended for operations below 400 m runway visual range (visibility of the lights along the runway). The standards also determine the airport equipment, which landing aids shall have emergency power and the maximum time for switching to emergency power. The emergency power source can be a power station completely independent of the airport's normal power supply, motor generator sets or battery-powered inverters.

Category I Most airports around the world intended for international traffic are in Category I. This category is intended for operations down to 60 m decision height (the height at which the pilot has to abort the approach if not able to continue visually) and 800 m runway visual range. During a power failure, the instrument landing aids must have emergency power within 10 s and the runway light within 15 s.

Category II Increasingly, intensive air traffic has created a demand to improve regularity (to minimize the time when an airport is closed due to bad weather). The less time an airport is closed, the smaller are the losses in landing and starting fees. It also gives negotiating power to attract flights, since

companies will select airports that are also available in bad weather. For this reason, there is now a trend to improve international airports to Category II. This is intended for operations down to 30 m decision height and 400 m runway visual range.

Some airport instruments require completely uninterrupted power supply while other instruments and runway lights must be able to switch to emergency power in a maximum of 1 s. The power demand of the systems requiring completely uninterrupted power is not so high, and is normally arranged by direct feeding from batteries. It is more difficult to arrange an emergency supply for the runway lights, which normally require hundreds of kilowatts.

Auxiliary supply

When the power required by a consuming device exceeds the output of the source, a local battery is necessary. Nickel–cadmium sintered plate sealed batteries, by their performance on discharge at high rates, have the perfect attributes for use to provide auxiliary supplies.

The battery plays a similar role to that of a capacitor used to store energy, but with the difference that it can discharge for considerably longer periods. Except during these periods of discharge, the battery is left on permanent charge.

For example, the case of an auxiliary supply for direction indicators on a lightweight motorcycle illustrates the possibilities of using uncontrolled rapid charge with cylindrical sealed nickel–cadmium batteries. When the generator on the motorcycle is a magdyno and when the engine speed is low, an auxiliary supply is necessary to supply the indicators. In this application only the charge presents a problem because the output of the magdyno depends on the engine speed and whether lights are on or not. In some machines the maximum charge current available may vary from 10 mA to 1 A. The use of a battery with a capacity of 0.5 A h gives a perfect solution because it can withstand overcharge at $2C_5$ A (1 A).

Another example of using sealed nickel–cadmium batteries as an auxiliary power source concerns the carrier wave equipment that permits the connection of a second telephone subscriber when only one line is available. The signal is modulated at the exchange, transmitted on the single line and demodulated at the receiver. The battery is kept on permanent charge for the telephone lines at a current of about $0.02C_5$ A. When the user is called, the battery supplies the energy necessary for the bell and the transmission. The daily load is small, a few milliampere hours, the source being a VBE type SAFT battery. A failure rate per operational hour of 10^{-8} would be typical for this type of SAFT battery.

Vehicle starting and high-power supplies

Nickel–cadmium sealed batteries are particularly well suited to starting small engines and high powers. Because of their very low internal resistance they can deliver very high rates of discharge for short periods. For example, the starting of light-weight motorcycles requires a breakaway current of 30–50 A followed by a discharge for several seconds at 20–25 A at a nominal 12 V. Recharge must be rapid in order to be sufficient, since runs are short and uncontrolled. All these requirements can be satisfactorily met by using a battery of 0.5 Ah capacity, with the following characteristics:

1. Voltage at end of starting at 20°C: 11 V.
2. Voltage at end of starting at −10°C: 9 V.
3. Uncontrolled rapid charge rate 0.8 A, which permits recharge after starting in less than 3 min.

The same type of application is the starting of a lawnmower, where the breakaway current may be as high as 75 A. The power necessary is supplied by a 1.2 Ah battery.

The nickel–cadmium battery is now the type used almost exclusively for engine starting in larger aircraft. Batteries with capacities up to 200 Ah are now available. Lead–acid batteries are used in the majority of small private aircraft and in some larger aircraft, e.g. the UK Harrier 'Jump Jet' and BAC Concorde.

33.2 Nickel–zinc secondary batteries

Nickel–zinc batteries may be the batteries of the future in applications such as utilities load levelling and electric vehicles, and indeed the nickel–zinc battery is now in commercial production in the USA.

33.2.1 Electric vehicles

In an effort to broaden the field use of the nickel–zinc system and to provide an early demonstration of the state of the art of this system in vehicle propulsion, Yardney are making available an unoptimized prototype battery module of 300 Ah nominal capacity, which can be assembled into a vehicle battery of the desired voltage; thus baseline performance can be established on which future performance can be projected. A specially designed charger is available, which maximizes the capacity maintenance and cycle life of this prototype battery.

The battery is 264 mm wide, 179 mm deep, 285 mm high and weighs 2.9 kg. At the $C/3$ discharge rate, the nominal voltage is 6.4 V and the nominal capacity is 300 Ah. The estimated cycle life is 200–400 cycles at 80% depth of discharge.

The module consists of four series-connected 300 Ah cells providing 6.4 V. The cells may be installed in any essentially upright orientation provided that the 179 mm × 285 mm face is constrained by the battery box or by adjacent modules. A typical 15-module arrangement, connected in series, would yield a 28.8 kWh ($C/3$) total energy content battery at a nominal 96 V.

The cell design uses electrochemically impregnated sintered nickel positive electrodes and rolled-bonded zinc oxide negative electrodes. The separator is a three-part system using proprietary Yardney separators. Factors affecting cycle life and capacity maintenance include depth of discharge, operating temperatures, vehicle design, battery mounting provisions, propulsion system design (that is, operating currents, pulse magnitudes, etc.) and frequency of maintenance.

The nickel–zinc battery is lighter and smaller for a given energy output and is capable of a higher specific power than either the lead–acid or nickel–iron systems and is consequently a prime candidate for a future electric vehicle power source. These features affect vehicle performance beneficially by both extending range and improving acceleration and operation at lower temperatures. The major limitation of the nickel–zinc battery system is its relatively short life. Extensive use of nickel–zinc batteries, particularly as electric vehicle power sources, will depend on significantly extending the deep discharge cycle life of full batteries beyond the current level of 100–300 cycles. The major problem with the nickel–zinc battery arises from the high solubility of zinc oxide, a discharge product of the negative electrode, in the battery's alkaline electrolyte. This results in zinc dendrite formation under charging conditions, as well as shape change and densification of the negative electrode on repeated charge/discharge cycles. In addition, active material may be lost from the negative electrodes through deposition of zinc oxide in the separators and negative electrodes of the battery. Each of these factors both reduces cycle life and is significantly influenced by the separator. Workers at W. R. Grace and Co., battery separator manufacturers, at Maryland, USA, have studied this problem of extending the deep discharge cycle life of nickel–zinc batteries and concluded that cycle life appears to be related to the separator's mass transport properties. Batteries containing separators with the lowest electrolytic resistivity and highest water permeability give the longest cycle life. The results favour the use of microporous separators with an average pore diameter of about 300×10^{-10} cm which provide the necessary mass transfer while retaining adequate dendrite penetration resistance.

For best battery performance, the Yardney nickel–zinc battery charger should be used. This battery charger is basically a voltage-controlled

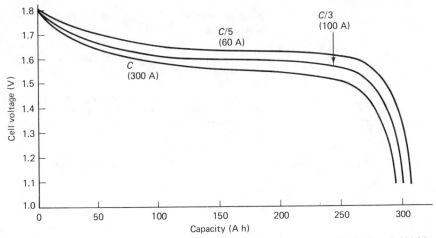

Figure 33.2 Projected performance at various discharge rates for a Yardney nickel–zinc cell, 300 A h capacity (Courtesy of Yardney)

current source with preregulation to minimize internal power dissipation and maximize efficiency. During the initial and final charge rate step, the unit is a constant-current charger. When charge is initiated, the battery is charged at a high rate (60 A) until the gas flow from a single pilot cell reaches a specific level, at which point the charger automatically switches over to a lower (20 A) final charging rate. The charge is continued at this rate until gas evolution reaches a second specific level, at which time the charge is automatically terminated. An SCR (thyristor) located at the output prevents both reverse charging from the battery and charger destruction from improper charger connection to the battery. The unit incorporates battery overvoltage protection. Charger dimensions for a 96 V battery charger are approximately 508 mm deep, 483 mm wide and 711 mm high; it weighs approximately 63.5 kg.

Typical curves of projected performance for the Yardney nickel–zinc battery as a function of discharge rate and temperature are shown in Figure 33.2. If the battery is to be operated in severe hot or cold climates, performance is affected (Figure 33.3) and a suitable cooling or heating system might be advisable to restore full performance as required by the user. The discharge and charge curves of the 75 A h battery used to make the above module are shown in Figures 33.4 and 33.5.

33.2.2 Load levelling

The installation of batteries on electric utility networks would permit the utilities to store energy generated at night by coal-burning or nuclear base-load plants. The battery could then discharge energy to the network during the day when demand is highest. This would reduce the need for gas- or

oil-burning turbine generators. The successful development of batteries for either application would reduce dependence on foreign sources of oil and also have beneficial effects on overall fuel costs.

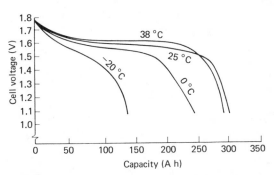

Figure 33.3 Projected effect of discharge temperature of a Yardney nickel–zinc cell, 300 A h capacity (Courtesy of Yardney)

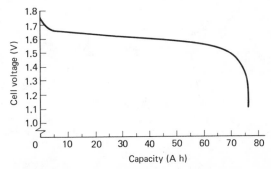

Figure 33.4 Typical discharge curve of a Yardney NZ EV3-XPI prototype bonded nickel–zinc cell, 75 A h nominal capacity, 25°C at 15 A discharge rate (Courtesy of Yardney)

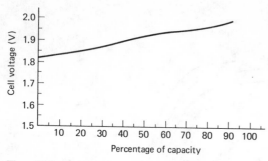

Figure 33.5 Typical charge curve (voltage *versus* percentage capacity) of a Yardney NZ EV6 (FIAT) nickel–zinc cell at 25°C and 25 A charge rate (Courtesy of Yardney)

33.3 Nickel–hydrogen secondary batteries

Potential future applications for this type of battery include electric traction, electric vehicle propulsion and utilities load levelling.

Eagle Picher (USA) are developing the newest and most advanced satellite power systems. This new battery is the nickel–hydrogen system, which offers more energy, longer life and greater reliability than the nickel–cadmium system. For further discussion of this type of battery, see Sections 4.19 and 5.1.

34

Silver batteries

Contents

34.1 Silver–zinc primary batteries

This type of battery is available from several suppliers. Varta, for example, produce a range of 1.55 V silver–zinc cells in the capacity range 45–175 mA h and a 6 V battery of capacity 170 mA h (Table 34.1). These are used mainly in calculators, hearing-aids and photographic applications.

Varta also produce a range of 1.5 V low-drain silver–zinc cells for electric and electronic watches with analogue or LCD display in the capacity range 50–180 mA h; and 1.5 V silver–zinc high-drain cells in the 38–190 mA h capacity range for use in solid-state watches with LED or LCD display with backlight.

Silver–zinc batteries are used, in addition, in photoelectric exposure devices, flashlights, instruments, reference voltage sources and missile telemetry, space launch vehicles, aerospace, portable equipment and military applications.

Other suppliers (Eable Picher, Silberkraft) produce remotely activated silver–zinc batteries for applications such as torpedoes, rockets, missiles and pyrotechnic devices, ignition and fuse applications, and emergency standby batteries.

Eagle Picher supply gas-generating pyrotechnic devices, which can be used for the activation of remotely activated batteries by transferring electrolyte from a reservoir to the battery cells. Eagle Picher have developed over 70 gas-generator configurations with capabilities ranging from 20 to 200 000 cm^3 (standard condition gas output), many of which have been used in a majority of missile/spacecraft systems.

The Chloride of Silver Dry Cell Battery Co. of Baltimore supply a silver–zinc battery with a mild acid electrolyte containing silver chloride as a depolarizer. This battery is recommended as a power unit in test instruments such as blasting galvanometers, circuit testers, volt-ohmmeters and clocks. Multi-cell batteries for B-voltages and bias are available for use in electronic circuits, electromedical instruments and other specialized low-drain uses where constant voltage and long life are important. The battery has an open-circuit voltage of 1.05 V. Under load, terminal voltage drops in proportion to load, but holds constant throughout the useful section of the discharge curve. Shelf life is several times longer than for ordinary dry cells at room temperature. However, the cells should not be stored below 0°C. Shelf life is 6 months under correct storage conditions and this may be extended by cold storage.

34.2 Silver–zinc secondary batteries

Some, now commonplace, applications for silver–zinc batteries are:

1. *Television and broadcasting* Electronic news gathering cameras, video tape recorders, portable lighting.
2. *Portable industrial instrumentation* Ultrasonic detectors, portable court recorders, portable thermocouple sensors, portable inventory devices, portable methane detectors, portable miners' lamps, portable crane control, portable oxygen mask testers, portable underground radios, portable floodlights, portable measuring and control instrument converters, emergency power supply of data-processing machines, flash and film lights in cameras.
3. *Medical instrumentation* Portable ambulance emergency transceivers, heart action recorders, portable organ transfer and storage devices, hand-held physicians' examining lights, portable cystoscopes, portable emergency operating-room lighting, portable defibrillators.
4. *Oil and gas pipeline and exploration* Power for instrumentation for examination devices, underwater sleds for external pipeline and

Table 34.1 Silver–zinc cells and batteries supplied by Varta

Varta type No.	Voltage (V)	Capacity (mA h)	Internal resistance (Ω)	Dimensions			IEC reference	Main application	Varta order No.
				Diameter (mm)	Height (mm)	Weight (g)			
V 8 GS	1.55	45	20.0	11.6	2.1	0.9	–	Calculators	4173
V 13 HS	1.55	75	4.2–6.5	7.9	5.4	1.1	SR 48	Hearing-aids	4012
V 10 GS	1.55	85	10.0	11.6	3.1	1.4	–	Calculators	4174
V 41 HS	1.55	120	3.2–5.0	11.6	4.2	1.8	SR43	Hearing-aids	4041
V 12 GS	1.55	130	6.0	11.6	4.2	1.9	SR 43	Calculators	4178
V 76 HS	1.55	175	3.5–5.5	11.6	5.4	2.4	SR 44	Hearing-aids	4076
V 76 PX	1.55	175	3.5–5.5	11.6	5.4	2.4	SR 44	Photography	4075
V 13 GS	1.55	175	3.5–5.5	11.6	5.4	2.4	SR 44	Calculators	4176
V 28 PX	6.00	170	15.0–22.0	13.0	25.8	12.8	–	Photography	4028

drilling rig base examination, down-hole instrumentation power, heated diving suits.

5. *Oceanographic exploration* Commercial submersibles, underwater seabed surface mining devices, underwater lighting portable cameras, buoys.
6. *Computers and microprocessors* Portable power for microprocessors and data banks, computerized portable weighing scales, portable paddles in stores for inventory of stock fed to computer.
7. *Vehicles* Power for electric vehicles.
8. *Hobbies* Remote-controlled model planes, ships, trains and vehicles.
9. *Aerospace and communications* Balloon flight-powering instrumentation, rocket instrumentation, stabilized aerostatic balloons (tethered), portable telephones in future computerized systems, portable heavy-duty radios, sonobuoy batteries, telemetry, energy storage for solar converters, launchers (satellites), planetary probes, energy systems, helicopter and aircraft batteries, portable receivers and transmitters, emergency systems for emergency lighting and power at airports.
10. *Military* Torpedoes, submarines, noiseless-running combat vehicles, rocket power supply, radio transmitters, night vision equipment.

34.2.1 Specialized applications

Some silver–zinc batteries are used in military applications; some other specialized applications are described below.

Space applications

Secondary (and high-rate primary) batteries were specially developed for the US Space Administration (NASA) Apollo ILM Saturn programme. A total of 24 secondary and primary silver–zinc batteries were used on each Saturn V vehicle. The Moon Buggy or Lunar Rover used for driving on the surface of the moon in 1971 was powered by two manually activated secondary silver–zinc batteries. This vehicle reached speeds of 5–6 miles/h. The Agena, a workhorse launch vehicle/satellite since 1959, is powered with silver–zinc batteries. When originally developed, in 1959, these batteries had a power density of 36 W h/kg. This has since been increased to 53 W h/kg.

Electric vehicles

Eagle Picher have developed an electric car, the Silver Eagle, utilizing 174 space-type secondary silver–zinc batteries as the power source. These tests were conducted in 1971 and 21 world speed records were broken, most of which stand today.

The Silver Eagle weighed 517 kg, including 118 kg of batteries. The vehicle was 409 cm long, 85 cm high, 129 cm wide and featured a wheel base of about 278 cm overall. Basically, the car was designed as a dragster. The engineering that went into the design of the car produced a very tangible result in that the unique speed control system was later to become part of golf or recreational vehicles manufactured by this company. Electric cars powered by silver–zinc batteries have achieved ranges of up to 350 miles in one discharge, making them a very attractive alternative to lead–acid batteries, which give a maximum range of about 70 miles.

A motorbike powered by a silver–zinc battery has broken the electric bike world speed record by achieving 165.3 miles/h. The battery, developed by Eagle Picher, weighed 40 kg, and had a nominal capacity of 100 A h and an energy content of 4680 W h.

Submarines

Yardney quote the following example of the use of silver–zinc batteries in submersibles in the place of batteries conventionally used in this application. A silver–zinc battery system designed to replace the lead–acid battery in the US Navy's turtle (DSV-3) would make the difference shown in Table 34.2.

Table 34.2

	Lead–acid battery	Silver–zinc battery
Volume (m^3)	0.78	0.78
Weight (kg)	1735	1643
Endurance at 1.5 knots (h)	10	31
Endurance at 2.5 knots (h)	1.9	6.07

The critical profit parameter in operating submersibles is the cost per hour while submerged. If the following assumptions are made, it would appear that, on cost considerations, the silver–zinc battery is more attractive than the lead–acid battery:

1. Cost of lead–acid battery, $15 000.
2. Cost of replacement of silver–zinc battery, $60 000.
3. Incremental cost of silver–zinc battery, $45 000.
4. Daily operating cost of the total system, $2200.
5. Daily incremental cost of silver–zinc battery (2 year life), $65.

With a lead–acid battery, a submersible can explore for 1.9 h at 2.5 knots for a distance of 4.75 miles. Thus, the cost per hour submerged is $1158. With a silver–zinc battery, submersibles could explore 6.07 h at 2.5 knots for a distance of 15.2 miles. Thus the cost per hour submerged is $373, and the potential saving is $785/h.

34.3 Silver–cadmium batteries

Applications for primary silver–cadmium batteries include oceanographic buoys, torpedoes, missiles, underwater scooters, portable field equipment balloons, sonar devices, planetary probes, and various civil and military applications.

Secondary batteries have a lower power density than silver–zinc batteries but have a greater cycle life. They are recommended, therefore, in applications where cycle life is paramount over total weight. Applications include underwater vehicles and torpedoes (rockets/missiles), target drones, flashlights, underwater scooters and portable electronic equipment, balloon experiments, oceanographic buoys, and various commercial applications (portable televisions, garden equipment, etc.).

35

Alkaline manganese primary batteries

The additional cost of the alkaline manganese construction, as opposed to the carbon–zinc type of battery, is usually not warranted for intermittent discharges, despite its improved performance. For example, comparing an HP2 carbon–zinc battery with an Mn1300 (IEC LR20) alkaline manganese cell on a five 2 h/day discharge to an end-voltage of 1.1 V, the Mn1300 gives twice the discharge life but costs three times as much as the HP2. Only on continuous high-rate discharges does the alkaline manganese system have an undoubted economic advantage. It is also recommended for low-temperature discharges.

In radios, alkaline manganese batteries usually last twice as long as standard carbon–zinc batteries. In battery-powered toys, alkaline manganese batteries last up to seven times as long as standard carbon–zinc batteries. Alkaline manganese batteries are excellent for photo-flash applications. In addition to high amperage, they have more energy than standard carbon–zinc photo-flash batteries. Some electronic flash units use transistor or vibrator circuits in a converter to change low-voltage d.c. into the high voltage necessary to charge the flash capacitor. The current drains involved strain the capabilities even of high-amperage photo-flash batteries. The alkaline manganese battery provides both a sustained short recycling time and two to three times as many flashes as carbon–zinc photo-flash or general-purpose batteries. This is because of the unusual cell construction, which provides a very low internal resistance, such that the battery delivers its energy faster than standard carbon–zinc types. The characteristics of these batteries make possible the development of equipment which up to this time had been thought impractical because of the lack of a suitable power source.

Alkaline manganese primary batteries are ideally suited for use in many types of battery-operated equipment, for example heavy-duty lighting, transistor radios (particularly heavy current drain), electric shavers, electronic photo-flash, movie cameras, photography, radio-controlled model planes and boats, glo-plug ignition (model planes), toys, cassette players and recorders, any high-drain heavy-discharge schedule use, alarm systems, roadside hazard lamps, radio location buoys, pocket calculators, hearing-aids, light meters, measuring instruments.

The choice of battery type from the wide variety available for a particular application undergoes distinct changes as new types of battery are developed over the years. Consider, for example, the use of primary batteries in cameras, which goes back two decades and is now accepted at all levels of photography. Modern cameras have numerous electrical functions, with loads ranging from the 20 min rate in built-in direct-fire flash bulbs to the 1 year plus rate found in exposure meters under dim lighting conditions. The old distinctions between meter, motor and flash batteries has become diffuse, with power now often supplied from a central source and subsequently stabilized to suit the different functions. The following camera functions may be electrically operated:

1. Film transport.
2. Light meter.
3. Aperture.
4. Shutter.
5. Focus.
6. Focal length.
7. Flash.
8. Low light warning.
9. Sound.

Table 35.1 shows that the number of camera applications has increased more than seven times since 1965. Note that alkaline manganese overtook

Table 35.1 Increase in number of applications of different types of primary battery in cameras since 1965

	Approximate launch year									
	Up to 1965	1966	1967	1969	1971	1972	1974	1976	1978	1980
New applications										
Alkaline manganese	173	82	42	107	172	135	312	313	385	447
Mercury–zinc	185	104	30	97	112	120	253	168	138	104
Silver oxide–zinc	0	3	0	6	7	21	46	68	118	160
Carbon–zinc	0	0	0	2	16	0	5	2	0	0
Total	358	189	72	212	307	276	616	551	641	711
Cumulative number										
Alkaline manganese	173	255	297	404	576	711	1023	1336	1721	2168
Mercury–zinc	185	289	319	416	528	648	901	1069	1207	1311
Silver oxide–zinc	0	3	3	9	16	37	83	151	269	429
Carbon–zinc	0	0	0	2	18	18	23	25	25	25
Total	358	547	619	831	1138	1414	2030	2581	3222	3933

Table 35.2 Applications growth rates for different types of primary battery in cameras since 1966

	1967	1969	1971	1972	1974	1976	1978	1980
Alkaline manganese	16.5	18.0	21.3	23.4	21.9	15.3	13.0	12.0
Mercury–zinc	10.4	15.2	13.5	22.7	19.5	9.3	5.5	4.0
Silver oxide–zinc	–	–	–	–	56.0	41.0	27.5	25.0

mercury batteries in terms of number of applications in 1971 and are now well ahead. Silver oxide batteries are not very significant, but are rapidly gaining strength.

Table 35.2 shows applications growth rate by system. Three points are of interest:

1. The growth rates for both alkaline and mercury batteries peaked between 1972 and 1974.
2. The growth rate of mercury battery applications is declining more rapidly than that of alkaline batteries.
3. The growth rate for silver oxide batteries, although declining, is significantly higher than for either mercury or alkaline batteries.

Figure 35.1 shows that 13 out of 14 of the growth rates listed in Table 35.2 are consistent with long-term effects, which means that it is possible to extrapolate the curves with a reasonable assurance of accuracy. The projected growth rates and the consequent new applications are added to Table 35.1, bringing the figures up to 1980.

Detailed analyses carried out by manufacturers forecast that, as far as camera applications are concerned, the popularity of alkaline batteries with camera designers will continue to increase, and so will the popularity of silver oxide. The corresponding decline in applications support for mercury batteries is already drastic, falling at roughly the same rate up to 1980, when only one in seven new camera applications required a mercury battery. One of the primary causes is not lack of battery performance but the developing worldwide anti-mercury lobby.

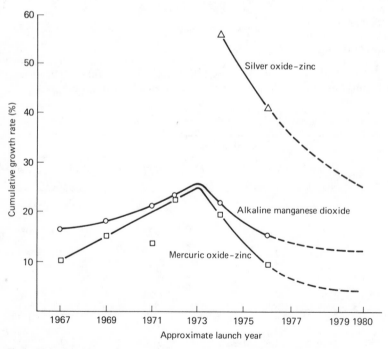

Figure 35.1 Cumulative growth rate of photographic applications of primary batteries, 1867–1980

36

Carbon–zinc primary batteries

Contents

These batteries retain the great majority of the primary battery market for portable-in-use applications. They are widely available in a variety of equivalent cylindrical sizes.

The earlier carbon–zinc system has the property of 'recovering' during rest periods in the discharge regimen. This property and the low cost resulting from high-volume automated manufacture make the earlier carbon–zinc systems ideal for the intermittent usage pattern found in most consumer appliances. Figure 36.1 shows the effect of increased current drain on service life of the SP2 (R20 size) battery. The discharge period is 5 h/day to an end-voltage of 0.9 V.

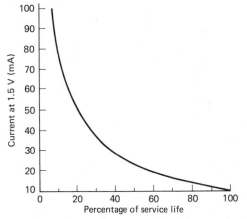

Figure 36.1 Effect of increased current drain on service life of SP2 (R20) Leclanché cells

For higher current drains, for example in motorized equipment, a high-power (HP) battery should be used. HP batteries contain electrolyte or chemically prepared manganese dioxides, which have enhanced discharge properties. Figure 36.2 compares SP11 and HP11 (R14 size) types on a 10 Ω 2 h/day discharge; the advantage of using the

high-power type is obvious. For low-rate discharges, the additional cost of the HP version is not warranted; Figure 36.3 compares the same cells on the 300 Ω 2 h/day discharge.

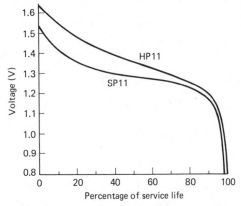

Figure 36.3 Low discharge effect. SP11 and HP11 (R14) alkaline manganese batteries: effect of using a high-power battery (HP11) on service life at high current drain (300 Ω, 2 h/day discharge)

Transistor applications require multiples of the unit cell voltage; the layer stack ('power pack') format provides a convenient battery in a smaller volume than the equivalent combination of round cells. Only one unit needs to be replaced, and connection is by simple non-reversible press studs ensuring correct polarity. Selection of the most economical layer stack battery should be made in conjunction with the battery manufacturer. Specification of a battery which is too small for the current drain can lead to unnecessarily frequent battery replacement. Figure 36.4 shows the comparative service lives of two 9 V power packs, PP9 and PP6, on a 450 Ω, 4 h/day discharge. Table 36.1 shows the most widely used batteries, in typical applications

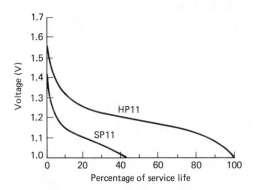

Figure 36.2 Heavy duty effect. SP11 and HP11 (R14) alkaline manganese batteries: beneficial effect of using a high-power battery (HP11) on service life at high current drain (10 Ω, 2 h/day discharge)

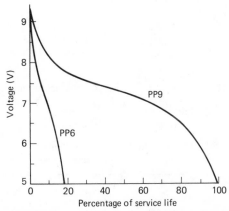

Figure 36.4 Size effect of Leclanché batteries: comparative service lives of two 9 V power packs, PP9 and PP6, on a 450 Ω, 4 h/day discharge

Table 36.1 Typical applications, load, voltage under load and amperage reading for testing of the most widely used carbon–zinc Leclanché cells and batteries

ANSI designation	Application	Load (Ω)	Voltage under load (% of applied voltage shown on cell or battery label)	Amperage (A) for usable cell or battery (test for an instant only)
AAA	Penlights	3.0	0–61% replace 61–71% weak 72–105% good	
AA	Toys, clocks, flashlights, movie cameras	2.25	0–61% replace 61–71% weak 72–105% good	
AA	Transistor radios	10.0	0–61% replace 61–71% weak 72–105% good	
C	Toys, clocks, flashlights, movie cameras	2.25	0–61% replace 61–71% weak 72–105% good	
C	Transistor radios	10.0	0–61% replace 61–71% weak 72–105% good	
D	Toys, clocks, flashlights, movie cameras	2.25	0–61% replace 61–71% weak 72–105% good	
D	Transistors	10.0	0–61% replace 61–71% weak 72–105% good	
N	Penlights	3.0	0–61% replace 61–71% weak 72–105% good	
6 V	Lanterns	5.0	0–61% replace 61–71% weak 72–105% good	
9 V	Transistor radios	250.0	0–61% replace 61–71% weak 72–105% good	
AAA	Photography	0.01 (entire resistance, lead-to-lead of tester)		1–3.8
AA	Photography	0.01 (entire resistance, lead-to-lead of tester)		3–6.5
C	Photography	0.01 (entire resistance, lead-to-lead of tester)		4–9
D	Photography	0.01 (entire resistance, lead-to-lead of tester)		10–13.5

and details of applicable loads, voltages and amperage readings for testing.

Other applications for carbon–zinc batteries of various voltages are as follows:

1. 1.5 V   Gas lighters, photographic, radio valves, transistorized clocks, calculators, lighting, electronic watches, hearing-aids, toys, transistor radios, glo-plug ignition, instruments, alarms.
2. 3 V   Lighting, radios, telephone.
3. 4.5 V   Alarms, emergency lighting, signalling, radio valves.
4. 6 V   Lanterns, roadside hazard, tape recorders, emergency lighting, ignition, road counting meters, marine handlamps, toys, instruments.
5. 7.5 V   Fences, emergency lighting, ignition.
6. 9 V   Calculators, frequency meters, transistor radios, emergency lighting.
7. 12 V   Alarms (intruder and fire), lighting, signalling, portable fluorescent lights, communication equipment, flashes, electric fence controllers, electric motors, portable televisions, toys.
8. 15–30 V   Photographic, wide-range oscillators, radiopaging, hearing-aids, portable applications, instrument operation, electronic cigarette lighters.
9. 60–90 V   Portable applications, lighting.
10. 225–500 V   Electronic flash, geiger counters, mass spectrometers.

36.1 Comparison of alkaline manganese and carbon–zinc cell drain rates

In general, devices using lights or motors will have high drain rates. The bigger the job to do (i.e. the brighter the light, the greater the load the motor is driving), the greater the drain rate will be. Another general point is that when any device is used continuously it will exhaust the energy of the cells much more rapidly, since the cells are deprived of a chance to rest and recover some of their energy.

Alkaline manganese dioxide cells have a superior performance over carbon–zinc cells at various drain rates. The benefits to be derived from alkaline batteries depend, of course, on the type of equipment and how it is used. Generally, the higher the drain rate of the equipment and the heavier the use, the greater the margin of superiority of alkaline cells. They are the ideal power source for calculators, photo-flash units, toys, flashlights, cassette recorders and other heavy-drain equipment. No matter what kind of battery-operated equipment is used, if it runs on regular carbon–zinc cells, it will always run longer on alkaline manganese cells.

Another point worth mentioning is the ability of alkaline manganese cells to deliver a constant level of power with high efficiency under continuous or heavy duty, high-drain conditions where carbon–zinc cells are unsatisfactory. In such applications as flashlights and cassette recorders, reduced power could result in dim lights and reduced tape speeds, respectively.

36.2 Drain characteristics of major consumer applications

36.2.1 Flashlights and lanterns

Because they obviously involve lightbulbs, these are an application with a relatively high drain rate, with active drains running anywhere in the range of 250–500 mA. Alkaline systems are proving especially suitable to lighting applications because of their long storage life and dependable performance.

When flashlights are used intermittently, as by a night watchman making his rounds and turning the light on occasionally to check a meter or a lock, the drain is the least severe and the alkaline cell would have only a small advantage. However, in situations of an emergency nature (power shortages, flat tyres, etc.) or when the light is used for longer periods (on night hikes or camping trips), the alkaline cell provides significantly greater performance. When an ordinary 2D cell flashlight is used continuously, the alkaline-powered light can last 18–20 h or more while ordinary carbon–zinc flashlight cells will only last 3–4 h.

Another area where alkaline cells have an even greater advantage is in the smaller penlight flashlights. The advantage for alkaline cells applies to smaller cells to an even greater degree. Also, since flashlights may be called on to do their job only once or twice a year, shelf life can be a very important factor to the user. Alkaline cells have distinct advantage in this respect and this is the reason why the smaller carbon–zinc flashlights have never become very popular.

36.2.2 Radios

Radios generally involve the lowest current drain of all the major categories of application. The drain of the more typical or popular radios runs from 10 to 20 mA. The bigger transoceanic or AM/FM radios that are becoming more and more popular can run higher than this, at about 50 mA. Also, if the volume of the radio is turned up very high, the drain will be higher. Again, the greater the power output, the greater the required drain rate. Of course, the life of cells in radios is dramatically affected by how the radio is played, but the generally lower drain rates will assure many more hours of use than are available for flashlights. Some radios can be played continuously for more than 100 h with alkaline batteries. However, in this application, the advan-

advantage for alkaline is a matter of degree, running from a low of about 2:1 to a high of about 4:1 over ordinary carbon–zinc radio cells.

36.2.3 Tape recorders

Tape recorders are becoming an increasingly important usage, and currently account for between 10 and 15% of the total replacement cell market. The sound reproduction in a tape recorder is, of course, much the same as in a radio, but the added requirement of a motor to turn the spindle for the tape adds quite significantly to the drain. As a result, the drain rate in tape recorders runs between 100 and 200 mA, which explains why they are a popular application for alkaline cells.

36.2.4 Toys

There is, of course, a wide variety of battery-powered toys and the newer electronic games. Generally speaking, the function of a cell in a toy or game is to provide the power for either a light or a motor that causes the toy to move in one way or another. For this reason, toys or games in general have a very high drain rate, running anywhere from 50 to 1000 mA (1 A). Many devices of this type would be at the high end of this scale, where the superiority of alkaline cells is most dramatic. At these drain rates they out-perform ordinary carbon–zinc cells by as much as 10:1.

36.2.5 Calculators

Some pocket calculators on the market today still employ LED (light emitting diode) display. This is the same basic circuitry or chip used in the new LED watches. A considerable amount of energy is required to power these displays when they are in use, and a relatively high drain rate results. The drain rate of the LED calculator runs from 30 to 80 mA.

Some of the newer calculators employ the LCD (liquid crystal display), which does not light up; therefore, the drain rates are dramatically lower than with LEDs. Drain rates on LCD-type calculators can run as low as 0.05 mA (50 µA). The drain rates in these new LCD-type calculators are so low that any type of cell would last for a considerable period of time. In fact, it is likely that shelf life may be a factor of equal importance to service life in these applications. Since smallness in size is the key with portable calculators, the smaller cells are generally involved. This will lead to the use of alkaline-type button cells or other new shapes. In addition, the new low-drain lithium cells are becoming a serious contender here because of their extremely long shelf life.

However, since the majority of calculators on the market or in use today employ LED display, they present ideal applications for alkaline cells. Con-

sumers are discovering this advantage very quickly as over 40% of the cells used in calculators are alkaline. The advantage of alkaline cells over carbon–zinc cells in these types of calculators runs as high as 6:1.

36.2.6 Cameras

In cameras, cells perform a wide variety of functions, including discharging flash units, motor powering, shutter control and film ejection. Obviously, the drain rate for these different functions varies significantly. The higher drain rates again apply to those applications involving motors or lights (flash). The drain rate in movie camera applications runs from 250 to 500 mA. Cameras with cell-operated flash require up to 750 mA. Electronic shutters require up to 90 mA.

While most applications in cameras have sufficient drain rates to justify alkaline (or mercury or silver) cells, there are other reasons why most camera manufacturers design their photographic equipment almost exclusively around this system. Their superiority in resisting extreme temperature variations is perhaps the most important of these, since cameras are obviously used extensively outdoors in all kinds of weather. The shelf life advantage is also important, as cameras are often left for long intervals between use.

In flash applications, the sharp demands in drain rate over very short time periods can only be handled by premium cells, as carbon–zinc cells simply could not respond to them.

36.2.7 Hearing-aids and watches

Unlike most conventional battery applications, hearing-aids and watches require miniature specialized mercury or silver oxide cells to provide the continuous power supply necessary. While the button cells used in hearing-aids and watches are similar in appearance, they differ in internal construction as they are designed to satisfy various power requirements at different drain rates and they are therefore not interchangeable.

LCD and quartz analogue watches (watches with a conventional face) have drain rates as low as 0.0025 mA (2.5 µA). Generally speaking, drain rates for hearing-aids are moderate, ranging from 0.15 mA to a high of 5 mA. In hearing-aids, as well as in LCD and quartz analogue watches, the drains are continuous. However, LED watches require drains up to 40 mA during the pulse. Thus, the requirements needed to power the displays in these watches are about eight times greater than in hearing-aids and 16 000 times greater than in LCD and quartz analogue watches. For this reason, a wide variety of mercury and silver oxide cells meets the growing demands of the various types of hearing-aids and watches.

37

Mercury batteries

Contents

37.1 Mercury–zinc primary batteries

Mercuric oxide batteries have been widely used as secondary standards of voltage because of the higher order of voltage maintenance and ability to withstand mechanical and electrical abuses (Figure 37.1). For use as reference sources in regulated

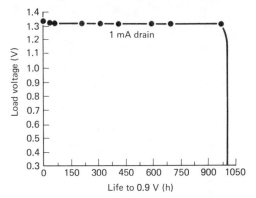

Figure 37.1 Mercury–zinc E1 cell: cell load voltage *versus* time at 1350 Ω at constant load at 21°C

power supplies, radiation detection meters, portable potentiometers, electronic computers, voltage recorders, scientific and military equipment, hearing-aids, electronic wrist watches and cameras, and similar equipment, desirable features are as follows:

1. *Voltage stability* versus *time* The uniform voltage of the mercury cell is due to the efficient nature of the cathode. Over long periods of time, regulation within 0.5–1.0% is maintained; for shorter periods, regulation of 0.1% may be realized.
2. *Short-circuit currents* Momentary short-circuits will cause no permanent damage with almost complete recovery of full open-circuit e.m.f. within minutes.
3. *Heavy load currents* Depending on the type of cell used, high drains without damage can be obtained; recovery to full open-circuit e.m.f. is rapid.

Miniature applications have become more important in recent years with the general acceptance of behind-the-ear hearing-aids and the advent of the electronic watch. High energy density per unit volume is the prime requirement for a battery in these products. The mercuric oxide–zinc, silver oxide–zinc, zinc–air and lithium-based systems appear to be likely contenders for this market. Although the last two types of battery have been produced in sizes suitable for miniature applications, they are not widely available in this format.

The mercuric oxide–zinc (mercury) cell for miniature applications is usually based on the familiar button construction using a compressed cathode of mercuric oxide and graphite (added for conductivity) in a plated steel can. The cell seal is supported by a cathode sleeve, on top of which is placed a synthetic separator and an electrolyte absorbing pad. The electrolyte is a solution of potassium hydroxide. The amalgamated zinc anode is added and the cell sealed with a polymeric gasket and a metal top cap.

The mercury cell has a low internal resistance and high cathode efficiency. Discharge characteristics are substantially flat, which is an obvious advantage for hearing-aid use. Capacity retention of the system on storage is good, with typically 95% of capacity retained for 10 months storage at 45°C and for 2 years at 20°C. Multi-cell batteries using the mercury system are available for applications requiring higher voltages, and some cylindrical sizes are produced. In general, the high cost of the system restricts it to those uses where space is at a premium or where voltage regulation is critical.

In hearing-aids, the current drain may be of the order of 1 mA for a total discharge of several days. The low-resistivity separator is able to retain the droplets of mercury which form as the cathode is discharged. In a watch, the batteries must perform adequately over a period of months or years, and the discharge pattern may be one of short periods of a drain of tens of milliamperes superimposed on a microampere continuous drain. Under these conditions a separator with carefully controlled properties is required to avoid possible mercury penetration and short-circuiting of the cell.

A general review of the types of application found for mercury–zinc batteries available from one particular supplier (Crompton-Parkinson) is found in Table 37.1, together with information on capacities available and battery weights. Voltages vary between 1.35 and 97.2 V and capacities vary up to 28 A h. These batteries are of the button cell or cylindrical design.

37.2 Mercury–cadmium primary batteries

Applications include mines, munitions, buoys, beacons, pyrotechnic igniters, missiles, sealed metering equipment, indicator lights and civil military electronic devices.

37.3 Mercury–indium–bismuth primary batteries

These have found applications in location beacons and memory core standby supply batteries.

Table 37.1 Applications of mercury–zinc primary batteries available from Crompton-Parkinson

Application	Voltage	Capacity range available (A h)	Approx. weight (g)
Photography	1.35–4.0	80–1000	4
	2.7–2.8	250–350	8
	5.4–5.6	100–500	30
Light meters	1.35–4.0	250	4
Flash units	5.4–5.6	110	7
Hearing-aids	1.35–4.0	16–2400	0.3–30
	2.7–2.8	1000	27
Watches	1.35–4.0	50–230	0.8–25
Instruments	1.35–4.0	45–14 000	0.8–174
	2.7–2.8	350	11
	4.05–4.2	350–1000	17–89
	5.4–5.6	500–2200	34–119
	6.75–7.0	250–2200	20–149
	8.1–8.4	1000–2200	81–179
	9.45–9.8	1000–3400	94–329
	12.6	500	76
	27	3400	950
	29.7	250	142
Detonators	1.35–4.0	500	–
Electric eyes	1.35–4.0	220–1000	2.5–12
	4.05–4.12	160–250	7–14
Radio receivers	1.35–4.0	500–2500	5–30
	4.05–4.2	1000–2200	40–90
	5.4–5.6	500	34
	6.75–7.0	180	12
	8.1–8.4	575–750	51–77
	9.45–9.8	215	24
Radio transmitters	97.2	1000	910
Transistor applications	2.7–2.8	1000	27
	4.05–4.2	500	25
	6.75–7.0	350	20
Voltage reference instruments	1.35–4.0	14 000	40
Radiation equipment	1.35–4.0	28 000	380
	6.75–7.0	250	21
	9.45–9.8	250	28
	10.8–11.2	250–3400	37–340
	16.2–16.8	250	51
Geophysical applications	2.7–2.8	2200	57
	47.25	1000	480
Paging devices	4.05–4.2	500	25
Fire alarms	5.4–5.6	2200–3400	128
	6.75–7.0	2400	170
TV tuners	6.75–7.0	500	42
Test equipment	6.75–7.0	2200–3400	170
Depth finders	8.1–8.4	3600	265
Garage door openers	9.45–9.8	160	28
Chart recorders	9.45–9.8	2400	255
Tape recorders	10.8–11.2	3600	400
Transceivers	12.6	750	100
	13.4	1000	–
	15.4	2200	–
	16.2–16.8	2200	370
Laser beam detectors	13.4		
Personal location beacons	13.4		
Solid-state relays	13.4		
Radiomicrophones	13.4		
Radiosondes	13.4		

38

Lithium primary batteries

Contents

Applications of some of the main types of lithium batteries are described below.

38.1 Lithium–sulphur dioxide

Because of their favourable characteristics these cells and batteries have found many applications in the military field and in the operation of electronic components, especially where high energy density, long shelf life and low weight are requirements. Portable radio transmitters, night vision devices, pyrotechnic ignition batteries, buoys, emergency power supply for laser devices, cameras, hand lamps, medical electronics, operation of micro-processors and signal lamps are some of the applications that have been found.

Applications for the lithium–sulphur dioxide reserve systems include underwater mine batteries and (active batteries) memory protection, manpack communications, life-support equipment, sono-buoys, space probes, missiles, mines, security systems, data buoys/stations, weather sondes and electronic counter measures. The non-reserve (active) systems are used for covert sensors, memory protection and weather sondes.

38.2 Lithium–vanadium pentoxide

There are two main applications for this type of cell. Reserve batteries are for applications requiring extremely long shelf and operational life (10–30 years) and relatively low discharge rates. The non-reserve types are used for applications such as covert sensors, implantable medical devices, watches, currency trackers, memory protectors and frangible weather sondes.

38.3 Lithium–thionyl chloride

Applications for lithium–thionyl chloride cells and active batteries include all those for which active lithium–sulphur dioxide cells and batteries are recommended, and also high-temperature applications.

Eagle Picher recommend their active lithium–thionyl chloride cells for micropower sources in advanced application areas, e.g. airborne instrumentation, undersea communications, mineral exploration, remote site monitoring safety controls, security, and space and/or defence systems.

Eagle Picher reserve batteries have found many applications in military and aerospace applications such as missiles and re-entry vehicles. They have also been used in mines.

38.4 Lithium–manganese dioxide

With the microprocessor involvement in all kinds of applications, the lithium–manganese dioxide battery is expected to become extensively used by the end of the 1980s. Possible applications include LCD watches, pocket calculators, CMOS memory protection, measuring instruments, electronic alarms and pace setters. The cells may either provide backup for memory retention in fixed installations or act as a side or part power source in portable equipment.

SAFT supply a range of 3 V button cells recommended for use in LCD watches, pocket calculators, CMOS memory protection measuring instruments and electronic alarms. These cells have an energy density of $400–500\,W\,h/dm^3$ (120–150 W h/kg) and an annual self-discharge rate of 2–3%. The capacities vary from 200 mA h (type LM 2425) through 90 mA h (type LM 2020) to 50 mA h (LM 2016). The temperature operating range is −20 to +50°C.

Varta also supply a range of 3 V lithium–manganese dioxide button cells ranging in capacity from 3 to 1400 mA h.

Lithium–manganese dioxide cells are a strong contender for CMOS data retention applications. The main drawback of lithium–manganese dioxide cells is that they are susceptible to damage from even small reverse currents and therefore require a more careful selection of isolation diodes. One maker cautions that the isolation diodes must have less than 10 nA leakage. Because of the tight specification on the isolation-diode leakage and the limited temperature specifications for the material, lithium–manganese dioxide is not an ideal choice for high-temperature applications.

38.5 Lithium–copper oxide

The lithium–copper oxide battery has two special characteristics. Its initial voltage is approximately 2 V, but when fully loaded it drops to a flat 1.5 V. Thus it can be used as a direct replacement for carbon and alkaline cells. (Most other lithium batteries have voltages from 2.8 to 3.5 V.) Lithium–copper oxide also has a wide temperature range, spanning −55 to +125°C. Although it is not the highest performance lithium combination, its voltage and temperature characteristics permit applications in specific niches.

The lithium–copper oxide battery has applications in power well lagging equipment, microprocessors, public address systems, high-temperature heat counters, CMOS memory protection, telex systems, high-temperature devices, and specialized industrial and military applications.

38.6 Lithium–silver chromate

These batteries are used in cardiac stimulators, other human implantable devices, high-reliability memory protection, inertial navigation systems and RAM backup power sources.

38.7 Lithium–lead bismuthate

These are used in analogue quartz watches and pulsing electronic devices.

38.8 Lithium–polycarbon monofluoride

These batteries are suitable for use in applications that require high-rate pulses of power, in low-current continuous drain applications and as a stable long-term reference voltage. They provide a high energy density when compared with other battery chemistries. These batteries are designed for applications requiring micropower sources in advanced application areas such as airborne instrumentation, undersea communications, mineral exploration, remote site monitoring and safety controls.

While lithium–iodine batteries suit low-current applications, the lithium–carbon monofluoride is becoming economically attractive for both low- and high-current requirements. The cost factor was improved when Panasonic convinced Eastman Kodak that this chemistry should be used in the mass-produced Kodak disc camera. In this camera, cylindrical cells, slightly smaller than AA size, produce the high 1.5 A pulses needed to recharge the flash and run the motor during automatic picture taking. This type of cell is not only being used in consumer products. Indeed, configured as small button- or coin-sized cells, lithium–carbon monofluoride batteries are being used by at least one computer manufacturer for CMOS-RAM data retention. The button cells are small enough to fit right on top of RAM dual-in-line packages and these button cells cost only 69–90 cents.

This lithium-based composition also has its disadvantages. Batteries are packaged with crimped rather than welded seals and are therefore subject to leakage. As a result, UL requirements dictate that a 1 kΩ series resistor be used to limit current if load short-circuiting does occur, or if an isolation diode fails to block forced charging from the main power source. However, according to Panasonic, the lithium–carbon monofluoride chemical combination forms a safe mixture; even if it does leak out past the seals, it is non-toxic. The wide distributions and use of the Kodak disc camera, using two UL-listed 1200 mA h cylindrical cells (Panasonic model BR2/3A) connected in series, is cited as proof that this chemistry will not pose regulatory problems for designers. Panasonic warn prospective users, however, that there can be regulatory restrictions for air shipment on passenger-carrying aircraft if the product has more than two cells connected in series.

38.9 Lithium solid electrolyte

These have a 15–20 year life, which is compatible with the design life of microprocessors, making it possible to incorporate these batteries as a permanent component of the equipment. The closely matched temperature characteristics of solid electrolyte batteries and CMOS logic circuits is a property that makes this type of battery ideally suited for memory protection applications. Basically, these batteries are suitable for any low-power long service life application.

38.10 Lithium–iodine

Many lithium cells are now available to the watch and calculator industry, but the designer should take care in selecting the right lithium cell for the application. Factors such as reliability, energy density, internal resistance and hermeticity should be considered. Since almost all lithium cells work best with a large area package, Catalyst Research Corporation propose standard button cell sizes of three diameters, 19, 23 and 27 mm, and a range of thicknesses from 1 to 3 mm in 0.5 mm steps. Many advantages will become apparent to the design engineer.

The lithium–iodine cell is an example of a highly reliable lithium power source, proven in the pacemaker industry. It is used in security, space and defence systems and, more recently, CMOS-RAM data retention applications. It is available in several sizes, suitable for watches (LCD and analogue stepping motor) and for low-current LCD calculators. Life spans as long as 10 years are possible in watches. In calculators, 5000 h operating times can be obtained over a 10 year period; for example, a 0.27 A h battery on a 50 mA drain would deliver 5000 h of service over a 10 year period.

More recently, Catalyst Research Corporation have issued a preliminary specification sheet for a 7.5 A h oil well logging battery (model 407220, Military specification D-cell). This is housed in a 38 mm × 57 mm circular container and is hermetically welded with a glass–metal compression seal. It has a voltage of 2.8 V (open circuit 2.8 V), a maximum operating temperature of 150°C and a self-discharge rate of about 5% in 10 years at 25°C. The prototype D-34 cell maintains its voltage of 2.8 V during 6 weeks continuous operation at 10 mA constant current at 130°C.

Also available is the model CRC Li D 2.8 D-type cell (2.8 V) with a capacity of 14 A h at 25°C and an operating temperature of −20 to +50°C. The battery weighs 80 g and has a height of 33.8 mm and a width of 57.1 mm. This cell suffers a capacity loss of less than 5% in 10 years at 25°C and can withstand substantial shock, vibration and short-circuiting at temperatures up to 150°C without venting, leaking, swelling or exploding. The cell continues operating at 2.8 V (25°C) for 18 months (1 mA) and 2.7–2.8 V for about 8 months (2 mA).

Of all the lithium compositions, lithium–iodine is considered by its inventor, Catalyst Research Corporation, to be a completely solid-state cell (although the electrolyte is purported to have a 'tar-like', rather than rock-hard, consistency).

The lithium–iodine combination has been used for most implanted cardiac pacemakers since 1972. It thus has a believable reliability history, at least at human body temperature. Because of its critical main use, the lithium–iodine cell comes in a hermetically sealed (laser-welded) nickel or steel container, which makes it more expensive than other types ($5–8 for the 200–350 mA h coin-shaped cells). Yet the high quality of the chemical processing and encapsulation makes it a good choice where very long shelf life (10 years and beyond) is the objective.

Besides its higher cost, the main disadvantage of the lithium–iodine system is that the current output is small; a maximum of only 50 μA is recommended at room temperature. Even with this current level, however, one of these cells has enough capacity to back up 65 kbytes of CMOS-RAM. In addition, these cells withstand a considerable amount of overloading, and even dead shorts, without suffering damage.

According to Catalyst Research Corporation, in the case of a continuous dead short, the battery cells are protected by the formation of a film, which raises the internal impedance, thereby limiting current. This self-limiting effect makes lithium–iodine safer than those lower impedance chemistries that heat up and, in the larger sizes, even explode when shorted.

Another important characteristic of a lithium–iodine cell is its sloping discharge curve over time; the voltage drops approximately 0.1 V every few years over a 10 year period for a current drain of a few microamperes. Many designers, however, might not consider this characteristic a disadvantage; it provides an easy means of monitoring the remaining life.

38.11 Comparison of lithium–iodine and nickel–cadmium cells in CMOS-RAM applications

Applications of CMOS-RAM (complementary metal oxide random access memory) are growing at a rapid pace. Manufacturers are using these memories with increasing frequency in industrial process controls and communications, energy management and security systems, consumer electronics, data loggers and computer peripherals such as teleprinters and word processing equipment, among others. Advantages of CMOS-RAM memories include relatively high data volumes, relatively fast reading and writing, easy reprogramming, low power requirements for memory operation and retention, comparatively low cost, and immunity to high noise.

Unfortunately, CMOS-RAM memories suffer one major drawback – they are susceptible to memory losses during power fluctuations or outages. This memory volatility dictates the need for some type of backup power source.

This could be accomplished by backing up the entire system with a large uninterruptible power supply which, in many cases, would be impractical because of the small size of many systems. A much more logical alternative is to selectively protect the individual memory chip or array of memory chips with an on-board battery power supply.

Selective battery backup of a CMOS-RAM provides numerous benefits, including space savings, easy inspection and servicing, long-life expectancy and reliability. Most importantly, a battery backed CMOS-RAM serves as a cost-effective non-volatile memory system.

The ideal battery for CMOS-RAM backup would be printed circuit board mountable and wave solderable, provide data retention of the life of the memory, retain data over a wide temperature range, not require recharging, not require replacement during memory life, be hermetically sealed to prevent leakage, have proved high quality and reliability and possess high safety characteristics. Nickel–cadmium and lithium-based batteries are two commercially available products which offer the most potential for a CMOS-RAM backup power source.

Selecting between nickel–cadmium and lithium batteries for CMOS-RAM backup depends on specific differences between the various battery systems in relation to the products being designed. Both systems have proved reliability, but there are fundamental differences which become significant when the two battery types are considered for use in battery backup applications. These differences include voltage generating capability, operating life, temperature range and recharging.

Nickel–cadmium batteries require more than one cell to provide power above 2.0 V, which is the common minimum data retention voltage of CMOS-RAM devices. In most cases, nickel–cadmium batteries also take up relatively large amounts of valuable board space. Lithium batteries, on the other hand, are capable of delivering more than 2.0 V of power in a single small unit.

Generally a lithium battery will last the life of the memory in which the battery is used. For long-term applications, nickel–cadmium batteries may need replacement in 3–5 years, causing service and battery hardware costs to become factors.

Most lithium systems operate over temperature ranges similar to those of nickel–cadmium systems

(−40 to +60°C). However, lithium systems, capable of operating throughout a temperature range of −55 to +125°C, are also available.

Nickel–cadmium batteries are secondary systems while lithium systems are usually primary devices. This means that, when nickel–cadmium batteries are used, a more complicated circuit is required. This translates to higher part, labour and board costs. In addition, limited nickel–cadmium battery capacity between charges can affect the duration of continuous data retention time. With a lithium cell, the total capacity may be used in one longer, continuous standby cycle.

In the past, nickel–cadmium batteries have been a favourite choice as a backup power source. Contrary to their popularity, they are far from the ideal battery for long-term memory backup because of their narrow operating temperatures, short replacement time, high replacement labour costs, recharging requirements and potential leakage problems. These shortcomings are not encountered with lithium-based batteries.

Lithium functions as the anode in the battery's chemical system. When considering using lithium batteries, it is crucial to take a look at the material used for the cathode. There are a number of cathode and depolarizer materials used in conjunction with the lithium metal anode to make up the generic term 'lithium batteries'. These materials, which include manganese dioxide, sulphur dioxide, carbon fluoride, thionyl chloride and lead iodide, greatly influence the properties and characteristics of lithium batteries (Table 38.1).

One currently available lithium system, which seems to meet most of the requirements for an ideal CMOS-RAM backup power source, is the lithium–iodine battery. The lithium–iodine battery has long life characteristics, and is designed for wave soldering and printed circuit board mounting.

When current is not being drawn from a lithium–iodine battery, as is often the case in memory backup applications, self-discharge takes place. Self-discharge is a natural process which consumes active battery materials and, hence, battery capacity. Capacity losses at 25°C vary from less than 5% over 10 years for lithium–iodine batteries to more than 25% per month for some secondary batteries.

A high equipment storage temperature will increase self-discharge and must be considered when selecting a battery for backup applications. Lithium–iodine batteries stored at an average temperature of 50°C or less will hold a charge for more than 10 years. On the other hand, a nickel–cadmium battery has a shelf life of about 1 month before recharging is required and is limited to 300–500 charge/recharge cycles (Table 38.2).

An advantage of the lithium–iodine battery is its temperature performance. Not only does the battery operate over a wide temperature range (−55 to +125°C), but it has a high impedance at low temperatures and low impedance at high temperatures. Therefore, as the power requirements of the memory chip decrease with a decrease in temperature, the lithium–iodine battery reacts similarly by providing less current to the chip and vice versa.

Table 38.1 Comparison of commercially available lithium batteries

System	Lithium–iodine P2VP*	Lithium–manganese dioxide	Lithium–sulphur dioxide	Lithium–carbon fluoride	Lithium–thionyl chloride	Lithium–lead iodide
Energy density (W h/cm^3)†	0.60	0.44	0.45	0.33	0.66	0.47
Seal	Hermetic	Crimp	Crimp or hermetic	Crimp	Crimp or hermetic	Hermetic
Voltage (V)	2.8	2.9	2.9	2.8	3.6	1.9
Electrolyte	Solid salt	Liquid organic	Liquid organic	Liquid organic	Liquid organic	Solid salt
Self-discharge rate	Low	Moderate	Moderate	Moderate	Moderate	Low
Internal resistance	High	Moderate	Moderate	Moderate	Moderate	High
Separator	Not required‡	Plastic	Plastic	Plastic	Plastic	LiI§
Relative life cycle costs (5+ years)	Low	Low to moderate	Low to moderate	Low to moderate	Low to moderate	Low to moderate

* Poly-2-vinylpyridine
† Energy density based on small sealed cells
‡ Chemical reaction of lithium anode and iodine. P2VP cathode forms its own self-healing lithium iodide separator
§ Added during fabrication process

Table 38.2 CMOS memory backup parameters for Nicad and lithium comparison

Item	Ideal	Nicad	Lithium*
Voltage (V)	2.0	1.2	2.8
Low current capability	Required	Yes	Yes
Shelf life	10 years	1 month†	95% of capacity retained after 10 years
Self-discharge	Low	High	Low
Operating life (years)	Life of memory	3–5‡	10
Energy density (W h/cm^3)	Higher the better	0.09	0.33 or better
Temperature range (°C)	Mil Spec	−40 to +60§	−55 to +125
Sealing	Hermetic	Hermetic/crimp	Hermetic/crimp
Recharge circuit	–	Required	Not required

* This is a summary of all commercially available lithium systems
† Between charges
‡ Exceptions claimed
§ Available from limited manufacturers, most are −40 to +60°C

Hazardous incidents have been experienced with some lithium systems, particularly those using sulphur dioxide and thionyl chloride cathodes. These incidents generally occur at later stages in battery life, under reverse current conditions, during voltage reversal and while operating at high temperatures. Safety incidents have not been experienced with lithium–iodine cells, which is why they power 90% of the cardiac pacemakers presently in use.

In theory, devising a workable memory–battery connecting circuit sounds simple, but in actual practice it is a major obstacle to adapting the battery backup concept. The battery, the memory, and other circuit component specifications must be considered when adapting a backup system. More specifically, additional consideration should be given to current drains and voltages of components, selection of RAM devices for minimal current and voltage drain properties, selection of diodes for optimal voltage drop and verification of data integrity after power-up. All of these factors must be considered in combination with battery specifications.

The connecting circuitry should constantly monitor system voltage to avoid causing errors in the software because of inadequate power. As long as the voltage remains above a specified level, usually around 2.9 V, the system's power retains the memory. It is when the voltage drops below this specified point that a smooth transition to data retention power is needed.

Recent research efforts have resulted in the development of an interface circuit that will dependably monitor voltage availability and switch between external power and battery backup as needed. The circuit and the lithium battery have been incorporated into module form. One module design even contains a 16K RAM. Modules are designed to be mounted on a printed circuit board, thus providing system designers with a black-box method of meeting the need for non-volatile memory.

39

Manganese dioxide–magnesium perchlorate primary batteries

Contents

Magnesium dry batteries are now finding widespread use in military communications equipment. With increasing use and production, the cost of these batteries has dropped significantly. For example, for the BA4386/U battery consisting of 16 long C-size cells, the cost dropped from $12.50 per battery to below $5.00 over 3 years. The equivalent zinc battery costs approximately $3.00.

Attempts to fabricate magnesium dry cell batteries using flat cell constructions have to date been unsuccessful. This is due to an expansion problem caused by the build-up of reaction products.

There are two main types of primary battery: the *non-reserve* type, which is ready for use off the shelf and requires no activation or pre-use conditioning, and the *reserve* type, which usually contains the active electrode materials but requires the addition of water or an electrolyte just before use.

39.1 Reserve batteries

Manganese dioxide–magnesium perchlorate reserve-type batteries are available from Eagle Picher.

The advantage of reserve manganese dioxide–magnesium perchlorate batteries is that they can be stored for years in an inactivated condition (that is, longer than the non-reserve type) and require activation with the electrolyte before use. However, unlike the non-reserve type, it is necessary, once they are activated, to use reserve batteries fairly soon, as they undergo self-discharge.

Reserve batteries are recommended for use in portable transceivers, locator beacons, electronic counter measures and expendable electronic equipment.

39.2 Non-reserve batteries

Recently, Marathon have announced a battery for military use. It is being used in equipment exposed to tropical conditions. The most important magnesium battery now in production by Marathon is the BA4386, which serves as the power source for the PRC 25 and PRC 77 transceiver sets, formerly powered exclusively by the BA386 carbon–zinc battery. The BA4386 battery consists of 18 CD-size cells in a series–parallel construction. It is used by the military not only because it has greater capacity per unit volume and weight than its carbon–zinc counterpart, but also because of its superior storage life both at normal and at higher temperatures.

Applications for non-reserve batteries include fire and burglar alarm systems, barricade lights, lantern and flasher applications, small d.c. motor applications, such as tape recorders, toys, etc., light packs for emergency slide crates in commercial aircraft, and emergency location transmitters on aircraft.

40

Zinc–air batteries

The zinc–air system, which attracted a great deal of investment in the late 1960s and early 1970s to make a consumer product in standard cylindrical sizes, suffered initially from four problems. It was difficult to produce air-breathing cathodes of consistent quality; the need to allow air into the cell led to electrolyte leakage; carbonation of the electrolyte occurred on long-term discharge; and during intermittent discharge oxygen ingress products caused wasteful corrosion of the active material.

Despite these initial difficulties, commercially available D and N cells were available in the UK in the early 1970s. The ECL D-size cell outperformed D-size nickel–cadmium cells under most conditions, except at sub-zero temperatures, in a military 'man pack' radio application. The cost of this zinc–air cell was about 25% of that of an equivalent nickel–cadmium cell and proved to be more economical. However, the nickel–cadmium cell had the advantage of being rechargeable. Primary zinc–air batter-

ies were used extensively during the Vietnam War. The N-size cell had been used successfully for the small pocket paging equipment market. The demise of zinc–air D and N cells rests on economic factors rather than on technical grounds. The picture for zinc–air button cells is quite different. Gould have manufactured them in the USA for several years for applications such as hearing-aids, watches, etc., and Gould and Berec have both been marketing them in the UK since 1980.

The design of a cell is shown in Figure 40.1, which shows a positive electrode made of a PTFE-laminated sandwich charcoal catalyser (air electrode) and a high-purity zinc high-surface-area anode. The separators are porous to ions and are made of several layers. These prevent the transfer of released reaction products and therefore ensure long-term discharge of very low intensity currents within short-circuits. The electrolyte is potassium hydroxide. This cell has a specific energy density of 650–800 mW h/cm^3 compared with 400–520, 350–430 and 200–300, respectively, for mercury-zinc, silver–zinc and alkaline manganese dioxide cells. It has an operational temperature range of -10 to $+60°C$ and a service voltage of 1.15–1.3 V. The cell has excellent shelf life because the air inlets are covered by a thin foil during storage, which keeps out the air needed to activate the cell. The discharge performance of the Varta zinc–air cell is compared with that of alkaline manganese–zinc, alkaline silver–zinc and alkaline mercury–zinc cells, shown in Figure 40.2.

Zinc–air cells also have applications in navigation aids. McGraw Edison, for example, supply the

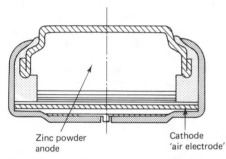

Zinc powder anode   Cathode 'air electrode'

Figure 40.1 Varta zinc–air cell type V4600 (Courtesy of Varta)

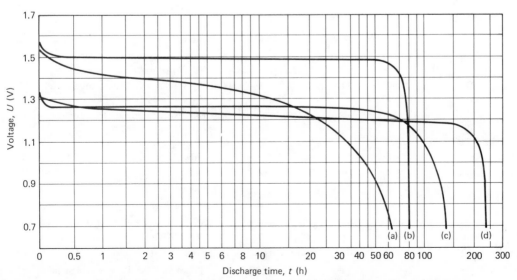

Figure 40.2 Varta cell discharge curves: (a) alkaline manganese–zinc, (b) alkaline silver oxide–zinc, (c) alkaline mercury oxide–zinc, (d) zinc–air (Courtesy of Varta)

Carbonaire range of batteries covering the following applications: short lights of all types (either flashed or continuously burned) such as reef lights and lights on fixed structures, offshore lights on oil platforms, fog horns, bell strikers, range lights, lights on barges under tow, bridge lights, pier and dock lights, obstruction lights, fish net marker lights, buoys (two standard types of 12 V disposable buoy battery packs, with capacities of 1100 and 3300 A h are available), portable warning lamps.

One outstanding advantage of Carbonaire batteries is that they may be activated with seawater. In navigation aid services, this means that batteries can be transported dry to the location. The weight to be transported is thus reduced to a minimum and the possibility of spillage of battery electrolyte in transit is completely eliminated.

In addition to this practical feature, the most desirable type of primary battery for navigation aids is one that combines large capacity with a minimum amount of labour required for setting up and replacing. While each of these advantages has been available previously in separate types of primary batteries, the Carbonaire is the first battery to combine them.

With available capacities of up to 2800 A h, only infrequent battery renewals are required. The result is a substantial saving in time and money, especially where the navigation aids are installed at remote, isolated or hard-to-reach locations. Another very pertinent operating characteristic of Carbonaire batteries is their narrow voltage range, which eliminates the necessity for adding or subtracting cells in an installation at any time during the service life of the battery.

McGraw Edison produce a 2.5 V, 1000 A h (type 2-S-J-I) and a 3.75 V, 1000 A h (type 3-S-J-I) battery and these, together with the 1.25 V, 2800 A h (type Y), can be used in various combinations to produce a variety of battery voltages, as shown in Table 40.1.

Some of the recommended applications of the range of zinc–air depolarized batteries (saline and alkaline electrolyte types) are illustrated in Tables 40.2 and 40.3.

Table 40.1 Voltage classification chart for Carbonaire batteries

Voltage required (V)	Types and number of Carbonaires recommended
2.5	$1 \times$ 2-S-J-I or $2 \times$ type Y
3.75	$1 \times$ 3-S-J-I or $3 \times$ type Y
6.0	$1 \times$ 2-S-J-I, $1 \times$ 3-S-J-I or $5 \times$ type Y
10.0	$4 \times$ 2-S-J-I or $8 \times$ type Y
12.0	$2 \times$ 2-S-J-I, $2 \times$ 3-S-J-I, $5 \times$ 2-S-J-I or $10 \times$ type Y

Table 40.2 Saline zinc–air depolarized batteries supplied by SAFT

Type	Rated capacity (A h)	Open-circuit voltage (V)	End-point voltage (V)	Average current (A)	Applications
AD 517-4	1000	1.45	0.8	0.80	Meteorology, railway signalling, marine navigation
AD 517	270	1.45	0.8	0.20	Railway signalling, telephone circuits, marine navigation, clocks
AD 524	145	1.45	0.8	0.10	Motorway emergency telephones, telephone circuits, clocks
AR 40	85	1.45	0.8	0.10	Telephone circuits, telecommunications, railway signalling
AD 538	80	1.45	0.8	0.09	Lift alarms, warning lamps, railway signalling
AD 539	45	1.45	0.8	0.04	Railway signalling
AD 542	25	1.45	0.8	0.025	Master clocks, railway signalling, measuring instruments
ADH 121	6	1.45	0.8	0.01	Security systems, clocks
AD2-519	130	2.90	1.6	0.10	Telephone circuits, railway signalling
AD2-522	70	2.90	1.6	0.08	Warning lamps

Table 40.3 Alkaline zinc–air depolarized batteries supplied by SAFT

Type	Rated capacity (A h)	Open-circuit voltage (V)	End-point voltage (V)	Average current (A)	Applications
AD 600	10 000	1.45	0.9	1.50	VHF/UHF relays, TV repeaters, radio guidance systems, radio telephone
AD 308 A	2 000	0.1	0.9	1.00	Railway signalling, meteorology, VHF/UHF relays, radio telephone, TV, marine navigation
AD 318 A	1 000	0.1	0.9	0.25	Railway signalling, telecommunications, telephone circuits, marine navigation
AD 304 J	800	0.1	0.9	3.00	Marine navigation
AD 319 J	350	0.1	0.9	0.30	Motorway emergency telephone, railway signalling
AD 810	2 200	0.1	0.9	1.5	Marine navigation, railway signalling, radio telephone
AD 820	1 200	0.1	0.9	1.2	Marine navigation, road traffic controls
AD 840	350	0.1	0.9	0.5	Roadside emergency warning lamps, railway signalling, marine navigation

41

High-temperature thermally activated primary batteries

These batteries find uses in fast activation applications such as firing squibs and explosive bolts. Other applications include electric matches, dimple motors and emergency starting power for military vehicles and aircraft.

Eagle Picher are supplying thermally activated batteries for a wide range of military and projectile applications. The latter include Maverick, Sidewinder, Dragon, Stringer Paveway, ACE, Viper, Korworan, Buzzard, Lance and ACE SII projectiles (calcium thermal cells); and Paveway, ASHT, AMRAAM and Hoe projectiles (lithium thermal cells). This Company also supplies cells for artillery fuses, mortar fuses, rocket fuses, proximity fuses and seat ejection devices in aircraft.

MSH supply thermally activated batteries for terminal guidance, actuation of control surfaces, telemetry, guidance and control systems, aircraft emergency power supplies, safety and arming power, and fuses for missiles, rockets, torpedoes, bombs, ammunition and missile/rocket firing units.

Thermally activated batteries are widely used as power sources in projectiles, rockets, bombs, mines, missiles, decoys, jammers and torpedoes.

42 Seawater-activated primary batteries

Silver chloride–magnesium batteries These are used in sonarbuoys, beacons, emergency equipment, torpedoes, flares, mines, pingers, balloons and life-jackets.

Silver chloride–zinc batteries These are used in balloons, buoys, military power systems, communications equipment and portable electronic devices.

Copper chloride–magnesium batteries These are used in balloons, beacons, flares, sonarbuoys, pingers and oceanographic equipment.

43

Electric vehicle secondary batteries

Contents

An effort to develop a practical electric vehicle is consuming a significant portion of the technical resources of the technically advanced nations. Its projected impact upon the environment is well known; an electric vehicle would contribute to a major reduction in emission pollutants, particularly in metropolitan areas. In addition, as natural energy resources become more scarce, it has been and will continue to be necessary to develop new energy sources and to centralize the use of old resources for greater efficiency. In virtually all of these new and changing energy systems, the conversion or usable output is in the form of electrical energy. The ability of an electric vehicle to use this future primary energy source directly may well overshadow its more familiar environmental benefits.

43.1 Lead–acid batteries

The starting point of power sources for electric vehicles was, of course, the lead–acid battery. Because of their relatively low power density, compared with other types of battery which are now becoming available, other types of batteries are now being considered for this application. The low power density of the lead–acid system puts severe limitations on the range that can be achieved between battery charges, although the restrictions regarding acceleration and maximum speed are not as limiting as they once were.

The lead–acid battery should not be dismissed lightly as a power source for electric vehicles. It is, in fact, the only type of battery being used in large-scale prototype vehicle trials. Chloride Batteries UK are currently conducting several such trials.

43.1.1 The Silent Carrier Programme

Silent Carrier is a 1.8 ton urban delivery van. The earlier versions of this vehicle had a maximum range of 35–40 miles and this has since been improved.

The maximum speed is 49 miles/h and the hill climbing ability is 1 in 8. Although the basic cost of early production models will be a little higher than that of vehicles with internal combustion engines, running costs will be lower. It is claimed that the maintenance costs of the battery-powered vehicle vary from 60% to as low as 30% of those of petrol- and diesel-powered vehicles.

43.1.2 The Silent Rider Programme

Silent Rider is an electric bus. It will carry 50 people at 40 miles/h over a 40 mile range on one charge. Trials are being conducted in collaboration with bus companies in two counties, Lancashire and Cheshire. Figures 43.1 and 43.2 show the layout of this bus. The 330 V battery in this vehicle weighs 4.4 tons, and consists of 165 cells, each with a capacity of 329 A h. Investigations by Chloride Batteries have shown that nearly 50% of buses on city centre journeys during peak periods cover less than 25 miles, with a further 40% covering no more than 40 miles. The shaded areas in Figure 43.3 show the potential for a battery-operated bus with a range of 40 miles.

43.2 Other power sources for vehicles

Types of batteries being developed or considered for vehicle traction include lithium–iron sulphide, nickel–iron, nickel–zinc, silver–zinc, sodium–sulphur and zinc–chlorine batteries. All of these are potentially capable of at least three times the rating of the best lead–acid systems. Some of these batteries (e.g. nickel systems) are seen as intermediate solutions to the problem of powering vehicles, whilst others (e.g. lithium systems) are seen as longer term solutions.

Present development of electric vehicles is limited to the non-availability of a low-cost battery system that offers an energy density much greater than the currently available 25 W h/kg. This energy density

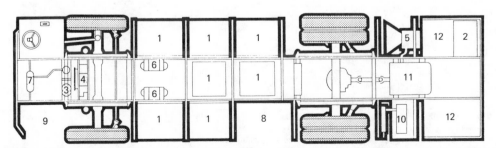

Figure 43.1 Layout of the Silent Rider battery-powered bus. 1, main 330 V battery; 2, auxiliary 24 V battery; 3, power steering pump and motor assembly; 4, air compressor and motor assembly; 5, motor cooling fan; 6, main air tanks; 7, auxiliary air tank; 8, exit; 9, entrance; 10, bus heater; 11, motor; 12, control panels (Courtesy of Chloride Batteries)

level restricts existing electric vehicles to narrow, short-range, intracity applications. On the other hand, electric vehicles capable of broad, intercity service will probably require batteries with energy densities in excess of 200 W h/kg and, although various groups in the battery industry are actively engaged in the development of such systems, the production of these batteries appears to be a number of years away. The only secondary batteries yet known with energy densities in excess of

Figure 43.2 Silent Rider Bus (Chloride Motive Power Group). Vehicles with predictable route patterns are particularly suitable for battery power (Courtesy of Chloride Batteries)

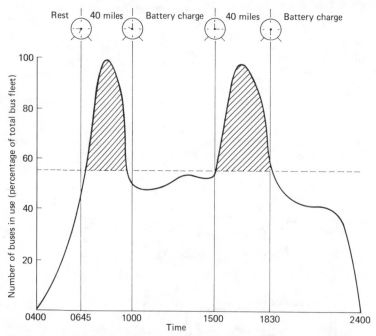

Figure 43.3 The potential for a battery-operated bus with a range of 40 miles. Nearly 50% of buses on city centre journeys during peak periods cover less than 25 miles, with a further 40% covering no more than 40 miles. The shaded areas of this graph clearly show the potential for a battery operated bus with a range of 40 miles (Courtesy of Chloride Batteries)

200 W h/kg are the sodium–sulphur, lithium–chlorine and lithium–sulphur systems, all of which are still only at the development stage. However, an intermediate energy density battery system could play a major role in electric vehicle development for the next decade. Specifically, a battery system with an energy density of 50–70 W h/kg could result in the development of an electric vehicle that would be suitable for a broad range of intercity applications. It is also possible that battery systems in this energy density range could find applications in hybrid electric vehicles for intercity service.

Eagle Picher, in a detailed study, have reached the conclusion that there are three attractive candidate batteries for powering medium-range vehicles, these being the nickel–hydrogen, nickel–iron and nickel–zinc systems.

Recent advances in nickel oxide electrode technology greatly increase confidence in the ability of the above systems to meet the desired requirements. New lightweight, high surface area electrode matrix materials should permit development of very thick, high capacity and high energy density electrodes at a lower cost. The new electrochemical impregnation process produces electrodes which are more dimensionally stable and thus capable of higher depths of discharge. Eagle Picher used a computer program to carry out a vehicle modelling study. Major inputs to the program include an assumed driving profile, cruising and limiting velocities, and frontal area, a rolling resistance factor and a power train efficiency factor (40% was assumed). For a specific vehicle gross weight and battery energy density, the computer program will predict through the generation and subsequent interpretation of least squares regression curves:

1. The total energy required per distance unit.
2. The energy required per acceleration.
3. The energy required per distance unit whilst cruising.
4. The time maximum power will be required during acceleration.
5. The peak power required during acceleration.

For example, assuming a gross vehicle weight of 1815 kg, a battery energy density of 60 W h/kg and the driving profile identified by schedule D of the SAE Electric Vehicle Test Procedure, a personal commuter car would offer a driving range of 160 km (100 miles) at an average velocity of 70 km/h (45 miles/h) with a payload of approximately 363 kg. Maximum power drain during acceleration would last approximately 25 s and the peak power density required during this period would be 59 W/kg.

These results agree well with a similar European study, in which it was reported that batteries with energy densities of 40–60 W h/kg with 2 h discharge times and power densities of 56 W/kg would permit electric vehicles to fulfil important intracity transportation requirements.

One of the conclusions of this study is that the nickel–hydrogen battery is one of the strongest candidate battery systems for the proposed electric vehicle application. Such systems are currently used in aerospace applications, offer energy densities of 50 W h/kg and have been successfully tested for thousands of cycles.

The nickel–hydrogen system is hermetically sealed and thus maintenance free, but it is not sensitive to operation mode as is the case with other hermetically sealed systems. Considerable abuse can be sustained with respect to both overcharge and overdischarge without system damage, a factor which may be very important considering the long series cell strings necessary to produce the high voltage required by the vehicle propulsion system.

The major problem associated with the nickel–hydrogen system is cost. Although used in very small quantities, one of the components of the negative electrode is a very expensive noble metal catalyst.

A lesser disadvantage of the nickel–hydrogen system is that one of the reactants is a gas (hydrogen). If this is stored under pressure, system energy volume is reduced and the gas may present a hazard in a passenger vehicle. Hydrogen gas has, however, been successfully stored in the form of a chemical hydride and this technique should be readily adaptable to the electric vehicle system.

43.2.1 The nickel–zinc battery system

The nickel–zinc system is a viable candidate for electric vehicle propulsion based on its high energy density and power density. The zinc electrode contributes a high electrode potential and a low equivalent weight, resulting in energy densities of 70 W h/kg presently obtainable in a configuration suitable for an electric vehicle application. Other features of the system that make it attractive for electric vehicles are:

1. The capability of accepting either constant-potential or constant-current charging with little overcharge required.
2. The potential of operation as a sealed, maintenance-free unit.
3. The ability to withstand considerable overdischarge without cell reversal, owing to the excess of zinc oxide active material.

Separator systems for nickel–zinc cells have extensively employed cellulosic materials. New materials of thinner cross-section and greater oxidation resistance are now being produced from polyethylene and polypropylene, which offer the advantages of decreased weight and volume, and

Table 43.1 Nickel battery systems cost and performance summary

	Nickel–hydrogen	*Nickel–iron*	*Nickel–zinc*
Energy density at 2 h rate (W h/kg)	60	50	70
Cycle life (100% depth of discharge) (cycles)	2000	2000	500
Power density (minimum) (W/kg)	100	100	150
Battery costs ($/kW h)	100–200	50–100	50–100
Costs/cycle ($/kW h)	0.05–0.10	0.03–0.05	0.10–0.20

increased resistance to oxidation degradation during overcharge. These materials, in conjunction with cellulosic films, or alone, can increase energy densities and life.

The utilization of the nickel–zinc system for electric vehicle propulsion is dependent on the realization of cycle life in excess of 500 deep cycles. Currently, cycle life is limited by the zinc electrode; specifically, the phenomenon known as zinc shape change. Work is currently being done at Eagle Picher which has demonstrated a capability to achieve 300 cycles at a 65% depth of discharge. This performance is based on an improved separator system, controlled charging methods, and incorporation of construction features designed to minimize zinc electrode shape change. Further improvements are anticipated in the zinc electrode based on construction features designed to control the concentration gradients of dissolved zinc ionic species within the cell.

An interesting cost trade-off for the nickel–zinc system is available concerning the nickel electrode. The ultimate design objective is, of course, to develop a deep cycling capability upwards to 1000 cycles. However, if a much lower cycle life can be tolerated (200–250 cycles), it is possible to construct cells with non-sintered nickel oxide electrodes at a lower cost.

Comparative cost and performance data for nickel–hydrogen, nickel–iron and nickel–zinc systems are given in Table 43.1.

43.2.2 Lithium aluminium alloy–iron sulphide batteries for vehicle propulsion, supplied by Eagle Picher

Eagle Picher entered into the development of the lithium–metal sulphide battery system for vehicle propulsion in 1975. The electrochemical system under study utilizes an anode of a lithium–aluminium alloy and a cathode of the metal sulphides. Iron sulphide (FeS) or iron disulphide (FeS$_2$) is currently most commonly utilized. Initial efforts were directed toward developing engineering-scale cells suitable for electric vehicle and utility load levelling applications. These cells have yielded energy densities in excess of 100 W h/kg and cycle

life in excess of 1000 cycles. Cells have been vibrated under road load conditions which simulate in-vehicle use with no problems encountered.

In 1979 Eagle Picher used this type of battery to power an electric van. A top speed of 45 miles/h and a range of 80 miles were achieved. Figure 43.4 shows the single cell voltage profiles for typical FeS and FeS$_2$ 4 h rate discharges.

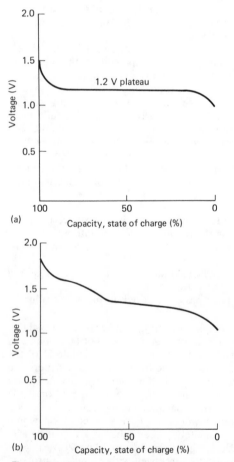

Figure 43.4 Lithium aluminium alloy–iron sulphide cells supplied by Eagle Picher. Typical single-cell voltage profiles for 4 h rate discharges for (a) FeS and (b) FeS$_2$ cells (Courtesy of Eagle Picher)

Feasibility of the lithium–metal sulphide battery technology has been demonstrated by Eagle Picher in engineering-scale cells and modules. Figure 43.5 shows a typical 10-cell module exhibiting a nominal 12 V e.m.f. curve and a capacity of 340 A h. The compact 12 V, 4 kW h package has been cycled in the insulated battery container in excess of 250 deep discharge cycles as testing continues. Ampere hour efficiencies are stable between 97 and 100%, while energy efficiency remains in excess of 85%. Low maintenance, high charge rates and minimal thermal management are characteristic of the system under a broad range of climatic conditions.

Table 43.2 lists the program goals Eagle Picher have defined for the lithium–metal sulphide system and the time frame for accomplishment.

Figure 43.5　Ten-cell lithium aluminium alloy–iron sulphide ten-cell module for vehicle propulsion (Courtesy of Eagle Picher)

Table 43.2 Program goals: lithium–metal sulphide systems

	1981	*1983*	*1986*	*1988*
Specific energy (W h/kg)				
Cell	95	120	150	200
Battery	70	100	130	160
Energy density (W h/l)				
Cell	320	400	525	650
Battery	100	200	300	375
Peak power (W/kg)				
Cell	150	185	200	250
Battery	120	120	160	200
Battery heat loss (W)	300	150	125	120
Life				
Deep discharges	250	500	1 000	1 000
Equivalent miles	25 000	60 000	150 000	200 000

Part 5

Battery charging

44 Introduction

Two principal methods of battery charging are used; constant-potential charging and constant-current charging. Many variants of these methods exist, such as the float charge regimen and the shallow charge regimen, which are modifications of the constant-potential method. Also, many methods of control of the charging process exist. These include charging for a particular time to a particular end-voltage, charging to a particular end-current, charging controlled by electrolyte temperature, and charging controlled by the internal pressure of the battery (in the case of sealed batteries). Another proposal, which has been tried out for charge control of sealed nickel–cadmium cells, is to incorporate an oxygen-consuming auxiliary electrode in the cell to consume oxygen gas as it is evolved towards the end of charge and on overcharge and thus prevent the build-up of excessive internal gas pressure. This, however, needs extensive modifications of the cell structure and consequently is expensive and increases the overall cost. The various methods used to charge different types of battery by these methods are discussed in Chapters 45–49.

45

Constant-potential charging

Contents

45.1 Standard CP charging

45.1.1 Calcium–lead alloy batteries

This is the preferred method for charging calcium–lead alloy grid low-maintenance batteries. By selecting an appropriate constant charging voltage and initial charging current, the battery can be safely recharged regardless of the depth of discharge in previous discharge cycles. The recommended initial charging currents for calcium–lead alloy low-maintenance batteries are listed in Table 45.1.

Table 45.1 Initial constant-potential charging currents for calcium–lead alloy low-maintenance batteries

Nominal capacity (A h)	Maximum initial charge current (A)
1.2	0.3
1.9	0.5
2.6	0.7
4.0	1.0
4.5	1.1
6.0	1.5
8.0	2.0
10.0	2.5
20.0	5.0

When carrying out constant-potential charges/discharges in cyclic use, the battery should be charged at a maximum of 2.5 V/cell. Charging current can be monitored and cut off when the voltage reaches the required value as indicated in Table 45.2.

Table 45.2 Charging current–charging voltage relationship: constant-potential charging of calcium–lead alloy low-maintenance batteries

Nominal capacity (A h)	Charging current (A)	
	At 2.40 V/cell	At 2.50 V/cell
1.2	10–30	30–60
1.9	20–40	60–120
2.6	30–60	80–160
4.0	40–80	110–220
4.5	50–100	130–250
6.0	60–120	180–360
8.0	80–160	250–500
10.0	100–200	300–600
20.0	200–400	600–1200

Figure 45.1 shows a typical charging curve for an approximately 10 A h lead–acid battery showing profiles of charge current as a function of charge time for two constant-potential voltages (2.3 and 2.5 V). The initial current resulting from constant

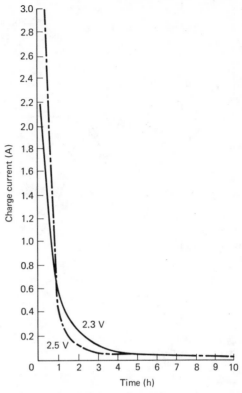

Figure 45.1 Constant-potential charging current *versus* time for a General Electric D cell at 23°C. Actual recharge time and current are circuit dependent. Charge CP at 23°C, discharge 250 mA at 23°C to 1.4 V (Courtesy of General Electric)

potential charging is high (2.2–3 A) and reduces to a relatively low current of less than 0.1 A as the cell approaches full charge after 4–5 h.

Figure 45.2 shows, in somewhat more detail, a constant-potential charging curve obtained by charging at 6.9 V a particular calcium–lead alloy low-maintenance battery (0–25C A) produced by Yuasa (Japan). The curve shows a plot against charging time of charging current, terminal voltage and charging volume. The charged capacity is indicated as a percentage against the discharged capacity in the previous cycle, i.e. percentage of charging volume. Normally, 100–130% recharge (charging volume) is required to charge a discharged battery fully. The initial charging current is limited to 0.25C A. The solid lines indicate 100% 10-h rate discharge in the preceding discharge cycle. The broken lines show their charging character after a recharge for 5 h (50%) at the same 10-h rate. The reduction in charge time required to attain an arbitrary 100% and 120% charging volume with increase in the constant potential used is clearly seen in the figures shown in Table 45.3.

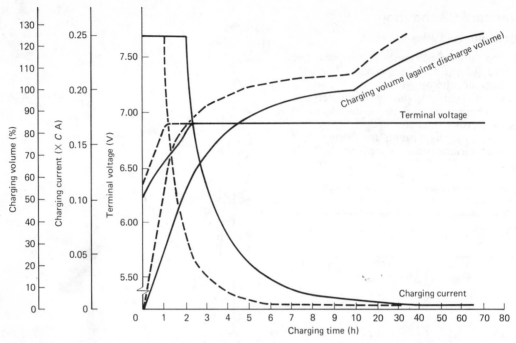

Figure 45.2 Charging characteristics of a Yuasa sealed lead–acid battery at 25°C. (*C* is the 20 h rated capacity. For further details, see text) (Courtesy of Yuasa)

Table 45.3 Reduction in charge time with increasing charging potential

Constant charging potential (V)	Hours to achieve a charging volume of	
	100%	*120%*
6.9	20	50
20	5	20
50	4	1

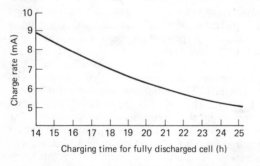

Figure 45.3 Sample discharge data for Eveready OB90 and OB90T sealed nickel–cadmium cells, 1.2 V, 90 A h capacity. Charge voltage 1.35–1.50 V (Courtesy of Union Carbide)

45.1.2 Nickel–cadmium batteries

Figure 45.3 shows a constant-potential charging curve for a 1.2 V, 90 μA h capacity nickel–cadmium cell. This curve demonstrates the relationship between charging rate and the time required to fully charge a fully discharged battery when a charge voltage of 1.35–1.5 V is used.

Constant-potential charge has always been used as a rapid-charge system for open and semi-open types of nickel–cadmium cell. This type of charge applied to sealed nickel–cadmium cells without precautions has almost always led to the destruction of the batteries. In fact, the voltage of sealed cells diminishes after the end of charge and the effect of this is to increase the charge current to proportions that are not acceptable to the batteries. It is possible, however, to use a simple means to charge sealed cells at constant potential by limiting the charge or overcharge current to the maximum permissible. It is sufficient to connect a simple resistance in the charge circuit.

45.1.3 CP chargers for calcium–lead alloy batteries

Figure 45.4 shows the charging circuits recommended by Yuasa for constant-potential charging of NP batteries. The constant-potential current-limited

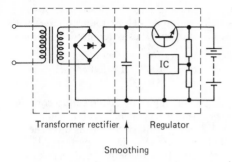

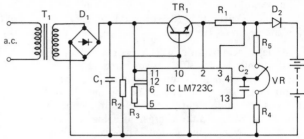

(a) Typical constant-voltage charging circuit

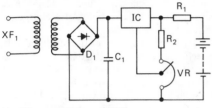

XF_1	6.3 V a.c. XMFR
D_1	1 A, 100 PIV
C_1	1000 μF, 50 W V d.c.
R_1	2.5 Ω, 5 W
R_2	2.5 kΩ, 1/2 W
VR	500 Ω, 1/2 W
IC	LM309 National Semiconductor

(c) Constant-voltage charging circuit comprising IC LM309IC

T_1	Transformer, d.c. 13 V (r.m.s.), 1–3 A (r.m.s.)
D_1, D_2	100 V, 1 A diode
C_1	50 V, 470 μF electrolytic condenser
TR_1	MJ2840 10 A 60 V 150 W (Motorola)
IC	LM723C (National Semiconductor)
R_1	4.7 Ω, 1/2 W 3
R_2	5.1 kΩ, 1/4 W
R_3	3.9 kΩ, 1/4 W
R_4	7.5 kΩ, 1/4 W
R_5	8.2 kΩ, 1/4 W
VR	2 kΩ
C_2	50 V, 1000 pF

(b) Constant-voltage current-limited charger comprising IC LM723C voltage regulator (for 12 V d.c. output 0.42 A max.)

Figure 45.4 Charging circuits for Yuasa sealed lead–acid batteries (Courtesy of Yuasa)

charger (Figure 45.4(b)) is available from National Semiconductors (IC LM723C Voltage Regulator). The Motorola MC 1723CL and the Fairchild MA 723DC equipment have equivalent performances.

45.2 Shallow cycle CP charging of lead–acid batteries

A battery is classified in this category when 5–50% of its rated capacity has been removed. Under this condition the constant potential and limiting current should be decreased from the values required for deep service. The charging conditions are:

1. Constant potential: 2.40–2.56 V/cell.
2. Limiting current: 10–15% of rated capacity.
3. Charge time: 10–18 h; should capacity appear to be decreasing, charge time should be increased periodically to 20–30 h.

Taper charging with constant potential, the simplest method of charging, is satisfactory for charging batteries in cyclic service. This method requires control of overcharging and of battery heating, both of which adversely affect battery life. In general, battery life is shorter than is obtained in the other charging methods discussed.

45.3 Deep cycle CP charging of lead–acid batteries

A battery is classified in this category when 50–100% of its rated capacity has been removed to the end-voltage specified by the manufacturer. Under this condition the constant potential and limiting current should be increased. The charging conditions are:

1. Constant potential: 2.45–2.50 V/cell.
2. Limiting current: 20–50% of rated capacity.
3. Recharge time: limited to 12–20 h; should capacity appear to be decreasing, charge time should be increased periodically to 24–30 h.

45.4 Float CP charging of lead–acid batteries

A battery on float service is one subjected to continuous charge and is called on to perform only in case of emergency. Under these conditions the battery is charged at a constant potential sufficient to maintain it at full charge. Charging conditions are:

1. Constant potential: 2.28–2.30 V/cell.
2. Limiting current: 1–20% of rated capacity.
3. Charge time: continuous.

All lead–acid batteries will charge most efficiently in the range of 15–30°C. When float charging is carried out under conditions where the battery temperature does not fall outside the limit 0–40°C there is no requirement for temperature compensation to be built into the charger to enhance charge efficiency. Many applications, however, require operation in environments from extreme cold to extreme heat. When batteries are likely to reach temperatures outside the 0–40°C range a temperature compensator built into the charger, operating on 4 mV/°C per cell based on 25°C is desirable. Constant-potential charge voltages for various battery temperatures and modes of service are shown in Figure 45.5.

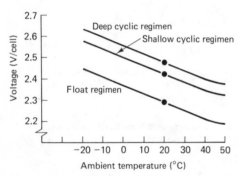

Figure 45.5 Charge voltage per cell *versus* temperature for an Eagle Picher Carefree rechargeable lead–acid battery (Courtesy of Eagle Picher)

45.5 Two-step cyclic voltage–float voltage CP charging

There are applications that require a lead–acid battery in standby service to be recharged rapidly after a discharge. Under this condition a two-step charger would be most efficient; that is, one having a capability to switch from a cyclic voltage to a float voltage. In these circumstances the following conditions are recommended:

1. Initial constant potential: 2.45–2.50 V/cell.
2. Float constant potential: 2.28–2.30 V/cell.
3. Limiting current: 20–40% of rated capacity.
4. Initial charge time: time required to reach 2.50 V/cell.
5. Float charge time: continuous.

Partially recombining sealed lead–acid batteries have the ability efficiently to accept a high-rate charge without a serious sacrifice in life. In applications where it is necessary to recharge in a

limited time, the battery can be restored to 90–95% of full capacity within a 1–3 h period. A charger with the following characteristics is required:

1. Constant potential: 2.50–2.55 V/cell.
2. Limiting current: 100% of rated capacity.
3. Charging time: 1–3 h.

If charging at the high rate is allowed to continue excessively beyond a 3 h period, greater than normal amounts of gassing will occur. It is not recommended that high-rate charging be performed in an airtight enclosure. It is recommended that battery compartments have adequate ventilation to prevent an accumulation of gases, which could reach dangerous proportions in the event of a charger or battery malfunction.

In general, the end of charge of the battery can be determined when the charge voltage has reached the level discussed above for the various modes of service, and the on-charge current has dropped to the approximate amounts shown in Table 45.4 (Eagle Picher Carefree Batteries). When the current has reached the value shown in the table, it can be removed from the charge source or switched to the float mode.

A charger two-step circuit for a 6 V battery is shown in Figure 45.6.

Table 45.4 Eagle Picher Carefree maintenance-free lead–acid batteries: current acceptance *versus* charge voltage

Batteries	Current (mA)
All 0.9 A h	<20
All 2.5 and 2.6 A h	<50
All 5 A h	<75
All 8 A h	<100
All 12 A h	<150
All 30 A h	<200

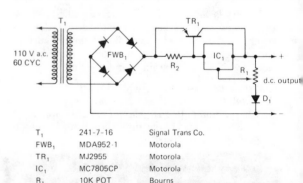

T_1	241-7-16	Signal Trans Co.
FWB_1	MDA952-1	Motorola
TR_1	MJ2955	Motorola
IC_1	MC7805CP	Motorola
R_1	10K POT	Bourns
R_2	15 Ω	Ohmite
D_1	F2	Sarkes-Tarzian

Figure 45.6 Charger circuit for 12, 15 and 30 A h Eagle Picher Carefree rechargeable lead–acid batteries. Note that TR1 should have an adequate heatsink (Courtesy of Eagle Picher)

46

Voltage-limited taper current charging of alkaline manganese dioxide batteries

The recommended method for charging alkaline secondary batteries is the use of voltage-limited taper current charging. This method offers the maximum cycle life and consequently the lowest battery operating costs. Cycle life can be as much as twice that obtained with other methods of charging. Although constant-current charging can be used for this type of battery, with consequent loss of cycle life, it is recommended that careful enquiries be made before such charging methods are contemplated.

Batteries should not be charged continuously, or float charged for extended periods, after the charge current has returned 120% of the ampere hours removed on previous discharge.

A rechargeable alkaline manganese dioxide battery must not be discharged completely. For best results, the rated capacity of the battery should not be exceeded on discharge. During deep discharge a secondary electrochemical reaction takes place. This reaction is not reversible and will seriously reduce the battery cycle life. To avoid short battery service, it is therefore desirable that the device using the battery contain some provision against excessive discharge of the battery.

The charger will replace the energy needed to recharge the battery, and a timer or ampere hour control is unnecessary to monitor the charging time. However, the battery should be removed from the charger after completion of the charge. Also, to prevent deep cell discharge, a timer will provide effective control over the energy withdrawn from the cell during any cycle.

With constant-current charging for this type of battery, a d.c. timer or ampere hour counter is needed to control both the discharge and charge cycles. A system of this sort would be adjustable to permit removal of the rated ampere hour capacity from the battery and thereby ensure maximum battery service life.

Voltage-limited taper current charging involves use of an inexpensive voltage regulator added to the basic constant-current type transformer–rectifier circuit with a current-limiting resistor between the battery and the regulated output voltage. This removes the burden of adjusting charge time to discharge time and provides automatic current control. Although the regulator circuit is somewhat more expensive than other common circuits, the initial cost is more than offset by 50–100% greater cycle life.

The basic characteristics of sealed alkaline manganese dioxide rechargeable cells when using voltage-limited taper current charging are given in Table 46.1 (based on a 6 h charge period).

A regulator providing voltage regulation of 2–3% at low current values is adequate. Poorer regulation when the battery is first placed on charge and the voltage is low is often satisfactory. The regulator output voltage is adjusted to the battery end-of-charge voltage. Correct battery voltage is determined by the number of cells and cell voltages given in Table 46.1. The current-limiting resistance (the total current-limiting resistance consists of the series resistance plus inherent resistance of the regulator circuit, which may vary with design and component selection) should limit the initial current to the 4–6 h rate at start of charge when the battery voltage is low. A typical starting voltage after withdrawal of rated capacity is 1.30–1.35 V/cell. The maximum recommended initial charge currents at these voltages are shown in Table 46.1.

The minimum voltage that a cell will reach at the end of the recommended discharge capacity is shown in Figure 46.1. This voltage is measured while the discharge current is flowing. The curves in Figure 46.1 represent cell behaviour at maximum discharge current rating to rated capacity and a minimum of 16 h recharge time. The initial closed-circuit and end-of-discharge voltage curves give the 'working' voltage range of the cell at various discharge levels. To prevent damage to the cell on charge cycle the charging potential should be limited to about 1.8 V/cell with a permissible maximum of 1.75 V/cell. The voltage rise on the charge and end-of-charge voltage value is dependent on the charge current character. As shown in Figure 46.1, the end-of-charge voltage will decrease slightly as cycle life progresses as the result of gradual change in the reversibility of the system.

Since the current tapers as the result of battery voltage increase, or is 'voltage responsive', the end-of-charge current will increase as cycling progresses. The current will equilibrate at some specific value but will never reach zero or shut-off since the voltage decreases as current is reduced as the result of internal resistance and chemical reaction. The battery voltage after charge will

Table 46.1 Basic characteristics of sealed alkaline manganese dioxide cells when using voltage-limited taper current charging

Cell size	Average operating voltage at maximum current (V)	Maximum initial charging current at 1.3 V/cell (A)	Current-limiting resistance (Ω/cell)	Source voltage limit (V/cell)
D	1.0–1.2	0.6	0.8	1.70–1.75
G	1.0–1.2	1.12	0.4	1.70–1.75

gradually decrease to a value of 1.5 V/cell when the charger is disconnected and the battery remains on open circuit. When changing from charge to discharge cycle with no time delay between, the battery voltage decay rate will vary with discharge current. At rates near 1 A, the battery voltage will

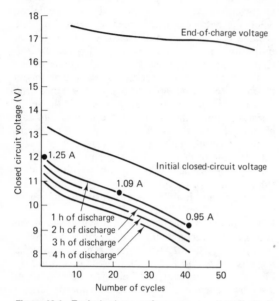

Figure 46.1 Typical voltage performance on voltage-limited taper current charge of an Eveready 561 15-V rechargeable alkaline manganese dioxide battery. Discharge cycle, 4 h through 9.6 Ω; charge cycle, voltage-limited taper current charging for 16 h, voltage limit 17.5 V, filtered regulated power supply, 1% regulation with 4.0 Ω limiting resistor (Courtesy of Union Carbide)

fall to the indicated initial closed-circuit values within 5 min.

Figure 46.2 shows a voltage-limited taper current charge circuit suitable for recharging a 15 V, 5 A h battery in 16 h. A tapered charging current is supplied to the battery through the series current-limiting resistor R_4, as battery voltage increases with charge. The performance of the circuit may vary with the components used. Often the replacement transistors, for example the SK 3009 replacement for 2N1304, may require reduction in value of R_3, for proper circuit operation. To determine the value of the series current-limiting resistor, R_4, and for initial circuit adjustment, a voltmeter (10 000 Ω/V or greater) is placed across R_3 (no battery connected) and R_3 adjusted to obtain the indicated source voltage limit. A voltage source equivalent to 1.3 V/cell is connected for the battery to be charged and R_4 is selected to limit the charge current to recommended values. The battery should be disconnected from the charger when the charge circuit is not energized to prevent discharge through resistance R_3.

The suggested linear taper current for the battery quoted above (Union Carbide Eveready 561) and the actual current characteristic obtained with the circuit of Figure 46.2 are shown in Figure 46.3. The current characteristic can be tailored by selection of components. The curves shown were obtained using the component values shown in Figure 46.2. The permissible deviation of charge current from that indicated suggests that linear taper varies somewhat with application and usage pattern. Where the application requires repetitive discharge to rated capacity, the charge current values should fall within

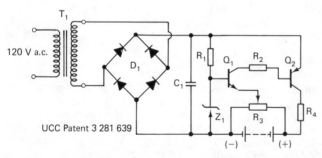

UCC Patent 3 281 639

T$_1$	Stancor RT 201 (input a.c. to terminals 1 and 7; connect terminals 3 and 6) (secondary connections to bridge 8 and 11; connect terminals 9 and 10)
D$_1$	2A50 Solitron devices, Inc. (or equivalent)
C$_1$	500 μF 50 V d.c.
Z$_1$	1/4 M12 Z (Motorola) in 963B TI or Int. Rec.
Q$_1$	2N1304 RCA, TI or equivalent
Q$_2$	2N1557 Motorola, 2N514 TI
R$_1$	4700 Ω, 1/2 W
R$_2$	390 Ω, 1/2 W
R$_3$	1000 Ω, 1 W
R$_4$	4 Ω, 10 W

Figure 46.2 Voltage-limited taper current charger circuit suitable for charging a 15 V, 5 A h Eveready 561 alkaline manganese dioxide battery in 16 h (Courtesy of Union Carbide)

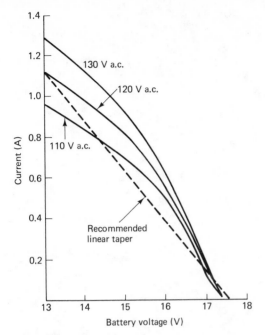

Figure 46.3 Suggested and actual linear taper currents obtained during 16 h voltage-limited taper current charging of 15 V, 5 A h Eveready 561 alkaline manganese dioxide battery (Courtesy of Union Carbide)

the shaded area of Figure 46.4, with charge time adjustment for circuit tolerance should operation fall within the 'minimum' area relative to linear character.

Figure 46.5 shows the battery voltage and charge current for a typical 16 h charge period at about mid-cycle life of a 15 V, 5 A h battery. The area under the charge current curve is the amount of charge in ampere hours; in this example it is approximately 6 A h.

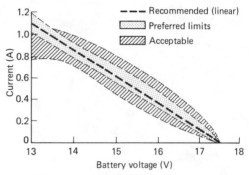

Figure 46.4 Permissible deviations of charge current of 16 h voltage-limited taper current charge of 15 V, 5 A h Eveready 561 alkaline manganese dioxide battery (Courtesy of Union Carbide)

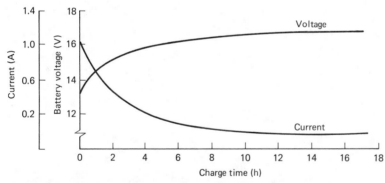

Figure 46.5 Voltage and current characteristics about mid-cycle life during 16 h voltage-limited taper current charge of a 15 V, 5 A h Eveready 561 alkaline manganese dioxide battery (Courtesy of Union Carbide)

47

Constant-current charging

Contents

When recharge time is critical and it is desirable to determine the amount of capacity restored to the battery, recharge with constant current is recommended. Constant-current charging is suitable when the discharged ampere hours taken out in the previous discharge cycle are known. It is necessary to monitor battery voltage or manually cut off the current at the end of charge to ensure a good service life for the battery. This type of charging requires regulation by means of an automatic timer. A normal recharge time of 20 h is recommended, whereby the constant-current charging rate is established to give an input of 110% of the previous output. If necessary, the charge time can be decreased, although it will require an increase in charging current such that

Amperes × Hours = 1.1 × Discharge output (A h)

Figure 47.1 shows a typical charging voltage profile as a function of percentage capacity returned, obtained at various constant-current charging rates between 25 and 500 mA, for a General Electric fully recombining lead–acid D cell.

47.1 Charge control and charge monitoring of sealed nickel–cadmium batteries

Many methods have been tried to determine the state of charge of a sealed nickel–cadmium battery, but none has yet been successful. Except in the case of complete discharge, neither cell condition nor state of charge can be determined by open-circuit voltage. Within a short time after charging it may be above 1.4 V; it will fall shortly thereafter to 1.35 V and continue to drop as the cell loses charge. To determine the capacity available it is necessary to discharge a fully charged battery at a constant current to 1 V and note the time taken to reach this fully discharged voltage. If the state of charge is unknown, a 12 h charge at the I_{10} rate will not harm the battery.

Sealed nickel–cadmium cells should not be charged in parallel unless each cell or series string of the parallel circuit has its own current-limiting resistor. Minor differences in internal resistance of the cells may result, after cycling, in extreme variation in their state of charge. This may lead to overcharge at excessive currents in some cells and undercharge in other cells.

Overcharge rates in excess of the I_{10} rate should be avoided; however, 100 h at the I_{10} rate will have no detrimental effect.

If cells or batteries are discharged below 1 V/cell, that is, overdischarged, and reverse polarity takes place, it is usually recommended that an extended charge of 24 h at the I_{10} rate be given to ensure that all the cells are charged to the same level.

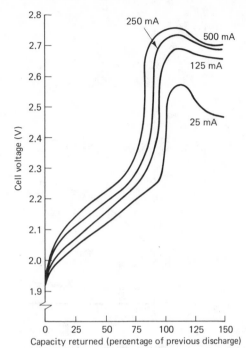

Figure 47.1 Typical voltage curves of a General Electric D cell charged at various constant-current rates at 23°C. Charge constant current at 23°C, discharge 250 mA at 23°C to 1.4 V (Courtesy of General Electric)

Sealed nickel–cadmium batteries and cells have a charge factor of 1.4; that is, 1.4 times the capacity withdrawn should be replaced. Therefore, at the 10 h rate ($0.1C_{10}$), 14 h are required to fully charge a discharged battery (1 V/cell). If a cell/battery has been overdischarged, a 24 h charge is recommended. This 24 h charge is also recommended for the first charge on a cell/battery that has been stored for prolonged periods. A further two cycles may be required to achieve the cell's rated capacity.

The maximum charge current to be considered for a particular method of charge is determined by the ability of the battery to accept this current on overcharge. These maximum currents differ with type of cell and battery and may even be different for various sizes of the same type. Such information is always available from the battery supplier.

When a nickel–cadmium battery is charged at a constant current, the cell voltage rises, as does the electrolyte temperature and the internal pressure of the cell. All three of these parameters offer a means of monitoring and controlling the battery charging operation. Figure 47.2 shows, in the case of charge at $0.3C_5$ (accelerated charge) and C_5A (rapid charge), the relationship between the state of charge (i.e. percentage of capacity charged), voltage, temperature and pressure.

The effects of these three parameters are discussed below.

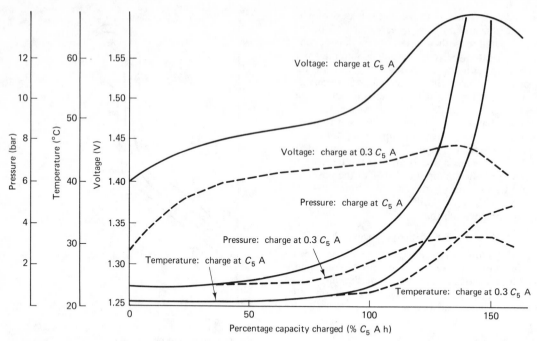

Figure 47.2 Figure 47.2 Effect of charge rate on pressure, temperature and voltage for two charge rates on sealed nickel–cadmium cells

47.1.1 On-charge voltage

To achieve charge control by the external sensing of the on-charge voltage of a nickel–cadmium cell or battery requires the use of auxiliary circuitry to shut off the charging current in response to a voltage sensor. The difficulty with the use of voltage sensing is that the voltage of many types of cell tends to change with negative cycling and may vary with temperature and charge rate.

47.1.2 Internal pressure

The internal pressure is very low during the greater part of the charge, and rises at the approach to full charge. At this moment nearly all the energy applied is producing oxygen. The pressure ceases to rise and reaches equilibrium when the rates of production and consumption of oxygen are equal. The pressure rises with the charge rate, falls when the temperature increases, and is different for each type of cell. The charge rate (and the maximum period of charge) to be considered for each type of cell corresponds to the equilibrium pressure that may be mechanically withstood by the cell. The equilibrium pressure must be lower than the pressure setting of the safety valve and must be less than that which will deform the cells, whether they are mounted in batteries or not.

Figure 47.3 shows the related characteristics of voltage, temperature and pressure on a rolled sintered cell when charged at $10 \times I_{10}$ starting from a discharged state. It may appear from these curves

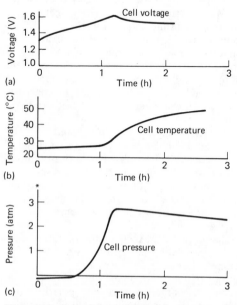

Figure 47.3 Related characteristics of voltage, temperature and pressure on charging of Varta sealed nickel–cadmium cells (Courtesy of Varta)

that pressure could be a suitable indicator for sensing the point of tripping out the charge (the tripping level). However, in practice, charge control based on pressure monitoring is not regarded favourably because of the technical and economic problems involved in fixing a pressure-sensitive device to a standard cell or in designing a special pressure-sensitive cell.

47.1.3 Electrolyte temperature

As shown in Figures 47.2 and 47.3, temperature is a suitable indicator of the state of charge of a battery. Temperature is practically constant during the greater part of the charge. When the fully charged state and overcharge are reached, the heat generated by the recombination reaction causes a rise of temperature. This stabilizes at a level that depends on the rate of charge and the thermal exchange characteristics of the cell or battery. The maximum rate of charge for the cell or battery corresponds to that at which the stabilized temperature is still acceptable; in general, this temperature should not exceed 60°C.

Cell voltage rises at the moment the charging ends; then the rise of temperature on overcharge has the effect of reducing the voltage. The increase of voltage at the end of charge is a function of the rate of charge and the temperature. This characteristic is developed in the system of controlled rapid charge.

One of the dangers of using electrolyte temperature as a means of controlling battery charging is that, if the battery is subjected to even short-term overcharge, the distinct temperature increase can cause physical damage to the cell and/or loss of cell capacity. High cell pressures are also a possible source of danger or damage.

Cell temperature monitoring has commonly been used as an indication of battery overcharge, the rise in temperature being fairly slow and concurrent with the rise in pressure. To achieve accurate temperature sensing it is necessary to have good thermal insulation between cells and normal ambient temperature during discharge. Without accurate control, cells may vent before the charge is terminated, with consequent dehydration and electrode imbalance.

In the sealed nickel–cadmium system most of the oxygen evolved is reduced at the cadmium–metal surface, with the consequent oxidation of the cadmium to cadmium hydroxide. This statement certainly applies for normal rates of charge. However, the rate of charge and temperature affect the recombination of oxygen. Provided the charge rate is sufficiently high, the oxygen evolved for a given temperature will not combine within the cadmium and the pressure will build up within the cell. While the pressure is rising, a pronounced rise in cell voltage will occur (approximately half the magnitude of that occurring in open cells). This voltage may be detected and used to terminate the charge.

More recently, however, cell temperature monitoring has been developed as a means of charge control as discussed below.

Sealed nickel–cadmium cells normally tend to warm up on overcharge in proportion to the rate of oxygen recombination at the negative electrode. Unfortunately, the ability of most commercially available nickel–cadmium cells to recombine oxygen gas is rather limited, as evidenced by the acceptance of the $C/10$ overcharge rate. The cells can withstand higher rates, but in overcharge at high rate only a fraction of the evolved oxygen will be recombined, resulting in heat generation. This change in temperature is gradual and difficult to detect. The excess of unrecombined oxygen will result in internal pressure build-up and cell venting. Excessive venting of cells may reduce useful capacity, and cause dehydration and electrode imbalance.

To accomplish charge termination safely and reliably through temperature sensing requires fast-acting, precise and expensive equipment at the lower charge rates. Because of the size, cost and complexity of such a system, the thermal sensing approach to overcharge control has previously been impractical for the consumer-oriented nickel–cadmium battery-powered portable appliances and devices. To overcome these problems, Union Carbide have developed a special fast-charge sealed nickel–cadmium battery design (the Eveready cell) and a temperature-sensing high-rate battery charger design.

The Eveready fast-charge cell has been specially designed to withstand high-rate overcharge (up to the 1 h rate) and thus to overcome these drawbacks. Nevertheless, Eveready cells and batteries should not be continually overcharged at high rates. Undesirable cell degradation or heat build-up damage will occur if batteries are left on high-rate charge.

Eveready sealed secondary nickel–cadmium cells and batteries are now widely used as a rechargeable power source in many different types of portable or cordless electrical appliance. Charging at the safe recommended $C/10$ rate has proved satisfactory for recharging the cells or batteries used in many of these appliances, such as toothbrushes, shavers, etc., where relatively long rest periods between uses are possible. However, there is now a demand for the use of sealed nickel–cadmium cells and batteries in other appliances, such as chain saws, electronic flashes, portable drills and professional hair clippers, where the rest periods between uses of the appliances are much shorter; consequently shorter recharging times, from about 3 h to about 1 h, that is, $C/3$ to $C/1$ rates, are required.

47.2 The Eveready fast-charge cell (nickel–cadmium batteries)

As the design of the cell and the charger are, in this case, intimately linked a description of the cell design will be given first.

47.2.1 Cell design

This cell has been designed to exhibit sufficient temperature rise to effect charge control without a significant change in operating pressures. The Eveready fast-charge cell series develop the desired temperature rise, and have the built-in ability to withstand short-term overcharge at rates to 1 h values without physical damage or loss in cell capacity. The cell construction is specifically designed to withstand overcharge at the 3 h rate without special control circuitry. Considerable heat can be generated within the cell, however, if overcharge is extended beyond a reasonable period of time. To prevent this heat from causing gradual cell degradation, it is recommended that the cell temperature should not exceed 46°C during this extended overcharge and that the cells be removed from the charger within 2 or 3 days of reaching full charge.

Before the introduction of this construction by Union Carbide, any cell overcharged at the 1 h rate would be permanently damaged. The Eveready fast-charge cell can withstand overcharge at these high rates long enough for the temperature rise to be sensed by simple control elements. This temperature rise is very pronounced, and provides a positive signal for charge control. As a result, the control element can be small, lightweight and inexpensive.

Sealed secondary nickel–cadmium cells have been manufactured for many years based on the so-called 'oxygen recombination' principle. The charge-accepting capacity of the negative electrode is made to exceed the charge-accepting capacity of the positive electrode. On charging, the positive electrode reaches a state of full charge before the negative electrode, and oxygen is evolved at the positive electrode. The oxygen gas reacts or combines with the active cadmium metal on the surfaces of the negative electrode. Thus, recombination of oxygen prevents the build-up of an excessive internal gas pressure.

In charging nickel–cadmium cells, an overcharge, that is, ampere hours input in excess of that previously removed upon discharge, must be provided to ensure that the cells have reached full charge. If overcharge is continued at too high a rate of charge current, the evolved oxygen gas may not fully recombine, and a build-up of excessive internal gas pressure may result. A safety resealable vent is provided to limit excessive build-up of pressure. In the Eveready design, proper selection of the electrolyte volume controls oxygen recombination pressure below the safety vent opening pressure. The safe charge rate for sealed secondary nickel–cadmium cells for extended charge periods has been established at the 10 h, or the $C/10$, rate; capacity (C) is the rated ampere hour capacity of the cell and 10 is the number of hours required at perfect charge efficiency to bring a completely discharged cell to full charge. At the 10 h rate and lower currents, an equilibrium condition is maintained in the cell and consequently there is no excessive build-up of internal gas pressure.

47.2.2 Charger design

The Eveready fast-charge cell exhibits a relatively sharp rise in temperature during high-rate overcharge. The particular type of thermal sensor to be used in combination with the cell or battery and the charger system is not critical. Probably the least expensive overall cell or battery control unit is provided by use of a simple snap-action thermostatic switch, which combines the temperature-sensing and circuit-switching functions in one small, inexpensive device that can be easily attached to the cell or battery.

A solid-state thermistor sensor may also be used. The thermistor is also relatively inexpensive and is even more compact, although it performs only the function of a sensor. Auxiliary circuitry and a switching means are required to cut off the charging current in response to the thermistor input. Among the commercially available types of thermistor, the positive temperature coefficient type is preferred because it changes resistance abruptly at a predetermined temperature. Auxiliary circuitry is therefore simplified without loss of reliability.

In constructing individual cell or battery units, it is not critical that the thermal sensor be placed or maintained in actual physical contact with the cell proper, although this is preferred. Individual cell units may be constructed with a small flat disc-type thermostatic switch welded in contact with the bottom of the cell. Similar battery units may be constructed with a small thermistor or bimetallic switch placed in the space between adjoining cells. Any arrangement is satisfactory provided the thermal sensor is well exposed to the heat generated by the individual cell or one or more cells of the battery. The use of extensive heat sinks, such as placing the entire battery in a water bath, is not recommended since this can prevent heat build-up, impede oxygen recombination within the cell, and lead to cell venting before sufficient heat rise occurs. The terminal leads from the thermal sensor may be connected by additional external contacts or may be brought out from the cell or battery unit and connected directly into the circuit. Where a sensor-switch device is used in a series-connected battery, it may be preferred to wire the switch internally between two series cells so that no additional external contacts are required. The practicality of this connection depends on discharge current value and sensor current rating. The advantage would be that the circuit would also open on discharge in case the battery becomes overheated for any reason. The charger circuit required for charging the individual cell or battery is not unique. A constant-current charger is recommended, and due regard should be paid to heat dissipation and wattage ratings of all components.

Figure 47.4 shows typical voltage and temperature characteristics of the Eveready CF1 fast charge cells when charged at the 1 h rate. Note that the temperature remains relatively constant until the cell approaches full charge. At this point, a steep rise in temperature is initiated and continues until the cut-off range of 43.3–48.9°C is reached.

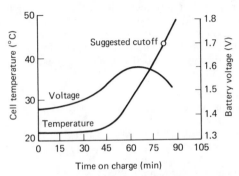

Figure 47.4 Typical voltage–temperature curves at 1 h charge rate for an Eveready sealed nickel–cadmium fast-charge CFI cell (Courtesy of Union Carbide)

Either mechanical or electronic methods can be used to terminate the charge at the desired temperature. A mechanical snap-action or bimetallic thermostat such as those used commercially in gas heaters, liquid heaters, percolators, fire alarm detectors, clothes dryers and the like has been used to terminate charge.

47.2.3 Fast-charge constant-current charger design

Three possible concepts using this type of device are shown in Figure 47.5.

In Figure 47.5(a):

1. Thermostat with non-lock-out features. The snap-action thermostat opens at 43.3–48.9°C and closes automatically at 35–40.6°C.
2. Negative battery terminal for discharge (2).
3. Positive battery terminal for both charge and discharge.

This is the least expensive means of control available. It will allow considerable overcharge if

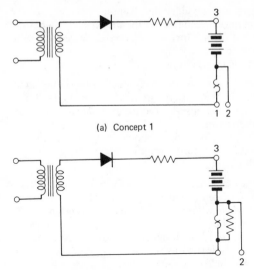

(a) Concept 1

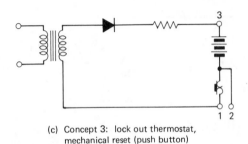

(b) Concept 2: lock out thermostat, automatic reset

(c) Concept 3: lock out thermostat, mechanical reset (push button)

Figure 47.5 Fast-charge concepts for Eveready sealed nickel–cadmium fast-charge cells (Courtesy of Union Carbide)

the battery is connected to the charger for long periods of time, since the battery will go back on charge as soon as the battery temperature drops below 35°C. One distinct advantage is that a battery that has reached a high temperature on discharge will automatically go on charge as soon as the battery cools down.

In Figure 47.5(b):

1. Thermostat with electrical lock-out. The snap-action thermostat opens at about 43.3–48.9°C.
2. Negative battery terminal for discharge (2).
3. Positive battery terminal for both charge and discharge.

The shunting of the charge current through a heating element in thermal contact with the thermostat holds the thermostat in the open position until the battery is disconnected from the charger. As a result, there will be no significant overcharge current if the battery is left connected to the charger for long periods. If some continuous charge current is desired, an appropriate resistor can be connected across the contacts of the thermostat as shown in the figure. A possible disadvantage would be the premature locking out of the thermostat if a battery still hot from a high-rate discharge were connected to the charger.

In Figure 47.5(c):

1. Thermostat with lock-out feature. The snap-action thermostat opens at about 43.3–48.9°C. There is a push-button for mechanically resetting thermostats.
2. Negative battery terminal for discharge (2).
3. Positive battery terminal for both charge and discharge.

The snap-action thermostat would cut off all charge currents. This concept might be useful in an application where high-rate charge was not normally needed, but would be useful at certain times. A charger could be designed to provide more moderate charging such as the $C/10$ rate, unless the user pushed the reset button when a fast charge was required (as shown in the figure). Again a heated

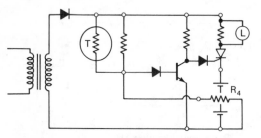

Figure 47.6 Fast-charge concept for Eveready sealed nickel–cadmium fast-charge cells. The circuit is designed to cut off charge at 43.3–48.9°C. The big advantage of this concept is the small size of the thermistor (R_4) contained in the battery pack. The circuit must be specifically designed for each application and its stability checked over the temperature range encountered by the battery in use (Courtesy of Union Carbide)

battery from high-rate discharge would necessitate a wait until the thermostat could be reset.

A second means of terminating the charge on the battery at the desired cut-off temperature is through the use of temperature-responsive electronic circuitry. In essence, this circuit will electronically terminate charge when the desired temperature is reached through the use of a sensing element and switching-type devices. A possible electronic circuit for this type of charge control is shown in Figure 47.6. The time to reach cut-off and the period of cycling 'on' and 'off' depend on charge rate, battery packaging and ambient temperature.

47.3 Types of constant-current charging

There are many constant-current methods that permit satisfactory charging of sealed nickel–cadmium batteries. These methods differ in the length of time required to charge, the complexity of the circuits used and the various techniques for charging with maximum safety and reliability. The following charging systems are worthy of mention (see Table 47.1):

1. Normal rate charge.
2. Accelerated charge.
3. Controlled rapid charge: timed, voltage–time control, voltage–temperature control.
4. Ultra-rapid charge.
5. Uncontrolled rapid charge.
6. Low-temperature charge.
7. Permanent or floating charge.

The various methods of charging of sealed nickel–cadmium cells and batteries are discussed briefly below.

Table 47.1 Summary of non-permanent methods of charge recommended for SAFT sealed nickel–cadmium batteries

Charge method	Duration	Application temperature (°C)	Battery type	Initial state before charge	Capacity recoverable (%) rated at 20°C	Continuous overcharge acceptance
Normal	14 h	5 to 40	VB–VR	Any	100	20 000 h
	14 h	5 to 40	VO sealed	Discharged	100	None
Accelerated	3 h	5 to 40	VR 0.1 to VR 1.8	Any	100	10 days
	7 h		VR 2.5 to VR 7			continuous
Controlled						
Rapid timed	50 min	10 to 40	VB	Discharged	80	None
Timed	1 h	10 to 40	VR 0.1 to VR 1.8	Any	80–100	None
Voltage–time control	1 h		VR (except VR 10)	Any	80–100	None
Voltage–temperature control	15–60 min	−20 to +40	VR	Any	80–100	None
Ultra-rapid	1–15 min	15 to 30	VR	Discharged	10–80	None
Uncontrolled rapid	30–60 min	0 to 30	VX	Any	100	Several days

47.3.1 Normal constant-current charging

This is the method recommended by all battery manufacturers for the charging of sealed nickel–cadmium batteries at the normal (i.e. unaccelerated) rate of charge.

It is permissible to use a nominal constant current, where the initial current (when the battery is discharged) is not critical, provided the I_{10} rate is not exceeded on overcharge.

Normal charge is carried out at a rate of $0.1C_5$ A or the recommended charging current at room temperature for 14–16 h and may, for example, be applied to all sealed cylindrical and button cells and batteries between 5 and 40°C, whatever the initial state of charge. Overcharge acceptance at this rate is great (more than 20 000 h). Sealed rectangular cells must be in the discharged state before being recharged for 14 h at $0.1C_5$ A. Below 10°C the rate of charge must be reduced.

The 10 h rate should not normally be exceeded unless overcharge is specifically to be prevented. During charging, the cell voltage rises from approximately 1.30 V to a maximum of approximately 1.45 V at near full charge. Finally, the voltage is stabilized at approximately 1.43 V. The decreasing voltage at the end of charging depends on a temperatue rise in the cell. It is caused by the reaction between oxygen and the negative electrode in the cell. This voltage characteristic, with a gradual and relatively moderate voltage increase followed by stabilizing at a level slightly less than maximum, is typical for sealed nickel–cadmium cells. Typical constant-current charging curves for a 0.4 A h nickel–cadmium cell are shown in Figure 47.7.

Instead of the normal charging current, both lower and higher currents can be used. The final voltage is dependent on the charging current and increases with rising current (Figure 47.7). These curves show typical final charging voltages in relation to charging current when the cells have been charged with $1.8C$ A h. Because of low charging efficiency at low charging currents, the final voltages in these curves corresponding to charging currents below approximately $0.1C$ A do not represent a full state of charge.

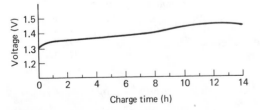

Figure 47.7 Charge voltage *versus* charge time at a constant current of 0.05 A at 25°C. Nife Jungner 0.4 A h sealed nickel–cadmium cells (Courtesy of Nife Jungner)

47.3.2 Accelerated constant-current charging

This method is applicable only to sealed nickel–cadmium cylindrical cells at cell temperatures between 10 and 40°C. 3 h at $0.4C_5$ A or 7 h at $0.2C_5$ A suffices to fully charge the battery according to cell type, with high overcharge acceptance which must be limited to about 10 days to avoid an adverse effect on cell life. For cylindrical cells the accelerated rates correspond to the maximum permissible on long-term overcharge for the cells in the temperature range 5–40°C. A few hours' overcharging at these currents will not damage the cells. The cells must not be fully charged with currents higher than $0.2C$ A. Completely discharged cells at normal temperature can be charged to approximately 80% of charge in 1 h with seven times the normal charging current shown, for example in the case of Nife Jungner sealed nickel–cadmium cells in Table 47.2. Continued charging to full state of charge must not be carried out with this high current, but the normal charging currents shown in Table 47.2 should be used. This requires several hours and thus charging to full charge with this method is not advantageous. In urgent situations, when full capacity is not required, it can be useful, however.

47.3.3 Controlled rapid constant-current charging

This is understood to imply a charge period of less than 1 h. (When the period is less than 15 min the charge is called ultrarapid; see below.) Controlled rapid charge implies that there is a system that interrupts the charge before the cells or battery reaches an overcharge state that is mechanically excessive (that is, could burst the case).

The methods of control recommended are suited to the operating temperature range and the ability of the nickel–cadmium cells or battery to accept or otherwise a degree of overcharge at high rate.

This method of charge should not be applied to cells and batteries that cannot accept overcharge at a high rate and which do not have safety valves that can operate in case of an excessive internal pressure.

Timed charge

Rapid charge for a limited time of cylindrical standard nickel–cadmium batteries (simplified 1 h rapid charge) is applicable to cylindrical standard batteries, differing from voltage–temperature controlled rapid charge which requires specially adapted batteries. According to whether the sizes of these cells will accept overcharge at a rate of C_5 A, a distinction is made between two methods of limited time rapid charge which may be applied to cells of an unknown initial state of charge. For example, SAFT VR 0.1 ⅓AA 0.1 A h cylindrical cells and

Table 47.2 Technical specification of Nife Jungner sealed nickel–cadmium cells

Cell type	KR 15/51	KR 23/43	KR 27/50	KR 35/62	KR 35/92
Nominal capacity (A h) (5 h rate C_5 at 25 ± 5°C to end-voltage 1 V)	0.40	1.1	1.6	3.6	5.6
Current on discharge (mA) ($0.2C_5$ A)	80	220	320	720	1120
Dimensions					
Diameter (mm)	14.5	22.7	26.0	33.2	33.2
Height (mm)	50.3	42.2	48.0	59.5	89.0
Weight, approx. (g)	20	45	65	140	210
Energy content (average discharge voltage with discharge at 5 h current, $0.2\,C$ A, is 1.2 V)					
W h/cell	0.48	1.32	1.92	4.32	6.72
W h/kg	24	29	30	31	32
W h/dm³	65	81	78	86	90
Internal resistance (m Ω)	90	27	22	13	11
Charging					
Normal charging current 14–16 h (mA)	50	120	200	400	600
Trickle charging minimum current (mA)	20	45	75	150	250
Temperature range	From 0 to 40°C, preferably normal room temperature				
Discharge					
Maximum recommended continuous discharge current (A)	1.5	5	8	15	20
Maximum recommended discharge current for 5 (A)	7	15	20	50	80
Temperature range	From −40 to +40°C				
Storage	In charged or discharge condition: use a cool storage place, preferably with temperature lower than 15°C				
Terminal arrangement:					
Cell type CF, drawing No.	90-51400	90-51405	90-52813	90-51411	90-51417
Cell type HH, drawing No.	90-51401	90-51406	90-52814	90-51412	90-51418
Cell type HB, drawing No.	90-51402	90-51407	90-52815	90-51413	90-51419
Batteries with even number of cells, drawing No.	–	90-52359	90-53536	90-52357	90-52355
Batteries with odd number of cells, drawing No.	–	90-52360	90-53537	90-52358	90-52356
Corresponding dry cell	R6 or AA	Sub C	R14 or C	R20 or D	–

batteries (Table 51.1) can withstand a double charge at a rate of C_5 A (0.1 A). The internal pressure does not reach the opening pressure of the safety valve and the temperature rise on overcharge is less than 20°C.

These batteries and cells may be charged at C_5 A, whatever their initial state of charge, and the chargers necessary are of a very simple design, a timer terminating the charge after 1 h.

Cylindrical cells and standard batteries such as SAFT VR 0.45 ½A (0.45 A h), VR 0.5 AA (0.5 A h), VR 1.2 RR (1.2 A h), and VR 1.8 C (1.8 A h) (see Table 51.1) cannot accept a double charge at the rate of C_5 A; the internal pressure and temperature on overcharge exceed permissible limits. An overcharge of limited duration is, however, possible in complete operational safety. These cells and batteries can accept an overcharge at C_5 A, 0.4 times their nominal capacity, without rises of internal pressure and temperature exceeding permissible operating limits. It is obviously possible that cells and batteries that are already fully charged might be put on charge, and a preliminary discharge

is therefore necessary. One hour chargers for these standard batteries and cells (available from SAFT) have the following features:

1. Preliminary discharge by connecting the battery to a fixed resistance calculated to discharge a maximum capacity of $0.6C_5$ A h.
2. Recharge in 1 h at C_5 A.

Rapid charge for limited time of cylindrical cells and standard batteries is applicable only in the temperature range 10–40°C. For cylindrical cells and standard batteries of capacities greater than 1.8 A h, no overcharge at a rate of C_5 A is acceptable, and it is necessary to consider a parameter other than time for the 1 h charge of cells and batteries having an unknown initial state of charge.

Voltage–time control and voltage–temperature control

Sealed nickel–cadmium cells and batteries on charge exhibit a voltage characteristic that passes

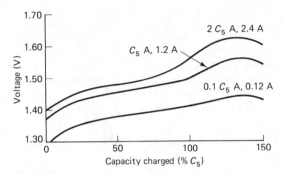

Figure 47.8 Cell voltage at various charging rates at 25°C for a SAFT VR 1.2 RR (1.2 A h) sealed nickel–cadmium cell (Courtesy of SAFT)

through a maximum at about the fully charged condition. Figure 47.8 shows that at very low rates of charge $(0.1C_5 A)$ the rise of voltage at the end of charge is very small; on the other hand at rapid rates $(C_5 A$ to $2C_5 A)$ it is quite large. It is generally accepted that a charge rate at least equal to $C_5 A$ is necessary for the end-of-charge voltage rise to be sufficiently sharp and reproducible and thus to serve as a signal to detect the end of charge. The on-charge voltage characteristics vary with the temperature and the charge rate, and show significant differences between sizes of cylindrical cells. For each cell, for a given rate of charge, it is possible to draw a theoretical cut-off voltage curve as a function of cell temperature. An example is shown in Figure 47.9(a) for a battery of ten SAFT VR 1.2 RR (1.2 A h) cylindrical cells (Table 51.1) charged at rates of C_5 and $2C_5 A$. At each temperature the selected cut-off voltages are defined to coincide approximately with the fully charged condition and to be clearly above the first voltage plateau and below the maximum end-of-charge voltage.

Depending on the operating temperature range, two methods of control (voltage–time and voltage–temperature) can be used. Control by voltage–time is used in applications where the operating temperature is of the order of 20°C (from 10 to 40°C). Chargers designed for this method of operation terminate the charge when the battery voltage reaches a fixed preset value. This value corresponds to the cut-off voltage at 40°C (the lowest cut-off voltage in the temperature range under consideration). A time-switch override is used to allow for the case where the battery temperature might exceed 40°C or the end-of-charge voltage might be abnormal.

This system of control does not require the use of special batteries and may be applied to all standard batteries including cylindrical types with capacities up to 7 A h charged at the $C_5 A$ rate. The time-switch terminates the charge in all cases after 1 h. The capacity recharged is close to the rated

capacity between 20 and 40°C and is less below this lower temperature.

In control by voltage–temperature, the termination of rapid charge is obtained from values of voltage that depend on the temperature of the battery. For economic reasons, it is not possible to follow precisely the theoretical characteristics of cut-off voltage as a function of temperature (Figure 47.9(a)). The characteristics may be linearized while retaining sufficient security for cut-off. Figure 47.9(b) shows the characteristics used in practice for a battery pack of ten SAFT 10 VR 1.2 RR 1.2 A h (i.e. 12 A h) cells (Table 51.1), charged at $C_5 A$ and $2C_5 A$.

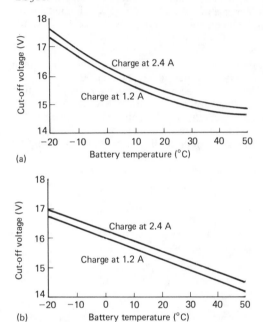

Figure 47.9 (a) End-of-charge cut-off voltage *versus* temperature: SAFT 10 VR 1.2 RR battery pack (ten 1.2 A h cells, 12 A h battery); fast charge at $C_5 A$ rate (1.2 A h) then $2C_5 A$ rate (2.4 A h). (b) Charger cut-off voltage *versus* temperature: SAFT 10 VR 1.2 RR battery pack (ten 1.2 A h cells; 12 A h battery); fast charge at $C_5 A$ rate (1.2 A h) then $2C_5 A$ rate (2.4 A h) (Courtesy of SAFT)

The capacity obtained with this system remains satisfactory in the temperature range of –20 to +40°C. Most rapid-charge systems with voltage–temperature control leave the batteries on low-rate charge after cut-off of the rapid charge. The object of this is to give 100% of capacity charged. To provide maximum security during charging, batteries can be equipped with thermostatic protection which interrupts this charge in case of irregularity in the battery or the charge system. This protection operates when the battery temperature reaches the abnormal value of 55–60°C. Batteries for rapid charge with voltage–temperature control must be adapted to this system because they must be equipped with a temperature sensor for voltage control and a thermostat for protection against overheating.

47.3.4 Ultra-rapid constant-current charging

By limiting the capacity charged to a certain proportion of the rated capacity of the cell or battery, it is possible to charge cylindrical and button nickel–cadmium sealed cells and batteries at very high rates and thus obtain a short-duration charge of the order of a few minutes. The amount of the capacity that may be charged is determined by the increase of internal pressure, which must always remain below the operating pressure of the valve for cylindrical cells and batteries and the deformation pressure for button cells and batteries. Ultra-rapid charge requires a prior discharge of the cells, which may be done in periods as short as 30 s.

Figure 47.10 shows the durations of ultra-rapid charge that are possible for a SAFT VR 0.5 AA (0.5 A h) cylindrical cell (Table 51.1), as well as the security limit corresponding to opening of the valves at 20°C. A method of charge derived from the

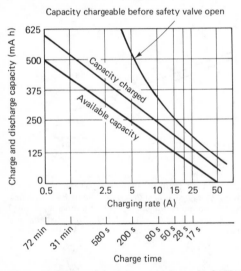

Capacity chargeable before safety valve open

Figure 47.10 Duration of ultra-rapid charge for SAFT VR 0.5 AA sealed cylindrical nickel–cadmium cells: charge duration *versus* charging rate (Courtesy of SAFT)

ultra-rapid is the one-shot charge for use in an application that requires only one single cycle, such as a military application where the battery is charged only once and is subsequently destroyed with the apparatus it supplies. For operational reasons, the battery must be charged in a few seconds, which is possible at very high rates if the capacity charged is limited to a few per cent of the rated capacity. This technique is applicable with complete safety over

the temperature range −40 to +60°C. In the case of one-shot charging, the available capacity is small, but the power available for short periods remains high. Thus, fully discharged button cells will accept between −40 and +60°C charge rates of up to $20C_5$ mA for a limited time. For example, SAFT VB 10 (0.1 A h) button cells (Table 51.1) charged for 5 s at 1 A have an available capacity of 2400 mA.

47.3.5 Uncontrolled rapid constant-current charging

This method is applicable to nickel–cadmium cylindrical cells and batteries such as the SAFT VO series (Table 51.1), which can withstand higher rates of charge and overcharge than standard cylindrical cells such as the SAFT VR range. This extra performance is made possible by the use in battery design of highly porous plates which improve oxygen consumption at the negative plate. A further consequence of this is that the available capacity per unit volume of the former type of cell is reduced to about 40–50% of that of the latter type. However, the power-to-volume ratio remains high. Overcharge at high rate may continue for several days without affecting the battery life.

The ability to overcharge at high rates permits the uncontrolled rapid charge of cells of unknown state of charge using chargers of simple design and relatively low cost. The times for uncontrolled rapid charge for cells that can stand higher rates of charge and overcharge (e.g. the SAFT VX range (Table 51.1)) are between 30 and 60 min in the temperature range 0–30°C.

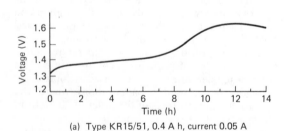

(a) Type KR15/51, 0.4 A h, current 0.05 A

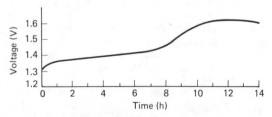

(b) Type KR35/92, 5.6 A h, current 0.6 A

Figure 47.11 Charge voltage *versus* charge time of Nife Jungner sealed nickel–cadmium cells at constant current and 0°C (Courtesy of Nife Jungner)

Table 47.3 Comparison of maximum charge rates in normal and continuous (float) charging at various temperatures recommended for SAFT VB series sealed nickel–cadmium button cells

Cell type		VB 4	VB 10	VB 22	VB 30	VB 60
Rated capacity (A h)		40	100	220	300	600
Nominal discharge voltage per cell (V)		1.20	1.20	1.20	1.20	1.20
Normal charge from −30 to +10°C, max. charge rates (mA) for charge time (h) in brackets	−30 (100)	0.4	1.0	2.2	3.0	6.0
	−20 (55)	0.8	2.0	4.0	6.0	12.0
	−10 (30)	1.5	4.0	8.0	11.0	21.0
	0 (20)	2.5	6.0	13.0	18.0	36.0
	+10 (14)	4.0	10.0	22.0	30.0	60.0
Continuous charge from −30 to +10°C, max. charge rates (mA) for charge time (h) in brackets	−30 (100)	0.2	0.5	1.0	1.5	3.0
	−20 (55)	0.4	1.0	2.0	3.0	6.0
	−10 (30)	0.8	2.0	4.0	5.0	11.0
	0 (20)	1.2	3.0	6.5	9.0	18.0
	+10 (14)	2.0	5.0	11.0	15.0	30.0

Occasional overcharge at these rates beyond the prescribed times is not detrimental to the cell. Rapid and ultra-rapid charge systems can be obtained for charging in times from a few seconds to 1 h

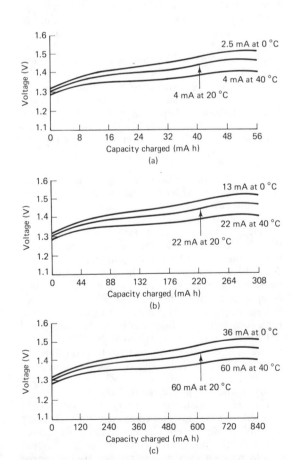

Figure 47.12 Voltage *versus* capacity for 'normal' charging of SAFT VB series sealed nickel–cadmium button cells. (a) UB4 $40C_5$ mA h; (b) UB22 $220C_5$ mA h; (c) VB60 $600C_5$ mA h (Courtesy of SAFT)

47.3.6 Low-temperature constant-current charging

Charging should preferably be done at normal temperature, but can be done between 0 and 40°C. Typical charging voltages are lower at a temperature higher than room temperature, and higher at temperatures lower than room temperature. Typical charging curves for nickel–cadmium cells at 0°C are shown in Figure 47.11. Manufacturers claim that occasional overcharging during a few weeks of cell life with normal recommended charging current will not damage the cells.

Table 47.3 compares the maximum charge rates (in milliamperes) for stated charge times (in hours) recommended for various charging temperatures between −30 and +10°C in the normal and the continuous (float) charging of a series of SAFT VB series (Table 51.1) sealed nickel–cadmium button cells.

Figure 47.12 presents a series of typical voltage *versus* capacity charged curves for various discharge rates for a series of button nickel–cadmium batteries (SAFT VB series, Table 51.1).

When the temperature of batteries is less than 5°C, it is necessary to limit the rates of charge and overcharge. The charge rate must be chosen so that the internal pressure on overcharge is less than the operating pressure of the valves in the case of cylindrical cells or the deformation pressure in the case of button and sealed rectangular cells. Figure 47.13 shows, for cylindrical and button cells, the maximum permissible rates on overcharge for short periods (a few days) and on permanent overcharge. A number of methods are used to vary the charge current as a function of temperature:

1. Fit into the battery a temperature sensor that operates on the charger to adjust the charge current to suit the battery temperature.
2. Charge at constant current and voltage limited to 1.55 V/cell, a voltage setting that will avoid all risk of internal overpressure at low temperature.
3. Use the charge current corresponding to the lowest operating temperature.

Also, in certain cases, for example when charging at low temperatures, constant-current charging can be combined with voltage limitation. With a normal charging current of approximately $0.1CA$ a fully charged state is reached after 14–16 h at room

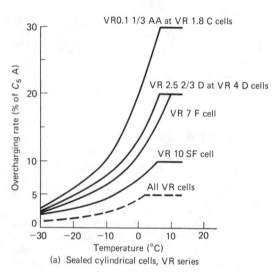

VR0.1 1/3 AA at VR 1.8 C cells

VR 2.5 2/3 D at VR 4 D cells

VR 7 F cell

VR 10 SF cell

All VR cells

(a) Sealed cylindrical cells, VR series

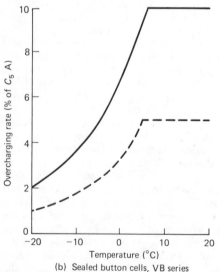

(b) Sealed button cells, VB series

Figure 47.13 Effect of temperature on optimum maximum permissible overcharge rates for short periods: SAFT VR and VB sealed nickel–cadmium cells: —, maximum rates for occasional overcharge; ----, maximum recommended permanent charge rates (Courtesy of SAFT)

temperature. A lower current, $0.04–0.05CA$, is often used as continuous charging current in different applications. The charging efficiency is then lower, because proportionally more energy is used to produce oxygen. The charging efficiency is also dependent on the temperature, and decreases with rising temperature.

47.3.7 Permanent or floating constant-current charging

In applications with permanent charge, the object of the charge current is to maintain an already fully charged nickel–cadmium battery in this state over a period of time and when necessary to recharge any capacity discharged at regular intervals or in a random manner. The first charge may be carried out either at the permanent rate if this is sufficient ($\geqslant 0.02C_5A$), or at a higher rate where the permanent charge value is too low for the effective recharge of a fully discharged battery. Rates vary from $0.005C$ mA to $0.05C$ mA. Where the appropriate rate is below $0.02C_5$ mA the battery should be given a normal charge ($0.1C_5$ mA for 14 h) when commissioning.

Trickle charge can be used in applications where the cells must always be kept fully charged. The minimum recommended current for trickle charging (Table 47.2) is chosen so that a full state of charge can be obtained when recharging with the same current. After charging for 24 h the state of charge is approximately 80%, and after 48 h approximately 95%. Full state of charge requires approximately 1 week.

In applications where the cells are mounted closely together with charging equipment or other heat-generating devices, the temperature increase can be considerable. The life of the cell is adversely affected by increased temperature. Thus the capacity will decrease to 75% of nominal when the cells are trickle charged with recommended minimum current for 1 year at 40°C. Therefore the cells must be shielded from heat radiation and the battery compartment must be well ventilated. The cell voltage decreases only slightly during discharge and for a completely discharged cell on open circuit will quickly be restored to approximately 1.2 V. The cell voltage can therefore not be used to determine the state of charge. Cells with an unknown state of charge should be considered discharged, but must not be recharged at the rapid-charge rate.

The nickel–cadmium battery can be trickle charged, but floating and constant-voltage charging are not recommended. For maximum performance in situations of long-term trickle charge, the current should be kept to a minimum. The trickle charge current required to keep the battery fully charged is approximately the 30–50 h rate plus whatever is necessary to compensate for any major withdrawals.

Lower trickle rates maintain the battery at some lower state of charge. If either floating or constant-voltage charging is mandatory, a thermal cut-out that senses battery temperature must be included in the charge circuitry. Battery overheating, which follows from any irregularity in the voltage–current control system, can be totally destructive to either battery or charger, or both.

Permanent charge with occasional discharge

This is usually the case where the nickel–cadmium battery supplies emergency power to apparatus in the event of a.c. supply failure. Permanent charge currents may be very low, from $0.005C_5$ to $0.02C_5$ A, when the depth of discharge foreseen is less than 50%. In this case, nickel–cadmium batteries are put into service after having been previously charged at a higher current. If the depth of discharge is great and if recharge must be obtained in a few hours, two solutions are possible:

1. Charge at a single rate, up to $0.05C_5$ A.
2. Charge by the two-rate method: fast charge at $0.1–0.3C_5$ A, then continue with a maintenance charge of $0.01C_5$ A.

Permanent charge with regular discharge

In this case the nickel–cadmium batteries are the power source for apparatus that requires a supply for more or less random periods, but for which the total capacity discharged in a given time is known. During the time when the batteries are not discharging they receive a continuous charge, the rate of which depends on the discharge capacity during a known period. Rates of charge may vary from $0.02C_5$ to $0.1C_5$ A. Telephones are an important application in which batteries operate on this principle; the batteries discharge when the telephones are in use and remain on charge when no communication is taking place; for telephones, charge currents vary from $0.006C_5$ to $0.02C_5$ A.

For example, for SAFT cylindrical VR series cells (Table 51.1), the following are the most appropriate charging methods for the constant-current charging of these particular types of battery at normal temperature:

1. *Minimum charge* $0.01C$ for 140 h irrespective of state of charge (all cylindrical SAFT VR types).
2. *Normal charge* $0.1C$ for 14 h irrespective of state of charge (all cylindrical SAFT VR types).
3. *Accelerated charge* Irrespective of state of charge, the following cylindrical cells can be charged within 3 h: SAFT VR 0.1 ⅓ AA (0.1 A h) to VR 2 C (2 A h). Irrespective of state of charge, the following cells can be charged within 7h: SAFT VR 2.5 (2.5 A h), VR 4 (4 A h) and VR 7 (7 A h) (Table 51.1).

4. *Uncontrolled rapid charge* Irrespective of state of charge, cylindrical VR series cells can be charged within 15–60 min.
5. *Ultra-rapid charge* 1–3 min. After discharge only cylindrical VR series cells type, SAFT VR 0.1 ⅓ A h (0.1 A h), VR 0.5 AA (0.5 A h), VR 0.7 ½ C (0.7 A h) and VR 1.2 RR (1.2 A h).
6. *Continuous overcharge current* The SAFT VR series of cylindrical cells are unaffected by indefinite charging at relatively high currents. Typically: VR 1.2 RR (1.2 A h) charged at 120 mA, VR 4 D (4 A h) charged at 400 mA and the VR 7 F (7 A h) charged at 700 mA.

47.4 Two-step constant-current charging

Because of the electrochemical processes involved, certain types of battery exhibit two distinct voltage steps when they are being charged by the constant-current method. The silver–zinc battery is a case in point. Figure 47.14 demonstrates such a charging curve for a Yardney silver–zinc battery, and it is seen that during the first 20% of the charge the potential across the electrodes is 1.6 V; thereafter for the remainder of the charge the potential increases to approximately 1.9 V.

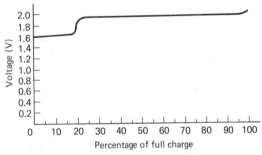

Figure 47.14 Typical charging curve at 20°C for high- and low-rate Yardney Silvercel silver–zinc batteries (Courtesy of Yardney)

47.5 Constant-current charger designs for normal-rate charging

47.5.1 SAFT charger designs for nickel–cadmium batteries

SAFT manufacture various types of charger for use with their sealed nickel–cadmium cells and batteries in the cylindrical (VR), button (VB) and rectangular (VO) series. The specifications of two of these chargers are:

- Model 4: 60 mA
  Model 5: 100 mA
  Model 7: 120 mA

- Charging capacity: CC 175; outputs available between 10 and 120 mA for ten cells in series or batteries up to 12 V.

  CC 175/4E; output available at 60 mA for 20 cells in series or batteries of up to 24 V.
- Dimensions: base 70 × 50 mm, height 41 mm.
- Weight: all models 0.25 kg.

SAFT series S charger

This constant-current charger is designed for use with rechargeable sintered plate nickel–cadmium cells primarily of the rectangular and button types. All models (see Table 47.4) are fully transistorized and short-circuit proof. Cells or batteries on charge are protected against a.c. supply failure. The a.c.

Table 47.4

Model	Charge rate (mA)	No. of cells	Case type*	Weight (kg)
S 202	100	10–20	B	0.9
S 401	400	10	B	1.0
S 402	400	20	B	1.0
S 70 IT	700	10	A	1.5

* Dimensions: Case type A: 100 mm high × 160 mm wide × 230 mm long;
 Case type B: 100 mm high × 120 mm wide × 230 mm long

supply input is via a 110 or 230 V selector and fuse. A 14 h adjustable electric timer is fitted to charger model S 70 IT.

SAFT CC 175 series charger

This is an encapsulated constant-current charger designed for use with button-type rechargeable sintered plate nickel–cadmium cells. The charger is fully transistorized and short-circuit proof. Cells and batteries on charge are protected against a.c. supply failure.

- AC supply input: 240 V, 50 Hz.
- Charging rates: Model 1, 10 mA.
  Model 2, 22 mA.
  Model 3, 30 mA.

47.5.2 Varta charger designs for nickel–cadmium batteries

Varta supply two types of universal constant-current charger for use with sealed nickel–cadmium batteries, the specifications of which are given below.

Varta type CC306C automatic

The charger will handle from one to 30 nickel–cadmium cells in series connection up to 6 A h capacity.

The rate of charge is adjustable over two ranges selected by the range switch 0–60 and 0–600 mA. A potentiometer in conjunction with a calibrated meter provides fine adjustment. The meter is calibrated 0–60 and 0–600 mA.

The circuitry of the charger is safe against short-circuit and reverse polarity connection.

The instrument incorporates a timer with the facility to provide preselected charging periods from 1 to 14 h with automatic cut-off, thus allowing the charger to be left unattended once the charging cycle has been started.

- AC supply input: 240 V a.c. 50 Hz/120 V a.c. 60 Hz by means of voltage selector mounted on the rear panel.
- AC supply cable: 2 m of three-core PVC covered.
- Charging rate: up to 600 mA over two ranges: 0–60 and 0–600 mA by means of panel-mounted range switch.
- Cell capacity: 1–30 cells in series.
- Controls (all on the front panel): meter calibrated 0–60 and 0–600 mA; potentiometer for current selection; manually reset timer with automatic cut-off.

Varta charger type CC106C OIK II

The output is preset up to ten nickel–cadmium cells connected in series, and the rate of charge is internally selected prior to dispatch. The circuit is designed such that the output terminals are 'floating', this gives the unit added versatility in as much as more than one can be added together, thus two units (with the same output) with output terminals in parallel (positive to positive and negative to negative) can give a capability of charging up to 12 A h batteries at 1.2 A; conversely two units connected in series (positive to negative, etc.) give a capability of charging up to twenty cells up to 6 A h.

Another feature of this low-cost design is the choice of 240 V a.c. or 120 V a.c. by means of internal tap.

Nominal constant current is derived from a transformer, full-wave silicon bridge rectifier and output limiting resistors.

- AC supply input: 240 V a.c. 50 Hz/120 V a.c. 60 Hz; a.c. input fuse.
- AC supply cable: 2 m three-core.
- Charging rate: up to 600 mA by means of internally selected resistors (present before dispatch).
- Maximum cell capacity: 10 cells up to 6 A h.
- Output: Belling Lee 4 mm socket/spade captive-type terminals on front panel.
- DC output fuse.

The final charge voltage of a nickel–cadmium battery can vary from approximately 1.4 to 1.5 V, but the effects of battery age, temperature and rate of charge will give variations outside this range. This voltage variation, together with the flat nature of the charge curve, makes it impossible to guarantee an overcharge rate by the constant-voltage method.

47.5.3 Nife Jungner designs for nickel–cadmium batteries

Nife Jungner recommend that their nickel–cadmium cells be charged with constant current and with the cells connected in series. In those cases where parallel batteries are used for capacity reasons, the charging equipment should be designed to give equal current to each battery. In effect, the parallel connection is eliminated during the charging sequence. Most applications require charging from a standard a.c. supply of 220 or 110 V. The charging equipment can be very simple, but should give a reasonably constant current. Figure 47.15 shows two examples of full-wave rectifiers recommended for charging sealed nickel–cadmium batteries.

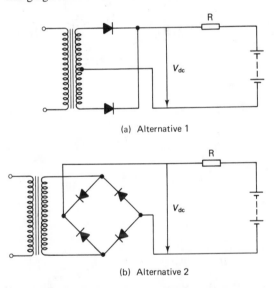

(a) Alternative 1

(b) Alternative 2

Figure 47.15 Typical charge circuits for Nife Jungner KR series cells. Voltage V_{dc} from the rectifier should be about three times the charging voltage of the battery; the resistance and power rating of resistor R should be adapted to limit the current to particular values (Courtesy of Nife Jungner)

Nife Jungner type NUC-612

Nife Jungner supply a constant-current charger type NUC-612 designed to charge from one to 12 of any of the range of KR series nickel–cadmium rechargeable cells, connected in series, at a constant current. The combined rotary on/off and selector switch is set to the type of cell to be charged, and the output voltage is then automatically adjusted to maintain a constant charging current according to the number of cells being charged. The charger is provided with a safety fuse and LED indicator, but any other control or ammeter is rendered unnecessary by the automatic system. The specification of this charger is given below.

- Supply voltage: Standard 220 V ± 10% single phase (other voltages can be provided for as an optional extra).
- Frequency: 50 Hz.
- Output current: KR 15/51, 0.05 A.
  KR 23/43, 0.12 A.
  KR 27/50, 0.20 A.
  KR 35/62, 0.40 A.
  KR 35/92, 0.60 A.
- Output voltage: 0–17.4 V.
- Dimensions: 170 × 150 × 60 mm overall (approx.).
- Weight: 1600 g (approx.).

47.6 Controlled rapid charger design for nickel–cadmium batteries

The Varta TSL charging system (Figure 47.16) charges the nickel–cadmium cell or battery with pulses of current with a voltage discriminator monitoring during the pauses between the pulses.

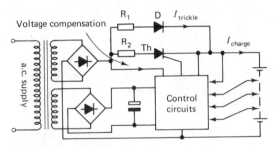

Figure 47.16 Varta TSL charging system for sealed nickel–cadmium cells (Courtesy of Varta)

With this method, battery interconnection and internal resistance potential differences do not influence the discriminator, only the cell potential being seen.

The reference voltage for the discriminator is matched to the temperature coefficient of the cells; that is, -4 mV/cell per °C. The Varta TSL system provides a safe and reliable method of fast charging sealed nickel–cadmium cells and batteries. Because of cell tolerances and variations in operating conditions, Varta do not recommend the use of more than ten cells per discriminator. For electrolyte temperatures up to 50°C it is recommended that the charge rate should not be less than the $10 \times I_{10}$ rate, and for electrolyte temperatures not exceeding 35°C the charge rate should not be less than $5 \times I_{10}$.

Provided these temperature recommendations are not exceeded, the charge terminating cell potential will be exhibited.

The battery tripping voltage occurs at approximately 80% of the battery's nominal capacity. The time taken to reach this point is dependent on the charge rate selected.

The temperature differential between charger and battery should not exceed ±10°C, which necessitates, for cyclic operation, an adequately ventilated battery. The Varta TSL system is most suitable for operation at elevated temperatures where problems normally occur because of poor charge acceptance. With the TSL system, charge acceptance is the same at 50°C as at 20°C. This system is of particular value in applications for standby supplies, since an interruption in the a.c. supply resets the charger, which then continues the charge to 80% capacity and then automatically switches over to trickle charge. This system gives similar battery life under normal operating conditions and superior battery life at extremes of temperature when compared with normal charging systems. If batteries have been stored for a prolonged period and are then connected to a fast charge system as described above, it will be found that the tripping voltage will be reached much sooner than is expected (possibly putting as little as 20% capacity into the battery). However, if the battery is left connected to the charger and the trickle charge fully charges the battery, when a further cycle of fast charge is started the battery will behave normally.

47.7 Transformer-type charger design (Union Carbide) for nickel–cadmium batteries

Published data for battery charging circuits, or d.c. power supplies with battery load, usually assume perfect transformers and rectifiers. Often these data do not provide for the series resistance element to limit current and prevent excessive changes in the charge current due to a.c. source voltage variations. The following design method and equations have been developed to overcome these limitations.

The common types of single-phase charging circuits and equations are shown in Figure 47.17. The terms of the design equations of Figure 47.17 will be taken in order for further definition. The value of E_{dc} or battery voltage varies with state of charge, temperature, type of nickel–cadmium cell construction and charge rate. These variations are less significant to constant-current operation than to other charge methods. The value of E_{dc} for a fully charged nickel–cadmium battery is between 1.35 and 1.45 V/cell for a 10 h charge rate at room temperature. However, 1.5 V/cell may be used for charger design calculations. The charging current,

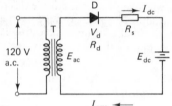

(a) Half-wave

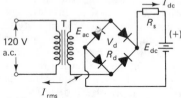

(b) Full-wave bridge

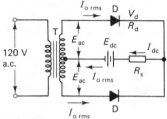

(c) Full-wave centre tap

$$E_{ac} = K(E_{dc} + nV_d) + MI_{dc}R \qquad (47.1)$$
$$R = R_s + R_t + nR_d \qquad (47.2)$$
$$I_{rms} = FI_{dc} \qquad (47.3)$$

$$E_{ac} = K(E_{dc} + 2nV_d) + \frac{M}{2}I_{dc}R \qquad (47.4)$$
$$R = R_s + R_t + 2nR_d \qquad (47.5)$$
$$I_{rms} = 0.707FI_{dc} \qquad (47.6)$$

$$E_{ac} = K(E_{dc} + nV_d) + \frac{M}{2}I_{dc}R \qquad (47.7)$$
$$R = R_s + R_t + nR_d \qquad (47.8)$$
$$I_{o\,rms} = \frac{F}{2}I_{dc} \qquad (47.9)$$
$$I_{o\,rms} = 0.707FI_{dc} \qquad (47.10)$$

Figure 47.17 Transformer-type charging circuits and equations for Eveready sealed nickel–cadmium cells and batteries. E_{dc} = battery voltage during charge, I_{dc} = average charging current (A), E_{ac} = open-circuit r.m.s. voltage of secondary winding, R = total circuit resistance, R_t = transformer winding resistance referred to secondary, R_d = rectifier dynamic resistance, R_s = series current-limiting resistor, V_d = rectifier forward threshold voltage, n = number of rectifier cells in series per leg, K = d.c. voltage equation factor (see Figure 47.18), M = d.c. current equation factor (see Figure 47.18), F = current from factor = ratio of I_{rms} to I_{dc} (see Figure 47.18). Rectifier materials and characteristics: germanium, R_d = 0, V_d = 0.35; silicon, R_d = 0, V_d = 0.80 (Courtesy of Union Carbide)

I_{dc}, is chosen to fit the specific application. The usage cycle, charge and discharge times of the particular device dictate the charge rate. The 10 h rate is the highest recommended charge current for the more common nickel–cadmium cells. To maintain minimum charge current change with line voltage variations, the ratio of E_{ac} to E_{dc} should be as large as possible. This, however, results in relatively high power losses and heat dissipation in the series current-limiting resistor. For practical reasons ratios of 1.5 to 2.5 are satisfactory, with the lower ratios being used for full-wave rectification in applications above 1 A.

Equations 47.2, 47.5 and 47.8 show that the value of R is the sum of three separate resistances. The value R_t, the resistance of the transformer, must be determined from manufacturers' specifications or by direct measurements of representative samples, and the value of R_d may be found in the caption to Figure 47.17. The value of the series limiting resistance, R_s, must be determined from the formulae in Figure 47.17. The purpose of R_t is to limit current. The value of R_s is normally high compared with the other resistances in the circuit, and in essence controls the current value since it constitutes the load on the power supply. The power dissipation in R_s varies as the square of r.m.s. current flow. For calculating the wattage rating of R_s, Equations 47.3, 47.6 and 47.10 give the relationship between I_{rms} and I_{dc}.

Typical values of rectifier forward threshold voltage, V_d, and rectifier dynamic resistance, R_d, for the design equations of Figure 47.17 are given in the caption to the figure. The current and peak inverse voltage rating of the rectifier must be adequate for desired circuit performance.

The equation factors K, M and F are functions of the current conduction angle. Their values are based on the ratio of E_{ac}, the open-circuit r.m.s. voltage of the transformer secondary, to the sum of the battery voltage and the forward threshold voltage of the rectifiers. The a.c./d.c. ratio must first be calculated from the formula, then the values of K, M and F can be read directly from Figure 47.18.

The half-wave rectification circuits (Equations 47.1, 47.2 and 47.3) are generally used only for low-current applications, of the order of 0.5 A or less. At higher currents, transformer efficiency is low and special core design is required because of the large direct current polarization effect.

The full-wave rectifiers may be bridge or centre-tap connection. The bridge connection is quite popular because of its flexibility, simplicity and use of a more economical transformer design. However, economics may dictate a choice between transformer cost and the total rectifier cost, the bridge connection requiring two additional diodes.

Other considerations in the selection of the charge circuit configuration are the usage cycle, charger

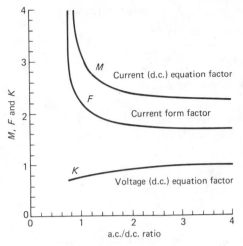

Half-wave and full-wave centre tap:

$$\text{a.c./d.c. ratio} = \frac{E_{ac}}{E_{dc} + nV_d} \qquad (47.11)$$

Full-wave bridge:

$$\text{a.c./d.c. ratio} = \frac{E_{ac}}{E_{dc} + 2nV_d} \qquad (47.12)$$

Figure 47.18 Design equation factors *versus* a.c./d.c. ratio (Courtesy of Union Carbide)

location, heat dissipation and transformer power ratings, overall size, cost, etc. The half-wave circuit produces additional heat in the transformer core material due to the saturating effect of the d.c. polarizing current. The physical dimensions of the transformer used with half-wave rectification may be larger than those of a transformer using the full-wave bridge connection for the same power rating. The following examples demonstrate the use of the equations, tables and graphs in the design of constant-current chargers.

Example 1

A circuit is required to charge two Union Carbide CH500 nickel–cadmium cells in series at the 10 h rate. This is assumed to be for experimental use, which does not justify acquiring a special transformer. The CH500 is a cylindrical high-rate cell rated at 0.5 A h capacity.

$$I = \frac{A\,h}{h} = \frac{0.5}{10} \quad 0.05\ A\ (I_{dc})$$

End-of-charge voltage at 10 h rate is estimated to be 1.45 V/cell and $2 \times 1.45 = 2.90 E_{dc}$.

Thus an a.c./d.c. ratio of about 2 is needed. A stock filament transformer with a secondary rating of 6.3 V, 0.6 A is selected (Stancor type P-6465 or equivalent). Measurements of the open-circuit output voltage and winding resistance give the following values.

With primary input voltage of 120 V a.c.,

Open-circuit secondary volts = $7.5\,V_{r.m.s.}$

Turns ratio, $N = \dfrac{V_p}{V_s} = \dfrac{120}{7.5} = 16$

Winding resistance (cold):

primary = $195\,\Omega$
secondary = $1.3\,\Omega$

Total resistance (cold) referred to secondary:

$$R_t = R_s + \frac{R_p}{N^2}$$

$$= 1.3 + \frac{195}{16^2} = 2.1\,\Omega$$

The hot resistance for a power transformer at rated output will be about 10% higher than the cold resistance. However, this transformer will be operating below rated capacity and a 5% increase is assumed. This gives a value

R_t (hot) = $2.1 \times 1.05 = 2.2\,\Omega$

Since the charging current is considered low, the half-wave circuit is chosen. A suitable rectifier is the low-cost IN2069 silicon type rated at 0.75 A d.c. and 200 peak inverse voltage (PIV). The diodes may be of any type having adequate current and PIV ratings. The PIV rating should be equal to or greater than the peak value of the power transformer secondary voltage plus the battery voltage, that is:

$$\sqrt{(2)}\,E_{ac} + E_{dc}$$

From Figure 47.17, $V_d = 0.80\,V$ and $R_d = 0$ for a silicon rectifier.

Equation 47.11 is used to obtain the a.c./d.c. ratio. Half-wave rectification was chosen, hence one diode was used and $n = 1$.

$$\frac{E_{ac}}{E_{dc} + nV_d} = \frac{7.5}{2.90 + (1 \times 0.8)} = 2.03$$

From Figure 47.18, $K = 0.92$, $M = 2.40$ and $F = 1.76$.

Rearranging Equation 47.1 to solve for R gives:

$$R = \frac{E_{ac} - K(E_{dc} + nV_d)}{MI_{dc}}$$

$$= \frac{7.5 - 0.92\,(2.90 + 1 \times 0.8)}{2.4 \times 0.05} = 34.1\,\Omega$$

and

$$R_s = R - R_t - nR_d$$

$$= 34.1 - 2.2 - 0 = 31.9\,\Omega$$

From Equation 47.3, the secondary r.m.s. current is

$I_{rms} = FI_{dc} = 1.76 \times 0.05 = 0.088\,A$

Example 2

A circuit is required to charge four CH1.2 nickel–cadmium cylindrical cells in series at the 10 h rate or I_{dc} of 0.12 A. E_{dc} will be about 6 V. The full-wave centre-tap circuit will be used for this design example. The specific application may dictate the best circuit configuration.

An a.c./d.c. ratio of 2 is assumed for the transformer secondary voltage. This requires two secondary windings of approximately 12 V ($2 \times 6 = 12$ V) or a 24 V centre-tapped secondary winding. For this first trial design, standard diodes will be used. The current rating should be selected to provide a safe temperature during operation. Heat sinks may or may not be required. The non-conducting rectifier will see a PIV equal to the peak value of the full secondary winding less the forward voltage drop of the conducting rectifier. Neglecting the rectifier drop the PIV will be

$2E_{ac}\sqrt{2} = 2 \times 12 \times \sqrt{2} = 33.94\,V$

Common diodes have ratings in excess of this value.

From Figure 47.17, $V_d = 0.8\,V$ and $R_d = 0$ for a silicon rectifier. From Equation 47.11, the a.c./d.c. ratio is

$$\frac{E_{ac}}{E_{dc} + nV_d} = \frac{12}{6 + (1 \times 0.8)} = 1.76$$

From Figure 47.18, $K = 0.90$, $M = 2.42$ and $F = 1.80$. Rearranging Equation 47.7 to solve for R gives:

$$R = \frac{E_{ac} - K(E_{dc} + nV_d)}{(M/2)I_{dc}}$$

$$= \frac{12 - 0.9(6 + 1 \times 0.8)}{(2.42/2)0.12} = 41\,\Omega$$

Rearranging Equation 47.8,

$$R_s = R - nR_d - R_t$$
$$= 41 - 1 \times 0 - R_t$$
$$= 41 - R_t\,\Omega$$

Total resistance, R_t, may be determined by the turns ratio and winding resistance (for the primary and half of the secondary) given in the transformer design data or by measurement of a sample. The measured value may be adjusted for operating temperature where warranted.

From Equation 47.10, the r.m.s. current in the centre leg is:

$$I_{rms} = 0.707FI_{dc}$$

$$= 0.707 \times 1.8 \times 0.12$$

$$= 0.153\,A$$

The series resistance power rating must handle I^2R or

$(0.153)^2(41 - R_t)$

The transformer secondary power is approximately

$$E_{ac} \times I_{rms} = 12 \times 0.153 = 1.84\,W$$

A better design may be provided by the full-wave bridge circuit. Although the number of diodes required is doubled, the transformer may be less expensive and somewhat smaller.

Example 3

A circuit is required to charge four CF1.2 nickel–cadmium cylindrical cells in series at the 1 h rate or I_{dc} of 1.2 A. E_{dc} will be about 6.4 V.

Again an a.c./d.c. ratio of 2 is indicated for the transformer secondary voltage. This means two secondary windings of approximately 12 V ($2 \times 6.4 = 12.8$ V) or a 24 V centre-tapped secondary winding. For this first trial design, standard diodes will be used. The current rating should be selected to provide safe temperature during operation. Heat sinks may or may not be required. The non-conducting rectifier will see a PIV equal to the peak value of the full secondary winding less the forward voltage drop of the conducting rectifier. Neglecting the rectifier drop, the PIV will be:

$$2E_{ac}\ \sqrt{2} = 2 \times 12 \times \sqrt{2}$$
$$= 34\,V$$

Common diodes have ratings in excess of this value.
From Figure 47.17, $V_d = 0.8$ V and $R_d = 0$ for a silicon rectifier. From Equation 47.11, the a.c./d.c. ratio is:

$$\frac{E_{ac}}{E_{dc} + nV_d} = \frac{12}{6.4 + (1 \times 0.8)} = 1.76$$

From Figure 47.18, $K = 0.90$, $M = 2.42$ and $F = 1.80$. Rearranging Equation 47.7 to solve for R gives:

$$R = \frac{E_{ac} - K(E_{dc} + nV_d)}{(M/2)I_{dc}}$$
$$= \frac{12 - 0.90(6.4 + 1 \times 0.8)}{(2.42/2)1.2} = 3.80\,\Omega$$

Rearranging Equation 47.8,

$$R = R - nR_d - R_t$$
$$= 3.9 - 1 \times 0 - R_t$$
$$= 3.9 - R_t\,\Omega$$

Total resistance, R_t, may be determined by the turns ratio and winding resistances (for the primary and half of the secondary) given in the transformer design data or by measurement of a sample. The measured value may be adjusted for operating temperature where warranted.

From Equation 47.10, the r.m.s. current in the centre leg is

$$I_{rms} = 0.707FI_{dc}$$
$$= 0.707 \times 1.8 \times 1.2 = 1.53\,A$$

The series resistance power rating must handle I^2R or

$$(1.53)^2\ (3.9 - R_t)$$

The transformer secondary power is approximately

$$E_{ac} \times I_{rms} = 12 \times 1.5 = 18\,W$$

A better design may be provided by the full-wave bridge circuit. Although the number of diodes required is doubled, the transformer may be less expensive and somewhat smaller.

47.8 Transformerless charge circuits for nickel–cadmium batteries

Though the transformer-type circuit offers versatility and desired isolation of battery terminals and a.c. power line there are other circuit configurations which are practical for special applications such as experimental batteries. The division or proportioning of the charge source voltage and current values may be obtained by reactive or resistive networks. The former has the advantage of being heatless. The

Table 47.5 Values for capacitor C in Figure 47.19(c)

Eveready battery number	Number of cells in series	Charge current (mA) (10 h rate)	Nominal capacitance (μF)
B20 or B20T	1 only	2	0.05
B50 or B50T	1–5	5	0.12
OB90 or OB90T	1–5	9	0.24
B150, B150T, CH150, CH150T	1–5	15	0.37
B225, B225T, BH225, BH225T, CH225, CH225T	1–5	22.5	0.56
CH450, CH450T	1–5	45	1.12
BH500, BH500T, CH500, CH500T	1–5	50	1.25

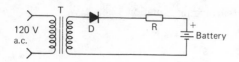

(a) Transformer and half-wave rectifier

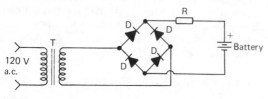

(b) Transformer and full-wave bridge rectifier

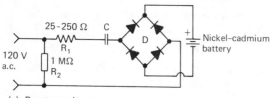

(c) Reactance charger

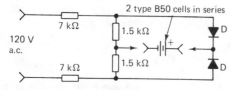

(d) Resistance charger D-IN2069 silicon diode (or equivalent)

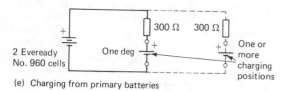

(e) Charging from primary batteries

Figure 47.19 Charger circuits for Eveready sealed nickel–cadmium cells and batteries (Courtesy of Union Carbide)

Union Carbide reactance-type charger of Figure 47.19(c) provides a constant current to nickel–cadmium batteries of one to five series-connected cells. This circuit has the advantages of simplicity, light weight, no heat, and can be used with variable numbers of cells. It is used quite widely in applications where the batteries and charger are built in and inaccessible to the user. This circuit usually takes the form of full-wave bridge rectification. The bridge circuit in Figure 47.19(b) will deliver about $40\,mA/\mu F$ (value of capacitor, C) to the battery (see Table 47.5 for suggested capacitor values for use with various Union Carbide Eveready nickel–cadmium batteries). Newer selenium bridge rectifiers and Mylar dielectric capacitors make practical charge currents possible in excess of $6.1\,mA/cm^3$.

The value of R_1 in Figure 47.19(c) is not critical since it is a current-limiting resistor to restrict initial surge when the circuit is energized. The value of R_1 may range from 25 to $250\,\Omega$. This resistor is often omitted when more recent improved components are used. The bleeder resistor, R_2, provides a discharge path for the capacitor, removing possible shock hazard across a.c. prongs when removed from the receptacle. For circuits having large capacitance values (higher current ratings) the rectifier may be burned out if the battery is removed when the circuit is connected to the 120 V a.c. supply. The resistance type charger shown in Figure 47.19(d) (see also Table 47.6) can also be used to provide a constant current to batteries of series connected cells. The circuit values shown will provide a charging current of approximately 4.5 mA for two B50 cells connected in series. This charger will not be damaged if it is connected to the a.c. power line with the batteries removed. By changing the resistor values, single cells or three to five cells in series could be charged. Higher currents requiring high wattage resistors would not be practical because of the excessive heat that would be produced. Small button cells, such as those used in hearing-aids, may be readily charged from flashlight cells or lantern batteries. Figure 47.19(e) (see also Table 47.6) is shown by way of example; this circuit will charge one B50 cell at approximately 5 mA. Additional cells may be charged as shown by the broken lines.

Table 47.6 Union Carbide sealed nickel–cadmium batteries and cells: resistance-type charger, circuit component values for charging

Eveready battery number	Description	Transformer T	Rectifier D	Charge current (mA) for 14 h at this current	Charge current (mA) for 4 h at this current	Circuit figure number	Resistance R to charge one cell (or battery) (Ω)	(W)	Resistance R to charge two cells in series (Ω)	(W)
BH1 (and BH1T)	Single cell	A	M	100		2.52d	22	1	15	1
CF1 (and CF1T)	Single cell	A	M		350	2.52d	4.7	2	2.7	1
CH1 (and CH1T)	Single cell	A	M	100		2.52d	22	1	15	1
CF1.2 (and CF1.2T)	Single cell	B or C	M		420	2.52d	4.7	2	3.3	2
CH1.2 (and CH1.2T)	Single cell	A	M	120		2.52d	18	1	15	1
CH1.2C	Single cell	A	M	120		2.52d	18	1	15	1
CH1.2D	Single cell	A	M	120		2.52d	18	1	15	1
CH1.5 (and CH1.5T)	Single cell	A	M	150		2.52d	15	2	10	1
CH1.8 (and CH1.8T)	Single cell	A	M	180		2.52d	11	2	7	1
CH2.2 (and CH2.2T)	Single cell	A	M	220		2.52d	8.2	2	5.6	2
CF4 (and CF4T)	Single cell	D or E	U		1400	2.52d	2.2	10	1.8	10
CH4 (and CH4T)	Single cell	B or C	M	400		2.52d	4.7	2	3.3	0.5
B20 (and B20T)	Single cell	A	M	2		2.52d	1200	0.5	1000	0.5
B50 (and B50T)	Single cell	A	M	5		2.52d	560	0.5	390	0.5
N64	Battery (12 CH 500 cells in series)	F or G	M	50		2.52d	120	0.5		
N65	Battery (11 CH1 cells in series)	F or G	M	100		2.52d	56	1		
N67	Battery (5 C900 cells in series)	F or G	M	90		2.52d	100	2		
N70	Battery (5 rectangular cells in series, 1.5 A h)	F or G	M	50		2.52d	68	3		
N86	Battery (10 CH1.2 cells in series)	F or G	M	120		2.52d	56	2		
N88	Battery (17 OB90 cells in series)	F	M	9		2.52d	930	0.5		
OB90 and (OB90T)	Single cell	A	M	9		2.52d	270	0.5	200	0.5
N91	Battery (5 CH1.2 cells in series)	F	M	120		2.52d	75	5		
B150 (and B150T)	Single cell	A	M	15		2.52d	180	0.5	120	0.5
CF150 (and CF150T)	Single cell	A	M		50	2.52d	47	1	33	1
CH150 (and CH150T)	Single cell	A	M	15		2.52d	150	0.5	120	0.5
B225 (and B225T)	Single cell	A	M	22.5		2.52d	120	0.5	100	0.5
BH225 (and BH225T)	Single cell	A	M	22.5		2.52d	100	0.5	82	0.5
CF225 (and CF225T)	Single cell	A	M		75	2.52d	33	1	22	1
CH225 (and CH225T)	Single cell	A	M	22.5		2.52d	100	0.5	82	0.5
CF450 (and CF450T)	Single cell	A	M		150	2.52d	15	1	10	1
CH450 (and CH450T)	Single cell	A	M	45		2.52d	56	1	39	1
BH500 (and BH500T)	Single cell	A	M	50		2.52d	47	1	33	1
CF500 (and CF500T)	Single cell	A	M		175	2.52d	12	3	8.2	3
CH500 (and CH500T)	Single cell	A	M	50		2.52d	47	1	33	1
1007	Battery (8 CH4 cells in series)	F or G	M	400		2.52d	15	5		

Table 47.6 (*Cont'd*)

Transformers				Rectifiers			
Designation	*Make and type No.*	*Primary*	*Secondary*	*Designation*	*Make and type No.*	*Primary*	*Secondary*
A	Stancor P-6465 (or equivalent)	120 V	6.3 V, 0.6 A	M	IN2069 Si diode	0.75 A, 200 V	PIV
B	Stancor P-6134 (or equivalent)	120 V	6.3 V, 1.2 A	S	IN420 Si diode	3 A, 100 V	PIV
C	Triad F-14X (or equivalent)	120 V	6.3 V, 1.2 A	U	Motorola MDA 970-2		
D	Stancor P-6466 (or equivalent)	120 V	6.3 V, 3.0 A		Si bridge (or equivalent)	4 A, 100 V	PIV
E	Triad F-16X (or equivalent)	120 V	6.3 V, 3.0 A				
F	Stancor P-6469 (or equivalent)	120 V	25.2 V, 1.0 A				
G	Triad F-45X (or equivalent)	120 V	24.0 V, 1.0 A				

48

Taper charging of lead–acid motive power batteries

Contents

Large battery installations such as one used by operators of fleets of electric vehicles, for example in distribution centres, require sophisticated battery charging systems to obtain the best possible performance from the batteries and to protect the large investment of cash involved. A good example of this type of charger for motive power batteries is the Spegel charger supplied by Chloride Legg UK and this equipment is described in some detail below.

Most chargers are designed so that, as the charging process progresses and the battery voltage rises during charge, the output current of the charger reduces. This charging characteristic is known as a 'taper' and is described by referring to the current obtained near the beginning and near the end of a typical charge.

The position chosen near the beginning of the charge is 2.1 V/cell, and near the end of the charge 2.6 V/cell. The values at these voltages are commonly expressed as ratios. Hence, a charger designed to give an output of 100 A at 2.1 V/cell and 50 A at 2.6 V/cell, is said to have a taper characteristic of 2:1. The steepness of a charger taper is important because it affects the charger's sensitivity to variations in mains voltage, the charging time required and the temperature rise of the battery, which must be controlled during charge.

A battery will accept a higher current at the start of charge than at the end. However, the higher the starting rate, the more expensive the charger is to make and buy. So, in practice, the steepness of the charger's taper is usually a compromise between economics and the battery's requirements for the value of the finishing current. In any event, the finishing current must be kept below specified values:

1. It is necessary to restrict the temperature rise in the battery, as excessive temperature can damage its plates and separators.
2. When the battery is between 75 and 80% charged, oxygen and hydrogen gases are released owing to chemical reaction within the cells. If the finishing current is too high, the excessive flow of these gases can dislodge particles of active material from the positive plates, resulting in reduced battery life.
3. Nevertheless, gassing is an essential part of the charging process as it mixes the electrolyte. Therefore, the finishing rate must be sufficiently high to allow this process, but not so high as to damage the positive plates as described in (2).

48.1 Types of charger

There are two basic types of charger.

48.1.1 Fully self-compensating chargers

These are becoming increasingly popular as they require no supervision and bring a greater flexibility to the battery charging operation.

When a battery on charge reaches a voltage of between 2.35 and 2.40 V/cell at the recommended charging rate, it is 75–80% charged. The time to termination of charge from this point is determined totally by the battery requirements and can vary over a wide range from say 1 to 6 h.

The self-compensating charger therefore monitors the battery voltage during the charging process and only when the stable voltage conditions achieved at the top of charge are reached will the controller terminate charge. A typical layout of such a charger is shown in Figure 48.1.

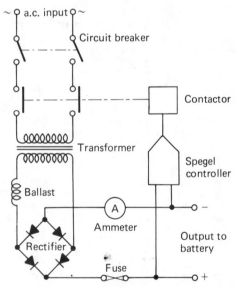

Figure 48.1 Layout of a Spegel continuous control charger (Courtesy of Chloride Batteries)

This seemingly complicated device is used instead of a voltage relay operating a switch at the end of charge because the top-of-charge voltage of a battery varies according to its age, condition and temperature. This would cause an old battery, which never reaches the tripping voltage, to go on receiving charge unnecessarily, or a new battery to reach the tripping voltage too early and not recieve the correct charge.

For each individual charge cycle, the battery itself determines exactly the amount of energy required, and the fully self-compensating charger purely responds to this demand without manual adjustment. The use of sophisticated electronic control to monitor battery condition permits recharge times for fully discharged batteries of 7.5 h to be achieved using a straightforward single-taper characteristic.

48.1.2 Pre-programmed chargers

When 10 h or more recharge time is available, a single-taper charger is used. This type of charger incorporates a voltage-sensing device set to operate when the battery voltage reaches 2.35 V/cell (42.3 V overall on an 18-cell battery). At this voltage a timing device is started which will terminate the charge after a predetermined but adjustable period. As stated previously, a discharged battery can absorb a high current during the early part of charge, but there is a limit to the safe current after the gassing point has been reached.

With a single-taper charger the current must have fallen to a safe value by the time the gassing point is reached, and must continue to fall as the gassing rate increases; thus, this type of charger is limited to a minimum charge time of 10 h.

When less than 10 h recharge time is available, a two-step taper charger is used, enabling battery manufacturers' recommendations to be met by using a higher current at the start of charge and a reduction to a lower level once the gassing point has been reached. As with the single-taper charger, a voltage-sensitive relay starts a timing device at the gassing point, but in this case it has a second function, which is to operate a contactor, which reduces the current to the required finishing rate. After 2.35 V/cell, both the single and two-step taper chargers have similar characteristics. In summary, the two-step taper charger has two separate taper characteristics in operation; the first at a high

Table 48.1 Range of Chloride Legg Spegel chargers

Battery ampere hour capacity (5 h rate) with time available for recharge														Starting current (A)
7½ h	8 h	8½ h	9 h	9½ h	10 h	10½ h	11 h	11½ h	12 h	12½ h	13 h	13½ h	14 h	
25	26	27	29	30	31	32	33	35	36	37	39	40	41	5
50	53	55	57	60	62	64	67	69	72	75	78	80	83	10
75	79	82	86	89	93	96	100	104	108	112	116	120	124	15
100	105	110	114	119	123	128	133	139	144	155	155	160	165	20
125	132	137	143	149	154	160	167	174	180	187	194	200	207	25
150	158	165	171	179	185	192	200	208	216	224	233	240	248	30
175	184	192	200	208	216	224	233	243	252	261	271	280	289	35
200	211	220	229	238	247	256	267	278	288	299	310	320	331	40
225	237	247	257	268	278	288	300	313	324	336	349	360	372	45
250	263	275	286	298	308	321	333	347	360	373	388	400	413	50
275	289	302	314	327	340	353	367	382	396	410	426	440	455	55
300	316	330	343	357	370	385	400	417	432	448	465	480	496	60
325	342	357	371	387	401	417	433	452	468	485	504	520	537	65
350	368	385	400	417	432	449	467	486	504	522	543	560	579	70
375	395	412	429	446	463	481	500	521	540	560	581	600	620	75
400	421	440	457	476	494	513	533	556	576	597	620	640	661	80
425	447	467	486	506	525	545	567	590	612	634	659	680	702	85
450	474	495	514	536	556	577	600	625	647	672	698	720	744	90
475	500	522	543	565	586	609	633	660	683	709	736	760	785	95
500	526	549	571	595	617	641	667	694	719	746	775	800	826	100
525	552	577	600	625	648	673	700	729	755	784	814	840	868	105
550	579	604	629	655	679	705	733	764	791	821	853	880	909	110
575	605	632	657	685	710	737	767	799	827	858	891	920	950	115
600	632	659	686	714	741	769	800	833	863	896	930	960	992	120
650	684	714	743	774	802	833	867	903	935	970	1008	1040	1074	130
700	737	769	800	833	864	897	933	972	1007	1045	1085	1120	1157	140
750	789	824	857	893	926	962	1000	1042	1079	1119	1163	1200	1240	150
800	842	879	914	953	988	1026	1067	1111	1151	1194	1240	1280	1322	160
850	895	934	971	1012	1049	1090	1133	1181	1223	1269	1318	1360	1405	170
900	947	989	1029	1071	1111	1154	1200	1250	1295	1343	1395	1440	1488	180
950	1000	1044	1086	1131	1173	1218	1267	1319	1367	1418	1473	1520	1570	190
1000	1053	1099	1143	1190	1235	1282	1333	1389	1439	1493	1550	1600	1653	200
1050	1105	1154	1200	1250	1296	1346	1400	1458	1511	1567	1628	1680	1736	210
1100	1158	1209	1257	1310	1358	1410	1467	1528	1583	1642	1705	1760	1818	220
1150	1211	1264	1314	1369	1420	1474	1533	1597	1655	1716	1783	1840	1901	230
1200	1263	1319	1371	1429	1481	1538	1600	1667	1727	1791	1860	1920	1983	240

Spegel chargers have a type reference in terms of the number of battery cells to be charged and the initial value of the charging current (starting current). For example, charger type S3P 36/95 indicates 36 cells and 95 A starting current, i.e. a 36-cell 500 A h battery will be charged in 8 h. Thus the correct size of charger will be determined by the number of cells, the ampere hour capacity and the time available for recharge

current rate, the second at a lower rate, brought in by a voltage sensitive relay, which at the same time starts a timing device.

48.2 Equalizing charge

Because each cell in a battery has its own characteristics, each requires a slightly different amoung of charge. Fully self-compensating chargers cater for this during the normal recharge function, therefore a separate 'equalizing' facility is not required.

With pre-programmed chargers an 'equalizing charge' is sometimes available via a switch that provides a continuous low current used to stabilize the voltage and relative density of the cells. If periodic equalizing charges are not given when pre-programmed chargers are being used, the life of the battery could be reduced. Frequency and extent of equalizing charges should be discussed with the battery manufacturer, who will assess the particular installation and make suitable recommendations.

48.3 How to choose the right charger

A charger must always be chosen by taking note of the battery and charger manufacturer's recommendations.

Table 48.1 illustrates the flexibility that can be achieved by using an electronically controlled self-compensating charger such as the Chloride Spegel. For example, a 36-cell 500 A h battery can be recharged in 12 h by a 36-cell Spegel charger with an output of 70 A. The same battery can be recharged in 8 h with a 36-cell Spegel charger with an output of 95 A.

On the other hand, a 36-cell Spegel charger with an output of 70 A can be used to recharge 36-cell batteries of capacities ranging from 350 to 579 A h.

Road vehicles are seldom needed to work more than 9–12 h/day, which leaves up to 12 h for recharging. The same is true of many industrial trucks. However, the demands of industry mean that trucks are sometimes needed to work longer periods. Sometimes even round-the-clock operation is required of them, with three shifts of drivers using the same vehicle.

In these circumstances, special charging methods have to be introduced. To overcome this problem, the truck can be supplied with one, or even two, spare batteries.

Two-battery operation

This enables one battery to be on charge while the other is at work on the truck. If this battery is recharged every 8 h (or once each shift) then obviously a charger capable of meeting this require-

ment must be used. Consequently, a two-step taper or Spegel charger would be required.

A battery that is continuously worked for 8 h, charged for the next 8 h and immediately worked again will have its life considerably reduced.

Batteries must be given a rest between charging and being discharged again, during which they can cool down. Where this kind of intense working is necessary, three-battery operation can be used.

Three-battery operation

This gives each battery the chance to rotate between working, recharging, and resting after charge. Charging the battery after every 8 h shift means that each battery has a 16 h period for recharging and resting.

48.4 Opportunity charging

Where an electric vehicle is being intensively worked, plugging into a Spegel charger during breaks and other brief down-periods can also help to extend the vehicle's effective working shift. However, too many opportunity-charging periods during a working shift may cause unacceptably high battery temperatures. This factor must be considered when planning an opportunity-charging schedule. Before the introduction of modern self-compensating electronically controlled chargers, opportunity charging was discouraged because of the high risk of battery overcharging. However, when used with the Chloride Spegel charger, the battery dictates its charge requirements and receives the minimum overcharge for its needs. A typical battery might be opportunity-charged for two or three periods of 30 min each during a working day, when, depending on the battery's state of charge when connected to the charger, this can boost the battery capacity by between 12 and 25%.

Table 48.2 shows that there is little benefit in charging the battery when it is 75% charged or

Table 48.2 Ampere hours replaced by opportunity charging related to state of charge

Battery state of charge when connected to charger (% charged)	Approximate percentage of ampere hour capacity restored per hour	
	8 h *Spegel* charger	12 h *Spegel* charger
95	6.0	4.5
85	8.0	6.5
75	10.0	8.0
65	13.5	11.0
50	15.0	12.0
25	17.0	12.5
0	17.5	13.0

above. Hence, opportunity charging should be carried out only when the battery is less than 75% charged, which corresponds to an electrolyte relative density of 1.240 or less. For example, when a 65% discharged battery of 400 A h capacity is connected to a 12 h Spegel charger for 1 h, it will receive 44 A h of charge (i.e. 11.0% of 400 A h).

48.4.1 Principles of operation of the Spegel charger

This charger operates on the well proven principle of continuously monitoring the on-charge battery voltage and terminating the charge only when the stabilized conditions occurring at the top of charge are detected. The charge duration is therefore totally dictated by the battery and, in practice, will vary considerably depending on actual requirements.

The detection system employs fully solid-state components, and enables continuous monitoring to be achieved without interruption of the charge cycle.

The system is designed to compensate for mains voltage variation so that the battery always receives the correct amount of charge irrespective of mains supply fluctuation or battery condition. This is illustrated in Figure 48.2.

A failsafe circuit is included, which is totally electrically independent of the main control and powered by the battery itself. This system will terminate charge after a fixed time from the commencement of gassing at 2.35 V/cell.

Following the connection of a battery and the switching on of the mains supply circuit breaker, a short delay is provided before the contact closes and current begins to flow. If the battery is inadvertently disconnected during the charging process without first switching off the mains supply circuit breaker, the supply contact within the charger will open. Electrical isolation of the exposed output terminals of the connector is therefore provided.

During the switch-on delay period the open circuit voltage of the battery is measured. The contact will be prevented from closing if the battery voltage lies outside the range of 1.8–3.4 V/cell. This will avoid damage if a battery of the incorrect number of cells is connected to the charger.

If the battery is not required for use it should be left connected to the charger after the main charging phase is complete, so that it receives periods of refreshing charge of 10 min during every 6 h to maintain its condition.

Spegel chargers are designed to provide a taper characteristic, which means that the d.c. output current falls as the battery voltage rises during the charge cycle. The rated output current is delivered at a cell voltage of 2.1 V at the start of charge, tapering to 37.5% of this value at a cell voltage of 2.6 V. Used in conjunction with Spegel control, this characteristic is suitable for all recognized makes of motive power lead–acid battery irrespective of age or condition.

When the charge time available has been determined (7.5–14 h) and the battery capacity required is known, the correct charger can be easily selected from the chart.

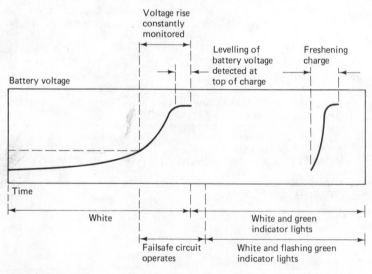

Figure 48.2 Principle of operation of the Spegel charger (Courtesy of Chloride Batteries)

49

Methods of charging large nickel–cadmium batteries

Contents

To charge a large vented nickel–cadmium battery there are normally three methods. The difference between these methods is caused by different voltage and current characteristics and results in a different charging time. The following abbreviations are used in this chapter:

W = decreasing current
I = constant current
U = constant voltage
O = automatic commutation
a = automatic end of charge

To bring nickel–cadmium batteries to the fully charged condition (100% C_5) they must be charged with the charging factor of 1.4 to compensate for the charge efficiency (Figure 49.1).

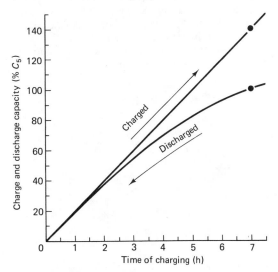

Figure 49.1 Graph of the charging efficiency of nickel–cadmium batteries

If there is a regular cycling of the battery, or after a deep discharge below the final discharge voltage, it is necessary to recharge with an increased charging factor of 1.6–1.8.

The battery is fully charged if the maximum voltage obtained does not increase over a period of 1–2 h.

The relative density of the electrolyte does not indicate the condition of the battery. The charging currents usually quoted are approximate values only as there is no theoretical limit. Compared with the lead–acid battery, it is not necessary to reduce the charging current if the gassing point has been reached. The true limit of the maximum value of the charging current is controlled by the electrolyte temperature, which must not exceed 45°C.

If charged with the I-characteristic, the charging current, I_L, is constant (Figure 49.2(a)). Towards the end of the charge period the process is

controlled manually (I) or automatically I(a) or by changing to trickle charge I(O) (Figure 49.2(b)). The charge method with the W, W(a), W(O) characteristic is the most used one for nickel–cadmium batteries of all types and kinds. The initial current drops during the charge to a final value (final charging current) whilst the battery voltage is increasing (Figure 49.2(c)). Towards the end of charge, the process is completed under manual control (W characteristic) or automatically (W(a) characteristic) (Figure 49.2(d)).

Using the IU characteristic, the charging current, I_L (initial charging current), is held constant until the fixed value of the charging voltage has been reached. Then the charging current, I_L, drops to a lower value (final charging current) whilst the charging voltage is kept constant (Figure 49.2(e)). The IU-charge is specially suited for charging batteries in parallel. This method of charging allows several similar batteries of identical voltage and capacity to be charged together irrespective of the individual state of charge.

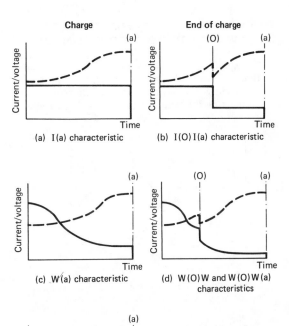

Figure 49.2 Methods of charging large nickel–cadmium batteries. —, current; ----, voltage. Charge with (a,b) constant current characteristic, (c,d) decreasing current characteristic, (e) constant current/constant voltage characteristics

49.1 Trickle charge/float charge

To compensate for losses for self-discharge, a trickle charge current of approx. 50–80 mA per 100 A h nominal capacity is essential. Depending on the type of battery and ambient temperature, the trickle charge voltage will be 1.35–1.45 V/cell. If charged in parallel using a constant potential of 1.40 V/cell, the trickle charge current, depending on the ambient temperature, will be 50–100 mA per 100 A h nominal capacity. If, in addition to the self-discharge, there is a low or temporary load connected to the battery, a float charge is essential. The float charge current should be from 120 to 150 mA per 100 A h nominal capacity. The cell voltage will be 1.4–1.45 V. The best float charge voltage should be found by experiment, the criterion being the loss of electrolyte consistent with having a fully charged battery. If there is a continuous trickle or float charge, it is essential to make an equalization charge with a charging current of I_5 every 3–6 months for 15 h or a boost charge depending on the characteristic of the charger to ensure full nominal capacity.

49.2 Charge/discharge operations on large vented nickel–cadmium batteries

1. *Charge/discharge operation* The charge and discharge of the battery is arranged separately by connecting the battery to the charger or by connecting the battery to the load (Figure 49.3).

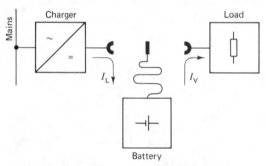

Figure 49.3 Charge/discharge operation of a large vented nickel–cadmium battery (Courtesy of Chloride Batteries)

2. *Switch-tripping operation* The battery is charged by the charger and kept in the fully charged condition. If necessary the battery is switched to the load (Figure 49.4).

Automatic recharge is often used in emergency lighting systems.

At first the battery is boost charged with the charging current, I_L, until the gassing point is reached. Then the final charge-timer takes over. This is operated by the automatic charge unit. The normal running time of the timer is 2–6 h. During

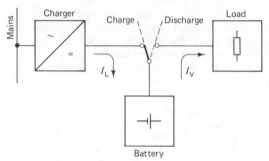

Figure 49.4 Switch-tripping operation of a large vented nickel–cadmium battery (Courtesy of Chloride Batteries)

the final charge, the boost charge current continues until the fully charged condition is reached. If the timer stops, the automatic charge unit connects the battery to the trickle charge. With the trickle charge current, the battery is maintained in the fully charged condition. The changeover to the load can be done manually or automatically (Figure 49.5).

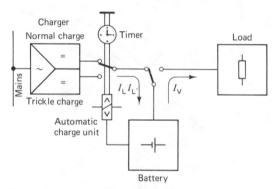

Figure 49.5 Automatic recharging of a large vented nickel–cadmium battery (Courtesy of Chloride Batteries)

The switch-tripping/continuous operation is another method used, e.g. in emergency lighting systems with continuous operation (maintained systems) (Figure 49.6). The load is supplied from

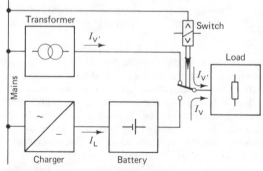

Figure 49.6 Auto switch-tripping/continuous operation recharging of a large vented nickel–cadmium battery (Courtesy of Chloride Batteries)

the mains via a transformer and the battery is charged from the charger with the charging current. If the mains fails, the load is switched over to the battery supply by the mains-failure switch. This method normally includes the automatic charge unit.

49.3 Standby operation

If the standby method is used, the charger/d.c. supply and the battery are continuously connected in parallel with the load (Figure 49.7). The load is

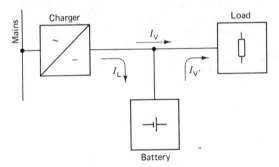

Figure 49.7 Standby operation of a large vented nickel–cadmium battery (Courtesy of Chloride Batteries)

d.c. and supplied by the current, I_V. The d.c. supply is also responsible for the charging of the battery with the charging current, I_L. If the mains fails, the battery takes over without any interruption to the load. If float operation is being used and the load current peaks exceed the output of the d.c. supply/charger, these peaks will be supplied from the battery.

With parallel operation and end cell switching, the load is supplied from the d.c. supply/charger with a voltage equal to 1.4 V/cell (Figure 49.8). With

a 20-cell battery, having a nominal voltage of 24 V means that a voltage of 28 V is available to ensure that the battery is fully charged by the charging current. In this mode of operation the end cell is bypassed. If the mains fails, the load is supplied without interruption directly from the battery with the current I_V.

With the mains reconnected, the battery will automatically experience initial boost charging at 1.55 V/cell. With a 20-cell battery having a nominal voltage of 24 V, this constitutes a charging voltage of 31 V.

During the automatic boost charge, the end cell group is connected in series with the load, V, in order to reduce the boost charge voltage from 1.55 to 1.4 V/cell.

At the end of the charge at 1.55 V/cell, switchover to 1.4 V takes place automatically by the action of a relay, and at the same time the end cell group is bypassed.

Some notes follow regarding precautions to be taken in battery charging rooms for vented nickel–cadmium batteries. Although these notes refer to the German Regulations (VDE 0510 and VDE 0100) they are applicable to any room in which charging of these batteries is carried out, and of course are similar to the UK and US regulations.

Rooms containing batteries up to 220 V nominal voltage are covered by VDE 0100 'Electric Installations'. Above 220 V these are 'Restricted Access Electric Installations'.

It is permitted to install nickel–cadmium batteries together with lead–acid batteries in the same room if there are provisions to ensure that the electrolyte spray of the lead–acid battery cannot contaminate the nickel–cadmium battery and vice versa.

The insulation resistance to earth or to a metal cubicle must be at least 1 mΩ when the battery is new. Batteries already in service must have a resistance of at least 50 Ω/V nominal voltage of the battery. The insulation resistance of the whole

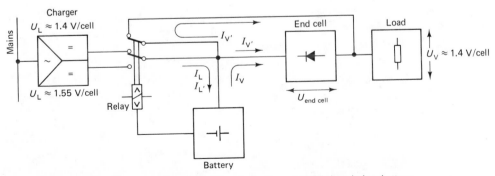

Figure 49.8 Parallel operation and end-cell switching of a large vented nickel–cadmium battery (Courtesy of Chloride Batteries)

battery must not be less than 1000 Ω. The height of the roof above the battery system must be at least 2 m. Doors have to open outwards. Electric tools normally causing sparks must be positioned at least 1 m from the battery. Heating radiators with a surface temperature of more than 200°C are not allowed. Doors, windows, floor covers and paints on walls and roofs have to be electrolyte resistant. Windows accessible from outside have to be locked and barred or a wire glass may be used.

The installation of batteries in cubicles and enclosures with convection venting is allowed up to a charging power of only 2 kW. The power is calculated from the nominal charging current of the battery. Battery cubicles or enclosures have to be provided with venting facilities.

If there is other electrical equipment in the room, it must be separated from the battery by a wall or barrier. Electric tools normally causing sparks are not allowed near the venting ducts of the battery.

49.4 Ventilation

Battery rooms and enclosures must be vented in a manner such that the hydrogen emitted on charging is diluted by natural or artificial ventilation. The dilution must be such that the mixture becomes non-explosive. The amount of hydrogen dissipated during charging a battery at ambient temperature of 0°C and a pressure of 1.013 bar is 0.42 l/A h per cell. A mixture of hydrogen gas and air is explosive if the amount of hydrogen gas exceeds 3.8%. The venting of the battery room should be in accordance with VDE 0510. The air necessary for venting battery cubicles or enclosures can be either by inlet or extraction provided the volume of the battery room contains at least 2.5 times the necessary air to be changed per hour to ensure that an explosive mixture does not occur. Battery rooms that comply with regulation VDE 0510 can be regarded as being explosion proof.

Part 6

Battery suppliers

50

Lead–acid (secondary) batteries

Contents

50.1 Motive power batteries

Motive power lead–acid batteries are the power source used in every kind of electric vehicle from road transport to fork lift trucks in industry. Many manufacturers throughout the world supply these types of batteries and it would be impossible to review all their products. Instead, the products of one major UK-based international producer, Chloride Motive Power, are considered in some detail. The applications of these types of batteries are considered in Part 4.

50.1.1 Chloride Motive Power

Chloride Motive Power produce five basic ranges of motive power batteries, the Classic series (three ranges), monobloc batteries and cells in hard rubber cases.

Three ranges of Classic batteries are available, the Classic batteries; the Classic 15 series which have at least 15% more capacity from the same cell dimensions than the Classic series; and the Classic 25 series which give 27% more capacity from the same cell weight and dimensions when compared with the Classic series. All the Classic batteries are housed in heat-sealed polypropylene containers and have tubular positive plate construction. The higher capacities per unit battery volume of the Classic 15 series are due to increases in the active material available in both negative and positive plates. The higher capacities per unit weight and volume of the Classic 25 series are due to a redistribution of the active material of the plates. Both the positive and negative have been redesigned so that, compared with the plates of a standard cell, there is 20% increase in usable area. The non-active components used in a cell have also been redesigned to optimize performance, with the result that a cell now has up to 27% more energy density in terms of watt hours per kilogram. Table 50.1 shows the relationship between electrolyte gravity and state of charge for the three Classic series of batteries.

Table 50.1 Electrolyte relative density related to state of charge

State of charge (% charged)	Approximate relative density (corrected to 15°C)		
	Classic cells	Classic 15 cells	Classic 25 cells
100 (fully charged)	1.280	1.295	1.295
75	1.240	1.250	1.240
50	1.200	1.210	1.200
25	1.160	1.160	1.150

In addition to the three Classic ranges (Tables 50.2–50.4 show selected examples), Chloride Motive Power also supply motive power monobloc batteries for fork lift truck and electric vehicle operation. Tubular plate monoblocs (6 and 12 V) are available in the 100–256 5-h capacity rate, 6 V flat plate monoblocs are available with 5 h capacities of 165 and 192. A range of motive power batteries housed in hard rubber containers with tubular positives are also available (Table 50.5 shows some selected examples).

Chloride Motive Power also offer two devices for the automatic topping up of cells. Such devices are important as accurate topping up will prolong battery life and remove the risk of abuse. Automatic topping up also assists, of course, in the sense that it is an important aspect of automation in battery charging and maintenance operations.

The Chloride Motive Power Battopper (Figures 50.1 and 50.2) is fitted with control logic to prevent overfilling and is self-powered by a maintenance-free battery. The system is continually monitored for faults. It is suitable for use for approximately 5 h between charges.

The Autofil device also automatically tops up the battery to the correct level. The device operates by controlling the topping up water flow by the formation of an air lock in the cell.

50.2 Standby power batteries

50.2.1 Chloride Industrial Batteries

Planté types YAP, YCP and YHP

These are designated as high-performance types. The range available is shown in Table 50.6 (types YAP and YCP) and Table 50.7 (Type YHP). A typical layout of a YHP battery is shown in Figure 50.3. Capacities up to 2000 A h per battery are available.

The nominal voltage is 2 V/cell, i.e. a nominal 110 V battery will have 55 cells. On discharge, the recommended final voltage at which the discharge should be terminated is shown in the discharge curves (Figure 50.4), i.e. the final voltage for the 3 h rate of discharge is 1.8 V. After the final voltage has been reached the voltage will fall away at an increasing rate.

The capacity that can be taken from a cell varies with the discharge rate as indicated in the capacity tables. Capacity is also affected by temperature.

Trickle charging is a method of keeping these cells in a fully charged condition by passing a small current through them. The correct trickle charge current does not allow the cell to gas and does not allow the density to fall over a period of time. It is normally in the region of 1 mA per ampere hour of the 10 h capacity for YAP cells and 0.3 × capacity at the 10 h rate + 70 mA for YCP cells. With this small

Table 50.2 Chloride Motive Power range of Classic motive power batteries

Design code	Capacity at 30°C (A h) 5 h	6 h	8 h	Cell dimensions (mm) Length	Width	Cell weight (kg) Dry	Filled	Charge current (A) at 2.1 V/cell 10–12 h*	8 h†	at 2.5 V/cell	Electrolyte quantity (litres)
Cell specifications 160 mm wide											
N13P HBKF-A / HBKF 7A / HBKF 25DA	57 / 228	59 / 235	62 / 246	Total height 240 mm over standard terminal take-off; cell box height 216 mm 61.5 / 205.5	158 / 158	4.3 / 14.7	5.5 / 19.1	8 / 32	10 / 45	4.5 / 19	0.9 / 3.4
N17P HXVF-A / HXVF 7A / HXVF 25DA	78 / 312	80 / 322	84 / 337	Total height 293 mm over standard terminal take-off; cell box height 260 mm 61.5 / 205.5	158 / 158	5.3 / 18.3	6.6 / 23.5	11 / 43	15 / 60	6.55 / 26	1.0 / 4.1
N23P HIMF-A / HIMF 7A / HIMF 25DA	110 / 438	113 / 451	118 / 473	Total height 357 mm over standard terminal take-off; cell box height 324 mm 61.5 / 205.5	158 / 158	6.8 / 23.7	8.5 / 30.3	15 / 61	20 / 80	9 / 37	1.3 / 5.1
N29P HILF-A / HILF 7A / HILF 25DA	141 / 564	145 / 581	152 / 610	Total height 435 mm over standard terminal take-off; cell box height 402 mm 61.5 / 205.5	158 / 158	8.3 / 29.1	10.5 / 37.9	20 / 61	25 / 110	12 / 47	1.7 / 6.9
N33P HTLF-A / HTLF 7A / HTLF 25DA	162 / 648	167 / 667	175 / 700	Total height 487 mm over standard terminal take-off; cell box height 454 mm 61.5 / 205.5	158 / 158	9.5 / 33.1	11.9 / 42.7	23 / 90	30 / 120	14 / 54	1.9 / 7.5
N38P HTHF-A / HTHF 7A / HTHF 25DA	189 / 756	195 / 779	204 / 816	Total height 547 mm over standard terminal take-off; cell box height 515 mm 61.5 / 205.5	158 / 158	10.7 / 38.0	13.4 / 48.7	26 / 105	35 / 140	16 / 63	2.1 / 8.3
N52P HTEF-A / HTEF 7A / HTEF 25DA	258 / 1032	266 / 1063	279 / 1115	Total height 728 mm over standard terminal take-off; cell box height 696 mm 61.5 / 205.5	158 / 158	14.3 / 49.1	18.3 / 64.9	36 / 144	50 / 200	22 / 86	3.1 / 12.3
Cell specifications 200 mm wide											
S24T HWAF-A / HWAF 7A / HWAF 21A	165 / 550	170 / 566	178 / 594	Total height 367 mm over standard terminal take-off; cell box height 328 mm 65 / 191	198.5 / 198.5	9.8 / 30.7	11.9 / 37.3	23 / 77	30 / 100	14 / 46	1.6 / 5.2
S29T HWBF-A / HWBF 7A / HWBF 21A	210 / 700	216 / 721	227 / 756	Total height 427 mm over standard terminal take-off; cell box height 388 mm 65 / 191	198.5 / 198.5	11.7 / 36.7	14.3 / 44.9	29 / 98	40 / 130	17 / 58	2.0 / 6.4
S34T HWCF-A / HWCF 7A / HWCF 21A	240 / 800	247 / 824	259 / 864	Total height 502 mm over standard terminal take-off; cell box height 463 mm 65 / 191	198.5 / 198.5	13.8 / 43.3	16.8 / 52.9	34 / 112	45 / 150	20 / 67	2.3 / 7.5
S40T HWEF-A / HWEF 7A / HWEF 21A	300 / 1000	309 / 1030	324 / 1080	Total height 588 mm over standard terminal take-off; cell box height 548 mm 65 / 191	198.5 / 198.5	15.8 / 51.0	19.4 / 62.3	42 / 140	50 / 190	25 / 83	2.8 / 8.8
S54T HWFF-B / HWFF 7B / HWFF 21B	360 / 1200	371 / 1236	389 / 1296	Total height 746 mm over standard terminal take-off; cell box height 707 mm 65 / 191	198.5 / 198.5	20.7 / 64.3	26.1 / 81.4	50 / 168	70 / 230	30 / 100	4.2 / 13.6

DA = Double pillar   * Taper characteristic of 2:1   † Taper characteristic of 2.67:1

Table 50.3 Chloride Motive Power range of Classic 15 motive power batteries

Total height over standard terminal take-off (mm)	Cell box height (mm)	Chloride type	5 h capacity at 30°C (A h)	Cell dimensions (mm)		Filled weight (kg)	Charge current (A) at 2.1 V/cell	
				length	Width		12 h*	8 h*
367	328	XWAF7	195	65	198.5	11.9	27	37
		XWAF21	650	191	198.5	37.3	91	123
427	388	XWBF7	240	65	198.5	14.3	34	46
		XWBF9	320	83	198.5	18.7	45	61
		XWBF11	400	101	198.5	23.5	56	76
		XWBF13	480	119	198.5	27.5	67	91
		XWBF15	560	137	198.5	32.7	78	106
		XWBF17	640	155	198.5	36.5	90	121
		XWBF19	720	173	198.5	40.6	101	136
		XWBF21	800	191	198.5	44.9	112	152
502	463	XWCF7	275	65	198.5	16.8	39	52
		XWCF21	920	191	198.5	52.9	130	175
588	548	XWEF7	345	65	198.5	19.4	48	66
		XWEF21	1150	191	198.5	62.3	160	220
746	707	XWFF7	435	65	198.5	26.3	61	83
		XWFF21	1450	191	198.5	81.7	203	275
435	402	XILF7	162	61.5	158	10.8	23	30
		XILF25	848	205.5	158	39.1	90	120
487	454	XTLF7	189	61.5	158	12.4	26	35
		XTLF25	756	205.5	158	44.7	105	140
547	515	XTHF7	219	61.5	158	14.2	30	40
		XTHF25	876	205.5	158	51.7	123	162
650	617	XTOF7	258	61.5	158	16.4	36	50
		XTOF25	1032	205.5	158	59.5	144	200
728	696	XTEF7	297	61.5	158	19.0	42	55
		XTEF25	1188	205.5	158	67.7	167	220
249	216	XBKF7	63	61.5	158	5.5	9	12
		XBKF25	252	205.5	158	19.1	35	48
293	260	XXVF7	86	61.5	158	6.6	12	16
		XXVF25	342	205.5	158	23.5	48	65
357	324	XIMF7	120	61.5	158	8.5	17	23
		XIMF25	480	203.5	158	30.3	67	91

* Taper characteristic 2.67:1 between 2.1 and 2.6 V/cell.

current flowing, the cell voltage will be approximately 2.25 V.

Float charging is keeping the voltage applied to the battery at 2.25 V/cell, i.e. constant voltage. This method is usually used where continuous and variable d.c. loads exist, and has the added advantage that some degree of recharge can be achieved without manual attention. A simple hydrometer reading indicates the state of charge.

A fully charged cell will have a specific gravity of between 1.205 and 1.215.

The ampere hour efficiency of the cells is 90%; therefore, on recharge, the amount of recharge required is equal to the discharge in ampere hours plus 11%. A YAP 9 cell discharged at the 1 h rate (18 A for 1 h) will require a minimum of 18 + 11%, i.e. 20 A h recharge. At the finishing rate of charge (2 A), this will take 10 h. It is possible to recharge in a shorter time by starting the charge at the 'starting rate', as given in the capacity tables, but this should be reduced to the finishing rate of charge when the voltage per cell reaches 2.3 V.

Table 50.4 Chloride Motive Power Classic 25 range of motive power batteries

Cell type	No. of plates	Standard cell equivalents		A h capacity at 5 h rate		% increase in capacity	Weight (kg)	Cell box dimensions (mm)			Overall height (mm)
		Hard rubber	HSP boxes	High energy	Standard cell			Height	Length	Width	
S26Y	9	WAF 7	HWAF 7	210	165	27	11.7	328	65	198.5	367
	17	WAF 13	HWAF 13	420	330	27	22.4	328	119	198.5	367
	25	WAF 19	HWAF 19	630	495	27	33.1	328	173	198.5	367
S32Y	9	WBF 7	HWBF 7	260	210	23	14.1	388	65	198.5	427
	17	WBF 13	HWBF 13	520	420	23	26.9	388	119	198.5	427
	25	WBF 19	HWBF 19	780	630	23	39.7	388	173	198.5	427
S36Y	9	WCF 7	HWCF 7	300	240	25	16.5	463	65	198.5	502
	17	WCF 13	HWCF 13	600	480	25	31.5	463	119	198.5	502
	25	WCF 19	HWCF 19	900	720	25	46.6	463	173	198.5	502
S45Y	9	WEF 7	HWEF 7	375	300	25	19.8	548	65	198.5	588
	17	WEF 13	HWEF 13	750	600	25	38.0	548	119	198.5	588
	25	WEF 19	HWEF 19	1125	900	25	56.1	548	173	588	588

Table 50.5 Chloride Motive Power range of motive power batteries in hard rubber containers

Overall height over standard terminal connectors (mm)	Type	Capacity at 30°C (A h)			Cell dimensions (mm)		Filled cell weight (kg)
		5 h	6 h	8 h	Length	Width	
245	BKF27A	247	255	267	224	160	20.6
	BKF29A	266	274	287	240	160	22.0
	BKF31A	285	294	308	255	160	23.5
289	XVF27A	338	348	365	224	160	25.4
	XVF39A	494	508	534	319	160	38.7
352	IMF27A	475	489	512	224	160	32.8
	IMF41A	730	752	788	335	160	49.5
430	ILF27A	611	629	660	224	160	40.9
	ILF41A	940	968	1016	335	160	62.4
483	TLF27A	702	723	758	224	160	45.8
	TLF41A	1080	1112	1166	335	160	70.4
543	THF27A	819	844	884	224	160	52.5
	THF41A	1260	1298	1360	335	160	79.2
740	MTF9	488	503	527	81	222	35.4
	MTF37	2196	2263	2372	310	227	153.8
363	WAF5A	110	113	119	47	200	8.7
428	WBF5A	140	144	151	47	200	10.2
498	WCF5A	160	165	172	47	200	11.9
578	WEF5A	200	206	216	47	200	14.2

Figure 50.1 Motive power Battoper automatic topping-up device (Courtesy of Chloride Batteries)

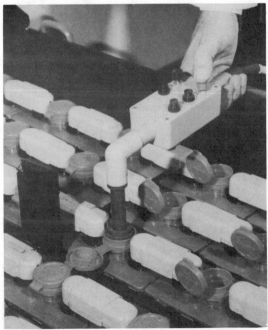

Figure 50.2 The Chloride Motive Power Battoper automatic topping-up device in use (Courtesy of Chloride Batteries)

Table 50.6 High-performance Planté-type lead–acid standby power batteries available from Chloride Industrial Batteries

	Final voltage (V)	Cell type														
		YAP 5	YAP 9	YAP 13	YAP 17	YAP 21	YCP 9	YCP 11	YCP 13	YCP 17	YCP 21	YCP 25	YCP 27	YCP 29	YCP 33	YCP 35
Capacity* at 15°C (A h) when discharged in																
10 h	1.85	15	30	45	60	75	100	125	150	200	250	300	325	350	400	425
3 h	1.80	12	24	36	47.5	60	80	99	119	159	199	239	258	278	318	338
1 h	1.75	9	18	27	36	45	60	75	90	120	150	180	195	210	240	255
Capacity* at 25°C (A h) when discharged in																
10 h	1.85	16	32	48	64	80	107	134	161	214	268	322	348	375	429	456
3 h	1.80	13	26	38.5	52	64	86	107	129	172	215	258	279	301	344	365
1 h	1.75	9.8	19.5	29.5	39	49	65	82	98	131	163	196	212	229	262	278
Charge current (A)																
Starting	–	2	4	6	8	10	14	17	20	28	34	42	46	48	56	60
Finishing	2.70	1	2	3	4	5	7	8.5	10	14	17	21	23	23	28	30
Weight of cell complete (kg)																
Filled		3.8	6.3	10.0	11.45	13.6	18.6	22.2	24.9	30.6	36.9	43.4	52.6	54.4	58.4	60.3
Acid only (relative density 1.210)		1.16	1.89	3.4	3.25	3.8	5.5	7.5	7.2	8.7	10.4	12.1	16.6	16.2	15.5	15.1
Approx. quantity of acid, relative density 1.210 (litres)		0.96	1.56	2.83	2.68	3.16	4.5	6.2	5.9	7.2	8.6	10	13.7	13.4	12.8	12.5
External dimensions of box (mm)																
Length†		76	114	190	190	228	134	172	172	210	248	286	362	362	362	362
Width		133	133	133	133	133	203	203	203	203	203	203	203	203	203	203
Height		212	212	212	212	212	349	349	349	349	349	349	349	349	349	349
Overall height of cell (mm)		260	260	260	260	260	423	423	423	423	423	423	423	423	423	423

* Larger capacity enclosed Planté cells are available in the YHP range (500–2000 A h)
† The length of a box or cell is measured at right angles to the plates. The width is measured parallel to the plates

Table 50.7 High-performance Planté-type lead–acid standby power storage batteries available from Chloride Industrial Batteries

	Final voltage (V)	Cell type															
		YHP 11	YHP 13	YHP 15	YHP 17	YHP 19	YHP 21	YHP 23	YHP 25	YHP 27	YHP 29	YHP 31	YHP 33	YHP 35	YHP 37	YHP 39	YHP 41
Capacity* at 15°C (A h) when discharged in																	
10 h	1.85	500	600	700	800	900	1000	1100	1200	1300	1400	1500	1600	1700	1800	1900	2000
3 h	1.80	406	487	568	649	730	811	892	973	1054	1135	1217	1298	1379	1460	1541	1622
1 h	1.75	300	360	420	480	540	600	660	720	780	840	900	960	1020	1080	1140	1200
Capacity* at 25°C (A h) when discharged in																	
10 h	1.85	536	643	750	858	965	1072	1179	1286	1394	1501	1608	1715	1822	1930	2037	2144
3 h	1.80	438	526	614	702	789	877	965	1052	1140	1228	1315	1403	1491	1578	1666	1754
1 h	1.75	327	392	458	523	589	654	719	785	850	915	981	1046	1112	1177	1242	1308
Charge current (A)	2.7																
Starting		70	84	98	112	126	140	154	168	182	196	210	224	238	252	266	280
Finishing		35	42	49	56	63	70	77	84	91	98	105	112	119	126	133	140
Weight of cell complete (kg)																	
Filled		95.2	106.2	133.5	144.5	155.5	179.3	190.4	218.0	229.0	240.1	268.3	279.2	290.2	318.2	329.2	340.2
Acid only, relative density 1.210		32.2	30.6	45.3	43.7	42.1	53.3	51.8	66.8	65.2	63.7	79.3	77.6	76.0	91.4	89.8	88.2
Approx. quantity of acid, relative density 1.210 (litres)		27.1	25.7	38.1	36.7	35.4	44.8	43.5	56.1	54.8	53.5	66.6	65.2	63.9	76.8	75.5	74.1

* Larger capacities can be obtained by connecting cells in parallel. Smaller capacities are available in the YAP range (15–75 A h) and YCP range (100–425 A h)

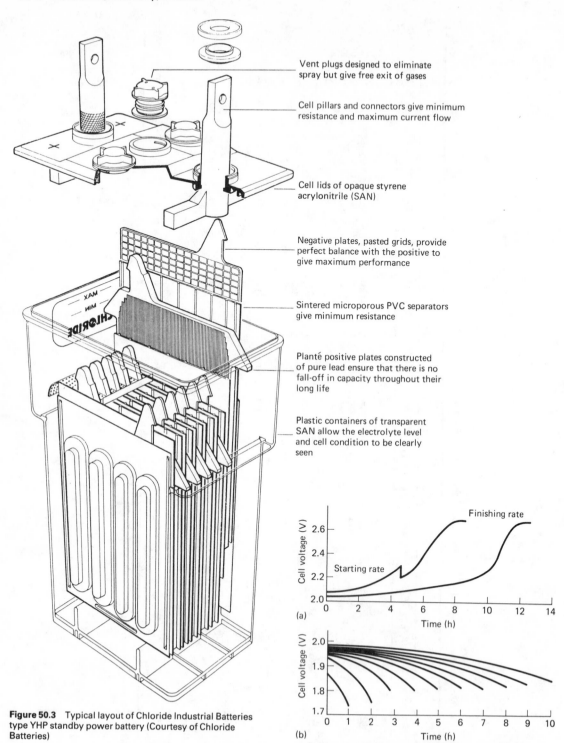

Vent plugs designed to eliminate spray but give free exit of gases

Cell pillars and connectors give minimum resistance and maximum current flow

Cell lids of opaque styrene acrylonitrile (SAN)

Negative plates, pasted grids, provide perfect balance with the positive to give maximum performance

Sintered microporous PVC separators give minimum resistance

Planté positive plates constructed of pure lead ensure that there is no fall-off in capacity throughout their long life

Plastic containers of transparent SAN allow the electrolyte level and cell condition to be clearly seen

Figure 50.3 Typical layout of Chloride Industrial Batteries type YHP standby power battery (Courtesy of Chloride Batteries)

Figure 50.4 High-performance Planté-type lead–acid standby batteries available from Chloride Industrial Batteries. (a) Typical recharge characteristics following discharge at the 3 h rate. (b) Discharge voltage characteristics at 15°C (Courtesy of Chloride Batteries)

Table 50.8 Super high-performance Planté type lead–acid standby batteries available from Chloride Industrial Batteries

Type YCP-H	13H	17H	23H	27H	31H	41H
Capacity at 15°C (A h) 10 h to 1.85 V	150	200	275	325	375	500
Charge current (A)						
Starting	20	28	38	45	52	70
Finishing	10	14	19	22	26	35
Weight (kg)						
Dry	16.7	20.9	28.1	32.7	38.8	51.4
Filled	21.7	27.5	36.0	42.3	50.2	66.1
Volume of electrolyte (relative density 1.240) (litres)	3.2	5.3	6.4	7.7	9.2	11.8
External box dimensions (mm)						
Length	134	172	210	248	286	362
Width	203	203	203	203	203	203
Height	349	349	349	349	349	349
Overall height (mm)	423	423	423	423	423	423

To calculate length of stands: double tier/single row = $C \times N/2 + 50$ mm, double tier/double row = $C \times N/4 + 50$ mm, where C = cell centre and N = cell numbers

On recharge, the voltage increases as the charge proceeds. The highest voltage reached with the finishing rate of charge flowing is approximately 2.70 V/cell. It is possible to recharge a cell by limiting the voltage of the charging equipment to a lower value than 2.70 V/cell, 2.4 V/cell being the minimum desirable value. This will result in an extended recharge period, as the battery will automatically limit the charge current irrespective of the charger output.

Chloride Industrial Batteries supply Super Planté type high-performance cells, type YCP-H, available in capacities up to 500 A h (Table 50.8).

This range of batteries is particularly suitable for any standby application calling for a good high-rate discharge characteristic. Such applications include electronic computer and telecommunications equipment, which requires a no-break 'clean' power supply free from voltage surges, frequency variations and other transients.

To achieve this the power supply is usually obtained through a no-break, static inverter or a no-break rotary set.

In either case a standby battery of proven reliability and maximum compactness is required, which is capable of supplying a high discharge current for a comparatively short period.

A large percentage of EHT switchgear is battery operated. Here again the requirement is for a good high-rate discharge characteristic, since the switchgear closing solenoids take a high current for a very short time.

The most reliable type of battery for engine starting is the Planté high-performance type. The standby engine is, of course, required for many emergency applications where maximum reliability is essential. These applications include emergency lighting, sprinkler systems, industrial processes, telecommunications, and any auxiliary power requirements.

High-performance flat-plate types

Chloride Industrial Batteries also supply high-performance flat-plate cell type standby batteries in the conventional vented (types FAP-H and FCP-H) and low-maintenance (types SAP, SBP and SEP) versions. Capacities up to 600 A h (vented, Table 50.9) and 1125 A h (low maintenance, Table 50.10) are available. Trickle and float charging methods are used as with the Planté types. A feature of both the vented and low-maintenance ranges is that a new plate alloy incorporating selenium is used in their manufacture. This new patented low-antimonial lead–selenium alloy has a substantially reduced antimony content, which reduces antimony migration and extends the battery life. The fine grain structure and strength of the alloy are maintained by the introduction of selenium to reduce intergranular corrosion.

The lead oxide used in the plates is made finer, producing a greater surface area with improved material utilization, giving higher ampere hour capacity per plate.

Table 50.9 High-performance flat-plate vented lead–acid standby power batteries available from Chloride Industrial Batteries

	Final voltage (V)	FAP 5-H	FAP 9-H	FAP 13-H	FAP 17-H	FAP 21-H	FAP 25-H	FAP 33-H	FAP 41-H	FCP 9-H	FCP 13-H	FCP 17-H	FCP 21-H	FCP 25-H	FCP 29-H	FCP 33-H	FCP 37-H	FCP 41-H
Capacity* at 20°C (A h) to BS 440 final voltage																		
10 h	1.85	16	32	48	64	80	96	128	160	120	180	240	300	360	420	480	540	600
3 h	1.80	12.3	24.6	36.9	49.2	61.5	73.8	98.4	123	92	138	184	230	276	322	368	414	460
1 h	1.75	8.4	16.8	25.2	33.6	42	50.4	67.2	84	62	93	124	155	186	217	248	279	310
Capacity* at 20°C (A h) emergency lighting ratings to final voltage 85% of nominal																		
3 h	1.70	13.6	27.2	40.8	54.4	68	81.6	108.8	136	102	153	204	255	306	357	408	459	510
2 h	1.70	12	24	36	48	60	72	96	120	90	135	180	225	270	315	360	405	450
1 h	1.70	8.8	17.6	26.4	35.2	44	52.8	70.4	88	66	99	132	165	198	231	264	297	330
Capacity* at 25°C (A h)																		
8 h	1.75	17	35	51	68	85	102	136	170	128	192	256	320	384	448	512	576	640
3 h	1.75	13.2	26.4	39.6	52.8	66	79.2	105.6	132	99	148.5	198	247.5	297	346.5	396	445.5	495
1 h	1.75	8.7	17.4	26.1	34.8	43.5	52.2	69.6	87	64	96	128	160	192	224	256	288	320
Charge current (A)																		
Starting	–	2	4	7	9	11	13	18	22	17	25	34	42	50	59	67	76	86
Finishing	2.70	1	2	3	4	5	6	9	11	8	12	17	21	25	29	33	38	42
Weight of cell complete (kg)																		
Filled		3.2	3.9	5.7	6.5	9.5	10.4	12.2	13.6	15.4	17.9	22.9	28.0	33.3	35.6	41.4	49.2	52.0
Acid only, relative density 1.240‡		1.3	1.2	2.0	2.0	3.6	3.5	3.5	4.0	5.8	5.2	7.3	9.0	10.7	10.1	12.2	16.3	16.0
Approx. quantity of acid, relative density 1.240‡ (litres)		1.1	1.0	1.6	1.5	2.8	2.8	2.7	3.1	4.68	4.19	5.89	7.26	8.63	8.15	9.83	13.15	12.90
External dimensions of box (mm)																		
Length†		76	76	114	114	190	190	190	228	134	134	172	210	248	248	286	362	362
Width		133	133	133	133	133	133	133	133	203	203	203	203	203	203	203	203	203
Height		212	212	212	212	212	212	212	212	349	349	349	349	349	349	349	349	349
Overall height of cell (mm)		253	253	253	253	253	253	253	253	423	423	423	423	423	423	423	423	423

Cell type

* Larger capacities can be obtained by connecting cells in parallel. Capacity levels applicable after 5 discharge/charge cycles
† The length of a box or cell is measured at right angles to the plates. The width is measured parallel to the plates
‡ 1.270 for FAP range

Table 50.10 High-performance flat-plate low-maintenance type lead–acid standby power batteries available from Chloride Industrial Batteries

Type	Nominal capacity (A h) at 25°C in 10 h	Width (mm)	Length (mm)	Height over pillar (mm)	Filled weight (kg)
SAP 5	50	206	103	389	12.0
SAP 7	75	206	103	389	13.6
SAP 9	100	206	103	389	14.8
SAP 13	150	206	145	389	19.5
SAP 17	200	206	145	389	22.7
SAP 21	250	206	187	389	28.3
SBP 17	320	206	166	505	35.4
SBP 21	400	210	254	505	47.1
SBP 27	520	210	254	505	54.3
SEP 21	625	210	210	694	63.8
SEP 23	688	210	210	694	66.5
SEP 25	750	191	233	694	74.3
SEP 27	813	191	233	694	77.0
SEP 29	875	191	233	694	79.7
SEP 31	938	240	270	648	90.0
SEP 33	1000	240	270	648	92.7
SEP 35	1062	240	270	648	95.4
SEP 37	1125	240	270	648	98.1

The low maintenance SAP, SBP and SEP ranges, in addition to being suitable for high-rate applications, are also suitable for low-rate applications such as SP telecommunications, where a low-maintenance standby power battery is required.

Uninterrupted power supply (UPS) systems are becoming employed increasingly with sophisticated computer and telecommunications equipment. Many kinds of batteries are available as standby within such systems, and the choice depends on the requirements in each case.

The SP range has been developed to meet the following specifications:

1. Flat plate design.
2. Life expectation of 10 years.
3. Minimum maintenance.
4. Transparent container for ease of inspection.

Table 50.11 LCP range small-size flat-plate lead–acid standby batteries supplied by Chloride Power Storage

	873 (LCP5)	874 (LCP7)	875 (LCP9)	876 (LCP13)	877 (LCP17)	878 (LCP21)
Nominal capacity at 20°C	20	30	40	60	80	100
Nominal charge current (A)	1.3	2.0	2.7	4.0	5.3	6.3
Trickle charge (mA)	10–20	15–30	20–40	30–60	40–80	50–100
Length (mm)	35	54	54	115	115	115
Width (mm)	129	129	129	129	129	129
Height (mm)	219	219	219	219	219	219
Cell centres (mm)						
face to face	38	57	57	117	117	117
edge to edge	–	–	–	132	132	132
Weight filled (kg)	1.79	2.58	2.92	5.10	5.70	6.36
Quantity of electrolyte (litres)	0.46	0.75	0.77	1.93	1.79	1.65

Voltage 12 V (6 cells) 6 V (3 cells)
Nominal voltage 2.0 V/cell
Floating voltage 2.20 to 2.25 V/cell
Charging voltage up to 2.70 V/cell
Final voltage 1.85 V at 10 h rate of discharge

Table 50.12 Range of flat plate type engine starting batteries supplied by Chloride Power Storage

	RSKA 62SA	RSKA 92SA	RSKA 154SA	RSKA 185SA	RSKB 118SA	RSKB 158SA	RSKB 236SA	RSKB 315SA
Number of plates	9	13	21	25	13	17	25	33
Capacity at 20°C (A h) at 5 h rate	62	92	154	185	118	158	236	315
Charge current (A)								
Starting	12	18	31	37	24	32	47	63
Finishing	4	6	10	12	8	11	15	21
Current for 30 s to 1.33 V p.c. at 0°C cell 25% discharged	360	534	893	1073	620	830	1239	1654
Cell weight complete (kg)								
Filled	7.7	10.7	17.2	20.2	12.7	16.8	23.8	30.8
Acid only	1.3	1.9	3.0	3.6	2.3	2.9	4.2	5.5
Approx. quantity of acid (litres)	1.05	1.50	2.36	2.82	1.77	2.27	3.32	4.36
Cell dimensions (mm)								
Length	62	86	135	160	86	111	160	208
Width	159	159	159	160	159	159	160	160
Height	283	283	283	283	321	321	321	321
One-cell tray*								
Length (mm)	102	126	175	200	126	152	200	246
Width (mm)	200	200	200	200	200	200	200	200
Layout	A	A	B	B	A	B	B	B
Centres (mm)	108	132	206	206	132	206	206	206
Weight (kg)	9.7	13	19.9	23.2	15.3	19.7	27.2	34.7
Four-cell tray*								
Length (mm)	164	212	–	–	212	–	–	–
Width (mm)	360	360	–	–	360	–	–	–
Layout	G	G	–	–	G	–	–	–
Centres (mm)	170	218	–	–	218	–	–	–
Weight (kg)	34.8	47.4	–	–	56	–	–	–
Tray height (mm)	272	272	272	272	310	310	310	310
Overall height (mm)	350	350	350	350	388	388	388	388
Height over end cell take-offs (mm)								

* Two-cell and three-cell versions are also produced
† Monobloc units in excess of 60 kg must be accommodated on stillages only

5. Good high-rate discharge characteristic.
6. High reliability.
7. Low floor area.
8. Minimum initial cost.

New separator technology has been incorporated into these flat-plate cells. The vented types incorporate separation between adjacent plates consisting of a glass wool mat and ribbed microporous PVC. The glass wool mats are fitted against the positive plates and act as an effective retainer of the positive active material. The low-resistance separators, which are inert chemically and have a high degree of porosity, provide a complete diaphragm between the plates but allow good circulation of the electrolyte and give mechanical support to the negative active material.

The low-maintenance types have separation between adjacent plates consisting of a glass wool mat and ribbed microporous phenol formaldehyde. The

| RK.144SA | RK.240SA | RK.288SA | RK.384SA | Monobloc† | | | | | | | | RK.192.M/4 |
				RSKA.77.M/4	RSKA.92.M/4	RSKA.123.M/4	RSKA.169.M/3	RSKB.158.M/3	RSKB.197.M/4	RSKB.217.M/4	RSKB.276.M/2	
13	21	25	33	11	13	17	23	17	21	23	29	17
144	240	288	384	77	92	123	169	158	197	217	276	192
29	48	58	77	15	18	25	34	32	39	43	55	38
9.5	16	19	26	5	6	8	11	11	13	14	18	13
713	1188	1426	1900	447	534	713	980	830	1034	1139	1449	950
14.5	23.4	27.4	35.4	38.2	45.0	60.0	54.5	52.9	84.8	99.0	57.1	81.0
2.5	4.1	4.8	6.4	6.5	7.6	9.8	11.1	8.8	14.4	16.7	10.0	14.0
2.00	3.23	3.82	5.00	5.10	6.0	7.7	8.7	6.9	11.3	13.1	7.8	11.0
86	135	160	208	170	194	248	160	359	296	305	391	248
159	159	160	160	340	343	343	483	184	343	327	185	343
356	356	356	356	288	288	288	281	325	326	319	326	360
126	175	200	246									
200	200	200	200									
A	B	B	B									
132	206	206	206									
17.3	26.8	31.1	39.6									
–	–	–	–									
–	–	–	–									
–	–	–	–									
–	–	–	–									
345	345	345	345									
423	423	423	423									
				340	340	340	334	378	378	372	378	411

glass wool mats are fitted against the positive plates and act as an effective retainer of the positive active material. The low-resistance separators, which are inert chemically and have a high degree of porosity, provide a complete diaphragm between the plates but allow good circulation of the electrolyte and give mechanical support to the active material.

Standard flat-plate types

Chloride Power Storage supply a lower capacity range LCP type (20–100 A h capacity) flat-plate type of standby power battery (Table 50.11) suitable for performance in small installations such as telephone exchanges, emergency lighting, laboratory equipment and security systems. This range also uses the low-antimony selenium-containing grid alloy.

Also available in the standard range is the R series of engine-starting standby batteries. These batteries are supplied as single cell containers with capacities up to 384 A h, or as monobloc containers with capacities up to 276 A h (Table 50.12).

Chloride also supply a range of marine batteries. These batteries are designed to fulfil a multitude of

Table 50.13 Batteries for marine applications

Basic construction	All-purpose marine battery (M series)	Engine starting battery (R series)	Buoy lighting battery MW 17 and MB 19
Positive plates	Tubular plates	Flat pasted plates	Flat pasted plates
Negative plates	Flat pasted plates	Flat pasted plates	Flat pasted plates
Separators	Microporous sleeves	Microporous sheet and glass wool mat	Microporous sheet and glass wool mat
Cell boxes	Resin rubber	Resin rubber	Resin rubber
Cell lids	Resin rubber	Resin rubber	Resin rubber
Collectors	Bolted inter-unit Burned inter-cell	Bolted	Wing nuts
Container	Wood tray holding 2–8 cells	Either resin rubber monoblocs or wood trays	Either metal containers or wood trays

duties at sea. They are vital for communications, engine starting, emergency lighting, sprinklers, navigation lights and many other ancillary services on ocean-going vessels. They are used also to provide power for communications and emergency services on smaller craft such as inshore trading vessels and pleasure boats.

There are slight variations in the construction of the three types of battery that Chloride offer for marine applications, to ensure that each battery matches the demands of its duty (Table 50.13).

A new addition to the Chloride Power Storage range of flat-plate standby batteries is the Powersafe series of low-maintenance cells. The batteries are housed in flame-retardant cases. In Powersafe cells the overcharge reaction is controlled using the principle of gas recombination. This means that the gases normally evolved during the flat/recharge operation in conventional lead–acid cells recombine to form water in the Powersafe cell. Consequently, for all practical purposes, Powersafe products lose no water during normal operation and therefore topping up is not required.

Gas-recombining cells contain a special highly porous glass microfibre separator through which oxygen evolved from the positive plate can diffuse to react at the negative electrode where it is electrochemically reduced to water.

Powersafe is designed specifically for standby power applications. Low internal resistance gives super high-rate performance and novel design features make Powersafe the ideal choice for a wide range of equipment including telecoms, uninterruptible power supplies, emergency lighting, power distribution, engine starting and a host of other standby duties. Powersafe has a design life of 10 years.

Only constant-potential chargers are recommended for use with Powersafe, and for efficient power consumption a current limit of $0.1C_3$ A, e.g. 8 A for the type 3VBII, is recommended.

For maximum service life the recommended float voltage is 2.27 V p.c. Where a higher float voltage is essential this is perfectly acceptable to Powersafe although this can be expected to reduce service life.

Powersafe will accept high currents for fast recharging (with normal voltage safeguards). For example, with constant potential recharge at 2.27 V p.c. and 30 A available current, a 3VBII battery would be 75% charged in 3 h.

Open-circuit losses of Powersafe from the fully charged condition are generally similar to those for Planté cells. As with Planté cells, the rate of loss is dependent on ambient temperature and it is therefore recommended that Powersafe batteries, which are delivered in fully charged condition, be stored at as low a temperature as possible and given a freshening charge every 6 months.

The product range is listed in Table 50.14.

Following the Powersafe range, in 1984 Chloride Power Storage introduced their Powerstore range of fully sealed high-performance long-life standby batteries (Table 50.15). These batteries also utilize recombination electrolyte technology to ensure that no topping up is required during life. They were introduced for use in the utilities and telecommunications markets.

Calcium–lead alloy positive and negative grids are used to ensure corrosion resistance and long life. In the smaller Powerstore sizes, the conventional intercell partition walls are replaced by high-density polyethylene envelopes into which individual cell groups are packed. This increases the volume of usable space available in the battery container and contributes to the high energy densities of the battery.

The Powerstore range covers capacity requirements from 3 to 240 A in 2, 4, 6 and 12 V monoblocs.

Table 50.14 Range of Powersafe low-maintenance lead–acid standby batteries available from Chloride Power Storage

Cell type	Voltage (V)	Capacity at 3 h rate (20°C/1.8 V p.c.)	Maximum dimensions (mm)			Weight (kg)
			Height	Length	Width	
VA17	2	48	208	106	117.5	5.2
2VB11	4	80	235	206	210	15.6
3VB9	6	64	235	206	210	18.8
3VB11	6	80	235	206	210	22.0
VC13	2	110	296	121	167.5	13.5
2VB13	4	90	224.5	204	204	17.6
2VB15	4	105	224.5	204	204	19.8
2VB17	4	120	224.5	204	204	22.0
VB26	2	180	224.5	204	204	17.6
VB30	2	210	224.5	204	204	19.8
VB34	2	240	224.5	204	204	22.0

Table 50.15 Range of low-maintenance standby Powerstore lead–acid batteries available from Chloride Power Storage

Nominal voltage (V)	Nominal capacity	Weight (g)	Internal resistance charged (mΩ)	Max. discharge current (A)	Operating temperature (°C)	Life (years)	Charging
6	20 h rate of 150 mA to 5.25 V 3.0 A h 10 h rate of 260 mA to 5.25 V 2.6 A h 5 h rate of 450 mA to 5.10 V 2.3 A h 1½ h rate of 1.15 V to 4.95 V 1.7 A h 1 h rate of 1.55 mA to 4.80 V 1.6 A h 30 min rate of 2.65 A to 4.65 V 1.3 A h	570	40	15	−10 to 50	5–8	Standby duty: maximum charging current 0.75 A, float voltage 6.75–6.9 V Primary power: maximum charging current 0.75 A, charge voltage 7.2 V
6	20 h rate of 225 mA to 5.25 V 4.5 A h 10 h rate of 440 mA to 5.25 V 4.0 A h 5 h rate of 675 mA to 5.10 V 3.4 A h 1½ h rate of 1.75 mA to 4.95 V 2.6 A h 1 h rate of 2.40 A to 4.80 V 2.4 A h 30 min rate of 4.2 A to 4.65 V 2.1 A h	630	35	22.5			Standby duty: maximum charging current 1.1 A, float voltage 6.75–6.9 V Primary power: maximum charging current 1.1 A, charge voltage 7.2 V
2	20 h rate of 290 mA to 1.75 V 5.8 A h 10 h rate of 500 mA to 1.75 V 5.0 A h 5 h rate of 8.75 mA to 1.70 V 4.4 A h						Standby duty: maximum charging current 1.5 A, float voltage 2.25–2.3 V

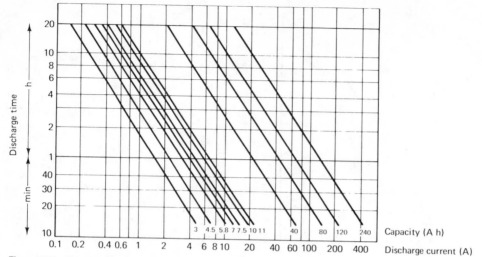

Figure 50.5 Chloride Power Storage Powerstore standby batteries: discharge current *versus* time (Courtesy of Chloride Batteries)

Powerstore batteries are designed for use at discharge rates from a few seconds to 20 h at temperatures from −10 to 50°C.

Figure 50.5 shows the relationship between discharge current and time for each Powerstore battery on continuous discharge at 25°C. The number on each line is the rated capacity (C). The battery capable of providing the required discharge current and time is the one whose discharge characteristic lies to the right of the point where current and time intersect.

In practice, the size may need to be different from this to take account of end-of-discharge voltage and ambient temperature. For example, an application requiring 10 A continuously for 30 min will be satisfied by a Powerstore battery with rated capacity (C) of not less than 10 A h. At this particular discharge rate it will deliver approximately $10 \times \frac{1}{2} = 5$ A h; half its rated capacity.

The curves in Figure 50.6 show the variation in voltage of a 2 V Powerstore battery at several discharge currents. The current is expressed as a multiple of the value of the rated capacity. For example, a 40 A h Powerstore delivering a continuous current of 8 A is discharging at $8/40 = 0.2C$ A.

For any discharge rate there is a minimum recommended discharge voltage. These minimum voltages lie on the broken line in Figure 50.6. If the system cannot tolerate end-of-discharge voltages as low as those in Figure 50.6, a larger size of Powerstore battery may be needed to supply the required current and time to this higher voltage. For example, a 10 A h Powerstore battery delivering 10 A is discharging at the $1C$ rate. This will continue for 30 min to 1.6 V/cell. However, if the system could not tolerate a final voltage of less than 1.9 V/cell, the 10 A h Powerstore battery discharging at the $1C$ rate would provide only 20 min of discharge time. To provide 30 min of discharge would require a Powerstore battery discharging at $0.8C$ when delivering 10 A, i.e. $10 + 2.5 = 12.5$ A h.

While Powerstore batteries will function over a wide temperature range, the available capacity varies with the temperature. The data presented in Figures 50.5 and 50.6 are for operation at 25°C. If the permanent discharge temperature is likely to be significantly different from 25°C, a different size of Powerstore battery may be needed to provide the same performance.

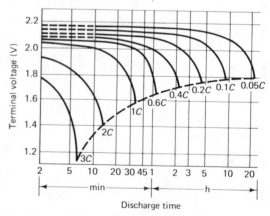

Figure 50.6 Chloride Power Storage Powerstore standby 2 V battery: voltage *versus* time for constant-current discharge at 25°C (Courtesy of Chloride Batteries)

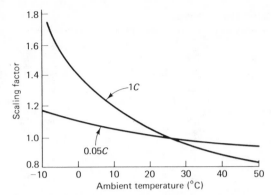

Figure 50.7 Chloride Power Storage standby batteries: capacity scaling factor *versus* discharge temperature (Courtesy of Chloride Batteries)

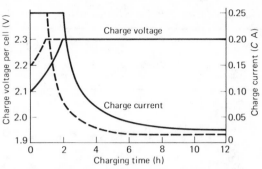

Figure 50.8 Electric Power Storage Powerstore standby batteries: typical charging characteristics (current limited) at 25°C. Charge voltage 2.3 V/cell, current limit 0.25C. —, 100% discharged; ----, 50% discharged (Courtesy of Chloride Batteries)

Figure 50.7 presents the scaling factor to be applied to the 25°C rated capacity to provide equivalent performance at different temperatures. The scaling factor is also affected by the discharge rate. For example, 10 A for 30 min to 1.60 V/cell are available from a Powerstore battery with a rated capacity of 10 A h at 25°C. The capacity would need to be increased to approximately $1.4 \times 10 = 14$ A h to provide 10 A for 30 min at 0°C on the $1C$ discharge curve.

Prolonged use at very high temperatures, while feasible, adversely affects the useful service life of all battery systems.

The performance and life of rechargeable batteries are significantly affected by the charging method. Current-limited constant-voltage charging is recommended for Powerstore batteries. The selection of the current limit and voltage depends on the required recharge time and battery temperature. To achieve the best performance and life from Powerstore batteries the following recommendations are made:

1. Limit the initial charge current to 0.25CA. Powerstore batteries will accept higher currents, but the small sacrifice in recharge time from using this current limit is more than compensated by increased battery life and reduced charger costs.
2. Avoid excessive overcharge. Using the recommended charging conditions will ensure the longest life.
3. Avoid prolonged undercharge. Keeping Powerstore batteries fully charged will help to prevent any permanent loss of capacity.

Figure 50.8 shows a typical current-limited constant-voltage charging characteristic for Powerstore batteries.

For standby applications, the recommended continuous float charge voltage is 2.25–2.30 V/cell at 25°C. This will keep Powerstore batteries in a fully charged, optimum working condition. The continuous float current will be in the region of 1–3 mA/A h. A higher float voltage serves no useful purpose and will merely lead to excessive overcharge, which could shorten the service life.

Where rapid recharge is required it is recommended that a charge voltage of 2.4 V/cell at 25°C should be used. However, charging at this voltage should be continued no longer than is necessary to restore full charge. The best way to ensure this is to monitor the charge current and either terminate the charge or switch to the float voltage of 2.25–2.30 V/cell when the charge current falls to 0.04–0.08C A. This represents the end-of-charge current for a battery being charged at 2.4 V.

The superior charge acceptance of Powerstore batteries means that they can be recharged rapidly. The time it takes will depend on the previous depth of discharge, the recharge voltage, the current limit and the temperature. Figure 50.9 shows the expected recharge times for a 4 V Powerstore at 25°C with current limits of 0.25C and 0.1CA from both 100% and 50% previous depth of discharge.

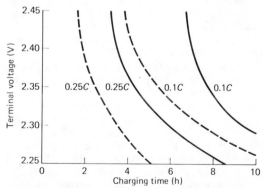

Figure 50.9 Chloride Power Storage Powersure standby batteries: recharge time to 90% capacity *versus* charging time for a 4 V type (all capacities) at 25°C. —, 100% discharged; ----, 50% discharged (Courtesy of Chloride Batteries)

Table 50.16 Tubular plate lead–acid standby batteries available from Chloride Power Storage

	Final voltage per cell (V)	6-BKF7SA	4-BKF7SA	3-BKF7SA	4-BKF9SA	3-BKF9SA	4-BKF11SA	3-BKF11SA	4-IMF7SA	3-IMF7SA	4-BKF13SA	3-BKF13SA	4-IMF9SA	4-IMF11SA	3-IMF11SA	1-IMF13SA	1-IMF15SA
Capacity (A h)																	
at 15°C																	
10 h	1.85	56	56	56	74	74	93	93	108	108	112	112	143	143	179	214	251
3 h	1.80	43	43	43	57	57	71	71	83	83	86	86	110	110	137	164	192
1 h	1.75	33	33	33	44	44	55	55	62	62	66	66	82	82	103	123	144
at 25°C																	
10 h	1.85	58	58	58	78	78	97	97	112	112	116	116	149	149	187	223	261
3 h	1.80	45	45	45	60	60	75	75	87	87	90	90	115	115	145	173	202
1 h	1.75	35	35	35	47	47	58	58	66	66	70	70	88	88	110	131	153
Charge current (A)																	
Start	–	11.5	11.5	11.5	15	15	19	19	22	22	23	23	29	29	37	44	51
Finish	2.75	4	4	4	5	5	6.5	6.5	7.5	7.5	7.5	7.5	9.5	9.5	12	14.5	17
Weight including container (kg)																	
Filled		36.5	24.5	18.6	31.1	23.5	36.9	27.8	37.0	28.7	43.4	32.7	46.4	35.6	43.6	17.6	20.2
Acid only		5.4	3.6	2.7	5.4	4.1	6.4	4.8	6.4	4.8	7.3	5.4	8.2	6.1	7.5	3.0	3.6
Approx. quantity of acid (litres)		4.3	2.9	2.2	4.3	3.2	5.0	3.8	5.0	3.8	5.7	4.3	6.4	4.8	5.9	2.4	2.9
External dimensions of wood tray (mm)																	
Length		411	283	219	343	264	407	312	283	219	471	360	343	264	312	138	154
Width		186	186	186	186	186	186	186	186	186	186	186	186	186	186	186	186
Height		227	227	227	227	227	227	227	335	335	227	227	335	335	335	335	335
Overall height of cell (mm)		279	279	279	279	279	279	279	387	387	279	279	387	387	387	387	387
Unit or cell centres (mm)		190	190	190	190	190	190	190	190	190	190	190	190	190	190	190	190

At a continuous float voltage of 2.30 V/cell, a recharge to 90% of full capacity after a full discharge will take less than 6 h if the current limit is 0.25C for 10 h at a current limit of 0.1C. In applications where the charge voltage can be varied, a faster recharge will be obtained by increasing the charging voltage to 2.4 V/cell. This should recharge a Powerstore battery to 90% capacity in 3–4 h from a fully discharged state. The recharge time from lower depths of discharge is correspondingly shorter.

Once fully charged at the higher voltage the battery should either be removed from charge or the charge voltage reduced to the continuous float level of 2.25–2.30 V/cell. The charge voltage should never be allowed to exceed 2.5 V/cell.

The voltage specified for constant-voltage charging refers to that which the charger should supply to a fully charged battery. The output characteristics of the charger are not critical provided this top-of-charge voltage can be controlled within the required range. All adjustments to charger output voltage should be made with the charger under load.

A fully discharged Powerstore battery will accept a high current for a short period at the beginning of charge. However, this high initial current does little in terms of charging the battery and is also expensive to provide in charger design. It is recommended that the charger output current is limited to a maximum of 0.25C A. This will both protect the battery and reduce charger costs.

The optimum charging voltage for Powerstore batteries varies with temperature. Where the battery temperature is likely to be consistently different from 25°C, Figure 50.10 should be used to select the required charging voltage. It is advisable to build this temperature compensation into the charger using a compensation rate of 4 mV/°C per cell based on 25°C. If a temperature sensor is used

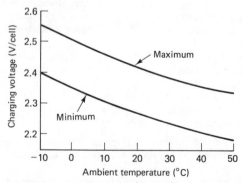

Figure 50.10 Chloride Power Storage Powerstore standby batteries: charging voltage *versus* temperature (Courtesy of Chloride Batteries)

THF 11SA	THF 13SA	THF 15SA	THF 17SA	THF 19SA	THF 21SA	THF 23SA	THF 25SA	THF 27SA	THF 29SA	THF 31SA	THF 33SA	THF 35SA	THF 37SA	THF 39SA	THF 41SA	MTF 23S	MTF 25S	MTF 37S
309	370	432	494	555	617	679	741	802	864	926	988	1050	1111	1173	1235	1315	1435	2152
237	283	331	378	425	473	520	568	614	662	709	757	804	851	899	946	1007	1099	1648
175	209	245	280	314	349	384	419	454	489	524	559	594	629	664	699	713	778	1166
321	386	450	514	578	643	707	771	835	900	964	1028	1092	1157	1221	1285	1369	1493	2240
248	299	348	398	447	498	547	597	646	697	746	796	845	896	945	995	1060	1156	1734
186	223	260	297	334	372	409	446	483	520	557	594	631	669	706	743	758	827	1241
63	75	88	101	113	126	139	151	164	176	189	202	214	227	239	252	268	293	439
21	25	29	34	38	42	46	50	54	59	63	67	71	76	80	84	89	97	146
24.3	28.5	32.9	37.1	41.3	45.6	49.6	53.8	57.9	62.4	66.7	71.1	75.3	79.5	83.7	88.0	101.4	109.8	160.3
4.3	5.2	6.1	6.8	7.7	8.9	9.5	10.4	11.3	12.3	13.2	13.8	14.7	15.7	16.6	17.5	19.5	21.3	32.2
3.4	4.1	4.8	5.4	6.1	7.0	7.5	8.2	8.9	9.6	10.3	10.9	11.6	12.3	13.0	13.7	15.3	16.7	25.2
122	138	154	172	187	203	219	235	251	267	283	298	314	330	346	362	222	233	337
186	186	186	188	188	188	188	188	188	188	188	188	188	188	188	188	253	253	256
526	526	526	526	526	526	526	526	526	526	526	526	526	526	526	526	707	707	707
580	580	580	580	580	580	580	580	580	580	580	580	580	580	580	580	770	770	770
125	190	190	190	190	190	190	190	190	190	190	190	190	190	190	190	260	260	260

this should be placed on the battery whenever possible.

Where Powerstore is to be operated between 15 and 25°C it is not generally necessary to provide temperature compensation.

Tubular plate types

Originally designed for electric road vehicles, the tubular plate cell can be used for a wide variety of standby power applications. This construction gives a very high electrical output per unit volume. Because the tubular plate design is compact it can be used particularly where space limitations are important. It will also stand up to vibration on work where movement of the equipment is involved, or where vibration occurs near the equipment.

The tubular plate construction is built up with antimonial lead spines held firmly in place by a rugged frame. The active material is held in firm contact with the spines by terylene tubes, minimizing shedding of the active material and prolonging battery life. The positive plates are matched by perfectly balanced pasted negative plates. They are designed to give the same commercial life as the positives, thus giving maximum economy in the life of the cell. Microporous envelopes surround each negative plate to give high porosity and low electrical resistance, and allow good circulation of the electrolyte. These combine to give efficient internal electrochemical performance.

Capacities up to 2152 A h are available in this range (Table 50.16). Batteries are also available in moulded monobloc containers with capacities up to 250 A h.

50.2.2 Standby power (stationary) batteries

Most of the major manufacturers of lead–acid batteries produce a range of these batteries. A typical range is that supplied by Varta. The Varta bloc type range is available in a wide range of capacities and is claimed to be maintenance-free for 3 years. These batteries can be used both for short-term operation, with bridging periods in the minutes ranges, and for long-term loads with battery bridging periods of the order of hours.

The Varta range is available in translucent plastic containers as 6 V modules with capacities from 12.5 to 150 A h, as 4 V modules with 175 and 200 A h, and as single cells with capacities from 250 to 2000 A h (Table 50.17 and Figure 50.11).

The bloc container and lid are of high impact resistant, temperature-resistant plastic and are welded electrolyte-tight together. In block batteries, the cells are connected to another inside the block

Table 50.17 Varta Bloc type lead–acid stationary batteries

No. of blocks or cells	Type designation	Type No.	Cell block external dimensions (mm)			Weight (kg)		Capacity (A h)			
			Length	Width	Height	Cell/block weight with acid	Weight of acid	10 h	8 h	3 h	1 h
6 V block 3 cells	Vb 611	160 1310 305	125	176	275	9.84	2.01	12.5	12.1	10.5	8
	Vb 612	160 1320 305	125	176	275	11.45	2.24	25	24.2	20.7	16
	Vb 613	160 1330 305	203	176	275	17.16	3.91	37.5	36.3	31.2	25
	Vb 614	160 1340 305	203	176	275	18.78	4.14	50	48.4	41.4	32
	Vb 615	160 1350 305	281	176	275	24.04	5.81	62.5	60.5	51.9	40
	Vb 616	160 1360 305	281	176	275	25.65	6.03	75	72.6	62.1	48
	Vb 624	160 2540 305	285	229	330	36.61	11.02	100	96.8	82.2	64
	Vb 625	160 2550 305	285	229	330	39.72	10.43	125	121	102.9	80
	Vb 626	160 2560 305	285	229	330	42.69	10.81	150	145	123.3	96
4 V block 2 cells	Vb 427	160 2570 205	248	229	330	34.65	9.21	175	169	144	112
	Vb 428	160 2580 205	248	229	330	36.34	9.47	200	194	164.4	128
Single cells	Vb 2305	190 5205 175	131	275	440	29.10	8.62	250	242	200	143
	Vb 2306	190 5206 175	131	275	440	31.20	8.62	300	290	240	172
	Vb 2307	190 5207 175	197	275	440	42.00	14.14	350	339	280	200
	Vb 2308	190 5208 175	197	275	440	44.20	14.15	400	387	320	229
	Vb 2309	190 5209 175	197	275	440	46.20	14.17	450	436	360	258
	Vb 2310	190 5210 175	242	275	440	54.50	17.81	500	484	400	287
	Vb 2311	190 5211 175	242	275	440	56.60	17.82	550	532	440	315
	Vb 2312	190 5212 175	242	275	440	58.80	17.83	600	581	480	344
	Vb 2407	191 0207 175	383	222	550	89.50	28.90	700	644	490	350
	Vb 2408	191 0208 175	383	222	550	94.70	28.30	800	736	560	400
	Vb 2409	191 0209 175	383	222	550	99.70	28.40	900	828	630	450
	Vb 2410	191 0210 175	383	222	550	104.90	27.70	1000	920	700	500
	Vb 2411	191 0211 175	383	307	550	131.60	41.90	1100	1012	770	550
	Vb 2412	191 0212 175	383	307	550	136.80	41.00	1200	1104	840	600
	Vb 2413	191 0213 175	383	307	550	141.70	40.40	1300	1196	910	650
	Vb 2414	191 0214 175	383	307	550	147.00	40.50	1400	1288	980	700
	Vb 2415	191 0215 175	383	307	550	152.10	39.80	1500	1380	1050	750
	Vb 2416	191 0216 175	383	392	550	178.80	54.00	1600	1472	1120	800
	Vb 2417	191 0217 175	383	392	550	184.00	53.20	1700	1564	1190	850
	Vb 2418	191 0218 175	383	392	550	189.10	52.50	1800	1656	1260	900
	Vb 2419	191 0219 175	383	392	550	194.20	52.00	1900	1748	1330	950
	Vb 2420	191 0220 175	383	392	550	199.40	51.90	2000	1840	1400	1000

container by connector technology patented by Varta. This results in optimum voltage conditions during discharge. In the case of single cells, this is attained through the use of copper cores in the terminals and connectors.

Figure 50.12 compares the discharge current intensity for a Varta bloc battery (Vb type) and a conventional stationary battery (OPzS type).

It can be seen that the bloc battery can provide a substantially higher utilization of capacity by the same discharge current or facilitate the use of a lower rated capacity by the same discharge currents.

Varta bloc batteries can be operated with a voltage of 2.23 V/cell without the necessity of an extra charging stage with 2.4 V/cell. This voltage ensures that the battery attains and maintains a fully charged state. Supplementary charges are therefore unnecessary.

In contrast to conventional lead–acid battery constructions, Varta bloc batteries ensure considerably higher energy throughput and also accept a low float charge current, and thereby ensure a minimum electrolyte decomposition of the water during the whole lifespan.

The quantity of electrolyte between the maximum and minimum level marks ensures maintenance-free operation for 3 years.

Because of the closed type of construction, together with the use of a safety vent plug system, a special battery room is not necessary. Installation of the block batteries and single cells up to type Vo 2312 is on steel step stands or steel tier stands. This

Capacity (A h)		Discharge current (A)						Final discharge voltage (V/cell)						1.75 V/cell	
$\frac{1}{3}$ h	$\frac{1}{6}$ h	10 h	8 h	3 h	1 h	$\frac{1}{3}$ h	$\frac{1}{6}$ h	10 h	8 h	3 h	1 h	$\frac{1}{3}$ h	$\frac{1}{6}$ h	Cap. 8 h (A h)	Disch. curr. 8 h (A)
5.5	4	1.25	1.51	3.5	8	16.5	24							12.8	1.6
11	8	2.5	3.03	6.9	16	33	48							25.6	3.2
16.5	12	3.75	4.54	10.4	24	49.5	72	1870	1865	1840	1800	1730	1665	38.4	4.8
22	16	5	6.05	13.8	32	66	96							51.2	6.4
27.5	20	6.25	7.56	17.3	40	82.5	120							64.0	8.0
33	24	7.5	9.08	20.7	48	99	144							76.8	9.6
44	32	10	12.1	27.4	64	132	192							104.0	13.0
55	40	12.5	15.1	34.3	80	165	240							129.6	16.2
66	48	15	18.1	41.1	96	198	288	1870	1865	1840	1790	1705	1630	156.0	19.5
77	56	17.5	21.1	48	112	231	336							181.6	22.7
88	64	20	24.2	54.8	128	264	384							208.0	26.0
92	70	25	30.3	66.7	143	277	420							256.0	32.0
111	84	30	36.3	80	172	333	504							307.2	38.4
129	98	35	42.4	93.3	200	388	588							358.4	44.8
148	112	40	48.4	106.7	229	444	672	1865	1855	1830	1785	1705	1620	409.6	51.2
166	126	45	54.5	120	258	499	756							460.8	57.6
185	140	50	60.5	133.3	287	555	840							512.0	64.0
203	154	55	66.5	146.7	315	610	924							563.2	70.4
222	168	60	72.6	160	344	666	1008							614.4	76.8
231	182	70	80.5	163.3	350	693	1092							689.6	86.2
264	208	80	92	186.6	400	792	1248							788.0	98.5
297	234	90	103.5	210	450	891	1404							880.0	110.0
330	260	100	115	233.3	500	990	1560							984.0	123.0
363	286	110	126.5	256.6	550	1089	1716							1080.0	135.0
396	312	120	138	280	600	1188	1872							1176.0	147.0
429	338	130	149.5	303.3	650	1287	2028	1860	1850	1845	1785	1685	1575	1280.0	160.0
462	364	140	161	326.6	700	1386	2184							1376.0	172.0
495	390	150	172.5	350	750	1485	2340							1472.0	184.0
528	416	160	184	373.3	800	1584	2496							1576.0	197.0
561	442	170	195.5	396.6	850	1683	2652							1672.0	209.0
594	468	180	207	420	900	1782	2808							1768.0	221.0
627	494	190	218.5	443.3	950	1881	2964							1864.0	233.0
660	520	200	230	446.6	1000	1980	3120							1968.0	246.0

type of assembly ensures ease of installation, which can be carried out by untrained personnel if necessary. Single cells of the types Vo 2407 and 2420 are installed on wooden base stands.

After filling with sulphuric acid, only a short commissioning charge is required and then the full performance capacity is attained. Even after several years' storage, a capacity of more than 80% is available immediately after filling.

50.3 Automotive batteries

Typical ranges of automotive batteries available from major manufacturers are exemplified in Tables

50.18 and 50.19. Table 50.18 illustrates the Motorcraft range of car, van and commercial batteries available from Ford Motors. These batteries are described either as low maintenance or as maintenance free, the latter employing recombination electrolyte technology (see Part 2). Reserve capacities of 46–210 min are available in the 10 V car/van battery range and of 180–600 min in the larger commercial 6 and 12 V batteries. Chloride Automotive Batteries supply two ranges of automotive batteries (Table 50.19). The Exide Punch Packer range is described as maintenance free with no topping up normally required. These batteries do have a facility for topping up with distilled water should component failure cause premature water loss.

Figure 50.11 Varta bloc standby power battery (Courtesy of Varta)

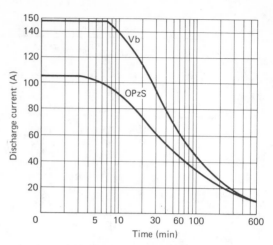

Figure 50.12 Varta bloc standby power battery discharge current depends on the bridging period E_{c-o} = 1.80 V per 100 A h (Courtesy of Varta)

(a)

(b)

Figure 50.13 Design features of Chloride Automotive Batteries low-maintenance TIR range of commercial vehicle batteries (Courtesy of Chloride Batteries)

The Torque Starter high-performance batteries (Table 50.19), which employ recombination electrolyte technology, are available in two sizes, designed in such a way that they will fit 92% of all cars. This has the advantage of enabling battery suppliers to carry less stock and thereby tie up less capital. In addition to zero maintenance, the recombination electrolyte principle of a fully sealed battery ensures that no acid spillage can occur and that there is no corrosion at the terminals or in the engine compartment. The Torque Starter battery delivers as much power in that vital first 5 s of starting as a conventional battery rated 20% more in cold cranking amps. It is also more rugged than conventional batteries, despite the fact that it is 25% lighter than a conventional battery.

Chloride Automotive Batteries also supply a range of standard (non-maintenance-free) commercial vehicle and van batteries (Tables 50.20 and 50.21) and the Exide TIR range of maintenance-free commercial vehicle batteries (Table 50.22).

The TIR batteries are available in three different starting performances: types 645 and 646 (350 A), 643, 644 and 649 (420 A), and 663, 664 and 665 (510 A).

TIR batteries are maintenance free. This means no topping up for life unless equipment malfunction causes excessive water loss, when the batteries can be topped up.

The TIR range of low-maintenance commercial vehicle batteries features several novel design features such as straight bar intercell construction and plate locking devices (Figure 50.13). The straight bar intercell connection minimizes voltage drop and increases the power available at the starter, meaning quicker engine starting and less wear on battery and starter. The plate locking device, together with the polypropylene container's specially strengthened walls, ensures that every TIR battery will meet arduous on-vehicle vibration

Table 50.18 Motorcraft batteries (Ford) specification data

Motorcraft number	Reserve capacity (min)	Cold start performance (A)			Bench charge rate (A)	Length (mm)	Width (mm)	Height (mm)	Layout	Terminals	Container	Approx. weight with acid (kg)
		BBS	SAE	AH								
Car/van batteries, 12 V Motorcraft Superstart Maintenance Free												
EMF 015	72	215	315	45	4.5	240	135	206	1	1	M012	12.5
EMF 016	72	205	315	45	4.0	236	135	206	0	4	M012	13.5
EMF 036	55	170	230	38	4.0	238	135	202	1	4	M012	12.2
EMF 037	60	180	250	40	4.0	240	135	206	0	4	M012	11.0
EMF 038	60	180	250	40	4.0	240	135	206	1	1	M012	11.0
EMF 048	65	185	290	40	4.0	226	135	225	0	1	M01	12.5
EMF 049	65	185	290	40	4.5	226	135	225	1	1	M01	12.5
EMF 063	65	210	310	41	4.5	207	175	175	0	1	M07	14.0
EMF 065	84	265	410	54	5.5	242	175	175	1	1	M07	16.0
EMF 067	115	300	480	66	6.6	306	175	190	0	1	M03	19.0
EMF 069	120	330	500	70	7.0	267	175	223	1	1	M06	20.0
EMF 074	100	260	430	60	6.5	265	175	206	1	1	N	20.0
EMF 077	60	195	335	44	4.4	210	175	190	1	4	M03	13.3
EMF 083	50	185	270	40	4.0	207	175	175	0	4	M03	11.0
EMF 085	60	195	360	40	4.0	251	175	175	1	1	N	12.5
EMF 091	80	220	340	48	5.0	251	175	175	0	4	M03	17.0
EMF 093	75	300	500	54	5.4	249	175	175	0	4	M03	15.5
EMF 097	110	385	610	63	6.3	306	175	175	0	4	M03	18.4
EMF 099	210	445	700	90	9.0	352	175	175	0	4	M03	24.0
Car/van batteries, 12 V Motorcraft Superstart Low Maintenance												
EBX 135	50	150	220	34	4.0	240	135	206	0	4	M012	10.5
EBX 136	46	150	210	40	4.0	240	135	206	0	1	M012	10.5
EBX 141	60	180	250	40	4.0	240	135	206	0	1	M012	11.0
EBX 154	65	185	290	40	4.0	226	135	225	1	3	M01	12.5
EBX 201	60	160	240	32	3.5	210	175	175	0	1	M07	11.0
EBX 206	60	160	240	32	3.5	210	175	175	1	1	M07	11.5
EBX 284	70	190	300	48	5.0	265	175	206	0	1	N	17.0
EBX 317	160	350	630	88	8.8	381	175	190	0	1	M03	24.2
Commercial batteries, 12 V Motorcraft Superstart Maintenance Free												
EMF 070	120	520	710	70	7.0	—	—	—	—	—	—	—
EMF 635	230	560	770	120	14.8	345	172	284	1	2	N	36.6
EMF 649	160	400	550	150	14.8	340	165	233	1	2	N	27.0
Commercial batteries, 12 V Motorcraft Superstart Low Maintenance												
EBX 222	210	450	620	120	12.0	508	207	208	3	1	N	43.0
EBX 243	180	430	650	96	9.5	345	173	231	0	1	N	29.5
EBX 244	180	430	650	96	9.5	345	173	231	1	1	N	29.5
EBX 323	220	430	590	110	11.0	513	189	225	4	1	N	40.0
EBX 325	340	550	760	180	18.0	518	291	242	3	1	N	56.0
EBX 329	250	590	810	143	14.3	513	223	225	3	1	N	44.0
EBX 333	220	390	550	120	12.0	356	252	233	6	1	N	42.0
EBX 347	220	510	700	12.0	12.5	345	173	282	0	1	N	34.5
EBX 348	220	510	700	12.0	12.5	345	173	282	1	1	N	34.5
EBX 355	220	550	760	125	13.0	345	173	282	0	1	N	37.0
EBX 356	220	550	760	125	13.0	345	173	282	1	1	N	37.0
Commercial batteries, 6 V Motorcraft Superstart Low Maintenance												
EBX 414	370	655	900	190	19.0	344	173	240	1	1	N	31.0
EBX 451	320	520	710	190	19.0	410	173	225	1	1	N	30.0
EBX 484	600	700	820	242	24.0	347	173	291	1	1	N	38.5
EBX 531	230	410	560	130	13.0	319	173	222	0	1	N	24.0

Table 50.19 Chloride Automotive Batteries Punch Packer and Torque Starter series of automotive batteries

Type No.	Starting performance (A)	Reserve capacity (min)	Bench charge rate (A)	Length (mm)	Width (mm)	Height (mm)	Weight with acid (kg)	Layout	Terminals	Container
				Maximum overall dimensions						
6 V car *										
404	216		6.5	173	173	185		0		B04
12 V maintenance free car Punch Packer †										
001	230	40	3.0	207	175	175	10.1	1	1	B03
007	270	50	3.5	207	175	175	10.8	0	1	B03
015	300	60	4.0	243	135	206	11.4	1	1	B12
016	300	60	4.0	243	135	206	11.4	0	4	B12
017	550	150	9.0	381	175	190	25.0	0	1	B03
029	540	120	7.0	271	175	205	19.9	1	1	B12
035	200	40	3.5	243	135	206	9.6	0	4	B12
036	200	40	3.5	243	135	206	9.6	1	1	B12
037	250	50	4.0	243	135	206	10.3	0	4	B12
038	250	50	4.0	243	135	206	10.3	1	1	B12
046	210	45	3.0	219	135	226	11.25	0	1	B01
047	210	45	3.0	219	135	226	11.25	1	1	B01
048	285	55	4.0	219	135	226	12.5	0	1	B01
049	285	55	4.0	219	135	226	12.5	1	1	B01
063	320	60	4.2	207	175	175	11.4	0	1	B03
065	420	75	5.0	243	175	175	14.2	0	1	B03
067	450	110	6.5	303	175	190	19.3	0	1	B14
068	540	120	7.0	256	175	205	19.9	0	1	N
069	540	120	7.0	271	175	205	19.9	1	1	B12
070	540	120	7.0	271	175	205	19.9	0	4	B12
071	360	85	5.5	271	177	222	16.5	1	1	B12
072	590	120	7.5	271	175	205	20.4	1	1	B12
073	360	85	5.5	271	177	222	16.5	0	1	B12
074	450	105	6.5	256	175	205	18.6	1	1	N
075	370	90	5.5	243	175	190	16.8	0	1	B14
077	320	60	4.2	207	175	175	11.4	1	1	B03
078	420	75	5.0	243	175	175	14.2	1	1	B03
083	270	50	3.5	207	175	175	10.8	0	4	B03
084	280	60	4.2	258	177	222	14.4	1	1	N
085	320	60	4.5	207	175	175	12.2	0	4	B03
091	360	85	5.5	258	177	222	16.5	1	1	N
092	420	75	5.0	243	175	175	14.2	0	4	B03
12 V standard car *										
154	132		3.0	198	133	227	12.5	1	3	N
895	125		2.5	185	125	176	9.0	0	4	N
Torque Starter †										
RE1	420			207	178	175	11.2			
RE2	310			187	140	181	8.1			

* Starting performance: SAE, 30 s at −18°C to 1.2 V p.c.
† Starting performance: IEC, 60 s at −18°C to 1.4 V p.c.

Table 50.20 Chloride Automotive Batteries standard (non-maintenance-free) range of commercial vehicle batteries

Type	Cranking performance (A)	Reserve capacity (min)*	Bench charge rate (A)	Maximum overall dimensions			Approx. weight with acid (kg)	Layout	Terminals	Container	European code equivalent
				Length (mm)	Width (mm)	Height (mm)					
Cranking performance: 60 s to 1.4 V/cell											
Standard (12 V)											
221	300	120	8.0	434	175	208	30.0	0	1	N	376
222	450	205	12.0	508	208	206	43.0	3	1	N	396
333	350	192	11.0	360	253	234	42.0	6	1	N	394
Standard (6 V)											
414	655	370	19.0	344	173	240	31.0	1	1	N	053
451	545	390	19.0	412	173	225	30.0	1	1	N	051
471	420	260	15.0	319	173	222	25.0	0	1	N	042
484	750	525	24.0	347	173	291	38.5	1	1	N	067
Double duty separation (12 V)											
321	540	270	14.3	513	224	220	40.7	3	1	N	404
323	540	270	14.3	513	224	220	40.7	4	1	N	405
324	725	375	21.0	518	290	240	54.3	4	1	N	417
327	430	215	11.0	513	192	220	34.4	3	1	N	380
Cranking performance: 90 s to 1.4 V/cell											
Double duty separation (12 V)											
602	186	105	7.5	378	173	221	28.5	1	1	N	449
612	225	140	9.0	440	173	221	34.0	1	1	N	451
Double duty separation (6 V)											
501	225	140	9.0	227	172	221	16.5	0	1	N	018
511	260	170	10.0	262	172	221	18.0	0	1	N	032
521	295	210	12.0	291	173	222	21.0	0	1	N	030
531	335	245	13.0	319	173	222	24.0	0	1	N	042
541	370	275	15.0	348	173	222	27.0	0	1	N	048
591	355	260	14.0	267	173	294	27.0	0	1	N	046
701	310	300	16.0	258	181	370	36.0	0	9	N	—
702	375	390	19.0	315	181	370	42.0	0	9	N	—
703	500	545	26.0	372	181	370	53.0	0	9	N	—
711	310	300	16.0	258	181	370	36.0	0	1	N	068
712	375	390	19.0	315	181	370	42.0	0	1	N	070
713	500	545	26.0	372	181	370	53.0	0	1	N	072
721	340	280	15.0	364	176	260	34.0	0	9	N	—
722	425	375	19.0	422	176	260	41.0	0	9	N	—
732	340	280	15.0	364	176	243	34.0	0	1	N	074
733	425	375	19.0	422	176	243	41.0	0	1	N	076

* The time taken for the voltage to drop to 1.75 V/cell when the battery is discharging at 25 A at 25°C.

Table 50.21 Chloride Automotive Batteries standard (non-maintenance-free) range of betteries for light vans

Type	Cranking performance* (A)	Bench change rate (A)	Maximum overall dimensions			Approx. weight (with acid) (kg)	Layout	Terminals	Container
			Length (mm)	Width (mm)	Height (mm)				
001	230	3.0	207	175	175	10.1	1	1	B03
016	300	4.0	243	135	206	11.4	0	4	B12
017	550	9.0	381	175	190	25.0	0	1	B03
037	250	4.0	243	135	206	10.3	0	1	B12
038	250	4.0	243	135	206	10.3	0	4	B12
049	285	4.0	219	135	226	12.5	1	1	B12
063	320	4.2	207	175	175	8.9	1	1	B01
067	450	6.5	303	175	190	19.3	0	1	B03
069	540	7.0	271	174	205	20.0	0	1	B14
070	540	7.0	271	174	205	20.0	1	1	B12
073	360	5.5	271	177	222	16.5	0	4	B12
074	450	6.5	256	174	205	18.6	0	1	B12
075	370	5.5	243	175	190	16.8	1	1	N
077	320	4.2	207	175	175	11.4	0	1	B14
078	420	5.0	243	175	175	14.2	1	1	B03
083	270	3.5	207	175	175	10.8	1	1	B03
085	320	4.5	207	175	175	12.2	0	4	B03
091	360	5.5	258	177	222	16.5	0	4	B03
092	420	5.0	242	175	175	14.2	1	1	N
154	132	3.0	198	133	227	12.5	0	4	B03
RE1	420		207	175	175	11.2	1	3	N
RE2	310		187	140	181	8.1			

* Cranking performance: 30 s to 1.2 V/cell (except type 154, which is 60 s to 1.4 V/cell)

conditions. This combination means that each cell's assembly is rigidly clamped in all directions to the container and lid. The battery vibrates as a whole, without over-stressing the intercell connectors, where a fracture would cause premature battery failure.

In the Exide range of low-maintenance bus batteries, low maintenance means 100 000 miles or 12 months between routine topping up of electrolyte. Battery design is such that battery topping up can be carried out as required, i.e. the batteries are not permanently sealed. These bus batteries are fitted with double duty separators. This type of separation is ideal where some discharge/charge cycling of batteries takes place, such as on public service vehicles. A discharge/charge regimen of operation can loosen the active material in the battery's plates. With ordinary single separation this active material will soon fall to the bottom of the cell, reducing capacity and shortening active life. With double separation a glass wool sheet is added to the conventional separator, which binds the active material in the plates into a tight cell assembly and prevents premature shedding of the active material. These design features lead to good starting performance and prolonged active life.

50.4 Sealed lead–acid batteries

As this type of battery is a relative newcomer, products available from the major suppliers are reviewed below.

Threee are two categories of sealed lead–acid cell. These are the non-recombining type, such as those manufactured by Sonnenschein and Crompton-Parkinson Ltd, and the fully recombining types, as manufactured by the General Electric Company and the Gates Rubber Company. The fully recombining types are also produced in the UK under licence by Chloride Gates Energy Ltd under the trade name Cyclon.

50.4.1 Sonnenschein

The lead–acid batteries referred to in Table 50.23 are some from the 'Dryfit' range supplied by Accumulatorenfabrik Sonnenschein, West Germany, although no doubt in many respects the comparisons will hold good for sealed lead–acid batteries produced by other suppliers.

The Dryfit battery has several design features, which contribute to its no-maintenance characteristic and its ability not to require overcharging to

Table 50.22 Chloride Automotive Batteries low-maintenance TIR 12 V range of commercial vehicle batteries

Type	Cranking performance* (A)	Reserve capacity (min)	Bench charge rate (A)	Maximum overall dimensions			Approx. weight (with acid) (kg)	Layout	Terminals	Container	European code equivalent
				Length (mm)	Width (mm)	Height (mm)					
643	420	170	9.0	345	173	233	28.1	0	1	N	246
644	420	170	9.0	345	173	233	28.1	1	1	N	247
645	350	150	7.5	345	173	233	27.1	0	1	N	246
646	350	150	7.5	345	173	233	27.1	1	1	N	247
647	510	220	12.0	345	173	284	36.9	0	1	N	388
648	510	220	12.0	345	173	284	36.9	1	1	N	389
649	420	170	9.0	345	173	233	28.1	1	8	N	247
655	550	220	12.0	345	173	284	36.9	0	1	N	388
656	550	220	12.0	345	173	284	36.9	1	1	N	389
663	510	170	9.0	345	173	233	29.0	0	1	N	246
664	510	170	9.0	345	173	233	29.0	1	1	N	247
665	510	170	9.0	345	173	233	29.0	1	8	N	247

* Cranking performance: 60 s to 1.4 V/cell

Table 50.23 Characteristics of sealed Dryfit batteries A300 series for standby operation

Type No. and type code*	Nominal voltage (V)	Nominal capacity for 10h discharge (A h)	Discharge current for 10h discharge (mA)	Weight (approx.) (g)	Length (mm)	Width (mm)	Height to top of lid (mm)	Max. height over contacts/cover (mm)	Power/weight ratio (W h/kg)	Power/volume ratio (W h/dm³)	Max. load (approx.) (A)‡	Interchangeable with‡
07 19 1172 00 A300-6V-1.0Ah-S	6	1.0	100	245	51.0	42.0	50.0	54.4	24.5	56.0	40	–
07 19 1182 00 A300-6V-1.1Ah-S	6	1.1	110	285	97.0	25.0	50.5	54.9	23.2	54.0	40	4 × R14/UM2/Baby/C
07 19 1185 00 A300-12V-1.1Ah-S	12	1.1	110	540	96.8	49.0	50.5	54.9	24.4	55.0	40	8 × R14/UM2/Baby/C
07 19 1262 00 A300-6V-2.0Ah-S	6	2.0	200	460	75.0	51.0	53.0	57.4	26.1	59.1	60	–
07 19 1302 00 A300-4V-3.0Ah-S	4	3.0	300	415	90.0	34.0	60.0	64.4	28.9	65.4	60	–
07 19 1312 00 A300-6V-3.0Ah-S	6	3.0	300	620	134.0	34.0	60.0	64.1	29.6	65.8	60	4 × R20/UM1/Mono/D
07 19 1432 00 A300-12V-5.7Ah-S	12	5.7	570	2225	151.2	65.0	94.0	98.4	30.7	74.0	80	–
07 19 1472 00 A300-6V-6.5Ah-S	6	6.5	650	1230	116.0	50.5	90.0	94.4	31.7	73.9	80	–
07 19 1502 00 A300-2V-9.5Ah-S	2	9.5	950	575	52.5	50.0	94.0	98.4	33.0	77.0	80	–
07 19 1523 00 A300-6V-9.5Ah-S	6	9.5	950	1710	151.2	50.0	94.0	98.4	33.3	80.2	80	–
07 19 1525 00 A300-12V-9.5Ah-S	12	9.5	950	3380	151.0	97.0	94.0	98.4	33.7	82.7	80	–

* The last letter of the type code indicates the type of terminals
† Only with mating contacts
‡ External dimensions of casings are chosen so that four, six or eight primary cells can be housed in the same volume – interchangeable, as shown, in accordance with DIN 408 66 and IEC Publication 86

maintain full capacity. In fact, no water needs to be added to these batteries during their entire life. If, because of improper charging or large variations in temperature, a gas pressure does build up, the safety valves ensure that this gas can escape immediately. Afterwards, the safety valves automatically shut off the electrolyte space from the outside atmosphere. Assuming that the operating and environmental conditions are satisfactory, Dryfit batteries are sealed and do not gas. They meet the requirement of the German Physikalische-Technische Bundesanstalt for use in hazardous areas containing combustible substances of all the explosive classes within the range of inflammability G1 to G5. Their method of construction also meets the requirements of VDE D171/1.69.

A 12 V, 36 A h Dryfit battery will operate for over 4 years at 14 V, 2.33 V/cell. Over this time, the battery loses only about 25 g of water and that virtually in the course of the first year. Dryfit batteries are available in two series, the A 300 with capacity range from 1 to 9.5 A h for standby operation and the A 200 series, capacity range 1–36 A h for high cyclic operations (Table 50.23).

Dryfit batteries may be charged and discharged over an ambient temperature range of −20 to +50°C. If continuous operation at one or more of the extreme values is required, a temperature sensor should be used to optimize charging. This can, however, be omitted if the battery is operated infrequently at extremes of temperature during the course of its life. Overcharging at high temperatures and charging, which cannot be completed in its usual time at low temperatures because of the lack of temperature compensation, affect the quality of the battery to only a minor degree. Short-term temperatures up to 75°C can be tolerated. No precautions need be taken against the lowest temperature, about −50°C, since the gelled electrolyte in the charged state cannot freeze solid.

Many types are severely limited by an allowable temperature range of 0–45°C for charging. Charging to a certain achievable percentage of full charge occurs in the freezing range and charging above 45°C is possible only for special types using sintered plate construction.

Dryfit batteries should, as far as possible, be stored in a fully charged condition. When stored at an average temperature of 20°C they need an additional charge after 16 months at the latest, and earlier when stored at higher temperatures. In general it is advisable to give an additional charge after storage periods corresponding to about 25% self-discharge.

The curves in Figure 50.14 give information regarding the variation of available capacity as a percentage of nominal capacity (withdrawable capacity) as a function of temperature over the range −30 to +50°C for three different loads with

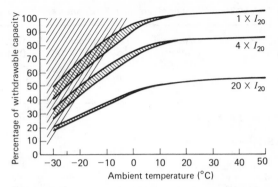

Figure 50.14 Percentage of withdrawable capacity of Dryfit batteries at different temperatures for three load examples. Upper edge of curve: charging at 20°C, discharging at quoted temperature. Lower edge of curve: charging and discharging at quoted temperature (Courtesy of Dryfit)

continuous discharge down to the relevant discharge cut-off voltage. For the values of the upper edges of the curves the batteries were charged at an ambient temperature of 20°C voltage limited to 2.3 V/cell. For the lower edge of the curves the batteries were charged at the indicated low ambient temperature, and hence under somewhat less favourable conditions. The curves show the behaviour of the batteries after a few charging cycles. When assessing both the upper and lower edges of the curves for the higher discharge currents, the dependence on the load must also be taken into account. (This is already included in Figure 50.14.) To prevent permanent damage to the capacity behaviour of the batteries at very low temperatures, the shaded areas in Figure 50.14 should be avoided.

50.4.2 Crompton-Parkinson

This company supplies rechargeable lead–acid maintenance-free batteries in the 1.5–4.5 A h capacity range (C_{10}). No electrolyte topping up is required during the battery life. On float charging within the recommended limits, the batteries are able to withstand overcharge for several years without significant deterioration. On deep discharge cycling duty, at least 200 cycles are normally obtained (80% capacity retained at the end of 200 cycles). On 30% partial discharge at least 1500 cycles are normally obtained. The batteries have a low internal resistance and consequently a high current delivery capability. They have a low level of self-discharge, and storage below 20°C would necessitate at most an annual recharge. The specification of the Crompton-Parkinson range of sealed lead–acid batteries is given in Table 50.24.

Figure 50.16 shows the discharge performance of this range of batteries.

Table 50.24 Specification of Crompton-Parkinson sealed lead–acid batteries

Type Ref.	Nominal voltage (V)	Nominal capacity (A h)		Width (mm)	Depth (mm)	Height (mm)	Weight (g)	Number of battery units per outer pack
		10 h rate	20 h rate					
SLA6-1.6	6	1.5	1.6	66	33	85	380	24
SLA6-3.2	6	3.0	3.2	66	33	125	660	24
SLA6-4.8	6	4.5	4.8	94	33	125	920	24
SLA6-6.4	6	6.0	6.4	91	49	115	1120	24
SLA6-8.0	6	7.5	8.0	115	49	115	1500	12
SLA12-1.6	12	1.5	1.6	66	66	85	760	12
SLA12-4.8	12	4.5	4.8	94	66	125	1840	12

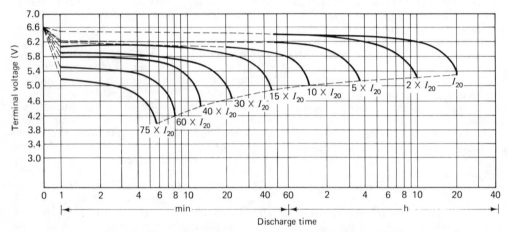

Figure 50.15 Discharge performance at 20°C of Crompton-Parkinson sealed lead–acid batteries. The curves show change in voltage at battery terminals with time during constant-current discharges at a number of different current rates. The current is given in multiples of I_{20} so that a single family of curves can be used for all the sizes of battery in the range. I_{20} is the current that will deliver the nominal battery capacity, C_{20}, in 20 h. Its value (A) is one-twentieth of the battery capacity ($C_{20}/20$). The lower end of each discharge curve indicates the recommended minimum voltage at the battery terminals for that particular rate of discharge; that is, the voltage at which the discharge should be discontinued (Courtesy of Crompton-Parkinson)

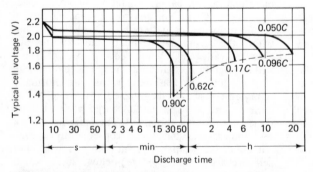

Figure 50.17 Typical voltage characteristics at various rates of discharge at 21°C of an Eagle Picher Carefree rechargeable lead–acid battery. To determine discharge rate of various batteries multiply rated capacity (C) by factor shown. For example, the rate at which an 8 A h battery must be discharged to yield a useful 10 h equals 0.096C or 0.096 × 8 A h = 0.77 A (Courtesy of Eagle Picher)

Table 50.25 Specification of Eagle Picher Carefree maintenance-free lead–acid batteries

Battery No.	Normal voltage (V)	Rated capacity at 20 h rate (A h)	Width (mm)	Depth (mm)	Height (mm)	Weight (kg)
CF6V1	2	0.9	50.8	41.9	50.0	0.25
CF12V1	12	0.9	83.8	50.8	50.0	0.50
CF12V1-L	12	0.9	102.1	41.9	50.0	0.50
CF18V1	18	0.9	124.4	50.8	50.0	0.77
CF12V1.5	12	1.5	178.3	33.7	61.0	0.86
CF2V2.5	2	2.6	58.4	26.4	66.0	0.21
CF6V2.5	6	2.6	80.5	58.7	66.0	0.63
CF8V2.5	8	2.6	105.7	58.7	66.0	0.86
CF10V2.5	10	2.6	133.4	58.7	66.0	1.04
CR12V2.5	12	2.6	116.8	79.2	66.0	1.27
CF6V2.6	6	2.6	134.6	34.3	60.2	0.63
CF13V2.6	12	2.6	134.6	69.1	60.2	1.27
CF2V5	2	5.2	59.1	26.9	100.3	0.35
CF6V5	6	5.2	80.8	58.7	100.3	1.09
CF8V5	8	5.2	107.9	58.7	100.3	1.45
CF10V5	10	5.2	134.6	58.7	100.3	1.77
CF12V5	12	5.2	117.3	80.8	100.3	2.13
CF12V5-L	12	5.2	155.4	61.4	100.1	2.18
CF12V5PP	12	5.2	161.0	66.3	167.6	2.27
CF8V6	8	6.0	84.8	54.6	165.1	1.77
CF6V8	6	8.0	152.4	50.8	95.2	1.63
CF12V8	12	8.0	152.4	102.4	95.2	3.22
CF6V15	6	15.0	91.4	84.8	164.8	2.90
CF12V15	12	15.0	183.1	84.8	164.8	5.81
CF12V20	12	22.0	165.3	124.7	165.8	7.35
CF2V30	2	28.0	84.8	54.5	165.1	1.72
CF6V30	6	28.0	158.7	85.1	165.1	5.22
CF6V30-L	6	28.0	255.0	54.5	165.1	5.22
CF12V30	12	28.0	170.7	158.7	165.1	10.44
CF12V30-L	12	28.0	255.0	109.5	165.1	10.44
CF6V40	6	44.0	165.3	124.7	165.8	7.49

50.4.3 Eagle Picher

Eagle Picher offer their Carefree range of recharge-able maintenance-free spillproof batteries in the capacity range 0.9–44 A h (Table 50.25). These batteries are equipped with self-sealing pressure relief valves in each cell to retard self-discharge, enhance recombination to water of hydrogen and oxygen gases released during self-discharge and ensure safety in the event of a malfunction.

Figure 50.16 illustrates typical voltage curves of Carefree rechargeable batteries at various rates of discharge. The effects of temperature on capacity and voltage are illustrated in Figures 50.17 and 50.18.

50.4.4 Yuasa

Yuasa manufacture a range of maintenance-free 6 and 12 V sealed lead–acid batteries with capacities up to 20 A h. The characteristics of these batteries are listed in Table 50.26.

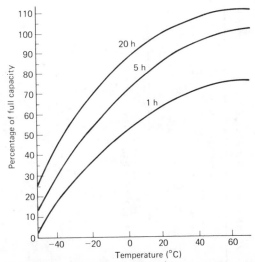

Figure 50.17 Capacity *versus* temperature at various rates of discharge for an Eagle Picher Carefree rechargeable lead–acid battery (Courtesy of Eagle Picher)

Table 50.26 Specification of a selection of Yuasa sealed lead–acid batteries

Type	Nominal capacity (A h)				Weight (kg)	Specific energy (W h/kg) 20 h rate	Internal resistance, charged battery (mΩ)	Max. discharge current, standard terminals (A)	Max. short-term discharge current (A)	Operating temperature range (°C)	Charge retention (%) at 20°C (shelf life)			Life expectancy	
	20 h rate	10 h rate	5 h rate	1 h rate							1 month	3 month	6 month	Standby use (years)	Cyclic use
NP1.2-6 (6 V)	60 mA to 5.25 V: 1.2	110 mA to 5.25 V: 1.1	200 mA to 5.10 V: 1.0	700 mA to 4.80 V: 0.7	0.34	21	45	40	45	−15 to +50 (charge) −20 to +60 (discharge)	97	91	85	3 to 5	100% DOD: 180 cycles; 50% DOD: 400 cycles; 30% DOD: 1200 cycles
NP2.6-6 (6 V)	130 mA to 5.25 V: 2.6	240 mA to 5.20 V: 2.4	440 mA to 5.10 V: 2.2	1600 mA to 4.80 V: 1.6	0.60	26	40	40	100	−15 to +50 (charge) −20 to +60 (discharge)	97	91	85	3 to 5	100% DOD: 180 cycles; 50% DOD: 400 cycles; 30% DOD: 1200 cycles
NP4.5-6 (6 V)	225 mA to 5.25 V: 4.5	418 mA to 5.25 V: 4.18	784 mA to 5.10 V: 3.92	2620 mA to 4.8 V: 2.62	1.11	24	35	40	150	−15 to +50 (charge) −20 to +60 (discharge)	97	91	85	3 to 5	100% DOD: 180 cycles; 50% DOD: 400 cycles; 30% DOD: 1200 cycles
NP8-6 (6 V)	400 mA to 5.25 V: 8.0	740 mA to 5.25 V: 7.4	1360 mA to 5.10 V: 5.1	4800 mA to 4.80 V: 4.8	1.65	29	20	40	300	−15 to +50 (charge) −20 to +60 (discharge)	97	91	85	3 to 5	100% DOD: 180 cycles; 50% DOD: 400 cycles; 30% DOD: 1200 cycles
NP1.9-12 (12 V)	95 mA to 10.5 V: 1.9	180 mA to 10.5 V: 1.8	320 mA to 10.2 V: 1.6	1100 mA to 9.6 V: 1.1	0.90	25	100	40	75	−15 to +50 (charge) −20 to +60 (discharge)	97	91	85	3 to 5	100% DOD: 180 cycles; 50% DOD: 400 cycles; 30% DOD: 1200 cycles
NP6-12 (12 V)	300 mA to 10.5 V: 6.0	560 mA to 10.5 V: 5.6	1020 mA to 10.2 V: 5.1	3600 mA to 9.6 V: 3.6	2.40	30	45	40	180	−15 to +50 (charge) −20 to +60 (discharge)	97	91	85	3 to 5	100% DOD: 180 cycles; 50% DOD: 400 cycles; 30% DOD: 1200 cycles

Table 50.27 Yuasa maintenance-free lead–acid batteries: capacity range 1.2–20 A h, discharge current at stipulated discharge rates (multiples of *C*)

20 h capacity (A h)	Discharge current (A)						
	$0.05C_{20}$	$0.1C_{20}$	$0.2C_{20}$	$0.4C_{20}$	$0.6C_{20}$	$1C_{20}$	$2C_{20}$
1.2	0.06	0.12	0.24	0.48	0.72	1.2	2.4
1.9	0.095	0.19	0.38	0.76	1.14	1.9	3.8
2.6	0.13	0.26	0.52	1.04	1.56	2.6	5.2
4.0	0.225	0.40	0.80	1.60	2.40	4.0	8.0
4.5	0.25	0.45	0.90	1.80	2.70	4.5	9.0
6.0	0.30	0.60	1.20	2.40	3.60	6.0	12.0
8.0	0.40	0.80	1.60	3.20	4.80	8.0	16.0
10.0	0.50	1.00	2.00	4.00	6.00	10.0	20.0
20.0	1.00	2.00	4.00	8.00	12.00	20.0	40.0

These batteries may be used in either cyclic or float services in a wide range of applications. The batteries incorporate a venting system to avoid excessive pressure build-up in the event of battery or charger malfunction. Because of the heavy-duty grids, 1000 cycles are claimed for these batteries and a normal life of 4–5 years is expected in float charge on standby applications. Tables 50.27 and 50.28 show the discharge current at stipulated charge rates from Yuasa batteries between 1.2 and 20 A h, 20 h rated capacities and the discharge capacity at various discharge rates for various types of battery.

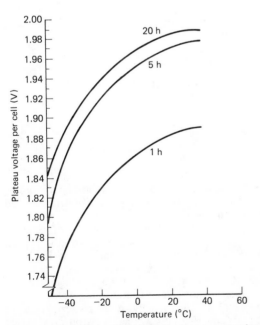

Figure 50.18 Plateau voltage *versus* temperature at various rates of discharge for an Eagle Picher Carefree rechargeable lead–acid battery (Courtesy of Eagle Picher)

Table 50.28 Yuasa maintenance-free lead–acid batteries: discharge capacity at various discharge rates

Type	Discharge capacity (A h)				
	20 h rate	10 h rate	5 h rate	3 h rate	1 h rate
NP1.2-6	1.2	1.1	1.0	0.9	0.7
NP2.6-6	2.6	2.4	2.2	2.0	1.6
NP4-6	4.0	3.7	3.4	3.1	2.4
NP4.5-6	4.5	4.2	3.8	3.5	2.7
NP6-6	6.0	5.6	5.1	4.6	3.6
NP8-6	8.0	7.4	6.8	6.2	4.8
NP10-6	10.0	9.3	8.5	7.7	6.0
NP1.9-12	1.9	1.8	1.6	1.5	1.1
NP2.6-12	2.6	2.4	2.2	2.0	1.6
NP6-12	6.0	5.6	5.1	4.6	3.6
NP20-12	20.0	18.6	17.0	15.4	12.0

50.4.5 SAFT

SAFT manufacture the Gelyte range of sealed rechargeable lead–acid batteries. These batteries are maintenance free and can be used in any position. They are available in capacities between 2.6 and 30 A h in 2, 4, 6, 12 and 24 V models (Table 50.29).

50.4.6 GEC/Chloride Gates (UK)

The fully recombining sealed lead–acid 2.00 V battery was developed by GEC. The first cylindrical cell in the range to be offered, and the only one for which details are currently available, is the D cell, which is rated at 2.5 A h at the 10 h rate. It is 61 mm in height and 34 mm in diameter. Capacities of 5.0 A h (x-type) and 25 A h (BC-type) are also available (see Table 50.30).

Table 50.29 Gelyte sealed lead–acid batteries available from SAFT

Model	Nominal voltage (V)	Nominal capacity (A h) 20 h *rate*	Length (mm)	Width (mm)	Height (mm)	Weight (kg)
SA-2300	2	30.0	46.3	58.4	144.0	1.90
SA-480	4	8.0	103.1	51.1	94.7	1.09
SA-626	6	2.6	134.6	35.1	59.7	0.64
SA-645	6	4.5	67.1	67.1	97.5	0.86
SA-645L	6	4.5	67.1	67.1	96.3	0.86
SA-660	6	6.0	70.6	70.6	99.7	1.04
SA-680	6	8.0	151.9	51.1	144.0	1.63
SA-690	6	9.0	96.3	58.4	144.0	1.68
SA-6100	6	10.0	96.3	58.4	145.0	1.90
SA-6180	6	18.0	192.5	58.4	145.0	3.36
SA-6200	6	20.0	192.5	58.4	145.0	3.81
SA-6270	6	27.0	175.3	96.3	145.0	5.03
SA-6300	6	30.0	175.3	96.3	145.0	5.71
SA-1226	12	2.6	134.9	35.1	125.7	1.27
SA-1260	12	6.0	141.2	70.6	97.8	2.09
SA-1290	12	9.0	192.5	58.4	145.0	3.36
SA-12100	12	10.0	192.5	58.4	145.0	3.81
SA-12180	12	18.0	192.5	116.8	145.0	6.71
SA-12200	12	20.0	192.5	116.8	145.0	7.62
SA-12270	12	27.0	192.5	175.2	195.0	10.07
SA-12300	12	30.0	192.5	175.2	145.0	11.43
SA-24100	24	10.0	192.5	116.8	145.0	7.62

Table 50.30 Fully recombining sealed lead–acid batteries from GEC

	D	X	BC
Nominal voltage (V)	2.0	2.0	2.0
Capacity rating (A h)			
10 h rate	2.5 (250 mA)	5.0 (500 mA)	25 (2.5 A)
1 h rate	2.8 (2.5 A)	3.2 (5 A)	20 (25 A)
Peak power rating (W)	135 (at 135 A)	200 (at 200 A)	600 (at 600 A)
Internal resistance ($\times 10^{-3}\,\Omega$)	10	6	2.2
Diameter (mm)	34.2	44.4	65.7
Height (mm)	67.6	80.6	176.5
Weight (kg)	0.181	1.369	1.67

The General Electric sealed lead–acid system offers a number of desirable discharge characteristics. The spiral plate design enhances long life in both float and cycle applications. Since the spiral configuration results in lower impedance, the cell may be discharged at higher rates. State of charge is conveniently determined by measuring open-circuit cell voltage after a short stabilizing period. A voltmeter is all that is required to measure the state of charge.

The total available capacity and general discharge characteristics on a D cell are described below. All discharge data were developed after stabilizing the performance by charging at standard conditions of 2.45 V/cell for 16–20 h at room temperature and discharging at 1.25 A to 1.4 V/cell.

Figure 50.19 illustrates the change of terminal voltage as a function of discharged capacity. Typical curves are shown for constant-current discharge rates at the standard conditions noted.

Both capacity and delivered voltage vary as inverse functions of discharge rate. At the lowest rates, voltage and capacity are the highest. A cut-off limit of the working voltage is arbitrarily set at 1.4 V. The voltage cut-off at that point is rapid, and little useful additional energy would be obtained

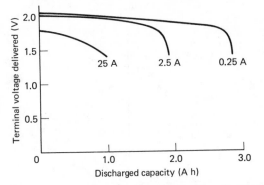

Figure 50.19 Typical discharge curves at various discharge rates (constant load) for a General Electric D cell under standard conditions: charge 2.45 V at 23°C for 16 h, discharge at 23°C to 1.4 V (Courtesy of General Electric)

50.5 Spillproof lead–acid batteries

Several manufacturers supply small lead–acid batteries which, although not sealed, are specifically designed to be spillproof. Thus, although these batteries will require topping up with electrolyte during their life, they have specific advantages in the field of portable battery-operated equipment.

50.5.1 Sonnenschein

Sonnenschein manufacture a wide range of spillproof accumulators designed for use with photoflash equipment and other portable units. These batteries are housed in a tough transparent plastics case and feature a labyrinth-like construction below the vent opening through which no acid will pass even with extreme changes of position and vibration. The vent does, however, permit the passage of gases during charging. Each individual cell has an indicator, based on floats, which accurately indicates the state of charge. The batteries at −40°C retain 38% of the capacity they possess at 20°C. These batteries are available in two versions, standard and permanent (see Table 50.31).

from the cell by attempting to use it below this voltage level. As the curves in Figure 50.19 indicate, higher drain rates aggravate the condition of concentration polarization (a decrease in electrolyte concentration near the active materials in the plate caused by lack of sufficient ion diffusion), resulting in the reduction of available voltage and capacity.

Table 50.31 Characteristics of Sonnenschein non-spill batteries

Type No.	Type designation	Nominal voltage (V)	20 h capacity (A h)	Max. charging current 2 V/cell (A)	Max. dimensions with terminals			Approx. weight (g)	
					Length (mm)	Width (mm)	Height (mm)	Without acid	With acid
Permanent batteries									
040 91511 00	2 Z3 F	4	2.0	0.20	42.0	58.0	66.5	256	305
040 91566 00	2 Bf3 G	4	3.5	0.35	42.0	58.0	102.0	390	465
040 91572 00	3 Bf3 S	6	3.5	0.35	62.0	58.0	100.5	570	685
040 91573 00	3 Bf3 K	6	3.5	0.35	62.0	58.0	98.5	570	685
040 91585 00	6 Bf3 SR	12	3.5	0.35	122.0	58.5	109.5	1140	1370
040 91706 00	1 B15 G	2	7.5	0.75	40.8	58.0	102.0	340	430
040 91727 00	3 B15 W/S	6	7.5	0.75	122.8	58.0	104.0	1050	1320
040 91742 00	3 T6 S	6	9.5	0.95	141.0	55.0	127.0	1700	2015
Standard batteries									
040 91011 00	2 Z3 F	4	2.0	0.20	42.0	58.0	65.5	256	305
040 91066 00	2 Bf3 G	4	3.5	0.35	42.0	58.0	102.0	390	465
040 91073 00	3 Bf3 K	6	3.5	0.35	62.0	58.0	98.5	570	685
040 91072 00	3 Bf3 S	6	3.5	0.35	62.0	58.0	105.5	570	685
040 91106 00	3 KS3 G	6	3.6	0.36	85.5	58.0	101.0	690	840
040 91126 00	1 KS4 G	2	4.8	0.48	35.0	58.0	101.0	265	330
040 91146 00	1 KS5 G	2	6.0	0.60	41.0	58.0	101.0	300	380
040 91156 00	2 KS5 G/A	4	6.0	0.60	82.0	58.0	101.0	600	765
040 91166 00	3 KS5 G/A	6	6.0	0.60	124.0	58.0	101.0	900	1150
040 91206 00	1 B15 G	2	7.5	0.75	40.8	58.0	102.0	340	430

Table 50.32 Specification of Varley non-spill lead–acid batteries

Type No.	Voltage (V)	Capacity (A h)					First formation charge (continuous) x h at y A	Normal charge (A) x h at y A	Freshening charge (A) x h at y A	Constant-potential taper charging* (V)	Weight (kg)
		20 h rate	10 h rate	5 h rate	1 h rate	½ h rate					
VPT 6.9/8	6	9.0	8.0	7.0	4.8	4.0	96 0.50 / 48 1.00	Std 16 0.75 / Max. 6 2.00	13 0.75	6.75 ± 0.9†	1.80
VPT 6.13/12	6	12.0	11.0	10.0	8.5	7.5	48 1.50 / 72 1.00	Std 12 1.50 / Max. 9 2.00	6 1.50	6.75 ± 0.9	2.25
VPT 6.15/20	6	20.0	19.0	17.0	14.0	11.0	48 1.00 / 24 2.00	Std 19 1.25 / Max. 10 2.50	16 1.25	6.75 ± 0.9	3.75
VPT 6.19/25	6	27.0	25.0	22.5	18.0	17.5	60 1.00 / 30 2.00	Std 17 2.00 / Max. 11 3.00	13 2.00	6.75 ± 0.9	3.75
VPT 6.15/30	6	28.0	25.5	23.0	18.0	16.0	90 1.00 / 60 1.50 / 30 3.00	Std 19 2.00 / Max. 10 4.00	16 1.25	6.75 ± 0.9	5.00
VPT 12.7/10	12	9.0	8.0	7.0	5.5	4.5	50 1.00 / 30 1.75	Std 10 1.25 / Max. 8 1.50	6 1.00	13.5 ± 0.18	3.85
F 12.19/25	12	27.0	25.0	25.5	18.0	17.5	60 1.00 / 30 2.00	Std 17 2.00 / Max. 11 3.00	13 2.00	13.5 ± 0.18	8.60
VPT 12.7/50	12	50.0	40.0	32.0	19.0	15.0	–	Std 16 4.00	6 4.00 (for 12 V system)	13.5 ± 0.18‡	2.60

* Recommended for rapid or float charging
† 13.5 ± 0.18 V for 12 V system
‡ 27.0 ± 0.36 V for 24 V system; 54.0 ± 0.72 for 48 V system

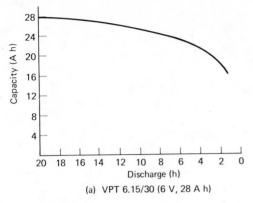

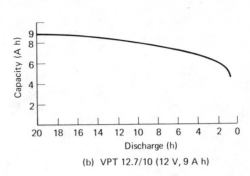

(a) VPT 6.15/30 (6 V, 28 A h)   (b) VPT 12.7/10 (12 V, 9 A h)

Figure 50.20 Discharge time *versus* discharge current for Varley 6 and 12 V non-spill lead–acid batteries (Courtesy of Varley)

50.5.2 Varley Dry Accumulators

Varley Dry Accumulators supply a range of non-spill 6 and 12 V lead–acid batteries in the capacity range 9–50 A h at the 2 h rate. These batteries are normally supplied filled and uncharged.

Figure 50.20 shows the effect of discharge current on discharge time for two batteries in the Varley range quoted in Table 50.32.

51

Nickel batteries

Contents

51.1 Nickel–cadmium (secondary) batteries

51.1.1 Sealed nickel–cadmium batteries

SAFT (Société des Accumulateurs Fixes et de Traction), a major manufacturer of alkaline batteries, produce a full range of sealed nickel–cadmium batteries, including open type (GPX) and semi-open type (VOK-VOKH-VHP), both of which are higher capacity types. SAFT also produce sealed cells and batteries and button cells (VB, VR and VY types), which are discussed below.

Details of the main types of sealed cell and battery manufactured by SAFT are given in Table 51.1.

The SAFT range consists of button and cylindrical types and the VR series. The button cells have very thin sintered positive and negative plates with a porous separator enclosed by two nickel-plated steel cups. Capacities are up to 500 mA h at 1.2 V for a size approximately 35 mm diameter × 10 mm thick with a weight of 100 g. Plastics-sleeved packs of button cells are available connected in series to give up to 12 V. Similar metal-clad packs are also available.

The cylindrical batteries are designed for a wide range of applications requiring high capacity and exceptionally long life. Cells are dimensionally similar to the principal conventional dry batteries. The sintered plate construction consists of a spirally wound plate assembly, which is tightly fitted into a steel container. All batteries have a special venting system, with automatic resealing should excess gas pressure occur. The design satisfies the requirements of BS 3456: Section 2.36: 1973 on all points, including safety on reverse charge. Capacities are up to 10 A h at 1.2 V for a size of 41.7 mm diameter × 90 mm height with a weight of 400 g. Series connected, plastics-coated modular packs are available to give up to 12 V at the above capacity.

The SAFT VR series of cells have been produced for applications requiring rapid charge and discharge. Up to 130 A peak discharge can be tolerated and the rapid charge time on a discharged battery can be as low as 15–60 min when using the appropriate charging equipment, which monitors cell temperature, voltage or time of charge. The maximum capacity available in this range at the moment is 10 A h. For uncontrolled rapid charge, SAFT have developed the VY range of cells.

SAFT

VB button cells These cells have the form of a button in various diameters and thicknesses, made up of a stack of disc-shaped sintered plates and separators held in two nickel-plated steel cups, one fitting into the other and pressed together with an insulating gasket. There is also a high-reliability version of the button cell with improved performance on high-rate discharge at very low temperatures (VBE cells). These have the same dimensions and capacities as the standard VB cells and are designed to withstand more severe operating conditions. They have an improved voltage characteristic at low temperatures and high discharge rates and higher mechanical strength and reliability.

There are five sizes of VB cell from 40 to 600 mA h available in VB and VBE versions (see Table 51.1).

VB button cells have a nominal discharge voltage of 1.2 V/cell. The rated capacity C_5 (Table 51.1) at 20°C is defined as the capacity in ampere hours obtained at the 5 h rate of discharge ($0.2C_5$ A) to an end-voltage of 1.10 V/cell after a full charge at $0.1C_5$ A for 14 h. Rates of charge and discharge are expressed in multiples of the rated capacity C_5, thus $0.2C_5$ A represents 20 mA for a cell having a rated capacity, C_5, of 100 mA h. $5C_5$ A represents 500 mA for the same cell.

The excellent temperature characteristic of the discharge of VB cells is illustrated in Figure 51.1.

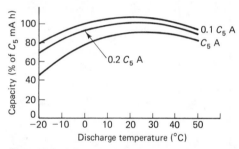

Figure 51.1 Influence of temperature on capacity of SAFT sealed nickel–cadmium sintered plate VB type button cells (Courtesy of SAFT)

The life of VB and VBE cells (Table 51.1) can extend over 10 years, but this is dependent on operating conditions and is principally influenced by the depth of discharge (discharged capacity/rated capacity) in a cycling application and by the number of ampere hours overcharged in a continuous charge application. The great reliability of VB and VBE cells makes their life comparable with that of most electronics components. During cycling approximately 500 cycles are obtained with 100% depth of discharge. After 5 years' continuous charge at $0.05C_5$ mA, the failure rate for VB cells is not greater than 5 per 1000. Under the same conditions, VBE cells offer ten times this reliability: a failure rate of 5 per 10 000.

A typical discharge curve for a VB button cell is given in Figure 51.2 (continuous discharge).

VB batteries are also available in standard plastics-cased button packs made of n identical VB button cells, where n is between 2 and 24, each of rated capacity C_5 mA h, connected in series and

Table 51.1 Sealed cells supplied by SAFT

Serial number	Type	Equivalent primary battery	Rated capacity C_5 (A h)	Max. dimensions (mm) Height	Diameter	Width	Depth	Max. weight (g)	Max. discharge rates (A) Continuous	Pulse	Max. power (W)
Cylindrical cells											
VR 1.2 V/cell	0.1 1/3AA		0.100	17.4	14.5			11	1.0	15.4	10
	0.45 1/2A		0.450	28.1	17.3			21	4.5	31	20
	0.5 AA	R6	0.500	50.2	14.5			24	5.0	24.5	16
	0.8 AF		0.800	49.8	16.17			34	8.0		
	1.2 RR		1.2	42.1	22.7			51	12	77	50
	2.0 C	R14	1.8	49.7	25.9			77	14	93	60
	2.5 2/3D		2.5	43.5	32.9			105	20	120	78
	4 D	R20	4	60.5	32.9			150	28	145	95
	7 F		7	91.3	32.9			240	35	170	110
	10 SF		10	89.2	41.7			400	80	200	130
VY	C5		1.150	42.1	22.7			50	12	90	66
VRE	1 RR		1.0	43.2	23.2			49	9	70	
	1.6 C		1.6	48.8	26.4			74	11	70	
	3.5 D		3.5	59.2	33.4			150	18	92	
	6 F		6.0	90.0	33.4			240	24	123	
Button cells (nominal discharge voltage 1.20 V)											
VBE	4		40	6	15.7			3.5	0.2	2.3	1.5
VB and VBE	10		100*	5.3	23.0			7	1	11	7
	22		220*	7.8	25.1			12	2.2	18.5	12
	30		300*	5.5	34.7			18	3	23	15
	60		600*	9.8	34.7			31	6	29	15
Rectangular cells											
VOS	4		4*	78.5		53	20.5	230	20	–	
	7		7*	112.5		53	20.5	360	35	–	
	10		10*	91		76.2	28.9	500	40	–	
	23		23*	165		76.2	28.9	1060	60	–	
	26		26*	185		76.2	28.9	1200	80	–	

* Rated capacity C mA h

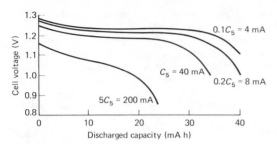

Figure 51.2 Discharge curve of a typical SAFT VB series nickel–cadmium button cell (continuous discharge at 20 ± 5°C). Maximum permissible continuous rate 200 mA (Courtesy of SAFT)

assembled in a rigid case, to make packs with discharge voltages between 2.4 V (2 cells) and 28.8 V (24 cells). Packs are available made up from VB 4, 10, 22, 30 and 60 cells (capacity 40, 100, 220, 300 and $600C_5$ mA h, see Table 51.1). The final assemblies have, of course, the same rated C_5 capacity as the individual cells but vary in discharge voltage.

Also available are environment-proof assemblies in which the interconnected cells are resin potted in a plastics container. An advantage of button cell packs is that these can operate as very compact low-voltage capacitors, giving much greater energy storage in a much smaller space. VB button cell packs can deliver very high currents, the maximum discharge rate being $10C_5$ mA. It should be ensured that the packs are not discharged to below 1 V/cell. Short-duration peak discharges may be as high as $50–100C_5$ mA at 0.8 V/cell.

VR cylindrical cells Nickel–cadmium cylindrical cells are designed for a wide range of applications requiring compact rechargeable batteries with high output capacity and exceptionally long life. Designed to be dimensionally similar to the principal cylindrical primary cells, of internationally standardized dimensions, the SAFT VR cells incorporate thin sintered plate construction. Complete spirally wound plate assemblies are tightly fitted into a steel container to form cells. The container is closed by a cover having an automatic self-reclosing safety valve for high pressure ($9–14 \times 10^5$ N/m²) should excess gas pressure be caused by maltreatment.

The cells satisfy BS 3456: Section 2.36: 1973 as regards safety on reverse charge. Contact between the plates and the cover or the case is through a special connection electrically welded to the edges of the plates. This method of assembly adopted by SAFT improves battery performance on charge and discharge. The range of VR cells available is from 100 mA h to 10 A h (Table 51.1). VR cells are characterized by a nominal voltage of 1.2 V/cell and their remarkable power (1.2 A h per cell) is, for example, capable of supplying a current of 77 A at

0.65 V for 0.3 s. The cylindrical cell is available in a high-reliability version (VRE). These cells have the cover welded directly on to the case, which improves the sealing. The plate group is also given supplementary support to increase the resistance to mechanical forces. Four sizes of VRE cell are available, covering the range 1–6 A h (Table 51.1). There is also a special version VRE cell available for missile applications.

VR cells when continuously discharged at normal temperatures supply currents of the order shown in Table 51.2. Short-duration peaks (maximum power) can be obtained. For example, for 300 ms at 0.65 V per cell, typical currents are given in Table 51.3.

Table 51.2 Continuous discharge

Cell type	Cell current (A)
VR 1.2 RR	12
VR 2 C	14
VR 4 D	28
VR 7 F	35
VR 10	80

Table 51.3 Short duration discharge peaks

Cell type	Cell current (A)
VR 1.2 RR	77
VR 2 C	93
VR 4 D	145
VR 7 F	170
VR 10	200

VR cells and batteries are designed for use in applications which require good high-rate discharge performance (continuous and peak). Due to their low internal resistance, higher discharge currents can be drawn and also substantially better voltages than apply in the case of cells available a few years ago. On continuous discharge, VR cells may be discharged at currents representing up to ten times their rated capacity (Table 51.4). In short-duration peak discharges currents may be drawn which are as much as 100 times the value of the rated capacity, with a cell voltage of 0.65 V/cell.

Table 51.5 shows the discharge performance for 0.3 s duration peaks at 20°C of various types of VR cell.

At low temperatures, VR cells and batteries give exceptional performance. Figure 51.3 shows a continuous peak discharge characteristic at +20 and −20°C for a 1.2 A h battery consisting of ten 1.2 A h VR 1.2 RR cells connected in series.

VR cells can also be obtained in standard battery packs of 2.4–28.8 V (Table 51.6) and extended

Table 51.4 Continuous discharge performance of VR cells at 20°C

Cell type	Rated capacity C_5 (A h)	Max. permissible current (A)	Capacity recoverable (A h) at max. current (end of discharge 0.9 V/cell)
VR 0.1 1/3AA	0.1	1.0	0.062
VR 0.45 1/2A	0.45	4.5	0.23
VR 0.5 AA	0.5	5.0	0.18
VR 0.8 AF	0.8	8.0	0.35
VR 1.2 RR	1.2	12.0	0.65
VR 2.0 C	1.8	14.0	0.87
VR 2.5 2/3 D	2.5	20.0	1.20
VR 4 D	4.0	28.0	1.90
VR 7 F	7.0	35.0	3.5
VR 10 SF	10.0	80.0	6.0

Table 51.5 Discharge performance for short-duration peaks (0.3 s) at 20 ± 5°C of VR cells

Cell type	Rated capacity C_5 (A h)	Max. power (W) at 0.65 V/cell	Current (A) at max. power 0.65 V/cell	Current (A) at 1 V/cell
VR 0.1 1/3AA	0.1	10	15.4	7.5
VR 0.45 1/2A	0.45	18	28	12
VR 0.5 AA	0.5	16	25	14
VR 0.8 AF	0.8			
VR 1.2 RR	1.2	50	72	32
VR 2.0 C	2.8	60	93	42
VR 2.5 2/3D	2.5	78	120	56
VR 4 D	4.0	95	195	70
VR 7 F	7.0	110	170	80
VR 10 SF	10.0	130	100	100

range modular packs of 12 and 24 V (Table 51.7), both in a range of capacities.

The VR series standard packs provide for applications requiring high performance on discharge at large currents either continuously or in short-duration peaks. The extended range modular packs are supplied in a plastics case in the range 0.5–10 A h at 12 V and 24 V. (Rated capacity is for discharge at the 5 h rate at 20°C after nominal charge and to a terminal voltage of 1.1 (C_5 A) V/cell.) The packs can be discharged steadily at up to the rated capacity and can be given a nominal charge at

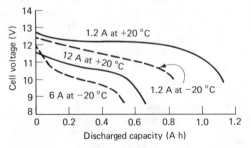

Figure 51.3 Continuous discharge characteristics of a SAFT 10 VR 1.2 RR, ten 1.2 A h cell nickel–cadmium battery (1.2 A h) at +20 and −20°C (Courtesy of SAFT)

constant current $0.1C_5$ for 14 h irrespective of their initial state of charge. Alternatively, the packs can be given a fast or accelerated charge. They will withstand a continuous charge of up to $0.05C_5$ and will operate in the temperature range −40 to +50°C. They may be stored indefinitely in any state of charge without deterioration in the above temperature range.

VY cylindrical cells The characteristics of VY cells (Table 51.1) are the same as those of the VR range. The VY cells have the novel conception of a plate that improves the recombination of oxygen produced on overcharge. They may thus be charged or overcharged at higher rates than those for VR cells. VY cells have a smaller capacity than the corresponding VR cells but the power characteristics are improved.

The full range of SAFT Nicad standard cylindrical cells ranging from 0.7 to 10 A h is listed in Table 51.8. These cells have a wide operating temperature range (−40 to 50°C) and an excellent cycle life (300–1000 cycles).

SAFT also supply a range of quick-charge cells (Table 51.9) and special low-resistance cells (Table 51.10). The quick-charge cells are designed for

Table 51.6 Characteristics and dimensions of SAFT VR series standard 6–12 V battery packs

Cell type	6V					9.6 V					12 V				
	Type No.	Length (mm)	Width (mm)	Height (mm)	Weight (g)	Type No.	Length (mm)	Width (mm)	Height (mm)	Weight (g)	Type No.	Length (mm)	Width (mm)	Height (mm)	Weight (g)
VR 0.45 1/2A 0.45 A h 4.5*	121.627	54	37	32	114	121.630	71	37	32	182	121.632	88	37	32	228
VR 0.5 AA 0.50 A h 5.0*	120.244	45	31	54	137	120.412	59.4	31	54	203	120.246	74	31	54	258
VR 1.2 RR 1.2 A h 12.0*	120.197	70	48	46	286	120.430	92	48	46	429	120.199	114	48	46	539
VR2 C 2.0 A h 14.0*	120.204	80	54	54	423	120.482	105	54	54	638	120.206	130	54	54	803
VR 2.5 2/3D 2.5 A h 20.0*	122.764	102	70	50	671	122.767	135	70	50	932	122.157	167	70	50	1155
VR4 D 4.0 A h 28.0*	120.210	102	70	67	852	120.440	135	70	67	1292	120.212	167	70	67	1617
VR7 F 7.0 A h 35.0*	120.216	102	70	98	1320	120.464	135	70	98	2024	120.185	167	70	98	2530
VR10 SF 10.0 A h 80.0*	120.221	130	88	96	2145	120.309	177	88	96	3333	120.223	212	88	96	4147

* Maximum current (A)

Table 51.7 Characteristics and dimensions of SAFT VR series extended range modular 12 and 24 V packs

Cell type	Battery voltage 12 V				Battery voltage 24 V			
	Length (mm)	Width (mm)	Height (mm)	Weight (g)	Length (mm)	Width (mm)	Height (mm)	Weight (g)
VR 0.4 AA, 0.5 A h	104	33	68	270	168	33	63	540
VR 1.2 RR, 1.2 A h	132	51	53	600	169	77	51	1200
VR2 C, 2 A h	147	57	58	870	188	84	53	1740
VR4 D, 4 A h	187	74	77	1600	177	145	81	3200
VR7 F, 7 A h	186	73	102	2403	177	145	112	4860
VR 10 SF, 10 A h	230	90	105	3920	–	–	–	–

Table 51.8 Standard Nicad cylindrical cells supplied by SAFT

Cell type		Maximum bare cell dimensions		Capacity (A h) rated at 20°C		Charge rates at 20°C				Bare cell weight (g)	Internal resistance (mΩ)	Vent type*
Model	Size	Height (mm)	Diameter (mm)	1C	0.2C	Continuous (A)	(h)	Maximum (A)	(h)			
0.070SC	1/2AAA	20.96	9.65	0.060	0.070	0.007	16	0.023	4.0	4.0	220.0	H
0.110SC	1/3AA	16.76	13.72	0.100	0.110	0.011	16	0.037	4.0	6.5	80.0	H
0.150SC	1/3A	16.13	15.06	0.140	0.150	0.015	16	0.030	8.0	7.9	100.0	H
0.180SC	AAA	44.45	9.65	0.160	0.180	0.018	16	0.036	8.0	9.0	120.0	H
0.225SC	1/3A$_f$	16.89	16.56	0.200	0.225	0.023	16	0.045	8.0	10.8	30.0	H
0.250SC	1/2AA	30.28	13.72	0.225	0.250	0.025	16	0.080	4.0	10.8	60.0	H
0.450SC	2/3AA	27.94	16.56	0.405	0.450	0.045	16	0.150	4.0	15.6	48.0	H
VR0.45	2/3A$_f$	28.09	17.04	0.405	0.450	0.045	16	0.180	3.5	21.0	24.0	R
0.500SC	AA	49.96	13.72	0.450	0.500	0.050	16	0.165	4.0	19.9	28.0	H
VR0.5	AA	50.20	14.30	0.450	0.500	0.050	16	0.200	3.5	24.1	26.0	R
VR0.5	AAL	49.20	14.30	0.450	0.500	0.050	16	0.200	3.5	24.1	26.0	R
0.600SC	A	48.77	15.06	0.530	0.600	0.600	16	0.120	8.0	24.0	26.0	H
0.750SC	1/2C	23.50	25.70	0.660	0.750	0.075	16	0.150	8.0	31.0	23.0	H
0.800SC	A$_f$	49.28	16.56	0.720	0.800	0.080	16	0.265	4.0	29.8	22.0	H
VR0.8	A$_f$	49.00	16.46	0.720	0.800	0.080	16	0.320	3.5	32.9	16.9	R
1.0SC	3/5C	29.21	25.70	0.880	1.000	0.100	16	0.200	4.0	39.7	20.0	H
1.2SC	C$_s$	41.96	22.25	1.060	1.200	0.120	16	0.240	8.0	45.4	15.0	H
1.2SC	C$_s$L	40.01	22.25	1.060	1.200	0.120	16	0.240	8.0	45.4	15.0	H
VR1.2	C$_s$	42.10	22.35	1.080	1.200	0.120	16	0.480	3.5	51.1	8.5	R
2.0SC	C	47.37	25.70	1.800	2.000	0.200	16	0.400	8.0	68.0	17.0	H
VR2	C	49.70	25.55	1.800	2.000	0.200	16	0.600	4.0	77.0	8.5	R
VR2	CL	47.70	25.55	1.800	2.000	0.200	16	0.600	4.0	77.0	8.5	R
2.2SC	1/2D	36.78	32.39	2.000	2.200	0.220	16	0.440	8.0	80.0	10.0	H
VR2.5	2/3D	43.51	32.54	2.300	2.500	0.250	16	0.500	8.0	105.0	6.0	R
4.0SC	D	58.88	32.39	3.500	4.000	0.400	16	0.800	8.0	130.0	9.0	H
VR4	D	60.50	32.55	3.500	4.000	0.400	16	0.800	8.0	150.0	5.0	R
VR4	DL	58.50	32.55	3.500	4.000	0.400	16	0.800	8.0	150.0	5.0	R
VRD4	D	60.50	32.55	3.500	4.000	0.400	16	NA	NA	140.0	5.0	R
VRD4	DL	58.50	32.55	3.500	4.000	0.400	16	NA	NA	140.0	5.0	R
7.0SC	F	89.30	32.39	6.300	7.000	0.700	16	1.400	8.0	210.0	7.0	H
VR7	F	91.30	32.55	6.300	7.000	0.700	16	1.400	8.0	240.0	4.4	R
VR7	FL	89.30	32.55	6.300	7.000	0.700	16	1.400	8.0	240.0	4.4	R
VRD7	F	91.30	32.55	6.300	7.000	0.700	16	NA	NA	230.0	5.5	R
VR10	SF	89.20	41.35	9.000	10.000	1.000	16	NA	NA	400.0	3.3	R

* H = hermetic; R = resealable

Table 51.9 Quick-charge sealed nickel−cadmium cells supplied by SAFT

Cell type		Maximum bare cell dimensions		Capacity (A h) rated at 20°C		Charge rates at 20°C				Bare cell weight (g)	Internal resistance (mΩ)	Vent type*
						Continuous		Maximum				
Model	Size	Height (mm)	Diameter (mm)	1C	0.2C	(A)	(h)	(A)	(h)			
0.070SC	1/2AAA	20.96	9.65	0.060	0.070	0.007	16	0.023	4.0	4.0	220.0	H
0.110SC	1/3AA	16.76	13.72	0.100	0.110	0.011	16	0.037	4.0	6.5	80.0	H
0.250SC	1/2AA	30.28	13.72	0.225	0.250	0.025	16	0.080	4.0	10.8	60.0	H
0.450SC	2/3A$_f$	27.94	16.56	0.405	0.450	0.045	16	0.150	4.0	15.6	48.0	H
VR0.45	2/3A$_f$	28.10	17.05	0.405	0.450	0.045	16	0.180	3.5	21.0	24.0	R
0.500SC	AA	49.96	13.72	0.450	0.500	0.050	16	0.165	4.0	19.0	28.0	H
VR0.5	AA	50.20	14.30	0.450	0.500	0.050	16	0.200	3.5	24.1	26.0	R
VR0.5	AAL	49.20	14.30	0.450	0.500	0.050	16	0.200	3.5	24.1	26.0	R
0.800SC	A$_f$	49.28	16.56	0.720	0.800	0.080	16	0.265	4.0	29.8	22.0	H
VR0.8	A$_f$	49.00	16.46	0.720	0.800	0.080	16	0.320	3.5	32.9	16.9	R
VR1.2	C$_s$	42.10	22.35	1.080	1.200	0.120	16	0.480	3.5	51.1	8.5	R
VR2	C	49.70	25.55	1.800	2.000	0.200	16	0.600	4.0	77.0	8.5	R
VR2	CL	47.70	25.55	1.800	2.000	0.200	16	0.600	4.0	77.0	8.5	R

* H = hermetic; R = resealable

Table 51.10 Low resistance sealed nickel−cadmium cells supplied by SAFT

Cell type		Maximum bare cell dimensions		Capacity (A h) rated at 20°C		Charge rates at 20°C				Bare cell weight (g)	Internal resistance (mΩ)	Vent type*
						Continuous		Maximum				
Model	Size	Height (mm)	Diameter (mm)	1C	0.2C	(A)	(h)	(A)	(min)			
VY1.15	C$_s$	42.10	22.35	1.040	1.150	1.000	1.5	4.000	15	51.0	7.0	R

*R = resealable

Table 51.11 Polytemp high-temperature nickel–cadmium cells available from SAFT

Model	Cell type Size	Maximum bare cell dimensions Height (mm)	Diameter (mm)	Capacity (A h) rated at 20°C 1C	0.2C	Charge rates at 20°C Continuous (A)	(h)	Maximum (A)	(h)	Bare cell weight (g)	Internal resistance (mΩ)	Vent type*
0.070PT	1/2AAA	20.96	9.65	0.060	0.070	0.004	24+	0.007	16	4.0	220.0	H
0.110PT	1/3AA	16.76	13.72	0.100	0.110	0.007	24+	0.011	16	6.6	80.0	H
0.150PT	1/3A	16.13	15.06	0.135	0.150	0.010	24+	0.015	16	7.9	100.0	H
0.180PT	AAA	44.45	9.65	0.160	0.180	0.012	24+	0.018	16	9.0	120.0	H
0.250PT	1/2AA	30.28	13.72	0.225	0.250	0.017	24+	0.025	16	10.8	60.0	H
0.450PT	2/3A$_f$	27.94	16.56	0.405	0.450	0.030	24+	0.045	16	15.6	48.0	H
0.500PT	AA	49.96	13.72	0.450	0.500	0.033	24+	0.050	16	19.9	28.0	H
VT0.5	AA	50.20	14.30	0.450	0.500	0.036	24+	0.050	16	24.1	32.5	R
0.600PT	A	48.77	15.06	0.540	0.600	0.040	24+	0.060	16	24.4	26.0	H
0.750PT	1/2C	23.50	25.70	0.675	0.750	0.050	24+	0.075	16	31.2	23.0	H
0.800PT	A$_f$	49.28	16.56	0.720	0.800	0.053	24+	0.080	16	29.8	22.0	H
1.0PT	3/5C	29.21	25.70	0.900	1.000	0.067	24+	0.100	16	39.7	20.0	H
1.2PT	C$_s$	41.96	22.25	1.080	1.200	0.080	24+	0.120	16	45.4	15.0	H
VT1.2	C$_s$	42.10	22.35	1.080	1.200	0.086	24+	0.120	16	51.1	12.1	R
2.0PT	C	47.37	25.70	1.600	1.800	0.120	24+	0.180	16	68.1	17.0	H
VT2	C	49.70	25.55	1.800	2.000	0.143	24+	0.200	16	77.0	11.2	R
2.2PT	1/2D	36.78	32.39	2.000	2.200	0.146	24+	0.220	16	79.5	10.0	H
4.0PT	D	58.88	32.39	3.500	4.000	0.267	24+	0.400	16	130.5	9.0	H
VT4	D	60.50	32.55	3.500	4.000	0.279	24+	0.400	16	150.4	7.0	R
VT4	DL	58.50	32.55	3.500	4.000	0.279	24+	0.400	16	150.4	7.0	H
7.0PT	F	89.10	32.39	6.300	7.000	0.467	24+	0.700	16	210.0	7.0	R
VT7	F	91.30	32.55	6.300	7.000	0.500	24+	0.700	16	241.0	5.5	R

* H = hermetic; R = resealable

applications where a full charge must be accomplished in 3.5–4 h. The low-resistance cells can be fully charged in as little as 15 min. They also demonstrate excellent voltage retention during very high rate discharging.

SAFT supply the Polytemp range of high-temperature nickel–cadmium cells (Table 51.11). The long life at high temperature of these cells makes them an ideal backup power source for many applications including computers and computer peripherals, emergency lighting and alarms, telephone and telegraph equipment, stationary and portable electronic instruments, and process control equipment.

Polytemp performance is the result of a combination of electrode design innovations with a polypropylene separator material. Polypropylene is much more resistant to deterioration than the nylon separator used in standard designs.

Table 51.12 illustrates the relative effect of continuous overcharge at the $0.1C$ rate on capacity over time (or life). The difference in life increases substantially as the continuous cell temperature is increased above 25°C.

Table 51.12 Percentage of life at 25°C

Cell temperature (°C)	Standard	Polytemp
25	100	100
35	40	75
45	20	60
55	10	45
65	1	30

Varta

Varta supply a range of sealed batteries and cells. Characteristics of some of these are detailed in Tables 51.13–51.18. These include button, cylindrical, and rectangular types, some of which have sintered electrodes. The nominal voltage of the cells is 1.2 V. The final voltage, when discharging at the 10 h rate, is 1.1 V. When discharging at higher rates, a lower voltage may be specified. Cells are normally recharged at the 10 h rate (I_{10}) for 14 h at room temperature. The end-of-charge voltage approximates to 1.35–1.45 V/cell. The stated maximum permissible temperature ranges for these batteries are as follows:

1. Charge: 0 to +45°C.
2. Discharge: −20 to +45°C.
3. Storage: −40 to +50°C.
4. Discharge and storage at +60°C permissible for a maximum of 24 h.

Varta batteries with sintered electrodes have a very low internal resistance and a high load capacity,

which result in an improved power to size and weight ratio. The capacity of their button-type cells is up to 3 A h at a voltage of 1.24 V. The size of the battery for this capacity is 50 mm diameter × 25 mm thick, weighing 135 g, and the permissible operating temperatures are −20 to +45°C. These cells are also available in welded and plastics-covered packs of up to ten cells.

Cylindrical batteries are suitable for heavy loads at steady voltages. They can be discharged continuously at 20 times the normal rate and 40 times the normal rate for short periods. Apart from ensuring the correct charging method, maintaining end-voltages and adhering to the temperature limits, no special care is needed. The rugged construction results in a battery that can withstand shocks and vibration. Capacities are up to 6 A h at 1.24 V, and the battery weighs 240 g and measures 35.5 mm diameter × 94 mm high. Permissible operating temperatures are from −40 to +45°C.

Rectangular or prismatic cells offer the highest capacities within the Varta range and in fact differ very little in construction from the heavy-duty nickel–cadmium type. Capacities are up to 23 A h at 1.22 V, and the cells weigh 1.39 kg and measure 51 × 91 × 125 mm. Permissible operating temperatures are the same as for the button type.

Single cells can be assembled into batteries, generally consisting of two to ten cells, depending on voltage requirements; exceptions are the 10DK and 20DK button cells which are available only as single cells. Batteries are available in stack formats, including insulating sleeve, plastics cassettes, plastics boxes and sheet steel cases.

The energy contents of the various types of sealed nickel–cadmium battery supplied by Varta are tabulated in Table 51.19.

In some instances Varta sealed nickel–cadmium batteries are interchangeable with dry cells (see Table 51.20). Each of the different types of Varta cell is commented on below.

Cylindrical cells with mass electrode, D series The 151 D cells are physically equivalent to primary cells and are suitable for either one- or two-cell operations. These cells are suitable for comparatively low rates of discharge where good charge retention is important. They are suitable for a continuous discharge up to ten times nominal discharge rate.

Cylindrical cells with sintered electrodes, RS series These cells are suitable for applications where a high rate of discharge, or operation at extremes of temperature, or permanent trickle charge or fast charging is required. The cells are constructed from rolled sintered positive and negative plates, separated by a highly porous separator which absorbs all the free electrolyte within the cell. A safety vent is

Table 51.13 Varta sealed nickel–cadmium cylindrical cells

Type	Type No.	Nominal capacity 5–10h rate (A h)	Discharge 10 h rate (mA)	Current 5 h rate (mA)	Charge current for 14 h (mA)	Max. recommended constant load (A)	Max. short-term load (A)	Internal resistance fully charged (μΩ)	Weight (g)	Diameter (mm)	Height (mm)	Remarks
Mass electrodes suitable for a continuous discharge up to 10 × nominal rate												
151D	3930150311	0.15	15	30	12.7	0.15	0.3	500	12	29	1.0	With contact cap and sleeving
180D	3930180011	0.18	18	36	10.0	–	–	–	10.5	44	1.0	With solder tags and sleeving
900D	3930090311	0.90	90	180	42.3	0.90	1.8	100	14.0	90	0.6	(Height measured without solder tags)
Sintered plate electrodes suitable for high-rate discharge and trickle charge												
100RS	3960100081	0.10	10	20	10.0	0.20	0.4	190	7.3	14.7	17.4	All cells can be provided with straight connectors
225RS	3960225041	0.225	22	44	22.0	0.45	0.9	82	11.3	14.7	25.0	
RS4	3964000051	4.00	400	800	400	8.0	16.0	10	147.0	33.5	61.0	
RS7	3967000051	7.00	700	1400	700	14.0	28.0	9	237.0	33.5	94.0	

Table 51.14 Varta sealed nickel–cadmium cylindrical cells

Type	Order No. for bare cell in shrink sleeve	Nominal capacity at $0.2C_5$ A (A h)	$0.1C_5$ A (I_{10}) (mA)	$0.2C_5$ A (I_5) (mA)	Charge current 14 h (mA)	Weight (g)	Dimensions (mm)	
							Diameter	Height
Sintered electrodes for continuous discharge rates up to $6C_5$ A ($60 \times I_{10}$)								
150RS*	05001 101 111	0.15	15	30	15	9	12–0.5	29–1.0
180RS	05003 101 111	0.18	18	36	18	10	10.5–1.0	44–1.0
800RS	50180 201 052	0.80	80	160	80	37	Max. 17.3	Max. 49.2
RS1	50210 101 052	1.0	100	200	100	41	23–1.0	35.7–1.0
Sintered electrodes for continuous discharge rates up to: $10C_5$ A ($100 \times I_{10}$), 750 RSH–RSH 1.8; $7C_5$ A ($70 \times I_{10}$), RSH 4; $6C_5$ A ($60 \times I_{10}$), RSH 7								
750RSH	50375 101 052	0.75	75	150	75	36	25.3–0.5	25–1.0
RSH 1.2	50412 101 052	1.2	120	240	120	50	23–1.0	42.2–1.0
RSH 1.8*	05014 101 111	1.8	180	360	180	67	26–1.0	49–1.0
RSH 4*	05020 101 111	4	400	800	400	147	33.5–1.0	61–1.0
RSH 7	50470 101 052	7	700	1400	700	237	33.5–1.0	94–1.0
Sintered electrodes for high-temperature applications up to 75°C								
100RST	50710 201 052	0.09	9	18	9	8	14.3–0.6	16.8–0.6
500RST	50750 101 052	0.5	50	100	50	24	14.5–0.5	49.5–1.0
RST 1.2	50812 101 052	1.2	120	240	120	50	23–1.0	42.2–1.0
RST 1.8	05814 101 111	1.8	180	360	180	67	26–1.0	49–1.0
RST 4	05820 101 111	4	400	800	400	147	33.5–1.0	61–1.0
RST 7	50870 101 052	7	700	1400	700	237	33.5–1.0	94–1.0

* These types are interchangeable with primary batteries of similar dimensions

Note: all characteristics assume cells are fully charged and discharged at 20°C

Table 51.15 Varta sealed nickel–cadmium rectangular cells

Type	Type No.	Nominal capacity 5–10 h rate (A h)	Discharge 10 h rate (A)	Current 5 h rate (A)	Charge current for 14 h (A)	Max. recommended constant load (A)	Max. short-term load (A)	Internal resistance fully charged (μΩ)	Weight (g)	Length (mm)	Height (mm)	Width (mm)
Mass electrodes suitable for a continuous discharge up to 10× nominal rate												
D23	3900023501	23.0 (10 h rate)	2.3	–	2.3	23	46	10	1390	51.0	91.0	125
Sintered electrodes suitable for high-rate discharge and trickle charge												
SD1.6	3941600001	1.6	0.16	0.32	0.16	3.2	6.4	16	115	16.8	41.4	64.1
SD2.6	3942600001	2.6	0.26	0.52	0.26	5.2	10.4	13	180	16.8	41.4	102
SD4	3944000001	4.0	0.40	0.80	0.40	–	–	–	260	24.2	41.4	102
SD7	3947000001	7.0	0.70	1.40	0.70	14.0	28.0	8	360	38.2	41.4	102
SD15	3940015001	15.0	1.50	3.00	1.50	30.0	60.0	2.5	780	30.0	77.0	126
SD2.4	52024101000	2.4	0.24	0.48	0.24				115	16.8	64	41.4
SD4.5	52045101000	4.5	0.45	0.40	0.45				180	16.8	100	41.4
SD10	52110101000	10	1.00	2.00	1.00				360	38.2	102	41.4
SD15	52115101000	15	1.50	3.00	1.50				780	30	126	77

All characteristics assume that cells are fully charged and discharged at 20°C

Table 51.16 Varta sealed nickel–cadmium cells: button cells with mass electrodes

Type	Type No.*	Nominal capacity 10h rate (mAh)	Discharge current 10h rate (mA)	Charge current for 14h (mA)	Max. recommended constant load (mA)	Max. short-term load (mA)	Internal resistance fully charged (μΩ)	Weight (g)	Diameter (mm)	Height (mm)	Remarks
Suitable for continuous discharge up to 10× nominal rate											
10DK	3910010001	10	1	1	1	1	4.9	0.8	7.65	5.2	Available only as single cells. Can be supplied with either straight or circular welded solder tags. 10DK and 50DK are available only with straight solder tags
50DK	3910050001	50	5	5	50	100	1.5	2.7	15.5	6.1	
90DKO	3910090001	90	9	9	90	180	0.6	5.0	oval†	6.0	
150DK	3910150001	150	15	15	150	300	0.5	9.1	25.1	6.7	
225DK	3910225001	225	22	22	225	450	0.465	11.8	25.1	8.8	
280DK	3910280001	280	28	28	–	–	–	16.5	34.35	5.3	
450DK	3910450001	450	45	45	450	900	0.25	32.4	43.1	7.9	
1000DK	3911000001	1000	100	100	1000	2000	0.11	57.0	50.3	10.0	
Suitable for higher rates of continuous discharge, up to 20× nominal rate											
160DKZ	3920160001	160	16	16	320	640	–	9.0	18.3	11.8	Available with either straight or circular welded solder tags except 160DKZ which is available only with straight welded tags
225DKZ	3920225001	225	22	22	450	900	210	13.0	25.1	9.1	
500DKZ	3920500001	500	50	50	1000	2000	80	26.0	34.4	10.0	
1000DKZ	3921000001	1000	100	100	2000	4000	50	57.0	50.3	10.0	

* Type numbers apply for cells with insulating sleeve or solder tag

† 90DKO dimensions: length 25.9–0.4, width 14.1–0.4

Note: all characteristics assume that cells are fully charged and discharged at 20°C

Table 51.17 Varta sealed nickel–cadmium batteries: miscellaneous types

Type	Type No.	Nominal voltage (V)	Voltage 10h rate (V)	Discharge current 10h rate (mA)	Nominal capacity at 10h rate (Ah)	Charge current for 14h (mA)	Max. recommended constant load (A)	Max. short-term load (A)	Weight (g)	Length (mm)	Width (mm)	Height (mm)
Batteries for pocket receivers, microphones, portable radiotelephones, calculators, etc.												
TR7/8	3520090760	9	8.5	9	0.09	9	0.09	0.18	42.5	26.4	15.1	49.0
6 V batteries with sintered electrodes in plastics cases; suitable for continuous discharge up to 10× nominal rate and trickle charge												
SML2, 5	3602500594	6.0	6.0	250	2.5	250	5.0	10.0	550	134	34.3	62.2
SM6	3606000594	6.0	6.0	600	6.0	600	12.0	24.0	1000	90.5	50.5	112.2
5/900D		6.0	6.0	90	0.9		0.9	1.8	220	37	37.0	97.0
10/RS1		12.0	12.0	100	1.0		2.0	4.0	475	119	50.0	42.0
3/SD1.6		3.6	3.6	160	1.6		3.2	6.4	360	52	43.0	67.5
3/SD2.6		3.6	3.6	260	2.6		5.2	10.4	570	52	43.0	110.0
5/SD2.6		6.0	6.0	260	2.6		5.2	10.4	980	67	59.0	106.0

Table 51.18 Varta sealed nickel–cadmium batteries: miscellaneous types

Type	Type No.	Nominal voltage (V)	Nominal capacity at $0.1C_5$ A (I_{10}) (A h)	$0.1C_5$ A (I_{10}) (mA)	Charging current 14 h (mA)	Weight (g)	Dimensions (mm)		
							Length	Width	Height
6 V batteries with sintered electrodes in plastic housing for continuous discharge rates up to $1C_5$ A ($10 \times I_{10}$)									
5 M 3	52530 101 063	6	3	300	300	550	134.5	34.3	62.2
5 M 6	52560 101 063	6	6	600	600	1000	90.5	50.5	112.5
Button cell batteries with mass plate electrodes in plastic encapsulation									
10/250DK	53025710057	12	0.25	25	25	135	54.8	28.8	54.9

All characteristics assume that cells are fully charged and discharged at 20°C

incorporated within each cell to enable gas, which may build up under fault conditions, to be released. Batteries containing RS cells are available either in standard formats or designed for specific applications.

Rectangular cells with mass electrodes, D23 type These cells are suitable for continuous discharge up to ten times the nominal discharge current. They are fitted with a safety vent which prevents excessive internal pressure damaging the cell case in fault conditions.

Table 51.19 Energy contents of Varta sealed nickel–cadmium batteries

	Volumetric energy density (W h/dm^3)	Energy density (W h/kg)
DK button cell (mass plate)	61	21
DKZ button cell (mass plate)	62	22
D round cell (mass plate)	69	22
RS round cell (sintered plate)	78	27
D rectangular (mass plate)	37	18
SD rectangular (sintered plate)	52	21

Rectangular cells with sintered electrodes, SD series These cells are suitable for applications where a high rate of discharge, or operation at extremes of temperature, or permanent trickle charge or fast discharge is required. The cells are constructed from cut sintered positive and negative plates, separated by a highly porous separator which absorbs all the free electrolyte within the cell. A safety vent is incorporated within each cell to enable gas, which may build up under fault conditions, to be released. All cells in this range have the cases connected to the positive electrodes. Cell cases are not insulated but intercell separators and nickel-plated connective links are available. The SD series are available made up in metal boxes.

Button cells with mass plate electrodes, DK and DKZ series Button cells are produced by Varta in two versions. DK-type cells are assembled using one positive and one negative electrode separated by porous insulating material. These cells are suitable for operation at normal temperatures and discharge rates of up to ten times the nominal discharge current. DKZ cells have two positive and two negative electrodes, which lower the internal resistance and therefore make the cell suitable for high rates of discharge (20 times nominal discharge

Table 51.20 Physically interchangeable cells/batteries

Varta Ni–Cd	IEC No.	USASI	Varta primary		Eveready	Mallory
			Zinc–carbon	Alkaline manganese		
151D	KR12/30	N	245	7245	D23	Mn9100
180RS	RO3	AAA	239	7239	U16	Mn2400
501RS	KR15/51	AA	280	7244	HP7	Mn1500
RS1.8K	KR27/50	C	281	7233	HP11	Mn1400
RS4K	KR35/62	D	282	7232	HP2	Mn1300
TR7/8	6F22	–	438	–	PP3	Mn1604

rate) and also for low-temperature operation. Button cells are suitable for applications where low cost and good packing density are required. The charge retention of mass plate button cells is superior to that of all other forms of sealed nickel–cadmium cell. All DK cells listed in Table 51.16, except the 10DK, can be made into stack batteries by connecting end on end, in sleeves, thereby ensuring a minimum internal resistance.

Varta supply a sealed sintered electrode nickel–cadmium cell (RSH and RSX type) for very high rates of discharge. Figure 51.4 compares the performance characteristics of the RS type of cell and the RSH type. A by-product of the lower internal resistance of the RSH cell is a reduction in its temperature on discharge.

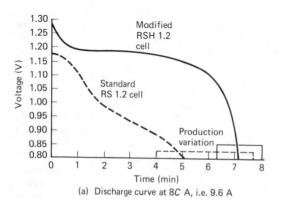

(a) Discharge curve at $8C$ A, i.e. 9.6 A

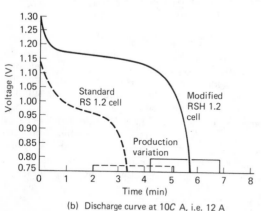

(b) Discharge curve at $10C$ A, i.e. 12 A

Figure 51.4 Comparison of discharge characteristics at 20°C: Varta RS1.2 and modified RSH1.2 sealed nickel–cadmium cells (Courtesy of Varta)

Nickel–cadmium sealed cells for CMOS-RAM memory protection Varta supply the Mempac range of button cells for this application and these cells are suitable for printed circuit board mounting. The advantages claimed for these cells are as follows:

1. *Rechargeable* The cells can be recharged many times or trickle charged permanently.

2. *Flat discharge curve* The cells remain at or above their nominal voltage (1.2 V/cell) for approximately 80% of their discharge time.
3. *Low self-discharge* Over 60% capacity remains after 6 months storage at +20°C.
4. *Low internal resistance* This ensures that the good voltage level of these cells is maintained during pulse discharges.
5. *Simple charging* Only a nominal constant current charge is required.
6. *Compact* Batteries, when board mounted, allow an 18 mm board spacing to be achieved as standard; 10 mm for special applications.
7. *Cost effective* A combination of the battery's low initial cost and its facile nature ensures minimum user expense.

Performance of Varta sealed nickel–cadmium batteries

Charge retention These batteries have a moderate self-discharge characteristic at normal temperatures. For applications requiring good charge retention, mass plate electrodes are preferred.

Effect of temperature on performance Extremes of temperature adversely affect the performance of sealed nickel–cadmium batteries, especially below 0°C. This reduction in performance at lower temperatures is mainly due to an increase in internal resistance, and therefore, because of their inherently lower internal resistance, batteries incorporating sintered cells (that is, RS and SD ranges) are normally recommended when lower operating temperatures are to be encountered. Figure 51.5 shows the relationship between the capacity and discharge voltage of rolled sintered cells up to 2 A h when discharged at 0 and −20°C after an initial charge at room temperature.

Information on the recommended conditions of temperature during charge, discharge and storage of Varta batteries is given in Table 51.21.

Discharge The available capacity from a cell is dependent on the discharge rate and is proportionally greater with a sintered plate cell. The maximum discharge current for Varta D and DK mass plate type cells is $10 \times I_{10}$, and for DKZ type mass plate and RS and SD types of cell with sintered electrodes it is $20 \times I_{10}$. In pulse load applications these rates can be doubled. Figure 51.6 shows the discharge at various rates between I_{10} and $50I_{10}$ of fully charged rolled sintered cells of up to 2 A h capacity. The voltage of a cell when discharged below 1 V/cell rapidly falls to zero (Figure 51.6(b) and (c)) and it is then possible under these circumstances for some cells in a battery to reverse their polarity. Provided that the discharge rage is at the I_{10} rate or less, this reversal will not harm a nickel–cadmium sealed cell,

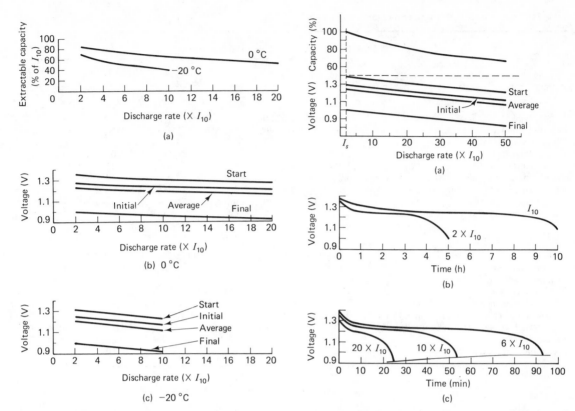

Figure 51.5 Capacity and voltage *versus* discharge rate for fully charged Varta sealed nickel–cadmium rolled sintered cells up to 2 A h (Courtesy of Varta)

Figure 51.6 Discharge curves at various rates of fully charged Varta sealed nickel–cadmium rolled sintered cells up to 2 A h. Starting voltage approx. 30 min after end of charge, initial voltage after 10% of discharge capacity, average voltage approx. 1.22 V, final voltage approx. 0.99 V. (b,c) Voltage during discharge below 1 V (Courtesy of Varta)

Table 51.21 Varta sealed nickel–cadmium batteries: recommended conditions of temperature during charge, discharge and storage

Charge temperature (°C)		Discharge temperature (°C)		Storage temperature (°C)	
Recommended	Possible	Recommended	Possible	Recommended	Possible
D, DK and DKZ cells with mass plate electrodes					
10 to 35	0 to 45	0 to 45	−20 to +50	0 to 45	−40 to +50
		At temperatures below 0°C maximum discharge rate $\leqslant 5C_{10}$ ($5 \times I_{10}$) for DK and $\leqslant 1C_{10}$ ($10 \times I_{10}$) for DKZ		Up to 60°C for a maximum of 24 h	
		Up to 60°C for a maximum of 24 h			
RS and SD cells with sintered electrodes					
10 to 35	−20 to +50	−20 to +45	−45 to +50	0 to 45	−45 to +50
At temperatures below 0°C charge current limited to $\frac{1}{3}I_{10}$ with a voltage limit of 1.55 V/cell		Up to 60°C for a maximum of 24 h		Up to 60°C for a maximum of 24 h	

Table 51.22 Full range of Varta Mempac memory protection cells

	Nominal capacity (mA h)					
	20	60	100	140	170	250
1.2 V *nominal battery voltage*						
Battery type	1/20 DK	1/60 DK	1/100 DKO	1/140 DK	1/170 DK	1/250 DK
Drawing number	DB 901/1	DB 892/1	DB 826/1	DB 862/1	DB 863/1	DB 864/1
Nominal voltage (V)	1.2	1.2	1.2	1.2	1.2	1.2
Final voltage (V)	1.0	1.0	1.0	1.0	1.0	1.0
End-of-charge on charge voltage (V)	1.45	1.45	1.45	1.45	1.45	1.45
Trickle charge rate (mA) $\pm 10\%$ at 1.45 V/cell	0.2	0.6	1.0	1.4	1.7	2.5
DC internal resistance (fully charged) (mΩ)	2200	1300	500	450	375	300
Battery length (mm)	5.35 − 0.3	6.2 − 0.4	6.0 − 0.25	6.1 − 0.6	6.7 − 0.6	8.8 − 0.6
Maximum battery diameter (mm)	11.5	15.5	Oval	25.1	26.0	26.0
6.0 V *nominal battery*						
Battery type	5/20 DK	5/60 DK	5/100 DKO	5/140 DK	5/170 DK	5/250 DK
Drawing number	DB 901/5	DB 892/5	DB 826/5	DB 862/5	DB 863/5	DB 864/5
Nominal voltage (V)	6.0	6.0	6.0	6.0	6.0	6.0
Final voltage (V)	5.0	5.0	5.0	5.0	5.0	5.0
End-of-charge on charge voltage (V)	7.25	7.25	7.25	7.25	7.25	7.25
Trickle charge rate (mA) $\pm 10\%$ at 1.45 V/cell	0.2	0.6	1.0	1.4	1.7	2.5
DC internal resistance (fully charged) (mΩ)	11000	6500	2500	2250	1875	1500
Battery length (mm)	28.8 − 2.0	31.5 − 2.0	30.4 − 1.25	30.9 − 3.0	33.9 − 3.0	44.4 − 3.0
Maximum battery diameter (mm)	12.0	15.8	Oval	26.0	26.0	26.0

Table 51.23 Varta Mempac S series of 100 mA h capacity cells for CMOS-RAM memory protection

Order No.	Nominal voltage (V)	Width (mm)	Length (mm)	Height without pins (mm)	Weight (g)
53010701012	1.2	17−0.4	42.4	10.5	9
53010702012	2.4	17−0.4	42.4	16.0	15
53010703012	3.6	22.2−0.4	40.3	16.0	21
53010704012	4.8	30.0−0.3	40.0	16.0	26

but for maximum life it should be avoided. Details of the 1.2 and 6 V batteries in the Mempac range of 100 mA h cells are given in Tables 51.22 and 51.23.

Most CMOS support systems require a battery backup capacity of a few milliampere hours. However, to ensure that the battery is cost effective, Varta recommend the use of the Mempac 100 DKO battery. This is rated at 100 mA h at the 10 h rate. However, at the discharge rates required for most memory support applications (possibly only a few hundred microamps), significantly greater capacity will be available.

Figure 51.7 shows voltage characteristics for the Mempac cell when discharged at the 10 h rate $0.1C_{10}$ A at $+20°$C. When discharged at rates less than $0.1C_{10}$ A, the voltage plateau will be extended. Mempac batteries may be discharged at lower rates than $0.1C_{10}$ A.

Figure 51.8 shows the voltage a memory may expect to see when using a three-cell 100 DKO Mempac to supply a load current of 20 μA. It can be

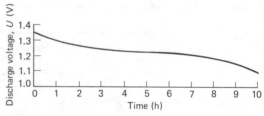

Figure 51.7 Discharge of a Varta Mempac 100 mA h memory protection cell at the 10 h rate, $0.1C_{10}$ A (Courtesy of Varta)

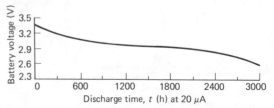

Figure 51.8 Discharge of a Varta Mempac 100 mA h memory protection cell to supply a load current of 20 μA (Courtesy of Varta)

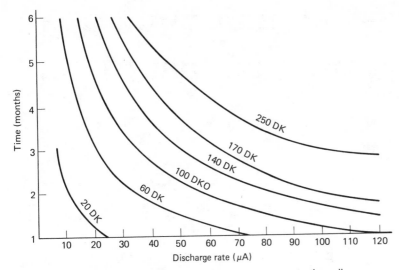

Figure 51.9 Effect of discharge rate of Varta Mempac memory protection cells on discharge duration (Courtesy of Varta)

seen from this curve that even at extremes of temperature, when the battery's negative temperature coefficient is taken into account, better than 2 V are still available, 2 V being the minimum voltage many volatile memories need in order to meet their performing specification.

To allow for variations in minimum operating voltages, the number of cells may be varied to suit the memory's requirements. Care should be taken to ensure that sufficient charging voltage is available to fully charge the battery.

Figure 51.9 shows the effect of discharge rate on discharge duration for a range of Mempac cells. The capacity required for memory support may be calculated by multiplying the discharge time by the discharge rate, adding sufficient capacity to allow for self-discharge losses.

When choosing a battery for memory support:

1. Allow for self-discharge when deciding the capacity required.
2. When selecting the number of cells, ensure that:
   (a) The minimum operating voltage of the chip can be achieved when the battery is fully discharged (1 V/cell).
   (b) Sufficient charge voltage is available to allow for the voltage drop across the current-limiting resistor and the battery to attain at least 1.45 V/cell.
3. If the cell is used in a purely standby mode, use the appropriate trickle charge rates.

Union Carbide

Union Carbide market their products under the trade name Eveready, manufacturing a range of sealed cylindrical and button-type nickel–cadmium batteries in the range 1.2–14.4 V.

In the Eveready design, gas evolution during charge and overcharge is prevented by the adoption of various design features. For the system to be overchargeable while sealed, the evolution of hydrogen must be prevented and provision is made for the reaction of oxygen within the cell container:

1. The battery is constructed with excess capacity in the cadmium electrode.
2. Starting with both electrodes fully discharged, charging the battery causes the positive electrode to reach full charge first, and it starts oxygen generation. Since the negative (cadmium) electrode has not reached full charge, hydrogen will not be generated.
3. The cell is designed so that the oxygen formed in the positive electrode can reach the metallic cadmium surface of the negative electrode, which it oxidizes directly.
4. Thus, in overcharge, the cadmium electrode is oxidized at a rate just sufficient to offset input energy, keeping the cell in equilibrium indefinitely. At this point of equilibrium the positive electrode is fully charged and the negative is somewhat less than fully charged.

When cells are connected in series and discharged completely, small cell capacity differences will cause one cell to reach complete discharge sooner than the remainder. The cell that reaches full discharge first will be driven into reverse by the others. When this happens in an ordinary nickel–cadmium sealed cell, oxygen will be evolved at the cadmium electrode and hydrogen at the nickel electrode. Gas pressure will increase as long as current is driven through the

Table 51.24 Union Carbide Eveready and Hercules nickel–cadmium cells and batteries: electrical characteristics of button and cylindrical types

Eveready or Hercules battery numbers	Voltage (V)	Capacity 1 h rate (mA h)	Current discharge 1 h rate (mA)	Capacity 10 h rate (mA h)	Current discharge 10 h rate (mA)	Charge for 4 h at (mA)	Charge for 14 h at (mA) (do not exceed)	Charging voltage (V)	Cut-off voltage 1 h rate (V)	Cut-off voltage 10 h rate (V)	Maximum dimensions				Weight (g)
											Diameter (mm)	Length (mm)	Width (mm)	Height (mm)	
B20	1.2			20	2		2	1.35 to 1.5		1.1	11.6			5.41	1.13
B20T	1.2			20	2		2	1.35 to 1.5		1.1	11.6			5.66	1.13
B50	1.2			50	5		5	1.35 to 1.5		1.1	15.5			5.97	3.4
B50T	1.2			50	5		5	1.35 to 1.5		1.1	15.5			6.22	3.4
CF4	1.2	4000	4000			1400		1.5 to 1.6	1.0		33.2			58.7	153.0
CF4T	1.2	4000	4000			1400		1.5 to 1.6	1.0		33.2			59.2	153.0
CH4	1.2	4000	4000				400	1.35 to 1.45	1.0		33.2			58.7	153.0
CH4T	1.2	4000	4000				400	1.35 to 1.45	1.0		33.2			59.2	153.0
2/B150	2.4			150	15		15	2.7 to 3.0		2.2	27.0			14.7	25.5
2/B150T	2.4			150	15		15	2.7 to 3.0		2.2	27.0			14.7	25.5
2/B225	2.4			225	22		22	2.7 to 3.0		2.2	27.0			18.7	28.4
2/B225T	2.4			225	22		22	2.7 to 3.0		2.2	27.0			18.7	28.4
2/BH500T	2.4			500	50		50	2.7 to 2.9		2.2	36.5			20.6	62.4
2/BH1	2.4			1000	100		100	2.7 to 2.9		2.2	52.4			20.6	113.0
2/BH1T	2.4			1000	100		100	2.7 to 2.9		2.2	52.4			20.6	113.0
3/B150	3.6			150	15		15	4.05 to 4.50		3.3	27.0			21.4	36.9
3/B150T	3.6			150	15		15	4.05 to 4.50		3.3	27.0			21.4	36.9
3/B225	3.6			225	22		22	4.05 to 4.50		3.3	27.0			27.4	39.7
3/BH500T	3.6			500	50		50	4.05 to 4.35		3.3	36.5			31.0	90.7
3/BH1	3.6			1000	100		100	4.05 to 4.35		3.3	52.4			31.0	176.0
3/BH1T	3.6			1000	100		100	4.05 to 4.35		3.3	52.4			31.0	176.0
4/B150	4.8			150	15		15	5.4 to 6.0		4.4	27.0			28.6	48.2
4/B151	4.8			150	15		15	5.4 to 6.0		4.4	27.0			28.6	48.2
4/B152	4.8			185	22		22	5.4 to 6.0		4.4	27.0			36.5	51.0
4/BH500T	4.8			500	50		50	5.4 to 5.8		4.4	36.5			40.9	116.0
4/BH1	4.8			1000	100		100	5.4 to 5.8		4.4	52.4			40.9	233.0
4/BH1T	4.8			1000	100		100	5.4 to 5.8		4.4	52.4			40.9	233.0
5/B150	6.0			150	15		15	6.75 to 7.50		5.5	27.0			35.7	59.5
5/B150T	6.0			150	15		15	6.75 to 7.50		5.5	27.0			35.7	59.5
5/B225	6.0	500	500	225	22		22	6.75 to 7.50		5.5	27.0			45.6	62.4
HS4172	6.0						50	6.75 to 7.25				40.6	34.2	50.8	111.0
N67	6.0			900	90		90	6.75 to 7.25		5.5		38.9	38.9	97.2	235.0
5/BH1	6.0			1000	100		100	6.75 to 7.25		5.5	52.4			51.2	298.0
5/BH1T	6.0			1000	100		100	6.75 to 7.25		5.5	52.4			51.2	298.0
N91	6.0	1200	1200				120	6.75 to 7.25				128.0	26.2	63.1	284.0
N70	6.0			1500	150		150	6.75 to 7.50		5.5		98.4	39.7	102.0	794.0
6/B150	7.2			150	15		15	8.1 to 9.0		6.6	27.0			42.9	70.9
6/B150T	7.2			150	15		15	8.1 to 9.0		6.6	27.0			42.9	70.9

Type														
6/B225	7.2	225		22	22			8.1 to 9.0	6.6	27.0			54.8	79.4
6/BH500	7.2	500		50	50			8.1 to 8.7	6.6	36.5			61.1	170.0
6/BH500T	7.2	500		50	50			8.1 to 8.7	6.6	36.5			61.1	170.0
6/BH1	7.2	1000		100	100			8.1 to 8.7	6.6	52.4			61.1	354.0
6BH1T	7.2	1000		100	100			8.1 to 8.7	6.6	52.4			61.1	354.0
N88	8.4	90		9	9			9.45 to 10.50	7.7	27.0	26.6	15.5	49.2	42.5
7/B150	8.4	150		15	15			9.45 to 10.50	7.7	27.0			49.6	82.2
7/B150T	8.4	150		15	15			9.45 to 10.50	7.7	27.0			49.6	82.2
7/BH500	8.4	500		50	50			9.45 to 10.15	7.7	36.5			71.4	198.0
7/BH500T	8.4	500		50	50			9.45 to 10.15	7.7	36.5			71.4	198.0
7/BH1	8.4	1000		100	100			9.45 to 10.15	7.7	52.4			71.4	414.0
7/BH1T	8.4	1000		100	100			9.45 to 10.15	7.7	52.4			71.4	414.0
8/B150	9.6	150		15	15			10.8 to 12.0	8.8	27.0			56.8	93.6
8/B150T	9.6	150		15	15			10.8 to 12.0	8.8	27.0			56.8	93.6
8/B225	9.6	225		22	22			10.8 to 12.0	8.8	27.0			72.6	102.0
8/BH1	9.6	1000		100	100			10.8 to 11.6	8.8	52.4			81.4	471.0
8/BH1T	9.6	1000		100	100			10.8 to 11.6	8.8	52.4			81.4	471.0
1007	9.6	4000	4000	400	400		175	10.8 to 11.6	8.8	52.4	163	81.8	79.4	1730.0
9/B150	10.8	150		15	15			12.15 to 13.50	9.9	27.0			63.9	108
9/B150T	10.8	150		15	15			12.15 to 13.50	9.9	27.0			63.9	108
9/B225	10.8	225		22	22			12.15 to 13.50	9.9	27.0			81.8	111
9/BH500	10.8	500		50	50			12.15 to 13.05	9.9	36.5			91.7	261
9/BH500T	10.8	500		50	50			12.15 to 13.05	9.9	36.5			91.7	261
9/BH1	10.8	1000		100	100			12.15 to 13.05	9.9	52.4			91.7	525
9/BH1T	10.8	1000		100	100			12.15 to 13.05	9.9	52.4			91.7	525
10/B150	12.0	150		15	15			13.5 to 15.0	11.0	27.0			70.6	119
10/B150T	12.0	150		15	15			13.5 to 15.0	11.0	27.0			70.6	119
10/B225	12.0	225		22	22			13.5 to 15.0	11.0	27.0			86.9	122
10/BH500T	12.0	500		50	50			13.5 to 14.5	11.0	36.5			102.0	289
HS4151	12.0	500	500			10		15.0 to 16.0			63.6	16.5	144.0	241
HS4153	12.0	500	500			10		13.5 to 14.5			63.6	16.5	144.0	255
10/BH1	12.0	1000		100	100			13.5 to 14.5	11.0	52.4			102.0	581
10/BH1T	12.0	1000		100	100			13.5 to 14.5	11.0	52.4			102.0	581
N86	12.0	1200	1200	120		10		13.5 to 14.5		36.5	128.0	51.6	62.3	567
N65	13.2	1000	1000			11		14.85 to 15.95			186.0	37.1	78.7	765
HS4280	13.2	1000	1000			11	75	14.85 to 15.95			186.0	37.1	78.7	765
HS4130	14.4	225	225			12		16.2 to 17.4			55.3	19.3	70.1	142
HS4080	14.4	225	225			12		18.0 to 19.2			55.3	19.3	70.1	142
HS4125	14.4	450	450			12		16.2 to 17.4			55.3	32.1	70.1	255
HS4068	14.4	500	500			12		16.2 to 17.4			73.0	33.7	71.1	340
N64	14.4	500	500			12		16.2 to 17.4			73.0	33.7	71.1	340
HS4073	14.4	4000	4000			12		16.2 to 17.4			84.9	66.7	203.0	2180

The capacity rating of Eveready nickel–cadmium cells and batteries is based on output in discharge at the 1 h rate to an end-voltage of 1 V/cell for all cylindrical cells (except CH1.8 and CH2.2) and at the 10 h rate to 1.1 V/cell for button cells (and CH1.8 and CH2.2). If current is withdrawn at faster rates than these standards, capacity is decreased

Table 51.25 Union Carbide sealed nickel–cadmium button cell stack assemblies (jacketed with plastics tubing)

Eveready battery number	Number of cells per stack	Voltage (V)	Terminals	Max. dimensions		Weight (g)
				Diameter (mm)	Height (mm)	
Assemblies of B150 cells: 150 mA h capacity (10 h rate)						
2/B150	2	2.4	flat	27	14.7	25.5
3/B150	3	3.6	flat	27	21.4	36.9
4/B150	4	4.8	flat	27	28.6	48.2
5/B150	5	6.0	flat	27	35.7	59.5
6/B150	6	7.2	flat	27	42.9	70.9
7/B150	7	8.4	flat	27	49.6	82.2
8/B150	8	9.6	flat	27	56.8	93.6
9/B150	9	10.8	flat	27	63.9	108.0
10/B150	10	12.0	flat	27	70.6	119.0
Assemblies of B225 cells: 225 mA h capacity (10 h rate)						
2/B225	2	2.4	flat	27	18.7	28.4
3/B225	3	3.6	flat	27	27.4	39.7
4/B225	4	4.8	flat	27	36.5	51.0
5/B225	5	6.0	flat	27	45.6	62.4
6/B225	6	7.2	flat	27	54.8	79.4
7/B225	7	8.4	flat	27	63.9	90.7
8/B225	8	9.6	flat	27	72.6	102.0
9/B225	9	10.8	flat	27	81.8	111.0
10/B225	10	12.0	flat	27	86.9	122.0
Assemblies of BH225 cells (high rate): 225 mA h capacity (10 h rate)						
2/BH225	2	2.4	flat	27	19.1	28.4
3/BH225	3	3.6	flat	27	28.2	42.5
4/BH225	4	4.8	flat	27	37.3	53.9
5/BH225	5	6.0	flat	27	46.4	68.0
6/BH225	6	7.2	flat	27	56.0	85.1
7/BH225	7	8.4	flat	27	65.1	96.4
8/BH225	8	9.6	flat	27	74.2	108.0
9/BH225	9	10.8	flat	27	83.3	122.0
10/BH225	10	12.0	flat	27	92.5	133.0
Assemblies of BH500 cells (high rate): 500 mA h capacity (10 h rate)						
2/BH500	2	2.4	flat	36.5	20.6	62.4
3/BH500	3	3.6	flat	36.5	31.0	90.7
4/BH500	4	4.8	flat	36.5	40.9	116.0
5/BH500	5	6.0	flat	36.5	51.2	142.0
6/BH500	6	7.2	flat	36.5	61.1	170.0
7/BH500	7	8.4	flat	36.5	71.4	198.0
8/BH500	8	9.6	flat	36.5	81.4	230.0
9/BH500	9	10.8	flat	36.5	91.7	261.0
10/BH500	10	12.0	flat	36.5	102.0	289.0
Assemblies of BH1 cells (high rate): 1 A h capacity (10 h rate)						
2/BH1	2	2.4	flat	52.4	20.6	113.0
3/BH1	3	3.6	flat	52.4	31.0	176.0
4/BH1	4	4.8	flat	52.4	40.9	233.0
5/BH1	5	6.0	flat	52.4	51.2	298.0
6/BH1	6	7.2	flat	52.4	61.1	354.0
7/BH1	7	8.4	flat	52.4	71.4	414.0
8/BH1	8	9.6	flat	52.4	81.4	471.0
9/BH1	9	10.8	flat	52.4	91.7	525.0
10/BH1	10	12.0	flat	52.4	102.0	581.0

cell and eventually it will either vent or burst. This condition is prevented in the sealed nickel–cadmium cells produced by some suppliers by special construction features. These include the use of a reducible material in the positive electrode, in addition to the nickel hydroxide, to suppress hydrogen evolution when the positive electrode expires. If cadmium oxide is used it is possible to prevent hydrogen formation and to react the oxygen formed at the negative electrode by the same basic process used to regulate pressure during overcharge. A cell is considered electrochemically protected against reversal of polarity if, after discharge at the 10 h rate down to 1.1 V, it may receive an additional 5 h discharge with the same current without being damaged or otherwise affected. This protection applies to all Eveready button nickel–cadmium cells except the B20.

Eveready cylindrical cells are protected against cell rupture (caused by gassing generated during polarity reversal) by a pressure-relief vent.

Eveready nickel–cadmium cells are available in two basic configurations: button and cylindrical. The capacity ranges are 20 mA h to 1 A h for the button cell and 150 mA h to 4 A h for the cylindrical.

Electrical and physical characteristics of some of the button and cylindrical nickel–cadmium cells and batteries that comprise the basic Eveready range are listed in Table 51.24. All of these cells may, of course, be assembled in series to make up batteries of various voltages. Table 51.24 shows the electrical and physical characteristics of 159 cells and batteries in the Union Carbide range. The items are arranged in ascending voltage categories varying from 1.2 to 14.4 V. Within any voltage category, batteries are arranged in ascending order of capacity. Union Carbide also supply assemblies of stacks of two to ten button cells (Table 51.25).

All Eveready high-rate cylindrical cells have a resealing pressure vent, except CH1.8 and CH2.2 (Table 51.24), which have a puncture-type failsafe venting mechanism. This vent permits the cell to release excess gas evolved if the cell is abused, for example. When the internal pressure has dropped to an acceptable level, the vent will reseal, permitting the cell to be recycled in the normal manner with little or no further loss of electrolyte or capacity.

Eveready sealed nickel–cadmium cells and batteries exhibit relatively constant discharge voltages. They can be recharged many times for long-lasting economical power. They are small convenient packages of high energy output, hermetically sealed in lead-resistant steel cases, and will operate in any position. The cells have very low internal resistance and impedance, and are rugged and highly resistant to shock and vibration.

The temperature range under which these cells may be operated is wide. Use at high temperatures, however, or charging at higher than recommended rates, or repeated discharge beyond the normal cut-offs, may be harmful. In the case of button cells, which do not contain a safety vent, charging at temperatures lower than those recommended may cause swelling or cell rupture.

Sealed nickel–cadmium cells should not be charged in parallel unless each cell or series string of the parallel circuit has its own current-limiting resistor. Minor differences in internal resistance of the cells may result, after cycling, in extreme variation in their states of charge. This may lead to overcharge at excessive currents in some cells and undercharge in others. Except in the case of complete discharge, neither cell condition nor state of charge can be determined by open-circuit voltage. Within a short time after charging it may be above 1.4 V; it will fall shortly thereafter to 1.35 V and continue to drop as the cell loses charge.

High-rate nickel–cadmium cells will deliver exceedingly high currents. If they are discharged continuously under short-circuit conditions, self-heating may cause irreparable damage. If the output is withdrawn in pulses spaced to limit the temperatures of a few critical areas in the cell to a safe figure, high currents can be utilized.

The heat problems vary somewhat from one cell type to another, but in most cases internal metal strip tab connectors overheat and/or the electrolyte boils. General overheating is normally easy to prevent because the outside temperature of the battery can be used to indicate when rest, for cooling, is required. In terms of cut-off temperature during discharge, it is acceptable to keep the battery always below 65°C. The overheated internal connectors are difficult to detect. This form of overheating takes place in a few seconds or less, and overall cell temperature may hardly be affected. It is thus advisable to withdraw no more ampere seconds per pulse, and to withdraw it at no greater average current per complete discharge, than recommended for the particular cell in question. In special cases, where cooling of the cell or battery is likely to be poor, or unusually good, special tests should be run to check the important temperatures before any duty cycle adjustment is made.

Output capacity in any discharge composed of pulses is difficult to predict accurately because there are an infinite number of combinations of current, 'on' time, rest time and end-voltage. Testing on a specific cycle is the simplest way to obtain a positive answer.

Self-discharge characteristics of various types of Eveready cell are illustrated in Figure 51.10, which shows a decline in percentage of rated capacity available over a 20 week period. (For identification of cells see Table 51.24.)

At elevated storage temperatures, self-discharge will be considerably higher than at room temperature. Union Carbide recommend that batteries be

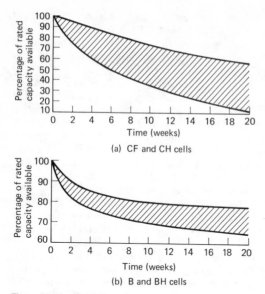

(a) CF and CH cells

(b) B and BH cells

Figure 51.10 Charge retention at 21°C of Eveready sealed nickel–cadmium cells (Courtesy of Union Carbide)

stored at 21°C or lower for this reason. When button cells (B and BH cells) have been stored for a long period (over 6 months), regardless of storage temperature, they should not immediately be charged, but should first be fully discharged and then charged once at half the normal rate; that is, 28 h at the 20 h rate. This procedure is not required for the CF and CH cells.

The overcharge capability of Eveready cylindrical nickel–cadmium cells is outstanding. The first discharge after the 2 year charge period yields a slightly reduced voltage curve and 65% capacity. The second cycle after 2 years' continuous overcharge provides essentially the same discharge curve as the initial one.

Figure 51.11 shows capacity maintenance *versus* months of continuous overcharge at the 20 h rate with periodic discharges every 3 months at the 1 h rate. The cells maintain 90% of their initial capacity after 2 years of this overcharge regimen. This

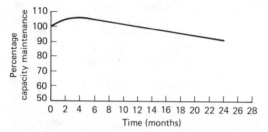

Figure 51.11 Capacity maintenance *versus* continuous overcharge of an Eveready sealed nickel–cadmium CH1.2 cylindrical cell discharged at 1C (1.2 A). First discharge every 3 months after continuous overcharge (Courtesy of Union Carbide)

pattern of use would occur where batteries are left on charge continuously and used one cycle only on an occasional basis. (These comments do not apply to the CH1.8 and CH2.2 types; see Table 51.21.)

Memory effect is that characteristic attributed to nickel–cadmium cells wherein the cell retains the characteristics of previous cycling. That is, after repeated shallow-depth discharges the cell will fail to provide a satisfactory full-depth discharge. Eveready cylindrical nickel–cadmium cells are particularly noted for their lack of memory effect. Figure 51.12 shows initial and subsequent cycles after repeated shallow discharges. The graph shows the initial discharge curve and the first and second discharge curves after 100 40% depth discharge cycles. Subsequent full-depth discharges yield nearly equal capacity to the initial curve at slightly reduced voltage levels. (These comments do not apply to the CH1.8 and CH2.2 types.

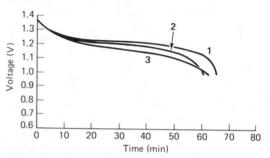

Figure 51.12 Initial and subsequent cycles of Eveready sealed nickel–cadmium cylindrical cells after shallow discharges: **1**, initial discharge; **2**, second full-depth discharge after 100 cycles at 40% depth discharge; **3**, first full-depth discharge after 100 cycles at 40% (C/5 (240 mA) for 2 h) depth discharge (Courtesy of Union Carbide)

Eveready sealed nickel–cadmium cells and batteries experience a relatively small charge of output capacity over a wide range of operating temperature. Charging, however, must be done in a much narrower temperature range. Temperature limits applicable to operation of the batteries are listed in Table 51.26.

Union Carbide have summarized the effects of low and high temperature on the storage, discharging and charging of Eveready cells and batteries as shown in Table 51.27.

Sealed nickel–cadmium cells have a high effective capacitance. Their impedance is so low that cells being, in effect, continuously overcharged make excellent ripple filters. Cell impedance is dependent on frequency and state of charge of the cell. It is lower for a charged cell than it is for a discharged cell. Values of impedance and resistance of Eveready cells are shown on the individual specification sheets for each cell (available from the manufacturer).

Table 51.26 Union Carbide sealed nickel–cadmium batteries: operating temperature limits

Cell type	Charge rate (h)	Charge temperature (°C)	Discharge temperature (°C)	Storage temperature (°C)
B	10	0 to 45	−20 to +45	−40 to +60
BH	10	0 to 45	−20 to +45	−40 to +60
CF	1 to 3	15.6 to 45	−20 to +45	−40 to +60
CF	10	0 to 45	−20 to +45	−40 to +60
CH	10	0 to 45	−20 to +45	−40 to +60

Table 51.27 Effect of high and low temperatures on storage, discharging and charging of Eveready nickel–cadmium cells and batteries

	Low temperature	High temperature
Storage (all types	at −40°C No detrimental effect. However, cells or batteries should be allowed to return to room temperature before charging	at 60°C No detrimental effect. However, self-discharge is more rapid starting at 32°C and increases as temperature is further elevated
Discharge (all types)	at −20°C No detrimental effect, but capacity will be reduced as shown by curves	at 45°C No detrimental effect
Charge		
CF and CH types (10 h rate)	at 0°C Cells or batteries should not be charged below 0°C at the 10 h rate	at 45°C Cells or batteries show charge acceptance of approximately 50%
CF types (1–3 h rate)	at 15.6°C Cells or batteries should not be charged below 15.6°C at the 1 h rate or below 10°C at the 3 h rate	at 45°C Cells or batteries show charge acceptance of approximately 90%
Button cells	at 0°C Cells or batteries should not be charged below 0°C at the 10 h rate	at 45°C Cells or batteries show charge acceptance of approximately 60%. Also possible detrimental effect on cycle life

Table 51.28 Cycle life

Cell	Estimated average number of cycles
Cylindrical	
CH types	1000 (CH1.8 and CH2.2: 300)
CF types	1000
Button	
B types	300
BH types	300

Cycle life of the nickel–cadmium sealed cell depends both on cell design and on the type of use to which it is subjected. Apart from violent abuse, the use factors that most seriously influence life expectancy are:

1. Amount of overcharge (excessive overcharge is undesirable).
2. Temperature of charge and overcharge (elevated or lowered temperature is undesirable).
3. End-point requirements regarding rate and capacity (increased cycle life will ordinarily be the result of a shallow discharge regimen).

Any treatment that causes a cell to vent itself is harmful. Frequent or extended venting of even properly valved cells eventually destroys them. In rating cycle life, the end of life for a sealed nickel–cadmium cell is considered to be when it no longer provides 80% of its rated capacity. The discharge currents used in determining the cycle lives listed below are the 10 h rate for button cells and the 1 h rate for cylindrical types. The charge current is terminated after return of approximately 140% of the capacity previously removed. If a cell can be considered to be satisfactory while delivering less than the arbitrary 80% end-point figure, cycle life will be greater than that listed in Table 51.28. The ratings are for 21.1°C performance.

The following information is supplied for each type of Eveready cell: discharge curves (Figure 51.13), capacity *versus* temperature (Figure 51.14), and charge rate *versus* charging time (Figure 51.15). Figures 51.16 and 51.17, respectively, show the capacities available at various discharge currents for the 0.5 A h Eveready 1.2 V cylindrical and button

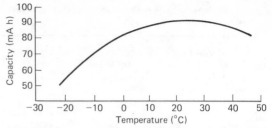

Figure 51.14 Sample capacity *versus* temperature curve for Eveready OB90 and OB90T sealed nickel–cadmium cells, 1.2 V, 20 µA h capacity, at 9 mA discharge current. The temperature ranges applicable are 0–45°C (charge), −20 to 45°C (discharge) and −40 to 60°C (storage) (Courtesy of Union Carbide)

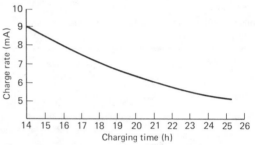

Figure 51.15 Sample discharge data for fully charged Eveready OB90 and OB90T sealed nickel–cadmium cells, 1.2 V, 90 µA h capacity. (Charge voltage 1.35–1.50 V) (Courtesy of Union Carbide)

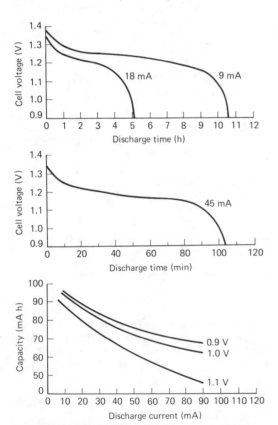

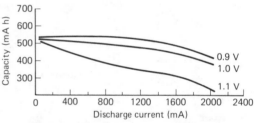

Figure 51.16 Typical capacity *versus* discharge current curves for Eveready CF500, CF500T, CH500 and CH500T sealed nickel–cadmium cylindrical cells, 1.2 V, 0.5 A h capacity (Courtesy of Union Carbide)

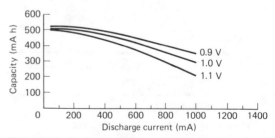

Figure 51.17 Typical capacity *versus* discharge current curves for Eveready BH500 and BH500T sealed nickel–cadmium button cells, 1.2 V, 0.5 A h capacity. Ranges of temperature applicable to operation of the cells are: charge 0 to 45°C, discharge −20 to +45°C, storage −40 to +60°C (Courtesy of Union Carbide)

Figure 51.13 Sample discharge data: average performance characteristics of Eveready OB90 and OB90T sealed nickel–cadmium cells, 1.2 V, 90 µA h capacity, at 21°C. Example: assume a 35 mA discharge to a 1.1 V end-point. The OB90 (or OB90T) cell has a capacity of 70 mA h. This provides a discharge time of 2 h (Courtesy of Union Carbide)

cells. Figure 51.18 shows the effect of temperature on the capacity for a 1 A h Eveready 1.2 V button cell. Table 51.29 shows the available types of battery, with the stated capacity, made up from Eveready cylindrical cells. Discharge curves and capacity *versus* discharge currents data for some of these batteries are given in Figure 51.19.

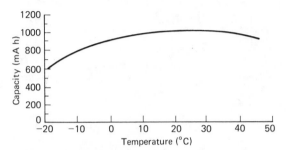

Figure 51.18   Effect of temperature on capacity of Eveready BH1 and BH1T sealed nickel–cadmium button cells, 1.2 V, 1 A h capacity, discharge current 100 mA (Courtesy of Union Carbide)

Nife Jungner SpA (Italy)

This company manufactures a range of rechargeable sealed nickel–cadmium cells in the nominal capacity range 0.4–5.6 A h. These are available with three different terminal arrangements; HH, HB and CF (Figure 51.20). Each cell has a nominal voltage of 1.25 V. A battery is assembled by series connection of a number of cells, which determine the battery voltage. The simplest way to connect a number of cells is to stack them on top of each other as in an ordinary flashlight (terminal arrangement CF, Figure 51.20). This method, however, is suitable only for low loads. For high loads, and when high reliability is of great importance, cells with soldered or spot-welded soldering tabs are recommended. Battery packs made up of between two and 20 cells are available.

The characteristics of the Nife Jungner KR series of cells (0.4–5.6 A h) are given in Table 51.30. These cells have an acceptably low rate of self-discharge. Self-discharge is very much dependent on the temperature, and increases rapidly with rising

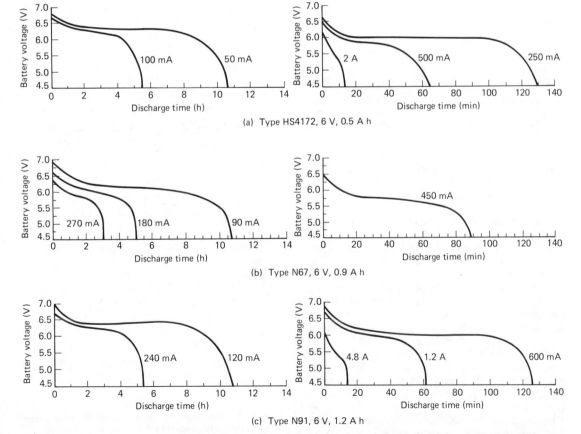

(a) Type HS4172, 6 V, 0.5 A h

(b) Type N67, 6 V, 0.9 A h

(c) Type N91, 6 V, 1.2 A h

Figure 51.19   Typical discharge curves at 20°C for Eveready sealed nickel–cadmium cylindrical batteries: see Table 51.29 (Courtesy of Union Carbide)

Table 51.29 Types of Union Carbide Eveready battery available

Voltage (V)	Number of 1.2 V cells in assembly	Type number								
		0.09 A h	0.225 A h	0.45 A h	0.5 A h	0.9 A h	1.0 A h	1.2 A h	1.5 A h	4.0 A h
6.0	5				HS4172	N67		N91	N70	
8.4	7	N88								
9.6	8									1007
12.0	10				HS4151 HS4153			N86		
13.2	11						N65 HS4280			
14.4	12		HS4130 HS4080	HS4125 HS4081	HS4068 N64					HS4073

Single 1.2 V cylindrical cells, available capacities (A h) 0.15, 0.225, 0.45, 0.50, 1.0, 1.2, 1.4, 1.8, 2.2 and 4.0

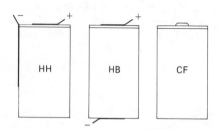

Figure 51.20 Terminal arrangements for Nife Jungner KR series sealed nickel–cadmium cells. The cells are available with three different terminal arrangements; HH, HB and CF. Types HH (head–head) and HB (head–base) are equipped with soldering tabs. Type CF (connection free) is equipped with a contact button on top, similar to dry cells (Courtesy of Nife Jungner)

temperature. At very low temperatures the self-discharge is small, and is almost zero at −20°C. These cells and batteries can be stored in the charged or discharged condition. When charged they should be handled with care to avoid short-circuits because the short-circuit current is very high. When storing for more than 6 months a storage temperature lower than 15°C is recommended by the manufacturers.

Figure 51.21 shows the effect on delivered capacity of KR cells of two of the factors that have most influence on capacity, namely current and cell temperature. The available capacity decreases at high currents and/or low temperatures. This is shown by the typical discharge curves for 25, 0 and −18°C in Figure 51.21. Maximum recommended continuous discharge current differs for different cell sizes. It ranges from 3.5 to 5CA. Considerably higher current can be used at short discharges (see Table 51.30).

Figure 51.22 shows the relationship between discharge current and capacity and final voltage for a KR 15/51 0.4 A h cell.

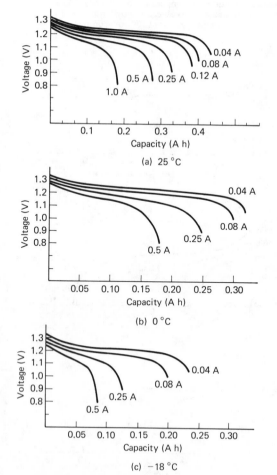

Figure 51.21 Discharge voltage *versus* capacity (at the stated temperature) of Nife Jungner KR 15/51 sealed nickel–cadmium cells, 0.4 A h capacity (Courtesy of Nife Jungner)

Table 51.30 NIFE Jungner sealed nickel–cadmium cells: technical specifications

Cell type	KR 15/51	KR 23/43	KR 27/50	KR 35/62	KR 35/92
Nominal capacity (A h) (5 h rate C_5 at $25 \pm 5°C$ to end-voltage 1 V)	0.40	1.1	1.6	3.6	5.6
Current on discharge (mA) ($0.2C_5$ A)	80	220	320	720	1120
Dimensions:					
Diameter (mm)	14.5	22.7	26	33.2	33.2
Height (mm)*	50.3	42.2	48	59.5	89
Weight, approx. (g)	20	45	65	140	210
Energy content (average discharge voltage with discharge at 5 h current), $0.2C$ A, is 1.2 V:					
W h/cell	0.48	1.32	1.92	4.32	6.72
W h/kg	24	29	30	31	32
W h/dm³	65	81	78	86	90
Internal resistance (mΩ)†	90	27	22	13	11
Charging:					
Normal charging current 14–16 h (m A)	50	120	200	400	600
Trickle charging minimum current (m A)‡	20	45	75	150	250
Temperature range	From 0 to 40°C, preferably normal room temperature				
Discharge:					
Maximum recommended continuous discharge current (A)	1.5	5	8	15	20
Maximum recommended discharge current for 5 (A)	7	15	20	50	80
Temperature range	From −40 to +40°C				
Storing	In charged or discharged condition: use a cool storage place, preferably with temperature lower than 15°C				
Terminal arrangement:					
Cell type CF, drawing No.	90-51400	90-51405	90-52813	90-51411	90-51417
Cell type HH, drawing No.	90-51401	90-51406	90-52814	90-51412	90-51418
Cell type HB, drawing No.	90-51402	90-51407	90-52815	90-51413	90-51419
Batteries with even number of cells, drawing No.	–	90-52359	90-53536	90-52357	90-52355
Batteries with odd number of cells, drawing No.	–	90-52360	90-53537	90-52358	90-52356
Corresponding dry cell	R6 or AA	Sub C	R14 or C	R20 or D	–

* Contact button is not included in the height with the exception of KR15/51. Add 2 mm for contact button. KR 15/51 always has a contact button even when soldering tabs are used
† Obtained from d.c. measurements of voltage and current after discharging 10% of nominal capacity from a fully charged cell
‡ Recommended tolerance for trickle charging current is +20 and −0%

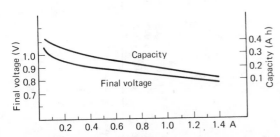

Figure 51.22 Capacity and final voltage *versus* discharge current for a Nife Jungner sealed nickel–cadmium cell type KR 15/51, 0.4 A h (Courtesy of Nife Jungner)

Marathon (USA)

Marathon supply a range of vented rechargeable 1.25 V sealed cells in the capacity range up to 63 A h, and also made-up batteries to customers' specifications (see Table 51.31). These batteries are equipped with a pressure relief vent. It is claimed by Marathon that these cells can be charged using a simple transformer–rectifier circuit, that they can safely be left on overcharge for months, if necessary, and that they have thousands of hours of life. Performance characteristics of these batteries are given in Table 51.28.

Table 51.31 Marathon sealed nickel–cadmium cells: physical and electrical characteristics

Marathon cell No.	Cell size	Overall dimensions (mm)						Weight (g)	Capacity (mA h)		Charge rate (mA) 14 h	Fast charge rate (mA) 4.5 h	Internal impedance at 1000 cycles (Ω)
		Diameter ± 0.010		Height ± 0.015									
				With positive button		Without positive button							
		With label and jacket	Without label and jacket	With label and jacket	Without label and jacket	With label and jacket	Without label and jacket		1 h rate	5 h rate			
S101	AA	14.4	14.0	49.1	19.1	—	—	22.4	450	500	50	125	0.021
S102	½C	25.8	25.4	25.4	25.4	23.1	22.8	33.6	700	800	80	200	0.017
S103	D	32.7	32.2	60.6	60.6	58.3	58.1	154.0	3800	4000	400	—	0.012
S104	C	25.8	25.4	49.0	49.0	47.6	47.6	72.8	1650	1800	190	450	0.016
S106	CD	25.8	25.4	77.6	77.6	75.3	75.0	112.0	2600	3000	300	750	0.014
S108	F	32.7	32.2	*	*	87.7	87.2	232.0	6300	6500	650	—	0.013
S113	0.875 × 1.625 (Sub C)	—	22.0	*	*	—	41.9	44.8	1150	1200	120	300	0.016
S126	0.625 × 0.6	—	15.9	*	*	—	15.5	8.4	140	160	20	40	0.052
S142	1 × 1.25	—	25.4	*	*	—	31.7	53.2	1200	1300	145	300	0.017
S143	1 × 1.14	—	25.4	*	*	—	28.9	47.6	1000	1150	100	250	0.017
S144	1.27 × 1.42	—	32.2	*	*	—	36.0	8.4	2100	2300	230	—	0.015
S147	½AA	—	14.0	*	*	—	31.6	14.8	250	285	30	65	0.025
S148	0.865 × 1.005	—	22.0	*	*	—	25.5	28.0	625	650	65	165	0.020
S150	0.645 × 1.095	—	16.3	*	*	27.9	27.8	18.8	390	450	45	115	0.024
S151	0.645 × 1.875	—	16.3	*	*	47.7	47.6	29.7	—	775	80	—	—

* Not available in this configuration

Table 51.32 Summary of GEC sealed nickel–cadmium cells (rechargeable)

Feature		Cylindrical cells	Rectangular and oval cells	Perma-cells
Range of capacity ratings (A h) at 1 h rate C		0.1 to 5.6	4.0	0.45 to 4.0
Outline	Top view			
	Side view			
Sealed construction		Yes	Yes	Yes
High-pressure resealing vent		Yes	Yes	Yes
Modular packaging efficiency		Good	Excellent	Good
High-rate discharge		Excellent	Excellent	Excellent
Low-rate discharge		Good	Good	Good
Maintenance-free		Yes	Yes	Yes
Position-sensitive		No	No	No
Constant-potential charge		No	No	No
Constant-current charge		Yes	Yes	Yes

General Electric Corporation (USA)

General Electric manufacture a very wide range of rechargeable sealed nickel–cadmium cells, including cylindrical, oval and rectangular cells, and also the Perma-cell, which is offered as an alternative to primary cells. Each of the three types of cell has its own typical applications (Table 51.32). Thus cylindrical, rectangular and oval cells are recommended for portable appliances and tools, photographic equipment, business machines, communications equipment, toys, hobby equipment, heating and lighting, while Perma-cells are recommended for calculators, tape recorders, toys, flashlights, photoflash equipment, etc. General Electric also supply special aerospace cells.

A range of extended-capability cells are available in three types. Goldtop cells are for normal charge rates and are capable of operating at cell temperatures up to 65°C. These cells are suitable for applications requiring long cell life at high ambient temperatures and/or high cell temperatures caused by continuous long-term overcharge. The second type of Goldtop cells are for use at quick charge rates where continuous overcharge rates of $0.3C$ are combined with high operating temperatures.

Power-Up-15 cells are supplied for use with Power-Up-15 battery charger systems that charge the cell or battery at fast charge rates of $1.0–4.0C$. These cells have voltage/temperature *versus* state of charge characteristics that permit effective control, using circuits that sense both battery temperature and voltage. The charge is thus terminated by either voltage or temperature, permitting reliable fast charging.

The available GEC range is given in Table 51.33. A discharge curve for a typical cell in this range is shown in Figure 51.23.

Oval cells are available at a rated capacity at the 1 h rate of 4 A h at 25°C. Their continuous overcharge rate is 400 mA maximum and 200 mA minimum, and their internal resistance is 5 μΩ. These cells are available either as a standard cell for normal charge rates (model GO4.0) or as a Goldtop cell for normal charge rates (model GO4.0ST). The rectangular cell VO-4 (model GR4.0) has a rated capacity of 4 A h at 25°C at the 1 h rate, maximum and minimum overcharge currents of 400 and 200 mA respectively, and an internal resistance of 4 μΩ.

The trade mark Perma-cell designates a rechargeable nickel–cadmium battery-charger system using sealed cylindrical cells, which consumers may use as alternatives to popular sizes of primary cell. Equipment for properly recharging sealed cells is not usually available where throwaway cells have been used, so General Electric provide the rechargeable cells as part of a whole Perma-cell system, which includes the proper charger for these nickel–cadmium batteries.

Table 51.33 GEC sealed nickel–cadmium cells (nominal voltage 1.25 V)

Type	Model No.	Application*	Rated capacity (mA h) at 25°C (1 h rate)	Continuous overcharge rate (mA)		Internal resistance (μΩ)	Weight (g)
				Max.	Min.		
1/3AA	GCF100ST	A	100	10	5	80	9.9
	XGCF100ST	B	100	10	5	80	9.9
	XKCF100ST	C	100	30	5	80	9.9
	XFCF100ST	D	100	10	5	80	9.9
1/3A	GCK150ST	A	150	15	8	10	11.3
	XGCK130ST	B	150	13	7	10	11.3
	XFCK130ST	C	150	45	7	10	11.3
1/3AF	GCL200ST	A	175	20	10	80	14.2
1/2AA	GCF250ST	A	250	25	13	60	17.0
	XGCF250ST	B	250	25	13	60	17.0
	XKCF250ST	C	250	75	13	60	17.0
	XFCF250ST	D	250	25	13	60	17.0
2/3AF	GCL450ST	A	400	40	20	35	22.7
AA (SD)	GCE450SB	A	450	45	23	35	21.3
AA	GCF500ST	A	500	50	25	28	28.3
	XGCF450ST	B	450	45	23	28	28.3
	XKCF450ST	C	450	150	23	28	28.3
	XFCF450ST	D	450	45	23	28	28.3
1/2Cs	GCR550ST	A	550	55	28	28	28.3
A	GCR0ST	A	600	60	30	26	28.3
	XGCK600ST	B	600	60	30	26	28.3
	XKCK600ST	C	600	180	30	26	28.3
	XFCK600ST	D	600	60	30	26	28.3
Cs	GCR10ST	A	1000	100	50	12	42.5
	XGCR10ST	B	1000	100	50	12	42.5
	XKCR10ST	C	1000	300	50	12	42.5
	XFCR10ST	D	1000	100	50	12	42.5
2/3C	GCT1.1ST	A	1000	110	55	22	45.4
	XGCT900ST	B	900	90	45	15	45.4
	XKCT900ST	C	900	270	45	15	45.4
	XFCT900ST	D	900	90	45	15	45.4
3/5C	GCT10ST	A	1000	100	50	17	45.4
Cs (1.2 A h)	GCR12ST	A	1200	120	60	13	42.5
C	GCT1.5SB	A	1500	150	75	10	65.2
	GCT1.5ST	A	1500	150	75	10	65.2
	XGCT1.5ST	B	1500	150	75	10	65.2
	XKCT1.5ST	C	1500	450	75	10	65.2
	XFCT1.5ST	D	1500	150	75	10	65.2
1/2D	GCW20ST	A	2000	200	100	8	90.7
	XGCW20ST	B	2000	200	100	8	90.7
	XKCW20ST	C	2000	600	100	8	90.7
	XFCW20ST	D	2000	200	100	8	90.7
D	GCW35SB	A	3500	350	175	5	147.0
	GCW35ST	A	3500	350	175	5	147.0
	XGCW35ST	B	3500	350	175	5	147.0
	XKCW35ST	C	3500	1000	175	5	147.0
	XFCW35ST	D	3500	350	175	5	147.0
F	XGCW56ST	B	5600	560	280	3	221.0

* A, standard cell for normal charge rate; B, Goldtop cell for normal charge rate; C, Goldtop cell for quick charge rate; D, Power-Up-15 cell (take fast charge rate up to predetermined fast charge cut-off point)

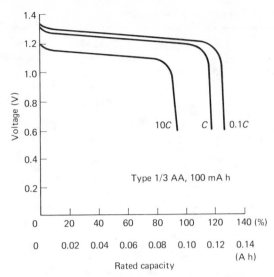

Figure 51.23 Typical discharge curve for General Electric sealed nickel–cadmium cells (Table 51.33) (Courtesy of General Electric)

Table 51.34 GEC charging equipment

Model No.	Cell size	Capacity (A h) 1 h rate	Maximum charge rate (mA)	GEC charger model
GC1	AA	0.5	50	BC-1
GC2	C	1.0	100	BC-1, BC-48B
GC3	D	1.2	100	BC-1, BC-48
GC4	D	3.5	350	BC-448

The Perma-cell system is a complete commercial battery-charger system. The most common sizes of consumer batteries (AA, C, and D) are available in the Perma-cell line. Typically these batteries can be recharged 1000 times or more with the Model BC-1, four-cell charger.

Three larger industrial chargers are available. Model BC-48 will recharge from one to 48 standard D-size model GC-3 Perma-cell batteries, rated 1.2 A h. Model BC-48B will recharge up to 48 C-size model CG-2 Perma-cells. Model BC-448 is designed to recharge from one to 48 heavy-duty model GC-4 Perma-cell batteries (Table 51.34).

The battery-charger systems are designed to provide the economy, convenience and long life of nickel–cadmium sealed cells for a wide variety of replaceable cell applications. These chargers operate under optimum conditions for safe operation (that is, avoidance of pressure and temperature increase). Details of three available types of GEC charger are given in Table 51.34.

Eagle Picher

This company supplies the Carefree II range of rechargeable sealed nickel–cadmium button cells. (The company also supplies sealed lead–acid batteries under the trade name of Carefree.) Three ranges are available. The ZA series (Table 51.35) has twin electrodes and is designed to produce maximum capacity with low internal resistance. The Carefree II nickel–cadmium batteries and cells provide a wide range of capacities (from 30 to 600 mA h) and the flexibility of stacking cells to produce the required voltage.

Table 51.35 Rechargeable Carefree II sealed nickel–cadmium button cells, ZA series, supplied by Eagle Picher

	ZA 30	ZA 35	ZA 60	ZA 110	ZA 180	ZA 200	ZA 280	ZA 600
Nominal voltage (V)	1.2	1.2	1.2	1.2	1.2	1.2	1.2	1.2
Nominal rating (at 5 h rate) (mA h)	30	35	60	110	180	200	280	600
when discharging down to 1.0 V at (mA)	6	7	12	22	36	40	56	120
Rating typical = 100% (mA h)	33	40	75	120	190	220	310	650
Discharge rate, max. (mA)	30	50	75	165	250	100	280	300
Surge power (max. 2 s) (A)	0.3	0.5	0.7	1.5	2	1	2	3
R_i (charged, $f = 50$ cycles), approx. (mΩ)	700	350	300	170	120	170	120	70
Charge rate, max. (mA)	3.0	3.5	6	11	18	20	28	60
Charge time (recommended) (h)	14–16	14–16	14–16	14–16	14–16	14–16	14–16	14–16
Charge voltage (V)	1.3–1.5	1.3–1.5	1.3–1.5	1.3–1.5	1.3–1.5	1.3–1.5	1.3–1.5	1.3–1.5
Current for trickle charge, max. (mA)	0.3	0.35	0.6	1.1	1.8	2.0	2.8	6.0
Cell diameter, d, approx. (mm)	11.5/9.5	15.5	15.5	22	25	22	25	35
Cell height, h, approx. (mm)	5.4	4	6	4.5	6	7.6	9	10
Cell weight, approx. (g)	2.0	2.50	4	6	9	10	13.5	30
Battery diameter, D, approx. (mm)	12	16	16	23	26	23	26	36

Up to 10 cells may be connected in series to form a battery

Table 51.36 Rechargeable Carefree II sealed nickel–cadmium button cells, Z2A series, supplied by Eagle Picher

	Z2A 35	Z2A 110	Z2A 180	Z2A 250	Z2A 300
Nominal voltage (V)	2.4	2.4	2.4	2.4	2.4
Nominal rating (at 5 h rate) (mA h)	35	110	180	250	300
when discharging down to 2.0 V at (mA)	7	22	36	50	60
Rating typical = 100% (mA h)	40	120	190	280	350
Discharge rate, max. (mA)	40	165	250	250	600
Surge power (max. 2 s) (A)	0.5	1.5	2	2	4
R_i (charged, f = 50 cycles), approx. (mΩ)	700	340	240	240	130
Charge rate, max. (mA)	3.5	11	18	25	30
Charge time (recommended) (h)	14–16	14–16	14–16	14–16	14–16
Charge voltage (V)	2.6–3.0	2.6–3.0	2.6–3.0	2.6–3.0	2.6–3.0
Current for trickle charge, max. (mA)	0.35	1.1	1.8	2.5	3.0
Cell diameter, d, approx. (mm)	15.5	22	25	25	35
Cell height, h, approx. (mm)	6.5	7.5	10	13.5	10
Cell weight, approx. (g)	4	10	14	22	32
Battery diameter, D, approx. (mm)	16	23	26	26	36

Up to 10 cells may be connected in series to form a battery

Table 51.37 Rechargeable Carefree II sealed nickel–cadmium button cells, DA and VA series, supplied by Eagle Picher

	DA 280	VA 600
Nominal voltage (V)	1.2	1.2
Nominal rating (at 5 h rate) (mA h)	280	600
when discharging down to 1.0 V at (mA)	56	120
Rating typical = 100% (mA h)	320	750
Discharge rate, max. (mA)	500	1500
Surge power (max. 2 s) (A)	3	8
R_i (charged, f = 50 cycles), approx. (mΩ)	80	30
Charge rate, max. (mA)	28	60
Charge time (recommended) (h)	14–16	14–16
Charge voltage (V)	1.3–1.5	1.3–1.5
Current for trickle charge, max. (mA)	2.8	6.0
Cell diameter, d, approx. (mm)	25	35
Cell height, h, approx. (mm)	9	10
Cell weight, approx. (g)	13.5	30.5
Battery diameter, D, approx. (mm)	26	36

Up to 10 cells may be connected in series to form a battery

The ZA series offers maximum performance when discharged over a long period of time at a low rate. The Z2A (Table 51.36) series has dual twin electrodes which, instead of the standard 1.2 V/cell has an e.m.f. of 2.4 V. This requires less space than individual single cells require. These twin cells can be series connected to provide multiple voltages. The DA and VA series of cells (Table 51.37) provide excellent performance under high-rate discharge or continuous overcharge conditions.

Eagle Picher also supply an NC series of rectangular transistor batteries (Table 51.38).

Typical discharge curves for the ZA, Z2A, DA and VA series of button cells are shown in Figure 51.24. Values shown are for a 10 h discharge of new cells at 20°C after a 16 h charge.

Eagle Picher also supply a range of higher capacity sealed nickel–cadmium cells and batteries (Tables 51.39 and 51.40). The RSN series of nominal 1.3 V cells have capacities in the range

Table 51.38 Rechargeable Carefree II sealed nickel–cadmium rectangular cells, NC series, supplied by Eagle Picher

	1604 NC	1605 NC
Nominal voltage (V)	7.2	8.4
Nominal rating (at 5 h rate) (mA h)	110	110
when discharging at (mA)	22	22
Rating typical = 100% (mA h)	120	120
Discharge rate, max. (mA)	165	165
Surge power (max. 2 s) (A)	1.5	1.5
R_i (charged, $f = 50$ cycles), approx. (mΩ)	1200	1200
Charge rate, max. (mA)	11	11
Charge time (recommended) (h)	14–16	14–16
Charge voltage (V)	9.1–10.5	9.1–10.5
Current for trickle charge, max. (mA)	1.1	1.1
Battery dimensions (length × width × height), approx. (mm)	26 × 16.5 × 48.5	26 × 16.5 × 48.5
Weight, approx. (g)	34	40

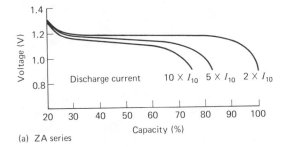

(a) ZA series

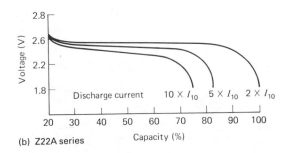

(b) Z22A series

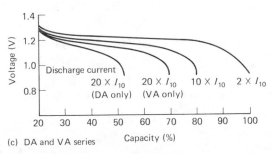

(c) DA and VA series

Figure 51.24 Discharge curves of Carefree II sealed nickel–cadmium button cells series 2A.

3–220 A h. Working voltages are between 1 and 1.25 V/cell. An SAR series of batteries containing between seven and thirty 1.3 V cells is also available. These cells are housed in stainless steel, magnesium and aluminium containers and are equipped with a Safety Pippet valve. Energy densities are 26–36 W h/kg as cells and 19.5–33 W h/kg as batteries. Operating temperature is in the range −18 to 52°C, and storage temperature in the range −18 to 32°C. A minimum shelf life of 5 years is claimed, and a cycle life of 5000. Self-discharge rates per day at various temperatures are given in Table 51.41. A typical cell discharge curve is shown in Figure 51.25. The effect of temperature on cell discharge characteristics is shown in Figure 51.26.

These cells have been used in applications such as orbiting spacecraft, undersea power, environmental test sheds and sealed ground power units.

Ever Ready (Berec) (UK)

This company supplies a range of sintered plate cylindrical cells and mass plate button cells (Tables 51.42 and 51.43). Eveready also supply batteries made up from standard single cell units to meet customers' requirements. These cells operate over a wide temperature range and in a variety of unusual environments such as gravity-free conditions or in a vacuum. The cells may be used or charged in any position. For sintered cells the permanent temperature range is −30 to +50°C on charge and discharge, and for mass plate button cells the temperatures are 0 to 45°C (charge) and −20 to +45°C (discharge). The cells have a good charge retention characteristic, retaining some 70% of capacity after 1 month's storage at 20°C. The charge retention characteristics

Table 51.39 Sealed nickel–cadmium cells available from Eagle Picher

Cell type	Nominal capacity (A h)	Length (mm)	Width (mm)	Height (mm)	Weight (kg)	Volume (cm^3)	Energy density	
							W h/kg	W h/dm^3
RSN 2	3.0	19.8	38.1	87.9	0.14	66	26.4	54
RSN 3	4.0	20.6	53.1	63.0	0.15	69	31.0	70
RSN 6	7.0	21.1	53.3	95.5	0.28	106	29.7	79
RSN 8	10	22.6	75.7	76.2	0.32	130	33.9	92
RSN 8A	10	22.6	75.7	69.3	0.27	119	44.7	101
RSN 9	10	21.1	53.3	114.8	0.36	127	33.4	94
RSN 10	11	22.6	75.7	83.8	0.38	143	35.0	92
RSN 12	14	22.6	75.7	115.8	0.52	198	32.1	85
RSN 12A	14	22.6	75.7	86.9	0.39	149	43.6	113
RSN 15	17	22.6	75.7	126.7	0.61	216	33.2	94
RSN 16	18	37.3	81.0	108.0	0.72	328	29.9	66
RSN 20	28	37.3	81.0	136.7	1.05	413	32.1	82
RSN 20A	23	22.6	75.7	125.2	0.59	215	46.6	129
RSN 20-3	25	22.6	75.7	177.8	0.91	282	33.0	106
RSN 22	26	27.2	81.0	154.9	0.91	341	34.3	92
RSN 30	33	22.6	75.7	190.5	1.09	326	36.3	121
RSN 36	40	37.3	81.0	162.1	1.27	488	37.6	98
RSN 45	50	37.3	81.0	193.5	1.64	585	36.7	102
RSN 55	60	37.3	81.0	215.1	1.82	652	39.6	110
RSN 60	65	33.0	127.0	158.8	2.05	667	38.1	117
RSN 100	110	84.8	81.0	193.5	3.32	1330	39.6	99
RSN 110	120	36.3	185.4	198.1	3.64	1333	39.6	108
RSN 200	220	36.3	254.0	266.7	6.82	2457	38.7	107

Table 51.40 Sealed nickel–cadmium batteries available from Eagle Picher

Battery type	Capacity (A h)	No. of cells	Length (mm)	Width (mm)	Height (mm)	Weight (kg)	Application
SAR-8013	36	20	478	216	178	32.7	USAF
SAR-8022-5	25	20	478	216	165	27.3	USAF
SAR-8054	45	22	566	229	208	46.8	USAF
SAR-8055-5	36	30	671	216	178	48.6	NASA
SAR-8055-19	36	30	671	216	178	55.9	NASA
SAR-8065	6	23	414	102	114	9.5	NASA
SAR-8079	110	24	615	381	264	104.5	NASA
SAR-8083	36	24	488	239	203	40.9	Navy
SAR-8087	15	24	343	183	135	16.8	USAF
SAR-8088	10	7	229	107	94	3.2	USAF
SAR-8091	60	22	592	229	254	54.5	USAF
SAR-8094	35	22	483	206	234	33.2	Space

Table 51.41 Self-discharge of Eagle Picher SAR batteries

Temperature (°C)	Self-discharge per day (%)
26	1.0
38	2.0
49	4.0
60	8.0
71	16.0

Table 51.42 Standard Ever Ready range of mass plate button cells

Ref. No.	Nominal capacity (mA h)	Diameter (mm)	Thickness (mm)	Weight (approx.) (g)
NCB22M	225	25.0	8.8	13
NCB50M	500	34.3	9.5	26
NCB100M	1000	50.3	10.0	57

Table 51.43 Standard Ever Ready range of sintered plate cylindrical cells

Size	Ref. No.	IEC designation	Nominal capacity* (A h)	Max. diameter (mm)	Max. height (mm)	Weight (approx.) (g)
1/3AA	NCC12		0.12	14.3	17.1	8
1/2AA	NCC24		0.24	14.3	28.1	14
AA	NCC50	KR15/51	0.50	14.3	50.0	25
Super AA	NCC60	KR16/51	0.60	15.6	50.0	30
RR	NCC120	KR23/43	1.20	22.8	42.7	50
C	NCC200	KR27/50	2.00	26.0	50.0	78
1/2D	NCC230		2.30	34.0	38.8	100
Dm	NCC350		3.50	32.5	57.5	146
D	NCC400	KR35/62	4.00	34.0	61.0	170
F	NCC700	KR35/92	7.00	34.0	91.0	255
–	NCC1000	KR44/91	10.00	42.0	91.4	410

* Defined as that which will be obtained when fully charged cells are discharged at such a rate as will bring them to an end-point of 1.0 V in 5 h

of mass plate cells are significantly better than those of sintered cells. Charge retention is improved at lower temperatures.

In general, and in agreement with the recommendations of other suppliers of sealed nickel–cadmium batteries, Ever Ready recommend constant-current charging methods for these cells. Constant-potential charging will, because of the very low internal resistance of these cells, lead to the drawing of very high currents, with the possibility of overheating or thermal runaway. These cells are also amenable to fast-charging methods and, when the application requires it, continuous charging can be used.

Yuasa (Japan)

The Yuasa hermetically sealed rechargeable nickel–cadmium alkaline batteries are largely produced in two types: the button type with pressed plates and the cylindrical type with sintered plates (Tables 51.44 and 51.45). These batteries permit overcharge and overdischarge without any deterioration in performance characteristics. Features claimed by Yuasa for their batteries include the following:

1. Electrolyte supply is unnecessary and the batteries are perfectly sealed.
2. All forms of quick charging are permitted.
3. Because the batteries are perfectly sealed they may be placed in any position to meet specific usage requirements.
4. The batteries can be used repeatedly, for as many as 300–3000 times (depending on the charging and discharging conditions).
5. They possess excellent electrical characteristics and one can obtain d.c. current of uninterrupted high efficiency because there is hardly any internal resistance.

6. The batteries are suitable for all emergency power requirements, when the battery is left in the float-charge state.
7. Permits overcharge and overdischarge.

The Yuasa button F cell is suitable for standard discharge requirements while the button FZ cell is suitable for high-rate discharge (for charge and discharge characteristics see Figure 51.27). The Yuasa RS-type cylindrical cell is for high-rate discharge applications, the NR-11 cell for high-temperature trickle charge requirements and the RSF cell for quick charging.

RS cells ensure steady and stable voltage even when internal resistance diminishes and discharge current increases. These cells can readily be used instead of dry cells.

NR-H cells maintain their rated capacity even under high-temperature conditions and are more suitable for continuous trickle charging at high temperatures. Consequently, these cells are recommended for emergency lighting and indicator lamps as an emergency standby power source. The cells have a high charging efficiency, even at high temperatures, and have excellent discharge characteristics. A comparison of the discharge performance of RC and NR-H cells is given in Figure 51.28.

RSF cells are well suited to quick charging due to the special plates used in their construction. Reduction in charging times of up to one-third are possible in comparison with RS cells. Charging methods are simple. These cells can be used as an alternative to dry cells, with the advantage of greater efficiency and more reliable high-rate discharge performance. The standard charging voltage characteristics of the RSF cell are shown in Figure 51.29. The basic data on Yuasa button and cylindrical cells are presented in Table 51.41, with button stack assemblies in Table 51.42.

Table 51.44 Basic data on Yuasa sealed nickel–cadmium cells

Type	JIS	Dry battery code	Nominal capacity (mA h)	Discharge voltage		Standard charging current (mA)	Dimensions		Weight (approx.) (g)
				Average (V)	Final (V)		Diameter (mm)	Height (mm)	
BUTTON TYPE									
For standard discharge									
10F			10 (0.1C discharge)	1.20	1.10	1	11.6	3.3	0.8
30F			30 (0.1C discharge)	1.20	1.10	3	15.6	4.8	2.0
50F			50 (0.1C discharge)	1.20	1.10	5	15.6	6.2	3.5
100F			100 (0.1C discharge)	1.20	1.10	10	25.2	6.3	9.0
150F			150 (0.1C discharge)	1.20	1.10	15	25.2	6.9	11.0
225F			225 (0.1C discharge)	1.20	1.10	22	25.2	8.9	12.5
For high-rate discharge									
225FZ			225 (0.2C discharge)	1.22	1.00	22	25.2	9.2	13.0
500FZ			500 (0.2C discharge)	1.22	1.00	50	34.5	9.9	26.0
CYLINDRICAL TYPE (ROUND TYPE)									
For high-rate discharge									
140RS			140 (0.2C discharge)	1.25	1.00	14	17.0	16.5	12
225RS	NR2/3AA		225 (0.2C discharge)	1.25	1.00	22	14.5	30.0	14
270RS	NR2/3AA		270 (0.2C discharge)	1.25	1.00	27	14.5	30.0	14.5
450RS	NR-AA	UM-3	450 (0.2C discharge)	1.25	1.00	45	14.5	50.0	25
500RS	NR-AA	UM-3	500 (0.2C discharge)	1.25	1.00	50	14.5	50.0	25
600RS			600 (0.2C discharge)	1.25	1.00	60	23.0	26.5	32
1000RS			1000 (0.2C discharge)	1.25	1.00	100	26.0	30.0	48
1200RS	NR-SC		1200 (0.2C discharge)	1.25	1.00	120	23.0	43.0	50
1500RS		UM-2	1500 (0.2C discharge)	1.25	1.00	150	26.0	50.0	70
1650RS	(NR-C)		1650 (0.2C discharge)	1.25	1.00	165	26.0	50.0	73
3500RS	(NR-D)	UM-1	3500 (0.2C discharge)	1.25	1.00	350	33.0	61.0	155
For quick charging									
450SF		UM-3	450 (0.2C discharge)	1.25	1.00	150	14.5	50.0	25
800SF			800 (0.2C discharge)	1.25	1.00	267	23.0	43.0	50
For high-temperature trickle charging									
NR-SCH	(NR-SCH)		1200 (0.2C discharge)	1.25	1.10	40	23.0	43.0	50
NR-CH	(NR-CH)	UM-2	1650 (0.2C discharge)	1.25	1.10	55	26.0	50.0	80
NR-DH	(NR-DH)	UM-1	3500 (0.2C discharge)	1.25	1.10	118	33.0	61.5	165

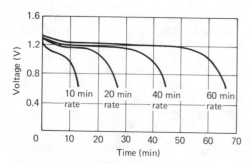

Figure 51.25 Typical cell discharge characteristics (26°C discharge) of Eagle Picher RSN series sealed nickel–cadmium cells (Courtesy of Eagle Picher)

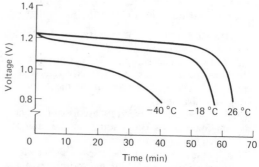

Figure 51.26 Effect of temperature on characteristics of cell discharge at 1 h rate of Eagle Picher RSN series sealed nickel–cadmium cells (Courtesy of Eagle Picher)

Table 51.45 Specification of Yuasa button stack assemblies made up of 2–10 cells of five different types

Battery number	No. of cells per stack	Voltage (V)	Diameter (mm)	Height (mm)	Weight (approx.) (g)
Assemblies of 50F cells, 50 mA h capacity (5 h rate)					
0/50F	2	2.4	15.5 ± 0.5	12 ± 0.5	7.0
3/50F	3	3.6	15.5 ± 0.5	18 ± 0.5	10.5
4/50F	4	4.8	15.5 ± 0.5	24 ± 1	14.0
5/50F	5	6.0	15.5 ± 0.5	30 ± 1	17.5
6/50F	6	7.2	15.5 ± 0.5	36 ± 2	21.0
7/50F	7	8.4	15.5 ± 0.5	42 ± 2	24.5
8/50F	8	9.6	15.5 ± 0.5	48 ± 2	28.0
9/50F	9	10.8	15.5 ± 0.5	54 ± 2	31.5
10/50F	10	12.0	15.5 ± 0.5	60 ± 2	35.0
Assemblies of 150F cells, 150 mA h capacity (5 h rate)					
2/150F	2	2.4	25.5 ± 0.5	14 ± 0.5	22
3/150F	3	3.6	25.5 ± 0.5	21 ± 1	33
4/150F	4	4.8	25.5 ± 0.5	27.5 ± 1	44
5/150F	5	6.0	25.5 ± 0.5	34 ± 1	55
6/150F	6	7.2	25.5 ± 0.5	41 ± 1.5	66
7/150F	7	8.4	25.5 ± 0.5	47.5 ± 1.5	77
8/150F	8	9.6	25.5 ± 0.5	54 ± 2	88
9/150F	9	10.8	25.5 ± 0.5	61 ± 2	99
10/150F	10	12.0	25.5 ± 0.5	67.5 ± 2	110
Assemblies of 225F cells, 225 mA h capacity (5 h rate)					
2/225F	2	2.4	25.5 ± 0.5	18 ± 1	25
3/225F	3	3.6	25.5 ± 0.5	27 ± 1	38
4/225F	4	4.8	25.5 ± 0.5	35.5 ± 1	50
5/225F	5	6.0	25.5 ± 0.5	44 ± 1.5	63
6/225F	6	7.2	25.5 ± 0.5	53 ± 1.5	75
7/225F	7	8.4	25.5 ± 0.5	61.5 ± 1.5	88
8/225F	8	9.6	25.5 ± 0.5	70.5 ± 2	100
9/225F	9	10.8	25.5 ± 0.5	79 ± 2	113
10/225F	10	12.0	25.5 ± 0.5	87.5 ± 2	125
Assemblies of 225FZ cells, 225 mA h capacity (5 h rate)					
2/225FZ	2	2.4	25.5 ± 0.5	18.5 ± 1	26
3/225FZ	3	3.6	25.5 ± 0.5	27.5 ± 1	39
4/225FZ	4	4.8	25.5 ± 0.5	36.5 ± 1	52
5/225FZ	5	6.0	25.5 ± 0.5	45 ± 1.5	65
6/225FZ	6	7.2	25.5 ± 0.5	54 ± 1.5	78
7/225FZ	7	8.4	25.5 ± 0.5	63 ± 2	91
8/225FZ	8	9.6	25.5 ± 0.5	72 ± 2	104
9/225FZ	9	10.8	25.5 ± 0.5	81 ± 2.5	117
10/225FZ	10	12.0	25.5 ± 0.5	90 ± 2.5	130
Assemblies of 500FZ cells, 500 mA h capacity (5 h rate)					
2/500FZ	2	2.4	35 ± 0.5	20.5 ± 1	53
3/500FZ	3	3.6	35 ± 0.5	30.5 ± 1	79
4/500FZ	4	4.8	35 ± 0.5	40.5 ± 1	105
5/500FZ	5	6.0	35 ± 0.5	50 ± 1.5	131
6/500FZ	6	7.2	35 ± 0.5	60 ± 1.5	157
7/500FZ	7	8.4	35 ± 0.5	70 ± 2	183
8/500FZ	8	9.6	35 ± 0.5	80 ± 2	209
9/500FZ	9	10.8	35 ± 0.5	90 ± 2.5	235
10/500FZ	10	12.0	35 ± 0.5	99.5 ± 2.5	261

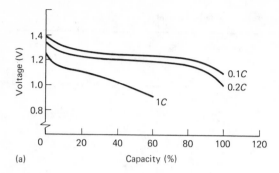

(a)

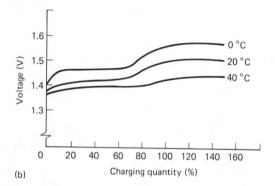

(b)

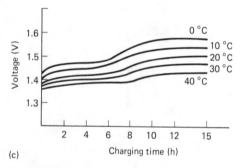

(c)

Figure 51.27 Standard discharge and charging characteristics for Yuasa F type sealed nickel–cadmium button cells. (a) Discharge after charge at 0.1C mA for 15 h, charge and discharge at 20–25°C. (b,c) Charging at 0.1C mA for 15 h (Courtesy of Yuasa)

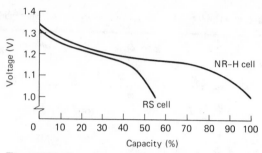

Figure 51.28 Temperature characteristic comparison of Yuasa RS and NR-H sealed nickel–cadmium cells at 45 ± 2°C: charging at 1/30C mA for 48 h, discharge at 1C mA (Courtesy of Yuasa)

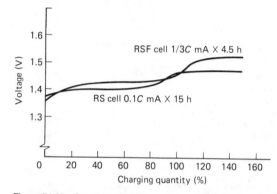

Figure 51.29 Standard charging voltage characteristics of a Yuasa RSF sealed nickel–cadmium battery at 20–25°C (Courtesy of Yuasa)

51.1.2 Unsealed (low-maintenance) nickel–cadmium batteries

Varta

Varta supply a non-sealed double cell (DTN) for portable lamps. This battery is available in a capacity range 4.5–11 A h (Table 51.46). It has a light weight, even voltage level, good capacity, long service life, lack of maintenance and low self-discharge rate.

Table 51.46 Varta DTN range of low-maintenance nickel–cadmium batteries

Battery type	Type No. 333	Nominal capacity in 5 h (A h)	Nominal discharge current in 5 h (A)	Medium discharge voltage (V)	Nominal charging current in 7 h (A)	Charging voltage (V)	Filled weight (kg)	Quantity of electrolyte (dm³)
DTN4.5K	2452 010	4.5	0.9	2.4	0.9		0.42	0.08
DTN6.5K	2652 010	6.5	1.3	2.4	1.3	Increasing	0.61	0.11
DTN7K	2702 010	7.0	1.4	2.4	1.4	from 2.7	0.75	0.14
DTN12K	3122 020	11.0	2.2	2.4	2.2	to 3.6	0.90	0.20

Table 51.47 Unibloc batteries: dependence of maintenance-free intervals on capacity and charging methods

Type	Capacity at 10 h rate (A h)	Maintenance interval (years)	
		Fully automatic operation without boost (float voltage 1.47 V/cell)	Close voltage application with 6-monthly boost of 1.65/1.70 V/cell (float voltage 1.42 V/cell)
5LP7	7.5	8	14
5LP7	12.5	5	9
5LP12	15.0	4	7.5
5LP27	27.0	2.5	5

Chloride Alkad (UK)

Chloride Alkad supply the Unibloc range of alkaline batteries, which are virtually maintenance free but unsealed. These range from 7.5 to 27 A h capacity. Unlike traditional alkaline batteries, which are built up from individual separately packaged cells, Unibloc starts off as a five 1.2 V cell battery in a lidded container. The Unibloc range is available in four basic capacities, each providing low-rate discharge performance:

1. Type SLP7, 7.5 A h.
2. Type SLP12, 12.5 A h.
3. Type SLP15, 15 A h.
4. Type SLP27, 27 A h.

The five cells are incorporated in a plastic case containing sufficient extra potassium hydroxide electrolyte to enable periodic topping up of the battery to extend battery life up to a maximum of 20 years (under ideal conditions), although 7–10 years would probably be the average. These batteries are recommended for applications such as switch-tripping, emergency lighting and alarm systems. For most applications, constant-voltage charging is recommended to minimize electrolyte loss, which is dependent on float charge voltage, the capacity of the battery and the frequency of boost charge. Table 51.47 shows how the maintenance interval across the Unibloc range of batteries is related to battery capacity and charging method. In many cases it should be possible comfortably to exceed these intervals up to a maximum of around 20 mainte-nance-free years for the lowest-capacity batteries on a constant 1.42 V/cell float charge without any boost charging.

The discharge performance of Unibloc batteries is summarized in Table 51.48.

Table 51.48 Chloride Alkad Unibloc battery performance to minimum voltage based on current and duration (fully discharged after 3 h open circuit)

V/cell		Discharge current (A) at 25°C													
		1 s	10 s	30 s	1 min	5 min	10 min	30 min	1 h	1.5 h	2 h	3 h	5 h	8 h	10 h
1.00	5LP7	18	14.0	13.0	12.0	9.2	8.0	6.1	5.1	4.2	3.5	2.6	1.6	1.00	0.83
	5LP12	29	23.0	22.0	19.0	14.8	13.0	10.0	8.3	6.8	5.6	4.1	2.6	1.70	1.33
	5LP15	33	26.0	25.0	22.0	17.3	15.1	11.5	9.6	7.9	6.5	4.8	3.0	1.90	1.55
	5LP27	62	49.0	47.0	42.0	32.4	28.4	22.6	18.1	14.9	12.3	9.1	5.7	3.60	2.92
1.05	5LP7	16	13.0	11.0	10.0	7.6	6.6	5.0	4.3	3.6	3.2	2.4	1.6	1.00	0.83
	5LP12	25	20.0	18.0	17.0	12.4	10.7	8.1	6.7	5.8	5.2	3.6	2.6	1.60	1.32
	5LP15	29	23.0	21.0	19.0	14.4	12.5	9.5	8.1	6.8	6.0	4.6	3.0	1.90	1.53
	5LP27	54	43.0	40.0	36.0	27.2	23.4	17.7	15.1	12.8	11.3	8.6	5.6	3.60	2.90
1.10	5LP7	14	11.0	9.1	8.0	6.2	5.3	4.1	3.7	2.9	2.6	2.2	1.5	1.00	0.81
	5LP12	22	17.0	14.7	13.0	10.0	8.6	6.5	6.0	4.7	4.2	3.5	2.5	1.60	1.31
	5LP15	25	20.0	17.2	15.1	11.7	10.0	7.6	7.0	5.4	4.9	4.1	2.9	1.90	1.53
	5LP27	47	36.0	32.2	28.4	22.0	18.9	14.3	13.2	10.2	9.2	7.7	5.4	3.50	2.88
1.14	5LP7	12	9.0	7.5	7.1	5.0	4.4	3.3	2.9	2.5	2.1	1.7	1.3	0.95	0.78
	5LP12	18	14.3	12.4	11.5	8.1	7.1	5.3	4.7	4.0	3.3	2.8	2.2	1.54	1.27
	5LP15	21	16.4	14.1	13.3	9.5	8.3	6.2	5.5	4.7	3.9	3.3	2.5	1.79	1.48
	5LP27	41	30.8	26.4	25.1	17.7	15.5	11.7	10.2	8.8	7.3	6.1	4.8	3.37	2.78

51.1.3 Large vented nickel–cadmium batteries

Although it is not possible to include all of the many suppliers of these batteries, a representative selection is covered below, including the various types of such batteries that are available for different applications.

Nife

Nife supply three versions of the vented nickel–cadmium battery, the H, M and L series, which overlap each other in performance.

The H cell (high-performance cell for short discharge times) has the thinnest plates and hence the greatest plate area per amount of active material.

The L cell (low-performance cell for long discharge times) has the thickest plates and hence the greatest amount of active material per plate area.

The M cell (medium-performance) has a plate thickness between that of the H and L cells. Power and energy are balanced between the two extremes of H and L.

Charging These batteries can be charged by all normal methods. Generally, batteries in parallel operation with charger and load are charged with constant voltage. In operations where the battery is charged separated from the load, charging with constant current or declining current is recommended. High-rate charge or overcharge will not damage the battery, but excessive charging will increase water consumption to some degree.

In constant-voltage charging (Figure 51.30) for continuous parallel operation with occasional battery discharge, the recommended charging voltages are:

1. Float charge:
   1.40–1.42 V cell for types H, M and L.
2. High-rate charge:
   1.55–1.65 V/cell for types H and M.
   1.55–1.70 V/cell for type L.

The minimum recommended charger rating for high-rate charging is:

1. $0.1C_5$ A for type H.
2. $0.05C_5$ A for types M and L.

Discontinuous parallel operation, for example for generator-charged starter batteries, normally uses short heavy-current discharge or other shallow discharge. The recommended charging voltage is:

1. 1.50–1.55 V/cell for types H and M.
2. 1.50–1.58 V/cell for type L.

In buffer operation, where the load temporarily exceeds the charger rating, the recommended charging voltage is:

1. 1.50–1.55 V/cell for type H.
2. 1.55–1.60 V/cell for types M and L.

The following maximum charging current may be used provided that the charger voltage limit is not higher than 1.72 V/cell:

1. $5C_5$ A for type H.
2. $2.5C_5$ A for type M.
3. $1.5C_5$ A for type L.

Constant-current charging currents (Figure 51.31) are:

1. Normal charging: $0.2C_5$ A for 7 h for types H, M and L.

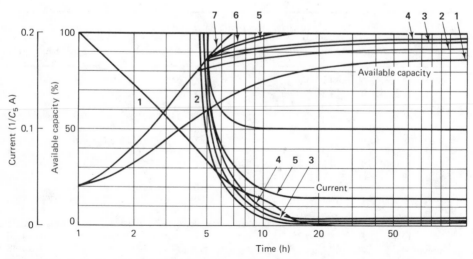

Figure 51.30 Typical constant-voltage recharge characteristics for M-type cells available from Nife. Charging voltage (V/cell): **1**, 1.40; **2**, 1.45; **3**, 1.50; **4**, 1.55; **5**, 1.60; **6**, 1.65; **7**, 1.70 (Courtesy of Nife Jungner)

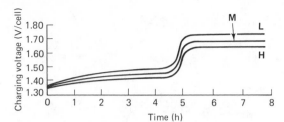

Figure 51.31 Typical constant current charge characteristics. Charging rate $0.2C_5$ A at 25°C (Courtesy of Nife Jungner)

2. Recommended fast recharging: $0.4C_5$ A for 2.5 h followed by $0.2C_5$ A for 2.5 h.
3. Minimum recommended charger rating for high-rate recharging: $0.1C_5$ A for type H; $0.05C_5$ A for types M and L.
4. Minimum recharging rate: $0.003C_5$ A.
5. Trickle charging rate: 0.001–$0.002C_5$ A. Adjust the current value to achieve a charging voltage level of 1.40–1.42 V/cell steady state.

Typical discharge performance data for some of the Nife range is shown in Table 51.49 (to 1.14 V end-voltage per cell) and Table 51.50 (to 0.83 V end-voltage per cell).

The tabulated discharge performance data and the rated capacities are valid for fully charged cells. For all cell types the rated capacity C_5 is defined as available ampere hours at 5 h discharge to an end-voltage of 1.0 V/cell. The rated capacity is no measure of performance. The performance depends on the battery construction. For example, an H-type cell at 15 min discharge can deliver about twice the discharge current compared to an L-type cell of equal rated capacity. It is therefore advisable to use the discharge performance tables to find the proper H, M and L type alternatives for a specific application. The final choice of cell type should be made by comparison of prices, dimensions, etc.

Typical discharge performance curves for the three types of battery are shown in Figure 51.32.

Varta

Varta produce a range of vented nickel–cadmium stationary batteries designed for areas of application with special operation conditions such as:

1. Ambient temperatures between 0 and −40°C.
2. Danger of deep discharge.
3. High mechanical stresses.
4. Limited space availability.
5. Possibility of being taken out of service.

They produce two ranges of batteries, the Special X and the Special M range, which differ from one another in the construction of the electrodes used.

The Special X range is for application with extreme electrical and physical conditions, and for peak current loads in the seconds and minutes ranges. At ambient temperatures of −20°C, these batteries yield 90% of their rated capacity.

The Special M range is for areas of application with battery bridging periods between 30 min and several hours.

The electrical characteristics of the product range available from Varta are shown in Table 51.51.

McGraw Edison

This company produces three ranges of batteries:

1. The CED range for low-rate discharge applications such as communications standby systems.
2. The ED and MED ranges for medium-rate discharges in applications, respectively, such as railway signalling, communications, rail transit (ED) and switch gear power operations (MED).
3. The HED range for high-rate discharge applications such as engine/turbine starting.

All the batteries in these ranges have a nominal voltage of 1.2 V/cell. The charging voltages on float charge and high-rate charge, respectively, are 1.42–1.47 and 1.52–1.60 V/cell and operating temperatures are in the range −40 to 45°C.

The CED range is used for communications, telecommunications, security and fire alarm systems, and low power control loads with long time periods. The electrical characteristics of this range are presented in Tables 51.52 and 51.53.

The ED range is used for railway signalling, communications and rail transit applications. The electrical characteristics of this range are presented in Table 51.54.

The MED range is used for uninterrupted power systems, high/low-rate duties, switchgear tripping and closing, emergency lighting power systems and solar applications. The electrical characteristics of this range are presented in Tables 51.55 and 51.56.

The HED range is used for engine starting, control power, uninterrupted power supply, switchgear tripping and closing, and pump and solenoid operation. The electrical characteristics of this range are presented in Tables 51.57–51.59.

FRIWO (West Germany)

This company manufactures two ranges. The T/TP or HK/HKP range is designed for normal discharge currents, i.e. capacity discharge of more than 1 h up to 3 h. The TS/TSP range is designed for high currents, i.e. a discharge of 1 h or less.

Each range has its advantages within its application limits. The T/TP and HK/HKP ranges offer a maximum capacity at low volume and low weight, whereas the TS/TSP range provides high-current short period discharges.

Table 51.49 Nife large vented range of nickel–cadmium batteries types H, M and L cell performance to end-voltage 1.14 V/cell available amps at 25°C fully charged

Cell type	C_5 A h	Hours							Minutes						Seconds			
		10	8	5	3	2	1.5	1	30	20	15	10	5	1	30	15	5	1
H 202	9	0.94	1.12	1.76	2.82	4.02	5.1	7.0	11.4	14.3	16.4	19.4	23.7	34.3	38.8	42.6	48.2	56
H 203	13	1.41	1.68	2.64	4.23	6.0	7.7	10.5	17.1	21.5	24.7	29.1	35.6	52	58	64	72	84
H 204	18	1.88	2.24	3.52	5.6	8.0	10.2	14.0	22.8	28.7	32.9	38.8	47.4	69	78	85	96	111
H 302	21	2.16	2.66	4.14	6.6	9.5	12.0	16.5	26.7	33.8	38.6	44.7	53	72	80	87	98	110
H 402	27	2.72	3.32	5.2	8.3	11.8	15.0	20.6	33.5	42.1	48.4	56	66	86	93	101	109	120
H 303	32	3.24	3.99	6.2	9.9	14.2	18.1	24.8	40.1	51	58	67	79	108	120	131	146	165
H 403	40	4.08	4.98	7.7	12.4	17.8	22.6	31.0	50	63	73	84	99	128	140	151	163	179
H 304	43	4.32	5.3	8.3	13.2	19.0	24.1	33.0	53	68	77	89	106	144	159	175	195	220
H 404	54	5.4	6.6	10.3	16.5	23.7	30.1	41.3	67	84	97	112	132	171	187	202	218	239
M 421	564	57	70	108	168	235	288	371	502	562	603	652	715	870	952	1020	1140	1290
M 327	581	58	72	111	173	242	296	382	515	579	621	672	760	966	1060	1160	1310	1490
M 328	603	60	74	115	180	251	307	396	535	601	644	697	788	1000	1100	1200	1360	1540
M 424	645	65	80	123	192	269	330	424	574	642	689	745	817	995	1090	1170	1310	1480
M 427	726	73	90	139	217	302	371	477	646	722	775	839	919	1120	1220	1320	1470	1660
M 428	753	76	93	144	225	313	384	495	670	749	804	870	953	1160	1270	1370	1530	1720
M 336	775	78	95	148	231	323	395	509	687	773	828	896	1010	1290	1420	1540	1740	1990
M 432	860	87	106	164	257	358	439	566	765	856	919	994	1090	1330	1450	1560	1740	1970
M 436	968	98	120	185	289	403	494	636	861	963	1030	1120	1230	1490	1630	1760	1960	2210
M 345	969	97	119	184	289	404	494	636	859	966	1040	1120	1270	1610	1770	1930	2180	2480
M 440	1070	108	133	205	321	448	549	707	957	1070	1150	1240	1360	1660	1810	1950	2180	2460
M 445	1210	122	149	231	361	504	618	796	1080	1200	1290	1400	1530	1870	2040	2190	2450	2770
L 101	2	0.19	0.97	1.48	2.18	2.88	3.43	4.19	5.3	5.8	6.1	6.6	7.5	9.9	11.0	12.2	13.7	15.8
L 201	15	1.58	1.93	2.96	4.35	5.6	6.8	8.4	10.7	11.7	12.4	13.1	14.7	18.7	20.5	22.3	25.0	28.8
L 202	30	3.16	3.86	5.9	8.7	11.5	13.7	16.8	21.4	23.4	24.7	26.1	29.5	37.4	41.0	44.6	50	58
L 301	34	3.47	4.24	6.5	9.6	12.6	15.0	18.4	23.5	25.8	27.2	28.8	31.5	37.3	40.3	43.1	47.1	53
L 401	44	4.52	5.5	8.5	12.5	16.4	19.6	24.0	30.1	31.8	32.6	33.5	35.8	42.1	45.2	48.3	53	60
L 203	46	4.74	5.8	8.9	13.1	17.3	20.6	25.1	32.0	35.1	37.1	39.2	44.2	56	62	67	75	86
L 204	60	6.3	7.7	11.8	17.4	23.0	27.4	33.5	42.8	46.8	49.6	62	59	75	82	89	100	115

Type	C1	C2	C3	C4	C5	C6	C7	C8	C9	C10	C11	C12	C13	C14	C15	C16	C17	C18
L 302	68	6.9	8.5	13.0	19.2	25.3	30.1	36.8	47.1	52	54	58	63	75	81	86	94	106
L 402	88	9.0	11.0	17.0	25.0	32.9	39.3	48.0	60	64	65	67	72	84	90	97	106	119
L 303	102	10.4	12.7	19.5	28.7	37.9	45.1	55	71	77	82	86	95	112	121	129	141	160
L 403	133	13.6	16.6	25.4	37.4	49.3	59	72	90	95	98	100	107	126	135	145	159	179
L 304	136	13.9	17.0	26.0	38.3	51	60	74	94	103	109	115	126	149	161	173	189	213
L 305	170	17.4	21.2	32.6	47.9	63	75	92	118	129	136	144	158	187	202	216	236	266
L 404	177	18.1	22.1	33.9	49.9	66	79	96	120	127	130	134	143	168	181	193	212	239
L 306	204	20.8	25.4	39.1	57	76	90	110	141	155	163	173	189	224	242	259	283	319
L 405	222	22.6	27.6	42.4	62	82	98	120	150	159	163	167	179	210	226	241	265	298
L 307	239	24.3	29.7	45.6	67	88	105	129	165	181	190	202	221	261	282	302	330	373
L 406	266	27.1	33.1	51	75	99	118	144	180	191	195	201	215	252	271	290	318	358
L 308	273	27.8	33.9	52	77	101	120	147	188	207	217	231	252	288	323	345	377	426
L 309	307	31.2	38.2	59	86	114	135	166	212	232	245	259	284	336	363	388	424	479
L 407	311	31.6	38.6	59	87	115	137	168	210	223	228	234	250	294	316	338	371	418
L 310	341	34.7	42.4	65	96	126	150	184	235	258	272	288	315	373	403	431	471	532
L 408	355	36.2	44.2	68	100	131	157	192	241	254	261	268	286	337	361	386	424	478
L 311	375	38.2	46.6	72	105	139	165	203	259	284	299	317	347	411	444	474	519	586
L 409	400	40.7	49.7	76	112	148	177	216	271	286	293	301	322	379	406	435	477	537
L 312	409	41.6	51	78	115	152	180	221	282	310	326	346	378	448	484	518	566	639
L 410	444	45.2	55	85	125	164	196	240	301	318	326	335	358	421	452	483	530	597
L 411	489	49.7	61	93	137	181	216	264	331	350	358	368	393	463	497	531	583	657
L 315	512	52	64	98	144	189	226	276	353	387	408	432	473	560	605	647	707	798
L 412	533	54	66	102	150	197	236	288	361	382	391	402	429	505	542	579	636	716
L 318	614	62	76	117	172	227	271	331	424	465	489	519	568	672	726	776	849	958
L 415	667	68	83	127	187	246	295	360	451	477	489	502	536	631	677	724	795	895
L 320	682	69	85	130	192	253	301	368	471	516	543	577	631	746	806	863	943	1060
L 418	800	81	99	153	225	296	354	432	541	573	586	602	644	757	813	869	955	1070
L 324	819	83	102	156	230	303	361	442	565	620	652	692	757	896	968	1040	1130	1280
L 325	853	87	106	163	239	316	376	460	588	645	679	721	788	933	1010	1080	1180	1330
L 420	889	90	110	170	250	329	393	480	601	636	652	669	715	841	903	996	1060	1190
L 330	1020	104	127	195	287	379	451	552	706	775	815	865	946	1120	1210	1290	1410	1600
L 424	1060	108	132	204	300	394	471	576	722	763	782	803	858	1010	1080	1160	1270	1430
L 425	1110	113	138	212	312	411	491	600	752	795	814	836	894	1050	1130	1210	1330	1490
L 430	1330	136	166	254	374	493	589	720	902	954	977	1000	1070	1260	1350	1450	1590	1790

Table 51.50 Nife large vented nickel–cadmium battery types H, M and L Cell performance to end-voltage 0.83 V/cell available amps at 25°C fully charged

Cell type	Minutes			Seconds				Cell type	Minutes			Seconds			
	10	5	1	30	15	5	1		10	5	1	30	15	5	1
H 202	33.8	55	119	136	151	170	197	M 412	1010	1360	1790	1930	2070	2260	2580
H 203	51	83	179	205	226	255	296	M 413	1090	1470	1940	2090	2240	2440	2790
H 204	68	111	239	273	301	340	394	M 414	1180	1580	2090	2250	2410	2630	3010
H 302	78	122	257	289	311	340	373	M 318	1210	1660	2290	2550	2750	3050	3340
H 402	97	153	303	333	357	381	421	M 415	1260	1700	2240	2410	2580	2820	3220
H 303	116	184	386	434	466	511	560	M 416	1350	1810	2390	2570	2760	3010	3440
H 403	146	229	454	500	535	572	632	M 321	1420	1930	2670	2970	3210	3550	3890
H 304	155	245	514	579	621	681	746	M 417	1430	1920	2540	2730	2930	3200	3650
H 404	194	306	606	667	713	763	842	M 418	1510	2040	2690	2890	3100	3380	3860
H 305	194	306	643	724	777	851	933	M 421	1770	2380	3140	3370	3620	3950	4510
M 303	202	276	381	424	459	508	556	L 418	1900	2240	2740	2940	3130	3370	3730
M 403	252	339	449	482	517	564	644	L 324	2080	2440	3160	3440	3670	3980	4350
M 304	270	368	508	566	612	677	742	L 325	2160	2540	3290	3590	3830	4140	4530
M 404	336	453	598	642	689	752	859	L 420	2110	2490	3040	3270	3470	3750	4140
M 405	420	566	748	803	861	940	1070	L 330	2600	3050	3950	4300	4590	4970	5430
M 406	504	679	879	963	1030	1130	1290	L 424	2540	2990	3650	3920	4170	4500	4970
M 407	588	792	1050	1120	1210	1320	1500	L 425	2640	3120	3800	4080	4340	4690	5170
M 408	673	905	1200	1280	1380	1500	1720	L 430	3170	3740	4560	4900	5210	5620	6210
M 409	757	1020	1350	1440	1550	1690	1930								
M 410	841	1130	1500	1610	1720	1880	2150								
M 411	925	1240	1650	1770	1890	2070	2360								
M 314	944	1290	1780	1980	2140	2370	2600								

Table 51.51 Varta large vented nickel–cadmium stationary batteries: electrical data

Type designation	Capacity (A h)				Discharge current (A)				Final discharge voltage (V)			
	5 h	1 h	½ h	¼ h	5 h	1 h	½ h	¼ h	5 h	1 h	½ h	¼ h
Special X												
Vs 410 X	30	28.5	27	24.75	6	28.5	54	99				
Vs 415 X	50	47.5	45	41.25	10	47.5	90	165				
Vs 420 X	65	61.8	58.5	53.75	13	61.8	117	215				
Vs 424 X	80	76	72	66	16	76	144	264				
Vs 431 X	100	95	90	82.5	20	95	180	330	1.10			
Vs 435 X	110	105	99	90.75	22	105	198	363		1.05		
Vs 443X	140	133	126	115.5	28	133	252	462			1.03	
Vs 451 X	165	157	149	136.25	33	157	298	545				0.90
Vs 461 X	200	190	180	165	40	190	360	660				
Special M												
Vs 105 M	45	43.2	40.5	37.8	9	14.4	20.25	37.8				
Vs 107 M	60	57.6	54	50.4	12	19.2	27	50.4				
Vs 110 M	85	81.6	76.5	71.4	17	27.2	38.25	71.4				
Vs 112 M	105	101	94.5	88.2	21	33.7	47.25	88.2	1.0			
Vs 115 M	130	125	117	109	26	41.7	58.5	109		0.98		
Vs 117 M	150	144	135	126	30	48	67.5	126			0.96	
Vs 121 M	185	178	167	155	37	59.3	83.5	155				0.91
Vs 125 M	215	206	194	181	43	68.7	97	181				
Vs 129 M	255	245	230	214	51	81.7	115	214				

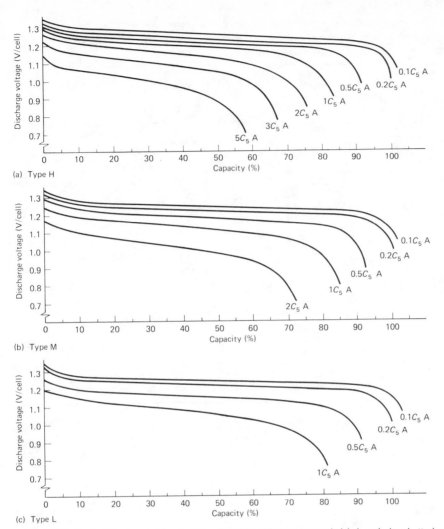

(a) Type H

(b) Type M

(c) Type L

Figure 51.32 Typical discharge characteristics at 25°C for Nife large vented nickel–cadmium batteries (Courtesy of Nife Jungner)

Table 51.52 Electrical performance data for McGraw Edison low-rate vented nickel–cadmium batteries, CED series

Cell type	Discharge current (A) to 1.14 V at 25°C							
	Nominal capacity (A h)	10 h	8 h	5 h	4 h	3 h	2 h	1½ h	1 h
CED-16	16	1.53	1.87	2.83	3.34	3.97	4.9	5.4	6.25
CED-32	32	3.07	3.74	5.7	6.5	7.7	9.4	10.6	12.3
CED-43	43	4.12	5.01	7.65	8.6	10.2	12.1	13.7	15.8
CED-58	58	5.5	6.75	9.1	10.2	11.45	13.25	14.5	16.6
CED-79	79	7.58	9.31	13.1	14.6	16.7	19.6	22	25
CED-100	100	9.6	11.8	16.9	18.8	22	26.1	29.2	33
CED-120	120	11.5	14.1	19.4	22.2	26.1	31.1	34.3	39.1
CED-150	150	14.5	17.7	24.1	27.6	32.2	38.9	42.7	48.8
CED-183	183	17.5	21.6	30	34	39.7	47.9	52.7	59.5
CED-216	216	20.8	25.5	36.2	41	47.6	57.1	63.1	72
CED-250	250	23.8	29.4	41.4	47.6	56.1	67.1	73	83

Table 51.53 Electrical performance data for McGraw Edison low-rate vented nickel–cadmium batteries, CED series

Cell type	Discharge current (A) to 1.00 V at 25°C								
	Nominal capacity (A h)	10 h	8 h	5 h	4 h	3 h	2 h	1½ h	1 h
CED-16	16	1.63	2.03	3.19	3.97	5.05	6.85	8	9.25
CED-32	32	3.28	4.06	6.4	7.8	9.85	13.86	16.8	21
CED-43	43	4.39	5.46	8.6	10.6	13.4	17.7	21.8	25.3
CED-58	58	5.92	7.35	11.5	13.7	16.3	21.5	25	30.5
CED-79	79	8.06	9.95	15.45	18.8	23	31.1	36.3	45
CED-100	100	10.2	12.55	20	25	31	40.5	48.8	60
CED-120	120	12.3	15.2	23.1	28.5	36.1	48.5	56.5	68
CED-150	150	15.3	18.9	29	36	45.5	63	74.1	88
CED-183	183	18.7	23.1	35.9	44	55.1	76.7	90.2	107.6
CED-216	216	22.1	27.3	40.5	50.7	65.1	90.5	106.5	127
CED-250	250	25.5	32.1	50.1	61.9	79	106	124	146

Table 51.54 Electrical performance data for McGraw Edison medium-rate large nickel–cadmium batteries, ED series

Cell type	Discharge period							
	8 h	5 h	3 h	60 min	30 min	15 min	5 min	1 min
Current (A) to 1.14 V/cell at 25°C								
ED-80	10	15.3	22.7	43	53	61	74	92
ED-120	15	22.3	35	62	76	89	113	138
ED-160	20	29.3	44	69	83	96	118	144
ED-240	30	44	66	110	132	152	184	222
ED-340	42.5	64	90	155	187	215	258	306
ED-400	50	74	105	178	214	247	298	360
Current (A) to 1.10 V/cell at 25°C								
ED-80	10.4	16	24.8	48	66	76	90	110
ED-120	15.7	24	38.2	75	92	108	134	164
ED-160	21	32	48.8	90	105	118	144	180
ED-240	31.5	48	72	128	152	178	220	274
ED-340	44.2	68	102	180	214	250	308	384
ED-400	52.3	80	119	211	251	294	352	445
Current (A) to 1.00 V/cell at 25°C								
ED-80	10.7	16.6	25.6	58	90	111	118	141
ED-120	16.1	25.2	39.5	91	130	153	178	212
ED-160	21.4	33.6	50.4	112	153	177	196	233
ED-240	32	51	74	165	230	273	321	369
ED-340	44.8	70.2	104	232	324	384	454	520
ED-400	52.7	83	122	272	384	450	534	610

Table 51.55 Electrical performance data for McGraw Edison medium-rate large nickel-cadmium batteries, MED series

Cell type	Nominal capacity (A h)	Discharge current (A) to 1.14 V at 25°C																
		Hours							Minutes						Seconds				
		8	5	4	3	2	1½	1	30	20	15	10	5	1½	60	30	10	5	1
MED-10	10	1.25	1.93	2.4	3	3.9	4.6	5.5	7	8	8.8	10	12	15	16	18	20.5	22	25
MED-24	24	3	4.63	5.7	7	9	10.5	12.3	15.3	16.7	18.5	20.8	24	31	33.5	38	45	48	55
MED-36	36	4.5	6.94	8.5	10.6	13.9	16.2	18.7	25	28	30	33	38	47	50	56	66	72	82
MED-48	48	6	9.2	11.4	14.2	19.2	22	25	34.5	40	43	47	54	64	68	74	84	91	102
MED-65	65	8.12	12.5	15.5	19.2	26	29.5	34	46	53	58	63	72	86	91	99	113	122	134
MED-80	80	10	15.4	19	23.6	32	36	42	55	63	69	76	87	106	112	122	139	150	165
MED-105	105	13.1	20.2	25	31	41	46	54.5	69	77	82	88	101	121	128	143	159	172	191
MED-126	126	15.8	24.3	30	37	49	55	65	80	89	94	104	120	143	152	166	183	197	215
MED-147	147	18.4	28.3	35	43.5	57	66	77	98	110	116	126	143	172	182	200	221	235	254
MED-168	168	21	32.4	40.2	50	66	75	87	112	125	133	144	163	199	208	226	250	267	290
MED-189	189	23.6	36.4	45.1	56	74	84	98	125	140	148	160	183	220	232	255	282	299	317
MED-210	210	26.3	40.5	50	62	82	93	110	138	154	164	178	202	243	258	280	310	330	350
MED-231	231	28.9	44.5	54.9	68	90	102	120	152	169	180	195	222	265	285	305	338	360	385
MED-240	240	29	45	55	70	90	106	124	150	162	172	185	213	254	273	292	324	345	368
MED-270	270	33	51	62	79	102	119	140	169	184	193	211	242	289	311	333	369	393	419
MED-300	300	37	57	70	88	115	132	155	188	206	219	236	271	324	349	373	414	441	470
MED-340	340	42	65	79	100	130	150	176	214	234	251	268	308	368	396	424	470	500	534
MED-360	360	45	69	85	105	140	159	186	226	251	267	288	330	394	424	454	504	536	572
MED-400	400	50	76	95	117	155	176	207	252	279	298	320	367	438	472	504	560	596	636

Table 51.56 Electrical performance data for McGraw Edison medium-rate large nickel-cadmium batteries, MED series

Cell type	Nominal capacity (Ah)	Discharge current (A) to 1.00 V at 25°C																	
		Hours							Minutes						Seconds				
		8	5	4	3	2	1½	1	30	20	15	10	5	1½	60	30	10	5	1
MED-10	10	1.31	2.06	2.6	3.3	4.8	6.1	8.5	12	14	15.5	17.5	21	26	27.5	30.5	35	37	42
MED-24	24	3.15	4.94	6.3	8	11.6	14.2	19.3	26.5	30.5	33	37	42	52	56	63.5	75	82	94
MED-36	36	4.72	7.41	9.4	11.9	17.5	22	29	40	46	50	56	65	78	84	95	112	123	140
MED-48	48	6.28	9.9	12.5	15.8	23.5	29.5	39	60	67	73	80	91	108	115	125	143	153	173
MED-65	65	8.5	13.3	16.9	21.4	32	40	53	81	91	99	108	123	147	155	169	193	207	228
MED-80	80	10.5	16.5	20.9	26.4	39	49	65	100	112	121	133	151	180	191	208	235	254	280
MED-105	105	13.7	21.6	27.5	35	50	64	84	114	129	136	151	172	204	217	244	270	288	325
MED-126	126	16.5	25.9	33	42	61	76	100	136	151	161	175	200	242	258	284	314	335	370
MED-147	147	19.3	30.2	38.1	48	69	89	119	168	190	201	219	247	297	312	343	378	400	435
MED-168	168	22.1	34.6	43.6	55	79	102	135	190	215	229	248	280	336	355	390	430	455	495
MED-189	189	24.7	38.9	49.6	63	90	114	153	213	240	257	276	311	370	395	435	480	510	540
MED-210	210	27.5	43.2	55	70	100	127	170	235	265	283	305	345	408	435	480	530	560	590
MED-231	231	30.3	47.5	60.5	77	110	140	185	258	288	308	330	375	440	475	515	575	615	650
MED-240	240	31.1	48.2	74	79.1	110	144	192	264	276	308	315	362	432	465	497	558	594	634
MED-270	270	35.1	54.5	84	89.2	125	162	216	297	314	346	359	412	492	529	566	635	676	722
MED-300	300	39.5	61.0	94	100.0	140	180	240	330	352	384	402	462	552	593	635	712	758	809
MED-340	340	44.2	69.0	107	113	160	204	272	374	400	438	457	524	626	674	720	809	861	918
MED-360	360	46.7	73	115	119	171	216	288	396	428	465	489	562	671	722	772	866	922	984
MED-400	400	52	81	128	133	190	240	320	440	476	518	544	624	746	802	858	963	1025	1093

Table 51.57 Electrical performance data for McGraw Edison high-rate large vented nickel-cadmium batteries, HED series

		Discharge current (A) to 1.14 V at 25°C																
		Hours						Minutes							Seconds			
Cell type	Nominal capacity (A h)	8	5	3	2	1½	1	30	20	15	10	5	1½	1	30	10	5	1
HED-10	10	1.24	1.97	3.2	4.6	6	8.3	12.6	16	17.3	21	26	36	39	45	51	55	65
HED-15	15	1.87	2.96	4.8	7	9	12.5	19	24	26.5	32	40	54	59	68	77	83	98
HED-20	20	2.5	4	6.4	9.3	12	16.7	25	30	33.5	38	47	63	67	75	87	95	107
HED-25	25	3.12	4.93	8	11.6	15	20.8	31	37	42	47	58	78	84	94	109	118	134
HED-30	30	3.75	5.92	9.6	13.9	18	25	37	44	50	56	70	94	101	113	131	142	161
HED-40	40	5	7.9	12.8	18.7	24	33	51	61	67	77	93	117	124	139	154	164	188
HED-46	46	5.75	9.1	15	21.5	28	38	59	70	78	89	108	135	144	161	179	191	219
HED-53	53	6.63	10.45	17.1	24.9	32	44	68	80	89	102	123	156	165	184	204	218	250
HED-60	60	7.5	11.85	19.3	28	36	50	76	90	100	115	138	175	185	207	229	245	281
HED-70	70	8.75	13.8	22.5	32.6	42	58	89	107	119	139	168	211	228	255	287	307	350
HED-80	80	10	15.7	25.8	37.3	48	67	102	122	136	159	190	233	253	283	319	341	389
HED-90	90	11.25	17.7	29	42	54	76	114	137	154	179	213	263	281	310	358	388	433
HED-100	100	12.5	19.6	31.8	46	60	83	125	153	170	196	238	295	315	350	390	413	465
HED-110	110	13.75	21.7	35.7	51.4	66	91	138	168	187	216	261	323	345	380	420	445	505
HED-120	120	15	23.7	39	56	72	100	150	183	204	235	285	350	375	415	460	485	550
HED-130	130	16.25	25.5	41.6	60	78	107	162	195	218	257	310	400	430	480	540	570	645
HED-140	140	17.5	27.4	44.8	64.5	81	115	175	210	235	275	332	420	455	505	575	610	690
HED-150	150	18.75	29.4	48	69	87	123	187	225	254	295	355	445	480	530	610	650	735
HED-160	160	20	31.3	51	74	93	131	200	240	265	307	378	475	510	570	640	680	780
HED-170	170	21.2	33.5	54.3	78.5	99	140	209	253	279	325	402	505	540	595	665	710	810
HED-180	180	22.5	35.3	57.5	83	105	148	222	268	296	344	425	535	570	630	700	750	860

Table 51.58 Electrical performance data for McGraw Edison high-rate large vented nickel-cadmium batteries, HED series

Cell type	Nominal capacity (A h)	Discharge current (A) to 1.00 V at 25°C																
		Hours						Minutes							Seconds			
		8	5	3	2	1½	1	30	20	15	10	5	1½	1	30	10	5	1
HED-10	10	1.3	2.03	3.4	5	6.5	9.3	18	22	26.7	34.5	47	62	67	76	87	95	112
HED-15	15	1.95	3.04	5.1	7.5	9.7	14	27	33	40	52	71	92	101	114	131	143	168
HED-20	20	2.6	4.07	6.8	10	13	18.7	34	45	52	63	82	106	115	129	151	161	184
HED-25	25	3.25	5.08	8.5	12.5	16.3	23.3	42	56	65	79	102	133	144	161	188	202	231
HED-30	30	3.9	6.1	10.2	15	19.5	28	51	67	78	94	122	162	173	193	226	242	277
HED-40	40	5.2	8.1	13.6	19.9	26	37	69	91	105	124	157	209	220	240	267	287	320
HED-46	46	6	9.4	15.6	22.9	30	42.5	80	105	121	143	181	240	256	279	311	334	367
HED-53	53	6.9	10.8	17.9	26.4	34.5	49	92	120	139	165	209	276	292	318	355	381	418
HED-60	60	7.8	12.2	20.2	29.8	39	55	104	135	156	185	235	312	328	358	399	428	470
HED-70	70	9.1	14.5	24.2	35	46	67	116	158	183	218	285	364	392	441	494	535	598
HED-80	80	10.4	16.5	27.6	40	52	76	132	180	209	249	325	404	435	490	548	594	664
HED-90	90	11.7	18.6	31	45	59	86	149	203	235	280	365	455	485	532	615	660	744
HED-100	100	13	20.7	35	50	65	95	165	216	250	300	390	500	530	590	660	700	795
HED-110	110	14.3	22.7	37.5	55	71	105	179	238	277	330	430	545	580	650	725	765	865
HED-120	120	15.6	24.7	41	60	78	114	195	260	305	360	470	595	638	710	790	835	945
HED-130	130	16.9	26.8	45	65	85	122	215	285	335	400	525	685	740	830	920	975	1110
HED-140	140	18.2	29	48.5	70	91	131	232	310	360	430	550	718	775	880	985	1045	1190
HED-150	150	19.5	31	51.5	75	98	140	248	330	385	460	580	760	820	920	1030	1105	1260
HED-160	160	20.8	33	56	80	104	150	263	345	400	475	620	810	890	1000	1110	1180	1340
HED-170	170	22.1	35.2	59.5	85	111	159	278	367	425	505	655	850	915	1030	1150	1230	1410
HED-180	180	23.4	37.3	63	90	117	168	295	389	450	535	695	900	970	1090	1220	1300	1490

Table 51.59 Electrical performance data for McGraw Edison high-rate large vented nickel–cadmium batteries, HED series

Cell type	Nominal capacity (A h)	Discharge current (A) to 0.65 V at 25°C						Discharge current (A) to 0.85 V at 25°C					
		90 s	60 s	30 s	10 s	5 s	1 s	90 s	60 s	30 s	10 s	5 s	1 s
HED-10	10	123	137	157	181	196	231	88	98	111	127	138	163
HED-15	15	185	206	236	271	294	346	132	147	167	191	207	245
HED-20	20	217	238	267	313	335	378	155	168	188	220	234	267
HED-25	25	271	298	333	392	417	472	193	210	235	275	293	333
HED-30	30	325	358	400	470	500	567	232	252	282	330	352	400
HED-40	40	442	459	493	546	579	661	309	320	344	380	408	470
HED-46	46	510	530	570	629	666	757	354	371	398	441	471	535
HED-53	53	587	610	655	725	768	864	408	425	457	507	542	615
HED-60	60	660	685	736	815	864	971	458	478	514	570	609	692
HED-70	70	732	810	900	1015	1102	1230	525	570	638	720	775	871
HED-80	80	813	899	999	1132	1223	1365	586	633	708	800	860	967
HED-90	90	905	990	1095	1261	1365	1520	660	699	771	892	965	1072
HED-100	100	1010	1075	1215	1350	1435	1610	715	765	860	960	1015	1150
HED-110	110	1115	1185	1340	1485	1575	1770	790	845	945	1050	1110	1250
HED-120	120	1215	1290	1460	1620	1715	1930	860	920	1030	1145	1210	1365
HED-130	130	1375	1515	1705	1890	2010	2290	990	1070	1195	1325	1415	1615
HED-140	140	1440	1590	1825	2020	2150	2450	1030	1120	1280	1420	1515	1730
HED-150	150	1540	1690	1900	2125	2270	2600	1090	1195	1355	1510	1610	1835
HED-160	160	1610	1810	2040	2270	2420	2770	1150	1280	1440	1600	1710	1960
HED-170	170	1710	1880	2110	2370	2530	2900	1220	1330	1500	1670	1780	2050
HED-180	180	1810	1980	2230	2500	2670	3060	1290	1410	1590	1770	1890	2170

The design features of the TS/TSP range are:

1. Compared with the T/TP or HK/HKP ranges, a 30% increase in plates resulting in increased active material surface.
2. Pin separators instead of corrugated perforated PVC separators.
3. An increased diameter for the terminal configuration, and larger cross-sectional areas for conductive sections of the plates. These features provide a low internal resistance, increasing high performance. Figure 51.33 indicates the enhanced voltage level of the TS/TSP range of cells compared with the general-purpose T/TP range. Figure 51.33 also indicates cells of the same capacity and discharge volt current.

Types T and TP These are storage batteries with pocket-type electrodes. The series T storage batteries, from 65 to 1250 A h, have a nickel-plated sheet steel case and are delivered in open wooden crates (from 65 to 380 A h) and as single cells (from 450 to 1250 A h). The 10–315 A h storage batteries, series TP, have a transparent plastic cell case, which makes it possible to control the electrolyte level from outside. They are available as compact blocks, portable blocks, and in open wooden crates. Types T and TP are universally designed for discharges, i.e. for power supplies, from more than 10 h down to short time discharges of 1 h with a favourable voltage performance. They can be used excellently

for charge/discharge operation, switch-tripping operation and standby parallel operation as well as for stabilizing operation. They can be charged with high charging currents within a short period of time. At low temperatures they provide favourable capacity characteristics. Deep discharges have no effect on the capacity.

Types HK and HKP Discharge data on types T and TP are given in Table 51.60. Types HK and HKP are storage batteries with pocket-type electrodes. The series HK storage batteries from 73 to 1390 A h have a nickel-plated sheet steel case and are delivered in open wooden crates (from 73 to 420 A h) and as single cells (from 500 to 1390 A h).

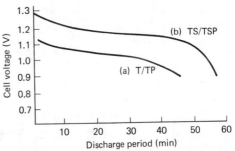

Figure 51.33 Schematic comparison between (a) T/TP and (b) TS/TSP cells; voltage–time characteristics when discharged with $5 \times I_5 = 1C$

Table 51.60 Discharge currents in amps (battery fully charged) at 20–25°C for FRIWO vented nickel–cadmium batteries types T and TP

Type of cell	Nominal capacity (C₅) down to 1.15 V/cell (A h)	Times of discharge										
		0.5 h	1 h	2 h	3 h	4 h	5 h	6 h	7 h	8 h	9 h	10 h
TP 10	10	7.00	5.00	4.00	3.00	2.40	2.00	1.70	1.50	1.30	1.15	1.05
TP 18	18	12.60	9.40	7.20	5.40	4.36	3.60	3.00	2.65	2.35	2.10	1.90
TP 24	24	16.80	12.50	9.60	7.20	5.78	4.80	4.10	3.50	3.10	2.80	2.55
TP 30	30	21.00	15.60	12.00	9.00	7.10	6.00	5.10	4.40	3.90	3.50	3.20
TP 40	40	28.00	20.80	16.00	12.00	9.50	8.00	6.80	5.90	5.20	4.70	4.25
TP 55	55	38.50	28.60	22.00	16.50	13.00	11.00	9.30	8.10	7.15	6.45	5.80
T 65* TP 65	65	45.50	33.80	26.00	19.50	15.40	13.00	11.00	9.50	8.45	7.60	6.90
T 75* TP 75	75	52.50	39.00	30.00	22.50	17.80	15.00	12.70	11.00	9.75	8.78	7.95
T 90* TP 90	90	63.00	46.80	36.00	27.00	21.30	18.00	15.30	13.20	11.70	10.50	9.55
T 110* TP 110	110	77.00	57.20	44.00	33.00	26.10	22.00	18.70	16.10	14.30	12.80	11.65
T 125* TP 125	125	87.50	65.00	50.00	37.50	29.60	25.00	21.20	18.30	16.25	14.60	13.25
T 140* TP 140	140	98.00	72.80	56.00	42.00	33.20	28.00	23.80	20.50	18.20	16.30	14.80
T 165* TP 165	165	115.50	85.80	66.00	49.50	39.10	33.00	28.00	24.20	21.45	19.30	17.85
T 185 /TP 185	185	129.50	96.20	74.00	55.50	43.90	37.00	31.40	27.20	24.00	21.65	19.60
T 200*/TP 200	200	140.00	104.00	80.00	60.00	47.40	40.00	34.00	29.40	26.00	23.40	21.20
T 230 /TP 230	230	161.00	119.60	92.00	69.00	54.50	46.00	39.10	33.80	29.90	26.90	24.30
T 275 /TP 275	275	192.50	143.00	110.00	82.50	65.20	55.00	46.70	40.40	35.70	32.10	29.10
T 315 /TP 315	315	220.50	163.80	126.00	94.50	74.70	63.00	53.50	46.30	40.90	36.80	33.30
T 380	380	266.00	197.60	152.00	114.00	90.00	76.00	64.60	55.80	49.40	44.40	40.20
T 450	450	315.00	234.00	180.00	135.00	106.70	90.00	76.50	66.15	58.50	52.60	47.70
T 520	520	364.00	270.40	208.00	156.00	123.30	104.00	88.40	76.45	67.60	60.80	55.10
T 600	600	420.00	312.00	240.00	180.00	142.20	120.00	102.00	88.20	78.00	70.20	63.60
T 750	750	525.00	390.00	300.00	225.00	177.80	150.00	127.50	110.20	97.50	87.70	79.50
T 1000	1000	700.00	520.00	400.00	300.00	237.00	200.00	170.00	147.00	130.00	117.00	106.00
T 1250	1250	875.00	650.00	500.00	375.00	296.30	250.00	212.50	183.70	162.50	146.20	132.50

* Cell types in steel sheet cases

Final discharge voltage 1.15 V/cell

The 11–355 A h storage batteries, series HKP, have a transparent plastic cell case, which makes it possible to control the electrolyte level from outside. They are available as compact blocks, portable blocks and in open wooden crates. Alternatively they can be delivered in enclosed wooden crates or in steel containers. Types HK and HKP are universally designed for discharges, i.e. for power supplies, from more than 10 h down to short-time discharges of 1 h with a favourable voltage performance. They can be used for charge/ discharge operation, switch-tripping operation and standby parallel operation as well as for stabilizing operation. They can be charged with high charging currents within a short period of time. Even at low temperatures they provide favourable capacity characteristics. Deep discharges have no effect on the capacity.

Discharge data on types HK and HKP are given in Table 51.61.

Types TS and TSP These are high-rate storage batteries fitted with pocket-type electrodes and are specially designed for high-rate and impulse discharge. The storage batteries of series TS from 150 to 650 A h have a nickel-plated sheet steel case and are delivered in open wooden crates (from 150 to 275 A h) and as single cells (from 300 to 650 A h). The 7.5–235 A h storage batteries, series TSP, have a transparent plastic cell case which makes it possible to control the electrolyte level from outside. They are available as compact blocks, portable blocks and in open wooden crates. Alternatively, they can be delivered in enclosed wooden crates or in steel containers. Types TS and TSP are specially designed for high-rate discharges and for use as starting batteries, i.e. for discharges from less than 1 h down to short-time discharges and high-rate/impulse loads with favourable voltage performance. They can be used for charge/discharge operation, switch-tripping operation and standby parallel operation as well as for stabilizing operation.

They can be charged with high charging currents within a short period of time. At low temperatures they provide favourable capacity characteristics. Deep discharges have no effect on the capacity.

Discharge data on types TS and TSP are given in Table 51.62.

51.1.4 Aircraft batteries

Another major application of vented nickel–cadmium batteries is aircraft batteries. These batteries are equipped with a battery temperature monitoring facility in order to comply with the requirements of the FAA. One or more thermal switches are installed in the battery to give advance warning and to prevent overheating of the battery, thereby achieving a greater safety of flight operation.

The thermal switches are connected to a connector on the battery container from which a cable leads to a warning lamp situated on the instrument panel of the aircraft.

This warning lamp lights up when the predetermined temperature limit for the thermal switch is reached, e.g. 65°C. When the warning lamp lights up, charging of the battery must be stopped. When the battery is cooled down the warning lamp goes out and the battery is ready for service.

Technical data on several of the Varta range of aircraft batteries are given in Table 51.63. Typical discharge data are given in Figures 51.34–51.36. The

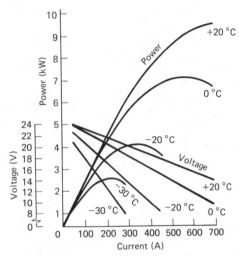

Figure 51.34 Varta vented nickel–cadmium battery type HI: dependence of initial voltage on current and temperature (Courtesy of Varta)

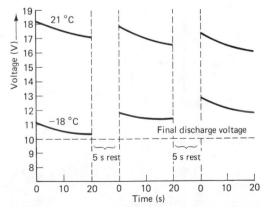

Figure 51.35 Varta vented nickel–cadmium battery type HI: start simulations at 20*C*A at 71°C and −18°C (Courtesy of Varta)

Table 51.61 Discharge currents in amps (battery fully charged) at 20–25°C for FRIWO vented nickel–cadmium batteries types HK and HKP

Type of cell	Nominal capacity (C_5) down to 1.0 V/cell (A h)	Times of discharge										
		0.5 h	1 h	2 h	3 h	4 h	5 h	6 h	7 h	8 h	9 h	10 h
HKP 1	11	7.00	5.00	4.00	3.00	2.40	2.00	1.70	1.50	1.30	1.15	1.05
HKP 2	19.5	12.60	9.40	7.20	5.40	4.36	3.60	3.00	2.65	2.35	2.10	1.90
HKP 2.7	26.5	16.80	12.50	9.60	7.20	5.78	4.80	4.10	3.50	3.10	2.80	2.55
HKP 3	33	21.00	15.60	12.00	9.00	7.10	6.00	5.10	4.40	3.90	3.50	3.20
HKP 5	44	28.00	20.80	16.00	12.00	9.50	8.00	6.80	5.90	5.20	4.70	4.25
HKP 6	61	38.50	28.60	22.00	16.50	13.00	11.00	9.30	8.10	7.15	6.45	5.80
HK 7* HKP 7	73	45.50	33.80	26.00	19.50	15.40	13.00	11.00	9.50	8.45	7.60	6.90
HK 9* HKP 9	83	52.50	39.00	30.00	22.50	17.80	15.00	12.70	11.00	9.75	8.78	7.95
HK 10* HKP 10	100	63.00	46.80	36.00	27.00	21.30	18.00	15.30	13.20	11.70	10.50	9.55
HK 12* HKP 12	122	77.00	57.20	44.00	33.00	26.10	22.00	18.70	16.10	14.30	12.80	11.65
HK 14* HKP 14	138	87.50	65.00	50.00	37.50	29.60	25.00	21.20	18.30	16.25	14.60	13.25
HK 16* HKP 16	155	98.00	72.80	56.00	42.00	33.20	28.00	23.80	20.50	18.20	16.30	14.80
HK 18* HKP 18	182	115.50	85.80	66.00	49.50	39.10	33.00	28.00	24.20	21.45	19.30	17.85
HK 20/ HKP 20	205	129.50	96.20	74.00	55.50	43.90	37.00	31.40	27.20	24.00	21.65	19.60
HK 23* HKP 23	220	140.00	104.00	80.00	60.00	47.40	40.00	34.00	29.40	26.00	23.40	21.20
HK 26/ HKP 26	255	161.00	119.60	92.00	69.00	54.50	46.00	39.10	33.80	29.90	26.90	24.30
HK 31/ HKP 31	305	192.50	143.00	110.00	82.50	65.20	55.00	46.70	40.40	35.70	32.10	29.10
HK 36/ HKP 36	350	220.50	163.80	126.00	94.50	74.70	63.00	53.50	46.30	40.90	36.80	33.30
HK 43	420	266.00	197.60	152.00	114.00	90.00	76.00	64.60	55.80	49.40	44.40	40.20
HK 51	500	315.00	234.00	180.00	135.00	106.70	90.00	76.50	66.15	58.50	52.60	47.70
HK 59	575	364.00	270.40	208.00	156.00	123.30	104.00	88.40	76.45	67.60	60.80	55.10
HK 67	665	420.00	312.00	240.00	180.00	142.20	120.00	102.00	88.20	78.00	70.20	63.60
HK 85	830	525.00	390.00	300.00	225.00	177.80	150.00	127.50	110.20	97.50	87.70	79.50
HK 115	1110	700.00	520.00	400.00	300.00	237.00	200.00	170.00	147.00	130.00	117.00	106.00
HK 140	1390	875.00	650.00	500.00	375.00	296.30	250.00	212.50	183.70	162.50	146.20	132.50

* Cell types in steel sheet cases

Final discharge voltage 1.15 V/cell

Table 51.62 Discharge currents in amps (battery fully charged) at 20–25°C for FRIWO vented nickel–cadmium batteries, types TS and TSP

Type of cell	Nominal capacity (C_5) down to 1.1 V/cell (A h)	Seconds					Minutes							Hours				
		1	5	10	20	30	1	3	5	10	20	30	45	1	1.5	2	3	5
TSP 7.5	7.5	25.5	24.0	21.0	19.5	18.7	17.5	14.7	14.2	12.7	10.1	9.2	7.5	6.25	4.5	3.5	2.40	1.45
TSP 10	10	34.0	32.0	28.0	26.0	25.0	23.4	19.6	19.0	17.0	13.5	12.3	10.0	8.35	6.0	4.7	3.20	1.95
TSP 14	14	47.6	44.8	39.2	36.4	35.0	32.7	27.5	26.6	23.8	18.9	17.2	14.0	11.70	8.4	6.6	4.50	2.75
TSP 20	20	68.0	64.0	56.0	52.0	50.0	46.8	39.2	38.0	34.0	26.9	24.5	20.0	16.70	12.0	9.4	6.45	3.95
TSP 30	30	102.0	96.0	84.0	78.0	75.0	70.0	59.0	57.0	51.0	40.5	36.8	30.0	25.00	18.0	14.1	9.70	5.95
TSP 40	40	136.0	128.0	112.0	104.0	100.0	94.0	79.0	76.0	68.0	54.0	49.0	40.0	33.40	24.0	18.9	12.90	7.95
TSP 55	55	187.0	176.0	154.0	143.0	138.0	129.0	108.0	105.0	94.0	74.0	67.5	55.0	45.90	33.0	25.9	17.80	10.90
TSP 65	65	221.0	208.0	182.0	169.0	163.0	152.0	128.0	124.0	111.0	88.0	79.8	65.0	54.30	39.0	30.7	21.00	12.90
TSP 80	80	272.0	256.0	224.0	208.0	200.0	187.0	157.0	152.0	136.0	108.0	98.0	80.0	66.80	48.0	37.7	25.90	16.90
TSP 100	100	340.0	320.0	280.0	260.0	250.0	234.0	196.0	190.0	170.0	135.0	123.0	100.0	83.50	60.0	47.2	32.50	19.80
TSP 125	125	425.0	400.0	350.0	325.0	313.0	292.0	245.0	238.0	213.0	169.0	153.0	125.0	104.00	76.0	59.0	40.70	24.80
TS/TSP 150	150	510.0	480.0	420.0	390.0	375.0	351.0	294.0	285.0	255.0	202.0	184.0	150.0	125.00	90.0	70.8	48.80	29.70
TS/TSP 185	185	629.0	592.0	518.0	481.0	463.0	433.0	363.0	352.0	315.0	249.0	227.0	185.0	154.00	111.0	87.3	60.30	36.70
TSP 200	200	680.0	640.0	560.0	520.0	500.0	468.0	392.0	380.0	340.0	270.0	246.0	200.0	167.00	120.0	94.4	65.10	39.60
TS/TSP 235	235	799.0	752.0	658.0	611.0	588.0	550.0	461.0	447.0	400.0	317.0	289.0	235.0	196.00	141.0	111.0	76.50	46.60
TS 275	275	935.0	880.0	770.0	715.0	688.0	644.0	539.0	523.0	468.0	371.0	338.0	275.0	230.00	165.0	130.0	89.60	54.50
TS 300	300	1020.0	960.0	840.0	780.0	750.0	702.0	588.0	570.0	510.0	404.0	368.0	300.0	250.00	180.0	141.0	97.70	59.40
TS 345	345	1173.0	1104.0	966.0	897.0	863.0	808.0	676.0	656.0	587.0	465.0	424.0	345.0	288.00	207.0	163.0	112.00	68.40
TS 370	370	1258.0	1184.0	1036.0	962.0	925.0	866.0	725.0	703.0	629.0	479.0	454.0	370.0	309.00	222.0	174.0	120.00	73.30
TS 415	415	1411.0	1328.0	1162.0	1079.0	1038.0	971.0	813.0	789.0	706.0	559.0	509.0	415.0	347.00	249.0	196.0	134.00	82.20
TS 460	460	1564.0	1472.0	1288.0	1196.0	1150.0	1076.0	902.0	874.0	782.0	620.0	565.0	460.0	384.00	276.0	217.0	149.00	91.10
TS 530	530	1802.0	1696.0	1484.0	1378.0	1325.0	1240.0	1039.0	1007.0	901.0	714.0	651.0	530.0	443.00	318.0	250.0	172.00	105.00
TS 575	575	1955.0	1840.0	1610.0	1495.0	1438.0	1346.0	1127.0	1093.0	978.0	775.0	706.0	575.0	480.00	345.0	271.0	187.00	113.00
TS 650	650	2210.0	2080.0	1820.0	1690.0	1625.0	1521.0	1274.0	1235.0	1105.0	876.0	798.0	650.0	543.00	390.0	307.0	211.00	128.00

Times of discharge

Final discharge voltage 1.15 V/cell

Table 51.63 Technical data for Varta vented nickel–cadmium aircraft batteries

Designation	F 19/04-1	F 19/15 H1	F 20/15 H1	F 19/25 H1	F 20/40 H1	F 20/40 H1T	F 20/40 H1WT	F 20/40 H1-E1	F 20/40 H1T-E1
Type No. 334	24 08 000	31 58 060	31 59 060	32 58 130	34 09 140	34 09 190	34 09 170	34 09 130	34 09 1301
Standards and specification	TL 6140-002 Lockheed ER 421487	DIN 29 833 A DIN 29 834 MS 24 496 MIL-B-26 220 C	DIN 29 833 B MIL-B-26 220 C	DIN 29 832 A	DIN 29 831 B DIN 29 834 MS 24 498 MIL-B-26 220 C	DIN 29 831 BS	DIN 29 831 BFG	DIN 29 998	Prl 56 101 A (E1) DIN 29 834
Connector		MS 18093		MS 18093	MS 18093				BAC 102
Number of cells	19	19	20	20	20	20	20	20	20
Rated voltage (V) (20°C)	22.8	22.8	24	24	24	24	24	24	24
Nominal capacity C_5 (A h) (20°C)	4	15	15	40	40	40	40	40	40
Dimensions (mm) — Length	226	198	198	254	254	254	254	362	362
Width	89	195	195	248	248	248	248	168	168
Height	134	196	196	262	262	262	262	267	267
Weight (kg max.)	4.75	15.5	16.3	24.5	36.3	36.4	36.6	37.5	37.7
Cell type	FP 4 H1	FP 15 H1	FP 15 H1	FP 40 H1	FP 40 H1	FP 40 H1	FP 40 H1	FP 40 H1	FP H40 H1
Type No. 374	25 03 000	31 53 040	31 53 040	34 03 040	34 03 040	34 03 040	34 03 040	34 03 040	34 03 040
Capacity 5 h rate	4	15	15	40	40	40	40	40	40
Capacity 1 h rate	3.7	13.5	13.5	38	38	38	38	38	38
Dimensions (mm) — Length	54	60	60	80	80	80	80	80	80
Width	16	27	27	35.5	35.5	35.5	35.5	35.5	35.5
Height	101	171	171	239	239	239	239	239	239
Weight (kg max.)	0.210	0.650	0.650	1.550	1.550	1.550	1.550	1.550	1.550
Constant current rate I_5 charge 6 h (A)	0.8	3	3	8	8	8	8	8	8
Constant current rate I_{10} charge 14 h (A)	0.4	1.5	1.5	4	4	4	4	4	4

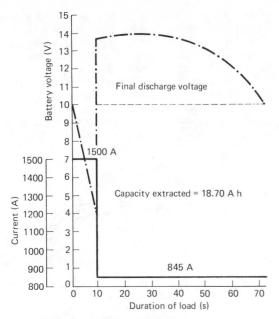

Figure 51.36 Varta vented nickel–cadmium battery type F20/40HI (24 V, 40 A h): effect of initial load on battery voltage at −12.2°C (Courtesy of Varta)

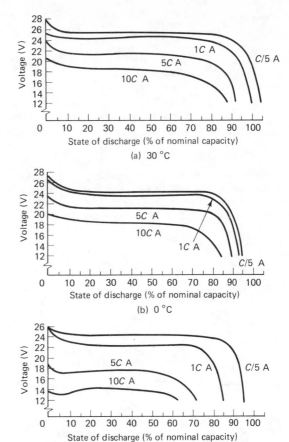

(a) 30 °C

(b) 0 °C

(c) −30 °C

Figure 51.37 Electrical performance of a Varta vented nickel–cadmium battery: typical discharge curves at different load temperatures for a battery of 20 cells. Initial state of charge: 100% of nominal capacity. Charging: 6 h at 0.2C A (25°C) (Courtesy of Varta)

effect of battery temperature in the range −30 to +30°C on discharge performance is shown in Figure 51.37.

Eagle Picher produce the VNC range of batteries for aircraft starting, aircraft emergency power and ground power units. Capacities up to 120 A h are available (see selected examples in Table 51.64). Cycling lives in excess of 5 years or 1000 cycles are claimed for these batteries. The working voltages are 1.0–1.25 V/cell with a voltage regulation that is

Table 51.64 VNC range of vented nickel–cadmium batteries supplied by Eagle Picher

Standard line cell types	Rated capacity (A h)*	Energy density		Weight (g)	Length (mm)	Width (mm)	Height (mm)	Military designation
		W h/kg	W h/dm³					
VNC 08	0.8*	17.6	32	57	18	29	59	
VNC 1.5	1.5*	19.8	35	91	17	29	102	
VNC 2A	2.0*	21.1	40	113	18	38	87	
VNC 2B	2.0*	22.9	47	99	17	29	102	
VNC 3	3.0*	18.9	37	190	13	49	153	
VNC 3.5	3.5	28.8	56	145	21	53	65	
VNC 5.5A	5.5*	25.7	49	255	24	55	102	BB 615/U
VNC 30	30.0*	30.4	64	119	27	81	255	
VNC 37	37.0*	30.1	67	1474	35	80	239	BB 600/U
VNC 42	42.0*	31.7	76	1588	35	80	239	
VNC 70	70.0*	33.9	82	2466	45	121	188	
VNC 120	120.0	37.0	88	3119	75	95	241	

* Based on 1 h discharge rate

Table 51.65

Temperature (°C)	Self-discharge (%/day)
27	0.5
37	1.1
49	2.5
60	5.0
71	10.0

uniform over 90% of the discharge. Operating temperatures are −40 to 74°C and storage temperatures −62 to 74°C when charged, and −54 to 38°C when discharged. Self-discharge rates are given in Table 51.65. Typical discharge characteristics are shown in Figure 51.38.

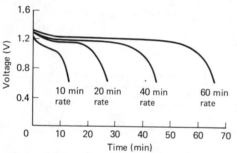

Figure 51.38 Typical cell discharge characteristics at 27°C of Eagle Picher VNC vented nickel–cadmium aircraft batteries (Courtesy of Eagle Picher)

51.2 Nickel–hydrogen and silver–hydrogen (secondary) batteries

At the time of writing, no commercial design of these batteries exists. However, Eagle Picher USA, and no doubt other companies, are conducting long-term development on such designs (as discussed elsewhere in this book) and it is hoped that these will eventually become viable contenders for non-space applications in addition to their well established space applications, where cost has not been an important consideration.

51.3 Nickel–zinc batteries

51.3.1 Yardney Corporation (USA)

Yardney USA pioneered the development of this type of battery and are still the main suppliers. The original application was electric traction, and a number of road vehicles are now operating powered by Yardney experimental batteries. Further details

of the Yardney traction battery assembled from 300 A h units are supplied elsewhere in this book. In a typical vehicle set-up, 15 of these modules are assembled in series to yield a 28.8 kW h energy content battery at a nominal 96 V.

Typical projected performance curves for the Yardney nickel–zinc battery as functions of discharge rate and temperature are shown in Figure 51.39. If the battery is to be operated in severe hot or cold climates, performance is affected (Figure

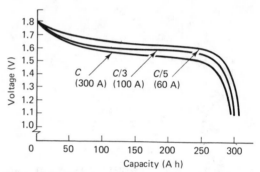

Figure 51.39 Projected performance at various discharge rates of a Yardney nickel–zinc cell of 300 A h capacity (Courtesy of Yardney)

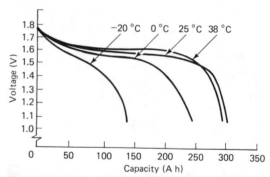

Figure 51.40 Projected effect of discharge temperature on a Yardney nickel–zinc cell of 300 A h capacity (Courtesy of Yardney)

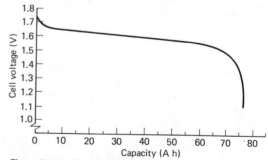

Figure 51.41 Typical discharge curve for a Yardney NZ EV3-XPI prototype bonded nickel–zinc cell of 75 A h nominal capacity, at 25°C and 15 A discharge rate (Courtesy of Yardney)

51.40) and a suitable cooling or heating system may be required. The discharge curve for the 96 V battery is shown in Figure 51.41. The characteristics of this battery are given Table 51.66.

51.3.2 Eagle Picher (USA)

This company offers a range of vented nickel–zinc batteries in the capacity range 2–35 A h. Unlike the Yardney battery, these are commercially available and are particularly recommended for field communication power supplies. They are recommended for use whenever one or more of the following characteristics are required:

1. High voltage (1.5–1.6 V working voltage per cell, the highest available in alkaline batteries).
2. Wide temperature tolerances (−39 to +81°C operating temperature, −53 to +73°C storage temperature).
3. Moderate cycle life (150 maximum cycles) and recycle flexibility.

Eagle Picher claim that with further improvements in design, coupled with a lowering of unit costs through mass production, the nickel–zinc battery holds promise of the lowest potential production cost per unit of energy supplied. The batteries have high-rate, low-temperature capabilities, and a flat discharge voltage for a large portion of the

Table 51.66 Characteristics of Yardney nickel–zinc batteries

Theoretical energy density	374 W h/kg
Practical energy density	44–77 W h/kg and 79–134 W h/dm^3 depending on cell model and conditions of use
Open-circuit cell voltage	1.75–1.8 V (i.e. the highest operating voltage of any of the alkaline batteries)
Nominal voltage under load	1.6 V
Voltage regulation	±5% at a fixed load and temperature limits within ±5°C
Recommended discharge cut-off voltage	1.0 V
Cycle life	Dependent on model and conditions of use; typical number of complete charge/discharge cycles obtainable:

Depth of discharge (%)	Approx. cycles
95–75	100–200
75–50	200–400
50–25	over 400

Wet life	Up to 3 years when manufacturer's recommended charge and discharge rates are used, and operating and storage temperatures do not exceed 37°C
Dry storage life	Up to 5 years
Gassing on discharge or stand	Slight
Operational temperature range	+73 to −21°C; down to −53°C with heaters; for optimum cell performance, from +56 to −45°C
Storage temperature range	Wet, +37 to −47°C; dry, +73 to −64°C
Operating attitude	In any position, although for optimum service, upright is recommended
Internal resistance	Very low; varies with cell model, temperature and rate of discharge
Resistance to mechanical stress	Excellent, extremely rugged, leakproof and spillproof. Construction similar to design of Yardney Silvercel batteries, which have met the stringent requirements of Spec. MIL-E-5272A. Can be packaged to meet the most severe requirements
Charging time	Can be fully recharged within 10–20 h, depending on requirements and type of cell
Charge retention	Up to 70% of nominal capacity after 6 months charged stand at room temperature

Table 51.67 Characteristics of Eagle Picher rechargeable nickel–zinc batteries

Cell type	Rated capacity (A h)	Working voltage (V)	Min. capacity (A h) at 30 min rate	Gravimetric energy density (W h/kg)	Volumetric energy density (W h/dm³)	Max. filled weight (g)	Overall volume (mm³) × 10³	Length (mm)	Width (mm)	Height (case only) (mm)
NZS 2.0	2.0	1.6	2.0	28.8	52	111	61	18.3	38.1	77.7
NZS 5.0	5.0	1.6	5.0	32.3	58	247	134	23.8	55.1	91.9
NZS 7.0	7.0	1.6	6.3	39.4	76	284	144	33.5	58.6	61.7
NZS13.0	13.0	1.6	11.7	75.5	96	398	213	33.5	58.6	96.2
NZS 20.0	20.0	1.6	18.0	53.7	111	596	284	33.5	58.6	132.0
NZS 25.0	25.0	1.6	25.0	43.3	88	923	446	35.3	79.0	149.0
NZS 30.0	30.0	1.6	27.0	60.3	126	795	374	35.3	79.0	178.0
Mono block NZS 5.5	5.5	1.6	5.0	51.7	71	682	487	85.3	58.6	85.3
Mono block NZS 35.0	35.0	1.6	31.5	53.2	107	409	2055	135.0	83.8	163.0

discharge, which is maintained over a wide current range, and they can be left fully discharged for long periods without deterioration. These batteries are claimed to have an excellent shelf life even when wet, and a charge retention of 60% for 30 days. Charging is non-critical and is carried out for 4 h at constant current until the voltage rises to 2.1 V, when charging is stopped. The batteries can be fully recharged with 10–15% overcharge.

The Eagle Picher range of nickel–zinc batteries is shown in Table 51.67. Typical discharge characteristics of a 5 A h nickel–zinc battery are shown in Figure 51.42.

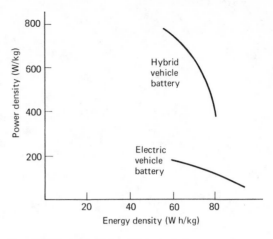

Figure 51.43 Eagle Picher nickel–zinc battery power/energy characteristics (Courtesy of Eagle Picher)

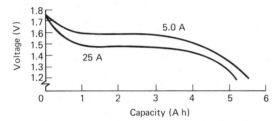

Figure 51.42 Typical discharge characteristics of an Eagle Picher NZS/5.0 (5 A h) nickel–zinc battery at 27°C (Courtesy of Eagle Picher)

More recently, this company has announced that it can now supply nickel–zinc batteries in the 2–850 A h size range. Energy densities are in the range 44–77 W h/kg (121–182 W h/dm^3), depending on the power density used in the particular application (Figure 51.43). Power capabilities of up to 800 W/kg and cycle lives of up to 500 cycles are achieved, depending on the depth of discharge and the operational energy density used in the application.

The discharge characteristics of a particular Eagle Picher nickel–zinc cell are shown in Figure 51.44.

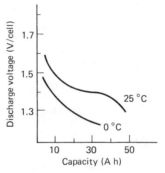

Figure 51.44 Eagle Picher nickel–zinc hybrid vehicle battery discharge characteristics at 500 A (Courtesy of Eagle Picher)

52

Silver batteries

Contents

52.1 Silver–zinc batteries

52.1.1 Primary batteries

Yardney

There are numerous applications in the aerospace field, particularly where a high energy density battery is required which can offer both fast activation and a long wet-stand charged life. It was to meet this demand that Yardney Electric launched the primary PM Silvercel battery.

This manually filled silver–zinc battery which, they claim, has the highest specific energy output of any battery, provides rapid activation and an activated stand time far longer than that of any other high-energy primary. Additionally, it offers a limited recyclability which allows performance testing before actual use. PM Silvercel batteries have surpassed the stringent mechanical requirements of missile applications and proved their ruggedness and reliability in the field. The following advantages are offered by the PM Silvercel battery system over alternative systems:

1. Up to 75% higher capacity.
2. Higher capacity with optimum life.
3. Higher voltages at peak pulses.

By carefully choosing the critical application requirement that must be met, whether it be capacity, life or voltage, and then optimizing these in relationship to each other, it is possible to satisfy the single most stringent demand. The characteristics of the PM cell are identical to those of the Yardney HR and LR Silvercel rechargeable secondary battery system (see below), with the exception that cycle life and wet life are somewhat limited.

Table 52.1 shows some of the types of primary AHP type silver–zinc batteries supplied by Yardney for various aerospace applications.

Eagle Picher manually activated high, medium and low rate batteries

Selected examples from the range of these manually activated batteries are listed in Table 52.2. Capacities between 0.5 and 410 A h are available. The high-rate series attains maximum performance at extremely high discharge rates for periods of 3–15 min. The medium-rate range is designed for discharge periods of 20–60 min, and the low-rate series develops maximum efficiency over 4–20 h.

Energy densities are in the range 110–220 W h/kg as cells and 77–176 W h/kg as batteries, or 214–457 W h/dm^3 as cells and 152–366 W h/dm^3 as batteries. The working voltage is 1.30–1.55 V. Voltage regulation is ±2% under fixed conditions; maximum voltage regulation is achieved at pulse rates of 100 ms or less. The battery operating temperatures are −40 to 54°C, or −54°C with a

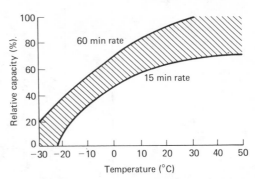

Figure 52.1 Effect of temperature on capacity of Eagle Picher manually operated silver–zinc cells (Courtesy of Eagle Picher)

heater. The storage temperature is −54 to 51°C when dry and −40 to 38°C when wet. Shelf life is 2–5 years when dry or 15–30 days when wet. Charge retention is 90–100% of nominal after 15 days wet stand.

Figure 52.1 shows the effect of temperature on capacity (as a percentage of nominal capacity) for silver–zinc cells discharged at the 60 min and the 15 min rates. Figure 52.2 shows typical discharge characteristics for high, medium and low rate cells.

Union Carbide

Data on the Eveready range of 1.5 V silver–zinc cells and their 6 V battery are reproduced in Table 52.3. Average service life data and other electrical characteristics are given in Table 52.4. Figure 52.3 shows discharge curves obtained for different silver–zinc cells in the Eveready range on a 16 h/day schedule at 21°C. Under these conditions silver–zinc cells are capable of rendering up to 120 h service, 1–2.5 mA drain at 625–1500 Ω load, whereas at lighter current drains (100–250 mA) and higher resistances (6500–15 000 Ω) up to 1100 h service are achievable (Figure 52.2).

Mallory

Data on the Mallory range of Duracell (1.5 V) silver–zinc cells and their 6 V battery are quoted in Table 52.5. Available capacities are in the range 36–250 mA h (compared with 35–210 mA h in the Union Carbide range). Typical discharge characteristics at stated current drains and resistances for a 1.5 V Duracell battery are reproduced in Figure 52.4.

Duracell Europe

Duracell Europe supply a range of 1.5 V silver–zinc button cells, as shown in Table 52.6.

Table 52.1 Characteristics and aerospace applications of some Yardney AHP-type primary silver–zinc batteries

Battery	Form	Activation system	Weight (kg)	Length (mm)	Width (mm)	Height (mm)	Volume (dm^3)	Energy (W h)	Energy density		Remarks	Applications
									(W h/kg)	(W h/dm^3)		
7 × 0.4 AHP	Cylindrical	Coil	0.3	51	51	51	0.102	4.2	14.1	40.4	EGLIN: percussion activated, cylinder cell block	Missile
40 × 1.0 AHP	Rectangular	Coil	4.4	229	114	106	2.79	70.0	15.9	25.6	ESI: multi-tapped, thin plate, moderate discharge rate, 6 h wet stand (total)	AT3K Polaris missile
19 × 1.3 AHP /20 × 9.5 AHP	Rectangular	Tank	13.0	289	134	178	6.95	470.0	36.8	67.1	SRAM: EMI filter, wet stand 60 min	Guidance hydraulics (SRA17)
20 × 1.8 AHP	Rectangular	Tank	9.0	254	159	127	5.11	181.0	20.0	35.4	TITAN III HYD: 5 h wet stand, high discharge rate, thin plate cell	FB hydraulics (Titan III)
10 × 2.2 AHP	Kidney	Coil	1.2				0.96	32.0	19.6	24.4	GAR: stainless steel case, nested-type cell cases, heater	Missile (Falcon)
20 × 3.15 AHP /21 × 4.8 AHP	Cylindrical	Tank	13.0	190	190	342	9.77	522.0	40.9	53.7	C3 ELEC: aluminium case, 6 h wet stand, low-rate discharge with pulses	C4 electronics
30 × 3.4 AHP /21 × 4.8 AHP	Cylindrical	Tank	13.0	190	190	342	9.77	501.0	38.3	51.2	C4 ELEC: 10 min and 6 h wet stand, high pulse discharge	

Table 52.2 Range of manually activated silver–zinc batteries available from Eagle Picher

Cell type	Rated capacity (A h)	Nominal capacity A h rates			Energy density		Weight (g)	Physical dimensions			
		3 min	6 min	15 min	W h/kg	W h/dm³		Length (mm)	Width (mm)	Height (mm)	Volume (cm³)
High-rate											
SZH 1.0	1.0	1.0	1.0	1.0	57	104	26				
SZH 1.6	1.6	1.0	1.6	1.6	66	110	37				
SZH 2.4	2.4	2.3	2.4	2.4	66	116	54				
SZH 38.0	38.0	35.0	37.0	38.0	73	159	780				
SZH 68.0	68.0	60.0	65.0	68.0	79	195	1290				
Medium-rate											
SZM 1.0	1.0	1.0	1.0	1.0	59	104	26				
SZM 1.8	1.8	1.8	1.8	1.8	68	128	40				
SZM 3.0	3.0	3.0	3.0	3.0	73	146	57				
SZM 58.5	58.5		50.0	58.5	103		850				
SZM 95.0	95.0		80.0	95.0	110	275	1318				
Low-rate											
SZL 1.7	1.7	1.7	1.7	1.7	84	171	31	10.9	26.9	51.6	14.8
SZL 2.8	2.8	2.8	2.8	2.8	88	201	48	12.4	30.7	57.2	21.3
SZL 4.5	4.5	4.5	4.5	4.5	92	220	74	14.2	35.1	63.2	31.1
SZL 105.0	105.0	100.0	105.0	105.0	158	458	1012	32.5	80.8	133.9	352.4
SZL 160.0	160.0		160.0	160.0	187	470	1332	37.3	92.7	150.9	522.8
SZL 410.0	410.0			410.0	209	561	3005	42.2	138.4	193.5	1122.7

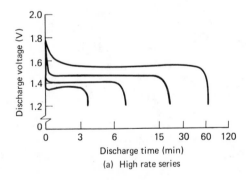

(a) High rate series

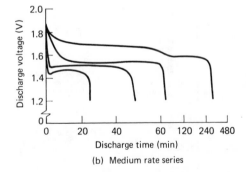

(b) Medium rate series

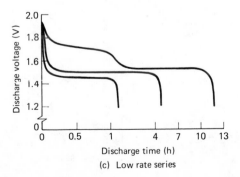

(c) Low rate series

Figure 52.2 Typical discharge characteristics at 27°C for high, medium and low rate Eagle Picher manually operated silver–zinc cells (Courtesy of Eagle Picher)

Table 52.3 Union Carbide Eveready primary silver–zinc cells and 6 V battery

Battery number (in order of increasing capacity)	Suggested current range (mA)	Service capacity (mA h)	Service capacity rated at (mA)	Maximum dimensions (mm)		Number and size of cells	Typical use	Weight of battery (g)	Terminals	Approximate volume of battery (cm³)	Voltage taps
				Diameter	Height						
1.5 V batteries											
384	0–0.06	35	0.10	7.87	3.63	1 × 7-31141	Watches	0.57	Flat contacts	0.13	–, +1.5
S312	0–5	38	1.00	7.87	3.56	1 × 6-31141	Transistor applications	0.57	Flat contacts	0.16	–, +1.5
S312E	0–5	38	1.00	7.87	3.56	1 × 6-31141	Hearing-aids	0.57	Flat contacts	0.16	–, +1.5
392	0–5	38	0.10	7.87	3.56	1 × 6-31142	Watches	0.57	Flat contacts	0.16	–, +1.5
309	0–0.1	60	0.10	7.87	5.33	1 × 7-31201	Watches	1.13	Flat contacts	0.21	–, +1.5
S13	0–5	75	1.07	7.87	5.33	1 × 6-31202	Transistor applications	1.13	Flat contacts	0.18	–, +1.5
S13E	0–5	75	1.07	7.87	5.33	1 × 6-31201	Hearing-aids	1.13	Flat contacts	0.18	–, +1.5
393	0–5	75	0.10	7.87	5.33	1 × 6-31202	Watches	1.13	Flat contacts	0.18	–, +1.5
301	0–0.1	100	0.10	11.60	4.06	1 × 7-45161	Watches	1.70	Flat contacts	0.38	–, +1.5
S41	0–10	120	1.60	11.60	4.19	1 × 6-45161	Microminiature lamps	1.70	Flat contacts	0.33	–, +1.5
S41E	0–10	120	1.60	11.60	4.19	1 × 6-45161	Hearing-aids	1.70	Flat contacts	0.33	–, +1.5
386	0–10	120	0.25	11.60	4.19	1 × 6-45163	Watches	1.70	Flat contacts	0.33	–, +1.5
303	0–0.24	165	0.25	11.60	5.59	1 × 7-45215	Watches	2.55	Flat contacts	0.51	–, +1.5
S76	0–10	190	2.56	11.60	5.36	1 × 6-45201	Transistor applications	2.30	Flat contacts	0.48	–, +1.5
S76E	0–10	190	2.56	11.60	5.36	1 × 6-45201	Hearing-aids	2.30	Flat contacts	0.48	–, +1.5
357	0–10	190	0.25	11.60	5.36	1 × 6-45202	Watches	2.30	Flat contacts	0.48	–, +1.5
355	0–10	210	0.25	15.50	4.83	1 × –	Watches	3.69	Flat contacts	0.90	–, +1.5
6 V battery											
544	0–10	190	2.50	13.00	25.20	4 × 6-45201	Electronic shutter and light meter	14.20	Flat contacts	2.80	–, +6

To provide the best contact to the terminals of silver oxide batteries, it is recommended that the device contacts be made of a spring material such as phosphor bronze or beryllium copper, which will maintain a contact force of at least 50 g for an extended period of time. The contacts should be plated with about 50.8 μm nickel (continuous) followed by a minimum of 5.08 μm gold. The reliability of the contact can be further increased by subdividing the main contact member into two, three or more individual points or prongs such as the tines of a fork

Table 52.4 Union Carbide Eveready silver–zinc cells

Eveready type No.	Voltage (V)	Suggested current range (mA)	Designation ANSI	Designation IEC	Average service capacity (mA h)	Schedule (h/day)	Starting drain	Load (Ω)	Cut-off voltage (V)	Estimated average service at 35°C (or 21°C if stated)	Impedance
544 (four 6-45201 cells)	6.0	0 to 10	–	–	190 to 3.6 V rated at 2.5 mA	24	2.5 mA	2400	3.6	60 (21°C)	–
355	1.5	0 to 10	–	–	210 to 1.3 V at 6500 Ω	24	246 μA	6500	0.9 / 1.3	915 / 905	Less than 4 Ω on open circuit at 1000 Hz
357	1.5	0 to 10	–	SR44	190 to 0.9 V at 6500 Ω	24	246 μA	6500	1.3	810	
S76E and S76	1.5	0 to 10	515	–	190 to 0.9 V rated at 2.56 mA	16 / 16	1.6 mA / 2.5 mA	1000 / 625	1.3 / 1.3	80 (21°C) / 117 (21°C)	3–12 over 40–5000 Hz at current drains quoted
303	1.5	0 to 0.24	WS16	SR47	165 to 1.3 V at 6500 Ω	24 / 24	17 μA / 17 μA / 246 μA / 246 μA	94000 / 94000 / 6500 / 6500	0.9 / 1.3 / 0.9 / 1.3	13 months / 12.8 months / 700 / 685	
386	1.5	0 to 10	WS10	SR43	120 to 0.9 V at 6500 Ω	24	240 μA	6500	1.3	510	
SHE and S41	1.5	0 to 10	S10	–	120 to 0.9 V rated at 1.6 mA	16 / 16	1.6 mA / 2.56 mA	100 / 625	1.3 / 1.3	80 (21°C) / 50 (21°C)	Open circuit 3–12 Ω over 40–5000 Hz at current drains quoted
SH301	1.5	0 to 100	WS10	SR43	100 to 1.3 V at 15000 Ω	24 / 24	10.6 μA / 10.6 μA / 107 μA / 107 μA	150000 / 150000 / 15000 / 15000	0.9 / 1.3 / 0.9 / 1.3	12.5 months estimated / 12 months estimated / 1050 / 1000	
393	1.5	0 to 5	WS5	–	75 to 0.9 V at 15000 Ω	24	104 μA	15000	1.3	725	
S13E and S13	1.5	0 to 5	S5	SR48	75 to 0.9 V rated at 1.07 mA	16 / 16	1.07 mA / 1.6 mA	1500 / 1000	1.3 / 1.3	74 (21°C) / 52 (21°C)	Open circuit 8–20 Ω over 40–5000 Hz at current drains quoted
309	1.5	0 to 0.1	WS5	SR48	60 to 1.3 V at 15000 Ω	24	107 μA / 5.9 μA	15000 / 270000	1.3 / 1.3	600 / 14 months estimated	Approximate open-circuit impedance at 1000 Hz, 35Ω
392	1.5	0 to 5	–	SR41	38 to 0.9 V at 15000 Ω	24	104 μA	15000	1.3	370	
S312E and S312	1.5	0 to 5	S4	SR41	38 to 0.9 V rated at 1 mA	16	1.0 mA / 1.5 mA	1500 / 1000	1.3 / 1.3	38 (21°C) / 22 (21°C)	
384	1.5	0 to 0.06	–	SR41	35 to 1.3 V at 15000 Ω	24	104 μA / 104 μA / 3.1 μA / 3.1 μA	15000 / 15000 / 500000 / 500000	0.9 / 1.3 / 0.9 / 1.3	370 / 360 / – / 15 months estimated	Open circuit at 1000 Hz, approximately 65 Ω

Table 52.5 Mallory Duracell silver–zinc cells and 6 V battery

Nominal voltage (V)	Nominal capacity (mA h)	Type	Maximum diameter (mm)	Maximum height (mm)	Weight (g)
1.50	36†	MS312H	7.75	3.56	0.57
1.50	38†	10L125	7.75	3.58	0.8
1.50	70†	10L122	11.56	2.79	1.0
1.50	70†	MS13H	7.75	5.33	1.0
1.50	75†	10L123	7.75	5.33	1.0
1.50	83*	WS12	11.56	3.43	1.4
1.50	85†	10L120	11.56	3.48	1.3
1.50	115*	WS11	11.56	4.19	1.7
1.50	120†	10L124	11.56	4.19	1.7
1.50	120†	MS41H	11.56	4.07	2.0
1.50		10L14	11.56	5.33	2.2
1.50	180†	MS76H	11.56	5.33	2.9
1.50	200*	WS14	11.56	5.59	2.2
1.50	250†	10L129	15.49	4.83	3.5
6.00	180	PX28	12.95	25.2	11.0

* Low-rate cell
† 10L125, 10L123, 10L120, 10L124, 10L129, 10L14 all for low-rate applications requiring high-rate pulses on demand

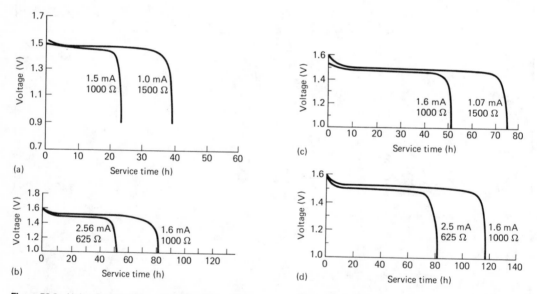

Figure 52.3 Union Carbide silver–zinc cell voltage discharge curves. Estimated hours of service at 21°C against voltage at stated starting drains (average drains in (a)) and loads and on a 16 h/day schedule. (a) S312 and S312E, 38 mA h to 0.9 V, rated at 1 mA; (b) S41 and S41E, 120 mA h to 0.9 V, rated at 1.6 mA; (c) S13E and S13 75 mA h to 0.9 V, rated at 1.07 mA; (d) S76E 190 mA h to 0.9 V, rated at 2.56 mA (Courtesy of Union Carbide)

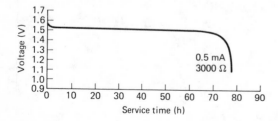

Figure 52.4 Typical discharge curves for a Duracell silver–zinc 10L125 (IEC designation SR41) 1.5 V cell, 38 mA h to 0.9 V on 30 000 Ω at 20°C: hours of service *versus* voltage (Courtesy of Duracell)

Table 52.6 Silver–zinc button cells supplied by Duracell Europe

Type No.	Capacity (mA h at 20°C)	Weight (g)	Volume (cm³)	Type No.	Capacity (mA h at 20°C)	Weight (g)	Volume (cm³)
D393	75 to 0.9 V on 20 kΩ	0.99	0.25	D391	35 to 0.9 V on 6.5 kΩ	0.8	0.23
D386	120 to 0.9 V on 15 kΩ	1.7	0.44	D392	38 to 0.9 V on 30 kΩ	0.8	0.16
10L14	130 to 0.9 V on 1.5 kΩ	2.0	0.55	D389	70 to 0.9 V on 13 kΩ	0.99	0.26

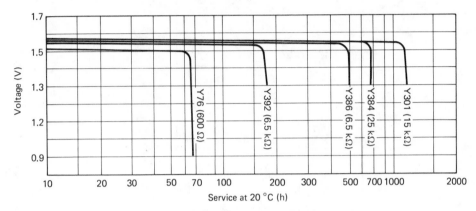

Figure 52.5 Yuasa silver–zinc cells: hours of service at 20°C *versus* voltage for various 1.5 V cells (Y301, Y384, Y386, Y392) and a 9 V battery (Y76) (Courtesy of Yuasa)

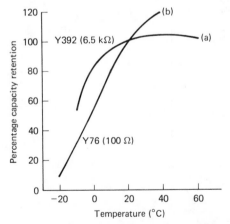

Figure 52.6 Effect of temperature on percentage capacity retention of Yuasa silver–zinc cells and batteries: (a) 1.5 V, 35 mA h cell on 6.5 kΩ load; (b) 9 V, 190 mA h battery on 100 Ω load (Courtesy of Yuasa)

Varta

Varta produce a range of 1.55 V silver–zinc cells in the capacity range 45–175 mA h and a 6 V battery of capacity 170 mA h (Table 52.7). These are used mainly in calculators, hearing-aids and photographic applications.

Varta also produce a range of 1.5 V low-drain silver–zinc cells for electric and electronic watches with analogue or LCD display; these are in the capacity range 50–180 mA h (Table 52.8).

Varta produce 1.5 V high-drain silver–zinc cells in the 38–190 mA h capacity range for use in solid-state watches with LED or LCD display with backlight (Table 52.9).

Yuasa

Yuasa also supply button design and cylindrical primary and rechargeable silver–zinc cells and batteries. Data in their range of 1.5 V watch cells are quoted in Table 52.10 and data on batteries of up to 9 V are quoted in Table 52.11. Service life data on Yuasa cells and batteries presented in Figures 52.5 and 52.6 show the effect of ambient temperature on percentage capacity retention of Yuasa silver–zinc cells and batteries.

Other suppliers

Other manufacturers' products equivalent to some of the Mallory silver–zinc Duracell range are quoted in Table 52.12.

Table 52.7 Silver–zinc cells and batteries supplied by Varta

Varta		Voltage (V)	Capacity (mA h)	Internal resistance (Ω)	Dimensions		Weight (g)	IEC reference	Main application
Type No.	Order No.				Diameter (mm)	Height (mm)			
V 8 GS	4173	1.55	45	20.0	11.6	2.1	0.9	–	Calculators
V 13 HS	4012	1.55	75	4.2–6.5	7.9	5.4	1.1	SR 48	Hearing-aids
V 10 GS	4174	1.55	85	10.0	11.6	3.1	1.4	–	Calculators
V 41 HS	4041	1.55	120	3.2–5.0	11.6	4.2	1.8	SR 43	Hearing-aids
V 12 GS	4178	1.55	130	6.0	11.6	4.2	1.9	SR 43	Calculators
V 76 HS	4076	1.55	175	3.5–5.5	11.6	5.4	2.4	SR 44	Hearing-aids
V 76 PX	4075	1.55	175	3.5–5.5	11.6	5.4	2.4	SR 44	Photography
V 13 GS	4176	1.55	175	3.5–5.5	11.6	5.4	2.4	SR 44	Calculators
V 28 PX	4028	6.00	170	15.0–22.0	13.0	25.8	12.8	–	Photography

Table 52.8 Varta low-drain silver–zinc cells for electric and electronic watches with analogue or LCD display

Type No.	Voltage (V)	Capacity (mA h)	Dimensions (mm)	Weight (g)	Shelf pack pieces	Replaces
521	1.5	180	11.6 × 5.4	2.33	10/100	303, RW 22/42, WS14, A, 9, SB-A9
523	1.5	55	9.5 × 2.6	0.75	10/100	395, 25, SB-Ap
524	1.5	67	9.5 × 3.6	1.04	10/100	394, 27, SB-A4
526	1.5	75	7.9 × 5.4	1.08	10/100	309, RW 28/48, 10L13, 70T, 16
527	1.5	50	7.9 × 3.6	0.69	10/100	384, RW 37/47, 10L15, 10, SB-A1
528	1.5	120	11.6 × 4.2	1.78	10/100	301, RW 24/44, WS11, D, 226, 1, SB-A8
529	1.5	110	11.6 × 3.6	1.49	10/100	344, RW 36, 10L120, 242, 12
531	1.5	15	6.8 × 2.15	0.33	10/100	364, SR620SW, SB-DG
532	1.5	20	7.9 × 2.1	0.40	10/100	362, RW 310, 19, SB-DK
533	1.5	45	11.6 × 2.1	0.90	10/100	381, RW 30, 23, SB-DS
543	1.5	85	11.6 × 3.1	1.32	10/100	390, RW 39, 10L122, 11/17, SB-BU
536	1.5	30	7.9 × 2.6	0.50	10/100	397, 26, SB-AL
537	1.5	30	9.5 × 2.1	0.61	10/100	371

Table 52.9 Varta silver–zinc high-drain cells for solid-state watches with LED or LCD display

Type No.	Voltage (V)	Capacity (mA h)	Dimensions (mm)	Weight (g)	Shelf pack pieces	Replaces
541	1.5	190	11.6 × 5.4	2.33	10/100	357, RW 22/42, 10L14, J, 228, 7, SB-B9
546	1.5	75	7.9 × 5.4	1.08	10/100	393, RW 28/48, 10L123, F, 255, 15
547	1.5	38	7.9 × 3.6	0.69	10/100	392, RW 27/47, 10L125, K, 247, 2, SB-B1
548	1.5	130	11.6 × 4.2	1.78	10/100	386, RW 24/44, 10L124, H, 260, 6, SB-B8
549	1.5	100	11.6 × 3.6	1.40	10/100	10L120, 14
553	1.5	45	11.6 × 2.1	0.90	10/100	391, RW 40, 10L130, 31, SB-BS
554	1.5	85	11.6 × 3.1	1.32	10/100	189, RW 49, 10L122, 17, SB-BU

Table 52.10 Yuasa 1.5 V watch silver–zinc cells

Number	Voltage (V)	Current range (mA)	Starting drain (mA)	Cut-off voltage (V)	Capacity (mA h)	Diameter (mm)	Height (mm)	Weight (g)
Y384 G33	1.5	0.1	0.06	1.3	35	7.9	3.6	0.7
Y392 G3	1.5	5.0	0.1	1.3	38	7.9	3.6	0.7
Y309 G55	1.5	0.2	0.1	1.3	60	7.9	5.4	1.1
Y393 G5	1.5	5.0	0.1	1.3	75	11.6	5.4	1.1
Y301 G12	1.5	0.25	0.1	1.3	100	11.6	4.2	1.7
Y386 G12	1.5	10.0	0.23	1.3	120	11.6	4.2	1.?
Y303 G514	1.5	0.5	0.23	1.3	165	11.6	5.6	2.?
Y357 G13	1.5	10.0	0.23	1.3	190	11.6	5.4	2.?
Y389	1.5	0.25	0.1	1.3	70	11.6	3.0	1.?

Table 52.11 Yuasa 1.5–9 V silver–zinc cells and batteries

Number	Voltage (V)	Current range (up to mA)	Starting drain (mA)	Cut-off voltage (V)	Capacity (mA h)	Diameter (mm)	Height (mm)	Weight (g)
Y312	1.5	5.0	1.0	0.9	38	7.9	3.6	0.7
Y13 G5	1.5	5.0	1.0	0.9	75	7.9	5.4	1.1
Y41 G12	1.5	10.0	1.6	0.9	120	11.6	4.2	1.7
Y76 G13	1.5	25.0	2.5	0.9	190	1.6	5.4	2.5
Y542 2G13	3.0	25.0	2.5	1.8	190	12.9	14.7	6.5
Y544 4G13	6.0	25.0	2.5	3.6	190	12.9	24.8	12.1
6Y76 6G13	9.0	25.0	2.5	5.4	190	13.4	34.7	17.5

Table 52.12 Equivalents in silver–zinc cells

Duracell (Mallory)	System	Nominal voltage (V)	Nominal capacity (mAh)	Height (mm)	Diameter (mm)	IEC	National	Ray-o-Vac	Ucar	Varta
WS11	Silver oxide	1.5	115	4.19	11.56	SR43	WS11	RW14, RW34	301	528
WS12	Silver oxide	1.5	83	3.43	11.56					529
10L14	Silver oxide	1.5	–	5.33	11.56	SR44	WL14	RW22, RW42	357	541
10L120	Silver oxide	1.5	85	3.48	11.56					549
10L122	Silver oxide	1.5	70	2.79	11.56					
10L123	Silver oxide	1.5	75	5.33	7.75	SR48	WS6	RW18, RW28 RW28, RW48	309, 393	546
10L124	Silver oxide	1.5	120	4.19	11.56	SR43	WL11	RW24, RW44	386	548
10L125	Silver oxide	1.5	38	3.58	7.75	SR41	WS1/WL1	RW27, RW37 RW47	384, 392	527, 547
10L129	Silver oxide	1.5	250	4.83	15.49			RW15, RW25 RW35	355	
WS14	Silver oxide	1.5	200	5.59	11.56		WS14	RW12, RW32	303 EPX 77	521
MS76H	Silver oxide	1.5	180							7301
MS13H	Silver oxide	1.5	70							7306
MS41H	Silver oxide	1.5	120							7308

52.1.2 Remotely activated silver oxide–zinc batteries

Silver–zinc primary batteries can be remotely activated for single use within seconds or fractions of a second even after long-term storage, by inserting the electrolyte under pressure. In addition to a unit containing dry charged plates they also contain an electrolyte vessel from which the cells are filled by electrical or mechanical means. Pile batteries are a new development of remotely activated primary batteries, which consist of bipolar electrodes that confer a very high density on the battery. Such batteries are available from Silberkraft (Table 52.13) and are recommended for use in aerospace, torpedoes and pyrotechnic ignition applications.

Eagle Picher also supply remotely activated primary silver–zinc batteries. In this cell design a vented system enhances the rapid clearing of the manifold and eliminates excessive intercell leakage paths. The battery has no moving parts. It has an extended shelf life. It can be activated directly into a load or in parallel with a mains system. The open-circuit voltage of this battery is 1.6–1.87 V with a working voltage of 1.20–1.55 V. The battery operating temperature is -29 to $+71°C$, and, with water assistance, -48 to $+71°C$. Units are available in the weight range 0.14–135 kg. Typical performance profiles at 27°C for the Eagle Picher remotely activated battery are shown in Figure 52.7. The dry storage charge retention characteristics of this battery are excellent.

Eagle Picher supply gas-generating pyrotechnic devices, which can be used for the activation of remotely activated batteries by transferring electrolyte from a reservoir to the battery cells. Eagle Picher have developed over 70 gas-generator configurations with capabilities ranging from 20 to 200 000 cm^3 (standard condition gas output), many of which have been used in a majority of missile/spacecraft systems.

The Chloride of Silver Dry Cell Battery Co. of Baltimore supply a silver–zinc battery with a mild

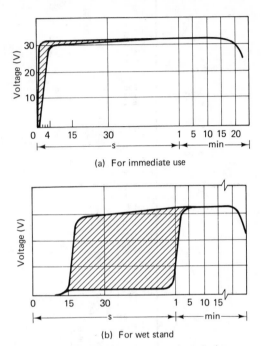

(a) For immediate use

(b) For wet stand

Figure 52.7 Eagle Picher remotely activated primary silver–zinc battery: typical performance profile at 27°C (Courtesy of Eagle Picher)

Table 52.13 Remotely activated primary silver–zinc batteries (Silberkraft)

	Voltage (V)	Discharge current	Discharge time (min)	Weight (kg)	Volume (cm^3)
Torpedo battery					
150PA/110/	210.0	480	13	400	285.0
20PA/29	28.0	40	45		
116PA/60/	156.0	480	6.5	196	124.3
20PA/7	28.0	40	25		
150PA 120	210.0	480	13	399	265.0
Rockets and missiles					
21PA9	30.5	35	5	14.5	9.5
20PA3, 5/4 PA3, 5	33.6	30	3	5.6	2.9
22PA30	31.5	260	5	33.0	21.5
Pyrotechnic ignition and fuse battery					
21PAB0, 19LS	30.0	1.5	0.18	0.042	0.02

Table 52.14 Characteristics of low-rate Yardney silver–zinc cells

Cell type	Nominal capacity (A h)	Charging rate (A)	Typical application data (at 21°C)								Max. filled weight (g)	Overall volume (mm³) × 10³	Overall height (mm)	Height less terminals (mm)	Width (mm)	Depth (mm)	Terminal threads
			10 h rate discharge			Power/ weight ratio* (W h/kg)	Power/ volume ratio† (W h/dm³)	1 h rate discharge									
			Discharge rate (A)	A h output*	Average voltage (V)			Discharge rate (A)	A h output*	Average voltage (V)							
LR-05	0.5	0.05	0.05	1.0	1.50	66	102	0.5	0.75	1.38	22	14.9	39.6	30.5	27.4	13.7	5-40
LR-1	1	0.07	0.10	2.0	1.50	95	150	1.0	1.75	1.35	31	19.3	51.3	42.9	27.4	13.7	5-40
LR-2	2	0.15	0.20	3.6	1.52	84	126	2.0	3.00	1.50	64	42.0	64.2	54.8	43.7	15.0	10-32
LR-3	3	0.20	0.30	4.0	1.50	70	126	3.0	3.50	1.46	84	47.5	72.6	63.2	43.7	15.0	10-32
LR-4	4	0.30	0.40	6.0	1.53	90	168	4.0	5.00	1.46	101	54.2	85.3	76.2	43.7	15.0	10-32
LR-5	5	0.30	0.50	7.5	1.52	88	134	5.0	6.50	1.44	126	78.4	73.9	63.2	52.8	20.0	10-32
LR-10	10	0.50	1.00	13.0	1.50	84	134	10.0	12.0	1.44	230	135.5	122.0	109.0	58.9	18.8	1/4.28
LR-15	15	1.00	1.50	20.0	1.52	107	198	15.0	18.0	1.40	280	149.7	125.0	107.0	58.7	20.3	5/16.24
LR-20	20	1.00	2.00	30.0	1.50	112	180	20.0	25.0	1.40	392	249.3	109.0	92.4	52.0	43.9	5/16.24
LR-2L	20	1.50	2.00	30.0	1.53	103	198	20.0	27.0	1.43	434	228.0	191.0	174.0	58.4	20.3	5/16.24
LR-40	40	2.00	4.00	46.0	1.52	108	186	40.0	43.0	1.40	644	374.0	180.0	162.0	82.5	25.1	1/4.28
LR-58	60	2.50	6.00	90.0	1.52	145	264	60.0	80.0	1.40	890	490.0	184.0	161.0	82.5	32.2	5/16.24
LR-60	60	3.00	6.00	75.0	1.52	121	204	60.0	55.0	1.36	812	475.0	114.0	98.0	69.3	59.9	5/16.24
LR-70	70	4.00	7.00	80.0	1.50	106	242	70.0	70.0	1.36	1120	526.0	159.0	141.0	92.4	35.8	5/16.24
LR-90	90	4.00	9.00	145.0	1.52	134	246	90.0	135.0	1.45	1599	815.0	179.0	162.0	82.8	54.8	5/16.24
LR-100	100	4.00	10.00	110.0	1.51	132	216	100.0	100.0	1.40	1232	756.0	102.0	113.0	87.3	70.6	5/16.24
LR-130	130	5.00	13.00	170.0	1.52	145	288	130.0	150.0	1.34	1758	882.0	166.0	149.0	83.2	63.5	7/16.20
LR-200	200	8.00	20.00	225.0	1.50	119	252	200.0	210.0	1.36	2940	1341.0	278.0	257.0	106.0	45.4	7/16.20
LR-525	525	30.00	60.00	640.0	1.52	95	198	Do not discharge at 1 h rate			10080	4854.0	428.0	393.0	106.0	106.0	3/4.16

* To a final voltage of 1.1 V
† Calculated using overall volume

acid electrolyte containing silver chloride as a depolarizer. This battery is recommended as a power unit in test instruments such as blasting galvanometers, circuit testers, volt-ohmmeters and clocks. Multi-cell batteries for B-voltages and bias are available for use in electronic circuits, electromedical instruments and other specialized low-drain uses where constant voltage and open life are important. The battery has an open-circuit voltage of 1.05 V. Under load, terminal voltage drops in proportion to load, but holds constant throughout the useful section of the discharge curve. Shelf life is several times longer than that of ordinary dry cells at room temperature. However, the cells should not be stored below 0°C. Shelf life is 6 months under correct storage conditions and this may be extended by cold storage.

52.1.3 Secondary batteries

Yardney

The extremely low internal resistance of silver–zinc batteries permits discharges at rates as high as 30 times the ampere hour capacity rating, and its flat voltage characteristic enables highest operational efficiency and dependability. Cells have been built with capacity ranging from 0.1 to 20000 A h. A recent advance using new separators patented by Yardney enabled their silver–zinc batteries to provide more than 400 cycles in special applications.

Yardney standard rechargeable silver–zinc batteries are available in two types; as a high-rate unit for applications where total energy must be delivered within 1 h, and as a low-rate unit where discharge rates are lower and are required for longer time periods.

The cells are leakproof and spillproof. The assembled cells can be enclosed in sturdy lightweight cases for special mechanical and environmental conditions. Batteries can also be designed and built to meet the most stringent requirements.

The data below indicate, in brief, the performance capabilities achieved with existing models of Yardney Silvercels showing the requirements which can be met.

1. *Vibration* 30–40 sinusoidal vibration.
2. *Random Gaussian vibration* 5–2000 Hz band, with as high as 60g r.m.s. equivalent in both non-operational and operational conditions: hard mounted unit.
3. *Shock* Tested up to 200g in all directions.
4. *Acceleration* 100g in all directions of sustained acceleration, except in the direction of the vent for vented cell systems.
5. *Thermal shock* From +70 to −50°C and repeated, with stabilization occurring at each temperature.

6. *Altitude* Successful activation and operation at a simulated altitude of 60 km for a period of 1 h without pressurization; with pressurization there is no limit.

The electrical characteristics of the full available range of 1.45–1.5 V Silvercels are given in Table 52.14 (low rate) and Table 52.15 (high rate). It will be noted that high-rate Silvercels are available in the capacity range 0.1–200 A h and the low-rate versions in the range 0.5–525 A h. High-rate cells are recommended where requirements demand maximum energy density and voltage regulation at high discharge rates through a cycle life period of 4–6 months. Low-rate cells are recommended where requirements demand maximum energy density and voltage regulation at low discharge rates through a cycle life period of 6–18 months.

Figure 52.8 shows the effect of discharge rate in terms of C nominal 1 h discharge rate on energy per unit weight and energy per unit volume for Silvercels in the 0.1–300 A h capacity range. The distinct improvement in the energy per unit weight or volume of silver–zinc over nickel–cadmium, nickel–iron and lead–acid battery systems is very apparent.

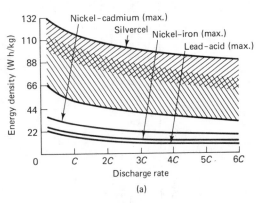

(a)

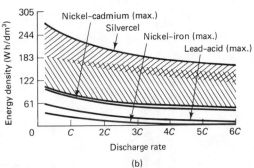

(b)

Figure 52.8 Effect of discharge rate on energy output: Yardney Silvercel rechargeable silver–zinc battery (▨ 21–300 A h capacity, ▧ 0.1–20 A h capacity) compared with other battery types (Courtesy of Yardney)

Table 52.15 Characteristics of high rate Yardney silvered zinc cells

Cell type	Nominal capacity (A h)	Charging rate (A)	Typical application data (at 21°C)													Max. filled weight (g)	Overall volume (mm³) ×10³	Overall height (mm)	Height less terminals (mm)	Width (mm)	Depth (mm)	Terminal threads
			60 min rate discharge					20 min rate discharge				10 min rate discharge										
			Discharge rate (A)	A h output*	Average voltage (V)	Power/weight ratio (W h/kg)	Power/volume ratio‡ (W h/dm³)	Discharge rate (A)	A h output	Average voltage (V)	Time limit† (min)	Discharge rate (A)	A h output	Average voltage (V)	Time limit† (min)							
HR-01	0.1	0.01	0.1	0.175	1.45	55	78	0.3	0.15	1.40	16	0.6	0.12	1.32	6	4.54	3.11	35.0	30.5	16.0	5.6	1-72
HR-02	0.2	0.02	0.2	0.22	1.48	50	72	0.6	0.21	1.35	15	1.2	0.17	1.25	5	6.53	4.43	49.2	42.9	16.0	5.6	1-72
HR-05	0.5	0.07	0.5	1.30	1.50	86	126	1.5	1.10	1.45	20	3.0	0.94	1.35	8	22.7	14.90	39.6	30.5	27.4	13.7	5-40
HR-1	1.0	0.10	1.0	1.75	1.50	84	132	3.0	1.60	1.44	20	6.0	1.40	1.35	8	31.2	19.30	51.3	42.9	27.4	13.7	5-40
HR-1.5	1.5	0.15	1.5	2.50	1.50	95	162	4.5	2.00	1.40	20	9.0	1.50	1.32	8	39.8	22.10	58.9	50.3	27.4	13.7	5-40
HR-2	2.0	0.20	2.0	3.30	1.51	73	114	6.0	3.00	1.46	20	12.0	2.40	1.40	8	68.1	42.00	64.2	54.9	43.7	15.0	10-32
HR-3	3.0	0.30	3.0	5.00	1.50	84	156	9.0	4.70	1.46	20	18.0	4.20	1.37	8	90.9	47.50	72.6	63.2	43.7	15.0	10-32
HR-4	4.0	0.30	4.0	6.00	1.50	86	162	12.0	5.40	1.49	20	24.0	4.80	1.40	8	105.1	54.30	85.3	72.6	43.7	15.0	10-32
HR-5	5.0	0.35	5.0	7.50	1.42	81	132	15.0	5.90	1.36	20	30.0	5.20	1.26	6	130.6	78.40	73.9	63.2	52.8	20.0	10-32
HR-10	10.0	0.75	10.0	12.00	1.44	75	126	30.0	10.50	1.33	18	60.0	8.00	1.26	5	232.9	135.40	122.0	109.0	58.9	18.8	1/4.28
HR-15	15.0	1.00	15.0	18.50	1.48	97	180	45.0	16.50	1.44	15	90.0	13.00	1.30	5	284.0	149.70	125.5	107.0	58.7	20.3	5/16.24
HR-18	18.0	2.00	18.0	20.00	1.48	79	130	55.0	18.00	1.40	15	100.0	16.00	1.30	5	372.0	249.30	178.0	159.0	53.8	20.6	5/16.24
HR-21	21.0	2.00	20.0	30.00	1.50	101	192	60.0	27.00	1.44	17	120.0	23.00	1.34	6	440.2	228.00	191.0	174.0	58.4	20.3	5/16.24
HR-40	40.0	3.00	40.0	49.00	1.45	90	186	120.0	43.50	1.36	18	240.0	39.00	1.20	5	710.0	373.90	180.0	162.0	82.5	25.1	5/16.24
HR-58	60.0	4.00	60.0	75.00	1.48	123	222	180.0	55.00	1.36	12	360.0	40.00	1.21	5	903.1	490.0	184.0	161.0	82.5	32.2	7/16.24
HR-60	60.0	4.00	60.0	74.00	1.47	117	228	180.0	61.00	1.39	12	360.0	46.00	1.28	5	937.2	476.8	114.0	98.0	69.3	59.9	5/16.24
HR-70	70.0	4.50	70.0	75.00	1.45	97	204	210.0	62.00	1.35	14	420.0	52.00	1.81	5	1136.0	526.4	159.0	141.0	92.4	71.4	5/16.24
HR-80	80.0	6.00	80.0	95.00	1.50	110	204	240.0	70.00	1.38	10	480.0	60.00	1.25	4	1363.0	685.5	159.0	141.0	71.4	44.4	5/16.24
HR-85	85.0	6.00	100.0	120.00	1.48	108	202	300.0	97.00	1.40	12	600.0	80.00	1.30	5	1647.0	187.0	216.0	198.0	71.4	46.0	5/16.24
HR-90	90.0	6.00	90.0	110.00	1.50	108	198	270.0	100.00	1.42	12	540.0	90.00	1.29	5	1534.0	815.1	240.0	220.0	82.8	54.8	3/8.24
HR-100	100.0	6.00	100.0	100.00	1.46	114	192	300.0	93.00	1.41	15	600.0	90.00	1.20	6	1278.0	756.0	179.0	162.0	87.3	70.6	5/16.24
HR-140	140.0	10.00	140.0	180.00	1.45	114	234	420.0	155.00	1.35	10	840	110.00	1.20	6	2272.0	1102.0	122.0	113.0	82.5	72.4	7/16.24
HR-160	160.0	10.00	160.0	200.00	1.45	150	312	480.0	170.00	1.32	20	600.0	140.00	1.26	10	1951.0	912.0	183.0	167.0	75.4	73.4	3/8.24
HR-200	200.0	10.00	100.0	290.00	1.47	143	312	200.0	250.00	1.40	60	380.0	170.00	1.25	20	2982.0	1341.0	278.0	257.0	106.0	45.4	3/8.24

* To a final voltage of 1.1 V
† For a maximum recyclability, observe discharge time specified in battery instructions
‡ Calculated using maximum volume

Table 52.16 Characteristics of Yardney Modular Silvercel Pac batteries

Module type	Nominal capacity (A h)	Charging rate (A)	Typical application data (at 21°C)						Max. filled weight (g)	Overall volume (mm³) × 10³	Volume less terminals (mm³) × 10³	Overall height (mm)	Height less terminals (mm)	Width (mm)	Depth (mm)	Terminal threads
			10 h rate discharge			1 h rate discharge										
			Discharge rate (A)	A h output	Average voltage (V)	Discharge rate (A)	A h output	Average voltage (V)								
LR3/3	3.0	15.0 *	0.3	3.3†	4.5	3.0	3.0†	4.2	162	129	118	73.1	66.8	47.5	37.3	6-32
LR3/5	5.0	0.25*	0.5	5.5†	4.5	5.0	5.1†	4.3	261	180	169	101.5	95.2	47.5	37.3	6-32
LR2/10	10.0	0.5 **	1.0	11.0‡	3.0	10.0	10.0‡	2.8	284	144	134	101.5	93.4	42.9	33.3	10-32
LR3/10	10.0	0.5 **	1.0	11.0†	4.5	10.0	10.0†	4.2	528	261	239	99.0	91.4	142.0	20.3	10-32
LR4/22	22.0	1.0 ***	2.0	24.0§	6.0	22.0	20.0§	5.6	1181	590	558	110.0	101.5	71.1	77.2	3/4.28
LR4/80	80.0	6.0	8.0	90.0§	6.1	55.0¶	80.0§	5.8	4317	2074	270	181.0	167.0	84.3	136.0	5/16.24

* Charge to 5.9–6.0 V/module
** Charge to 3.95–4.0 V/module
*** Charge to 7.9–8.0 V/module
† To a final voltage of 3.3 V
‡ To a final voltage of 2.2 V
§ To a final voltage of 4.4 V
¶ Discharge for 85 min at 55 A rate

Figure 52.9 shows the charging curve for both types of Silvercel. Figure 52.10 shows the typical effect of temperature on the energy obtained per unit weight for Silvercels operated without heaters. As would be expected, energy per unit weight decreases somewhat with a lowering of battery temperature.

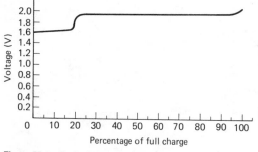

Figure 52.9 Typical charging curve at 20°C for high- and low-rate Yardney Silvercel silver–zinc batteries (Courtesy of Yardney)

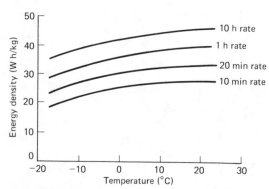

Figure 52.10 Effect of battery temperature on energy per unit volume for a Yardney Silvercel rechargeable silver–zinc battery operated without heaters (Courtesy of Yardney)

Yardney also supply modular battery packs of Silvercels; the Silvercel Pac. These are available in nominal capacities between 3 and 80 A h (Table 52.16). These batteries are between 2.8 and 15.8 V and have equivalent energy densities of about 100 W h/kg, which exceeds the energy of a nickel–cadmium battery by four times and a lead–acid battery by five times. The silver–zinc batteries are, of course, more expensive than nickel–cadmium or lead–acid batteries, but for many applications the space and weight savings they offer would make them a very attractive alternative.

Silberkraft

Silberkraft manufacture silver–zinc cells and batteries in West Germany under licence to Yardney. They offer a range of silver–zinc cells and batteries in the capacity range 0.1–750 A h, covering several types of application which meet the NATO AQAPI specification. These types include the VG 95284 range of rechargeable secondary cells suitable for the propulsion of torpedoes, submarines, combat vehicles, rockets and emergency power for radios (Table 52.17). In addition, standard secondary silver–zinc batteries and cells for general use are available (Table 52.18).

Characteristic data for the cells and batteries in Table 52.18 are as follows:

1. Type S for a 2–10 h discharge, electrodes not charged.
2. Type SHV for a discharge time of 2 h down to a few minutes, with dry charged electrodes: discharge of cell can be performed in 24 h after filling with electrolyte.
3. Cycle life: Type S 80–120 cycles in filled condition 1–2 years. Type SHV 5–50 cycles, depending on load and ambient temperature, 0.5–1 year in filled condition.

Table 52.17 Characteristics of Silbercraft secondary silver–zinc cells and batteries meeting NATO Standard VG No. 95284

	Nominal capacity (A h)	Nominal voltage (V)	Filled weight (kg)	Length (mm)	Width (mm)	Height (mm)
For propulsion batteries for submarines, torpedoes, combat vehicles, rockets, emergency power supplied for radio transmitters						
02100 SHV	20	1.5	0.380	80	85	90
04200 SHV	40	1.5	0.770	80	35	178
6500 SHU	60	1.5	1.005	80	35	184
09000 SHU	90	1.5	1.540	83	55	179
012000 SHU	120	1.5	1.860	80	68	184
HR 36DC	36	1.5	0.890	82	32	152
Propulsion battery for torpedo U/K 53	90	90.0	125.000	762	406	402

Table 52.18 Characteristics of Silberkraft standard secondary silver–zinc cells

Type	Nominal capacity (A h)	Mean discharge (V)	Charging current (A)	Weight filled (g)	Length (mm)	Width (mm)	Height including terminals (mm)
050 S	0.5	1.50	0.05	25	28	16	38
050 SHV	0.5	1.50	0.07	25	28	16	38
0100 S	1.0	1.50	0.07	35	28	16	52
0100 SHV	1.0	1.50	0.10	37	28	16	52
0300 S	3.0	1.50	0.20	70	35	21	62
0300 SHV	3.0	1.50	0.30	70	35	21	62
0500 S	5.0	1.50	0.30	105	35	21	82
0500 SHV	5.0	1.45	0.35	105	35	21	82
0700 S	7.0	1.50	0.50	150	35	21	117
0700 SHV	7.0	1.48	0.60	150	35	21	117
0800 S	8.0	1.50	0.60	160	43	23	100
0800 SHV	8.0	1.48	0.70	160	43	23	100
01000 S	10.0	1.50	0.70	215	50	23	114
01000 SHV	10.0	1.46	0.80	215	50	23	114
01500 S	15.0	1.50	1.00	310	43	43	105
01500 SHV	15.0	1.50	1.30	310	59	21	125
02000 S	20.0	1.50	1.40	440	62	43	105
02000 SHV	20.0	1.50	2.00	380	80	35	90
04000 S	40.0	1.50	2.00	750	83	25	180
04000 SHV	40.0	1.50	4.00	770	80	35	178
05000 SHV	50.0	1.50	4.00	900	77	35	170
06000 S	60.0	1.50	4.00	910	83	32	184
06000 SHV	60.0	1.48	5.00	1005	80	35	185
09000 S	90.0	1.50	4.00	1540	83	55	179
09000 SHV	90.0	1.50	6.00	1540	83	55	179
9200 S	120.0	1.50	10.00	1920	80	68	185
9200 SHV	120.0	1.50	10.00	1920	80	68	185

4. Open-circuit voltage: $U_o = 1.82–1.86$ V.
5. Voltage under nominal load: $U_e = 1.5$ V.
6. Storage time: 5 years in dry condition.
7. Storage temperature: -60 to $+80°C$ dry, -40 to $+75°C$ wet.
8. Operating temperature: -20 to $+75°C$.
9. Self-discharge: 15% per year.
10. Gas generation: very low under normal operating conditions.
11. Internal resistance: at the milliohm level depending on capacity, temperature and discharge rate.
12. The cells are short-circuit proof; the magnitude of the currents under short-circuit are eight times the nominal capacity (for S cells) and 80 times the nominal capacity (for SHV cells).
13. Charge temperatures: when cell voltage (U_2) is 2.00–2.05, charging may be given during battery operation by either constant-current or constant-voltage methods.

Eagle Picher

Eagle Picher supply ranges of low-rate and high-rate silver–zinc cells and batteries in the rated capacity range 0.8–320 A h. The characteristics of some batteries in this range are given in Table 52.19. The cases of these batteries are low-pressure vented and non-spill. Battery heaters are available for operation at $-39°C$ and below. Although the theoretical energy density of the silver–zinc battery is about 440 W h/kg, the actual energy density of practical cells is in the range 55–209 W h/kg as cells and 37–114 W h/kg as batteries (see Table 52.19). The practical volumetric energy density of Eagle Picher batteries is 80–415 W h/dm^3 as cells and 55–262 W h/dm^3 as batteries. The cells have a working voltage of 1.55–1.0 V and require voltage regulation of ±0.1 V for 90% of discharge. Operating and storage temperatures, respectively, are -39 to $+73°C$ and -53 to $+73°C$. The batteries have good thermal, vibration and acceleration resistance and can be used in any position at any altitude. Eagle Picher claim a dry shelf life of 2.5 years and a wet shelf life of up to 1 year at 21°C. In excess of 500 cycles are claimed, with a charge retention of up to 1 year at 26°C. Either constant-current or constant-potential (1.96–1.98 V) charging methods can be used and 16 h is a typical recharge time. No electrolyte maintenance is normally required.

Table 52.19 Characteristics of Eagle Picher rechargeable silver–zinc batteries

Cell type	Rated capacity (A h)	Nominal capacity (A h) at given rates			Power/ weight ratio (W h/kg)	Power/ volume ratio (W h/dm³)	Weight (g)	Depth (mm)	Width (mm)	Height (mm)	Volume (mm³) × 10³
		4 h	10 h	20 h							
Low rate											
SZLR 0.8	0.8	0.8	0.8	0.8	53	78	22.7	10.92	26.92	40.64	13.27
SZLR 1.5	1.5	1.5	1.5	1.5	66	102	34.0	12.45	30.73	46.99	21.30
SZLR 3.0	3.0	3.0	3.0	3.0	84	144	53.9	14.22	35.05	53.34	31.14
SZLR 70.0	70.0	65.0	70.0	70.0	156	294	671.8	32.51	80.77	121.41	352.32
SZLR 115.0	115.0	110.0	110.0	115.0	172	324	997.9	37.34	92.71	139.45	522.75
SZLR 320.0	320.0	–	290.0	320.0	176	426	2732.9	42.16	138.43	181.10	1124.15
		15 min	30 min	60 min							
High rate											
SZHR 0.8	0.8	0.7	0.7	0.8	53	78	22.7	10.92	26.92	40.64	13.27
SZHR 1.5	1.5	1.4	1.5	1.5	66	112	34.0	12.45	30.73	46.99	21.30
SZHR 2.8	2.8	2.6	2.7	2.8	77	132	53.9	14.22	35.05	53.34	31.14
SZHR 65.0	65.0	–	50.0	65.0	152	270	640.7	32.51	80.77	121.41	352.32
SZHR 100.0	100.0	–	80.0	100.0	158	282	952.5	37.34	92.71	139.45	522.75
SZHR 290.0	290.0	–	–	290.0	165	438	2608.1	42.16	138.43	181.10	1124.15

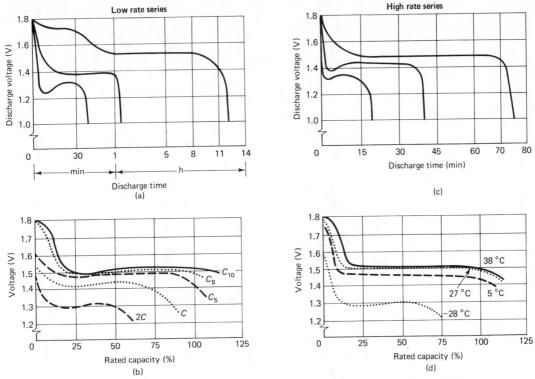

Figure 52.11 Characteristics of Eagle Picher high- and low-rate rechargeable silver–zinc batteries. (a–c) 27°C (Courtesy of Eagle Picher)

Figure 52.11 shows typical capacity and voltage characteristics for high- and low-rate Eagle Picher cells. Discharge characteristics at various rates and the effect of temperature on discharge voltages are also shown.

52.2 Silver–cadmium (secondary) batteries

Yardney

The Yardney Silcad silver–cadmium battery combines in one system the high energy and excellent space and weight characteristics of the silver–zinc battery with the long life, low-rate characteristics with some resistance to overcharge characteristic of the nickel–cadmium batteries. It provides high efficiency, an extended shelf life in charged or uncharged conditions, level voltage and mechanical ruggedness. Energy densities are two to three times greater than for a comparable nickel–cadmium battery.

The electrical characteristics of these batteries are given in Table 52.20. Figure 52.12 shows typical charge and discharge curves for a Silcad battery.

Yardney also supply a range of Silcad batteries made up into modular packs (Table 52.21) giving a versatile compact power source for industrial and commercial battery uses.

Silberkraft

Silberkraft supply silver–cadmium batteries in the 0.1–300 A h range providing a voltage of approximately 1.4 or 1.1 V. These have a high cyclic life and a slightly lower energy density and voltage than silver–zinc batteries. They are also suitable for use at low discharge rates. These batteries have been used in satellites for up to 6000 cycles with a depth of discharge of 50%.

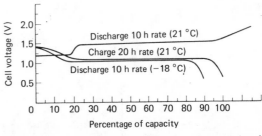

Figure 52.12 Typical charge and discharge characteristics of a Silcad silver–cadmium cell (Courtesy of Eagle Picher)

Table 52.20 Characteristics of Yardney Silcad rechargeable silver–cadmium batteries

Cell type	Nominal capacity (A h)	Charging rate (A)	Max. continuous discharge rate (A)	Typical application data (at 21°C)						Max. filled weight (g)	Overall volume (mm³) × 10³	Volume less terminals (mm³) × 10³	Overall height (mm)	Height less terminals (mm)	Width (mm)	Depth (mm)	Terminal threads
				Discharge rate (A)	Discharge time (h)	A h output	Average voltage (V)	Power/weight ratio (W h/kg)	Power/volume ratio (W h/dm³)								
YS-01	0.1	0.007	0.1	0.05	2.70	0.14	1.08	29	46	5.11	3.11	2.79	35.0	30.5	16.0	5.6	1-72
YS-05	0.5	0.035	0.5	0.05	15.00	0.75	1.10	40	55	21.30	14.20	11.50	39.3	30.5	27.4	13.7	5-40
YS-1	1.0	0.070	1.0	0.10	15.00	1.50	1.10	48	84	34.10	19.30	16.10	51.3	42.9	27.4	13.7	5-40
YS-2	2.0	0.140	2.0	0.20	14.00	2.80	1.10	48	72	65.30	42.10	35.90	64.2	54.8	43.7	15.0	10-32
YS-3	3.0	0.200	3.0	0.30	13.30	4.00	1.10	48	90	90.90	47.90	41.00	72.6	63.2	43.7	15.0	10-32
YS-5	5.0	0.300	5.0	0.50	14.60	7.30	1.08	55	96	142.00	78.30	67.10	73.9	63.2	52.8	20.0	10-32
YS-10	10.0	0.500	10.0	1.00	11.50	11.50	1.10	48	90	261.00	135.30	120.50	122.0	109.0	58.9	18.7	1/4.28
YS-18	18.0	1.000	18.0	2.00	10.00	20.00	1.10	59	108	369.00	198.40	178.70	178.0	159.0	54.1	20.6	5/16.24
YS-20	20.0	1.500	60.0	2.00	1.30	26.00	1.08	64	114	429.00	249.30	211.50	109.0	92.4	43.9	52.0	5/16.24
YS-40	40.0	3.000	40.0	4.00	12.00	48.00	1.10	70	138	746.00	372.30	337.80	179.0	161.5	82.5	25.1	5/16.24
YS-60	60.0	3.000	60.0	6.00	1.25	75.00	1.10	68	168	1207.00	475.50	406.70	114.0	98.0	69.3	59.9	5/16.24
YS-70	70.0	4.000	70.0	7.00	11.50	80.00	1.10	75	162	1193.00	526.40	467.40	159.0	141.0	92.4	35.8	3/8.24
YS-85	85.0	5.000	100.0	7.00	14.30	100.00	1.08	62	132	1732.00	793.80	713.40	240.0	216.0	71.4	45.9	5/16.24
YS-100	100.0	4.000	100.0	10.00	11.50	115.00	1.08	84	162	1505.00	754.40	700.30	122.0	113.0	70.6	87.3	5/16.24
YS-150	150.0	8.000	120.0	30.00	5.50	165.00	1.06	56	132	3095.00	1309.00	1215.00	272.0	227.0	106.0	45.2	3/8.24
YS-300	300.0	20.000	150.0	30.00	12.00	360.00	1.07	75	132	5197.00	2148.00	2017.00	432.0	419.0	106.0	25.4	3/8.24

Table 52.21 Characteristics of Yardney rechargeable modular Silvercel Pac batteries

Module type	Nominal capacity (A h)	Charging rate (A)	Typical application data (at 21°C)						Max. filled weight (g)	Overall volume (mm³) × 10³	Volume less terminals (mm³) × 10³	Overall height	Height less terminals (mm)	Width (mm)	Depth (mm)	Terminal threads
			10 h rate discharge			1 h rate discharge										
			Discharge rate (A)	A h output	Average voltage (V)	Discharge rate (A)	A h output	Average voltage (V)								
BD-3/2 High rate	2.0†	0.2	0.2	2.4	3.3	2.0	2.1	3.1	15.9	130	118	73.1	66.8	47.5	37.3	6-32
BD-3/3 Low rate	3.0	0.15	0.3	3.2	3.2	3.0	3.1	3.0	16.7	130	118	73.1	66.8	47.5	37.3	6-32
BD-3/4 High rate	4.0	0.4	0.4	4.5	3.3	4.0	4.1	3.1	27.2	180	168	101.0	95.2	47.5	37.3	6-32
BD-3/5 Low rate	5.0	0.2	0.5	5.5	3.25	5.0	5.3	3.0	29.5	180	168	101.0	95.2	47.5	37.3	6-32
BD-8/5.5 Low rate	5.5	0.4‡	5.5	6.0*	8.4	5.5	5.0*	8.0	105.6	640	640	100.0	100.0	79.2	80.5	Plug
BD-3/10 Low rate	10.0	0.5	1.0	11.0	3.2	10.0	10.0	3.1	48.8	261	230	99.0	91.4	142.0	20.3	10-32

* To a final voltage of 2.7 V
† Charge to 4.65–4.76 V/module
‡ Charge to 12.4–12.7 V/module

Table 52.22 Characteristics of Eagle Picher rechargeable high-rate and low-rate silver–cadmium batteries

Cell type	Rated capacity (A h)	Nominal capacity (A h)			Gravimetric energy density (W h/kg)	Volumetric energy density (W h/dm³)	Max. filled weight (g)	Depth (mm)	Width (mm)	Height (mm)	Overall volume (mm³) × 10³
Low rate		1 h rate	4 h rate	10 h rate							
SCLR 0.7	0.7	0.7	0.7	0.7	33	51.0	22.7	11.4	26.9	40.6	14.8
SCLR 1.1	1.1	1.1	1.1	1.1	33	55.8	36.9	12.4	30.7	47.0	21
SCLR 2.2	2.2	2.2	2.2	2.2	40	78	59.6	14.2	35.0	53.3	31
SCLR 45.0	45.0	35.0	40.0	45.0	64	138	781	32.5	80.8	121.0	352
SCLR 70.0	70.0	50.0	65.0	70.0	68	144	1136	37.3	92.7	139.0	523
SCLR 175.0	175.0	NA*	150.0	175.0	73	168	2655	42.1	138.0	181.0	1125
High rate		15 min rate	30 min rate	60 min rate							
SCHR 0.5	0.5	0.5	0.5	0.5	24	36	22.7	11.4	26.9	40.6	14.8
SCHR 1.0	1.0	1.0	1.0	1.0	31	48	37	12.4	30.7	47.0	21
SCHR 2.0	2.0	2.0	2.0	2.0	37	72	60	14.2	35.0	53.3	31
SCHR 17.0	17.0	16.5	16.8	17.0	44	78	434	28.7	71.4	107.0	244
SCHR 31.0	31.0	28.0	30.0	31.0	51	96	690	32.5	80.8	121.0	352
SCHR 55.0	55.0	28.0	40.0	55.0	57	114	1045	37.3	92.7	139.0	523

* Not applicable.

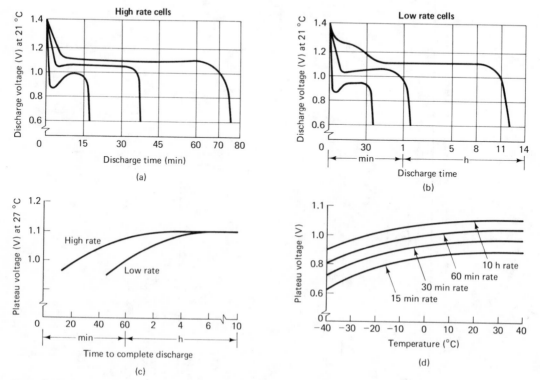

Figure 52.13 Characteristics of Eagle Picher high- and low-rate rechargeable silver–cadmium batteries (Courtesy of Eagle Picher)

Eagle Picher

Eagle Picher supply silver–cadmium batteries in the nominal capacity range 0.5–175 A h. The electrical characteristics of some batteries in this range are given in Table 52.22. The batteries are available in high-rate and low-rate versions.

The major application for the high-rate series is for service conditions where maximum energy density and voltage regulation at high rates of discharge are required throughout a cycle life of 6–18 months. The system also has an exceptional ability to retain its charge during long periods of wet storage.

The major application for the low-rate series is for service conditions in which maximum energy density and voltage regulation at low rates of discharge are required throughout a cycle life of 12–36 months. The system also has an exceptional ability to retain its charge during long periods of wet storage.

The cell cases are fabricated in styrene–acrylonitrile, nylon or acrylonitrile–butadiene–styrene and the battery cases in stainless steel, magnesium, titanium, fibreglass or polyvinyl chloride. Optional battery heaters are available. Although the theoretical energy density of the silver–cadmium battery is 310 W h/kg, the actual energy density of practical Eagle Picher batteries is in the range 24–73 W h/kg as cells and 18–44 W h/kg as batteries.

The practical volumetric energy density is 36–171 W h/dm^3 as cells and 24–122 W h/dm^3 as batteries. The batteries have a working voltage of 0.9–1.12 V/cell and require voltage regulation of ±5% under fixed conditions. The operating temperature is −40 to +71°C, or to −54°C with a heater. Batteries can be stored at −54 to +51°C dry or −40 to +38°C wet. A shelf life of 2.5 years when dry and 1.3 years when wet is to be expected. Above 90% charge retention is claimed after 6 months at 21°C. Up to 3000 cycles are claimed. Constant-current charging is recommended, with a charge rate of one-tenth nominal capacity to 1.65 V/cell or 1.55 V/cell minimum. Alternatively, constant-potential charging can be adopted with an input of 1.65 ± 0.02 V. With constant potential charging, the current is limited to one-fifth nominal capacity charging to the trickle charge rate.

Figure 52.13 shows typical capacity and voltage characteristics for high- and low-rate silver–cadmium batteries. Discharge characteristics at various rates and the effect of temperature on discharge voltage are also shown.

53

Alkaline manganese dioxide batteries

Contents

Table 53.1 Union Carbide Eveready alkaline manganese cells and batteries: estimated service time at 21°C

Type	Suggested current range (mA)	Designation ANSI	IEC	Voltage (V)	Weight (g)	Volume (cm³)	Number of cells	Schedule (h/day)	Starting drain (mA)	Load (Ω)	Service time (h) at cut-off voltage	Height (mm)	Length (mm)
EPX825 (Button)	0 to 10	–	–	1.5	6.9	2.0	1	Continuous	10.0	150	0.8 V 17 1.2 V 7	5.79	22.9
E90 (Cylinder)	0 to 85	–	–	1.5	9.9	2.8	1	16 4	6.0 37.5	250 40	0.75 V – 0.9 V 120 21 180	11.7	30.2
E189 (Cylinder)	0 to 85	–	–	1.5	11.3	4.4	1	16 4	10.0 12.0	150 250	0.9 V 83 63	27.6	14.3
537 (Cylinder)	0 to 10	–	–	6.0	14.2	3.3	4	Continuous	2.5	2400	3.6 V 31		
539 (Cylinder)	0 to 100	–	–	6.0	31.0	16.4	4	4 ½	18.0 60.0 180.0	333 100 33.3	3.3 V 30.5 4.6 V 18.3 8.7 4 2.3 0.6		
520 (Rectangular)	0 to 1300	–	–	6.0	1130.0	883.0	4	24	30.0 3000.0	200 2	3.2 V 775 4.0 V 560 4.4 V 505 5 2.8 0.6		
522 (Rectangular)	0 to 100	–	–	9.0	45.0	21.0	6	4 ½	18.0 60.0	500 150	4.2 V 33 5.4 V 28 9 9		

53.1 Primary batteries

These batteries are available from numerous suppliers. Details are given below of the products available from some of the major suppliers.

Union Carbide

The range available is shown in Table 53.1, with information on the estimated service time at 21°C. The effect of operating temperature on internal resistance of these batteries (EP3 1.5 V) is shown in Table 53.2.

Table 53.2 Union Carbide Eveready 1.5 V (EP3) alkaline manganese cell: effect of temperature on internal resistance

Temperature (°C)	Internal resistance (Ω)
21	0.18
0	0.24
−6.7	0.31
−17.8	0.46
−28.9	0.74
−40	1.14

Mallory

Details of the full Mallory Duracell range together with details of service life and impedance of some cells in this range are given in Table 53.3 The capacities of these batteries lie between 300 and 10 000 mA h. Figure 53.1 shows typical discharge characteristics for a 1.5 V Duracell cell, and Figure 53.2 shows the effect of ohmic load on service life at 20°C.

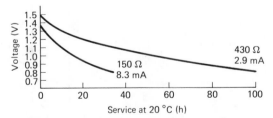

Figure 53.1 Typical discharge characteristic of a Mallory Duracell PX825 300 mA h alkaline manganese 1.5 V dry cell: current drain at 1.25 V (Courtesy of Mallory)

Various battery manufacturers supply what are obviously the same alkaline manganese batteries under their own designation. The situation regarding Mallory, Varta and Berec is listed in Table 53.4.

Crompton Vidor and several other manufacturers also supply alkaline manganese batteries.

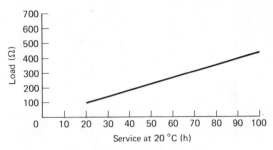

Figure 53.2 Mallory Duracell PX825 300 mA h alkaline manganese 1.5 V dry cell: a typical performance on constant resistance load. End-point 0.8 V (Courtesy of Mallory)

Chloride Automotive

Chloride supply a range of primary alkaline manganese batteries (Chloride Gold range) in the range 1.5–9 V. The batteries have a shelf life of up to 2 years and are claimed to have an operational life of up to five times longer than carbon–zinc batteries of the same size. The batteries available in the Chloride Gold range are shown in Table 53.5.

Varta

Varta supply several ranges of alkaline manganese primary batteries and button cells. The button cells are 1.5 V and cover the capacity range 18–300 mA h (Table 53.6). The cylindrical cells cover the voltage range 1.5–30 V and the capacity range 40–600 mA h (Table 53.7). The main applications of these batteries are in photography, hearing-aids and pocket calculators.

Varta also supply 1.5 V cylindrical alkaline manganese calculator batteries (type No. 4103, 800 mA h, and type No. 4106, 1800 mA h), and an E-block 9 V calculator battery (type No. 4122); 1.5 V cylindrical alkaline manganese batteries of capacities 600 mA h (type Nos. 4001, 4003); a range of cylindrical alkaline manganese Super Energy batteries, 1.5 V (type Nos. 4001 (size N), 4003 (size AAA), 4006 (size AA), 4014 (size C), 4020 (size D)), and an E-block 9 V battery (type No. 4022); alkaline manganese high-performance heavy duty batteries, 1.5 V (type Nos. 2606 (size AA), 2614 (size C) and 2620 (size D)) and a 9 V battery (type No. 3022). Standard long-life batteries are also available: 1.5 V (type Nos. 1006 (size AA), 1014 (size C) and 1020 (size D)), 3 V (type No. 1010) and 4.5 V (type No. 1012).

Duracell Europe

Duracell Europe supply various types of alkaline manganese battery (Table 53.8).

In addition, Duracell Europe supply low-capacity 1.5 V cells, 55–100 mA h capacity (types LR43,

Table 53.3 Mallory Duracell alkaline manganese dry cells and batteries

Type	IEC designation	Voltage (V)	Weight (g)	Volume (cm³)	Number of cells	Capacity (mAh) at 20°C	Service life (h) at 20°C Starting drain (mA)	Load (Ω)	Cut-off voltage and hours service		Height (mm)	Width (mm)	Impedance (Ω)	
PX825 (button)	LR53	1.5	7	2.4	1	300 to 0.8 V on 430 Ω	8.3 2.9	150 430	0.8 V 32 100	1.0 V 15 50	5.9	23	–	
MN9100 (cylinder)	LR1	1.5	9.6	3.1	1	700 to 0.8 V on 83 Ω	50 15 10	25 83 125	0.8 V 13 47 68	1.0 V 10 39 60	30.2	4.5	0.7 at 10 Hz 0.6 at 1000 Hz 0.4 at 100 000 Hz	Average of 15% of discharge on 83 Ω at 20°C
MN2400 (cylinder)	LR03	1.5	13	3.6	1	800 to 0.8 V on 83 Ω	50 25 15	25 60 83	0.8 V 18 35 57	1.0 V 13 28 48	44.5	10.5	0.8 at 10 Hz 0.6 at 1000 Hz 0.4 at 100 000 Hz	Average of 15% of discharge on 83 Ω at 20°C
7K67 (rectangular)	–	6.0	34	15	4	500 to 3.2 V on 250 Ω	50 25 10	100 200 500	3.2 V 10 20 50	4.0 V 8 16 42	48.3 × 35.6 ×	9.0	3.2 at 10 Hz 1.3 at 1000 Hz 0.9 at 100 000 Hz	Average of 15% of discharge on 300 Ω at 20°C
5K65 (rectangular)	–	9.0	64	21.3	6	500 to 4.8 V on 375 Ω	50 25 10	150 300 750	4.8 V 10 20 50	6.0 V 8 16 42	125.7 × 19.3 ×	7.0	3.5 at 10 Hz 2.0 at 1000 Hz 1.4 at 100 000 Hz	Average of 15% of discharge on 500 Ω at 20°C
5K69 (rectangular)	–	9.0	63	21.5	6	500 to 4.8 V on 375 Ω	50 25 10	150 300 750	4.8 V 10 20 50	6.0 V 8 16 42	48.2 × 51.0 ×	9.0	3.5 at 10 Hz 2.0 at 1000 Hz 1.4 at 100 000 Hz	Average of 15% of discharge on 500 Ω at 20°C
MN1604 (rectangular)	–	9.0	51	21	6	500 to 4.8 V on 750 Ω	50 25 10	150 300 750	4.8 V 10 20 50	6.0 V 8 18 42	48.5 × 26.5 × 17.5		3.5 at 10 Hz 2.0 at 1000 Hz 1.4 at 100 000 Hz	Average of 15% of discharge on 500 Ω at 20°C

Outline

A   B

Table 53.4

Voltage (V)	Mallory Duracell	Varta	Berec
1.5	PX825, button	–	PX825
	MN1500	7244	MN1500
	MN1300	7232	MN1300
	MN2400	7239	MN2400
	MN1400	7233	MN1400
	MN9100	–	MN9100
3	PX24, 7301	–	
4.5	PX21, 7251	–	
	PX19, 7251	–	

Table 53.5 Chloride Gold alkaline manganese batteries

Chloride type	Code	ASA	Japanese code	IEC	Voltage (V)
MN 1300	Mono	D	AM 1	LR 20	1.5
MN 1400	Baby	C	AM 2	LR 14	1.5
MN 1500	Migron	AA	AM 3	LR 6	1.5
MN 2400	Micro	AAA	AM 4	LR 03	1.5
MN 9100	Lady	N	AM 5	LR 1	1.5
MN 1604	–	–	006 P	6LF 22	9.0

Table 53.6 Varta alkaline manganese button cells and batteries

Varta Type No.	Order No.	Voltage (V)	Capacity* (mA h)	IEC reference	Main application
V 8 GA	4273	1.5	18	–	Calculators
V 10 GA	4274	1.5	45	–	Calculators
V 12 GA	4278	1.5	55	–	Calculators
V 13 GA	4276	1.5	80	LR 44	Calculators
V 625 U	4262	1.5	170	LR 9	Photography
V 825 PX	4825	1.5	300	LR 53	Photography
V 24 PX	4024	3.0	600	–	Photography
V 21 PX	4021	4.5	600	–	Photography
V 19 PX	4019	4.5	600	–	Photography
V 4034 PX	4034	6.0	100	–	Photography
V 74 PX	4074	15.0	40	10 LR 54	Cameras, light meters, lighting, measuring instruments
V 72 PX	4072	22.5	80	15 LR 43	External light meters, measuring instruments
V 73 PX	4073	30.0	100	20 LR 44	External light meters, measuring instruments

* Mean values at a discharge voltage of 0.8 V/cell

Table 53.7 Varta alkaline manganese cylindrical and rectangular cells and batteries

Varta type No.		Order No.	Capacity (mA h)	Nominal voltage (V)	Size	Dimensions (mm)		
						Diameter/ length	Width	Height
4001		4001	550	1.5	N/LR1/AM5	12.0		30.2
4003		4003	850	1.5	AAA/LR03/AM4	10.5		44.5
4006		4006	1600	1.5	AA/LR6/AM3	14.5		50.5
4014	Energy	4014	6000	1.5	C/LR14/AM2	26.2		50.0
4018	2000	4018	450	6	(7K67)	48.26	35.56	9.02
4020		4020	13 000	1.5	D/LR20/AM1	34.2		61.5
4022		4022	450	9	6LR61/6AM6	26.5	17.5	48.5
V1500PX		4206	1700	1.5	AA/LR6/AM3	14.5		50.5
V2400PX		4203	850	1.5	AAA/LR03/AM4	10.5		44.5

Table 53.8 Duracell alkaline manganese batteries

Voltage (V)		Type	Capacity at 20°C (mA h)
1.5	Button	LR 54	55
1.5	Cylindrical	LR 43	85
1.5	Button	LR 44	100
1.5	Cylindrical	MN 1300	10 000
1.5	Cylindrical	MN 1400	5 500
1.5	Cylindrical	MN 1500	1 800
9	Rectangular	MN 1604	500
1.5	Rectangular	MN 2400	800
1.5	Cylindrical	MN 9100	650
9	Rectangular	5K 65	400
9	Rectangular	5K 69	400–500
6	Rectangular	7K 67	400
3	Cylindrical	9K 62	400
4.5	Cylindrical	PX 21	580
4.5	Flat	MN 1203	4 400
6	Rectangular	ID 9080	20 000
6	Rectangular	ID 9150	20 000
6	Rectangular	ID 9180	40 000
12	Rectangular	ID 9260	20 000

LR44 and LR54) and also 6–12 V batteries in the 20 000–40 000 mA h capacity range (types ID 9080, ID 9150, ID 9180 and ID 9260).

53.2 Secondary batteries

Union Carbide

Certain manufacturers supply rechargeable alkaline manganese cells and batteries; for example, Union Carbide offer the Eveready rechargeable alkaline battery range. These batteries use a unique electrochemical system, are maintenance free, hermetically sealed and will operate in any position. They have been designed for electronic and electrical applications where low initial costs and low operating costs, compared with nickel–cadmium cells, are of paramount importance.

Present types of Eveready battery available from Union Carbide include 4.5 and 7.5 V batteries using D-size cells, and 6, 13.5 and 15 V batteries composed of G-size cells. Specifications are listed in Tables 53.9 and 53.10 and average performance characteristics are shown in Figure 53.3.

The discharge characteristics of the rechargeable alkaline manganese dioxide battery are similar to those exhibited by primary batteries; the battery voltage decreases slowly as energy is withdrawn from the battery. The shape of this discharge curve changes slightly as the battery is repeatedly discharged and charged. The total voltage drop for a given energy withdrawal increases as the number of discharge/charge cycles increases. Coupled with this is the fact that the available energy per cell lessens with each discharge/charge cycle even though the open-circuit voltage remains quite constant. When a rechargeable battery of the alkaline manganese dioxide type is discharged at the maximum rate for a period of time to remove the rated ampere hour capacity and then recharged for the recommended period of time, the complete discharge/charge cycle can be repeated many times before the battery voltage will drop below 0.9 V/cell in any discharge period, depending on cell size. Decreasing either the discharge current or the total ampere hour withdrawal, or both, will increase the cycle life of the battery by a significant percentage. Conversely, if the power demands are increased to a point where they exceed the rated battery capacity, the cycle life will decrease more quickly than the increase in power demand. During the early part of its cycle life there is a very large power reserve in the Eveready rechargeable alkaline manganese dioxide battery. In these early cycles, the battery terminal voltages may measure 1.0–1.2 V/cell after the battery has delivered its rated ampere hour capacity. If it is discharged beyond its rated capacity, however, total battery cycle life will be reduced. Nevertheless, this reserve power can be used in situations where maximum total battery life can be sacrificed for immediate power. During the latter part of cycle life, there is little or no reserve power and the terminal voltage of the battery will fall to between 1.0 and 0.9 V/cell at end of discharge.

These charge-retention characteristics also apply during the charge/discharge cycling of the battery. With this type of battery it is necessary to discharge it to its rated capacity before it will be capable of standing any overcharge.

The recommended method for charging alkaline secondary batteries is the use of voltage-limited taper current charging. This method offers the maximum cycle life and consequently the lowest battery operating costs. Cycle life can be as much as twice that obtained with other methods of charging. Although constant-current charging can be used with consequent loss of cycle life, it is recommended that careful enquiries be made before such charging methods are contemplated.

Batteries should not be charged continuously after the charge current has returned 120% of the ampere hours removed on previous discharge or float charged for extended periods.

A rechargeable alkaline manganese dioxide battery must not be discharged completely. For best results, the rated capacity of the battery should not be exceeded on discharge. During deep discharge a secondary electrochemical reaction takes place. This reaction is not reversible and will seriously reduce the battery cycle life. To avoid possible complaints of short battery service, it is therefore desirable that

Table 53.9 Characteristics of Eveready rechargeable alkaline manganese dioxide batteries

Eveready battery number	Voltage (V)	Number and size of cells	Capacity (Ah)	Maximum recommended discharge current (A)	Weight (g)	Volume (mm^3) $\times 10^3$	Diameter (mm)	Length (mm)	Width (mm)	Height (mm)	Terminals
563	4.5	3 D	2.5	0.625	425	166	34.5			182	Flat
565	6.0	4 G	5.0	1.250	1130	476		70.6	70.6	136	Socket
560	7.5	5 D	2.5	0.625	709	646		67.5	38.9	182	Socket
564	13.5	9 G	5.0	1.250	2500	2245		211.0	71.4	149	Socket
561	15.0	10 G	5.0	1.250	2720	2245		211.0	71.4	149	Socket

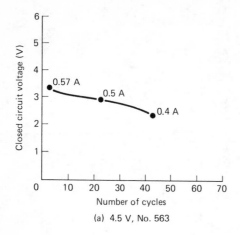

(a) 4.5 V, No. 563

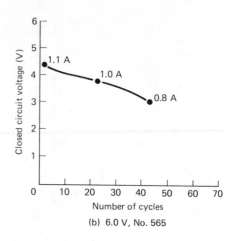

(b) 6.0 V, No. 565

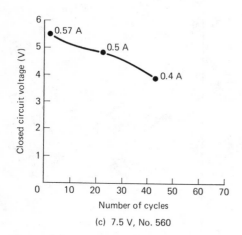

(c) 7.5 V, No. 560

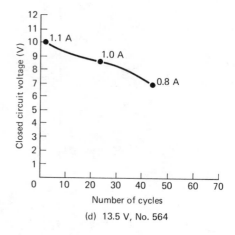

(d) 13.5 V, No. 564

(e) 15.0 V, No. 561

	Discharge cycle		Charge cycle*		
Figure	Time (h)	Through resistance (Ω)	Time (h)	Voltage limit (V)	Limiting resistance (Ω)
(a)	4	5.76	16	5.25	2.4
(b)	4	3.84	16	7	1.6
(c)	4	9.6	16	8.75	4.0
(d)	4	8.64	16	15.75	3.6
(e)	4	9.6	16	17.5	4.0

*Voltage limited taper current charge, filtered regulated power supply with 1% regulation

Figure 53.3 Average performance characteristics, voltage *versus* cycles at 21.1°C, of Eveready rechargeable alkaline manganese dioxide batteries, 4.5–15 V (Courtesy of Union Carbide)

Table 53.10 Characteristics of cells in Eveready rechargeable alkaline to manganese dioxide batteries

Cell size	Voltage (V)	Average operating voltage (V)	Rated A h capacity	Maximum recommended discharge current (A)
D	1.5	1.0–1.2	2.5	0.625
G	1.5	1.0–1.2	5.0	1.250

the device using the battery contain some provision against excessive discharge of the battery.

Where voltage-limited taper current charging is used, the charger will replace the energy needed to recharge the battery, and a timer or ampere hour control is unnecessary to monitor the charging time. However, the battery should be removed from the charger after completion of the charge. Also, to prevent deep cell discharge, a timer would provide effective control over the energy withdrawn from the cell during any cycle.

With constant-current charging, a d.c. timer or ampere hour counter is needed to control both the discharge and charge cycles. A system of this sort would be adjustable to permit removal of the rated ampere hour capacity from the battery and thereby ensure maximum battery service life.

Voltage-limited taper current charging involves use of an inexpensive voltage regulator added to the basic constant-current type transformer–rectifier circuit with a current-limiting resistor between the battery and the regulated output voltage, which removes the burden of adjusting charge time to discharge time and provides automatic current control. Although the regulator circuit is somewhat more expensive than other common circuits the initial cost is more than offset by 50–100% greater cycle life.

The basic characteristics of sealed alkaline manganese dioxide rechargeable cells when using voltage-limited taper current charging are given in Table 53.11 (based on a 6 h charge period).

A regulator providing voltage regulation of 2 or

3% at low current values is adequate. Poorer regulation when the battery is first placed on charge and the voltage is low is often satisfactory. The regulator output voltage is adjusted to the battery end-of-charge voltage. Correct battery voltage is determined by number of cells and cell voltages given above. The current-limiting resistance (the total current-limiting resistance consists of the series resistance plus inherent resistance of the regulator circuit, which may vary with design and component selection) should limit the initial current to the 4–6 h rate at start of charge when battery voltage is low. Typical starting voltage after withdrawal of rated capacity is 1.30–1.35 V/cell. The maximum recommended initial charge currents at these voltages are shown in Table 53.11.

The minimum voltage that a cell will reach at the end of the recommended discharge capacity is shown in Figure 53.4. This voltage is measured while the discharge current is flowing. The curves in Figure 53.4 represent cell behaviour at maximum discharge current rating to rated capacity and a minimum of 16 h recharge time. The initial closed-circuit voltage and end-of-discharge voltage curves give the 'working' voltage range of the cell at various discharge levels. To prevent damage to the cell on charge cycle the charging potential should be limited to about 1.7 V/cell with a maximum permissible of 1.75 V/cell. The voltage rise on the charge and end-of-charge voltage value is dependent on charge current character. As shown in Figure 53.4, the

Table 53.11 Sealed alkaline manganese dioxide cells under voltage-limited taper current charging

Cell size	Average operating voltage at maximum current (V)	Maximum initial charging current at 1.3 V/cell (A)	Current limiting resistance (Ω/cell)	Source voltage limit (V/cell)
D	1.0–1.2	0.6	0.8	1.70–1.75
G	1.0–1.2	1.12	0.4	1.70–1.75

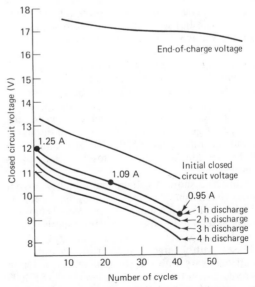

Figure 53.4 Typical voltage performance of an Eveready 561 15 V rechargeable alkaline manganese dioxide battery on voltage-limited taper current charge. (Discharge cycle 4 h through 9.6 Ω. Charge cycle 16 h, voltage limit 17.5 V, filtered regulated power supply 1% regulation with 4.0 Ω limiting resistor) (Courtesy of Union Carbide)

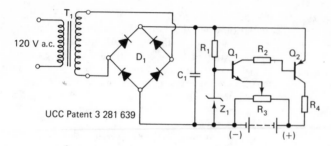

UCC Patent 3 281 639

T_1 Stancor RT 201 (input a.c. to terminals 1 and 7; connect
    terminals 3 and 6) (secondary connections to bridge 8 and
    11; connect terminals 9 and 10)
D_1 2A50 Solitron Devices, Inc. (or equivalent)
C_1 500 μF, 50 V d.c.
Z_1 1/4 M12 Z (Motorola) in 963B TI or int. rec.
Q_1 2N1304 RCA, TI or equivalent
Q_2 2N1557 Motorola, 2N514 TI
R_1 4700 Ω, 0.5 W
R_2 390 Ω, 0.5 W
R_3 1000 Ω, 1 W
R_4 4 Ω, 10 W

Figure 53.5   Voltage-limited taper current charger circuit
suitable for charging a 15 V, 5 A h Eveready 561 alkaline
manganese dioxide battery in 16 h (Courtesy of Union
Carbide)

end-of-charge voltage will decrease slightly as cycle
life progresses, as the result of gradual change in the
reversibility of the system.

Since the current tapers as the result of battery
voltage increase or is 'voltage responsive', the
end-of-charge current will increase as cycling
progresses. The current will equilibrate at some
specific value but will never reach zero or shut-off
since the voltage decreases as current is reduced as
the result of internal resistance and chemical
reaction. The battery voltage after charge will
gradually decrease to the value of 1.5 V/cell when
the charger is disconnected and the battery remains
on open circuit. When changing from charge to
discharge cycle with no time delay between, the
battery voltage decay rate will vary with discharge
current. At rates near 1 A the battery voltage will
fall to the indicated initial closed-circuit values
within 5 min.

Figure 53.5 shows a voltage-limited taper current
charge circuit suitable for recharging an Eveready
561 battery (15 V, 5 A h in 16 h). A tapered charging
current is supplied to the battery through the series
current-limiting resistor R_4 as battery voltage
increases with charge. The performance of the
circuit may vary with the components used. Often
the replacement transistors, for example the SK3009
replacement for 2NI304, may require a reduction in
value of R_3 for proper circuit operation. To
determine the value of the series current-limiting
resistor, R_4, and for initial circuit adjustment, a
voltmeter (10 000 Ω or greater per volt) is placed
across R_3 (no battery connected) and R_3 adjusted to
obtain the indicated source voltage limit. A voltage
source equivalent to 1.3 V/cell for the battery to be
charged is connected and R_4 selected to limit the
charge current to recommended values. The battery
should be disconnected from the charger when the

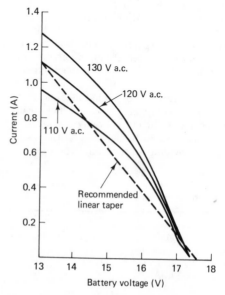

Figure 53.6   Suggested and actual linear taper currents
obtained during 16 h voltage-limited taper current charging of
a 15 V, 5 A h Eveready 561 alkaline manganese dioxide battery
(Courtesy of Union Carbide)

charge circuit is not energized, to prevent discharge through resistance R_3.

The suggested linear taper current for the 561 battery and actual current characteristic obtained with the circuit of Figure 53.5 are shown in Figure 53.6. The current characteristic can be tailored by selection of components. The curves shown were obtained with the use of component values shown in Figure 53.5. The permissible deviation of charge current from that indicated suggests that linear taper varies somewhat with application and usage pattern. Where the application requires repetitive discharge to rated capacity, the charge current values should fall within the shaded area of Figure 53.7, with charge time adjustment for circuit tolerance should operation fall in the 'minimum' area relative to linear character.

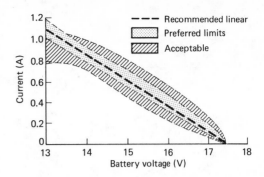

Figure 53.7 Permissible deviations of charge current of 16 h voltage-limited taper current charge of 15 V, 5 A h Eveready 561 alkaline manganese dioxide battery (Courtesy of Union Carbide)

54

Carbon–zinc batteries (primary) and carbon–zinc chloride batteries

Union Carbide

Union Carbide produce a vast range of carbon–zinc dry cells and batteries (see Table 54.1). This range covers from single 1.5 V cells with current ranges between 0.06 and 1500 mA to a multi-cell 510 V battery with a current range of 0–2.5 mA; typical applications for these various types are given in Table 54.1.

Average service life data for a selection of these various types of cells and batteries are listed in Table 54.2 (1.5–9 V) and Table 54.3 (12–510 V) for different test schedules.

Carbon–zinc cells (1.5 V) come in a variety of standard sizes covered by the ANSI designation. The type numbers of Union Carbide Eveready 1.5 V cells available under each ANSI designation are listed in Table 54.4 (see also Table 54.1).

Eveready transistor batteries (see Table 54.1) exhibit exceptional current drain carrying capabilities. Battery impedance is low and essentially constant even to low-voltage cut-offs. Shelf life characteristics are excellent. Battery capacity is nearly constant at withdrawal rates of 2–6 h/day. The construction used in Eveready transistor batteries perform well in the following abuse tests:

1. Continuous short-circuit.
2. Continuous discharge.
3. High-temperature storage.
4. High-temperature high-humidity storage.
5. Shock and vibration.
6. Temperature cycling from -28.9 to $+60°C$.

The ampere hour capacity of a carbon–zinc dry battery is not a fixed value. It varies with current drain, operating schedule, cut-off voltage, temperature and storage period of the battery before use. Approximate service capacity for each cell used in the batteries shown in the specification listing for carbon–zinc batteries is given in Table 54.5 for two different current drains. The values given are for fresh batteries at 21°C; the operating schedule is 2 h/day; and the cut-off voltage is 0.8 V per 1.5 V cell for all of the cells, which is consistent with the normal cut-off voltages of the cells in service. The data in Table 54.5 are based on starting drains and fixed resistance tests. From the voltage of the battery and the total number of cells, the number of 1.5 V cells in series and in parallel may be determined. Service capacity is given for single 1.5 V cells. If a battery uses cells in parallel, dividing the current drain which the battery is supplying by the number of parallel strings of cells and looking up this value of current will give the service life.

Table 54.6 shows the most widely used batteries, in typical applications, and details of applicable loads, voltages and amperage readings for testing.

Ever Ready (Berec) (UK)

The Eveready range of carbon–zinc dry cells and batteries covering 1.5–90 V supplied is given in Table 54.7.

Crompton Vidor (UK)

Crompton Vidor supply dry cells and batteries suitable for the operation of road warning lights, lighting, radio, burglar alarm and fire alarm systems. Other examples are the V22, a 15 V layer stack battery, which was uniquely designed by Vidor to power the transformer/capacitor-generated spark in an electronic cigarette lighter, and the L120-1196, a high-voltage unit in use with mass spectrometers and wherever a stable-high-voltage low-noise power source is required.

The L12-1470 12 V battery weighs approximately 1.14 kg and is recommended for intruder and fire alarm systems, portable fluorescent lighting, small d.c. motors, portable television receivers and toys.

The HP992 Hi-Power 6 V battery was made specially for the alarm industry, having:

1. High power capability.
2. Screw terminals with insulated nuts for ease and flexibility of connection.
3. All plastics construction, which is resistant to leakage and ingress of moisture.

Applications include alarm systems, such as door alarm locks, road counting meters, marine handlamps, toys and laboratory power sources.

The L12-150 12 V battery weighs approximately 1.05 kg and is stated to have good shelf life and an internal enclosed cell block resistant to ingress of moisture, and can be used for longer periods at higher current drains. Applications include intruder and fire alarm systems, portable fluorescent lighting, small d.c. motors, portable television receivers and toys.

Varta (West Germany)

Varta produce the range of carbon–zinc cells and batteries shown in Tables 54.8 and 54.9. They also supply 1.5 V zinc chloride cells (Super Dry 280, 281 and 282). Their cell type 246 is a 1.5 V carbon–zinc button cell.

Yuasa (Japan)

Yuasa supply two ranges of carbon–zinc dry cells, the Golden G Hi Capacity series and the Standard Diamond series. Table 54.10 gives the range available, with Union Carbide, Ever Ready (Berec) and Ray-o-Vac equivalents.

Table 54.1 Union Carbide carbon–zinc Eveready Leclanché dry cells and batteries

Battery number in order of increasing service capacity	Suggested current range (mA)	Typical use	Maximum dimensions (mm)				Cells		Battery weight (g)	Terminals	Approx. volume of battery body (cm³)	Voltage taps
			Diameter	Length	Width	Height	Number	Size				
1.5 V batteries												
201	0 to 0.06	Electronic watch	11.3			3.3	1	103	1.39	Flat contact	0.31	−, +1.5
E340E	0 to 20	Hearing-aid	11.3			30.0	1	N	6.24	Flat contact	2.62	−, +1.5
904	0 to 20	Flashlight, toys	11.3			30.0	1	N	6.24	Flat contact	2.62	−, +1.5
812	0 to 20	Photo-flash	10.3			44.5	1	AAA	8.50	Flat contact	3.28	−, +1.5
912	0 to 20	Flashlight	10.3			44.5	1	AAA	8.50	Flat contact	3.28	−, +1.5
815	0 to 25	Photo-flash	14.3			50.0	1	AA	17.00	Flat contact	7.87	−, +1.5
HS15	0 to 25	Flashlight	14.3			50.0	1	AA	14.80	Flat contact	7.87	−, +1.5
915	0 to 25	Flashlight	14.3			50.0	1	AA	14.80	Flat contact	7.87	−, +1.5
1015	0 to 25	Transistor radio	14.3			50.0	1	AA	17.00	Flat contact	7.87	−, +1.5
1215	0 to 25	General-purpose (heavy duty)	14.3			50.0	1	AA	15.80	Flat contact	7.87	−, +1.5
T35	0 to 1	Electric clock	26.2			49.2	1	C	39.7	Flat contact	24.9	−, +1.5
835	0 to 80	Photo-flash	26.2			49.2	1	C	39.7	Flat contact	24.9	−, +1.5
HS35	0 to 80	Flashlight	26.2			49.2	1	C	41.0	Flat contact	24.9	−, +1.5
935	0 to 80	Flashlight	26.2			49.2	1	C	41.0	Flat contact	24.9	−, +1.5
1035	0 to 80	Transistor radio	26.2			49.2	1	C	39.7	Flat contact	24.9	−, +1.5
1235	0 to 80	General-purpose	26.2			49.2	1	C	45.6	Flat contact	24.9	−, +1.5
T50	0 to 1	Electric clock	34.1			61.1	1	D	93.6	Flat contact	52.0	−, +1.5
850	0 to 150	Photo-flash	34.1			61.1	1	D	93.6	Flat contact	52.0	−, +1.5
HS50	0 to 150	Flashlight	34.1			61.1	1	D	85.1	Flat contact	52.0	−, +1.5
950	0 to 150	Flashlight	34.1			61.1	1	D	85.1	Flat contact	52.0	−, +1.5
1050	0 to 150	Transistor radio	34.1			61.1	1	D	85.1	Flat contact	52.0	−, +1.5
HS150	0 to 150	Industrial flashlight	34.1			61.1	1	D	85.1	Flat contact	52.0	−, +1.5
1150	0 to 150	Industrial flashlight	34.1			61.1	1	D	85.1	Flat contact	52.3	−, +1.5
1250	0 to 150	General-purpose (heavy duty)	34.1			61.1	1	D	93.6	Flat contact	52.0	−, +1.5
W353	0 to 500	Portable A		65.9	34.9	106.0	2	F	354.0	Socket	243.0	−, +1.5
711	0 to 500	Glo-plug ignition		66.7	33.3	102.0	2	F	340.0	Knurled nut and screw	203.0	−, +1.5
742	0 to 1000	Portable A		66.7	66.7	97.6	4	F	624.0	Socket	434.0	−, +1.5
735	0 to 1000	Instruments, hobbies		66.7	66.7	110.0	4	F	652.0	Knurled nut and screw	434.0	−, +1.5
HS6	0 to 1500	Industrial general-purpose	66.7			168.0	4	6	907.0	Knurled nut and screw	480.0	−, +1.5
IF6	0 to 1500	General-purpose	66.7			172.0	6	6	907.0	Fahnestock	480.0	−, +1.5
IS6	0 to 1500	General-purpose	66.7			168.0	6	6	907.0	Knurled nut and screw	480.0	−, +1.5
IS6T	0 to 1500	General-purpose	66.7			166.0	6	6	907.0	One brass knurl, one wire	480.0	−, +1.5
EA6F	0 to 1500	General-purpose	66.7			172.0	1	6	1020.0	Fahnestock	480.0	−, +1.5
EA6	0 to 1500	Alarm	66.7			166.0	1	6	964.0	Brass knurls	480.0	−, +1.5
EA6FT	0 to 1500	Alarm	66.7			172.0	1	6	964.0	One Fahnestock, one wire	480.0	−, +1.5
EA6ST	0 to 1500	Alarm	66.7			166.0	1	6	964.0	One brass knurl, one wire	480.0	−, +1.5
3 V batteries												
750	0 to 250	Flat flashlight		31.0	15.9	67.5	2	AA	56.7	Spring	26.6	−, +3
W356	0 to 500	Radio		66.7	66.7	116.0	4	F	624.0	Knurled nut and screw	459.0	−, +3
W357	0 to 1000	Telephone		96.0	68.3	148.0	8	F	1250.0	Fahnestock	895.0	−, +3
4.5 V batteries												
703	0 to 50	Miscellaneous		61.9	21.4	77.8	3	B	142.0	Spring	83.6	−, +4.5
781	0 to 50	Portable C		61.9	21.4	76.2	3	B	142.0	Knurled nut and screw	83.6	−, +4.5
714	0 to 200	Miscellaneous		101.0	34.1	95.3	3	55	383.0	Knurled nut and screw	277.0	−, +4.5
736	0 to 250	Portable A		100.0	33.3	102.0	3	F	454.0	Socket	339.0	−, +4.5

Type	Range	Application								Terminals		Polarity
6 V batteries												
724	0 to 25	Portable A		31.0	31.0	59.5	4	AA	70.9	Flat contact	55.7	–, +6
509	0 to 250	Lantern		66.7	66.7	112.0	4	F	609.0	Coil springs	434.0	–, +6
510F	0 to 250	Miscellaneous		66.7	66.7	112.0	4	F	567.0	Fahnestock	434.0	–, +6
HS10S	0 to 250	Lighting, miscellaneous		66.7	66.7	110.0	4	F	652.0	Knurled nut and screw	434.0	–, +6
510S	0 to 250	Lighting, miscellaneous		66.7	66.7	110.0	4	F	652.0	Knurled nut and screw	434.0	–, +6
744	0 to 250	Portable A		66.7	66.7	97.6	4	F	624.0	Socket	434.0	–, +6
HS90	0 to 250	Industrial lantern		66.7	66.7	112.0	4	F	638.0	Coil springs	434.0	–, +6
1209	0 to 250	Lantern		66.7	66.7	112.0	4	F	595.0	Coil springs	447.0	–, +6
2744N	0 to 250	Barricade flasher, miscellaneous		66.7	66.7	97.6	4	F	652.0	Socket	447.0	–, +6
2745N	0 to 250	Barricade flasher, lantern		66.7	66.7	112.0	4	F	652.0	Coil springs	447.0	–, +6
2746N	0 to 250	Lighting, miscellaneous		66.7	66.7	110.0	4	F	652.0	Knurled nut and screw	447.0	–, +6
HS31	0 to 500	Lighting		136.0	72.2	125.0	8	F	1470.0	Insulated screw	1139.0	–, +6
731	0 to 500	Lighting		136.0	72.2	125.0	8	F	1470.0	Insulated screw	1139.0	–, +6
706	0 to 1000	Emergency lighting		136.0	71.4	164.0	16	F	2580.0	Insulated screw	2245.0	–, +6
1461	0 to 1500	Ignition, lighting		265.0	69.1	183.0	4	6	4200.0	Knurled nut and screw	3261.0	–, +6
7.5 V batteries												
773	0 to 50	Portable C	19.0	99.2	21.4	76.2	5	B	225.0	Five knurled nut and screw, one wire	135.0	+, –1.5, –3, –4.5, –6, –7.5
717	0 to 70	Portable A		54.8	49.2	77.0	5	172	227.0	Socket	213.0	–, +7.5
715	0 to 1000	Emergency lighting		184.0	103.0	164.0	20	F	3460.0	Insulated screw	2835.0	–, +7.5
1562	0 to 1500	Ignition, lighting		199.0	126.0	183.0	5	6	5100.0	Knurled nut and screw	4474.0	–, +7.5
9 V batteries												
206	0 to 7	Transistor radio		26.2	16.7	50.8	6	109	34.0	Flat contact negative recessed	13.9	–, +9
216	0 to 15	Transistor radio		26.2	17.5	48.4	6	117	42.5	Miniature snap	19.7	–, +9
1222	0 to 15	Transistor radio (heavy duty)		26.2	18.7	49.2	6	118	42.5	Miniature snap	21.0	–, +9
226	0 to 9	Transistor radio		25.4		49.2	6	127	56.7	Snap	21.5	–, +9
2709N	0 to 16	Transistor radio		35.7	34.1	119.0	12	117	70.9	Snap	74.9	–, +9
246	0 to 15	Transistor radio		35.7	46.0	69.9	6	148	128.0	Miniature snap	82.3	–, +9
266	0 to 20	Transistor radio		46.0	51.6	61.9	6	165	126.0	Snap	126.0	–, +9
276	0 to 30	Transistor radio		65.1		80.2	6	175	425.0	Snap	267.0	–, +9
2356N	0 to 80	Transistor radio		55.6	29.4	159.0	54	117 & 118	312.0	Snap	288.0	–, +9
716	0 to 80	Emergency lighting		218.0	103.0	164.0	24	F	3860.0	Knurled nut and screw	3359.0	–, +9
12 V batteries												
228	0 to 9	Communications equipment		25.4		61.9	8	127	65.2	Snap	27.5	–, +12
732	0 to 250	Lighting		136.0	72.2	125.0	8	F	1420.0	Insulated screw	1139.0	–, +12
2780N	0 to 250	Flasher		136.0	72.2	106.0	8	F	1360.0	Socket	1034.0	–, +12
1463	0 to 600	Fence controllers		265.0	69.1	183.0	16	G	3490.0	Knurled nut and screw	3261.0	–, +12
1862	0 to 1500	Barricade flasher		260.0	133.0	168.0	8	6	9530.0	Socket	5834.0	–, +12
15 V batteries												
504	0 to 1.5	Radio paging		15.9	15.1	34.9	10	105	17.0	Flat contact	8.36	–, +15
411	0 to 2.5	Radio paging		26.2	15.9	36.9	10	112	26.9	Flat contact	15.6	–, +15
417	0 to 6	Radio paging		33.3	24.6	39.7	10	132	51.0	Flat contact	30.6	–, +15
22.5 V batteries												
505	0 to 1.5	Radio paging B-C flash		15.9	15.1	50.2	15	105	25.5	Flat contact	12.1	–, +22.5
412	0 to 2.5	Radio paging B-C flash		26.2	15.9	50.8	15	112	36.9	Flat contact	21.5	–, +22.5
420	0 to 6	Radio paging		33.3	24.6	55.6	15	132	70.9	Flat contact	43.6	–, +22.5
425P	0 to 10	Hearing-aid		34.5	26.2	100.0	15	135	116.0	Socket	91.0	–, +22.5
763	0 to 40	Portable B or C		88.9	53.2	77.8	15	161	397.0	Knurled nut and screw	303.0	–, +22.5
778	0 to 50	Portable C		102.0	61.9	79.4	15	B	567.0	Fahnestock	477.0	+, –3, –4.5, –6, –9, –10.5, –16.5, –22.5

Table 54.1 (Continued)

Battery number in order of increasing service capacity	Suggested current range (mA)	Typical use	Maximum dimensions (mm)				Cells		Battery weight (g)	Terminals	Approx. volume of battery body (cm³)	Voltage taps
			Diameter	Length	Width	Height	Number	Size				
30 V battery												
413	0 to 25	Radio paging		26.2	15.9	65.1	20	112	454.0	Flat contact	26.1	−, +30
45 V batteries												
415	0 to 4	Portable B		26.6	15.9	93.7	30	112	70.9	Miniature snap	38.5	−, +45
460	0 to 6	Photographic		48.0	34.9	61.7	30	132	150.0	Flat contact	100.0	−, +45
455	0 to 10	Portable B		67.5	25.4	93.7	30	135	221.0	Snap	157.0	−, +45
738	0 to 25	Portable B		76.2	58.7	105.0	30	AA	539.0	Socket	469.0	−, +22.5, +45
482	0 to 40	Portable B		91.3	46.8	139.0	30	165	851.0	Socket	597.0	−, +45
762S	0 to 70	Instruments, portable B		104.0	65.1	138.0	30	172	1250.0	Knurled nut and screw	872.0	−, +22.5, +45
484	0 to 70	Portable B		100.0	64.3	135.0	30	172	1420.0	Socket	875.0	−, +45
487	0 to 80	B		130.0	52.4	184.0	30	175	1870.0	Socket	1256.0	−, +22.5, +45
63 V battery												
477	0 to 8	Portable B		48.8	27.0	138.0	42	118	244.0	Snap	179.0	−, +63
67.5 V batteries												
416	0 to 3	Portable B		33.7	25.0	88.9	46	114	113.0	Miniature snap	72.9	−, +67.5
457	0 to 6	Portable B		71.4	34.9	63.5	45	132	216.0	Snap	155.0	−, +67.5
467	0 to 10	Portable B		71.4	34.9	94.1	45	135	340.0	Snap	233.0	−, +67.5
69 V battery												
646	0 to 30	Lighting		65.1	54.0	206.0	46	155	1450.0	Flat contact	721.0	−, +69
90 V batteries												
479	0 to 8	Portable B		48.8	27.0	190.0	60	118	340.0	Snap	246.0	−, +90
490	0 to 10	Portable B		94.5	34.9	94.1	60	135	425.0	Snap	308.0	−, +90
225 V battery												
489	0 to 10	Electronic flash		110.0	68.3	106.0	152	135	1.25	Socket	810.0	−, +225
240 V battery												
491	0 to 2.5	Electronic flash		65.9	33.3	114.0	160	112	369.0	Flat recessed and socket	246.0	−, +240
300 V battery												
493	0 to 2.5	Geiger counters		68.3	56.4	99.2	200	112	411.0	Pin jacks	384.0	−, +300
450 V battery												
496	0 to 10	Electronic flash		172.0	76.2	127.0	300	135	2380.0	Socket	1655.0	−, +450
510 V battery												
497	0 to 2.5	Electronic flash		76.2	40.5	143.0	336	112	737.0	Flat recessed	441.0	−, +180, +510

Table 54.2 Estimated average service life at 21°C for a selection of Union Carbide Eveready carbon–zinc Leclanché dry cells and batteries, voltage range 1.5–9 V

Eveready battery number	Battery voltage (V)	Schedule (h/day)	Starting drain (mA)	Load (Ω)	Service time (h) at cut-off voltage	Suggested current range (mA)
1215	1.5	4	37.5	40.0	0.75 V: 33, 0.9 V: 31	
	1.5	4	60.0	25.0	0.75 V: 20, 0.9 V: 18	
1235	1.5	4	37.5	40.0	0.75 V: 97, 0.9 V: 91	0–25
	1.5	4	60.0	25.0	0.75 V: 59, 0.9 V: 55	
H550 1050	1.5	4	60.0	25.0	0.75 V: 113, 0.9 V: 100	
	1.5	4	18.0	83.3	0.75 V: 440, 0.9 V: 400	
750	3.0	2	5.0	600.0	1.6 V: 265, 2.0 V: 240, 2.4 V: 145	0–25
	3.0	2	50.0	60.0	1.6 V: 14, 2.0 V: 10, 2.4 V: 3.8	
703 and 781	4.5	2	5.0	900.0	2.4 V: 420, 3.0 V: 340, 3.6 V: 250	0–25
	4.5	2	100.0	45.0	2.4 V: 14.5, 3.0 V: 9, 3.6 V: 4.5	
510S, HS105, 510F, 509	6	Discharge 100 ms/s, 24 h/day, 7 days/week, 10% 'on' time Safety flasher test	—	Load is RCR (resistor–capacitor–resistor) network. Components chosen to provide initial peak pulse of 5000 mA (at 6 V) in 20 ms and remaining at this level for remainder of 100 ms pulse (to simulate lamp characteristics)	3.6 V: 825	0–250
2744N, 2745N, 2746N	6	Discharge 100 ms/s, 24 h/day, 7 days/week, 10% 'on' time Safety flasher test	—	Load is RCR network. Components chosen to provide initial peak pulse of 500 mA (at 6 V) in 20 ms and remaining at this level for remainder of 100 ms pulse (to simulate lamp characteristics)	875	0–250
1209	6	Discharge 100 ms/s, 24 h/day, 7 days/week, 10% 'on' time Safety flasher test	—	Load is RCR network. Components chosen to provide initial peak pulse of 500 mA (at 6 V) in 20 ms and remaining at this level for remainder of 100 ms pulse (to simulate lamp characteristics)	975	0–250
773	7.5	2	5.0	1500.0	4.0 V: 420, 5.0 V: 340, 6.0 V: 250	0–50
	7.5	2	100.0	75.0	4.0 V: 14.5, 5.0 V: 9, 6.0 V: 4.5	
717	7.5	2	10.0	750.0	4.0 V: 485, 5.0 V: 350, 6.0 V: 240	0–70
	7.5	2	80.0	94.0	4.0 V: 50, 5.0 V: 32, 6.0 V: 10	
266	9	4	0.5	18.0	4.8 V: 2400, 6.0 V: 1950, 7.2 V: 1400	0–20
	9	4	80.0	112.5	4.8 V: 14.5, 6.0 V: 8, 7.2 V: 1.3	
276	9	2	5	1800	4.8 V: 920, 6.0 V: 800, 7.2 V: 600	0–30
	9	2	50	180	4.8 V: 110, 6.0 V: 80, 7.2 V: 38	
2356N	9	4	15	600	4.8 V: 305, 6 V: 290	0–80
	9	4	36	250	4.8 V: 125, 6 V: 115	
716	9	2	80	112.5	4.8 V: 490, 6.0 V: 410, 7.2 V: 310	0–1000
	9	2	1200	7.5	4.8 V: 23, 6.0 V: 18, 7.2 V: 6.8	

Table 54.3 Estimated average service life at 21°C for a selection of Union Carbide Eveready carbon–zinc dry batteries, voltage range 12–510 V

Eveready battery number	Battery voltage (V)	Schedule (h/day)	Starting drain (mA)	Load (Ω)	Service time (h) at cut-off voltage			Suggested current range (mA)
228	12	2	12.0	1000	5.6 V 59	7.2 V 52	8.8 V 37	0–90
1463	12	24	20.0	600	4 V 2300	7.2 V 1750		0–6000
504	15 15	2 2	1.0 5.0	150 000 3 000	8 V 1010 11.4	10 V 930 7	12 V 820 3.8	0.1–1.5
417	15 15	2 2	0.5 20.0	30 000 750	8 V 720 7.5	10 V 630 4.5	12 V 460 0.8	0–6
778	22.5	2	5.0 100.0	4 500 225	12 V 420 14.5	15 V 340 9	18 V 250 4.5	0–50
412	22.5	2 2	0.1 5.0	225 000 4 500	12 V 1300 24	15 V 1210 18	18 V 1100 5	0–2.5
413	30	4 4	0.2 5.0	150 000 6 000	16 V 760 23	20 V 660 19	24 V 570 13	0–2.5
484 and 762S	45	2 2	10.0 80.0	4 500 562	24 V 485 50	30 V 350 32	36 V 240 10	0–70
415	45	2 2	0.1 5.0	450 000 9 000	24 V 1300 24	30 V 1210 18	36 V 1100 5	0–4
477	43	4	8.4	7 500	45 V 49			0–8
416	67.5	2 2	3.37 6.75	20 000 10 000	30 V 57 25	45 V 50 16	51 V 42 9	0–3
467	67.5	2 2	0.5 35.0	135 000 1 929	36 V 920 7	45 V 850 5.4	54 V 730 0.8	0–10
646	69	4 4	19.0 78.0	3 630 885	36.9 V 195 20	46 V 140 15	55.2 V 105 7.5	0–30
479	90	4	9.0	10 000	48 V 60	60 V 51		0–8
490	90	2	0.5 35.0	180 000 2 570	48 V 920 7	60 V 850 5.4	72 V 730 0.8	0–10
489	225	2	0.5 35.0	450 000 6 430	121.6 V 920 7	152 V 850 5.4	182.4 V 730 0.8	0–10
491	240	2	0.1 5.0	2.4×10^6 48 000	128 V 1300 24	160 V 1210 18	192 V 1100 5	0–2.5
493	300	2	0.1 5.0	3×10^6 60 000	160 V 1300 24	200 V 1210 18	240 V 1100 5	0–2.5
496	450	2	0.5 35.0	900 000 129 000	240 V 920 7	300 V 850 5.4	360 V 730 0.8	0–10
497	510	2	0.1 5.0	5.1×10^6 102 000	269 V 1300 24	336 V 1210 18	403 V 1100 5	0–2.5

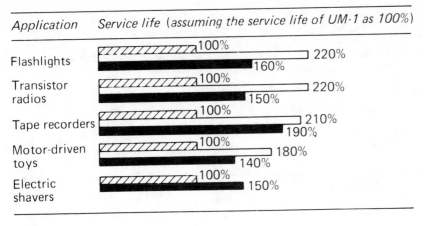

Application *Service life (assuming the service life of UM-1 as 100%)*

Flashlights — 100% / 220% / 160%

Transistor radios — 100% / 220% / 150%

Tape recorders — 100% / 210% / 190%

Motor-driven toys — 100% / 180% / 140%

Electric shavers — 100% / 100% / 150%

UM-1

SUM-1G, continuous use

SUM-1G, intermittent use

Figure 54.1 Comparison of service life between Yuasa standard and Hi Capacity carbon–zinc cells of the same dimensions (Courtesy of Yuasa)

Table 54.4 Type numbers of Union Carbide carbon–zinc Leclanché dry cells available under various ANSI designations

Union Carbide type number	ANSI designation	Voltage (V)	Suggested current range (A)
HS6, 1F6, 1S6, 1S6T, EA6F, EA6, EA6FT, EA6ST	6	1.5	0–1.50
HS150, HS50, 1150, 8500001, T50, 1050, 950, 1250*	D	1.5	0–0.15
HS35, 935, T35, 835, 1035, 1235*	C	1.5	0–0.08
HS15, 815, 915, 1015, 1215*	AA	1.5	0–0.025
904	N	1.5	0–0.02
201	WO	1.5	0–0.06
812, 912	AAA	1.5	0–0.02
E340E	N	1.5	0–0.02

* Zinc chloride cells

Table 54.5 Service capacity of Union Carbide carbon–zinc Leclanché cells

Cell	Starting drain (mA)	Service capacity (h)	Cell	Starting drain (mA)	Service capacity (h)
N	1.5 / 15.0	320 / 20	114	0.7 / 7.0	300 / 25
AAA	2.0 / 20.0	350 / 21	117	0.8 / 8.0	475 / 49
AA	3.0 / 30.0	450 / 32	118	0.8 / 8.0	525 / 54
B	5.0 / 50.0	420 / 32	127	1.0 / 10.0	475 / 72
C	5.0 / 50.0	520 / 53	132	1.3 / 13.0	275 / 16
D	10.0 / 100.0	525 / 57	135	1.3 / 13.0	550 / 52
F	15.0 / 150.0	630 / 60	148	2.0 / 20.0	610 / 60
G	15.0 / 150.0	950 / 78	155	5.0 / 30.0	620 / 74
6	50.0 / 500.0	750 / 95	161	3.0 / 30.0	500 / 55
55	15.0 / 150.0	635 / 61	165	3.0 / 30.0	770 / 90
105	0.4 / 4.0	210 / 8	172	5.0 / 50.0	780 / 90
109	0.6 / 6.0	710 / 75	175	5.0 / 50.0	1000 / 110
112	0.7 / 7.0	210 / 12			

Figure 54.1 shows a comparison of service life under various conditions between Yuasa's UM-I standard cell and their SUM-IG Hi Capacity cell of the same dimensions. The improvement in life of the Hi Capacity cell is very apparent in both continuous and intermittent use applications.

The carbon–zinc batteries discussed above are representative of the various types available. There are, of course, many other suppliers of this type of battery and they are listed under Suppliers of primary batteries.

Table 54.6 The most widely used carbon–zinc Leclanché cells and batteries: typical applications, load, voltage under load, amperage reading for testing

ANSI designation	Application	Load (Ω)	Voltage under load (% of applied voltage shown on cell or battery label)	Amperage (A) for usable cell or battery (test for an instant only)
AAA	Penlights	3.0	0–61% replace; 61–71% weak; 72–105% good	
AA	Toys, clocks, flashlights, movie cameras	2.25	0–61% replace; 61–71% weak; 72–105% good	
AA	Transistor radios	10.0	0–61% replace; 61–71% weak; 72–105% good	
C	Toys, clocks, flashlights, movie cameras	2.25	0–61% replace; 61–71% weak; 72–105% good	
C	Transistor radios	10.0	0–61% replace; 61–71% weak; 72–105% good	
D	Toys, clocks, flashlights, movie cameras	2.25	0–61% replace; 61–71% weak; 72–105% good	
D	Transistors	10.0	0–61% replace; 61–71% weak; 72–105% good	
N	Penlights	3.0	0–61% replace; 61–71% weak; 72–105% good	
6 V	Lanterns	5.0	0–61% replace; 61–71% weak; 72–105% good	
9 V	Transistor radios	250.0	0–61% replace; 61–71% weak; 72–105% good	
AAA	Photo	0.01 (entire resistance, lead-to-lead of tester)		1–3.8
AA	Photo	0.01 (entire resistance, lead-to-lead of tester)		3–6.5
C	Photo	0.01 (entire resistance, lead-to-lead of tester)		4–9
D	Photo	0.01 (entire resistance, lead-to-lead of tester)		10–13.5

Table 54.7 Ever Ready (Berec) carbon–zinc dry cells and batteries

Catalogue number	Voltage (V)	Diameter (mm)	Height (mm)	Length (mm)	Width (mm)	Applications
High-power cells						
HP2	1.5	34.2	61.8			Gas lighter
HP11	1.5	26.2	50.0			Gas lighter
HP7	1.5	14.5	50.5			
HP16	1.5	10.5	45.0			
HP992	6.0		109.0	67.0	67.0	
HP1	12.0		136.5	133.5	683.0	
HP16	1.5	45.0	10.5			Photographic
HP1	12.0		136.5	133.5	68.3	Alarms, lighting, signalling
Power pack batteries						
PP1	6.0		55.6	65.1	55.6	
PP3-P	9.0		48.5	26.5	17.5	
PP3	9.0		48.5	26.5	17.5	
PP4	9.0	25.5	50.0			
PP6	9.0		70.0	36.0	34.5	
PP7	9.0		61.9	46.0	46.0	
PP8	6.0		200.8	65.1	51.6	
PP9	9.0		81.0	66.0	52.0	
PP10	9.0		226.0	66.0	66.0	
PP11	4.5 + 4.5 (in series or parallel externally)		91.3	65.1	52.4	
Photographic batteries						
B121	15.0		37.0	27.0	16.0	
B122	22.5		51.0	27.0	16.0	
B123	30.0		65.0	27.0	16.0	
B154	15.0		35.0	16.0	16.0	
B155	22.5		51.0	16.0	16.0	
Alarm, emergency lighting and signalling batteries						
AD28	4.5		106.0	102.0	35.0	
126	4.5		91.3	103.2	34.9	
D23	1.5	12.0	30.1			
Valve radio batteries						
B101	67.5		95.0	71.0	35.0	
B126	90.0		100.0	71.0	50.0	
AD4	1.5		103.0	67.0	67.0	
AD28	4.5		106.0	102.0	35.0	
AD55	1.5		38.1	79.4	66.7	
Lantern batteries						
P5996	6.0		102.0	67.0	67.0	
991	6.0		125.4	135.7	72.2	
Fencer batteries						
R1585	7.5		196.8	198.4	96.8	
A0300	7.2	182.4	193.7			
PP8	6.0		200.8	65.1	51.6	
Transistorized clock battery						
C11	1.5	26.2	50.0			
Calculator batteries						
C7	1.5	14.5				
PP3-C	9.0		48.5	26.5	17.5	
Gas lighter battery						
U14	1.5	51.2	109.5			
Lighting batteries						
SP2	1.5	34.2	61.8			
No. 8	3.0	21.5	74.0			
SP11	1.5	26.2	50.0			
800	3.0		81.0	68.3	36.5	
U16	1.5	10.5	45.0			
1289	4.5		67.0	62.0	22.0	

Table 54.8 Varta carbon–zinc dry cells and batteries

Designation	Voltage	Equivalent Eveready size	Designation	Voltage	Equivalent Eveready size
Single power (SP)			245	1.5	D23
211	1.5	U2	434	6.0	PX23
212*	1.5	SP2	74	15.0	B154
213*	1.5	SP11	72*	22.5	B122
214	1.5	U11	28	9.0	PP6
251	1.5	U12	29*	9.0	PP4
201	4.5	1289	438*	9.0	PP3
259	3.0	No. 8	439*	9.0	PP9
430 (PJ430)	6.0	996 (PJ996)	489	9.0	PP7
High power (HP)			*High drain*		
232*	1.5	HP2	222*	1.5	–
233*	1.5	HP11	236	1.5	–
244	1.5	HP7	239	1.5	HP16

* Steel clad

Table 54.9 Varta carbon–zinc cells and batteries

Varta type order No.		Capacity (mA h)	Nominal voltage (V)	Size	Dimensions (mm)		
					Diameter/ length	Width	Height
1006		900	1.5	AA/R6/UM-3	14.5		50.5
1012	Standard	1800	4.5	3R12	62.0	22.0	67.0
1014		2200	1.5	C/R14/UM-2	26.2		50.0
1020		4500	1.5	D/R20/UM-1	34.2		61.5
2006, 2606		1000	1.5	AA/R6/SUM-3	14.5		50.5
2010	Super	925	3.0	2R10	21.8		74.6
2012	High	2000	4.5	3R12	62.0	22.0	67.0
2014, 2614	Performance	2650	1.5	C/R14/SUM-2	26.2		50.0
2020, 2620		6000	1.5	D/R20/SUM-1	34.2		61.5
2022, 2622		300	9.0	6F22	26.5	17.5	48.5
3006		1100	1.5	AA/R6/SUM-3	14.5		50.5
3012		2300	4.5	3R12	62.0	22.0	67.0
3014	Super Dry	8000	1.5	C/R14/SUM-2	26.2		50.5
3020		350	1.5	D/R20/SUM-1	34.2		61.5
3022		350	9.0	6F22	26.5	17.5	48.5

Table 54.10 Yuasa carbon–zinc dry cells

Type	Nominal voltage (V)	Dimensions (mm)		Jacket material	IEC	Equivalent batteries		
		Diameter	Height			Ray-o-Vac	Eveready (Union Carbide)	Ever Ready (Berec)
Standard Diamond series								
UM-1	1.5	33	60	Metal	R20	2D	950	U2, PLU2
UM-2	1.5	26	49	Metal	R14	1C	935	U11
UM-3	1.5	14	50	Paper	R6	7R	915	U12
006P	9.0	17	49	Metal	6F22	1604	216	PP3
Hi capacity Golden G series								
SUM-1G	1.5	33	60	Metal	R20	13	1050	HP2, PP12
SUM-2G	1.5	26	49	Metal	R14	14	1035	HP11, PP13
SUM-3G	1.5	14	50	Metal	R6	15	1015	H97, PP14
SUM-4G	1.5	10	44	Paper	R03	400	912	U16

55

Mercury batteries

Contents

55.1 Mercury–zinc (primary) batteries

Union Carbide and Mallory are two of the major producers of mercury–zinc cells and batteries. There are, of course, other important producers such as Berec (UK), Crompton-Parkinson (Hawker Siddeley) (UK) and Varta (West Germany). Table 55.1 tabulates the type numbers of equivalent cells, which meet a given International Electrochemical Commission (IEC) designation as produced by Union Carbide, Mallory, Berec and Varta. In many instances equivalent cells are available from more than one manufacturer. In such instances comparative cost quotations would be of interest to the intending purchaser.

Both Union Carbide and Mallory use a vented construction, which renders their products leak-proof and free from bulging under all normal working conditions. Self-venting occurs automatically if, for example, excessive gas is produced within the cell under sustained short-circuit conditions. Both manufacturers assemble cells in a wide variety of structures to provide batteries of varying size, voltage and capacity. A wide range of purpose-built batteries can also be obtained. For maximum reliability, inter-cell connection is by spot-welded strips of nickel-plated steel. Outer containers are made from materials ranging from cardboard to metal.

Union Carbide

These batteries can be stored for periods of up to 3 years; it is not, however, good practice to store them for long periods.

Mercuric oxide batteries are available with two formulations designed for different field usage. In general the 1.35 V cells (that is, 100% mercuric oxide depolarization) or batteries using these cells are recommended for voltage reference sources (that is, high degree of voltage stability) and for use in applications where higher than normal temperatures may be encountered (and for instrumentation and scientific applications). The 1.4 V cells (a mixture of manganese dioxide and mercuric oxide depolarization), or batteries made up of these cells, are used for commercial applications of all types other than those just mentioned. The 1.4 V cells or batteries should be used for long-term continuous low-drain applications if a very flat voltage characteristic is not needed.

Stabilized battery characteristics are unaffected by high temperatures. The mercuric oxide cell has good high-temperature characteristics. It can be used up to 65–70°C, and operation at 145°C is possible for a few hours. In general, mercuric oxide batteries do not perform well at low temperatures. However, recent developments have produced several popular cell sizes, which have good low-temperature characteristics. For the mercuric oxide batteries not in this group, there is a severe loss of capacity at about 4.4°C, and near 0°C the mercuric oxide cell gives very little service except at light current drains.

The life curve (Figure 55.1) shows typical performance of the cylindrical mercuric oxide battery, E12, over the temperature range of −23 to +71°C for current drains encountered in many applications. Successful operation at temperatures above 120°C for short periods has been reported, but it is recommended that 70°C is not exceeded. Voltage depression is slight at low temperatures when drains are 100 μA or less, in the large cell types, or when intermittent drains of this order are used with smaller cells.

Union Carbide supply a range of Eveready mercury–zinc cells and batteries with voltages between 1.35 and 97.2 V and capacities from 16 to 28 A h. These batteries are of the button cell or cylindrical design and the electrical and dimensional specifications together with information on typical

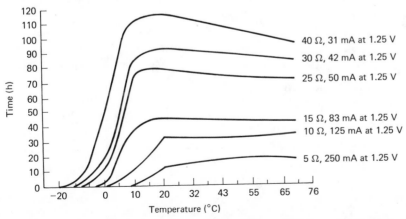

Figure 55.1 Mercury–zinc cell life *versus* temperature (E12 type, 3.6 A h capacity) (Courtesy of Union Carbide)

Table 55.1 Some equivalent mercury–zinc cells from different manufacturers

IEC Designation	Union Carbide Eveready			Mallory Duracell			Varta		Berec Ever Ready	
	No.	Voltage (V)	Capacity (mA h)	No.	Voltage (V)	Capacity (mA h)	No.	Voltage (V)	No.	Voltage (V)
MR41	325	1.35	50							
MR08	E400N	1.35	80							
	E400	1.40	80							
NR48				RM13H	1.40	85	7106	1.40	RM13H	1.40
MR48	E138, E13E	1.40	85	WH6	1.35	95				
	323	1.35	95							
NR08				RM575H	1.40	100	7109	1.40	RM575H	1.40
MR42	343	1.35	120	WH12NM	1.35	110				
NR43				RM41	1.40	150				
MR43	354	1.35	150							
MR07	EPX675	1.35	220	PX675	1.35	210	7001	1.35		
NR07				MP675H	1.40	210	7103		MP6754	1.40
				SP675	1.40	230				
MR44	313	1.35	220	WH3	1.35	220				
MR9	E625	1.40	350							
	E630	1.40	350							
	E625N	1.35	250							
	EPX625	1.35	250	PX625	1.35	250	7002	1.35	PX625	1.35
	EPX13	1.35	250	RM625N	1.35	350			RM625N	1.35
MR52				PX640	1.35	500				
NR52				RM640H	1.40	500				
MR1	E401N	1.35	800							
	E401, E401N	1.40	800							
MR50				RM1N	1.35	1000			RM1N	1.35
				PX1	1.35	1000			PX1	1.35
MR7	EPX1	1.35	1000							
	E1	1.40	1000							
NR50				RM1	1.40	1000				
NR1				MP401	1.40	1000				
MR17	E3	1.35	2200							
NR6				ZM9C	1.40	2400			ZM9C	1.40
MR6	E502E	1.40	2400							
	E9	1.40	2500							
	E502	1.35	2400							
MR51				RM12R	1.35	3600				
				RM675H	1.40		7101	1.40	RM675H	1.40
				RM625			7102	1.40		
				RM312H			7107	1.40		
				RM41H			7108	1.40		
				PX825			7201	1.50		
									RM1H	1.40
									RM401H	1.40

Table 55.2 Union Carbide Eveready zinc–mercuric oxide cells, 1.35 and 1.40 V

Eveready number	Voltage (V)	Suggested current range (mA)	Designation ANSI	IEC	Average service capacity (mAh)	Schedule (h/day)	Starting drain (mA)	Load (Ω)	Cut-off voltage (V)	Service life (h)	Average impedance at 1000 Hz (Ω)
E212E	1.35	0–1	M4	–	16 to 0.9 V at 0.9 mA	24	0.9000 1.1000	1 500 1 250	0.9 0.9	20 16	
325	1.35	0–0.1	WM5	MR41	50 to 1.2 V at 13 000 Ω load	24 24	0.1040 0.0054	13 000 25 000	1.2 1.2	540 14 months	65
388	1.35	0–0.07	–	–	65 to 1.2 V at 20 000 Ω load	24	0.0675	20 000	0.9 1.2	1000 980	
387	1.35	0–0.1	–	–	80 to 0.9 V at 13 000 Ω load	24	0.1040	13 000	1.0	800	
E502E	1.40	0–200	M55	MR6	2400 to 9 V at 48 mA	24 24 24 24	17.0000 17.0000 54.0000 54.0000	83 83 26 26	0.9 1.1 0.9 1.1	174 154 51.2 42.7	
E9	1.40	0–200	M55	MR6	2500 to 9 V at 50 mA	24 24 24 24	16.9000 16.9000 56.0000 56.0000	83 83 25 25	0.9 1.1 0.9 1.1	180 175 55 52	
E12	1.40	0–250	M70	–	3800 to 9 V at 83 mA	24 24 24 24	44.0000 44.0000 140.0000 140.0000	32 32 10 10	0.9 1.1 0.9 1.1	100 91 33 28	
E42	1.40	0–1000	M100	–	14 000 to 9 V at 280 mA	24 24 24 24	318.0000 318.0000 1120.0000 1120.0000	4.4 4.4 1.25 1.25	0.9 1.1 0.9 1.1	60 29 7.5 1	

Table 55.3 Range of Union Carbide Eveready mercury–zinc batteries

Battery No. (in order of increasing A h capacity)	Suggested current range (mA)	Service capacity (mA h)	Service capacity (A h) rated at (mA)	Maximum dimensions (mm)				Number of cells
				Diameter	Length	Width	Height	
1.35 and 4.0 V batteries								
E212E	0–1.0	16	0.75	5.72			3.30	1
E312	0–5.0	45	0.85	7.87			3.56	1
E312E	0–5.0	45	0.85	7.87			3.56	1
E42	0–1000	14000	280.00	30.40			60.70	1
42N	0–1000	14000	280.00	30.40			60.70	1
E302651	0–1000	14000	250.00	33.30			61.90	1
E302157	0–2000	28000	500.00		63.50	34.90	108.00	2
2.7 and 2.8 V batteries								
EPX14	0–5	250	5.00	16.90			15.40	2
E152	0–30	350	5.00	12.30			28.90	2
E132	0–100	1000	20.00	16.80			32.90	2
E132N	0–100	1000	20.00	16.80			32.90	2
E302702	0–60	2200	42.00	28.60			41.30	2
4.05 and 4.2 V batteries								
EPX29	0–1	160	1.00	12.10			17.30	3
EPX25	0–5	250	5.00	16.80			21.50	3
E233	0–60	2200	42.00	26.00			50.60	3
E233N	0–60	2200	42.00	26.00			50.60	3
5.4 and 5.6 V batteries								
EPX23	0–10	110	2.00	15.20			20.00	4
EPX32	0–50	500	10.00	17.20			44.90	4
E164	0–50	500	10.00	16.80			44.30	4
E302351	0–60	2200	42.00	27.00			66.70	4
E302904	0–80	3400	63.00	34.10			74.60	4
6.75 and 7 V batteries								
E175	0–10	180	2.20	12.60			27.80	5
E115N	0–20	250	5.00	16.80			33.50	5
16.2 and 16.8 V batteries								
E302362	0–20	250	5.00	19.10			80.20	12
E303314	0–250	2200	50.00		72.40	71.40	33.50	12
27 V battery								
E302580	0–80	3400	63.00		69.10	69.10	99.20	20
29.7 V battery								
E302579	0–20	250	5.00		35.70	19.10	81.00	22
47.25 V battery								
E302465	0–100	1000	30.00	54.00			95.30	35
97.2 V battery								
E302462	0–100	1000	30.00		71.40	38.90	163.00	72

Cell size (g)	Application	Weight (cm^3)	Terminals taps	Volume	Voltage
212	Hearing-aids	0.28	Flat contact	0.08	−, +1.35
4-31141	Instruments	0.85	Flat contact	0.13	−, +1.40
4-31141	Hearing-aids	0.85	Flat contact	0.13	−, +1.40
42D	Instruments	167.00	Flat contact	47.50	−, +1.40
42D	Instruments (voltage reference)	167.00	Flat contact	47.50	−, +1.35
42D	Instruments	174.00	Special leads	47.50	−, +1.35
42D	Radiation equipment	383.00	Socket	220.00	−, +1.35
625	Photographic (with cadmium sulphide cell)	8.50	Flat contacts	3.28	−, +2.70
450	Instruments	11.30	Flat contacts	3.10	−, +2.80
1	Transistor applications, hearing-aids	26.70	Flat contacts	6.50	−, +2.80
1	Transistor applications	26.70	Flat contacts	6.50	−, +2.80
3	Geophysical	56.70	Wire leads	21.30	−, +2.70
675	Electric eye (movie camera)	7.25	Flat contact	1.81	−, +4.05
625	Electric eye (movie camera)	13.90	Flat contact	4.59	−, +4.05
3	Radio	89.30	Flat contact negative recessed	26.20	−, +4.20
3	Instruments	89.30	Flat contact	26.20	−, +4.05
575	B–C flash	7.25	Flat contact	3.44	−, +5.60
640	Camera	31.20	Flat contact	10.00	−, +5.60
640	Radio	34.00	Flat contact	10.00	−, +5.60
3	Fire alarm	128.00	Flat contact	37.70	−, +5.40
4	Meter	170.00	Wire lead	59.00	−, +5.40
675	Radio	11.90	Flat contact negative recessed	3.60	−, +7.00
625	Instruments	20.00	Flat contact	7.20	−, +6.75
625	Radiation equipment transceiver	51.00	Wire lead	18.50	−, +16.20
15 AA low temperature cell		363.00	Recessed flat contact	156.00	−, +16.80
4	Instruments	953.00	Socket	472.00	−, +27.00
625	Test equipment	142.00	Wire lead	47.50	−, +29.70
1	Geophysical	482.00	Wire lead	198.00	−, +47.25
1	Transmitter B	907.00	Wire lead	420.00	−, +97.20

usages for some of these batteries are given in Table 55.3. Further information on the electrical specification including capacity and estimated hours service at 21°C under various conditions for a range of these cells is given in Table 55.2 (1.35 and 1.4 V cells) and Table 55.4 (2.7 to 12.6 V batteries). Typical discharge curves at 21°C for a typical Eveready cell are reproduced in Figure 55.2.

Table 55.4 Union Carbide Eveready zinc–mercuric oxide cells, 2.7 and 12.6 V

Eveready number	Voltage (V)	Suggested current range (mA)	Average service capacity (mA h)	Estimated average hours service at 21°C				
				Schedule (h/day)	Starting drain (mA)	Load (Ω)	Cut-off voltage (V)	Service life (h)
EPX14	2.7	0–5	250 to 1.8 V rated at 5 mA	24	0.98	2750	1.8	280
					5.40	500	1.8	58
E152	2.8	0–30	350 to 1.8 V rated at 7 mA	24	5.60	500	1.8	75.8
					5.60	500	2.2	74.7
					22.40	125	1.8	12.8
					22.40	125	2.2	10.5
E133N	4.05	0–100	1000 to 2.7 V rated at 20 mA	24	22.50	180	2.7	52.6
					22.50	180	3.3	42
					37.50	108	2.7	31.7
					37.50		3.3	18
E163	4.2	0–50	500 to 2.7 V rated at 10 mA	24	11.20	375	2.7	50.8
					11.20	375	3.3	46.5
					28.00	150	2.7	22
					28.00	150	3.3	8.1
E164N	5.4	0–50	500 to 3.6 V rated at 10 mA	24	10.80	500	3.6	51
					10.80	500	4.4	46
					27.00	200	3.6	15.8
					27.00	200	4.4	7.5
EPX23	5.6	0–10	110 to 3.6 V rated at 2 mA	24	1.70	3340	3.6	66.3
					1.70	3340	4.4	61.7
					5.60	1000	3.6	16.5
					5.60	1000	4.4	15
E135N	6.75	0–100	1000 to 4.5 V rated at 20 mA	24	22.50	300	4.5	52.6
					22.50	300	5.5	42
					37.50	180	4.5	31.7
					37.50	180	5.5	18
E175	7.0	0–10	180 to 4.5 V rated at 2.2 mA	24	2.30	3000	4.5	87
					2.30	3000	5.5	86
E136N	8.1	0–100	1000 to 5.4 V rated at 20 mA	24	22.50	360	5.4	52.6
					22.50	360	6.6	42
					37.50	216	5.4	31.7
					37.50	216	6.6	18
E146X	8.4	0–30	575 to 5.4 V rated at 15 mA	24	11.00	750	5.4	60
					11.00	750	6.6	44
					28.00	300	5.4	19
					28.00	300	6.6	5
E177	9.8	0–10	215 to 6.3 V rated at 2.05 mA	2	5.40	1800	4.2	44
					5.40	1800	6.6	42
				4	9.80	1000	4.2	25
					9.80	1000	6.6	21
E289	12.6	0–50	750 to 8.1 V rated at 15 mA	24	17.50	720	0.9	51.5
					17.50	720	1.1	41.5
					56.00	225	0.9	12
					56.00	225	1.1	4.6

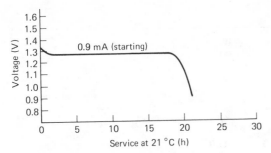

Figure 55.2 Union Carbide 212E mercury–zinc cell, 1.35 V (16 mA h to 0.9 V at 0.9 mA, load 1500 Ω). Voltage discharge curve, estimated average service at 21°C *versus* voltage (Courtesy of Union Carbide)

Mallory

The voltage, capacity and physical characteristics of some of the Mallory Duracell range of mercury–zinc cells and batteries are shown in Tables 55.5 and 55.6. Battery types 304116 (12.6 V) and TR431 (11.2 V) (Table 55.6) are recommended as smoke alarm centre batteries. Mallory recommend that, for the best contact to the terminals of mercuric oxide batteries, the device contacts be made of a spring

Table 55.5 Mallory Duracell single-cell mercury–zinc cells

Nominal voltage (V)	Nominal capacity (mA h)	Type	Max. diameter (mm)	Max. height (mm)	Weight (g)
1.35	45	RM312	7.75	3.56	0.6
1.35	50*	WH1	7.80	3.60	0.8
1.35	62*	10R10	8.84	3.30	0.9
1.35	80*	PX400	11.60	3.43	1.1
1.35	80*	W2	11.56	3.55	1.4
1.35	95*	WH6	7.75	5.33	1.2
1.35	110*	WH12NM	11.58	3.60	1.4
1.35	120*	WH8	11.56	3.84	1.6
1.35	210*	PX675	11.60	5.40	2.6
1.35	220*	WH3	11.56	5.40	2.9
1.35	250	PX625	15.60	6.20	4.0
1.35	13 000	RM2550R	65.00	13.97	178.0
1.35	14 000	RM42R	34.00	60.70	167.0
1.40	16	RM212H	5.59	3.10	0.32
1.40	75	RM400H	11.60	3.43	1.1
1.40	85	RM13H	7.87	5.34	1.0
1.40	100	RM575H	11.60	3.50	1.4
1.40	210	MP675H	11.60	5.40	2.6
1.40	230	SP675	11.60	5.40	2.6
1.40	500	RM640H	15.90	11.20	7.9
1.40	600	RM660	17.40	7.57	7.6
1.40	800	RM401	12.00	29.00	11.0
1.40	1000	MP401	12.00	30.20	13.0
1.40	1000	RM1	16.40	16.80	12.0
1.40	2400	ZM9C	14.50	50.00	30.0
1.40	3600*	ZM12	16.26	49.71	40.0

* Low-rate cell

material such as phorphor-bronze or beryllium copper, which will maintain a contact force of at least 50g for an extended period of time. The contact should be plated with about 50.8 μm nickel (continuous) followed by a minimum of 0.508 μm gold. The reliability of the contact can be further increased by subdividing the main contact member into two, three or more individual points or prongs such as the tines of a fork.

Voltage discharge curves for representative cells and batteries in the Duracell range are shown in Figure 55.3 These curves show the estimated average service life at 20°C to be expected at various current drains on constant-resistance load for various Duracell 1.35 and 1.4 V cells, for cells with nominal capacities in the range 80–85 mA h.

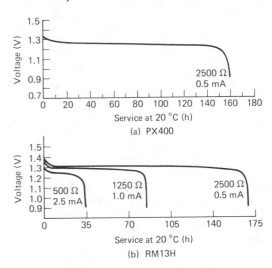

Figure 55.3 Mallory Duracell mercury–zinc cells and batteries, voltage discharge curves: estimated average service at 21°C *versus* voltage. Current drains at 1.25 V. (a) 1.35 V, 80 mA h to 0.9 V on 25 000 Ω at 20°C, (b) 1.4 V, 85 mA h to 0.9 V on 1250 Ω at 20°C, impedance (average at 15% discharge) 7.6 Ω at 10 Hz, 7.0 Ω at 1000 Hz, 6.0 Ω at 100 000 Hz (Courtesy of Mallory)

Mallory produce special mercury cells based on the wound-anode construction principle, which are particularly suitable for use at lower temperatures than those recommended for normal mercury cells. Thus Mallory types RM1438, RM1450R and RM2550R operate efficiently at a temperature 15°C lower than the standard mercury cells.

Duracell Europe

Duracell Europe supply a selection of the mercury–zinc batteries produced by Mallory under the same type number. These are listed in Table 55.7, together with the table numbers from which details concerning these batteries can be obtained.

Table 55.6 Mallory Duracell mercury–zinc multi-cell batteries

Nominal voltage (V)	Nominal capacity (mA h)	Type	Maximum dimensions (mm)		Weight (g)	Outline (see below)	Cell type	Number of cells
			Diameter	Height				
2.70	250	PX14	16.40	15.40	8.3	A	PX625	2
2.70	350	TR112N	16.76	15.49	8.2	A	RM625N	2
2.70	350	TR152R	12.70	28.90	11.0	A	RM450R	2
2.70	500	TR162R	16.81	22.53	17.0	A	RM640R	2
2.70	1000	TR132N	17.10	33.52	27.0	A	RM1N	2
2.70	2200	TR232R	26.04	34.04	60.0	A	RM3R	2
4.05	210*	PX29	11.71	16.51	7.7	A	PX675	3
4.05	250	PX25	16.76	21.46	14.0	A	PX625	3
4.05	350	TR113N	16.76	21.46	14.0	A	RM625N	3
4.05	350	TR153R	12.70	43.38	17.0	A	RM450R	3
4.05	500	TR163R	16.81	22.53	26.0	A	RM640R	3
4.05	1000	TR133N	17.10	50.04	41.0	A	RM1N	3
4.05	2200	TR233R	26.04	50.67	89.0	A	RM3R	3
4.20	225	7R31	17.02 × 11.43 × 40.64		14.0			
5.40	350	TR114N	16.76	27.43	16.0	A	RM625N	4
5.40	350	TR154R	12.70	57.89	23.0	A	RM450R	4
5.40	500	TR164R	16.81	44.90	34.0	A	RM640R	4
5.40	1000	TR134N	17.10	66.68	54.0	A	RM1N	4
5.40	2200	TR234R	26.04	67.56	119.0	A	RM3R	4
5.60	100	PX23	15.20	20.00	8.0	A	RM575H	4
5.60	150	PX27	12.70	20.50	9.2	A	RM41	4
5.60	175	7H34	13.00	25.40	11.0	A		
5.60	500	TR164	16.81	44.90	34.0	A	RM640H	4
6.75	350	TR115N	16.76	33.53	20.0	A	RM625N	5
6.75	500	TR165R	16.81	56.13	43.0	A	RM640R	5
6.75	1000	TR135N	17.10	83.31	67.0	A	RM1N	5
6.75	2200	TR235R	26.04	84.46	149.0	A	RM3R	5
7.00	210	TR175	12.70	27.18	12.0	A	MP675H	5
8.10	1000	TR136N	17.10	99.95	81.0	A	RM1N	6
8.40	575	TR146X	26.20 × 16.76 × 44.50		53.0			
8.40	600	TR126	19.20	50.80	50.0	A	RM660	6
8.40	750	TR286	25.40	49.20	77.0	B		
9.45	1000	TR137N	17.10	116.59	95.0	A	RM1N	7
9.80	210	TR177	12.95	48.31	19.0	B	MP675H	7
10.80	350	TR118N	16.76	51.94	32.0	A	RM625N	8
11.20	210	TR178	12.70	43.31	19.0	A	MP675H	8
11.20	1000	TR431	26.16	81.26	116.0	B		
12.60	100	TR149	14.99	35.00	17.0	A	RM575H	9
12.60	750	TR289	25.40	61.11	102.0	B		
12.60	900	304116	25.16	81.28	127.0	B		
Recent additions								
2.80	240	TR172	12.70	16.60	4.5			2
4.20	240	TR173	12.70	16.60	4.1			3
5.60	240	TR174	12.70	22.00	9.4			4
8.10	350	TR116N	16.80	39.74	24			6
9.45	350	TR117N	16.80	45.84	28			7

Outline

A   B

---

* Low-rate battery

Crompton-Parkinson

This company, under the trade name of Vidor, supplies a 13.4 V mercury–zinc battery (K13-4-1499) in which the cells are encapsulated in epoxy resin to

Table 55.7 Mercury–zinc batteries supplied by Duracell (See also Table 55.5)

Type No.	Voltage (V)	Capacity at 20°C (mA h)
RM1N	1.35	1000
RM13H	1.4	85
MP401	1.4	1000
PX625	1.35	250
RM625N	1.35	350
PX640	1.35	500
MP675H	1.4	210
SP675	1.4	240
PX23	5.6	80
PX27	5.6	150
TR132N	2.7	1000
TR133N	4.05	1000
TR134N	5.4	1000
TR162	2.7	500
TR163	4.05	500
TR164	5.4	500

produce a solid block, resistant to all forms of damage. It is recommended for use in electronic equipment, transceivers and rocket sondes and as a secondary standard voltage source. This battery is $59 \times 40 \times 59$ mm and weighs 285 g. The battery terminal open voltage is 13.56 V at 0°C, 13.57 V at 20°C and 13.58 V at 30°C. At a constant discharge of 10 mA the battery has a life of approximately 2 years at temperatures between -10 and $+20°C$, decreasing towards 1 year at temperatures outside this range. Typical discharge characteristics for this cell at temperatures between -10 and $+20°C$ and on constant resistance loads are shown in Figure 55.4. The discharge characteristics are little affected by whether the duty is continuous or intermittent.

Varta

Varta produce a wide range of mercuric oxide batteries and button cells in the voltage range 1.35–5.6 V and the capacity range 40–1000 mA h (Table 55.8). These batteries are mainly used in photographic and hearing-aid applications.

In addition, Varta produce a range of low-drain cells for electric and electronic watches with analogue display (Table 55.9).

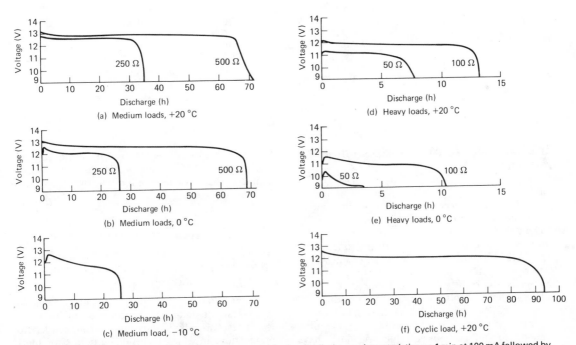

Figure 55.4 Vidor K13-4-1499 mercury–zinc dry battery: typical cyclic discharge characteristics on 1 min at 100 mA followed by 9 min at 10 mA, at temperatures between -10 and $+20°C$ (constant resistance load). Typical voltages at the end of each 100 mA cycle have been plotted (Courtesy of Vidor)

Table 55.8 Mercury–zinc button cells and batteries supplied by Varta

Type No.	Voltage (V)	Capacity* (mA h)	Internal resistance (Ω)	Dimensions (mm)		Weight (g)	IEC reference	Main application	Order No.
				Diameter	Height				
V 400 PX	1.35	80	–	11.6	3.6	1.1	–	Photography	4400
V 675 PX	1.35	230	2.5 to 4.5	11.6	5.4	2.7	MR 07	Photography	4677
V 625 PX	1.35	350	1.5 to 2.2	16.0	6.2	4.6	MR 9	Photography	4625
V 640 PX	1.35	500	–	15.9	11.2	7.9	MR 52	Photography	4640
V 1 PX	1.35	1000		16.4	16.8	12.0	MR 50	Photography	4002
V 312 HM	1.40	40	9.0 to 14.0	7.9	3.6	0.78	NR 41	Hearing-aids	4312
V 13 HM	1.40	95	7.0 to 12.0	7.9	5.4	1.2	NR 48	Hearing-aids	4013
V 575 HM	1.40	100	3.4 to 5.0	11.6	3.6	1.5	NR 08	Hearing-aids	4575
V 41 HM	1.40	150	2.5 to 4.0	11.6	4.2	1.8	NR 43	Hearing-aids	4042
V 675 HM	1.40	180	3.6 to 5.4	11.6	5.4	2.6	NR 07	Hearing-aids	4676
V 675 HP	1.40	250	1.8 to 3.5	11.6	5.4	2.6	NR 07	Hearing-aids	4675
V 625 HM	1.40	350	1.8 to 2.7	16.0	6.2	4.6	NR 9	Hearing-aids	4627
V 14 PX	2.70	350	3.0 to 4.5	16.8	16.0	10.0	–	Photography	4015
V 164 PX	5.40	500	–	16.8	44.9	34.0	–	Photography	4164
V 23 PX	5.60	100	14.0 to 20.0	15.2	20.0	12.0	–	Photography	4023
V 27 PX	5.60	150	10.0 to 16.0	12.85	20.5	10.2	–	Photography	4027

* Mean values at a discharge voltage of 0.9 V/cell

Table 55.9 Varta low-drain mercury–zinc cells for electric and electronic watches with analogue displays

Type No.	Voltage (V)	Capacity (mA h)	Dimensions (mm)	Weight (g)	Shelf pack pieces	Replaces
501	1.35	230	11.6 × 5.4	2.69	10/100	313, RW 52, WH3, 13
502	1.35	90	11.5 × 3.6	1.05	10/100	387, RW 51, W2, 214
503	1.35	70	8.5 × 3.15	0.92	10/100	388, 10 R 10, 221
506	1.35	95	7.9 × 5.4	1.24	10/100	323, RW 58, WH6, C, 8, SB-C3
507	1.35	50	7.9 × 3.6	0.78	10/100	325, RW 52, WH1, 6UDC, 5, SB-C1
508	1.35	150	11.6 × 4.2	2.05	10/100	354, RW 54, 10 R 11, 4, SB-C8
509	1.35	110	11.6 × 3.6	1.63	10/100	343, RW 56, WH8, B, 218, 3

55.2 Mercury–zinc cardiac pacemaker batteries

The first pacemaker batteries were produced in about 1973, and since then many improvements have been made in their design.

The type I battery has a nominal voltage of 1.35 V and a nominal capacity of 1000 mA h defined for a load of 1250 Ω, with a cut-off voltage of 0.9 V. For lighter loads, tests on early production batteries have given the results shown in Table 55.6.

The values in Table 55.6 are to be considered as minima. It is clear that reducing the drain does not result in a proportional increase of life. To meet the demand for longer service life, Mallory have developed a new type RM 2 cell. Its construction differs from the earlier one in a number of respects.

Its nominal capacity is 1800 mA h, and a capacity realization of 93% over 28 months' discharge is claimed. The service life of this battery is likely to be less than 5 years.

55.3 Other types of mercury battery

Mercuric oxide–indium–bismuth and mercury–cadmium (primary) batteries supplied by Crompton-Parkinson

Crompton-Parkinson supply mercuric oxide–indium–bismuth and mercuric oxide–cadmium batteries. These are alkaline systems recommended for applications in which high reliability in particularly onerous long-term storage and use conditions is a prime requirement.

Operating characteristics for a 0.6 A h mercury–cadmium cell are shown in Figure 55.5.

Mercury–cadmium batteries supplied by Mallory

Mallory supply a mercury–cadmium battery in their Duracell range. This battery (No. 304116) is of

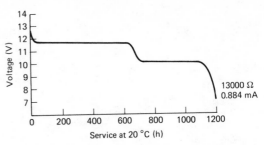

Figure 55.6 Mallory No. 304116 12.6 V mercuric oxide–cadmium battery: two-stage discharge curve. Current drain at 11.5 V (Courtesy of Mallory)

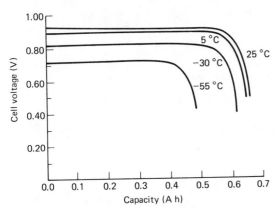

Figure 55.5 Operating characteristics for 0.6 A h mercury oxide–cadmium cell at various temperatures. Hourly discharge rate is 2% of capacity (i.e. $C/50$) (Courtesy of Mallory)

12.6 V with a capacity of 6.30 mA h to an end-point of 10.3 V or a capacity of 900 mA h to an end-point of 7.0 V, both on 13 000 Ω at 20°C. This cylindrical battery weighs 237 g and occupies a volume of 40.5 cm^3. Typical two-step discharge characteristics for this battery are shown in Figure 55.6.

Mercury–cadmium batteries supplied by Varta

Varta supply 0.9 V, 3200 and 2000 mA h mercury–cadmium cells (types 4910 and 4911 respectively).

56

Lithium batteries

Contents

56.1 Lithium–vanadium pentoxide (primary) batteries

The excellent storage characteristics of this type of battery are illustrated in Figure 56.1. Honeywell can supply small lithium–vanadium pentoxide cells with capacities up to 100 mA h, glass ampoule reserve

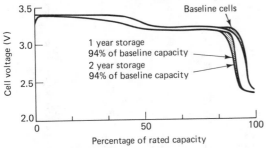

Figure 56.1 Honeywell lithium–vanadium pentoxide cell: discharge performance after 1 and 2 years storage at 24°C, current density = 0.4 mA/cm². Average voltage: baseline, 3.26 V; storage, 3.21 V (Courtesy of Honeywell)

cells with capacities up to 500 mA h, cylindrical button, prismatic and flexible cell configurations, and a range of active cells. Details of some of these cells are given in Table 56.1.

Honeywell have recently developed a high-integrity crimp-sealed lithium–vanadium pentoxide cell. It has an open-circuit voltage of 3.42 V and a voltage under load of more than 3.0 V. The G3093 model has a rated capacity of 100 mA h, and the

G3094 of 100 mA h, both under the rated load. The cells have energy densities of 180–300 W h/dm³ and 132 W h/kg. The cells are 29 mm in diameter, 25–50 mm thick and weigh 6–8 g. The operating temperature range is −29 to +49°C and the cells have a projected storage capability of more than 20 years. Both reserve and non-reserve type lithium–vanadium pentoxide batteries are available from Honeywell.

56.2 Lithium–sulphur dioxide (primary) batteries

Honeywell

Honeywell use two types of cell construction: jelly-roll electrodes in crimp-sealed or hermetically sealed cylindrical cells, and large 20–100 A h 12 V flat-plate electrodes in large reserve batteries. It is a relatively high-pressure system and cells must have safety vents to avoid explosion in the event of accidental incineration. Standard cylindrical cells are available in low, medium and high rate configurations. Honeywell estimate that these batteries should store for 12 years at room temperature. A typical energy density is 420 W h/dm³ or 260 W h/kg.

The lithium–sulphur dioxide system is versatile and relatively inexpensive. The battery has excellent storage characteristics. The batteries can be supplied either as reserve batteries with capacities between 20 and 100 A h or as active (non-reserve) batteries in the 0.7–20 A h range (Table 56.2).

Table 56.1 Characteristics of Honeywell lithium – vanadium pentoxide cells

P/N description	Diameter or width (mm)	Length (mm)	Thickness (mm)	Volume (cm³)	Weight (g)	Capacity (A h)	Rated load (mA)
Reserve cells							
G2659 glass ampoule	15.2	25.4	–	5.1	10	0.18*	0.25
G2664†	25.4	18.4	–	9.5	23	0.75*	50.0
G2665 glass ampoule	15.2	55.9	–	11.1	25	0.50*	5.3
G2666 glass ampoule	12.7	20.3	–	2.6	6	0.10*	1.0
Representative active cells							
G2682 cylinder DD	34.0	128.0	–	108.8	287	30.0	15.0
G3005 cylinder A	16.0	71.1	–	12.5	38.5	4.00	10.0
G2655 ½ in button	51.7	12.7	–	32.3	79	3.50*	250.0
G2692 ¼ in button	50.8	6.3	–	–	–	1.20*	10.0
G2679 prismatic	20.0	50.0	9.9	10.0	28	1.00	10.0
G3003 flex cell	26.2	63.5	1.65	–	–	0.12	10.0
G3025 0.625 button	15.9	11.0	–	2.3	6	0.20*	1.0
G3004 flex cell	38.1	25.4	1.65	1.6	4	0.12	10.0

* To a 2.5 V cut-off
† Uses a throwaway activator; parameters given are for cell less activator
Capacities given are to a 2 V cut-off at rated load at room temperature. Those rated to a 2.5 V cut-off will deliver 50–100% more capacity under reduced load when discharged to 2 V

Table 56.2 Honeywell lithium–sulphur dioxide batteries

P/N description	Diameter or width (mm)	Length (mm)	Thickness (mm)	Volume (cm³)	Weight (g)	Capacity (A h)	Rated load (mA/cm²)	Open-circuit voltage (V)
Reserve batteries (metal bellows)								
20 A h	97.7	152.0	–	–	–	20.0	0.1	12.0
60 A h	157.5	152.0	–	–	–	60.0	0.1	12.0
100 A h	20.3	152.0	–	–	–	100.0	0.1	12.0
Active cells and batteries								
G3033 SO₂ DD cell	39.0	129.0	–	107.4	185	20.0	1000	2.9
G3008 SO₂ D cell	39.0	60.9	–	52.9	94	10.0	350	2.9
G3058 SO₂ half D cell	39.0	30.5	–	25.7	48	4.8	175	2.9
G2686 SO₂ CC cell	26.1	99.8	–	49.8	85	7.6	700	2.9
G3039 SO₂ long C cell	26.1	60.9	–	30.2	55	4.8	350	2.9
G3010 SO₂ C cell	26.1	49.8	–	24.4	45	4.6	250	2.9
G3024 SO₂ C cell	26.1	49.8	–	24.4	45	3.7	350	2.9
G3029 SO₂ A cell	16.0	49.8	–	9.5	18	1.7	125	2.9
G3018 SO₂ short A cell	16.0	33.0	–	6.4	13	1.2	70	2.9
G3028 SO₂ AA cell	13.4	49.7	–	7.4	14	1.2	100	2.9
G3041 SO₂ short AA cell	13.4	25.4	–	3.6	9	0.7	15	2.9
G3042 SO₂ 9 V battery	26.1	48.5	16.7	19.5	50	0.7	15	8.7

Mallory

Mallory supply hermetically sealed lithium–sulphur dioxide organic electrolyte cells in the capacity range 1.1–10 A h with a nominal voltage of 3.00 V. Further details are given in Table 56.3.

Discharge and performance on constant resistance load curves for this range of batteries are given, respectively, in Figures 56.2 and 56.3.

Mallory claim that their lithium cells offer gravimetric energy densities of up to 330 W h/kg, nearly three times that of mercury cells and four times that of alkaline manganese cells. The volumetric energy density is 50% greater than that of alkaline manganese cells.

The manufacturers also claim that their batteries operate efficiently between −40 and +70°C. They retain 75% initial capacity after 5 years storage at 21°C.

In 1975 Mallory introduced their new glass-to-metal hermetic seal to reduce sulphur dioxide losses from the cell and consequently reduce capacity loss on storage and inhibit corrosion effects. Such cells have been stored for 1.75 years at 21°C and 4 months at 72°C without leakage. The excellent high-rate properties of the lithium sulphur dioxide

Table 56.3 Specification of Mallory lithium–sulphur dioxide primary cells

Type	Nominal capacity to 2 V at 21°C (A h)	Maximum diameter (mm)	Maximum height (mm)	Weight (g)	Volume (cm³)
L032	1.1 on 120 Ω	16.5	34.3	12	7.0
L028	3.9 on 36 Ω	24.4	49.5	43	23.1
L029	4.4 on 32 Ω	25.9	49.5	48	26.1
L027	5.8 on 24 Ω	25.9	59.9	58	31.5
L030	7.5 on 18 Ω	25.9	59.7	62	42.0
L026	10.0 on 14 Ω	33.8	59.7	85	53.4
Recent additions					
L0285HX	27	24.6	61.3	38	
L0295X	3.5	26.0	51.3	40	
L0305H	4.3	28.7	60.4	62	
L032S	0.85	15.9	35.2	12	

Because of the high-rate capability of the lithium-sulphur dioxide system, precautions must be taken to prevent a short-circuit of the cells. Mallory supply this system only in complete assemblies using an internal fuse or other form of protection

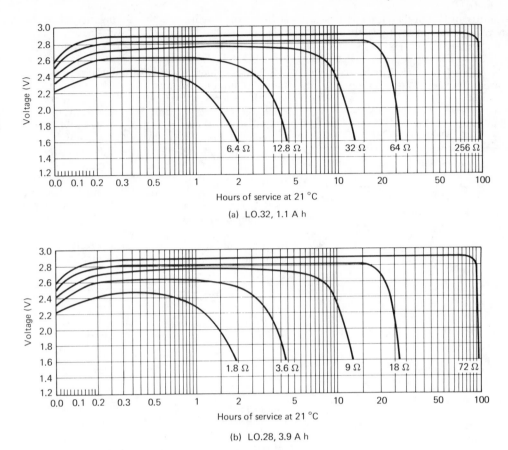

Figure 56.2 Mallory primary lithium–sulphur dioxide cells: typical discharge characteristics (Courtesy of Mallory)

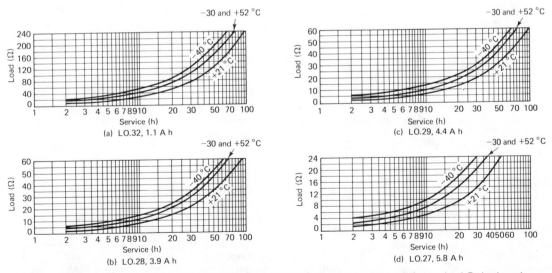

Figure 56.3 Mallory primary lithium–sulphur dioxide cells: typical performance on constant-resistance load. End-point voltage 2.0 V (Courtesy of Mallory)

system have been confirmed in the hermetic D-size cell. Results covering the current range 10 mA to 1 A and the temperature range +60 to −23°C are quoted in Table 56.4. Full capacity is realized at 24°C regardless of load, and at 10 mA load full capacity is realized regardless of temperature.

Table 56.4 High-rate discharge performance of Mallory lithium-sulphur dioxide hermetic D-size cell (test performed on resistive loads after 6 months' storage at 20°C)

Nominal discharge current	Capacity (A h) at various discharge temperatures				
	−23°C	0°C	24°C	49°C	60°C
1 A	3.7	5.8	9.7	6.9	6.8
250 mA	7.1	8.3	10.0	8.2	8.8
50 mA	9.7	9.7	10.1	8.9	9.5
10 mA	10.1	10.3	10.2	10.5	10.5

The estimated current and temperature limits for the hermetic D-size cell for 80, 90 and 100% capacity realization are shown in Figure 56.4.

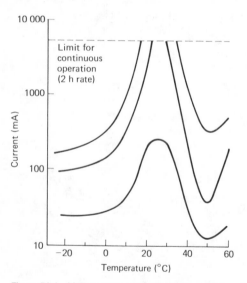

Figure 56.4 Mallory hermetic D-size lithium-sulphur dioxide cell: current and temperature limits for various capacity realizations (Courtesy of Mallory)

Silberkraft

Data on various types of lithium-sulphur dioxide batteries in the 500–30 000 mA capacity range available from Silberkraft are given in Tables 56.5 and 56.6.

These cells and batteries are claimed to have a particularly good shelf life. Silberkraft claim that the low self-discharge rate of Eterncells gives them a

shelf life of 10 years, and that during this period the cells are ready for immediate use. These products conform to the NATO temperature requirements and also to MIL standards.

Silberkraft have recently introduced a revised G range of lithium-sulphur dioxide batteries, which are characterized by improved low-temperature dischargeability (see Table 56.7).

Figure 56.5 shows typical discharge curves for a lithium-sulphur dioxide 3 V cell supplied by Silberkraft. Load-capacity and load-voltage plots for the same battery are reproduced in Figure 56.6. The G06 and G32 cells (see Table 56.7) and the G03 cell are particularly recommended for power supply for CMOS backup applications. With these cells, simply by diode separation from the operating voltage, a constant voltage of not greater than 2.2 V is able to maintain the memory circuit. The reverse current from the cell may amount to 30 μA maximum.

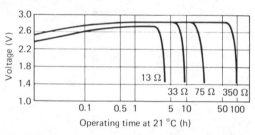

Figure 56.5 Silberkraft type 440 (renumbered 932) lithium-sulphur dioxide 3 V, 0.8 A h cell typical discharge curves (Courtesy of Silberkraft)

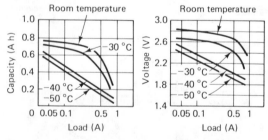

Figure 56.6 Silberkraft type 440 (renumbered 932) lithium-sulphur dioxide 3 V, 0.8 A h cell load capacity and load voltage curves (same battery as in Figure 56.5) (Courtesy of Silberkraft)

Duracell International

Duracell International produce a range of lithium-sulphur dioxide primary batteries. This company manufactures lithium-sulphur dioxide cells in a variety of cylindrical cell sizes, ranging in capacity from 0.45 to 21 A h. Larger cells are under development but currently not available on a commercial basis. A number of the cells are manufactured in standard ANSI (American Nation-

Table 56.5 Types and applications of primary lithium–sulphur dioxide batteries obtainable from Silberkraft

Type	Nominal voltage (V)	Dimensions (mm)	Weight (g)	Application
BA5590/U	24 (2 × 12)	63 × 112 × 127	1025	XM-8 VINSON MX-9331 wire-line chemical alarm
BA5568/U	12	28 dia. × 77.5	85	AN/PRC-90
BA5598/U	14.4/3	120 × 92 × 54	680	AN/PRC 77/KY-38 AN/GRA-114/AN/PPS-15
BA584	14.4/3	242 × 92 × 54	1300	AN/PRC 77/AN/PRC 25
BA5100/U	6	39 dia. × 60	85	Night vision devices
BA5567/U	3	26.5 dia. × 21	20	Night vision devices
BA5585/U	4	112 × 89 × 62.5	795	Furture development
BA5842/U	6	70 × 37 × 180	568	Raft light
BA5574/U	6	–	46	SDU 5/E
BA5841/U	18	53.5 dia. × 143	285	AN/RTIO
SK660Li	28	242 × 92 × 54	1300	PRC 660
PE3.440Li	8.4	62 × 39 × 18	75	Battery for pyrotechnic ignition

Table 56.6 Cylindrical types of lithium–sulphur dioxide batteries (0.52–30 A capacity range) supplied by Silberkraft

Type	Nominal capacity (mA h)	Nominal current (mA)	Dimensions (mm) Diameter	Height	Weight (g)	Nominal voltage (V)	Comments
400-5	525	11	14.5	25.0	7.5	2.8	½ AA size
440	1 000	15	16.5	34.0	11.5	2.8	
Li440 FS	1 000	12	23.0	43.0	22.0	2.8	
400 AA	1 200	24	14.5	51.5	13.0	2.8	AA size IEC R 6 (Mignon)
440-2	2 000	75	16.5	58.5	20.0	2.8	
660-2	2 400	50	25.0	41.0	33.0	2.8	¾ C size
660-C	3 000	100	25.0	51.5	40.0	2.8	C size, IEC R 14 (Baby)
660-3	3 800	125	25.0	60.5	48.0	2.8	1¼ C size
550-D	8 000	250	34.0	61.0	83.0	2.8	D size, IEC R 20 (Mono)
660-50	9 000	250	25.0	130.0	95.0	2.8	
660-4	10 000	250	42.0	50.5	100.0	2.8	
660-5	25 000	1000	42.0	115.0	228.0	2.8	
660-5A	30 000	1000	42.0	140.0	280.0	2.8	
400-9	525	11	48.5 × 26.51 × 17.5		25.0	8.4	9 V Transistor, IEC 6 F 22

Table 56.7 Revised G range Silberkraft lithium–sulphur dioxide cells

Type New	Previous	Rated capacity (A h)	Rated current (mA)	Maximum dimensions (mm) Diameter	Height	Weight (g)	Rated voltage (V)	Notes on design
*G03	*400-5S	*0.35	*14	*14.5	*25	*7	*2.8	*½ AA size
*G32	*440 S	*0.8	*30	*17	*35	*14	*2.8	
*G06	*400 S	*1.0	*45	*14.5	*50.5	*14	*2.8	*AA size, IEC R6 (Mignon)
*G50	660-2 S	2.4	100	26	42	38	2.8	¾ C size
*G52	*660 S	*3.0	*125	*26.2	*50	*46	*2.8	*C size, IEC R14 (Baby)
*G20	*550 S	*7.0	*300	*34.2	*61.5	*85	*2.8	*D size, IEC R 20 (Mono)
G70	1500-2 S	8.5	350	39	53	90	2.8	
G58	660-4 S	9.5	400	42	51	105	2.8	
G60	660-5 S	22.5	900	42	115	228	2.8	
*G62	*660-5 AS	*27.5	*1100	*42	*140	*290	*2.8	

al Standards Institute) cell sizes in dimensions of popular conventional zinc primary cells. While these single cells may be physically interchangeable, they are not electrically interchangeable because of the high cell voltage of the lithium cell (3.0 V for lithium, 1.5 V for conventional zinc cells).

Table 56.8 lists the cells currently available and gives their major physical and electrical characteristics. The cells are classified in two categories:

Table 56.8 Duracell lithium–sulphur dioxide cells

Duracell cell type	Part No.	Rated capacity* (A h)	Weight (g)	Diameter (mm)	Height (mm)
LO34S	331018	0.95	11.75	25.4	19.0
LO37S	331050	0.45	6.5	13.7	24.0
LO32S	331035	0.90	12.0	16.3	34.5
LO36SX	331051	0.65	13.0	24.0	18.4
LO30SK	331079	5.4	62.0	29.1	59.6
LO45SX	331074	6.3	75.0	31.5	57.1
LO26SX	331061	8.0	85.0	33.8	59.6
LO25SX	331075	8.5	95.0	38.7	49.8
LO50SX	331036	21.0	207.0	38.7	114.0
LO27SHX	331078	3.6	47.0	25.7	59.6
LO30SH	331034	4.3	62.0	29.1	59.6
LO26SH	331029	6.4	85.0	33.8	59.6

*Rated capacity: S and SK cells at 30 h rate, SH and SHX cells at 2 h rate

standard cells (Mallory S designation) and high-rate cells (Mallory SH designation). The standard cell is optimized to deliver high-energy output over a wide range of discharge loads and temperatures. The high-rate cell is designed with longer and thinner electrodes than the standard cell and delivers more service at a high discharge rate (higher than the 10 h rate) and at low temperatures. At lower discharge rates, the service life of the high-rate cell is less than that delivered by the standard cell.

In addition, Duracell manufacture a lithium-limited (or balanced) cell (designation SX). The cell is designed with a stoichiometric ratio of lithium to sulphur dioxide of the order of 1:1 rather than the excess of lithium used in the other designs. The lithium-limited feature ensures the presence of sulphur dioxide throughout the life of the cell to protect the lithium from chemically reacting with the other cell components. This design has been found successfully to withstand extended reverse discharge below 0 V at rated loads. In addition, these cells do not produce the toxic chemicals that form when standard cells are fully discharged, thus simplifying disposal procedures. The lithium-limited cell does, however, deliver lower capacity compared with the standard cell at low discharge rate (below the 5 h rate).

56.3 Lithium–thionyl chloride batteries

Honeywell

The types of lithium–thionyl chloride reserve and active (non-reserve) cells currently available from Honeywell are shown in Tables 56.9 and 56.10. These include cells with capacities up to 17 000 A h.

Table 56.9 Types of lithium–thionyl chloride reserve and active cells available from Honeywell

P/N description	Diameter or width (mm)	Length (mm)	Height (mm)	Capacity (A h)	Rated load (A)
Reserve					
G2659B1	16.0	25.4	–	0.36	0.025
Active					
G3013C	16.0	30.7	–	1.3	0.80
G3037	91.0	109.0	137	381.0	3.00
G3038	58.4	106.0	127	160.0	50.00
G3059	92.2	107.0	156	517.6	3.00

Cell construction techniques similar to those of lithium–sulphur dioxide and vanadium pentoxide cells

SAFT

SAFT supply the LS range of lithium–thionyl chloride cells and batteries (see Table 56.11). The batteries are available in 3.65 and 7.3 V with capacities between 0.5 and 18 A h. The discharge curves exhibit a very flat voltage plateau. The cells are manufactured without any initial internal gas pressure and, because the discharge reaction generates only a limited amount of gas, the need for venting is eliminated. Cells operate between −55 and +71°C, and a shelf life of up to 10 years can be expected because of the negligibly low self-discharge rate of these cells. The energy density of the SAFT lithium–thionyl chloride cell is 800 W h/dm^3 compared to 400 W h/dm^3 for zinc–mercury, 200 W h/dm^3 for carbon–zinc and 300 W h/dm^3 for alkaline manganese dioxide. (The corresponding W h/kg data for the four types of cells are 420, 100, 80 and 100.)

Typical discharge curves for the SAFT LS210, 3.5 V, 500 mA h cell are shown in Figure 56.7. Figure 56.8 shows the capacity–current drain relationship for the same battery.

Figure 56.9 shows the effect of shelf storage temperatures between 25 and 70°C on percentage capacity retention: 80% of capacity is retained after 10 years at 25°C or about 1 year at 70°C.

SAFT lithium–thionyl chloride cells exhibit a highly stable service voltage and a far greater capacity than conventional zinc–manganese dioxide and carbon–zinc cells of the same size and configuration.

Table 56.10 Honeywell hermetically sealed lithium–thionyl chloride cells

Type	Designation	Open-circuit voltage (V)	Voltage under load	Rated capacity (Ah)	Energy density Wh/cm³	Wh/kg	Diameter or width (mm)	Length (mm)	Height (mm)	Weight (g)	Volume (cm³)	Operational temperature range (°C)	Projected storage capability at 23°C (years)
Active	G3013C	3.63	(40 Ω load) 3.4	1.6 to 2 V cut-off into an 82 mA load at 23°C	0.8	301	16	30.7		18	6.7	−40 to 74	>5
	G3102	3.62	3.4	16 under rated load (17 Ah) limited capacity	0.94	418	29.9	36.3	53.6	130	58	−40 to 41	>15
	G3103	3.62	3.4		0.94	418	29.9	36.3	63.0	130	58	−40 to 41	>15
	G3066	3.62	>3.0	17 000 under rated load >50 000 Wh/cell	1.1	635	381	381	381	91 000	55 300	−40 to 41	>15
Active spacecraft A battery	G3038	3.63	(0.2 Ω load) 3.3	166 to 2 V cut-off into average 18 A load at 23°C 50 A pulse capacity	0.7	334	58	107	130	1639	808	0 to 60 short voltage delay of <200 ms after 1 month at 60°C	>5
Active spacecraft B battery	G3037	3.63	(9 Ω load) 3.6	400 to 2 V cut-off into average 0.4 A load at 23°C 15 A pulse capability	1.1	592	91	107	140	2429	1362	−40 to 60	>5
Active spacecraft B battery	G3059	3.63	3.6	500 A to 2 V cut-off into average 0.4 A load at 23°C 15 A pulse capability	1.2	642	91	107	158	2803	1536	−40 to 60	>5

Table 56.11 Characteristics of lithium–thionyl chloride cylindrical cells supplied by SAFT

Type	Nominal capacity (mAh) at 20°C	Open-current voltage (V)	Nominal voltage (V)	Maximum recommended current drain rate (mA)	Weight (g)	Operating temperature range (°C)	Storage temperature range (°C)	Diameter (mm)	Height (mm)
LS 210	500 to 2 V at 1 mA drain	3.67	3.5 at 1 mA drain	3	11	−40 to 75	−55 to 75	21.2	9.6
LS 3	850 to 2 V at 1 mA drain	3.67	3.5 at 1 mA drain	50	8.5	−40 to 75	−55 to 75	14.5	25.0
LSH 274	1000 to 2 V at 15 mA drain	3.67	3.5 at 15 mA drain	500	25	−40 to 75	−55 to 75	26.2	18.7
LS 120	1100 to 2 V at 2 mA drain	3.67	3.5 at 2 mA drain	50	10.5	−40 to 75	−55 to 75	12.5	42.0
LS 6/LS6	1800 to 2 V	3.67	3.5 at 5 mA drain	100	15.5	−40 to 75	−55 to 75	14.5	50.5
LSH 14	5000 to 2.5 V	3.67	3.5	600	56	−40 to 75	−55 to 75	26.2	50.0
LSH 14.2	9000 to 2.5 V	3.67	3.5	1500	105	−40 to 75	−55 to 75	26.2	100.0
LSH 20	10 000 to 2.5 V	3.67	3.5	4000	120	−40 to 75	−55 to 75	33.5	61.5
LSH 20.2	18 000 to 2.5 V	3.67	3.5	2000	180	−40 to 75	−55 to 75	33.5	123.0
LS 622	1100 to 5 V at 2 mA drain	7.34	7.0	50	30	−40 to 75	−55 to 75	26 × 17 (rectangular)	49.0

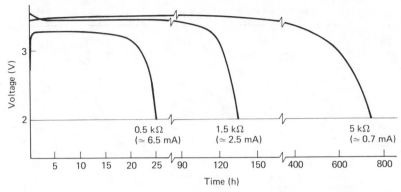

Figure 56.7 Discharge curve at 20°C for a SAFT LS210, 3.5 V, 500 mA h lithium–thionyl chloride cell (Courtesy of SAFT)

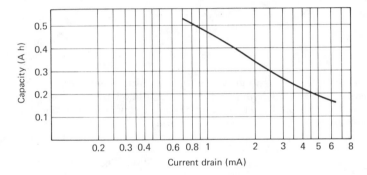

Figure 56.8 SAFT LS210, 3.5 V, 500 mA h lithium–thionyl chloride cell: capacity *versus* current drain chart at 20°C, discharge to 2 V (Courtesy of SAFT)

Safety characteristics Figure 56.10 shows a curve obtained for a SAFT type h S622 1100 mA h, 7.34 V battery short-circuited at room temperature. The total short-circuit results in only a moderate temperature (40°C) with no adverse effects.

Eagle Picher

The range of lithium–thionyl chloride Keeper II batteries available from Eagle Picher is tabulated in Table 56.12. These comprise 3.5 V cells in the capacity range 750–1600 mA h and 7 V batteries with capacities of 750 and 1600 mA h. Operating temperature ranges are exceptionally good with this type of battery; from −40 to 93°C or, in some cases, from −40 to 125°C.

Reserve batteries Gas-activated lithium–thionyl chloride reserve batteries are now supplied by Eagle Picher. It will be recalled that Honeywell supply lithium–sulphur dioxide and lithium–thionyl chloride gas-activated reserve batteries. Some of the advantages claimed for this battery design concept are as follows:

1. The batteries provide long-term inactive storage followed by instantaneous high power delivery.

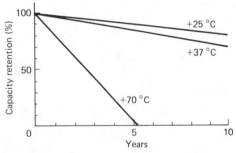

Figure 56.9 SAFT LSL6 3.5 V, 1800 mA h lithium–thionyl chloride battery. Effect of shelf life in storage on percentage capacity retention (Courtesy of SAFT)

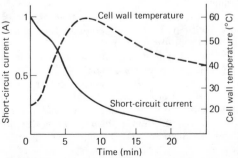

Figure 56.10 Thermal behaviour on total short-circuit of a SAFT type LS622 1100 mA h, 7.34 V battery (Courtesy of SAFT)

Table 56.12 Lithium–thionyl chloride hermetically sealed cells and batteries available from Eagle Picher (Keeper II batteries)

Type	Open circuit voltage (V)	On load voltage (V)	Capacity at 350 h rate (mA h)	Length (mm)	Width (mm)	Thickness (mm)	Weight (g)	Operating temperature range (°C)
LTC-7PMS*	7.30	7.00	750	38.1	30.5	8.4	19.0	−40 to 125
LTC-7PMP†	3.65	3.50	1500	38.1	30.5	8.4	19.0	−40 to 125
LTC-7PN	3.65	3.50	750	25.4	16.5	6.35	6.8	−40 to 125
LTC-7P	3.65	3.50	750	30.5	17.8	8.4	9.0	−40 to 125
LTC-12P	3.65	3.50	1200	25.4	13.5	13.5	12.0	−40 to 93
LTC-16P	3.65	3.50	1600	36.5	13.5	13.5	16.0	−40 to 93
LTC-16P9	7.30	7.00	1600	45.7	28.6	17.5	25.0	−40 to 93
LTC-20P	3.65	3.50	1600	49.2	13.5	13.5	24.0	−40 to 93

* Two LTC-7PN cells in series
† Two LTC-7PN cells in parallel

2. Electrolyte storage design allows minimal capacity loss, requires no maintenance and is insensitive to temperature fluctuations.
3. Increased safety is inherent in reserve battery design.
4. The design allows use of features that improve power capability.
5. Reserve batteries prevent the operation of the device until desired.
6. Higher energy density is available than with any other electrochemical system.
7. Costs are lower because no expensive raw materials are used and because the higher cell voltage allows the use of fewer cells and brings reduced labour costs.
8. There are longer activated life and stand capability because there is no inherent electrochemical process that limits activated life.
9. Voltage delays are eliminated by the use of reserve configuration.
10. Batteries can be activated and operated in any orientation.
11. Both prismatic and cylindrical shapes are available.
12. Cells can be configured so that moulded single-cell cavities nest together to form a cell stack with an integral manifold.
13. The activation input signals can be electrical, mechanical or pressure signals.
14. The electrolyte reservoirs can be spheres, cylinders with pistons or coil tubing.
15. Low-temperature operation can be achieved with electrical heaters, using external power, or chemical heaters that require no external power.
16. Multi-section or multi-tap cell stacks are available.

17. Options include removable electrolyte reservoirs and designs that allow reservoirs and cell stacks to be physically separated and mounted in different locations.
18. Batteries can be designed to be sealed, vented or delay vented.
19. The reserve configuration eliminates most safety concerns associated with lithium–thionyl chloride active batteries.
20. Design emphasis is placed on preventing inadvertent activation due to mechanical failure or abusive conditions.
21. No-load conditions can be safely tolerated by lithium–thionyl chloride reserve batteries.
22. Hundreds of batteries have been tested by Eagle Picher and no safety incidents have arisen.
23. Electrodes can be designed for the specific application; cathode or anode limited.
24. Internal parasitic loads can be included to safely discharge the battery in the event that the discharge cycle is not completed.
25. Electrical and thermal fuses can be used to prevent accidental battery activation.
26. Removable electrolyte reservoir and valves between the reservoir and cell stack can be used to prevent inadvertent battery activation.

Eagle Picher develop and manufacture reserve batteries for a wide range of both military and aerospace applications. Eagle Picher lithium–thionyl chloride reserve batteries have found applications in mines, missiles, and re-entry vehicles where high-power, high energy density and multi-section, multi-tap configurations are required.

Table 56.13 Lithium–manganese dioxide cells and batteries supplied by Duracell

Duracell type No.	IEC type No.	Nominal voltage (V)	Rated capacity (mA h)	Weight (g)	Diameter (mm)	Height (mm)	
Low-rate cells							
DL-1220	CR-1220	3	30	0.8	12.5	2.0	175 h on 15 kΩ to 1.4 V at 21°C
DL-1620	CR-1620	3	50	1.2	16.0	2.0	106 h on 5.6 kΩ to 1.4 V at 21°C
DL-2016	CR-2016	3	50	2.0	20.0	1.6	106 h on 5.6 kΩ to 1.4 V at 21°C
DL-2025	CR-2025	3	120	2.5	20.0	2.5	143 h on 2.7 kΩ to 1.4 V at 21°C
DL-2420	CR-2420	3	120	3.0	24.5	2.0	118 h on 2.7 kΩ to 1.4 V at 21°C
DL-2032	CR-2032	3	170	3.0	20.0	3.2	194 h on 2.7 kΩ to 1.4 V at 21°C
DL-2430	CR-2430	3	200	4.0	24.5	3.0	228 h on 2.7 kΩ to 1.4 V at 21°C
High-rate cells							
DL-2016H	CR-2016H	3	50	2.0	20.0	1.6	
DL-2025H	CR-2025H	3	100	2.4	20.0	2.5	
DL-2420H	CR-2420H	3	100	3.0	24.5	2.0	
DL-2032H	CR-2032H	3	130	2.8	20.0	3.2	
DL-2430H	CR-2430H	3	160	4.0	24.5	3.0	
Cylindrical cells							
DL-1/3N	CR-1/3N	3	160	3.0	11.6	10.8	59 h on 1 kΩ to 1.4 V at 21°C
DL-2N	CR-2N	3	1000	13.0	12.0	60.0	20 h on 25 kΩ to 1.4 V at 21°C
Batteries							
PX-28L	2-CR-1/3N	6	160	8.8	13.0	25.2	59 h on 2 kΩ to 2.8 V at 21°C

Ratings: low-rate cell, *C*/200 h rate; high-rate cell, *C*/30 h rate; cylindrical cell, *C*/30 h rate

56.4 Lithium–manganese dioxide batteries

Duracell

These batteries are claimed to perform well at temperatures between −20 and +50°C. In some designs, up to 70% of the rated 200°C capacity is delivered at 20°C. The shelf life is limited by the amount of solvent diffusing through the seal. Negligible capacity loss occurs after 3 years storage at 20°C. Projections from this suggest that 85% of the initial capacity will be available after storage for 6 years at 20°C.

The cell types available from Duracell are listed in Table 56.13. The cells are manufactured in a variety of button and cylindrical cell designs ranging in capacity from 30 mA h to 1 A h.

Discharge characteristics of Duracell cylindrical and low-rate button lithium–manganese dioxide cells are shown, respectively, in Figures 56.11 and 56.12.

SAFT

SAFT supply a range of 3 V button cells. These cells have an energy density of 400–500 W h/dm^3 (120–150 W h/kg) and an annual self-discharge rate of 2–3% at 20°C. The capacities range from 200 mA h (type LM 2425) through 90 mA h (type LM 2020) to 50 mA h (type LM 2016). Operating

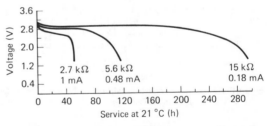

Figure 56.11 Discharge characteristics of a Duracell DL 1/3N, 3 V, 160 mA h cylindrical lithium–manganese dioxide cell. Current at 2.7 V (Courtesy of Duracell)

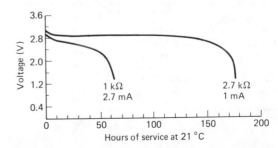

Figure 56.12 Discharge characteristics of a Duracell DL1620, 3 V, 50 mA h low-rate lithium–manganese dioxide button cell. Current at 2.7 V (Courtesy of Duracell)

temperatures are, again, in the range $-20°C$ to $+50°C$.

SAFT have recently introduced a battery to match the requirement of the microelectronics industry for a battery to provide standby power for RAM memory applications and other integrated circuits. This battery (Memoguards LH and LF) has similar

characteristics (3.0 V nominal voltage, 200 mA capacity) to the LM2425. However, as shown in Table 56.14, the Memoguard battery has superior operating and storage temperature characteristics.

The Memoguard battery has a life expectancy of more than 10 years at room temperature and a failure rate comparable with those of other electronic components. The electrical characteristics of these batteries are shown in Figure 56.13.

Varta

Varta supply 3 V button cells ranging in capacity from 3 to 1400 mA h (see Tables 56.15 and 56.16).

Table 56.14 Operating and storage temperatures of SAFT lithium–manganese dioxide cells

	LM 2425	Memoguard LH and LF
Operating temperature range (°C)	−20 to 50	−40 to 70
Storage temperature range (°C)	−20 to 50	−40 to 70

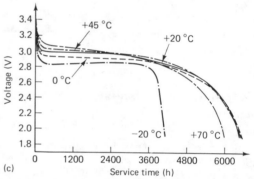

Figure 56.13 Electrical characteristics of SAFT Memoguard LH and LF and of type LM 2425 lithium–manganese dioxide button cells, 3 VV 2.00 mA h. (a,b) Discharge to 2.0 V. (c) Discharge under 90 kΩ (≈30 μA) (Courtesy of SAFT)

Table 56.15 Lithium–manganese dioxide batteries supplied by Varta

Type	Order No.	Diameter (mm)	Height (mm)	Weight (g)	Nominal voltage (V)	Capacity (mA h)
Cylindrical						
CR-⅓N	6131 140 401	11.6	10.8	3.0	3	160
CR-2N	6201 140 401	12.0	60	13.0	3	1000
CR-2NP	6202 140 401	12.0		13.0	3	1400
2CR-⅓N	6231 140 401	13.0		8.8	6	160
CR 1220	6220 140 401	12.5	2.0	0.8	3	30
CR 1620	6620 140 401	16.0	2.0	1.2	3	50
Low-rate button						
CR 2016	6016 140 401	20.0	1.6	2.0	3	60
CR 2025	6025 140 401	20.0	2.5	2.5	3	120
CR 2032	6032 140 401	20.0	3.2	3.0	3	170
CR 2420	6420 140 401	24.5	2.0	3.0	3	120
CR 2430	6430 140 401	24.5	3.0	4.0	3	200

Table 56.16 Lithium–manganese dioxide batteries supplied by Varta

Product	Manufacturer	Model	Batteries			Battery life
			Type	Sanyo No.	Varta order No.	
Pocket calculators	Sanyo	CX-1270, CX-1272, CX-1254, CX-7232M	CR 1620		6620	2000–8000 h
		CX-1251, CX-7231M, CX-8181L	CR 2016	LF-¼V	6016	2000–5000 h
		CX-7211T, CX-8178L, CX-8183L, CX-8185LR, CX-1212, CX-1221, CX-7215T, CX-1252, CX-7214, CX-1214	CR 2032	LF-½V	6032	2000–10 000 h
		CX-8176L, CX-8176LM, CX-8179L, CX-8179LG, CX-8180TL, LX-8182L, LX-7250H, CZ-8114, CZ-0125L, CZ-1201, CZ-1203	CR2430	LF-½W	6430	1000–5000 h
	Sharp	EL-8150, EL-8154, EL-1116, EL-819	CR 2016	LF-¼V	6016	1000 h
		EL-8062	CR 2025		6025	1 year
	Giko	LCD-480	CR 2016	LF-¼V	6016	3000 h
	Towa	V-006LC	CR 2016	LF-¼V	6016	3000 h
	Sanitron	LC-906	CR 2016	LF-¼V	6016	1200 h
		LC-7, LC-7T	CR 2016	LF-¼V		4000 h
	Canon	LC-61T, LC-6T, LC-5T, LC-6, LC-51S, LC-51, LC-52, R-Q1, F-53, F-54, F-72	CR 2032	LF-½V	6032	2000–4000 h
	Commodore	LC-5	CR 2420	LF-⅓W	6420	2000 h
		F-52, F-62	CR 2430	LF-½W	6430	2000 h
		CIL-100	CR2032	LF-½V	6032	2000 h
		CQ-alarm	CR 2430	LF-½W	6430	2000 h
	Toshiba	LC-851	CR 2420	LF-⅓W	6420	2000 h

Table 56.16 (Continued)

Product	Manufacturer	Model	Batteries		Varta order No.	Battery life
			Type	Sanyo No.		
Watches	Texas Instruments	Starburst	CR 2032	LF-½V	6032	5 years
	Sanyo	DO 371-11 GBC, DO 371-11 SBB	CR 2420	LF-⅓W	6420	5 years
Clock radio	Sanyo	RPM-6800	CR 2032	LF-½V	6032	2 years
Radio with cassette recorder	Sanyo	MR-A 110	CR 2032	LF-½V	6032	2 years
Fishing equipment	Novel	E-L 121	CR 333	LR 333	6333	20 h
	Novel	E-L 101	CR 772	LR-bH	6772	20 h
	Daiwa	LR 10, LR 20, LR 30, LR 40, LR 40S, LR 50, LR 60	CR 333	LR 333	6333	20 h
	Olympic	EF-504	CR 772	LR-bH	6772	20 h
	Yuasa	Lithium No.1, No. 3, No. 6	CR 772	LR-bH	6772	20 h
Alarm systems	Glanville Reid		CR 772	LR-bH	6772	
	Meta system		CR 772	LR-bH	6772	
	Telonics		CR 772	LR-bH	6772	
	Taouer		CR 772	LR-bH	6772	
Pace setter	Maging	Jogger's Pace Setter	CR 2430	LF-½W	6430	1000 h

Figure 56.14 shows discharge curves for one of the Varta series carried out at discharge temperatures between −20 and 50°C.

Figure 56.15 shows the discharge performance at 20°C obtained for the same cell when the discharge is carried out in the impulse mode, consisting of a 1.25 s discharge followed by a 5 min rest period.

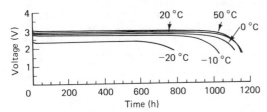

Figure 56.14 Effect of discharge temperature on discharge performance of a Varta CR 2430, 3 V, 200 mA h low-rate lithium–manganese dioxide button cell under 15 kΩ load (Courtesy of Varta)

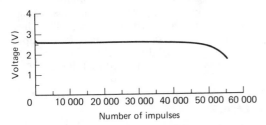

Figure 56.15 Impulse discharges of Varta CR 2430, 3 V, 200 mA h low-rate lithium–manganese dioxide button cells at 20°C. Discharge 1.25 s on, 5.0 min off; load 220 Ω (Courtesy of Varta)

56.5 Lithium–silver chromate batteries

SAFT supply a range of these 3.1 V cells with capacities in the range 130–3100 mA h (cylindrical cells) and 2090–2450 mA h (rectangular cells) (Table 56.17).

Table 56.17 Lithium–silver chromate cells supplied by SAFT

Cell type	Diameter (mm)	Height (mm)	Capacity (mA h)	Energy density (W h/dm³)
LI 114	11.6	5.4	130	550–750
LI 210	21.0	9.2	810	550–750
LI 273	27.3	7.9	1220	550–750
LI335/6	35.5	6.0	1750	550–750
LI335/10	35.5	6.0	3100	550–750
LIR 123	45 × 8	23.0	2090	550–750
LIR 132	45 × 8	32.0	2450	550–750

The special features of these cells are a high reliability (better than 0.7×10^{-8}), indication of end of life through a second plateau voltage, a discharge capability up to 100 μA and low internal impedance.

Varta also supply a lithium chromate cell, type ER½AA, which is a 3 V cylindrical cell with a capacity of 900 mA h. It weighs 10 g, and is of 14.5 mm diameter × 25 mm height. These cells are designed for 10 year life applications where voltage level is critical. Typical voltages in excess of 3.5 V/cell are claimed. A typical discharge curve for this cell is shown in Figure 56.16.

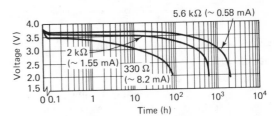

Figure 56.16 Discharge characteristics of Varta type R ½ AA, 3 V, 900 mA h cylindrical lithium chromate cell (Courtesy of Varta)

56.6 Lithium–copper oxide batteries

SAFT claim a 10 year shelf life and an ability to operate at temperatures between −20°C and +50°C for these batteries, and a volumetric capacity of 750 A h/dm³. The self-discharge rate is 2–3% per annum at 20°C.

SAFT supply battery packs containing groups of cells connected in series/parallel. Cylindrical 1.5 V batteries are available in 3.6 A h (type LC01), 1.6 A h (type LC02), 0.5 A h (type LC07), 3.3 A h (type LC06), 10 A h (type LCH14) and 20 A h (type LCH20). Further details concerning these batteries are supplied in Table 56.18.

Typical discharge curves for the LC07 are shown in Figure 56.17.

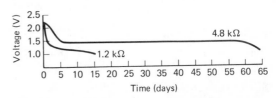

Figure 56.17 SAFT lithium–copper oxide cells, LC07, 1.5 V, 0.4 A h. Typical discharge curve at 20°C and stated ohmic loads (Courtesy of SAFT)

Table 56.18 Characteristics of lithium–copper oxide cells supplied by Varta

Type	Nominal capacity (A h)*	Open circuit voltage (V)	Nominal voltage (V)†	Maximum recommended drain current (mA)‡	Weight (g)	Diameter (mm)	Height (mm)	Operating temperature range (°C)	Storage temperature range (°C)
LC07	0.4	2.4	1.5	8	4.5	14.0	12.5	−20 to 55	−20 to 55
LC02	1.4	2.4	1.5	20	7.3	14.0	25.0	−20 to 55	−20 to 55
LC06	3.3	2.4	1.5	40	17.4	14.0	50.0	−40 to 70	−40 to 70
LCH14	10.0	2.4	1.5	600	56.0	25.5	50.0	−40 to 60	−40 to 60
LCH20	20.0	2.4	1.5	1000	110.0	34.2	61.5	−40 to 60	−40 to 60

* to 0.9 V at 1 mA drain at 20°C
† at 1 mA drain at 20°C
‡ to recover 75% of nominal capacity in continuous discharge at 20°C

56.7 Lithium–lead bismuthate batteries

SAFT

SAFT supply cells with a nominal voltage of 1.5 V. These cells have an energy density of 400–500 W h/dm^3 (90–150 W h/kg). The annual self-discharge rate is 2–3% at 20°C. An operating temperature of −10 to +45°C is claimed. The cells have a discharge pulse capability up to 500 μA.

SAFT supply 1.5 V button cells at three capacities: 185 mA h (type LP1154), 110 mA h (type LP1136) and 50 mA h (type LP1121).

Varta

Varta also supply two 1.5 V lithium–lead bismuthate button cells; type DR 991, with a capacity of 35 mA h, and type DR 926 with a capacity of 48 mA h.

56.8 Lithium–copper oxyphosphate cells

The cylindrical cells supplied by SAFT have an open-circuit voltage of 2.8 V and a nominal voltage of 2.8 V, as opposed to the 1.5 V nominal voltage available from lithium–copper oxide cells. These cells are designed to operate at temperatures up to 175°C with high reliability. Details of the available range of these cells are supplied in Table 56.19.

56.9 Lithium–polycarbon monofluoride batteries

Eagle Picher supply four types of this battery, as shown in Table 56.20.

The LDFS-5PN and LDFS-5P cells are supplied hermetically sealed in a stainless steel case. The LDFS-5PMS 6 V battery consists of two LDFS-5PN

Table 56.19 SAFT LCP series lithium–copper oxyphosphate cells

	LCP 6	LCP 6HT	LCP 14HT
Nominal capacity	2.3 A h to 1 V at 1 mA drain at 20°C	1.9 A h to 1 V at 15 mA drain at 150°C	5 A h to 1 V at 40 mA drain at 150°C
Open circuit voltage (V)	2.8	2.8	2.8
Nominal voltage	2.4 V at 1 mA drain at 20°C	2.4 V at 15 mA drain at 150°C	2.4 V at 40 mA drain at 150°C
Max. recommended drain current	40 mA to recover 75% of nominal capacity in continuous discharge at 20°C	50 mA to recover 75% of nominal capacity in continuous discharge at 150°C	120 mA to recover 75% of nominal capacity in continuous discharge at 150°C
Weight (g)	17	18	55
Diameter (mm)	14	14	25.5
Height (mm)	50	50.5	50
Operating temperature (°C)	−40 to 70	0 to 175	0 to 175
Storage temperature (°C)	−40 to 70	−40 to 70	−40 to 70

Table 56.20

Battery No.	Working voltage (V)	Capacity at 350 h rate (mA h)	Length (mm)	Width (mm)	Thickness (mm)	Weight (g)
LDFS-5PN	2.8	500	25.4	16.5	6.35	6.8
LDFS-5P	2.8	500	30.5	17.8	8.4	9.0
LDFS-5PMP	2.8	1000	38.1	30.5	8.4	19.0
LDFS-5PMS	5.6	500	38.1	30.5	8.4	19.0

cells connected in series hermetically sealed in a plastic module. The LDFS-5PMP 3 V battery consists of two LDFS-5PN cells connected in parallel in a hermetically sealed plastic module.

All cells in the series have an operating temperature range of −73 to 93°C; storage temperatures up to 150°C are permitted. Operating lives of up to 10 years are claimed for these batteries. Voltage characteristics remain flat throughout battery life. The battery design is non-pressurized, allowing for its use in non-vented applications.

Figure 56.18 shows discharge curves for LDFS-5PN cells at various discharge rates between 25 μA and 3 mA at 75, 20 and −40°C.

Figure 56.19 compares the capacity retention at 20°C and 75°C of lithium–polycarbon monofluoride batteries (type LDFS) with that of thionyl chloride batteries (type LTC), both produced by Eagle Picher. Figure 56.20 compares the effect of discharge rate on projected battery life at 25°C for the same two types of battery, which show a very similar performance in this respect.

These micropower source batteries are suitable for use in applications that require high-rate pulses of power, low-current continuous drain applications, or a stable long-term performance voltage. They provide a high energy density. They are applicable in airborne instrumentation, undersea communications, mineral exploration, remote site monitoring, safety controls, security, space, defence and CMOS-RAM retention applications.

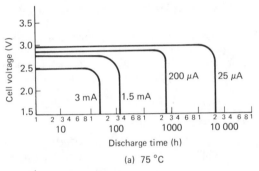

(a) 75 °C

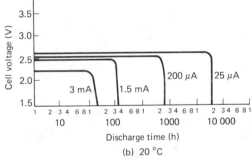

(b) 20 °C

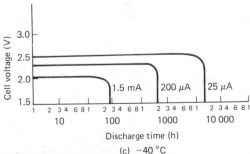

(c) −40 °C

Figure 56.18 Discharge characteristics of Eagle Picher lithium–polycarbon monofluoride cells, LDFS-5PN series, 2.8 V, 500 mA h (Courtesy of Eagle Picher)

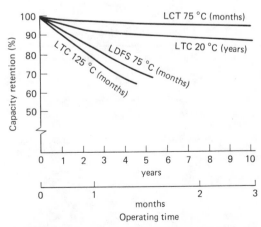

Figure 56.19 Capacity retention of Eagle Picher lithium–polycarbon monofluoride (LDFS) and lithium–thionyl chloride (LTC) batteries (Courtesy of Eagle Picher)

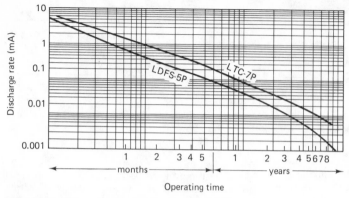

Figure 56.20 Effect of discharge rate on projected operating life at 25°C for Eagle Picher lithium–polycarbon monofluoride (LDFS) and lithium–thionyl chloride (LTC) batteries (Courtesy of Eagle Picher)

56.10 Lithium solid electrolyte batteries

Duracell supply these batteries, which are basically a 2 V system capable of supplying currents up to a few hundred microamps at room temperature. An ability to withstand discharge temperatures up to 125°C without deterioration in performance is claimed for these batteries. The cell may be stored at temperatures up to 200°C without serious losses. A projected storage life of 15–20 years at 20°C is believed to be possible. Duracell solid electrolyte batteries are currently available in a button or circular disc configuration, with a nominal 25.4 mm diameter, and rated at 350 mA h.

Details of two particular types of Duracell lithium solid electrolyte batteries are given in Table 56.21.

56.11 Lithium–iodine batteries

Since 1972, Catalyst Research Corporation has been the supplier of lithium–iodine batteries for cardiac pacemakers and other medical implantable devices and, more recently, for other applications of these batteries.

56.11.1 Pacemaker batteries

A full specification of the Model 802/23 battery is given in Table 56.22. Projected electrical performance of various types of battery over 20 years is given in Figure 56.21.

The 900 Series Lithiode lithium–iodine implantable medical cell manufactured by Catalyst Re-

Table 56.21 Duracell solid electrolyte batteries

	Duracell type No.	
	305127	305159
Nominal voltage (V)	2.0	4.0
No. of cells	1.0	2.0
Rated capacity (mA h)*	28.9 ± 0.13	29.7 ± 0.13, -0.05
Dimensions		
Diameter (mm)	2.54 ± 0.25, -0.13	5.84 ± 1.8
Height (mm)		
Volume (cm³)	1.44	4.04
Weight (g)	7.25	15.85

* Rated at 1 µA discharge at 21°C

Table 56.22 Characteristics of Catalyst Research Corporation model 802/23 solid lithium–iodine battery

Nominal size	23.0 mm × 45.1 mm × 13.5 mm
Volume	11.2 cm³
Weight	30 g
Density	2.7 g/cm³
Lithium area	17.1 cm³
Voltage	2.8 V under no load
Recommended currents	1–50 µA at 38°C
Nominal capacity	2.3 A h
Energy	6.0 W h
Energy density	530 W h/dm³; 200 W h/kg
Self-discharge	10% in 10 years
Seal	Heliarc welded with glass/metal hermetic seals less than 4.6×10^{-8} maximum helium leak
Insulation resistance	$>10^{10}\ \Omega$ from pin to case
Storage temperature	50°C, brief excursions to 60°C

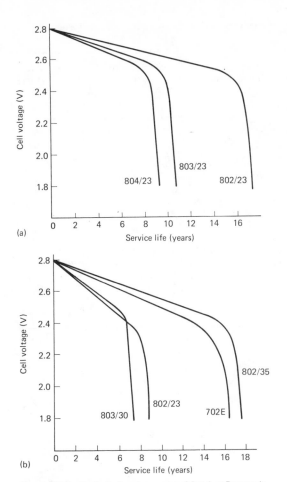

(a)

(b)

Figure 56.21 Projected performance of Catalyst Research Corporation Lithiode TM lithium–iodine cells based on 5 years test data. (a) Voltage under 190 kΩ load, (b) voltage under 90 kΩ load (Courtesy of Catalyst Research Corporation)

search represents the latest innovation in lithium–iodine batteries. The innovation is in the cell's internal construction. Instead of using a cooked-and-cast cathode, as employed in the earlier 700 and 800 series of Catalyst Research lithium–iodine cells, the 900 series utilizes a unique pelletized cathode.

Because of this pelletized cathode the 900 series features:

1. Higher energy density.
2. Higher deliverable capacity.
3. Lower internal impedance.
4. Higher current drains.
5. Exacting quality control.

All components of the cell are solid, including the cathode. This precludes cathode migration to the anode current collector, a possible latent failure mode in other lithium–iodine cells.

These improved internal components are encased in a corrosion-resistant nickel or stainless steel container. The 900 series battery case is electrically positive. Each battery is hermetically sealed by metal-to-metal fusion welding and high-quality ceramic feedthroughs.

During manufacture, these batteries are subjected to environmental testing, consisting of the following levels of stress:

1. Shock: 2000g, ½sine, 0.5 ms, 6 times.
2. Vibration: MIL-STD-810, random 20–2000 Hz, 16.9g r.m.s., 6 h.
3. Bump: 36g, 5 ms, 4000 times.
4. Temperature cycle: −40 to 60°C, 10 cycles, 6 h/cycle.
5. High temperature: 60°C, 8 weeks.

The full range of Catalyst Research Series 900 pacemaker lithium–iodine cells is shown in Table 56.23. A diagrammatic representation of one of the cells in this series (Model 901) is shown in Figure 56.22. Projected performance curves for a single anode cell in this range are shown in Figure 56.23. Included in the series are two cells (types 920 and 921) having double, as opposed to single, anodes. Figure 56.24 illustrates the increase in power for a double anode cell (type 920) as compared with a single anode cell of comparable size (type 909). Approximately 80% of the capacity of type 920 is achieved below a 2 kΩ level of resistance, and more than 90% of capacity is achieved below a 5 kΩ level of resistance. Figure 56.25 illustrates the resistance–capacity relationship for these two cells.

56.11.2 Other applications

Figure 56.26 gives performance characteristics of the S-23P-15 button cell. The largest button cell in the range is the D27P-20 (Table 56.24). This performance is shown in Figure 56.27. This has a capacity of 0.26 A h and has applications in digital watches and calculators; 460 mA h (type 2736) and 870 mA h (type 3740) button cells are also available. A 7.5 A h (type 3740) oil well logging battery model 407220 D cell is also available. This is housed in a 38 mm × 57 mm circular container and is hermetically welded with a glass–metal compression seal. It has a voltage of 2.8 V (open-circuit 2.8 V), a maximum operating temperature of 150°C and a self-discharge rate of about 5% in 10 years at 25°C. The prototype D 34 cell maintains its voltage of 2.8 V during 6 weeks of continuous operation at 10 mA constant current at 130°C.

Also available is the model CRC Li D 2.8 D-type cell (2.8 V) with a capacity of 14 A h at 25°C and an operating temperature of −20 to +50°C. The battery weighs 80 g and has a height of 33.8 mm and a width of 57.1 mm. This cell suffers a capacity loss of less than 5% in 10 years at 25°C and can withstand

Table 56.23 Catalyst research 900 series lithium–iodine pacemaker batteries electrical and dimensional characteristics (based on 2.6 V)

Model	Capacity (A h)	Energy (W h)	Energy density		Self-discharge (% in 10 years)	Storage temperature (°C)	Volume (cm³)	Weight (g)	Thickness (mm)	Height (mm)	Width (mm)
			W h/cm³	W h/g							
901	2.6	6.5	0.88	0.25	<10	−40 to 50	7.39	26	9.2	23.0	45.0
902	2.1	5.2	0.86	0.24	<10	−40 to 50	6.0	22	9.2	22.5	45.0
903	1.6	3.9	0.87	0.24	<10	−40 to 50	4.5	16	9.1	15.0	44.9
904	3.8	9.8	0.98	0.30	<10	−40 to 50	10.0	33	13.4	22.2	45.0
905	2.0	5.2	0.93	0.25	<10	−40 to 50	5.6	21	9.2	18.0	45.0
906	1.8	4.42	0.88	0.23	<5	−40 to 52	5.05	19	6.6	19.0	45.2
908	1.9	4.68	0.88	0.26	<5	−40 to 52	5.34	19	9.1	15.5	44.9
909	3.1	7.8	0.95	0.28	<5	−40 to 52	8.17	28	9.4	22.9	44.9
910	1.8	4.6	0.92	0.24	<5	−40 to 52	4.95	18	6.6	22.3	45.2
911	2.3	5.72	0.90	0.25	<5	−40 to 52	6.35	23	6.6	27.2	45.2
912	2.0	5.1	0.94	0.26	<5	−40 to 52	5.4	20	7.8	30.6	27.3
914	1.6	4.1	0.94	0.26	<5	−40 to 52	4.3	16	7.8	25.0	27.3
915	2.2	5.6	1.00	0.27	<5	−40 to 52	5.7	21	8.6	26.0	33.0
916	2.3	5.9	0.98	0.25	<5	−40 to 52	6.0	23	8.8	22.0	45.0
917	2.6	6.9	0.98	0.28	<5	−40 to 52	7.1	25	8.6	30.6	33.0
920	2.8	7.2	0.91	0.28	<10	−40 to 52	7.9	26	9.1*	22.9	44.9*

* Of flange

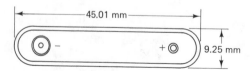

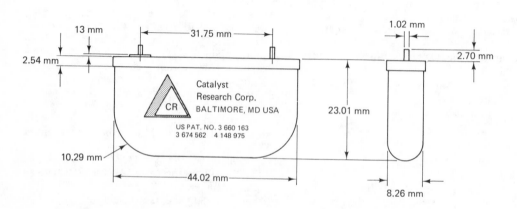

Figure 56.22 Configuration of Eagle Picher lithium–iodine cells (Courtesy of Eagle Picher)

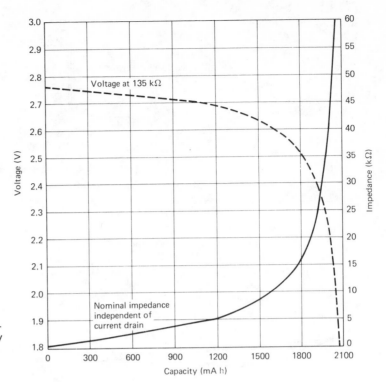

Figure 56.23 Catalyst Research lithium–iodine 2.8 V cells. Voltage *versus* capacity relationship. Model 902, 2.8 V, 2.1 A h (single anode) (Courtesy of Catalyst Research Corporation)

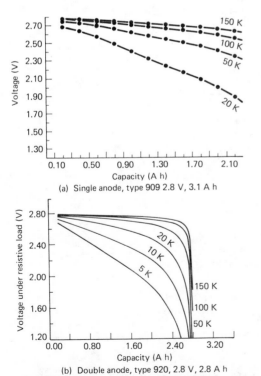

(a) Single anode, type 909 2.8 V, 3.1 A h

(b) Double anode, type 920, 2.8 V, 2.8 A h

Figure 56.24 Voltage-capacity relationship for Catalyst Research lithium iodine 2.8 V cells (a) types 909 (single anode). and (b) 920 (double anode) (Courtesy of Catalyst Research Corporation)

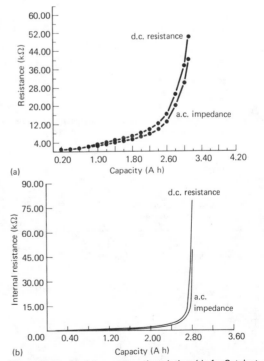

(a)

(b)

Figure 56.25 Resistance–capacity relationship for Catalyst Research lithium iodine 2.8 V cells (a) types 909 (single anode) and (b) 920 (double anode) (Courtesy of Catalyst Research Corporation)

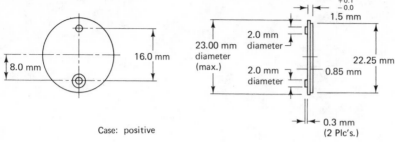

Case: positive

Weight: 4.25 g

Voltage: 2.8 V, no load

Current: 0–20 μA at 25 °C pulsing allowable
up to several hundred microamperes

Nominal capacity: 100 mA h

Self discharge (measured by microcalorimetry):
0.01 A h maximum in 10 years at 25 °C

Construction: Laser welded 304 stainless steel case,
glass to metal terminal seal

Operating temperature: 0–50°C nominal, −10 °C at
reduced drain rates

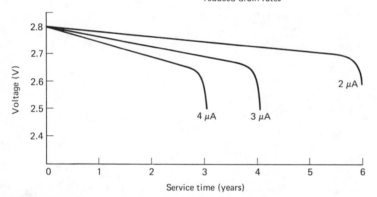

Figure 56.26 Projected performance of Catalyst Research Corporation Model S23P-1.5 lithium–iodine cell (0.1 A h) at 25°C (Courtesy of Catalyst Research Corporation)

Table 56.24 Catalyst Research Corporation non-medical CRC range of lithium–iodine batteries

Type	Diameter (mm)	Thickness (mm)	Capacity (mA h)
SP19-10	19	3.0	110*
SP19-20	19	3.0	120
S23P-15	23	1.5	120*
S27P-15	27	1.5	140*
S27P-20	27	3.0	250
DP27P-20	27	2.0	260†

* Single sided, only one face of lithium exposed to iodine
† Double sided, hence can support higher currents owing to increased surface area

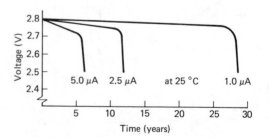

Figure 56.27 Projected performance of Catalyst Research Corporation D27P-20 cell (23 mm diameter × 2 mm thick, 250 mA h), based on 5 years' test data (Courtesy of Catalyst Research Corporation)

Table 56.25 Types of non-medical CMOS-RAM 2.8 V lithium–iodine cell available from Catalyst Research Corporation

Type No.	Voltage (V)	Capacity (mA h)	Current capability at 25°C (A)	Self-discharge (A h) in 10 years at 25°C	Operating temperature (°C)	Diameter (mm)	Height (mm)	Weight (g)
11620	2.8	40	100			11.6	2.0	2.4
1935	2.8	200 above 2 V at 600 kΩ at 25°C	15	0.01	−55 to 70	19	3.5	4.8
2736	2.8	350 above 2 V at 500 kΩ at 25°C	15	0.02	−55 to 125	27	3.6	7.1
3440	2.8	700 above 2 V at 250 kΩ at 25°C	30	0.04	−55 to 125	34	4.0	17.8

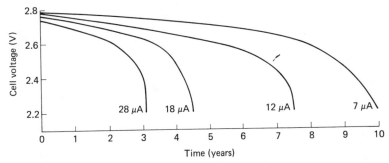

Figure 56.28 Projected discharge curves at 25°C of Lithiode lithium–iodine cell (type 3440, 2.8 V, 700 mA h) supplies by Catalyst Research (Courtesy of Catalyst Research Corporation)

substantial shock, vibration and short-circuiting at temperatures up to 150°C without venting, leaking, swelling or exploding. The cell continues operating at 2.8 V (25°C) for 18 months (1 mA) and 2.7–2.8 V for about 8 months (2 mA).

Catalyst Research Corporation supply the Lithiode range of disc-shaped lithium–iodine power sources with PC bound mounting for CMOS-RAM as standby power sources and for reference voltage sources and other low-current applications where the cell is the main power source.

These cells can be recharged at currents up to 10 μA at room temperature. This is particularly useful in memory backup applications where a small current (about 1 μA) can be used to trickle charge the cell whenever main power is available. Then, when the cell is called upon in the standby mode, it will be in a fully charged condition.

Four types of 2.8 V Lithiode cell are supplied by Catalyst Research (Table 56.25). Figure 56.28 shows the discharge curve of one of these cells. The dependence of maximum current on cell temperature for the same cell is shown in Figure 56.29.

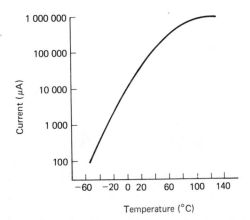

Figure 56.29 Lithiode lithium–iodine cell (type 3440, 2.8 V, 700 mA h) supplied by Catalyst Research Corporation. Dependence of projected maximum continuous current of fresh cells on cell temperature (Courtesy of Catalyst Research Corporation)

57

Manganese dioxide–magnesium perchlorate (primary) batteries

Contents

These batteries are available in two types: the non-reserve type ready for use off the shelf requiring no activation; and the reserve type, which usually contains the active electrode materials but requires the addition of water or an electrolyte just before use.

57.1 Reserve-type batteries

Eagle Picher manganese dioxide–magnesium perchlorate batteries of this type meet the requirements of SCS ELCD–2010/0001. The batteries are non-hazardous, being vented to atmosphere, and non-explosive; the electrolyte is far less corrosive than that used in conventional alkaline batteries.

The theoretical energy density is 242 W/kg against a practical energy density of 110 W h/kg (130 W h/dm^3) as cells and 88 W h/kg (120 W h/dm^3) as batteries. The open-circuit voltage is 2.0 V and the nominal working voltage is 1.55 V/cell. The output range available is 2–300 W with rated capacities between 3 and 120 A h.

The operating and storage temperatures, respectively, are −54 to +18°C and −68 to +18°C.

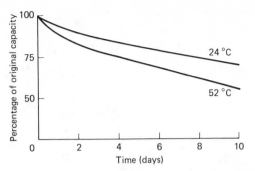

Figure 57.3 Capacity retention of an Eagle Picher reserve primary manganese dioxide–magnesium perchlorate battery in activated state to 1.0 V/cell end-voltage (Courtesy of Eagle Picher)

Batteries are available with volumes and weights of 4 cm^3 upwards and 28 g upwards. A shelf life of 3–5 years is claimed.

Figure 57.1 shows the voltage–time curves (43 Ω load) when a 92 g Eagle Picher manganese dioxide–magnesium perchlorate reserve battery is activated by addition of electrolyte at −53, −40 and +24°C. Even at −54°C a near maximum voltage is achieved within 80 s. The subsequent discharge curves at −54 and +24°C are shown in Figure 57.2. Figure 57.3 shows the capacity retention at 24 and 52°C.

57.2 Non-reserve batteries

Marathon manufactures the BA4386 battery, which serves as the power source for the PRC 25 and PRC 77 transceiver sets. This battery consists of 10 CD-size cells in series–parallel construction.

57.2.1 The CD cell

The electrical and physical characteristics of the CD cell are shown in Table 57.1. The discharge curve at

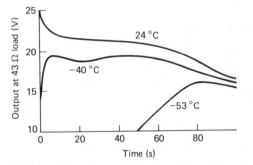

Figure 57.1 Voltage–time curves for an Eagle Picher reserve primary manganese dioxide–magnesium perchlorate battery on activation at various temperatures (Courtesy of Eagle Picher)

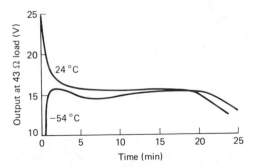

Figure 57.2 Discharge curves for an Eagle Picher reserve primary manganese dioxide–magnesium perchlorate battery (6.6 × 6.6 × 2.5 cm, 94 g) after activation at −54 and +24°C (Courtesy of Eagle Picher)

Table 57.1 Electrical and physical characteristics of Marathon magnesium CD cell

Electrical	
Nominal operating voltage	1.5 V
Nominal capacity	6 A h
Energy density	
per unit weight	132 W h/kg
per unit volume	210 W h/dm^3
Shelf life	
at 20°C	5 years
at 71°C	3 months
Physical	
Maximum height	83.8 mm
Maximum diameter	22.8 mm
Case material	magnesium alloy
Maximum weight	58.1 g

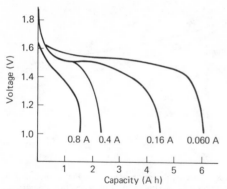

Figure 57.4 Marathon CD-size magnesium cell constant-current discharge curves at 23°C (Courtesy of Marathon)

24°C is shown in Figure 57.4, and Table 57.2 gives discharges at various rates at 24°C. The battery delivers its maximum capacity when it is continuously discharged for a period of about 100 h. When the continuous time of discharge exceeds 200 h, a capacity decrease occurs due to the self-discharge process of the magnesium anode.

Table 57.2 Marathon CD cell capacity *versus* discharge rate at 24°C

Discharge rate (A)	Life (h)	Capacity (A h)	Capacity as % of 100 h rate
0.060	100	6.0	100
0.16	28	4.5	75
0.40	5.5	2.2	38
0.80	1.9	1.5	25

As well as the CD size, Marathon also supply the A, C, D and F sizes. At the 50 h rate, their capacities are estimated to be 0.6 A h (A cell), 3.0 A h (C cell), 6.0 A h (D cell) and 12.0 A h (F cell).

57.2.2 The BA4386.18 CD battery

Marathon CD cells may be assembled, as required, in any series or series–parallel construction. Cells are connected together by means of soldered leads. The terminals of the battery may vary, depending on the requirement. The outer packaging of the battery depends on the intended use and may vary from a simple cardboard container to a metal enclosure. It is customary to pot the cells of a multi-cell battery with an asphalt compound to prevent movement and to ensure a mechanically strong package. Proper design of the battery package enables it to meet the most rigid vibration and shock tests as well as conditions of high humidity.

The electrical and physical characteristics of this battery are given in Table 57.3.

Table 57.3 Electrical and physical characteristics of Marathon magnesium battery type BA4386

Electrical
The BA4386 battery consists of two sections, both of which are made up of CD-size cells: section A1 yields 3 V and section A2 yields 14.1 V

	Section A1	Section A2
Nominal voltage	3 V*	14.4 V
Load	6.75 Ω, 2 min; open-circuit voltage, 18 min	14.2 Ω, 2 min; 291 Ω, 18 min
End-voltage	2.12 V	10 V
Performance	60 h at room temperature to 55 h after 7 days' storage at 71°C	

Physical

Maximum height	53.97 mm
Maximum width	92.07 mm
Maximum length	24.13 mm
Weight	1.36 kg
Number of cells	18†
Battery case material	Electrolyte-resistant cardboard
Terminal	Five-hole socket
Cell size	CD

* With dropping resistor in series
† 18 cells in series–parallel arrangement (section A1 plus section A2)

Field experience has shown that shelf life is an outstanding characteristic of Marathon batteries. After 18 months storage at room temperature, magnesium batteries have yielded 90% of their initial capacity; 90 days at 55°C has shown that the battery still retains 85% of its initial capacity; and after storage at 71°C for 30 days the capacity retention is 85%. The shelf life of BA4386 batteries is demonstrated in Figure 57.5 for conditions in which the batteries were discharged under transmit–receive conditions after storage at temperatures between 21 and 71°C.

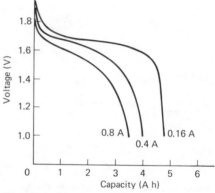

Figure 57.5 Marathon CD-size magnesium BA4356 battery constant-current discharge curve at 71°C (Courtesy of Marathon)

Under intermittent conditions of use the realizable capacity of the magnesium battery depends on factors such as the magnitude of off-time period, the rate of discharge, the frequency of discharge intervals and the ambient temperature. The effect of each of these and their interrelationship makes it difficult to predict the capacity of the battery unless the mode of application is completely defined. In general, intermittent usage at heavy discharge drains is a favourable condition for optimum performance. This battery performs better at higher operating temperatures than at 21°C, especially at the higher drains. This is readily seen by comparing the family of curves in Figure 57.5 (discharged at 71°C) with those of Figure 57.4 (discharged at 23°C). Its performance diminishes when discharged at lower temperatures. However, the battery operates at temperatures as low as −18°C, especially at light drain rates.

58 Magnesium–organic electrolyte batteries

Magnesium perchlorate-*m*-dinitrobenzene batteries, supplied by Marathon, are still in an early stage of development. A limited range is available from Marathon, and other suppliers are carrying out development work. One of the interests in this system is its amazingly high theoretical energy density of 1900 W h/kg, although, to date, practically realized energy densities fall considerably short of this value. Another interesting feature of this type of cell is its very flat discharge curve.

59

Zinc–air (primary) cells

Gould and Berec

Gould and Berec market zinc–air button cells in the UK.

Varta

Varta also produce this type of cell, an example being the 1.4 V hearing-aid battery (type V4600) with a capacity of 300 A h. This cell has a specific energy density of 650–800 mW h/cm³. Its operational temperature range is −10 to +60°C.

The Varta V675HPA hearing-aid battery has a capacity of 260 mA h. Varta also produce rectangular zinc–air batteries in the capacities, 85 A h (types 441, 459, 445, 461 and 465) and 53 A h (type 446).

Duracell

Duracell produce a 400 mA h, 1.4 V hearing-aid button cell (type DA675). The storage life is estimated at 2.5 years in the sealed state. When the seal is broken the cell has a usable life of up to 3 months. Typical discharge characteristics for the DA675 button cell are shown in Figure 59.1. This cell has a life of 195 h on 625 Ω load at 20°C. It weighs 1.63 g and has a volume of 0.56 cm³.

Eagle Picher

Eagle Picher supply a range of primary expendable zinc–air batteries in the capacity range 4–50 A h at 25 V, equivalent to 95–1250 W (0.7–5.5 kg). It is claimed that larger sizes with watt hour efficiencies up to 330 W h/kg are feasible with the larger types. Eagle Picher supply primary expendable one-shot batteries and also mechanically rechargeable types (anode replacement), but do not supply electrically rechargeable types on the grounds that these have not yet been developed to a satisfactorily high standard.

Mechanically rechargeable secondary batteries are available from Eagle Picher in the 88–220 W h/kg capacity range. Their output is 5–150 A h; between 50 and 125 cycles are to be expected. On particular discharges of 11–0.75 A service, some 50 cycles life is to be expected. Storage life in the inactivated state is about 5 years, and in the activated state it is 36 h. Some specific characteristics of this type of battery are listed in Tables 59.1 and 59.2.

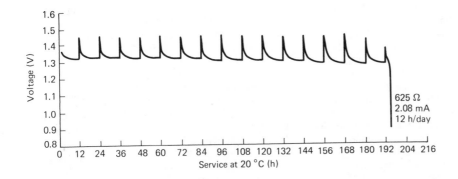

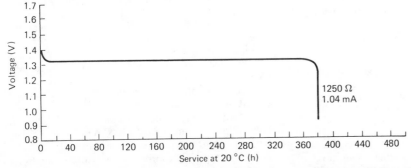

Figure 59.1 Typical discharge characteristics of a Duracell type 675 zinc–air button cell. Current at 1.3 V (Courtesy of Duracell)

Table 59.1 Characteristics of Eagle Picher mechanically rechargeable secondary batteries

Load at steady state (A)	Temperature (°C)	Nominal rating (A h)
1	21	30
1	−40	24
2.5	−40	20

Table 59.2 Characteristics of Eagle Picher mechanically rechargeable secondary batteries

Load in service	Load ratio	Nominal rating (A h)
3 A to 1.6 A	1:9	27.5
11 A to 0.75 A	1:9	25.0
11 A to 0.75 A	1:9	10–12 at −40°C

McGraw Edison

McGraw Edison supply the Carbonaire range of zinc–air cells and batteries listed below.

Types ST-22 and ST-33 (navigational aids) These are, respectively, two- and three-cell types, housed in transparent acrylic plastics cases and lids. The batteries as supplied are contained in plastics bags to prevent deterioration of the zinc anode during storage by ingress of air and moisture. The batteries are activated simply by topping up with water. Individual cells can be connected in series or in parallel with external connections. Normally ST Carbonaires require no routine maintenance or inspection for the first year. Should the service period run well over a year, an annual visual inspection of battery solution levels is desirable. The transparent case permits the user to see when each battery is ready for replacement.

McGraw Edison supply 1100–3300 A h versions of these batteries with voltages between 6 and 22.5 V.

Type Y This consists of a single 2800 A h, 1.25 V cell sealed in a moulded hard rubber case. This battery will give up to 3 years' active life in many services and will operate at temperatures down to −18°C electrolyte temperature. It is recommended for use in aids to navigational services.

Type BY-301 This 3000 A h, 1.25 V cell in a hard rubber case is designed for the operation of navigational signal aids, both visible and audible. The size and shape of the battery will permit it to

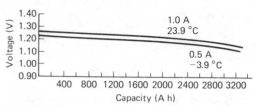

Figure 59.2 Voltage curves: McGraw Edison zinc–air batteries, type TR (Courtesy of McGraw Edison)

replace lead storage cells now widely used in this application. Special disposable battery racks are currently available for buoys with 559 or 610 mm battery pockets. These racks are furnished completed with the proper number of BY Carbonaires and the entire unit is prewired at the factory.

The problem of electrolyte spillage is eliminated by the use of a spillproof valve, which is applied after the battery is activated. Depending on the electrical energy requirements, one set of BY-301 Carbonaires will give up to 2 years' continuous service.

Type TR This 3300 A h, 1.25 V cell is housed in a hard rubber case. It is designed for the operation of railway d.c. track circuits. The TR cell has the straight-line voltage characteristic so essential to satisfactory track circuit operation. Figure 59.2 illustrates this narrow voltage range throughout the entire service life of the TR Carbonaire, using the maximum recommended continuous current of 1.0 A at 23.9°C and 0.500 A at −3.9°C battery solution temperatures.

Another advantage of using the TR Carbonaire is the ease with which it can be carried. Using the specially designed Edison Carbonaire Carrier, two batteries can be carried at once in a natural carrying position that minimizes accidental electrolyte spillage from activated batteries.

Type F This is a twin single cell (mounted in series) 1300 A h, 2.5 V battery intended for railway signalling services. This battery has an extremely uniform voltage. In general, the F Crystalaire can be continuously discharged to full rated capacity at any rate up to 0.150 A. Under favourable operating conditions a higher continuous current than 0.050 A may be sustained. In very low temperature ranges, maximum recommended continuous current may vary considerably, depending on actual service conditions encountered.

McGraw Edison also produce 2.5 V, 1000 A h (type 2-S-J-I) and 3.75 V, 1000 A h (type 3-S-J-I) batteries and these, together with the 1.25 V, 2800 A h (type Y), can be used in various combinations to produce a variety of battery voltages, as shown in Table 59.3.

Table 59.3 Voltage classification chart for Carbonaire batteries

Voltage required (V)	Types and number of Carbonaires recommended
2.5	1 × 2-S-J-I or 2 × type Y*
3.75	1 × 3-S-J-I or 3 × type Y*
6.0	1 × 2-S-J-I*, 1 × 3-S-J-I or 5 × type Y*
10.0	4 × 2-S-J-K* or 8 × type Y*
12.0	2 × 2-S-J-I*, 2 × 3-S-J-I* or 5 × 3-S-J-I* or 10 × type Y*

* Batteries to be connected in series

McGraw Edison also supply Carbonaire buoy battery packs. All are rated at 12 V. Type ST-1010 consists of two 2-cell and two 3-cell Carbonaire batteries connected in series and has a capacity of 1000 A h. Type ST-2010 consists of ten 2-cell parallel-connected batteries. These ten batteries are connected in series to provide a pack having a capacity of 2000 A h. Type ST-3010 consists of ten 3-cell parallel-connected batteries. These ten batteries are connected in series to provide a pack with a capacity of 3000 A h.

One outstanding advantage of Carbonaire batteries is that they may be activated with seawater. In navigational aid services, this means that batteries can be transported dry to the location. The weight to be transported is thus reduced to a minimum and the possibility of spillage in transit is eliminated.

SAFT zinc–air depolarized primary batteries

SAFT supply two main types of air depolarized batteries, saline and alkaline (liquid and jellified). Both use the electrochemical couple zinc–oxygen.

Saline batteries are suited to applications in which a low constant current is required (of the order of tens of milliamps) for several years. They can also be discharged at a higher regimen for a shorter period (up to 12 A).

Alkaline batteries, either liquid or jellified, are suitable for installations in which the requirement demands a continuous discharge (up to 4 A) and a long service life. This system is fitted for use in extreme climatic conditions (polar and tropical).

Air depolarized batteries can be readily assembled in series and/or parallel to obtain appropriate voltages, currents and capacities.

It is advisable to place air depolarized batteries in a well ventilated area. Sheltered containers are recommended in cold or hot climates. In order that the batteries function correctly, allow 1 litre of air per battery and per discharged ampere hour. Air depolarized batteries must be stored and used in an upright position.

Further characteristics of SAFT air depolarized batteries are given in Table 59.4. Electrical characteristics of some of these batteries are tabulated in Table 59.5. Discharge data for some saline and alkaline types of SAFT zinc–air depolarized cells are given in Figures 59.3–59.8. The applications of this type of battery are discussed in Part 4.

Table 59.4 Characteristics of SAFT air depolarized zinc–air batteries

Characteristics	Saline batteries	Alkaline batteries	
		Liquid	Jellified
Electrochemical couple	Zinc–oxygen	Zinc–oxygen	Zinc–oxygen
Electrolyte	A set solution of ammonium chloride	Potassium solution	Jellified potassium
	Ready for use	By adding of water or electrolyte	Ready for use
Electrolyte temperature range (°C)	−5 to +50	−40 to +55	−30 to +40
Capacity range (A h)	6–1000	350–10 000	350–2200
(A)	0.010–0.800	0.250–1.500	0.600–1.500
(A h/kg)	120	150	220
(A h/dm^3)	135	200	240
Storage under temperature conditions (years)	2	3	1

Table 59.5 Electrical characteristics at 20°C of air depolarized batteries available from SAFT

Types	Open circuit voltage (V)	Rated capacity (Ah)	End-point voltage (V)	Average current (A)	Resistance (Ω)	Maximum continuous discharge (A)	Intermittent discharge, rate ≤1 min (A)	Pulse discharge, rate ≤1s (A)	Maximum hourly capacity, intermittent or pulse (Ah)	Length (mm)	Width (mm)	Height with terminals (mm)	Diameter (mm)	Weight (kg)
Saline batteries														
AD 517.4	1.45	1000	0.8	0.800	0.800	1.200	4.000	12.000	0.600	212	212	195	–	9.2
AD 517	1.45	270	0.8	0.200	5	0.300	1.000	3.000	0.150	106	106	195	–	2.3
AD 524	1.45	145	0.8	0.100	10	0.200	0.500	1.500	0.100	86	86	195	–	1.5
AD 2-519	2.90	130	1.6	0.100	20	0.150	0.500	1.500	0.070	106	110	195	–	2.45
AD 2-522	2.90	70	1.6	0.080	20	0.100	0.200	0.500	0.050	89	87	185	–	1.75
Alkaline batteries														
Liquid														
AD 600	1.45	10 000	0.9	1.500	0.8	2.000	3.500	4.500	1.100	306	306	520	–	65*
AD308A_M	1.45	2000	0.9	1.000	1.1	1.500	2.000	2.500	0.800	210	210	310	–	14*
Jellified														
AD 810	1.45	2200	0.9	1.500	0.75	2.000	3.500	4.600	1.000	–	–	285	212	10.1†
AD 820	1.45	1200	0.9	1.200	1	2.000	3.000	4.000	1.000	200	104	216	–	5
AD 840	1.45	350	0.9	0.500	2.5	1.000	1.500	2.000	0.500	85	85	200	–	1.7

* After filling with water
† Cylindrical

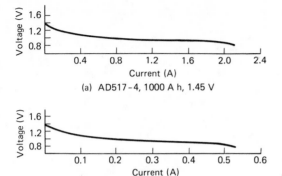

(a) AD517 - 4, 1000 A h, 1.45 V

(b) AD517, 270 A h, 1.45 V

Figure 59.3 SAFT zinc–air depolarized saline batteries: average voltage *versus* discharge rate at 20°C (Courtesy of SAFT)

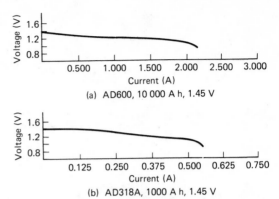

(a) AD600, 10 000 A h, 1.45 V

(b) AD318A, 1000 A h, 1.45 V

Figure 59.6 SAFT zinc–air depolarized alkaline batteries: average voltage *versus* discharge rate at 20°C (Courtesy of SAFT)

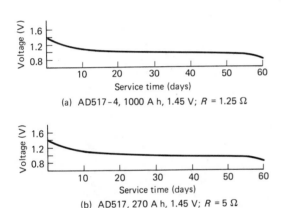

(a) AD517 - 4, 1000 A h, 1.45 V; $R = 1.25\ \Omega$

(b) AD517, 270 A h, 1.45 V; $R = 5\ \Omega$

Figure 59.4 SAFT zinc–air depolarized saline batteries: discharge curve on continuous load at 20°C (Courtesy of SAFT)

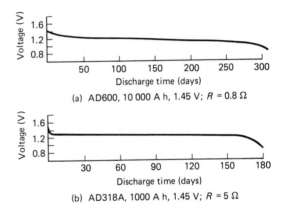

(a) AD600, 10 000 A h, 1.45 V; $R = 0.8\ \Omega$

(b) AD318A, 1000 A h, 1.45 V; $R = 5\ \Omega$

Figure 59.7 Discharge curve of a SAFT zinc–air depolarized alkaline battery on continuous load at 20°C (Courtesy of SAFT)

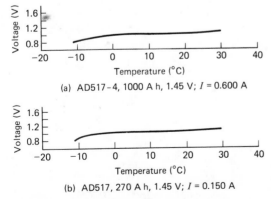

(a) AD517 - 4, 1000 A h, 1.45 V; $I = 0.600$ A

(b) AD517, 270 A h, 1.45 V; $I = 0.150$ A

Figure 59.5 SAFT zinc–air depolarized saline batteries: average voltage *versus* temperature (Courtesy of SAFT)

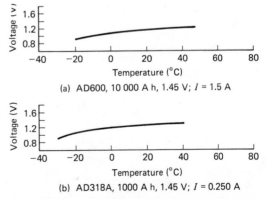

(a) AD600, 10 000 A h, 1.45 V; $I = 1.5$ A

(b) AD318A, 1000 A h, 1.45 V; $I = 0.250$ A

Figure 59.8 SAFT zinc–air depolarized alkaline batteries: average voltage *versus* temperature (Courtesy of SAFT)

60

Thermally activated batteries

Catalyst Research Corporation

Table 60.1 shows current–life parameters for a wide variety of general-purpose thermal batteries produced by Catalyst Research Corporation. The height of the batteries is determined by the power required of the battery. The same power output can be achieved in a number of case diameters, by using different intercell connections (series, parallel or series–parallel) to provide the desired voltage–current parameters. A thermal battery is a low-impedance device; much higher currents than indicated may be drawn for shorter periods of time.

Catalyst Research Corporation also supply fast-activation thermal batteries. These are to meet the needs of applications requiring battery power to fire squibs, electric matches, dimple motors, explosive bolts, etc., after battery activation.

Table 60.2 presents size and performance data for some typical units supplied in the three categories of fast-start batteries: A, high current pulse; B, medium life; and C, high current. These may be used to ripple fire squibs under $20\,\Omega$ load, $50\,ms$ on, $50\,ms$ off, $0.7\,\Omega$ load.

Catalyst Research Corporation have also designed and developed a high-efficiency electrochemical system for use in high current density applications. This Power Cell system has the flexibility for use in both short- and long-life battery requirements. Available power output is approximately 25% greater than that of conventional thermal battery systems. Performance data on this battery are given in Table 60.3.

Catalyst Research Corporation also produce a high-voltage thermal battery designed to meet the need for high-voltage low-current power supplies. Typical specifications for such units are shown in Table 60.4.

Table 60.1 General-purpose thermal batteries from Catalyst Research Corporation

Nominal case diameter (mm)	Current (A) for life shown		
	60 s	120 s	180 s
31.7	0.9	0.3	0.1
38.1	1.5	0.5	0.3
41.1	1.9	0.6	0.4
50.8	2.8	0.9	0.5
57.1	3.8	1.3	0.7
76.2	5.7	2.0	1.3
95.2	9.2	3.0	1.9

Table 60.2 Fast-activation thermal batteries from Catalyst Research Corporation

Type	Nominal diameter (mm)	Nominal height (mm)	Nominal voltage (V)	Activation current (A)	Activation time (s)	Life (s)
A	12.7	19.0	10–20	5	0.01	0.5
A	41.1	38.1	20–32	40	0.03	5.0
A	41.1	33.2	12–19	17–19	0.20	0.5
A	41.1	69.8	24–42	15–25	0.10	5.0
B	31.7	31.7	12–16	1.0	0.05	30.0
C	88.9	165.1	28–36	3–4	0.50	105.0

Table 60.3 'Power Cell' thermal batteries from Catalyst Research Corporation

Diameter (mm)	Height (mm)	Voltage (V)	Current (A)	Life (min)	Average power output (W)
35.5	58.9	25 ± 4	0.7	5.0	—
53.3	76.2	50 ± 7	15.6	0.6	780
53.3	76.2	45 ± 5	45.0	0.25	2025
88.9	88.9	28 ± 4	5.7	15.0	144
88.9	149.8	28 ± 4	16.7	11.5	426*
152.9	101.6	28 ± 4	20.0	14.0	492*

* Operated at 25°C

Table 60.4 High-voltage thermal batteries from Catalyst Research Corporation

Diameter (mm)	Height (mm)	Voltage output (V)	Current (A)	Life (s)
45.7	31.7	140 ± 4	0.014	30
		1.4 ± 0.1	0.40	30
		8.2 ± 0.3	0.07	30
55.3	55.3	276 ± 13	0.001	30
		56 ± 8	0.030	30
76.2	86.8	356 ± 24	0.120	30

Catalyst Research Corporation thermal batteries are capable of operation at temperatures between -54 and $+73°C$, at accelerations up to $2500g$ lateral, under a shock of up to $200g$ and vibration of $30g$, and also at spin rates of $400\,rev/s$.

Activation of the battery by ignition or an internal heat source can be achieved by several methods:

1. *Electrical* High- and low-energy matches and squibs with wire bridges are available. Firing energies of two of the units are shown in Table 60.5.
2. *Mechanical* An M42G primer similar to that in a .38 calibre bullet, with a modified charge, has been used extensively in thermal batteries. Spring or explosive-driven firing pins are used to deliver the $1800\,cm\,g$ of energy required for primer firing. A second method of mechanical initiation uses a self-contained inertial starter, which is activated by forces of $300–600g$ imposed on the battery.

Table 60.5 Firing energies

	No-fire	*All-fire*
Match	0.25 A	0.50 A
	5 s	50 ms
Igniter	1 A	3.5 A
	5 min	20 ms

Eagle Picher

Eagle Picher are major suppliers of thermal batteries. The batteries are available in a variety of configurations. Sizes range from 1.25 mm diameter × 12.5 mm length to multi-cell battery packs of up to 99 mm diameter × 127 mm length. Fast activation and high amperage pulse power are available from the smaller units. Regulated power up to 4 min duration is available from larger units. Battery voltages from 2.5 to 450 V are available from existing designs.

The batteries have an open-circuit voltage of 2.3–3.0 V/cell and voltages available of 2.3–500 V. Voltage regulation is within 10% of nominal under constant resistive load. The battery operation range covers the interval 1–300 s. A typical discharge curve for an Eagle Picher thermal battery is shown in Figure 60.1. Performance and other characteristics of these batteries are:

1. High-rate capacity: 10 A for 60 s.
2. Low-rate capacity: 250 mA for 300 s.
3. Activation time: 100 ms to 3.5 s.
4. Operating temperature: -54 to $+93°C$.
5. Storage temperature: -54 to $+74°C$.
6. Storage life: up to 10 years at the recommended storage temperature.

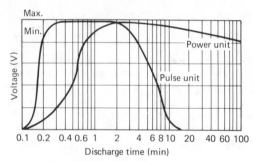

Figure 60.1 Typical discharge curves of Eagle Picher thermal batteries. (Courtesy of Eagle Picher)

7. Vibration: withstands $300g$ all areas.
8. Shock: withstands $3000g$ longitudinal.
9. Acceleration: withstands $450g$ longitudinal.

The characteristics of some of the wide range of thermal batteries available from Eagle Picher are shown in Table 60.6.

The EAP-6273 12 V battery based on the calcium chromate–calcium system was specifically designed to provide emergency starting power for military vehicles and aircraft. The basic design of this battery is eight parallel cell stacks of 152.4 mm diameter. The cell area of 181 cm^2/cell for the eight sections results in 1449 cm^2 of active area. At a discharge rate of 0.7 A/cm^2, the EAP-6273 could produce 1000 A. This rate is well within the normal operating limits of the standard calcium chromate–calcium system. The cell is fabricated in one piece using the standard two-layer (DEB) configuration and is 152.4 mm in diameter, 2.2 mm thick, and contains a 12.7 mm centre hole.

The header is of standard 304 stainless steel 9.5 mm thick. Because of the high current, four terminals for each negative and positive lead are used. At the 800 A level, each terminal would carry 200 A. The battery is connected externally by paralleling the four positive and the negative terminals at the load connections.

The battery case is assembled from 321 stainless steel welding tubing with a 9.5 mm steel plate welded in the bottom. The burst strength of this container was calculated to be in excess of 454 kg.

Because of the total height of the cell stacks (approximately 356 mm) the stack was assembled in two separate sections, each containing four parallel stacks of cells. These are wrapped and inserted into the case separately. The case is lined with 12 mm of thermoflex and 12 mm of asbestos in the bottom. Each battery section is wrapped with 3.8 mm of glass cloth. Filler asbestos for the top of the battery and the header insulation are added to obtain a closing pressure of 18–22 kg. The total battery weighs 20.4 kg, and is 406.4 mm tall and 177.8 mm in diameter.

Table 60.6 Characteristics of Eagle Picher thermal batteries

EAP-6266: 101.6 mm diameter, five-cell stacks
Voltage	27 ± 3 V
Life	at 71°C, 221 s
	at 24°C, 366 s
	at −54°C, 317 s
Load	33 A nominal
Size	121 mm diameter × 193 mm long
Weight	5.2 kg
Watt hours	89.1 (= 17.2 W h/kg)

EAP-6230: 101.6 mm diameter five-cell stacks, DEB type
Voltage	28.5 ± 3.5 V
Life	at 38°C, 24 min
	at 5°C, 16 min
Load	4.3 A nominal
Size	122 mm diameter × 104 mm long
Weight	2.86 kg
Watt hours	48.3 (= 16.9 W h/kg)

EAP-6251: 76 mm diameter, parallel stacks
Voltage	27 ± 3 V
Life	at 71°C, 110 s
	at 24°C, 200 s
	at −40°C, 140 s
Load	23 A nominal
Size	94 mm diameter × 145 mm long
Weight	3 kg
Watt hours	31 (= 10.4 W h/kg)

EAP-6261: 76 mm diameter, parallel stacks
Voltage	$+33 \pm 5.5$ V
	-33 ± 5.5 V
Life	at 71°C, 170 s
	at 24°C, 200 s
	at −56°C, 135 s
Load	4.2 A each section
Size	89 mm diameter × 117 mm long
Weight	1.9 kg
Watt hours	20.4 = 8.1 W h/kg (138.6 × 2 sections)

EAP-6273
Voltage	12 ± 3 V
Life	at 24°C, 350 s
Load	500 A nominal
Size	177.8 mm diameter × 406 mm long
Weight	20.4 kg
Watt hours	583.3 = 28.6 W h/kg

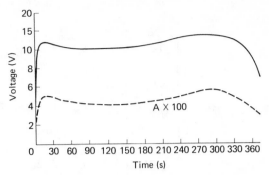

Figure 60.2 Performance of Eagle Picher EAP-6273 thermal battery under 0.024 Ω load (Courtesy of Eagle Picher)

The load desired is $0.024 \, \Omega$, which is capable of dissipating 25 kW. A 15.8 mm outside diameter stainless steel pipe is cut to length and welded to two copper bars. Connections are made to allow water to flow through the pipe. A 1000 A current shunt is bolted into series with one copper bar to provide current-monitoring capabilities.

The performance of the EAP-6273 is shown in Figure 60.2. Essentially, the battery performs as a 12 V battery delivering 550 A. Additional current could be drawn on if necessary, as would be required for an electric motor starter. However, the power developed would be capable of starting any 12 V vehicle requiring up to 800 A of peak current loads.

Mine Safety Appliances

As well as supplying a range of conventional thermal batteries varying from the well established calcium chromate closed cell to the pellet variety, Mine Safety Appliances also manufacture the range of new lithium-based thermal batteries discussed below. The batteries are cylindrical, in the form of hermetically sealed metal canisters. Terminals providing connections to cells are located in glass or ceramic insulating seals. The batteries are now in production and are capable of operating at electrode current densities of up to 2.0 A/cm², with long battery durations. Battery results have demonstrated a capability of achieving 100 W h/kg compared with approximately 20 W h/kg in existing designs of conventional calcium batteries, such as the Ca/LiCl–KCl and CaCrO$_4$/Fe systems.

Some of the cylindrical calcium chromate cells available from MSA are shown in Table 60.7. These batteries are claimed to have vibration, shock and acceleration resistances that make them suitable for use in a military environment. The batteries are stored in a hermetically sealed metal container. Batteries can be stored for periods in excess of 15 years at normal ambient temperatures without self-discharge or deterioration. The batteries operate well over a temperature range of −40 to 70°C; in some designs, −55 to 110°C can be tolerated without deterioration in performance. The batteries can be orientated in almost any direction with reference to movement.

When activated, the molten salt electrolyte ensures that the battery has a very low internal impedance. In the inactive state the solid electrolyte is non-conductive and the battery has an extremely high impedance; thus, full missile circuit testing can be carried out with the circuitry connected to the battery.

The high impedance before initiation also allows the use of ship or aircraft external power supplies to

Table 60.7 Technical specifications of cylindrical calcium–calcium chromate types of thermal battery supplied by MSA

MSA part No.	Output (A)	Maximum voltage (V)	Minimum voltage (V)	Activation time (s)	Duration (s)	Operating temperature (°C)	Weight (g)	Height (mm)	Diameter (mm)
270072	1.27 at 23 V 1.24 at 6 V	30.5 8.3	23.0 6.0	1.5	15	−30 to 70	213	98	32
270046	0.033 at 42 V 0.002 at 26 V	64.0 289.0	42.0 263.0	5	33	−10 to 50	304	57	51
270086	1 at 12.25 V	15.75	12.25	1.8	27	−10 to 60	132	153	79
270085	2A pulses at 5 V	13.0	5.0	0.35	23.5	−40 to 105	100	51	32
270036-2	11.06 at 25 V	28.7	25.0	1.8	140	−20 to 60	2800	231	76
290050	2.7 at 27 V	33.0	27.0	<1.9	>93	0 to 40	879	66	76
290070	1 at 21.6 V	26.4	21.6	<2 to 21.6 V	>41.6 to 21.6 V	−10 to 50	400	68	51
270095	12 at 22 V	33.0	22.0	<4 to 30.75 V	>60 to 22 V	−17 to 34	1240	116	76
270028	43.5 at 17 V	23.0	17.0	1 to 17 V	70 to 17 V	−40 to 60	95	34	31
279608	5 at 27 V	35.0	27.0	<2.2 to 27 V	>41.5	−37 to 85	1457	116	76
270027	0.147 at 22 V	31.0	22.0	1 s to 22 V	70 s to 22 V	−40 to 60	142	47	38
290027	3.45 at 26 V 3.05 at 23 V	32.0	26.0 23.0	2	155	−20 to 55	2041	186	76
279611	0.2 at 4 V	15.0	4.0	0.4	10	−39 to 83	54	22	31

Initiator type: electrical

Table 60.8 Technical specifications of cylindrical lithium–iron disulphide types of thermal batteries supplied by MSA

MSA part No.	Ouput (A)	Maximum voltage (V)	Minimum voltage (V)	Activation time (s)	Duration	Initiator type	Operating temperature (°C)	Weight (g)	Height (mm)	Diameter (mm)
279694	2	22	16	<0.3	>60 s	Electrical	−40 to 70	70	40	25
279649	7	32	24	<5	>20 min	Electrical	−23 to 6	3500	114	114
279655	2	27	21	<0.6	>50 s	Fuse	−55 to 75	275	68	14.9
279678	2	22	16	<0.3	>60 s	Electrical	−40 to 70	100	37	31
279682	1.25	26.6	19	<1.5	>50 s	Percussion	−40 to 70	230	65	38
	1.9	26.6	19							
	3	11.2	8							
279687	35.5	75	45	<10	>8 min	Electrical	0 to 30	3800	234	98
279693	3.5	84	52	<1	>180 s	Electrical	−30 to 60	470	100	14.3
279657	6 A at 28 V	39.5	28	<2.5	>600 s	Electrical	−20 to 40	2000	137	89
279651	23 A at 26 V	37.5	26	<2.5	>400 s	Electrical	−20 to 55	3400	210	89
279683	5	32.5	26	<2.5	>100 s	Electrical	−25 to 75	1200	85	77
279659	50 A at 25 V	32	25	<1	>180 s	Electrical	−40 to 70	3100	165	89

be connected to the supply terminals before and during activation. Diode protection may, however, be necessary to prevent excess current being drawn from the external supply after initiation of the thermal battery.

These features are ideal for devices that must have instant readiness and incorporate completely automatic electrical check systems during test or firing sequences.

Mine Safety Appliances have also designed a system with lithium as the active anode material and ferrous sulphide as the cathode, which shows significant improvements in performance over the conventional systems (Table 60.8). The lithium–iron sulphide system is based on a pile-type construction, which allows the battery voltage to be easily adjusted, by varying the number of series cells, to meet different application requirements.

Quite large sized thermal batteries of this type are being produced; one battery weighs 3.8 kg and consists of 30 series cells housed in a container 89 mm in diameter by 236 mm long. This battery shows smooth discharges, free from electrical disturbance, for average current densities varying from 0.3 to 1.86 A/cm^2. To 80% of the peak voltage, durations ranged from 3 to 21 min. It is also noteworthy that the cell capacity is very constant at 275 A minimum over this range of discharge rates, although there are indications of some reduction at the highest current density. For this capacity of

lithium–iron sulphide cell, the current density performance is at least 15 times better than with calcium–calcium chromate systems for discharge densities of over 2–3 min, and analysis of the discharge curves also shows that the energy density achieved by the cell (including pyrotechnic) is 113 W h/kg, and 63 W h/kg for the battery. The volumetric energy density from the battery is 113 W h/dm^3.

A further battery design gave a duration of 7 min, a short activation time and a low internal resistance for good voltage regulation for a wide range of current drains. It uses two parallel stacks of 75 mm nominal diameter cells, housed in a container 89 mm in diameter and 190 mm long. The battery weight is 2.9 kg. At an average discharge current of 20 A (0.23 A/cm^2 electrode current density), 8 min discharge duration was obtained to 80% of the maximum voltage for both conditioning temperatures. The battery internal resistance remains constant for the discharge duration to 80% of mean voltage, with a value of 0.09–0.11 Ω. The batteries tested after initial conditioning at +50°C activated in approximately 1.6 s giving 12.8 A at 25.5 V, and those conditioned at −30°C activated in approximately 3.0 s. The battery design used anodes with excess coulombic capacity (380 A minimum), and considerable improvement in activation time is possible if these are matched to the rated cathode capacity of 200 A minimum.

61 Zinc–chlorine batteries

Energy Development Associates, Michigan, are the only known supplier of this type of battery. Details are given in Part 2.

62 Sodium–sulphur batteries

Details of this type of battery, which has not yet reached commercial realization, are given in Part 2.

63

Water-activated batteries

Contents

63.1 McMurdo Instruments magnesium–silver chloride seawater batteries

McMurdo Instruments Ltd supply a range of magnesium–silver chloride cells (Aquacells), which are activated by seawater and have obvious applications in marine and aircraft safety lights, buoys and flashing beacons. These batteries have a long storage life provided they are kept in a sealed condition. Once unsealed and immersed in water they must be considered expended. They cannot be recharged.

An example of this type of cell is the L18A divers' underwater searchlight battery. This cell has dimensions $103 \times 35 \times 25$ mm, weighs 95 g and has an output of 4 V for 1.5 h at a 1 A loading. The L37 submarine escape-hatch lighting battery is a 24 V emergency supply system designed to operate at pressures up to 2756 kPa. The battery has dimensions $305 \times 254 \times 72$ mm, weighs 6.6 kg and will operate two 24 V, 36 W lamps for 8 h. The L43A distress beacon (13.5 V) battery for ships and aircraft has dimensions $175 \times 51 \times 125$ mm, weighs 110 g and has a 90 h discharge time. McMurdo also supply a range of Aquacells for life-rafts and life-jackets (Table 63.1).

63.2 SAFT magnesium–silver chloride batteries

These cells have an energy density of 30–120 W h/kg (power density 1200 W/kg) and 40–250 W h/dm^3 (volumetric power density 2000 W/dm^3). Operating temperatures are between -20 and $+60°C$ and cells up to 150 kW are obtainable. Designs in the range 1–250 V are available. The cells are activated by immersion in water or saline water and durations are between several seconds and 20 days. The cells are used to operate sonobuoys, beacons, torpedoes, flares, mines, pingers, balloons and life-jackets.

The types available are listed in Table 63.2.

A typical discharge curve for the AM 34 type (1.5 V, 7.5 A h) is shown in Figure 63.1. The two curves contrast the cell performance in water at low salinity (15 g sodium chloride per litre) and low temperature ($0°C$) with the performance obtained in warm ($30°C$) seawater (36 g sodium chloride per litre).

Figure 63.2 shows the effects of repeated reimmersions of the cell in water. No adverse effects of repeated immersions on cell performance are apparent.

Table 63.1 Magnesium–silver chloride cells for life-jackets and life-rafts from McMurdo Instruments

Designation	Application	Dimensions (mm)	Weight (g)	Power output	Duration (h)
L8.1	Life-jacket	$93 \times 26 \times 12$	34	1.5 V, 0.3 W	8
M8M	Life-jacket	$93 \times 26 \times 12$	34	1.5 V, 0.165 W	20
				or 1.5 V, 0.25 W	14
L12B	Life-jacket and life-raft	$110 \times 30 \times 22$	80	3 V, 1 W	6
				or 3 V, 0.5 W	12
				or 3 V, 0.25 W	24
L50	Life-raft	$117 \times 66 \times 36$	120	3 V, 0.5 W	12

Table 63.2 SAFT magnesium–silver chloride batteries

Type No.	Nominal voltage (V)	Nominal capacity (A h)	End-of-discharge voltage (V)	Average current (A)	Maximum continuous current (A)	Length (mm)	Width (mm)	Thickness (mm)	Weight (g)
AM 18U	3	3	2.4	0.400	–	122	74	42	135
AM 99F	3	2	2.4	0.400	0.500	94	32	24	100
AM 99	3	2	2.4	0.400	0.500	94	32	24	70
AM 14, AM14T	1.5	1.6	1.2	0.200	0.300	93	26	12	32
AM 34	1.5	7.5	1.2	0.250	0.900	92	66	19	100
AM 12, AM 125	1.5	2.4	1.2	0.250	0.500	93	26	12	35

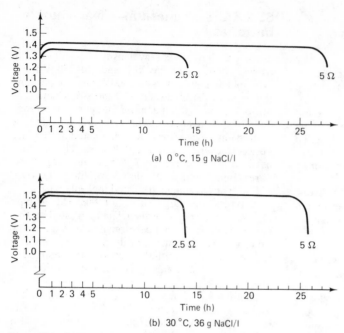

(a) 0 °C, 15 g NaCl/l

(b) 30 °C, 36 g NaCl/l

Figure 63.1 SAFT AM34, 1.5 V, 7.5 A magnesium–silver chloride seawater-activated batteries: influence of salinity and temperature on cell discharge performance (Courtesy of SAFT)

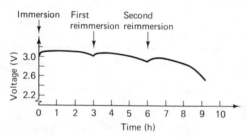

Figure 63.2 SAFT AM18, 3.0 V, 30 A h magnesium–silver chloride seawater-activated batteries: effect of repeated immersions in seawater (32 g NaCl/l) on discharge performance at 20°C (Courtesy of SAFT)

63.3 SAFT zinc–silver chloride batteries

These cells have an energy density of 15–60 W h/kg (power density 1000 W/kg) and 20–150 W h/dm³ (volumetric power density 2400 W/dm³). Operating temperatures are between −30 and +60°C and cells up to 2500 W are obtainable. The cells are water or seawater activated, have a long dry shelf life, high performance and a long discharge time capability at low rates (up to 300 days). Designs are available in the range 1–50 V. The cells are used in balloons, oceanographic buoys, military power systems, communications equipment and portable electronic devices.

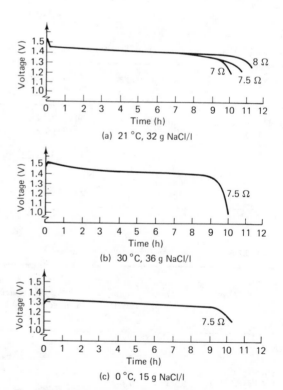

(a) 21 °C, 32 g NaCl/l

(b) 30 °C, 36 g NaCl/l

(c) 0 °C, 15 g NaCl/l

Figure 63.3 SAFT CIM 120, 1.4 V, 1.8 A h magnesium–copper iodide seawater-activated primary batteries: discharge curves (Courtesy of SAFT)

Table 63.3 SAFT magnesium–copper iodide primary batteries

Type No.	Nominal voltage (V)	Nominal capacity (A h)	End-of-discharge voltage (V)	Average current (A)	Maximum continuous current (A)	Length (mm)	Width (mm)	Thickness (mm)	Weight (g)
CIM 990	2.8	1.5	2.4	0.150	0.200	94	32	24	80
CIM 120	1.4	1.8	1.2	0.180	0.200	93	26	12	40

63.4 SAFT magnesium–cuprous chloride batteries

These cells have an energy density of 20–90 W h/kg and 18–150 W h/dm^3. They operate in the temperature range -20 to $+60°$C and are available with up to 50 W output. Designs in the 1–100 V range are available. These cells are water or seawater activated. They have durations from seconds to many hours and are used in balloons, beacons, flares, sonobuoys, pingers and oceanographic equipment.

63.5 SAFT magnesium–copper iodide seawater-energized primary batteries

The types available are listed in Table 63.3. Typical discharge curves at different cell temperatures and salinities are shown in Figure 63.3.

Suppliers of primary and secondary batteries

Belgium

Chloride Belgium NV, Groenstraat 31, Mortsel 2510
  Primary batteries, zinc–alkaline manganese dioxide; secondary batteries, nickel–cadmium, lead–acid.
Varta SA, Rue Uyttenhove 49–51, B-1090 Brussels
  Primary batteries, zinc–alkaline manganese dioxide, mercury–zinc, carbon–zinc, Leclanché, magnesium types, lithium types, silver oxide–zinc, zinc chloride Leclanché, zinc–air.

Canada

Chloride Industries Battery Ltd, 7480 Bath Road, Mississauga, L4T 1L2 Ontario
  Primary batteries, zinc–alkaline manganese dioxide; secondary batteries, nickel–cadmium.
Varta Batteries Ltd, 4 Lansing Square, Suite 237, Willowdale, Ontario, M2J 1T1
  Primary batteries, zinc–alkaline manganese dioxide, mercury–zinc, carbon–zinc Leclanché, magnesium types, lithium types, silver oxide–zinc, zinc chloride Leclanché, zinc–air.

Denmark

Chloride Scandinavia AS, Literbuen 9–11, 2740 Skovlunde
  Primary batteries, zinc–alkaline manganese dioxide; secondary batteries, nickel–cadmium, lead–acid.

France

Chloride France SA, 18 Place de la Seine, Zone Industrielle de la Silic, 94563 Rungis, BP Silic 194
  Primary batteries, zinc–alkaline manganese dioxide; secondary batteries, nickel–cadmium, lead–acid.
Duracell France Technical Division, 5 Rue Chantecoq, F-92801 Puteaux
  Primary batteries, mercury–zinc, silver–zinc, lithium solid electrolyte types.
La Pile Leclanché, Department Piles de SAFT, 156 Avenue de Metz, Romainville 93239
  Primary batteries, zinc–alkaline manganese dioxide, carbon–zinc Leclanché, lithium types (lithium–lead bismuthate, lithium–silver chromate).
SAFT Société des Accumulateurs, Fixes et de Traction, 156 Avenue de Metz, 93230 Romainville
  Secondary batteries, nickel–hydrogen, cuprous chloride.

Israel

Tadiron Israel Electronics Industries Ltd, Derech Hashalon, Tel Aviv 6100
  Primary batteries, carbon–zinc Leclanché, magnesium types, lithium types, silver oxide–zinc; secondary batteries, nickel–cadmium, silver–zinc, silver–cadmium, sealed lead–acid.

Italy

Duracell Italia Technical Division, Via Arrivabene 14, 20158 Milan
  Primary batteries, mercury–zinc, silver–zinc, lithium solid electrolyte types.
FAR, Fabbriche Accumulatori Riunite, Stabilimento Tudor, Via Martiri della Liberta 20, Milan 20066
  Secondary batteries, nickel–cadmium.
Nife Italia SpA, Viale Cembrano 11, Casella Postale 1283, 16148 Genoa
  Alkaline, nickel–cadmium.
Varta SpA, Via Tertulliano 70, 20137 Milan
  Primary batteries, zinc–alkaline manganese dioxide, mercury–zinc, carbon–zinc Leclanché, magnesium types, lithium types, silver oxide–zinc, zinc chloride Leclanché, zinc–air; secondary batteries, alkaline, nickel–iron, nickel–cadmium, silver–zinc, silver–cadmium.

Japan

Shinkobe Electric Machine Co. Ltd, PO Box 218, Mitsui Building, Shinjuku-ku, Tokyo 160
  Secondary batteries, nickel–cadmium.
Varta Batterie AG, Room 810, 8F Shuwa Kioichu, TBR Bldg, No. 7, 5-chome, Kohi-machi, Chiyoda-ku, Tokyo 102
  Primary batteries, zinc–alkaline manganese dioxide, mercury–zinc, carbon–zinc Leclanché, magnesium types, lithium types, silver oxide–zinc, zinc chloride Leclanché, zinc–air.
Yuasa Battery Co. Ltd, 6-6 Josai-cho, Takatsuki-shi, Osaka-fu 569; also International Division, 12-112 Chome, Higashi-Shinbashi Minako-ku, Tokyo 105
  Primary batteries, carbon–zinc Leclanché, silver oxide–zinc; secondary batteries, nickel–iron, nickel–cadmium, silver–zinc, silver–cadmium, sealed lead–acid.

Kenya

Chloride Exide Kenya Ltd, PO Box No. 14242, Nairobi
  Primary batteries, zinc–alkaline manganese dioxide; secondary batteries, nickel–cadmium.

Mexico

Varta SA, Apartado Postal 7355, Mexico IDF, Mexico 9 DF
Primary batteries, zinc–alkaline manganese dioxide, mercury–zinc, carbon–zinc Leclanché, magnesium types, lithium types, silver oxide–zinc, zinc chloride Leclanché, zinc–air.

The Netherlands

Accumulatorefabriek Varta NV, Molensingel 19, Venlo
Primary batteries, zinc–alkaline manganese dioxide, mercury–zinc, carbon–zinc Leclanché, magnesium types, lithium types, silver oxide–zinc, zinc chloride Leclanché, zinc–air; secondary batteries, alkaline, nickel–iron, nickel–cadmium, silver–zinc, sealed lead–acid.

Bataafsche Accufabriek BV, Bovendijk 317, Postbus 10 125, Rotterdam 3004
Primary batteries, zinc–alkaline manganese dioxide, mercury–zinc, carbon–zinc Leclanché, magnesium types, lithium types, silver oxide–zinc, zinc chloride Leclanché, zinc–air.

Chloride Batterijen BV, Postbus 117, Produktiestraat, Vlaardigen
Primary batteries, zinc–alkaline manganese dioxide; secondary batteries, nickel–cadmium, lead–acid.

New Zealand

Chloride Batteries New Zealand Ltd, PO Box No. 36-026, Morewa
Primary batteries, zinc–alkaline manganese dioxide; secondary batteries, nickel–cadmium.

Norway

Norsk Chloride A/S, Ostre Aker vei 203, Oslo 9, Norway
Primary batteries, zinc–alkaline manganese; secondary batteries, lead–acid, nickel–cadmium.

Portugal

Accumuladores Autosil SARL, Avenida 24 de Julho 26-B, Lisbon
Secondary batteries, alkaline.

South Africa

Raylite Battery Pty Ltd, Cor Apex and Detroit Streets, Apex Ind. Sites, PO Box 166, Benoni 1500, Transvaal
Secondary batteries, nickel–iron, nickel–cadmium.

Spain

Nife Espana SA, Hermosilla IIF, Madrid 9
Secondary batteries, nickel–cadmium.

Sweden

Nordiska Akkumulatorfabriker, NOACK AB, Kommendorsgatan 16, Box 5317, S-10246 Stockholm
Primary batteries, zinc–alkaline manganese dioxide, mercury–zinc, carbon–zinc Leclanché, magnesium types, lithium types, silver oxide–zinc, zinc chloride Leclanché, zinc–air.

Nife Jungner AB, 5572 01 Oskarshamm
Secondary batteries, alkaline.

Chloride Branza A13, Lautvornsgaten 8, Box 260, 651 07, Karlstad
Primary batteries, zinc–alkaline manganese dioxide; secondary batteries, lead–acid, nickel–cadmium.

United Kingdom

Chloride Automotive Batteries Ltd, Chequers Lane, Dagenham, Essex RM9 6PX
Primary batteries, zinc–alkaline manganese dioxide.

Chloride Power Storage, PO Box 5, Clifton Junction, Swinton, Manchester M27 2LR
Secondary batteries, alkaline, nickel–cadmium, lead–acid.

Duracell UK Technical Division, Duracell House, Gatwick Road, Crawley RH10 2PA
Primary batteries, mercury–zinc, silver–zinc, lithium solid electrolyte types.

Ever Ready Holdings Ltd (Berec), Ever Ready House, 1255 High Road, Whetstone, London N20 0EJ; also Ever Ready House, Wishart Street, Glasgow E1; Ever Ready House, Dewhurst Street, Manchester 8; Ever Ready House, Coneygree Industrial Estate, Burntree, Tipton, Staffs.; Ever Ready House, Narroways Road, Bristol 2; Ever Ready House, Station Approach, North Coulsdon, Surrey; Ever Ready Co. (Holdings) Ltd, Central Laboratories, St Ann's Road, London N15 3TJ
Primary batteries, zinc–alkaline manganese dioxide, mercury–zinc, carbon–zinc Leclanché.

Hawker Siddeley Electric Export Ltd, PO Box 72, Crompton House, Aldwych, London WC2B 4JH; also Crompton-Parkinson Vidor Works, River Drive, South Shields, Tyne and Wear NE33 2TR; 32 Duke Street, London SW1
Primary batteries, zinc–alkaline manganese dioxide, carbon–zinc Leclanché.

Radenite Batteries Ltd, 21 Bayton Road, Exhall, Coventry CV7 9EL, UK
Secondary batteries, lead–acid.

McMurdo Instrument Co. Ltd, Rodney Road, Portsmouth PO4 8SS
Primary batteries, silver oxide–zinc seawater batteries, copper-type seawater batteries.

Mallory Batteries Ltd, Gatwick Road, Crawley RH10 2PA
Primary batteries, zinc–alkaline manganese dioxide, mercury–zinc, carbon–zinc Leclanché, magnesium types, lithium types, silver oxide–zinc; secondary batteries, alkaline, nickel–iron.

Mine Safety Appliances Co. Ltd, Blairtummock Road, Queenlie Industrial Estate, Glasgow G33 4BT.
Primary batteries, thermal batteries, lithium–iodine.

Ray-o-Vac International Corporation, Westminster House, 97 St Mary Street, Cardiff
Primary batteries, carbon–zinc.

Sogea (SAFT) Batteries Ltd, Castle Works, Station Road, Hampton, Middlesex
Primary batteries, zinc–alkaline manganese dioxide, carbon–zinc Leclanché, lithium types (lithium–lead bismuthate, lithium–silver chromate); secondary batteries, nickel–hydrogen.

Varley Dry Accumulators Ltd, Alfreds Way, Barking, Essex, IG11 0TB (wholly owned subsidiary of Ever Ready Co. (Holdings) Ltd)
Secondary batteries, lead–acid.

Varta Batteries Ltd, Varta House, Hanger Lane, Ealing, London W5 1EH; also Hermitage Street, Crewkerne, Somerset TA18 8EY
Primary batteries, zinc–alkaline manganese dioxide, mercury–zinc, carbon–zinc Leclanché, magnesium types, lithium types, silver oxide–zinc, zinc chloride Leclanché, zinc–air; secondary batteries, sealed lead–acid, sealed nickel–cadmium.

United States

ACR Electronics Inc., 3901 North 29th Avenue, Hollywood 33020, Florida
Primary batteries, magnesium types, copper types.

Atlantic Battery Corporation Inc., 80–86 Elm Street, PO Box No. 172, Watertown 02172, Massachusetts
Secondary batteries, nickel–cadmium.

Battery Service Co., 6500 N 91st Plaza, Omaha 68122, Nebraska
Secondary batteries, nickel–cadmium.

Bright Star Industries, 600 Getty Avenue, Clifton 07015, New Jersey
Primary batteries, carbon–zinc Leclanché; secondary batteries, nickel–iron.

Burgess Inc., Freepost 61032, Illinois
Primary batteries, zinc–alkaline manganese dioxide, mercury–zinc, carbon–zinc Leclanché.

Catalyst Research Corporation, 1421 Clark View Road, Baltimore 21209, Maryland
Primary batteries, lithium–iodine, silver oxide–zinc, thermal batteries.

Chloride Inc., 7 Mallard Lane, North Haven 06473, Connecticut
Primary batteries, zinc–alkaline manganese dioxide.

Chloride of Silver Drycell Battery Co., 500–504 North Paca Street, Baltimore 21201, Maryland
Primary batteries, silver oxide–zinc.

Chloride Systems, PO Box 484, 118 Quinnipiac Avenue, North Haven 06473, Connecticut
Secondary batteries, nickel–cadmium.

Dual Life Inc., Simm Lane, PO Box 468, Newtown 06470, Connecticut
Secondary batteries, nickel–cadmium.

Duracell International Inc. Lithium Systems Division, 175 Clearbrook Road, Elmsford 10523, New York
Primary batteries, mercury–zinc, silver–zinc, lithium solid electrolyte types.

Eagle Picher Inc., 580 Walnut Street, Cincinnati 45202, Ohio; also Eagle Picher Electronics Division, PO Box 47·C' and Porter Street, Joplin 64801, Missouri
Primary batteries, magnesium perchlorate–manganese dioxide remote or manually activated types, silver–zinc, remotely activated silver–zinc, silver seawater batteries, copper seawater types, thermal batteries; secondary batteries, nickel–cadmium, silver–zinc, silver–cadmium, nickel–zinc.

Energy Development Associates, 110 W Whitcomb Avenue, Madison Heights, Michigan 48071
Secondary batteries, zinc–chlorine.

Ford Motor Co., PO Box 600, Wixom, Michigan 48096
Secondary batteries, lead–acid.

General Electric Corporation, PO Box 992, Gainesville 32601, Florida
Secondary batteries, nickel–cadmium, sealed lead–acid.

Gould Inc. Portable Battery Division, 931 Vandalia Street, St Paul 55114, Minnesota
Secondary batteries, nickel–cadmium.

W. R. Grace & Co., 7379 Route 32, Columbia, Maryland 21044
Battery separator manufacturers.

Hawaiian Pacific Battery Corporation, 250N Clark Avenue, Pomona 91767, California; also Hawaiian Pacific Battery Corporation, 1803 Kahai Street, Honolulu 96819, Hawaii
Primary batteries, zinc–alkaline manganese dioxide, mercury–zinc, carbon–zinc Leclanché, magnesium types; secondary batteries, nickel–cadmium.

Honeywell Inc. Power Sources Center, 104 Rock Road, Horsham 19044, Pennsylvania
Primary batteries, lithium types.

McGraw Edison Corporation, Edison Battery Division, PO Box 28, Bloomfield 07003, New Jersey
Primary batteries, zinc–air; secondary batteries, nickel–iron, nickel–cadmium.

Mallory Battery Corporation, South Broadway and Sunnyside Lane, Tarrytown 10591, New York
Primary batteries, zinc–alkaline manganese dioxide, mercury–zinc, carbon–zinc Leclanché, magnesium types, lithium types, silver oxide–zinc.

Marathon Battery Co., PO Box No. 8233, Waco, Texas
Secondary batteries, nickel–cadmium.

Molecular Energy Corporation, 132T Floral Avenue, Murray Hill 07974, New Jersey
Secondary batteries, silver–zinc, silver–cadmium.

Nife Inc., PO Box 100, George Washington Highway, Lincoln 02865, Rhode Island
Secondary batteries, alkaline, nickel–cadmium.

SAFT Corporation of America, 50 Rockefeller Plaza, New York 10020, New York
Primary batteries, zinc–alkaline manganese dioxide, carbon–zinc Leclanché, lithium types (lithium–lead bismuthate, lithium–silver chromate); secondary batteries, nickel–cadmium.

Shigoto Industries Inc., 350 Fifth Avenue, New York 10000, New York
Primary batteries, zinc–alkaline manganese dioxide, mercury–zinc, carbon–zinc Leclanché, magnesium types, copper types; secondary batteries, nickel–cadmium, silver–zinc.

Sonotone Corporation Battery Division, Saw Mill River Road, Elmsford 10523, New York
Secondary batteries, nickel–cadmium.

Union Carbide Corporation Consumer Products, 270 Park Avenue, New York 10017, New York
Secondary batteries, nickel–cadmium, silver–oxide.

Wisco Division ESB Inc., 1222 18th Street, Racine 53403, Wisconsin
Secondary batteries, silver–zinc, lead chloride, cuprous chloride.

Union Carbide Inc., 270 Park Avenue, New York 10017, New York
Primary batteries, zinc–alkaline manganese dioxide, mercury–zinc, carbon–zinc Leclanché, silver oxide–zinc, zinc chloride Leclanché; secondary batteries, alkaline, nickel–cadmium.

Varta Batteries Inc., 85 Executive Boulevard, Cross Westchester Executive Park, Elmsford 10523, New York
Primary batteries, zinc–alkaline manganese dioxide, mercury–zinc, carbon–zinc Leclanché, magnesium types, lithium types, silver oxide–zinc, copper types, zinc chloride Leclanché; secondary batteries, alkaline, nickel–iron, nickel–cadmium, silver–zinc, sealed lead–acid.

Yardney Electric Corporation, 82 Mechanic Street, Pawcatuck 02891, Connecticut
Primary batteries, magnesium types, lithium types; secondary batteries, silver–zinc, silver–cadmium.

Yardney Electric Corporation, Power Sources Division, 3850 Olive Street, Denver 80207, Colorado
Secondary batteries, nickel–hydrogen, silver–zinc, silver–cadmium, nickel–zinc.

West Germany

Akkumulatorenfabrik Sonnenschein GmbH, 64–70 Budingen
Secondary batteries, sealed lead–acid.

Chloride Batteries GmbH, Otto Hahn Strasse 1, 6084 Gerscheim
Secondary batteries, nickel–cadmium.

Diamon Werke GmbH, Sellerstrasse 14, 1000 Berlin 65; also 5000 Koln 50, Ossendorf; Kohlstrasse 37, Cologne 36; Postlach 3 000 420 and Selterstrasse 14, 1000 Berlin 65
Primary batteries, zinc–alkaline manganese dioxide; secondary batteries, nickel–cadmium.

Duracell Deutschland Technical Division, D-5020 Frechen, Hermann-Seger-Strasse 13
Primary batteries, mercury–zinc, silver–zinc, lithium solid electrolyte types.

Friemann Wolf Silberkraft, Leichtakkumulatoren, 41 Duisburg 1, Meidericher Strasse 6–8, Postschliesstfach 100703
Primary batteries, silver–zinc and silver chloride–magnesium seawater batteries; secondary batteries, nickel–cadmium, silver–zinc, silver–cadmium.

Gottfried Hagen AG, Rolshoverstrasse 95/101, Postfach 91 01-10-D5
Secondary batteries, nickel–cadmium.

Nife Jungner GmbH, Nicholstrasse 2–6, 1000 Berlin 46; also Blucherstrasse 33–35, 2000 Hamburg; Benrotherstrasse 7, 4000 Dusseldorf 13; Ostwaltstrasse 48, 6000 Frankfurt; Nockherstrasse 50, 8100 Munich 90
Secondary batteries, nickel–cadmium, alkaline.

Varta Batterie AG, Leineufer 51, Postfach 210 540, 3000 Hanover 21; also Postfach 63, 7090 Ellwargen, Jagst
Primary batteries, zinc–alkaline manganese dioxide, mercury–zinc, carbon–zinc Leclanché, magnesium types, lithium types, silver oxide–zinc, zinc chloride Leclanché, zinc–air; secondary batteries, alkaline, nickel–iron, nickel–cadmium, silver–zinc, sealed lead–acid.

Yugoslavia

Jugometal, Deligradska 28/IX, Belgrade 11001
Secondary batteries, nickel–cadmium.

Zambia

Chloride Zambia, PO Box 1892, Kitwe
Primary batteries, zinc–alkaline manganese dioxide; secondary batteries, nickel–cadmium.

Glossary

Activated stand life The period of time, at a specified temperature, that a secondary cell can be stored in the charged condition before its capacity falls below a specified level.

Activation The process of making a cell without electrolyte functional, either by introducing an electrolyte or by immersing the cell into an electrolyte. Activation of thermally activated reserve cells involves heating to melt a solid.

Active materials The substances of a positive or negative plate that react to produce current.

Ampere hours Product of current in amperes multiplied by time current is flowing; capacity of a cell or battery is usually expressed in ampere hours.

Anode An electrode at which an oxidation reaction (loss of electrons) occurs; in secondary cells either electrode may become the anode, depending on the direction of current flow.

Available capacity *See* Capacity.

Baffle A non-conductive barrier inserted in a vented cell above the plate pack and used as a reference for electrolyte level; it also protects the top edges of the plates from objects dropped into the vent cap hole.

Battery One or more cells connected to form one unit and having provisions for external connections.

Battery case Box or enclosure which contains the cells, associated connectors and hardware.

C_p Effective parallel capacitance.

C **rate** Discharge or charge current rate in amperes, numerically equal to rated capacity of a cell in ampere hours, rated at the 1 h discharge rate.

C_n **rate** Charge or discharge current numerically equal to capacity, rated at the n h discharge rate.

XC_n **rate** Charge or discharge current numerically equal to X times the capacity, rated at the n h discharge rate.

Cadmium electrode *See* Negative plate.

Capacity Ampere hours available from a fully charged cell or battery.

Capacity, available Total capacity that may be obtained at defined charge and discharge rates and their associated environmental conditions.

Capacity, deliverable *See* Capacity, available.

Capacity, functional loss of Reduction in cell capacity due to non-standard charging or discharging parameters such as cell temperature, current, and discharge voltage cut-off. *See also* Failure, function.

Capacity, nameplate *See* Capacity, rated.

Capacity, nominal *See* Capacity, rated.

Capacity, permanent loss of Reduction in cell capacity from 'as new' value, under standard rating conditions; not recoverable by reconditioning. *See also* Failure, permanent.

Capacity, rated A designation by the battery manufacturer which helps identify a particular cell model and also provides an approximation of capacity; usually expressed in ampere hours at a given discharge current.

Capacity, residual Capacity remaining at a particular point in time after any set of operating conditions, usually including a partial discharge or long rest.

Capacity, restorable *See* Capacity, temporary loss of.

Capacity, temporary loss of Reduction in cell capacity that is recovered when cell is subjected to several reconditioning cycles. *See also* Failure, reversible.

Capacity, useful *See* Capacity, available.

Capacity reconditioning *See* Reconditioning.

Cathode An electrode at which a reduction reaction (gain of electrons) occurs; in secondary cells either electrode may become the cathode, depending on the direction of current flow.

Cell Electrochemical device, composed of positive and negative plates, separator and electrolyte, that is capable of storing electrical energy; when encased in a container and fitted with terminals, it is the basic 'building block' of a battery.

Cell case *See* Container.

Cell reversal Reversing of polarity of terminals of a cell in a multi-cell battery due to overdischarge.

Charge Return of electrical energy to a battery.

Charge, state of Residual capacity of a cell expressed in terms of the fully charged capacity.

Charge rate The current at which a secondary cell or battery is charged, expressed as a function of the cell or battery's rated capacity; for example, the 20 h charge rate of a 10 A h cell would be equal to $C/20 = 10/20 = 0.5\,A$.

Charge retention The tendency of a charged cell to resist self-discharge.

Charger Device capable of supplying electrical energy to a battery.

Connector Electrical conductor which joins individual cells together in a battery.

Constant current Charging method in which current does not change appreciably in magnitude, regardless of battery voltage or temperature.

Constant potential Charging method that applies a fixed voltage to a cell; often abbreviated to CP.

Container Cell enclosure in which plates, separator and electrolyte are held; it consists of the cell jar and cover, which are permanently joined.

Contaminant Undesirable element, usually in the electrolyte, that reduces the capacity of the cell; in vented cells, contaminants can be introduced by use of tap water or operation without the vent cap.

Contraction The shrinkage of active material.

Corrosion The oxidation of a metal electrode.

Coulometer Electrochemical or electronic device, capable of integrating current–time, used for charge control or measurement.

Counter electromotive force A voltage opposite to the applied voltage, also referred to as back e.m.f.

Current density The amount of current per unit area passing from one plate to another (A/cm^2).

Cut-off voltage Voltage at which a discharge or charge is terminated.

Cycle One sequence of discharge and recharge. Deep cycling requires that all the energy to an end-voltage established for each system is drained from the battery on each discharge; in shallow cycling the energy is partially drained in each discharge, that is, discharge up to 50% depth of capacity.

Cycle life In a secondary storage battery, the number of cycles the battery may experience before its capacity falls to a point considered a failure. *See also* Failure.

Cylindrical cell A cylinder-shaped sealed wound cell containing a high-pressure safety vent.

Dead band The range of temperatures between the point at which the thermostat opens and the point where it recloses (resets). If the temperature first exceeds the point at which the switch opens and then drops below this point, the switch remains open within the dead band until the temperature falls below the reset point.

Deep cycling *See* Reconditioning.

Deep discharge The condition in which a cell is discharged to 0.5 V or less at low rate, that is, to below cut-off voltage or withdrawal of at least 80% of rated capacity.

Deionized water Water that has been freed of ions by treatment with ion-exchange resins.

Depolarizer A material to offset a polarizing effect of an electrode.

Depth of discharge (DOD) The percentage of rated capacity to which a cell or battery is discharged; for example, if 5 A h of capacity is discharged from a 100 A h battery, the depth of discharge is 5%.

Discharge Withdrawal of electrical energy to end-point voltage before the cell or battery is recharged.

Discharge rate *See* Rate.

Discharge voltage The voltage of a battery during discharge.

Discharging The withdrawing of electrical energy from a battery or cell.

Distilled water Water that has been freed of ions by a process of vaporization and subsequent condensation.

Drain Withdrawal of current from a cell or battery.

Dry Indication that the electrolyte in a cell is immobilized, being either in the form of a paste or gel or absorbed in the separator material, and/or that a cell may have insufficient electrolyte.

Dry, charged An electrochemical system in which the electrodes are in a charged state, ready to be activated by the introduction of electrolyte.

Dry, uncharged Indicates that a cell must be activated and receive initial formation cycles before it can be brought into use.

Dump timed charge (DTC) A charging method in which the cell is first discharged at a rate–time combination equal to or greater than the charge rate–time combination which immediately follows.

Duty cycle The condition and usage to which a battery is subjected during operation, consisting of charge, overcharge, rest and discharge.

E_0 *See* Equivalent no-load voltage.

Effective internal resistance, R_e The apparent opposition to current flow within a battery that manifests itself as a drop in battery voltage proportional to the discharge current; its value depends on battery design, state of charge, temperature and age.

Electrode Conducting body and the active materials in which the electrochemical reaction occurs.

Electrode potential The voltage developed by a single plate, either positive or negative; the algebraic difference in voltage of any two electrodes equals the cell voltage.

Electroformation The conversion of the material in both the positive and negative plates to their respective active material.

Electrolyte Fluid used in a cell as a medium for movement of ions.

End-of-charge voltage The voltage of the battery at termination of a charge but before the charge is stopped.

End-of-discharge voltage The voltage of the battery at termination of a discharge but before the discharge is stopped.

End-point voltage Voltage below which connected equipment will not operate, or below which operation of a cell is not recommended; frequently interchangeable with 'cut-off voltage'.

Energy Output capability, ampere hours capacity times average closed-circuit discharge voltage, expressed as watt hours.

Energy density A figure of merit in common use with batteries, expressing the stored energy as a function of the weight or volume (W h/kg or W h/dm^3); rate dependent.

Entrainment Process whereby gases generated in the cell carry electrolyte out through the vent cap.

Environmental conditions External circumstances to which a cell or battery may be subjected, such as ambient temperature, humidity, shock, vibration and altitude.

Equalization *See* Reconditioning.

Equivalent circuit   A circuit presented to simulate the electrical behaviour of a cell.

Equivalent internal resistance   *See* Effective internal resistance.

Equivalent no-load voltage, E_0 The numerical value of the source voltage in the equivalent circuit.

Fading   The long-range loss of capacity with use.

Failure   The condition in which a battery is unable to perform satisfactorily.

Failure, function   Condition in which the battery has caused the end-use device to fail to function at the performance level expected.

Failure, permanent   A non-reversible condition that makes the battery essentially useless. There is no recognized standard method of rating or defining exactly when a cell has reached the end of its life; however, the end of useful life is defined as that condition when the cell cannot perform above 50% of its rated ampere hour capacity with all charging and discharging done at room temperature.

Failure, reversible   A failure condition that may be corrected through the application of certain electrical procedures.

Failure, temporary   *See* Failure, reversible.

Fast-charge battery   A battery that can be charged at the fast-charge rate and that gives a suitable signal which can be used to terminate the fast-charge current without damage to the battery.

Fast charging   Rapid return of energy to a battery at the C rate or greater.

Float charging   The use condition of a storage battery wherein charge is maintained by a continuous, long-term constant-potential charge.

Flooded cell   *See* Vented cell.

Flooding   Filling of pores of a porous electrode with electrolyte solution, thereby minimizing access of gases to electrode surface.

Form factor   Battery configurations that may be created by interconnecting cells in various arrangements.

Function failure   *See* Failure, function.

Gassing   The liberation of hydrogen and/or oxygen gases from a cell.

Grid   The metallic part of a battery plate which retains the active material.

Group   An assembly of positive or negative plates which fit into a cell.

High-resistance short   *See* Short-circuit.

High-rate discharge   Withdrawal of large currents in short intervals of time, usually at a rate less than 1 h.

I_{mp} *See* Maximum-power discharge current.

Impedance   An a.c. circuit's apparent opposition to current; consists of reactance and ohmic resistance. For the equivalent phenomenon in a d.c. battery, *see* Effective internal resistance.

Initial drain   Current that the cell or battery supplies when first placed on load.

Intermittent short   *See* Short, intermittent.

Internal impedance   The opposition of a cell or battery to an alternating current of a particular frequency, usually 60 Hz. (Note: the internal impedance may vary with the state of charge.)

Internal resistance   Opposition to direct current flow in a cell; this can be measured by dividing the difference between the open-circuit voltage and the voltage at a particular current, by that current. (Note: the internal resistance may vary with the current and state of charge.)

Jar   The bottom portion of the cell container which mates with the cell cover.

KOH   *See* Potassium hydroxide.

Life   The duration of satisfactory performance, measured as usage in years or as the number of charge/discharge cycles.

Load   The current drain through a fixed resistance to which the battery will be subjected.

Low-rate discharge   Withdrawal of small currents for long periods of time, usually longer than 1 h.

Low-resistance short   *See* Short, low-resistance.

Maintenance   The care and procedures necessary to keep a battery in a usable condition, such as reconditioning and water addition to electrolyte of a vented cell.

Manufacturing variations   Differences in performance characteristics between products of the same design, attributable to process deviations within expected tolerances.

Maximum-power discharge current, I_{mp} The discharge rate at which the terminal voltage is equal to one-half of E_0 and at which maximum power (energy rate) is transferred to the external load.

Memory effect   A phenomenon in which a nickel–cadmium battery, operated in successive cycles of identical depth of discharge, temporarily renders the rest of its capacity inaccessible at normal voltage levels.

Mid-point voltage   The battery voltage at the halfway point in the discharge between the fully charged state and the fully discharged state of a cell.

Negative electrode   *See* Negative plate.

Negative plate   The plate that has an electrical potential below that of the other plate during normal cell operation.

Nickel electrode   *See* Positive plate.
Nominal capacity   *See* Capacity, rated.
Nominal voltage   The mid-point voltage observed across a battery during discharge at a selected rate, usually at the $0.2C$ or $0.1C$ rate.

On-charge voltage   Voltage of a cell while on charge.
Open-circuit voltage   The no-load voltage of a cell or battery measured with a high-impedance voltmeter at about room temperature ($21°C$). It is also implied that, in the case of a primary, the cell is freshly manufactured, and, for a secondary, fully charged.
Overcharge current   The charging current flowing to the battery after all the active material has been converted into a dischargeable state.
Overcharging   Continuing charge after the battery has accepted its maximum amount of charge. In a vented cell, a result will be decomposition of water in the electrolyte into hydrogen and oxygen gases. In a sealed cell, a result will be increased cell temperature.
Oxidation   The release of electrons, by the cell's active material, to the external circuit. During discharge, active material at the negative electrode is oxidized.
Oxygen recombination   The electrochemical process in which oxygen generated at the positive plate during overcharge is reacted (reduced) with water at the negative plate at the same time, generating heat.

Parallel   Electrical term used to describe the interconnection of batteries in which all the like terminals are connected together.
Permacell   A registered trademark of the General Electric Corporation which designates a rechargeable nickel–cadmium battery-charger system using sealed cylindrical cells; offered as an alternative to popular sizes of primary cell.
Permanent failure   *See* Failure, permanent.
Peroxide voltage   Initial voltage observed when current load is first applied to a fully charged cell; this voltage is normally higher than the plateau voltage.
Plaque   A porous body of sintered metal used as a current collector and holder of electrode active materials.
Plate   Common term for an electrode.
Plateau voltage   Level portion of discharge curve, varying with discharge rate.
Polarization   The increased internal resistance of a cell which occurs during discharge.
Polarity   Electrical term used to denote the relative voltage relationship between two electrodes.
Positive electrode   *See* Positive plate.

Positive plate   The plate that has an electrical potential higher than that of the other plate during normal cell operation.
Potassium hydroxide   A chemical compound, symbol KOH, which, when mixed with pure water in the correct proportions, is the electrolyte solution used in nickel–cadmium cells and other alkaline cell systems.
Powerup-15 battery   The General Electric trademark for a nickel–cadmium battery which permits charging at rates up to the $4C$ rate to at least 90% of rated capacity in 15 min at room temperature. As the battery reaches full charge, the fast-charge rate is switched by the charger to a lower, continuous overcharge rate by sensing either the battery voltage or temperature.
Primary cell   A cell designed to be used only once, then discarded. It is not capable of being returned to its original charge state by the application of current.

Quick charge   Charging rate that ranges from $0.2C$ to $0.5C$ rate.
Quick-charge battery   A battery that can be charged fully in 3–5 h by a simple, constant-current charger and is capable of continuous overcharge at this quick-charge rate.

R_e *See* Effective internal resistance.
Rate   Amount of charge or discharge current, frequently expressed as a fraction or multiple of the 1 h rate, C.
Rated capacity   *See* Capacity, rated.
Rating   *See* Capacity.
Recharge   Return of electrical energy to a battery.
Rechargeable   Capable of being recharged; refers to secondary cells or batteries.
Recombination   The chemical reaction of gases at the electrodes to form a non-gaseous product.
Reconditioning   Maintenance procedure consisting of deep discharge, short and constant-current charge used to correct cell imbalance which may have been acquired during battery use.
Reduction   The gain of electrons; in a cell, refers to the inward flow of electrons to the active material.
Relative density   The ratio of the mass of a given volume of a substance to the mass of an equal volume of water at a temperature of $4°C$.
Resealable   In a cell, refers to a safety vent that is capable of closing after each pressure release, in contrast to the non-resealable one-shot vent.
Residual capacity   *See* Capacity, residual.
Resistance   *See* Effective internal resistance.
Reversal   *See* Cell reversal.
Reversible failure   *See* Failure, reversible.
Reversible reaction   A chemical change that takes place in either direction, as in the reversible reaction for charging or discharging a secondary battery.

Sealed cell A cell that is free from routine maintenance and can be operated without regard to position; all reactants are retained within the container.

Secondary battery A system that is capable of repeated use by using chemical reactions that are reversible; that is, the discharged energy may be restored by supplying electrical current to recharge the cell.

Self-discharge The spontaneous decomposition of battery materials from charged to discharged state. That rate at which a primary or secondary battery or cell loses service capacity when standing idle.

Semipermeable membrane A porous film that will pass selected ions.

Separator Material which provides separation and electrolyte storage between plates of opposite polarity.

Series Electrical term used to describe the interconnection of cells or batteries in such a manner that the positive terminal of an individual cell is connected to the negative terminal of the next cell.

Shallow discharge A discharge of a battery equal to only a small part of its total capacity.

Shedding The loss of active material from a plate during cycling.

Shelf life A measure of the active life of a battery in storage.

Short, high-resistance A short measuring $20\,\Omega$ or more.

Short, intermittent A condition where the open-circuit voltage of a battery is unstable when mechanically shocked; an intermittent short can be either low or high resistance.

Short, low-resistance A short measuring less than $20\,\Omega$.

Short-circuit The condition in a battery where two plates of opposite polarity make electrical contact with each other.

Sintered-plate electrode A sturdy plate formed by sintering (agglomerating metallic powders by heating). The resulting electrode has a porous structure with a large surface area containing active materials which are accessible to be charged and discharged at high rates.

Slow charge 'Overnight' return of energy to a battery at $0.05-0.1C$ rates.

Soak time That time required for the electrolyte to be absorbed into the active materials after activation.

Spalling The flaking off of active material from a plate or electrode.

Split rate charge A charging method in which the battery is charged at a high rate and then automatically reduced to a lower charge rate as the battery approaches full charge.

Stack Flat plates assembled in a parallel orientation.

Standby charge A low overcharge current rate, of the order of $0.01-0.03C$, applied continuously to a vented-cell battery to maintain its capacity in a ready-to-discharge state. Often called 'trickle' charging.

Starved cell A cell containing little or no free fluid electrolyte solution; this enables gases to reach electrode surfaces readily, and permits relatively high rates of gas recombination. *See* Sealed cell.

State of charge Residual capacity of a cell expressed in terms of fully charged capacity.

Tab A battery terminal, often containing a hole for wire connection.

Temperature, ambient The average temperature of the battery's surroundings.

Temperature, cell The average temperature of the battery's components.

Temperature cut-off (TCO) A method of switching the charge current flowing to a battery from fast charge to topping charge by means of a control circuit in the charger that is activated by battery temperature.

Temporary failure *See* Failure, reversible.

Terminals The external part of a cell or battery from which current can be drained.

Thermal runaway A condition whereby a battery on constant-potential charge at elevated temperature will destroy itself through internal heat generation, which is caused by high overcharge currents in constant-potential charging.

Topping charge A reduced rate charge that completes (tops) the charge of a cell and can be continued in overcharge without damaging the cell.

Trickle charge *See* Standby charge.

Trough voltage The instantaneous open-circuit voltage of a battery that is charged by a pulsating current.

Unactivated storage life The period of time, under specified conditions of temperature and environment, in which a dry-charged cell can stand before deteriorating below a specified capacity.

Undercharging Applying less than the amount of current required to recharge a battery.

Vent A normally sealed mechanism that allows the controlled escape of gases from within a cell.

Vent cap A device that contains a vent and fits on to the top of a cell; normally used on vented cells to allow access for electrolyte level adjustment.

Vented cell A heavy-duty cell design in which the vent operates at low pressure during the normal duty cycle to expel gases generated in overcharge. A vented cell plate pack contains flat plates separated by a gas barrier and separator, completely immersed in electrolyte. Often called a 'flooded' cell.

Voltage cut-off (VCO)   A method of switching the charge current flowing to a battery from fast charge to topping charge by means of a control circuit in the charger that is activated by battery voltage.

Voltage limit   In a charge-controlled battery, the limit beyond which battery potential is not permitted to rise.

Voltage–temperature cut-off (VTCO)   A method of switching the charge current flowing to a battery from fast charge to topping charge rate by means of a control circuit in the charger that is activated by either battery voltage or temperature.

Wet   Indicates that the liquid electrolyte in a cell is free flowing.

Wet charged stand   Period of time a battery can stand in the wet charged condition while retaining a large percentage of its capacity.

Wet formed discharge   Indicates that a cell is activated, formed and discharged. The cell need only be recharged for use.

Wet shelf life   Period of time a battery can stand in the wet discharged condition before losing charging capability.

Wound   The interior cell construction in which plates are coiled into a spiral.

Yield   The energy output of a battery.

Battery Standards

The International Electrochemical Commission (IEC) have prepared a battery standards specification: *EEC Publication 86: Primary Cells and Batteries*, Parts 1 and 2, 1975. The relevant British Standard is BS 397. Both are available from the British Standards Institution, 2 Park Street, London W1A 2BS.

The American National Standard, ANSI C18.1 is available from ANSI, 1430 Broadway, New York 10018, NY, USA.

The Japanese Standards Association issues Standard J15 8501 *Dry Cells and Batteries*, J15 8508 *Mercury Cells and Batteries*, J15 8509 *Alkaline-Manganese Dioxide Cells and Batteries*, J15 8510 *Silver Oxide Cells and Batteries*, and J15 8511 *Alkaline Primary Cells and Batteries*. All are available through Japanese Embassies or direct from the Japanese Standards Association, 1–24 Atasak 4 Chome, Minato-ku, Tokyo, Japan.

British Standards

AU 118 1965, *Storage, Shipment and Maintenance of Lead Acid Batteries for Motor Vehicles*. 4 pages. Gr 2 (£7.00).

BS 1335: 1968, *Air Depolarized Primary Cells – Dimensions of Containers*. Dimensions, materials of electrodes for dry, wet, inert cells and batteries. Details of electrolyte for wet cells. Performance, test methods. 20 page A5 size. Gr 4 (£10.50).

BS 5932: 1980, *Specification for Sealed Nickel–Cadmium Cylindrical Rechargeable Single Cells*. Performance requirements, cell designation, terminations, marking, dimensions, charging requirements and methods of test for low and high rate of discharge at normal and low temperature, charge retention, overcharge, life cycle, mechanical and after storage. AMD 3418, May 1980. 4 pages. Gr 3 (£9.50).

BS 6260: 1982, *Specification for Open Nickel–Cadmium Prismatic Rechargeable Single Cells*. Specifies characteristics, technical requirements and test methods to be used for the type testing of open nickel–cadmium prismatic rechargeable single cells. AMD 4386, October 1983. (Gr 1), 4 pages. Gr 3 (£13.70).

BS 397: Part 1: 1985, *Specification for General Requirements*. Applies to primary cells and batteries based on any electrochemical system, and aims to ensure the electrical and physical interchangeability of products from different manufacturers, to limit the number of battery types and to define a standard of quality and provide guidance for its assessment. 20 pages. Gr 8 (£31.00).

BS 397: Part 2: 1985, *Specification Sheets*. Gives dimensions, drawings (where applicable), discharge conditions and applications of primary batteries approved for a standardization by the IEC. 137 pages. Gr 11 (£44.00).

BS 397: *Primary Batteries* Part 3: 1985. Specification for batteries not included in Parts 1 and 2. Specifications for batteries manufactured in the UK which are not standardized internationally and therefore are not included in Parts 1 and 2 of this standard. Gives dimensions, discharge conditions and drawings and terminal details where applicable. No current standard is superseded. 46 pages. Gr 7 (£22.00).

BS 6115, *Sealed Nickel Cadmium Prismatic Rechargeable Single Cell Batteries*.

IEC Standards

Primary batteries: applies to dry primary cell batteries with the object of enumerating the types corresponding to the most current needs, defining their characteristics, ensuring their interchangeability and limiting their number.

86-1 (1982) Part 1: *General*. (Fifth edition) 47 pages. Applies to primary cells and batteries on any electrochemical system. Specifies nomenclature, dimensions, terminals, marking, test conditions and service output requirements for each battery in order to:

1. Ensure interchangeability.
2. Limit the number of types.
3. Define a standard of quality. (£60.83)

Amendment No. 1 (1984), 5 pages.

86-2 (1982) Part 2: *Specification sheets*. (Fifth Edition) 144 pages. Specifies dimensions together with outline drawings of batteries, conditions and minimum duration of discharges and applications. Defines three new coding letters: S = standard, C = high capacity, P = high power. Includes new accelerated application tests for clocks and automatic cameras. This is a loose-leaf publication and supplements, containing new and revised sheets, are issued from time to time. (£112.50).

86-2A (1984) *First Supplement to Publication 86-2 (1982)*. 62 pages. Specifies dimensions together with outline drawings of batteries, conditions and minimum duration of discharges and applications. Defines three new coding letters: S = standard, C = high capacity, P = high power. This is a loose-leaf publication, and supplements containing new and revised sheets, to be inserted in Publication 86-2, are issued from time to time.

95: *Lead Acid Starter Batteries*. Applies to lead–acid accumulator batteries with a rated voltage of 6 or 12 V, used primarily as a source of starting and ignition current for internal combustion engines and also for the auxiliary installations of internal combustion engine vehicles.

95-1 (1980) Part 1: *General Requirements and Methods of Test.* (Fourth edition) 29 pages. Defines international specifications for several groups of batteries according to the general type of application and the climatic conditions. (£27.50).

95-2 (1984) Part 2: *Dimensions of Batteries and Dimensions and Marking of Terminals.* (Third edition) 23 pages. Applies to lead–acid batteries used for starting, lighting and ignition of passenger automobiles and light commercial vehicles with a nominal voltage of 12 V fastened to the vehicles by means of ledges on the long sides of the battery case. Two alternative admissible means are specified in Section 3. It specifies:

1. The main dimensions of starter batteries of four standard series.
2. The location of the positive and negative terminals with respect to the fastening system.
3. The dimensions of tapered terminals of starter batteries.
4. The marking of the polarity.

(£26.66).

254: *Lead Acid Traction Batteries.*

254-1 (1983) Part 1: *General Requirements and Methods of Test.* (First edition) 19 pages. Applies to lead–acid traction batteries intended for installation in electric traction vehicles or materials handling equipment and lays down, in general terms, their main characteristics and methods for their testing. (£24.16).

254-2 (1973) Part 2: *Dimensions of Traction Battery Cells.* (First edition) 10 pages. Gives the outer maximum dimensions (overall) of all traction battery cells. (£13.33).

254-2A (1974) First Supplement. 2 pages. Gives additional dimensions for extra-low cells.

285 (1983) *Sealed Nickel–Cadmium Cylindrical Rechargeable Single Cells.* (Second edition) 25 pages. Specifies tests and requirements for sealed nickel–cadmium cylindrical rechargeable single cells, suitable for use in any composition. (£28.33).

509 (1976) *Sealed Nickel–Cadmium Button Rechargeable Single Cells.* (First edition) 11 pages. Applies to elements suitable for use in any position, in electric or electronic appliances. Specifies the designation system, the electrical characteristics, marking and dimensions. (£15.83).

622 (1978) *Sealed Nickel–Cadmium Prismatic Rechargeable Single Cells.* (First edition) 11 pages. Establishes a system of designation of the cells. Defines the electrical characteristics and the relevant test methods, as well as the marking of the cells. (£13.33).

623 (1978) *Open Nickel–Cadmium Prismatic Rechargeable Cells.* (Second edition) 21 pages.

Gives the designation system of the cells, dimensions, electrical and mechanical tests as well as the marking applicable. (£13.33).

428 (1973) *Standard Cells* (First edition) 23 pages. Applies to two kinds of standard cells used as electromotive force references, namely saturated and unsaturated standard cells, and deals with test conditions relating to certification and requirements for their electrical and mechanical characteristics. (£22.50).

The following is an extract from *The IEC System of Battery Designation*, prepared by IEC Technical Committee No. 35.

Introduction

The last decade has seen a large expansion in the range of portable battery-operated consumer goods; often these are sold and used in countries other than those in which they were manufactured. During the same period the range of electrochemical systems available for battery use has multiplied and even within a system different application grades have appeared. These changes when coupled with the increase in travel between countries have led to the user being confused as to the correct type of replacement battery to purchase.

The battery purchaser requires a simple, unambiguous symbol which defines internationally the important characteristics of the battery. These characteristics are:

(a) Dimensions and terminals (physical interchangeability).
(b) Voltage and electrical performance (electrical interchangeability).

At the present time there are three possible sources of such information.

The battery manufacturer publishes literature describing his products and sometimes includes recommendations for the battery to be used in specific equipment. In the nature of things this information is not complete.

The equipment manufacturer may include in the equipment a label stating the specific type of battery to be used, or the size required. Often a more detailed list of approved batteries is given in the equipment instruction manual but again this is likely to be incomplete. Furthermore, the information may lead to frustration when the equipment originates in a different country from that in which the batteries are purchased.

Finally assistance may be given at the sales point by experienced and qualified technicians, although more usually the sales assistant is non-technical. It is apparent that the casual purchaser is rarely able to derive adequate benefit from the sources described above. In recognition of this, several attempts have been made to establish systems at a national level.

It is the aim of the IEC system to overcome the deficiencies described and the limitations imposed by national boundaries.

Examples

In the IEC system fully described later and in Publication 86, a series of numbers and letters are used to give a unique reference to each size of battery, its terminals, voltage and electrochemical system. The following examples illustrate the use of the IEC system.

(a) R6 (commonly used in small transistor radios), Figure 1 The letter R indicates that the battery is cylindrical and combined with the number 6 signifies a particular size with closely defined terminals. The absence of another letter shows that the electrochemical system is manganese dioxide–ammonium chloride, zinc chloride–zinc. This system has a nominal voltage of 1.5 V and since no preceding number is given the battery voltage is identical with that of the electrochemical system, namely 1.5 V.

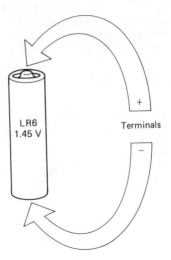

Figure 2

(c) 3R12 (commonly used in pocket lamps), Figure 3 The complete symbol defines the size, terminals and voltage of the battery. As in Example (a) the electrochemical system is manganese dioxide/ammonium chloride, zinc chloride/zinc but the presence of the number 3 denotes a battery voltage of three times the nominal voltage of the system, i.e. 4.5 V.

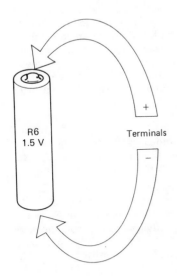

Figure 1

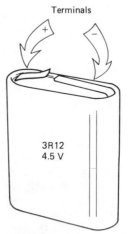

Figure 3

(b) LR6 (commonly used in cine-cameras), Figure 2 R6, as in the preceding example, defines the same shape, dimensions and terminals. The additional letter L indicates that the electrochemical system is manganese dioxide/alkali metal hydroxide/zinc having a nominal voltage of 1.45 V.

(d) 6F22 (commonly used in pocket size transistor radios), Figure 4 As in Example (c) the complete symbol defines the size, terminals and voltage of the battery. The letter F denotes a flat cell construction which is convenient for compact multicell batteries. As in the preceding example the absence of an additional letter shows that the electrochemical

system is manganese dioxide/ammonium chloride, zinc chloride/zinc. The number 6 indicates that the battery voltage is six times 1.5 V, i.e. 9 V.

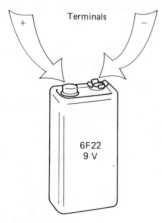

Figure 4

(e) MR07 (commonly used in hearing-aids), Figure 5 The new feature in this example is the letter M which denotes that the electrochemical system is mercuric oxide/alkali metal hydroxide/zinc of 1.35 V nominal, which in this case is also the battery voltage.

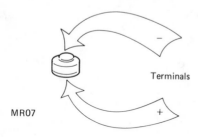

Figure 5

(f) MR44 (commonly used in watches), Figure 6 This battery is identical in all respects with the MR07 except that a closely controlled and precisely defined profile is mandatory. This ensures satisfactory fitment in the battery compartment of watches. For this reason the battery is defined in conjunction with a profile gauge and is given a different number. The IEC system thus defines the battery construction, its precise dimensions including profile and contact arrangements, the electrochemical system used and the battery voltage.

Whilst at first sight such a system appears to be extremely complex and perhaps difficult to understand, it does not present any difficulties to battery users since all they need to know is the IEC number.

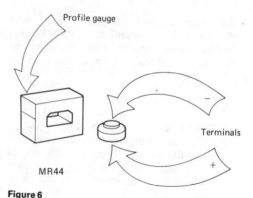

Figure 6

For example, a battery which is familiar to all users of pocket size transistor radios and is known by such different references as PP3, 216, R0603, 438. 333, 006P 1604, TR1, 410, LPI-9, Tiber, 485, TM146 can be simply identified by the symbol 6F22. Another instance is the popular size of a lighting battery (approximately 34 mm diameter × 61 mm height) which is identified in the IEC system as the R20, whereas at present it is variously referred to as the 212, SP2, 950, UMI, D, 250, 211, 734, 6TIL, LPV 2, 152, 209 and 70 in different parts of the world.

Since the appearance of the IEC designation upon a product is a claim by the manufacturer that the product complies with the requirements of the relevant standard, which includes a minimum standard for electrical performance, the purchaser has the additional guarantee that batteries will not only fit the equipment but that they will also provide reasonable service.

The major factor controlling the price and electrical performance of batteries is the electrochemical system employed. The IEC recommendations include at present six electrochemical systems (Table 1), whose principal characteristics are given below:

(a) The manganese dioxide/ammonium chloride, zinc chloride/zinc system has the lowest prime cost. It is very efficient under conditions of intermittent use at low and medium current drains. For this reason transistor radios and torches are almost exclusively used with batteries of this system.

(b) The air or oxygen/ammonium, zinc chloride/zinc system (letter reference A) is another low price system, suitable for applications requiring a high ampere-hour capacity under low current drain conditions, where the volume is unrestricted. Railway signalling and electric fence equipment employ this system.

(c) The manganese dioxide/alkali metal hydroxide/zinc system (letter reference L) has a medium

Table 1

Letter	Positive electrode	Electrolyte	Negative electrode	Nominal voltage (V)
–	Manganese dioxide	Ammonium chloride, zinc chloride	Zinc	1.5
A	Airo or oxygen	Ammonium, zinc chloride	Zinc	1.4
L	Manganese dioxide	Alkali metal hydroxide	Zinc	1.45
M	Mercuric oxide	Alkali metal hydroxide	Zinc	1.35
N	Mercuric oxide with minor addition of manganese dioxide	Alkali metal hydroxide	Zinc	1.4
S	Argentous oxide	Alkali metal hydroxide	Zinc	1.55

Note: the letter K relates to nickel–cadmium secondary batteries for which recommendations are given in IEC Publication 285

prime cost and is most suitable for applications requiring a medium current drain under semi-continuous conditions. Its capacity per unit volume is higher than that of the manganese dioxide/ammonium chloride, zinc chloride/zinc system. It is most commonly used in specialized motor-driven equipment.

(d) The mercuric oxide/alkali metal hydroxide/zinc system (letter reference M) has a high prime cost, a very high capacity per unit volume, and an almost constant voltage discharge characteristic. It finds applications in miniature electronic equipment, e.g. hearing-aids.

(e) Mercuric oxide with minor addition of manganese dioxide/alkali metal hydroxide/zinc system (letter reference N) is similar in all respects to the M system described above but has a slightly lower capacity per unit volume and a slightly lower cost.

(f) The argentous oxide/alkali metal hydroxide/zinc system (letter reference S) has a very high prime cost. It operates at the highest current drains and voltages of any of the systems under discussion and is used in miniature equipment.

Within each system (particularly the manganese dioxide/ammonium chloride, zinc chloride/zinc) differing levels of electrical performance are possible, dependent upon the quality and quantity of the active materials used and therefore on prime cost. IEC is considering how best to indicate these differences in performance level.

For the IEC system to be successful it is essential that all countries support its use through the medium of their National Standards. Many countries have already done so.

It is regrettable that at present some battery manufacturers do not use the IEC system on their products. It is hoped that the manufacturers will revise their policy and participate in this International cooperation.

The IEC battery nomenclature defines unambiguously the physical dimensions, polarity and type of terminals, electrochemical system and the nominal voltage.

Cells

A cell is designated by a capital letter followed by a number. The letters F, R and S define flat (layer-built), round and square (or rectangular) cells respectively.

This letter, together with the following number provide the reference to the appropriate entry in a table of approximate dimensions.

Each designation is related to a specific set of dimensions. There may however be more than one designation for any one set of dimensions. These cases arise due to differences of detail in profile or polarity.

At the present time numbers are allocated sequentially.

Electrochemical system

With the exception of the manganese dioxide/ammonium chloride, zinc chloride/zinc system, the letter R, F or S is preceded by an additional letter which denotes the electrochemical system, as shown below.

Batteries

If the battery contains only a single cell, the cell designation is used.

In other cases each section of the battery is designated separately, each designation being separated by an oblique stroke.

If a section contains more than one cell in series, their number precedes the cell designation.

If cells are connected in parallel, the number of parallel groups succeeds the cell designation, being attached to it by a hyphen.

Examples:

R20   A battery consisting of a single R20 cell.

3R12   A battery consisting of three R12 cells connected in series.

R25-2   A battery consisting of two R25 cells connected in parallel.

3R20-2   A battery consisting of two parallel groups of cells, each group containing three R20 cells in series.

60F20
2R14-2   A battery of two sections in which one section contains 60 F20 cells in series and the other section contains two parallel groups, each of two R14 cells in series.

Battery journals, trade organizations and conferences

Battery journals

Journal of the Electrochemical Society (monthly), Electrochemical Society Inc., Box 2071, Princeton, New Jersey 0854, USA.

Battery Council International, 111 East Wacker Drive, Chicago, Illinois 60601, USA publish an annual year book.

Independent Battery Manufacturers Association Inc., 100 Larchwood Drive, Largo, Florida 33540, USA publish an annual SLI9 Buyers Guide (Battery main listing).

Electroanalytical Chemistry and Interfacial Chemistry, Elsevier Sequoia SA, Lausanne, Switzerland.

Journal of Membrane Science (battery separators), Elsevier Science Publishers BV, Amsterdam, The Netherlands.

Trade organizations concerned with batteries

Battery Council International, 111 East Wacker Drive, Chicago, Illinois 60601 (annual meeting).

Electrochemical Society, Inc., Box 2071, Princeton, NJ 08540, USA (meetings twice annually).

Canadian Battery Manufacturers Association, One Yonge Street, Suite 1400, Toronto, Ontario, M5E IJP, Canada.

Independent Battery Manufacturers Association, Inc., 100 Larchwood Drive, Largo, Florida 33540 (have an annual meeting).

Arthur D. Little, USA and UK. Periodic surveys of lead–acid and small batteries.

Conferences

UK Brighton Power Sources Symposium, Brighton, UK, every 2 years. Secretary Miss S. M. Borner, International Power Sources Symposium Committee, PO Box 17, Leatherhead, Surrey, KT22-9QB, UK. Reprints of papers published by Pergamon Press, Oxford/New York and Oriel Press, Newcastle upon Tyne, UK Editor D. H. Collins.

France

NATO Conference on Materials for Advanced Batteries, Aussois, France, 1979.

Murphy, D. W., Broadhead, J. and Steel, B. C. H. (eds) (1980) *Materials for Advanced Batteries.* Plenum Press, New York. (NATO conference series. VI. Materials science: vol. 2) Proceedings of a NATO conference on Materials for Advanced Batteries, held September 9–14, 1979 in Aussois, France.

Canada

Symposium on Zinc–Silver Oxide Batteries, Montreal, 1968.

USA

Fleischer, A. and Lander, J. J. (eds) (1971) *Zinc–Silver Oxide Batteries.* Wiley, New York. (The Electrochemical Society series). Held during the 1968 fall meeting of the Electrochemical Society and co-sponsored by the Battery Division of the Society and the Air Force Aero Propulsion Laboratory. Bibliography, pp. 505–526.

Independent Battery Manufacturers Association, 100 Larchwood Drive, Largo, Florida 33540, USA, Annual Battery Convention.

Battery Council International, 111 East Wacker Drive, Chicago, Illinois 60601, Annual Battery Convention.

Bibliography

Arrance, F. C., Greve, R. and Rosa, A. (1968) *Inorganic Separator for a High Temperature Silver–Zinc Battery.* National Aeronautics and Space Administration, Washington, DC. For sale by the Clearinghouse for Federal Scientific and Technical Information, Springfield, Va. NASA contractor report, NASA CR-965, prepared under contract No. NAS 3-7639 by Douglas Aircraft Company, Newport Beach, Calif. for Lewis Research Center, National Aeronautics and Space Administration.

Austin, F. E. (1873) *Examples in Battery Engineering.* Hanover, NH.

Bagotskii, V. S. and Skundin, A. M. (1980) *Chemical Power Sources* (trans. O. Giebov and V. Kisin), Academic Press, London.

Bagshaw, N. E. (1983) *Batteries on Ships* (Battery Applications Series), Wiley, Chichester.

Barak, M. (ed.) (1980) *Electrochemical Power Sources: Primary and Secondary Batteries* (IEE Energy Series, No. 1), IEE, London.

Barak, M. (ed.) (1980) *Electrochemical Power Sources: Primary and Secondary Batteries* (IEE Energy Series: 1) Peter Peregrinus/Institution of Electrical Engineers, Stevenage.

Dry Batteries (1970) Key Note Publishers.

Batteries and Bulbs (1971) (Macdonald Starters Series), Macdonald.

Battery Powered Road Transportation World Guide (1970) McGraw-Hill, Maidenhead.

Battery Powered Vehicles: Electrical Equipment in Market Analysis and Forecast for Western Europe (1972) Electrical Research.

Battery Powered Vehicles: Electric Vehicles in Western Europe (1970) Electrical Research.

Batteries Selection and Application of Conference Proceedings (1978) Electrical Research.

Bechtel National Inc. (1982) *Photovoltaic Power Systems: Handbook for Battery Energy Storage.* Includes solar energy information.

Benjamin, P. (1893) *The Voltaic Cell, Its Construction and Its Capacity*, 1st edn, Wiley, New York. Includes bibliography of the voltaic cell.

Bode, H. (1977) *Lead Acid Batteries* (trans. H. Bode and O. Kordesch), Electrochemical Society, USA/Wiley International, New York.

Bogenschütz, A. (1968) *Fachwörterbuch für Batterien und Energie-Direktumwandlung* (Dictionary of Batteries and Energy Transformation) Brandstetter. A German–English dictionary.

Bogenschütz, A. (ed.) (1968) *Technical Dictionary for Batteries and Direct Energy Conservations*, Brandstetter/Heyden.

Bolen, M. N. and Weil, B. H. (1948) *Literature Search on Dry Cell Technology with Special Reference to Manganese Dioxide and Methods for its Synthesis* (Georgia State Engineering Experiment Station, Atlanta Special Report No. 27) State Engineering Experiment Station, Georgia Institute of Technology, Atlanta, Georgia. Sponsored by the Battery Branch, Signal Corps Engineering Laboratories, Fort Monmouth, NJ.

Bottone, S. R. (1902) *Galvanic Batteries: Their Theory, Construction and Use, Comprising Primary, Single and Double Fluid Cells, Secondary and Gas Batteries* (Whittaker's Library of Arts, Sciences and Industry) Whittaker, London.

Bubier, E. T. (1910) *How to Make Electric Batteries*, 3rd edn, Bubier Publishing Co., Lynn, Mass. Earlier editions are entitled *How to Make Electric Batteries at Home.*

Burgess Battery Co. (1958) *Burgess Engineering Manual: Complete Data on Dry Batteries for the Design Engineer*, Burgess Battery Co., Freeport, Ill.

Cahoon, N. and Heise, G. W. (1976) *The Primary Batteries*, Vol. 2 (Electrochemical Society Series), Wiley Interscience, New York.

Carhart, H. S. (1891) *Primary Batteries*, Allyn and Bacon, Boston.

Carhart, H. S. (1920) *Thermo-electromotive Force in Electric Cells, the Thermo-electromotive Force between a Metal and a Solution of one of its Salts*, D. Van Nostrand Co., New York.

Codd, M. A. (1929) *Practical Primary Cells*, Pitman, London.

Collins, D. H. (1977) *Batteries International Symposium*, 4th edn, Pergamon, Oxford.

Cooper, W. R. (1917) *Primary Batteries: Their Theory, Construction and Use* (The Electrician Series), The Electrician Printing and Publishing Co., New York.

Crane, O. and Dale, A. (eds) (1975) *Aviation Maintenance Publishers Aircraft Batteries Lead Acid and Nickel Cadmium* (Aviation Technician Training Series), Aviation Maintenance Publishers.

Crompton, T. R. (1983) *Small Batteries, Primary Cells*, Vol. 2, Halsted Press, New York.

Dart, H. F. (1921) *Generation of Electromotive Force*, International Textbook Company, Scranton, Pa.

Dixon, M. (1977) *Batteries, Bulbs and Circuits* (Let's Do Science Series), Edward Arnold, London.

DOE Technical Information Centre (1978) *Electric Batteries: A Bibliography*, Department of the Environment, London.

Electric Storage Battery Co. (1914) *Manual of 'Exide' Batteries in Electric Vehicles*, The Electric Storage Battery Co., Philadelphia, Pa.

Electric Vehicle Council (1980) *World Guide to Battery Powered Road Transportation*, Edison Electric.

Falk, S. U. and Salkind, A. J. (1969) *Alkaline Storage Batteries* (Electrochemical Society Series), Wiley Interscience, New York.

Frost, O. and Sullivan, F. (1985) *Industrial Batteries and Fuel Cells (Europe)*.

Gabano, J. B. (ed.) (1983) *Lithium Batteries*, Academic Press, London.

Graham, R. W. (1978) *Primary Batteries – Recent Advances* (Chemical Technology Review 105. Energy Technology Review 25) Noyes Data Corporation, Park Ridge, NJ.

Graham, R. W. (ed.) (1980) *Rechargeable Batteries: Advances Since 1977* (Energy Technology Review Set, No. 55; Chemical Technology Review Set, No. 160), Noyes Data Corporation, Park Ridge, NJ.

Graham, R. W. (ed.) (1981) *Electrochemical Cell Technology: Advances Since 1977* (Chemical Technology Review Set, No. 191), Noyes Data Corporation, Park Ridge, NJ.

Grevich, J. D. (1972) *Testing Procedures for Automotive AC and DC Charging Systems*, McGraw-Hill, Maidenhead.

Gross, S. (ed.) (1978) *Battery Design and Optimization Symposium of 1978, Proceedings* (Electrochemical Society Proceedings Series, Vol. 79.1).

Guthe, K. E. (1903) *Laboratory Exercise with Primary and Storage Cells*, G. Wahr, Ann Arbor, Mich.

Harrison, N. J. (1962) *RCA Battery Manual for Industrial and Consumer-Product Applications*. Radio Corporation of America Electron Tube Division.

Hehner, N. E. (1976) *Storage Battery Manufacturing Manual*, 2nd edn, IBMA Publications.

Heise, G. W. and Cahoon, N. C. (eds) (1971) *The Primary Battery*, Vol. 1 (Electrochemical Society Series) Wiley, New York. Includes bibliography.

Himy, A. (1985) *Silver–Zinc Battery, Phenomena and Design Principles*, Vantage.

Jasinski, R. (1967) *High-Energy Batteries*, Plenum Press, New York.

Jones, B. E. (ed.) *Electric Primary Batteries: a Practical Guide to their Construction and Use*, Cassell, London.

Kordesch, K. V. (ed.) (1974) *Batteries*, Vol. 1, *Manganese Dioxide*, Marcel Dekker, New York. Includes bibliographical references.

Kordesch, K. V. (ed.) (1977) *Batteries: Lead Acid Batteries and Electric Vehicles*, Vol. 2, *Batteries*, Marcel Dekker, New York.

Lincoln, N. E. S. (1945) *Primary and Storage Batteries* (The Essential Modern Electrical Series) Duell, Sloan and Pearce, New York.

Linden, D. (1983) *Handbook of Batteries and Fuel Cells*, McGraw-Hill, Maidenhead.

Mantell, C. L. (1970) *Batteries and Energy Systems*, McGraw-Hill, New York. Includes bibliographies.

Mantell, C. L. (1983) *Batteries and Energy Systems*, 2nd edn, McGraw-Hill, Maidenhead.

Marsh, K. (ed.) (1981) *Battery Book One: Lead Acid Traction Batteries*, Curtis Instruments.

Merrell, J. P. (1875) *Lecture on Galvanic Batteries and Electrical Machines, as used in Torpedo Operations*, US Torpedo Station, Newport, RI.

Mueller, G. A. (1973) *The Gould Battery Handbook*, 1st edn, Gould, Mendota Heights, Minn.

Murphy, D. W. *et al.* (eds) (1981) *Materials for Advanced Batteries*, Vol. 2 (NATO Conference Series VI – Materials Science) Plenum Press, New York.

National Electrical Manufacturers (1984) *Safety Recommendations for Lead Acid Industrial Storage Batteries for Railway and Marine Starting Application*, National Electrical Manufacturers.

National Electrical Manufacturers (1984) *Safety Recommendations for Lead Acid Industrial Storage Batteries Used for Motive Power Service*, National Electrical Manufacturers.

Niaudet, A. (1890) *Elementary Treatise on Electric Batteries*, 6th edn (trans. L. M. Fishback) Wiley, New York.

Nicholas, K. F. (1914) *Bright Way, the Use and Abuse*, Punton-Reed Publishing Co., Kansas City, Mo.

Richards, T. W., Williams, J. H. and Garrod-Thomas, R. N. (1909) *Electrochemical Investigation of Liquid Amalgams of Thallium, Indium, Tin, Zinc, Cadmium, Lead, Copper and Lithium*. Carnegie Institution of Washington, Washington, DC.

Ruben, S. (1978) *The Evolution of Electric Batteries on Response to Industrial Needs*, Dorrance, Philadelphia.

Sanders, D. R. *et al.* (eds) (1984) *Radial Keratotomy*, Slack Inc.

Sandia National Laboratories (1986) *Design and Development of a Sealed 100-A h Nickel Hydrogen Battery*. Includes solar energy information.

Schneider, N. H. (ed.) (1904) *Dry Batteries – How to Make and Use Them* (The Model Library, Vol. 1, No. 3). Spon and Chamberlain, New York. Carnegie Institution of Washington Publication No. 118. Contributions from the Chemical Laboratory of Harvard College. Gives full detailed instructions for the manufacture of dry cells of any shape and size, especially adapted for automobile, launch and gas engine work, by a dry battery expert.

Schneider, N. H. (1905) *Modern Primary Batteries: Their Construction, Use and Maintenance*, Spon and Chamberlain, New York. Includes batteries for telephones, telegraphs, motors, electric lights, induction coils, and for all experimental work.

Shaw, M., Paez, O. A. and Ludwig, F. A. (1969) *Electrochemical Characterization of Nonaqueous Systems for Secondary Battery Application*, National Aeronautics and Space Administration, Washington, DC. For sale by the Clearinghouse for Federal Scientific and Technical Information, Springfield, Va. NASA contractor report, NASA CR-1434, prepared under contract No. NAS 3-8509 by Whittaker Corporation, San Diego, Calif. for Lewis Research Center.

Sittig, M. (1970) *Battery Materials* (Electronics Materials Review, No. 10) Noyes Data Corporation, Park Ridge, NJ. Based on US patent literature since 1960.

Small Batteries Secondary Cells, Vol. 1, Halsted Press, New York.

The Battery Market (Industrial Equipment and Supplies Set) Business Trends.

Tucker, A. E. (ed.) (1984) *Cylinders and Accumulators*, Vol. G, *Fluid Power Standards*, National Fluid Power.

Union Carbide Chemicals Company (1963) *Eveready Battery Applications and Engineering Data*, Union Carbide, New York.

Union Carbide Corporation, Union Carbide Consumer Products Company (1965) *Eveready Battery Applications and Engineering Data*, Union Carbide, New York.

Varta Sealed Nickel Cadmium Batteries (1982) *Batterie*, VDI, W. Germany/Heyden.

Vinal, G. W. (1950) *Primary Batteries*, Wiley, New York.

Vinal, G. W. (1955) *Storage Batteries: A General Treatise on the Physics and Chemistry of Secondary Batteries and Their Engineering Applications*, 4th edn, Wiley Interscience, New York.

Vincent, C. A. (1976) *Modern Batteries*, Edward Arnold, London.

Waterford, V. (1985) *The Complete Battery Book*, TAB Books.

White, R. (ed.) (1984) *Electrochemical Cell Design*, Plenum Press, New York.

Index